Advances in Experimental Medicine and Biology

Volume 1477

Advances in Experimental Medicine and Biology provides a platform for scientific contributions in the main disciplines of the biomedicine and the life sciences. This series publishes thematic volumes on contemporary research in the areas of microbiology, immunology, neurosciences, biochemistry, biomedical engineering, genetics, physiology, and cancer research. Covering emerging topics and techniques in basic and clinical science, it brings together clinicians and researchers from various fields.

Advances in Experimental Medicine and Biology has been publishing exceptional works in the field for over 40 years, and is indexed in SCOPUS, Medline (PubMed), EMBASE, Reaxys, EMBiology, the Chemical Abstracts Service (CAS), and Pathway Studio.

2023 CiteScore: 5.9

Daniela Tropea • Emanuela Giacometti
Editors

Brain-Body Connections

Bidirectional Communication Between the Brain and Body Systems

Springer

Editors
Daniela Tropea
Trinity Translational Medicine Institute
St James Hospital
Dublin, Ireland

FutureNeuro Research Ireland Centre
Trinity College
Dublin, Ireland

Trinity College Institute of Neuroscience
Dublin, Ireland

Emanuela Giacometti
Amsterdam, The Netherlands

ISSN 0065-2598 ISSN 2214-8019 (electronic)
Advances in Experimental Medicine and Biology
ISBN 978-3-031-89524-1 ISBN 978-3-031-89525-8 (eBook)
https://doi.org/10.1007/978-3-031-89525-8

This Springer imprint is published by the registered company Springer Nature Switzerland AG
The registered company address is: Gewerbestrasse 11, 6330 Cham, Switzerland

To Snaiha, and her great love for learning.

Preface

The human body is made up of multiple systems such as the nervous, endocrine, cardiovascular, and others. Although it is considered the "supreme controller" of all other body systems, the brain does not work in isolation. The systems are interconnected and communicate with each other.

In this book, researchers from different fields explore the cellular and molecular systems that regulate such communications.

The role/function of the brain has been discussed by several scientists and thinkers. In 335 BC, the Greek philosopher Aristotle declared that the brain was simply a cooling system that kept the most important organ: the heart, from overheating. Around 170 AD, the Roman physician Galen suggested that the brain's four ventricles (fluid-filled cavities) were the seat of complex thought and determined personality and bodily functions. This was one of the first suggestions that the brain was the residence place of our memory, personality, and thoughts (although not in the ventricles).

For a long time the brain has been considered the "master controller" of the body, because it is the region of the human nervous system responsible for processing sensory information, coordinating and executing motor functions, and managing thought, memory, and emotions. It send and receives signals through a vast network of neurons, allowing us to respond to our environment, make decisions, and perform complex tasks. Quoting a neurology text book from 1997 "In essence then, it is the brain that makes us what we are, and indeed it can be argued that all together the organs are there simply to support the brain. .... The brain is one of the latest great frontiers on the biological sciences, and the unravelling of its mysteries is comparable in complexity and intellectual challenge to the search for elementary particles of matter or the effort to explore space" (from *The Neuron: Cell and Molecular Biology*, 2nd Edition by Irwin B. Levitan, Leonard K. Kaczmarek).

On the other hand, considering the brain in isolation can lead to a narrow understanding of its functions. For instance, the brain works closely with the endocrine system to regulate hormones, which in turn affect brain function as well as other important organs in the body. Another example is the cardiovascular system, where proper blood flow is essential for delivering nutrients and oxygen to brain cells. Isolating the brain from these connections can result in an incomplete picture of its operations and the factors that influence its health. It's important to consider the brain as part of the larger system (body) and in the context of the individual's health and environment. An additional factor to consider is how the brain uses metabolic information from organs to drive physiological changes and behavior, although the types of information conveyed to the brain, and how the brain processes such information are not yet well understood. In this book we cover many examples of the interactions between the brain and different systems and organs.

The book is organized in 12 chapters covering different examples of brain-body connections, but it is not exhaustive. Chapters are broadly organized by topic and the relevant links between chapters are indicated when a common topic is discussed.

Summary Overview of the Chapters

The book starts with the system that isolates the brain from the body to protect it, but also to organize the selective transport of oxygen, glucose, neurotransmitters, and other molecules: the blood brain barrier (BBB). The BBB is a highly specialized structure that is crucial for brain homeostasis and is in turn affected by "body" factors such as sex and age as well as environmental factors such as stress. In pathological conditions BBB disruption can contribute to disease such as Alzheimer, Multiple Sclerosis, and Epilepsia. *Matthew Campbell and Angela Garcia* discuss the basic structure and transport mechanisms that allow selected molecules to cross the BBB. Synchronization of brain-body responses to external stimuli is a great example of how the brain and the body interact. Norepinephrine and Epinephrine are molecules that can cross the BBB and that serve as both neurotransmitters and hormones, regulating a wide array of physiological functions. They control, respectively, central and peripheral adrenergic systems. Although there seems to be no direct link between the adrenergic and noradrenergic systems their feedback loops allow them to synchronize brain and body response to stimuli. *Mriganka Sur and colleagues* explore adrenergic systems in the brain and body and share evidence that noradrenaline (NE) plays a crucial role in linking neural control to physiological functions.

Released from sympathetic nerve terminals and the adrenal medulla in response to stress, NE also stimulates hepatic glucose production by activating glycogen phosphorylase and enhancing gluconeogenic enzyme activity. Glucose metabolism controls the interplay between brain function, obesity, type 2 diabetes mellitus, and other metabolic disorders. As a result, obesity-driven brain dysfunction may complicate the treatment of diseases characterized by aberrant glucose control (including type 2 diabetes mellitus). In addition, obesity is associated with a light inflammatory state and has been linked to neurodegenerative diseases like multiple sclerosis, Alzheimer's, and Parkinson's disease. *Susanne La Fleur and colleagues* discuss how the brain senses glucose and regulates its levels, how the brain is communicating with the liver and pancreas to regulate glucose metabolism, and what the clinical implications are of dysfunctional glucose regulation.

Insulin-like growth factor I (IGF-I) is a protein with significant structural homology with insulin. IGF1 is produced in the liver and in the brain and crosses the BBB. In the brain it has a neuroprotective effect and in both brain and body it acts to enhance insulin sensitivity. It is in fact used for ~~diabetic~~

patients with severe insulin resistance. Brain or body IGF1 deficiencies lead to a disrupted BBB and altered cerebrovascular microenvironment. *Daniela Tropea and colleagues* discuss about IGF1 and its different roles within the brain and the body, and suggest that it may act as a sensor between body energy levels and brain states. Therapeutically, IGF-1 is used as a diagnostic and prognostic marker, and a potential treatment Dwarfism, Rett Syndrome, Autism, Neuropathy, and neurodegenerative conditions like Alzheimer's Disease. Metabolic dysfunction is a unifying pathology underlying many of these disorders. Another important mechanism controlling energy levels is the oxidative metabolism in the mitochondria. This topic is presented by *Jill Glausier* who explains how neuronal mitochondria and synaptic functioning are vulnerable to perturbations within different organ systems. The effects of mitochondria's dysfunctions in schizophrenia, Parkinson's disease, metabolic disorders, and normal ageing processes are also discussed.

The brain and gut are connected through the nervous system, hormones, and the immune system. This connection is known as the brain-gut axis. The gut-brain connection is complex and bidirectional. Key players in this connection include the enteric nervous system, the vagal nerve, and the gut microbiota. Millions of nerves and neurons run between gut and brain, and neurotransmitters and other chemicals produced in the gut also affect the brain. NE is a key player in the communication between gut and brain via the vagal nerve. The enteric nervous system, also referred to as the "second brain," has the same type of neurons and neurotransmitters found in your central nervous system. This connection between the brain and gut affects digestion, mood, and the way we think. *John Cryan and colleagues* describe the multifaceted bidirectional communication system connecting gut microbial activity with central nervous system processes through immune pathways, metabolic by-products, and neural circuits like the vagal nerve. The vagal nerve is a direct physical link between the periphery and the brain, transmitting signals related to various bodily functions including heart rate, digestion, and immune response, arousal and stress responses.

Stress plays a role in the brain-gut axis, with implications for multiple prevalent health conditions that are characterized by chronic gastrointestinal symptoms. Although the afferent and efferent pathways linking the gut and the brain are modulated by stress, the fields of neurogastroenterology and psychoneuroendocrinology/psychoneuroimmunology remain only loosely connected. Functional gastrointestinal disorders, for example, are a common conditions modulated by stress. Patients show increased levels of chronic stress and high comorbidity with affective disorders, especially symptoms of anxiety and depression. Furthermore, psychological factors demonstrably contribute to or even explain alterations in brain function or structure observed in patients with Irritable Bowel syndrome. When our environments change in extreme ways and for prolonged periods of time, a different set of hormonal stress responses are recruited. These chronic stress responses produce adaptive changes but can also drive maladaptations. *Ki Goosens and colleagues* review the peripheral hormones that are recruited as part of the acute stress response and describe their adaptive impact on brain and peripheral function. They discuss the endocrine factors that underlie stress-induced vulnerability in rodent models of post-traumatic stress disorder (PTSD), Major Depressive Disorder (MDD), and obesity. Stress has long been known to affect eating behaviors in humans. Generally, acute stress results in decreased eating whereas chronic stress results in increased eating. Peptides associated with the brain-gut axis are traditionally known for their ability to control feeding and appetite but have over the last years been attributed to an additional role in reward regulation. An extensive number of gut-brain peptides have been identified, and among them ghrelin and glucagon-like peptide-1 (GLP-1) appear to be central players. This gut–brain signaling has been suggested to influence addiction-related behaviors through two main mechanisms. The first involves a condition known as "leaky gut," where the molecules contained in the gut are released to impact structures outside the digestive system. The second mechanism is related to the molecules that gut microbes produce, which could influence addiction by directly influencing brain functioning. Therefore, gut dysbiosis and its subsequent impact on GABA, serotonin, dopamine, and tryptophan could increase the susceptibility to addiction, and increase the severity of withdrawal symptoms in people with a disrupted gut microbiome. *Elisabeth Jerlhag Holm* explains that appetite regulatory peptides could be candidates for novel treatments for Alcohol use disorder (AUD). The author in this chapter describes the crosstalk and the downstream mechanisms that link ghrelin and GLP to addiction processes.

Functional studies performed during the past decade have shown that most rare and common variants associated with neurodevelopmental disorders (NDDs) affect genes that have a role in a few conserved pathways such as protein synthesis, epigenetic regulation, and synaptic signaling. Often, genes involved in NDDs have a broad effect on brain development, but they are also expressed in other parts of the body. One such example is *MECP2*, a gene coding for a chromatin remodeler and transcriptional regulator (MeCP2) that when mutated leads to an NDD called Rett Syndrome. However, the influence of MeCP2 deficiencies extends beyond neural boundaries, impacting a broad spectrum of other tissues and organ systems. *Rudolf Jaenisch and colleagues* describe the Rett Syndrome (RTT) disorder and explore the potential pathways involved in the brain-body circuit and their impact on the development of RTT. These insights not only enhance our understanding of the multisystem involvement in RTT but also reveal the necessity for using a more comprehensive approach to investigating the systemic courses of neurological disorders. Indeed, in genetic NDDs, the mutations are present in multiple tissues beyond the brain. Peripheral symptoms of NDDs frequently correlate with metabolic disorders, endocrine dysregulation, and immune system abnormalities. For instance, multiple hormone dysfunctions within the endocrine system are commonly seen in autism. Depression is often associated with hyperactivity of the hypothalamic-pituitary-adrenal axis and systemic immune activation, with a notable correlation between depression severity and increased pro-inflammatory cytokines. The HPA axis is an important signaling cross-road that is discussed in many chapters in this book.

Stress responses can be adaptive, but when experienced repeatedly or over a protracted period of time, stress is strongly associated with multiple types of disease, including posttraumatic stress disorder, depression, obesity, and diabetes.

In recent years, the interplay between the immune and nervous systems has been well described in a wide array of pathological cases: from autoimmune diseases such as multiple sclerosis to more classical cognitive conditions of dysfunction like Alzheimer disease. Beyond this link, it is now recognized that immune cells also actively contribute to homeostatic processes in the nervous system. Brain residents' innate immune cells—microglia—have the capacity to sculpt neural circuitry and coordinate copious and diverse neurodevelopmental processes. *Yuri Bozzi and colleagues* describe how the accumulation of dysfunctional immune cell types contributes to the pathogenesis of NDDs with a special emphasis on the connection between immune dysfunction, inflammation, and oxidative stress and introduces the role of the immune system in brain development and function. In fact, our understanding of how the brain and immune system interact has changed substantially over the past decade, acknowledging the existence of a complex brain-immune network that involves multiple peripheral immunological players. This brain-immune relationship is important during development, adulthood, and aging, as well as during various pathologies.

Environmental conditions, stress, and health of the body of the mother can greatly influence brain development in her child both during gestation but somehow also later in their life. It is now well-established that adverse prenatal environments, such as exposure to maternal stress, viral infections, drugs, and toxins, can disrupt normal brain development and have lasting consequences for the structure and function of the brain. Moreover, prenatal adversity can significantly increase the risk of developing mental disease in later life, including schizophrenia, depression, anxiety, and autism, among others. *Natalia Duque-Wilckens and Taylor Doolittle* explain a novel neuro-immune mechanism mediated by Mast cells (MCs). The chapter provides an overview of the role of maternal-fetal immune interactions in healthy and altered brain development and examines the link between immune alterations, prenatal chemical exposures, and the onset of neuropsychiatric disorders in later stages of life.

Understanding, diagnosing, and treating brain diseases require analysis of cell function and communication. Since their discovery, just over 30 years ago, microRNAs (miRNAs) have emerged as a major class of non-protein coding genes with critical roles in orchestrating gene expression and communication between cells. The brain is the most miRNA-rich organ in the body, and miRNAs regulate every step during brain development. After brain development is complete, miRNAs continue to enforce the gene expression programs. Expression of miRNAs is disturbed in brain disease, leading to under- and over-suppression of gene pathways and even the re-direction of miRNAs to new target.

miRNAs can efficiently regulate gene expression. Recent studies show that miRNAs exist not only in cells but also in a variety of body fluids. Characterization of the composition of such extracellular microRNAs indicates some have CNS origins. This has led to the idea that microRNAs have extracellular functions in paracrine and long-distance signaling, including the connection between the brain and rest of the body, which makes it possible to imagine they play a key role in connecting the body and the brain. David Henshall explains how MicroRNAs fine-tune brain and body communication in health and disease. Levels of circulating microRNAs have been reported to be altered in many brain diseases, including epilepsy, indicating potential use as diagnostic biomarkers

Our understanding of the systems interconnecting the brain and the different body systems is continuously evolving. The study of these connections in their physiological context and the effects of the environment could offer additional insights on the regulatory function of the brain, and might have applications in diagnostics and therapeutics.

Dublin, Ireland — Daniela Tropea
Amsterdam, The Netherlands — Emanuela Giacometti

Contents

Understanding the Blood-Brain Barrier: From Physiology to Pathology

1

Angela Garcia-Gallardo and Matthew Campbell

Abstract

The blood-brain barrier (BBB) is a highly specialized structure that separates the circulating blood from the brain, ensuring protection, homeostasis, and optimal functioning of the central nervous system (CNS). Cerebral endothelial cells, pericytes, astrocytes, and basement membrane form the BBB. The interaction of these elements with neurons and microglia is called the neurovascular unit (NVU). Each component plays a fundamental role in maintaining the integrity and functioning of the BBB.

The BBB regulates the transport of molecules between the blood and the brain, and vice-versa, selectively allowing the entry of essential nutrients while preventing the passage of potentially harmful substances, such as toxins and pathogens. This selective permeability is crucial for maintaining an optimal environment for neuronal function and protecting the brain from potential damage.

BBB integrity is essential for brain health, and its disruption is a hallmark of many neurological disorders, neurodegenerative diseases, and even aging. BBB disruption has been described in pathologies such as multiple sclerosis, where it facilitates immune cell migration and demyelination. In epilepsy, BBB dysfunction has been considered as contributing factor to the initiation and progression of seizures. In Alzheimer's disease, BBB's permeability contributes to neuroinflammation and neuronal damage. Age-related BBB breakdown can lead to toxins, pathogens, or immune cell invasion, which can result in neuronal dysfunction and potentially, neurodegeneration.

In this chapter, we will discuss the structure and functions of the BBB, sex-related physiological differences, the impact of conditions like temperature and stress on it, and how its disruption affects brain homeostasis and contributes to disease.

Keywords

Blood-brain barrier · Tight junctions · Claudin-5 · Neurology

A. Garcia-Gallardo (✉)
Department of Neurology, St James's Hospital, Dublin, Ireland

Department of Genetics, Smurfit Institute of Genetics, Trinity College Dublin, Dublin, Ireland
e-mail: garciaa2@tcd.ie

M. Campbell
Department of Genetics, Smurfit Institute of Genetics, Trinity College Dublin, Dublin, Ireland

D. Tropea, E. Giacometti (eds.), *Brain-Body Connections*, Advances in Experimental Medicine and Biology 1477, https://doi.org/10.1007/978-3-031-89525-8_1

Abbreviations

ABC	ATP-Binding cassette family of transporters
AD	Alzheimer disease
AJ	Adherens junctions
ALS	Amyotrophic lateral sclerosis
ANXA1	Annexin A1
APOE4	Apolipoprotein E-ε4
APP	Amyloid precursor protein
AQP4	Aquaporin-4
ATP	Adenosine triphosphate
BBB	Blood-brain barrier
BM	Basement membrane
BMEC	Brain microvascular endothelial cells
CAMS	Soluble cell adhesion molecules
CNS	Central nervous system
CSF	Cerebrospinal fluid
DCE-MRI	Dynamic contrast-enhanced magnetic resonance imaging
DNA	Deoxyribonucleic acid
EC	Endothelial cells
ECM	Extracellular matrix
GFAP	Glial fibrillary acidic protein
GLUT-1	Glucose transporter-1
ICAM-1	Intercellular adhesion molecule-1
IFN-γ	Interferon-gamma
IL-1	Interleukin-1
IL-6	Interleukin 6
JAM (-A, -B, -C)	Junctional adhesion molecules -A,-B,-C
LRP-1	Lipoprotein receptor-related protein 1
MMP (−2, −9)	Matrix metalloproteinase (−2, −9)
MS	Multiple sclerosis
NO	Nitric oxide
NVU	Neurovascular unit
PECAM-I	Platelet endothelial cell adhesion molecule-I
PP-MS	Primary progressive multiple sclerosis
RAGE	Receptor for advanced glycation end products
RNS	Reactive nitrogen species
ROS	Reactive oxygen species
RR-MS	Relapsing-remitting multiple sclerosis
S100B	S100 calcium-binding protein B
SE	Status epilepticus
SP-MS	Secondary progressive multiple sclerosis
sVCAM-I	Soluble vascular cell adhesion molecule-I
TBI	Traumatic brain injury
TEER	Transendothelial electrical resistance
TEM	Trans-endothelial migration
THT	Therapeutic hypothermia therapy
TJ	Tight junctions
TLE	Temporal lobe epilepsy
TNF-α	Tumour necrosis factor-alpha
VCAM-1	Vascular cell adhesion molecule-1
VE-Cadherin	Vascular endothelial-cadherin
VEGF	Vascular endothelial growth factor
ZO (−1, −2)	Zonula occludens (−1, −2)

1.1 Introduction

The brain is a complex and sophisticated organ with dynamic functions that make it the most metabolically demanding organ in the body (Schaeffer and Iadecola 2021). Despite representing only 2% of the total body weight, the human brain consumes approximately 20% of the total oxygen and glucose available in the body (Kisler et al. 2017; Dion-Albert et al. 2023).

Given its metabolic needs, the brain possesses more than 600 km of vessels that provide nutrients, remove waste and toxins, and maintain the homeostasis of the nervous tissue, among other functions (Dion-Albert et al. 2023; Schlageter et al. 1999). Such vessels are distributed in a manner that ensures a correlation of almost one capillary per neuron ratio (Schlageter et al. 1999), allowing a direct response to the changing blood supply needs of the neurons. This phenomenon is known as neurovascular coupling (Alahmari 2021).

These brain vessels possess unique properties that distinguish them from the rest of the circulatory system and isolate the neural tissue, creating a multilayered barrier capable of strict influx/efflux control, known as the blood-brain barrier (BBB) (Obermeier et al. 2013).

1.2 The Discovery

The existence of the BBB was first documented by Humphrey Ridley in the book Anatomy of the Brain, published in 1695, after observing that cerebral vessels were less permeable to beeswax and mercury compared to other bodily tissues (Ridley 1695). Almost two centuries later, in 1885, Paul Ehrlich and his student, Edwin Goldman, revived this concept through their experiment findings. Ehrlich described that "after intravenous dye injection, the brain and spinal cord remained white as snow while peripheral organs were readily stained," (Ehrlich 1885). demonstrating the ability of the CNS to exclude certain molecules.

In 1900, Max Lewandowski confirmed these findings, concluding that "brain capillaries must prevent certain molecules from entering the brain," a concept that he called "bluthirnschranke," the German word for blood-brain barrier (Lewandowsky 1900). Finally, in 1967, with the invention of electron microscopy, Thomas S. Reese and Morris J. Karnovsky first described a morphological correlation to these barrier properties, identifying the presence of tight junctions connecting the overlapping endothelial cells (Reese and Karnovsky 1967). From then on, BBB knowledge continues to grow, fed by an endless number of completed and ongoing research.

1.3 The Blood-Brain Barrier

The blood-brain barrier (BBB), present in most vertebrates and not exclusive to humans (Sweeney et al. 2018), is a dynamic, tightly controlled, and highly selective microvascular system with unique properties restricted to the Central Nervous System (CNS) circulation (Obermeier et al. 2013). It consists of cellular and non-cellular vascular and neural components that support the endothelial cells (EC) in their role of creating a structural, transport, and metabolic barrier that limits the entry of foreign and unwanted molecules and cells, safeguards nutrient distribution, and controls the neuronal extracellular environment (Alahmari 2021; Obermeier et al. 2013), maintaining a balanced microenvironment that protects the neurons (Zhao et al. 2021).

Acting as the primary gatekeepers, ECs interact dynamically with pericytes, astrocytes, basement membranes, microglial cells, and neurons. This collective interaction forms the neurovascular unit (NVU) and constitutes the building block of the BBB (Zhao et al. 2021).

1.3.1 BBB Function

Over the past century, as research in the field has expanded, the understanding of the BBB has evolved from a "brick wall" concept to an active, robust, regulatory, and dynamic multicomponent unit that adapts to meet CNS and systemic demands under physiological and pathological conditions (Lampron et al. 2013). BBB modulation is influenced by the crosstalk between NVU components, vascular and immune elements, as well as neurons. Recent studies using single-cell sequencing have revealed vascular changes occurring in response to neuronal activity (Hrvatin et al. 2018; Nishijima et al. 2010).

The BBB maintains CNS homeostasis by regulating the influx and efflux of ions, oxygen, and nutrients. It tightly controls molecular and cellular migration, preventing nonmeaningful crosstalk (Abbott et al. 2006; Bernacki et al. 2008), and restricting the entry of blood cells, plasma components, toxins, and pathogens that could damage the brain environment (Harati et al. 2012; Zhao et al. 2015). Additionally, the BBB regulates the bidirectional water flow between the bloodstream and the extracellular space, ensuring proper volume balance (Rasmussen et al. 2022).

It has been proposed that there may be heterogeneity in the BBB across different functional regions of the brain, such as the cortex, hippocampus, white matter, and basal ganglia. This differentiation could contribute to the specialized functions of each region, though there is currently insufficient data to fully support this hypothesis to date (Profaci et al. 2020).

1.4 Neurovascular Unit Components

1.4.1 Endothelial Cells (ECs)

In the CNS, ECs create a continuous lining of highly specialized barrier cells, where the smallest capillary circumference is formed by a single EC (Profaci et al. 2020). The brain ECs are simple squamous epithelial cells characterized by a unique phenotype that makes them different from their peers in the rest of the circulatory system (Aird 2007). They are non-fenestrated, rich in mitochondria cells, with luminal/abluminal polarization, abundant junctional adhesion molecules (JAMs), and tight junctions (TJ) with high electrical resistance that regulate the paracellular movement of ions, macromolecules, and polar solutes. These cells also have specific transporters and receptors controlling transcellular molecule flow and exhibit a notably low level of transcytotic vesicle activity (Michalicova et al. 2020).

The luminal aspect of the ECs is covered by glycocalyx, a dense coat composed of glycoproteins and glycolipids that restricts the interaction of ECs with large molecules (Pillinger and Kam 2017), but that can be permeated by smaller solutes like fluorescein (376 daltons, Da) and Alexa Fluor (643 Da) (Kutuzov et al. 2018). As expected, glycocalyx degradation has a negative impact on BBB integrity, being associated with more severe BBB leakage (DellaValle et al. 2019; Zhu et al. 2018). On the contralateral aspect of the ECs lies the basal lamina, or basement membrane (BM), a matrix of laminins, fibronectin, collagens, tenascin, and proteoglycans encasing the ECs and pericytes that acts as a selective binding interface for molecules and cell migration (Del Zoppo et al. 2006).

CNS EC's tight junctions also hold special features when compared with the ones in peripheral vessels. They are more abundant and 50–100 times closer to each other, resulting in an extremely high transendothelial electrical resistance (TEER) (Nagasawa et al. 2006).

Despite being considered the core element of the BBB, ECs' unique characteristics are the result of the regulation carried out by the intricate molecular crosstalk network between the other cells in the NVU (Alvarez et al. 2013), required in the BBB formation during embryonic development and maintenance of its integrity in adulthood (Keaney and Campbell 2015).

1.4.2 Pericytes

Pericytes are multifunctional perivascular cells that wrap their finger-like projections around ECs at the capillary level (Dore-Duffy 2008). As they play a significant role in the stability and integrity of the NVU and therefore the BBB, the pericyte concentration and coverage over the endothelial walls of the CNS are much higher than in the peripheral circulation (Winkler et al. 2011). The pericyte-to-EC ratio has been shown to correlate with cerebral blood flow and barrier integrity directly (Delaney and Campbell 2017).

Within the NVU, pericytes are located between the ECs and the BM, surrounded by astrocytic end-feet, microglia, and neuron (Fig 1.1). Pericytes place themselves at the ECs junctions and face the luminal aspect of astrocytes' end-feet (Bertrand et al. 2019). This positioning allows them to perform their fundamental role in BBB development, growth, stability, and maintenance (Daneman and Prat 2015) as they participate in the regulation of angiogenesis, "barriergenesis," vascular remodeling, and vascular tone (Winkler et al. 2011; Sweeney et al. 2016).

Pericytes control the formation of TJ and transendothelial vesicle traffic (Dyrna et al. 2013), guide astrocytic foot processes to vessel walls, and mediate the polarization of astrocytic end-feet (Delaney and Campbell 2017). Pericytes are key modulators of extracellular matrix protein secretion and levels, have a high phagocytic

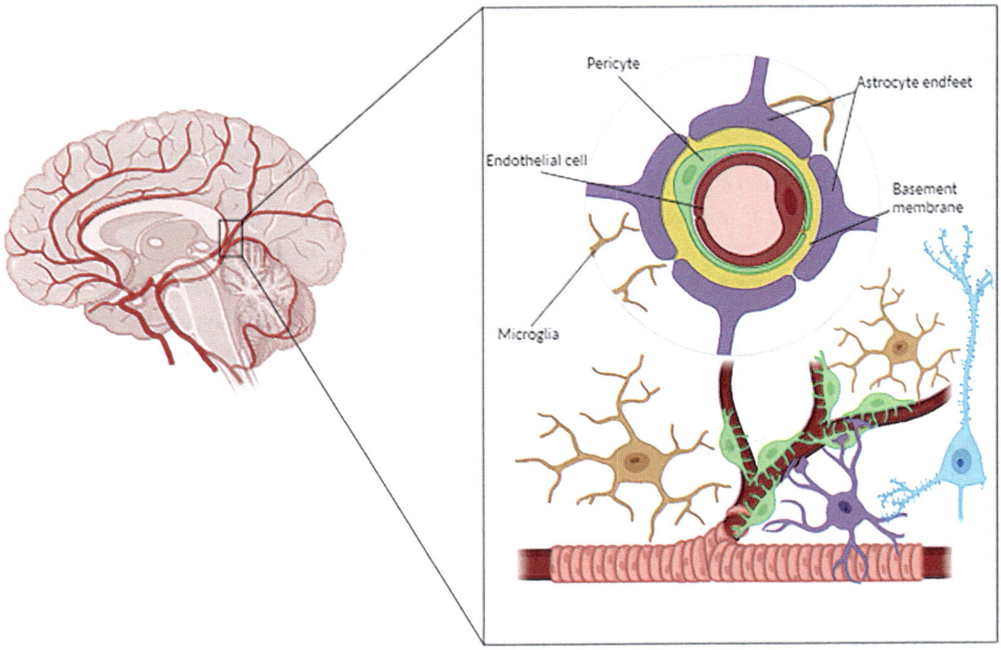

Fig. 1.1 Components of the Neurovascular Unit. (Illustration created using BioRender.com)

activity for clearing harmful foreign molecules (Sweeney et al. 2016), and actively suppress genes increasing vascular permeability or infiltration of blood-derived leukocytes (Daneman and Prat 2015; Dyrna et al. 2013).

Along with aging, there is pericyte loss and a subsequent reduction of TJ proteins (Dyrna et al. 2013). This leads to increased vascular permeability due to paracellular leakage that might contribute to the accumulation of toxic molecules in the vascular walls and CNS extracellular space, interfering with microvascular perfusion and resulting in neurodegeneration (Winkler et al. 2011; Bell et al. 2010).

1.4.3 Astrocytes

Astrocytes, a star-shaped subtype of glial cells, are the most abundant cell type in the vertebrate CNS (Vasile et al. 2017). Their strategic position between neurons and the other cells in the NVU allows them to protect the neuronal microenvironment homeostasis (Dyrna et al. 2013). Astrocytes interact with ECs through polarized end-feet projections that collectively ensheath the abluminal aspect of the cerebral microvessels (Alvarez et al. 2011a). This interaction is fundamental to providing endothelial signaling to the capillary segment, modulating blood flow and regulating local immune responses through the downregulation of proinflammatory molecules in ECs (Dyrna et al. 2013).

Astrocytes are crucial elements within the NVU, both for preserving BBB integrity and ensuring neuronal well-being. They release molecules involved in enhancing and maintaining the tightness and optimal function of the barrier (Michinaga and Koyama 2019), as well as in repairing and reducing BBB dysfunction after a brain injury (Zlokovic 2008). Along with the ECs, they synchronize metabolite levels with cerebral blood flow, and regulate brain water content through Aquaporin-4 (AQP4), the dominant perivascular water channel located in the astrocytes' endfeet (Alvarez et al. 2011a; Zlokovic 2008).

In their neuroprotective role, astrocytes guide the migration of developing neurons (Alahmari 2021), promote synapse formation (Vasile et al. 2017), control neurotransmitter concentrations and their metabolism, regulate ion concentrations, and maintain the balance of the neural microenvironment (Delaney and Campbell 2017). Despite not conducting electric signals like the neurons, astrocytes enfold synapses, modulate synaptic transmission, and ensure the continued function of neurons (Vasile et al. 2017).

1.4.4 Basement Membrane (BM)

The BBB basement membrane is a dense, sheet-like extracellular matrix (ECM) composed of proteins secreted by ECs, pericytes, and astrocytes (Xu et al. 2018). It is constructed from multiple types of highly cross-linked glycoproteins, such as laminins 4 and 5, collagen IV, fibronectin, nidogen, perlecan, and proteoglycans, among other molecules. The BM consists of two layers: an inner vascular layer produced by ECs and pericytes, and an outer glial one, synthesized by astrocytes. These layers are fused at the capillary level but separate on larger vessels (Sorokin 2010).

The BM plays a critical role in BBB permeability, not only by serving as a physical barrier and scaffold (Xu et al. 2018) but also due to its dynamic properties. It displays selective permeability to solutes and cells, modulates cell-signaling and communication among NVU components (Yao et al. 2014), helps to regulate pericyte differentiation, guides the migration and proliferation of cells involved in barrier repair, participates in neuroimmune surveillance (Engelhardt and Ransohoff 2012), and contributes to leukocyte recruitment during neuroinflammation (Sixt et al. 2001).

The BM's metabolism and degradation are mediated by matrix metalloproteinases (MMPs), which are fundamental enzymes for BBB integrity. In several neurological conditions, MMP induction has been shown to directly compromise the BBB, increasing permeability and facilitating leukocyte infiltration (Rempe et al. 2016).

Over time, the elasticity and thickness of the BM change, along with its susceptibility to degradation, thereby increasing the BBB's vulnerability during aging and especially in response to CNS damage (Delaney and Campbell 2017).

1.4.5 Microglia

Microglia are another type of glial cell distributed throughout the CNS, where they play a vital role in innate immune response (Al-Onaizi et al. 2020). Derived from yolk sac's hematopoietic progenitor cells, they migrate to the CNS during fetal development and mature there (Ginhoux and Prinz 2015). Microglia have a highly ramified appearance and constitute approximately 5–20% of all the glial cell population (Ginhoux et al. 2013).

Located in the proximity to ECs, microglia conduct immune surveillance and act as the CNS's main line of defense after crossing the BBB, phagocytizing harmful particles (Streit 2005). As part of the NVU, they also contribute to the resealing of the BBB after a vascular injury (Fernández-López et al. 2014), by stimulating TJ expression, and they assist in the repair of injured brain tissue (Michell-Robinson et al. 2015). Activated microglia can exist in two states: those activated through the M1 pathway release proinflammatory cytokines, whereas those activated through the M2 pathway are involved in tissue repair and activation of neurotrophic pathways (Guo et al. 2022).

Emerging research on BBB breakdown and neuroinflammation in CNS pathologies, such as multiple sclerosis and Alzheimer disease, suggests that microglial activation may be both a cause and a consequence of BBB dysfunction (Guo et al. 2022).

1.5 BBB Main Molecules

CNS endothelial cells (ECs) feature highly specialized paracellular and transcellular transporters, along with enzymes that degrade unwanted substances in the blood that establish contact with the BBB.

The paracellular gap functional integrity is controlled by multiprotein complexes located between neighboring ECs and is in charge of interconnecting them structurally and functionally. Disruption of ECs junctional proteins can compromise BBB integrity, allowing harmful substances to enter the brain and potentially causing neuroinflammation or toxicity (Alvarez et al. 2013; Yasuda and Mochizuki 2010).

1.5.1 Tight Junctions (TJ)

TJ protein complexes are constituted by an ensemble of multiple families of endothelial-specific transmembrane proteins connected to the ECs' cytoskeleton through anchoring cytoplasmic molecules. These TJs are essential for CNS homeostasis, as their interactions modulate the paracellular movement of molecules and ions, support the maintenance of ECs polarity (Burek et al. 2019), increase the trans-endothelial electrical resistance (TEER) (Liebner et al. 2018), and participate in the cell signaling process within NVU components (Haseloff et al. 2015). TJ-associated proteins can alter the BBB permeability under both normal and pathological conditions (Fig. 1.1).

Claudins The claudin family is composed of 24 transmembrane proteins with a molecular weight of 20–27 kDa (Hawkins and Davis 2005). Depending on the specific requirements of the tissue, cells expressing claudins often display multiple isoforms of these proteins, enabling the formation of TJs of varying sizes (Dyrna et al. 2013). In the CNS, claudins are considered the keystone of TJs and the primary factor determining the paracellular tightness between ECs (Yang et al. 2020).

The claudins expressed in CNS ECs are predominantly claudins-5 and -12 (Cardoso et al. 2010). Although claudins-1 and -3 have been shown to have an important role in sealing the BBB (Brunner et al. 2020), claudin-5 is by far the most abundant and dominant TJ component in the CNS, expressed at least 500-fold more than other family members (Nitta et al. 2003). Its paracellular sealing function is the result of its association with other proteins on the adjacent EC's surface, creating a cell-to-cell adhesion mechanism that reduces the intercellular distance to almost zero, thereby helping to maintain the integrity of the vessel structure (Greene et al. 2019).

Claudin-5 is necessary for the embryonic development of TJs and the maintenance of BBB integrity throughout life. A congenital claudin-5 deficit causes early postnatal brain swelling and death (Engelhardt and Liebner 2014). Downregulation or altered expression of claudin-5 after BBB formation is associated with increased BBB permeability, infiltration of immune cells to the CNS, and neuroinflammation (Greene et al. 2019).

Occludin This 65 kDa transmembrane protein was the first TJ protein discovered (Furuse et al. 1993). It is extensively expressed in CNS's endothelium (Hirase et al. 1997) and, together with claudins and Zonula Occludens-1 (ZO-1), is a staple component of the TJ. Occludin participates in the modulation of ECs' paracellular permeability (Sakakibara et al. 1997), not only through its junctional activity (Huber et al. 2001) but also by regulating the formation, maintenance, and functioning of other TJs (Yang et al. 2020). Additionally, occludin contributes to the regulation of EC differentiation and signal transduction (Erickson et al. 2007). Unlike with claudin-5, congenital occludin deficiency is compatible with life.

Junctional Adhesion Molecules (JAM-A,-B,-C) JAMs are transmembrane protein members of the immunoglobulin superfamily with a molecular weight of 40 kDa (Martìn-Padura et al. 1998). JAM-As are localized at endothelial and epithelial cell gaps (Martìn-Padura et al. 1998), whereas JAM-B and JAM-C are exclusively present in ECs (Aurrand-Lions et al. 2001). As part of the TJs, JAMs participate in cell-to-cell adhesion, as well as in the development and preservation of other components of TJs (Pimentel et al. 2020).

JAM proteins are also implicated in establishing the apical-basal EC axis and play a prime role in the regulation of immune cell traffic through the BBB by modulating the leukocyte-EC interaction, leukocyte adhesion, and migration (Johnson-Léger and Imhof 2003).

Zonula Occludens (ZO) The ZO family are membrane-associated intracytoplasmic proteins with a molecular mass ranging from 130 to 220 kDa (Stevenson et al. 1986). These molecules function as scaffolding elements, providing the structural foundation for the formation of functional, stable, and lasting TJs. ZO proteins link and bind to various TJ components, anchoring them to the EC's actin cytoskeleton (Fanning et al. 1998). ZO proteins also play important roles in cellular signal regulation, such as cell proliferation and motility (Gumbiner et al. 1991).

There are three types of ZO proteins: ZO-1, ZO-2, and ZO-3. ZO-1 was the first described cytoplasmic component identified as an anchor for TJs. It interacts with claudin-5, occludin, and the JAM family (Shin and Margolis 2006). ZO-1 has also been found to associate with other junctional proteins, such as those in adherens junctions, and gap junctions (Itoh et al. 1993). The dissociation of ZO-1 from their junctional proteins directly correlates with increased BBB permeability (Fischer et al. 2002). Recent studies suggested that in the absence of ZO-1, ZO-2 can compensate, forming morphologically normal TJ (Umeda et al. 2004).

1.5.2 Adherens Junctions (AJ)

AJs are protein complexes located on the abluminal side of ECs, below TJs. Structurally similar to TJs, AJs consist of transmembrane proteins, called cadherins, and intracytoplasmic catenins (Guo et al. 2007).

Cadherins are responsible for cell-to-cell and cell-to-matrix adhesion. In the CNS endothelium, vascular endothelial (VE)-cadherin is the predominant AJ protein (Stamatovic et al. 2016). Catenins, including p120, Β-catenin, and plakoglobin act as scaffolding proteins, linking the transmembrane portion of the AJ complex to ZO-1 and the actin cytoskeleton (Stamatovic et al. 2016). In addition to their structural role, catenins also participate in regulating cellular signaling processes (Guo et al. 2007). The cross-talk between components of AJs and TJs is essential for establishing and regulating the paracellular barrier (Keaney and Campbell 2015).

1.5.3 Transporters

Due to its high selectivity, only a limited number of molecules can freely cross the BBB. Features such as a strong affinity for hydrogen bond formation, flexible rotatable bonds, and an extensive polar surface area generally prevent particles from entering the CNS (Bodor and Buchwald 2002). Passive diffusion is restricted to water, gases, particles with less than 10 hydrogen bonds (Patching 2017), and lipophilic solutes smaller than 180 Da (Liu et al. 2004).

Hydrophilic nutrients and metabolites with structures that do not permeate the BBB require transcellular transport systems. Essential molecules, like glucose, fatty acids, and amino acids, move across the BBB through carrier-mediated transport mechanisms (Miller 2015). CNS endothelium is densely populated with transporters. By relying on active transport, the BBB maintains strict regulation over molecular traffic. Transporters can be divided into two distinct classes: solute carriers and efflux transporters (Ghersi-Egea et al. 2015; Bouldin and Krigman 1975).

Paracellular Transport It involves the movement of small, dissolved molecules through the gaps between ECs, driven by a concentration gradient from the bloodstream to the brain tissue. TJs, acting as gatekeepers of the intercellular space, limit the passage of hydrophilic molecules larger than 2–4 Å (Table 1.1) (Chen and Liu 2012).

Transcellular Transport This process refers to the transit of solutes across the ECs. Gaseous molecules such as oxygen, carbon dioxide, nitro-

Table 1.1 Paracellular gap structural components

Category	Description	Function	Proteins	Features
Tight junctions (TJ)	Protein complexes located in the endothelial cells' paracellular gap, forming a seal between cells	Regulate paracellular permeability and maintain barrier integrity	Claudins	Essential for forming the tight seal in junctions Claudin-5 is critical for BBB integrity, ensuring selective permeability
			Occludin	Regulate paracellular permeability and interact with signaling pathways
			Junctional adhesion molecules	JAM-A, -B, -C. Facilitate TJ assembly and regulate leukocyte migration
Scaffolding proteins	Proteins anchoring TJs to the cytoskeleton and maintaining its structural integrity	Support junctional complexes and coordinate cell signaling	Zonula Occludens (ZO)	ZO-1, ZO-2, ZO-3, interact with transmembrane TJ proteins, ensuring structural integrity
Adherens junctions (AJ)	Protein complexes mediating cell-cell adhesion and communication	Facilitate cellular adhesion and contribute to mechanical strength	Cadherins	Calcium-dependent adhesion proteins, like VE-cadherin are critical for endothelial barrier integrity. They mediate homophilic cell-cell adhesion.
			Catenins	p120, Β-catenin, and Plakoglobin link cadherins to actin cytoskeleton, stabilizing AJs and participating in signaling pathways
Cytoskeletal	Cytoskeletal elements providing structural support	Stabilize cellular junctions and maintain cytoskeletal dynamics	Actin filaments	Provide mechanical support and interact with adhesion proteins to stabilize cellular junctions Play a role in cytoskeletal rearrangement and barrier regulation

gen, and alcohol traverse the BBB through this mechanism (Verscheijden et al. 2019). Although small lipophilic particles, such as steroids, can move freely through EC, their traffic is regulated by efflux pumps, the ATP-Binding Cassette Family (ABC) of transporters (Hindle et al. 2017). These pumps attach to and remove excessive, unwanted, or potentially harmful liposoluble substrates, both xenobiotics and endogenous, back into the bloodstream (Campos-Bedolla et al. 2014).

Transporter Protein Pathway (Carriers) This active transport mechanism involves both solute carriers and ABC transporters. Solute carriers, expressed in the EC membranes, facilitate a bidirectional movement of metabolites, such as glucose, amino acids, ions, and fatty acids, across the EC and between the bloodstream and the CNS (Shimizu et al. 2008).

ABC transporters, primarily located on the luminal aspect of the EC and also present on pericytes and astrocytic endfeet, actively move a broad range of substrates against their concentration gradients, returning them to the systemic circulation (Patel and Patel 2017).

Transcytosis ECs employ multiple specific and unspecific transcytotic mechanisms to facilitate the transport of other hydrophilic macromolecules across the BBB and into the CNS (Zhang et al. 2020). The transcytosis pathway comprises three key steps: endocytosis, intracellular vesicular trafficking, during which lysosomal degradation is bypassed, a unique feature of the BBB's endothelium, and exocytosis. This vesicle-mediated process includes receptor-mediated and adsorptive-mediated transcytosis (Miller 2015).

- *Receptor-mediated transcytosis:* In this process, specific receptors on the luminal surface of EC bind to ligands, initiating internalization and subsequent transport across the cell to the abluminal side (Pardridge 2005). This mechanism is responsible for the entry of peptides and large molecules, like insulin, transferrin, and low-density lipoproteins, into the brain (Ding et al. 2020).
- Receptor-mediated transcytosis is the main mechanism that allows Insulin-like Growth Factor 1 (IGF-1) to cross the BBB. This book talks more in detail about how IGF-1 plays a vital role in the brain and body, acting homeostatically, to promote growth and repair. *See the Chap. 4 "Insulin like Growth Factor-1 (IGF1) as brain-body connector."*
- *Adsorptive-mediated transcytosis:* This pathway facilitates the movement of macromolecules and charged nanoparticles across the BBB via electrostatic interactions between the ligands and EC's glycocalyx. This method supports the binding and uptake of cationic molecules, lipids, albumin, and nanoparticles on the luminal surface of ECs (Pardridge 2005; Ding et al. 2020).

Cell-Mediated Transport In this pathway, molecules such as drugs cross the BBB encapsulated in liposomes within peripheral blood leukocytes as they migrate into the CNS, both under physiological and pathological conditions (Ransohoff and Engelhardt 2012).

Trans-Endothelial Leukocyte Migration The CNS is considered an immune-privileged organ due to minimal leukocyte infiltration under healthy conditions, as a result of the modulatory role played by the BBB. Immune surveillance occurs continuously at the blood-CSF interfaces, including the meninges and choroid plexus (Greenwood et al. 1995).

During immunopathological conditions, regional leukocyte infiltration into the CNS occurs in response to danger signals expressed by the NVU components. The trans-endothelial migration (TEM) of immune cells across the BBB is a dynamic process initiated by the activation of ECs, leading to the release of proinflammatory molecules (Carman and Martinelli 2015), and upregulation of cell adhesion molecules on their luminal surfaces. This process is followed by leukocyte tethering, rolling, crawling, arrest, and diapedesis across the ECs (Winger et al. 2014). Although its precise mechanism remains unclear, it is known to involve TJ remodeling and reorganization of the endothelial cytoskeleton (Carman and Martinelli 2015).

1.6 BBB Dysfunction

As a dynamic system composed of multiple interacting components, even a minor change in one of BBB's elements can lead to altered function and the loss of neural homeostasis (Profaci et al. 2020). BBB dysfunction may result from structural damage, such as direct destruction of ECs, or from pathological processes including inflammation, hypoxia, excitotoxicity, lipid peroxidation, and calcium-mediated injury (Sun et al. 2019). The compromise in BBB's integrity triggers ion dysregulation, edema, neuroinflammation, and neurotoxicity, which elevate intracranial pressure and cause cellular dysfunction and degeneration, particularly in neurons (Shen et al. 2019).

BBB disruption is associated with CNS diseases such as Alzheimer disease (AD), multiple sclerosis (MS), stroke, epilepsy, encephalitis, traumatic brain injury (TBI), and amyotrophic lateral sclerosis (ALS) (Profaci et al. 2020). Also for neurodevelopmental disorders, it has been suggested a role for the disregulated passage of immune cells across the BBB, *as mentioned by Pangrazzi and colleagues in the Chap. 10* "Brain-resident immune cells in neurodevelopmental disorders." In some of these conditions, BBB impairment has been identified as an early and significant factor in disease pathogenesis, while in other conditions, its role remains controversial and subject to ongoing investigation. Additionally, BBB compromise can result from physiological processes like aging and menopause, making the

CNS more vulnerable to external insults (Delaney and Campbell 2017).

Factors like age, sex, temperature, and stress can promote BBB disruption. Initial damage is often mediated by cytokines and inflammatory molecules, including tumor necrosis factor-alpha (TNF-α), interferon-gamma (IFN-γ), interleukin-1 (IL-1), interleukin 6 (IL-6), reactive oxygen and nitrogen species (ROS and RNS), MMPs, bradykinin, substance P, thrombin, and endothelin-1 (Almutairi et al. 2016). These agents target different cellular and molecular components of the BBB, leading to structural and functional compromise (Zhao et al. 2021).

Following the BBB disruption, paracellular and transcellular permeability increase driven by dysregulation of TJ, ECs' cytoskeleton, transporters, transcytosis mechanism, and gene expression. Oxidative damage occurs to proteins, lipids, and DNA (Pun et al. 2009). At the same time, immune cells infiltrate the CNS due to the upregulation of adhesion molecules and pro-inflammatory mediators, after a cascade of aberrant cell signaling (Kvichansky et al. 2019). All this causes fluid extravasation, neuroinflammation, and neuronal dysfunction, further exacerbating the damage caused by the initial insult (Takeshita and Ransohoff 2012).

The mechanisms behind BBB failure and its involvement in the onset, progression, and of diseases remain poorly understood but under investigation.

1.6.1 BBB Biomarkers

Quantification and analysis of BBB-associated molecules, along with advanced imaging techniques, are nowadays used as biomarkers to assess BBB integrity and functioning (Hussain et al. 2021). The discovery and understanding of these biomarkers have progressed with advances in molecular biology, immunoassays, and neuroimaging, establishing them as reliable indicators of BBB status (Table 1.2) (Hussain et al. 2021). These biomarkers hold potential for understanding, diagnosing, and monitoring CNS diseases where BBB disruption is a hallmark, such as Alzheimer disease, multiple sclerosis, and epilepsy (this includes, in the latter case, microRNAs as epilepsy biomarkers covered *in Henshall's Chap. 12 "MicroRNAs fine-tune brain and body communication in health and disease")*. Here are some of the main serum biomarkers of BBB dysfunction:

Brain Microvascular Endothelial Cells (BMEC) Any injury to the BBB can cause ECs to exfoliate and circulate in the bloodstream. As these cells are key components of the BBB, the quantification of circulating BMECs in the blood may serve as an indicator of the extent of BBB degradation (Huang et al. 2013).

Cellular Fibronectin (c-Fn) ECs produce and release c-Fn, which is essential for maintaining BM stability. When there is a BM rupture, c-Fn is released into the bloodstream, promoting leukocyte migration to the damaged area (Miao and Liao 2014). Since c-Fn is primarily located within ECs and along the luminal side of the BM, elevated plasma c-Fn levels serve as a marker of endothelial damage at the BBB (Li et al. 2018).

Soluble Cell Adhesion Molecules (CAMS) Soluble intercellular adhesion molecule-I (sICAM-I), soluble vascular cell adhesion molecule-I (sVCAM-I), platelet endothelial cell adhesion molecule (PECAM-I), endothelial-leukocyte adhesion molecule 1 (E-selectin) and P-selectin, are involved in leukocyte adhesion and migration through the BBB. These molecules are upregulated in the event of BBB compromise, indicating enhanced immune cell trafficking across the barrier and neuroinflammation (D'Ambrosio et al. 2015).

S100B This calcium-binding protein participates in cell differentiation, migration, and cognitive processes such as learning and memory (Delaney and Campbell 2017). It is primarily produced by astrocyte endfeet, and its levels rise significantly following a BBB injury, as it is released into the bloodstream and CSF (Kadry et al. 2020).

Table 1.2 BBB biomarkers

Biomarker	Description	Change under pathological conditions
Brain microvascular endothelial cells (BMEC)	Key components of the BBB; injury leads to exfoliation and circulation in the bloodstream	Increased circulating BMECs indicate the extent of BBB degradation
Cellular fibronectin (c-Fn)	Produced and released by endothelial cells; essential for basement membrane, stability and promotes leukocyte migration upon BM rupture	Elevated plasma c-Fn levels signify endothelial damage at the BBB
Soluble cell adhesion molecules sICAM-I, sVCAM-I, PECAM-I, E-selectin, P-selectin	Facilitates leukocyte adhesion and migration through the BBB during inflammation	Elevated levels are markers of increased leukocyte trafficking and BBB integrity compromise
S100B	Calcium-binding protein involved in cell differentiation, migration, and cognitive processes; primarily produced by astrocyte endfeet	Increased levels in serum and CSF indicate BBB injury and astrocyte activation or damage
Glial fibrillary acidic protein (GFAP)	Cytoskeletal protein expressed by astrocytes; serves as a marker of astrocytic activation or injury	Elevated levels in serum or CSF reflect astrocyte damage and BBB disruption
Vascular endothelial growth factor (VEGF)	Signaling protein involved in angiogenesis, vascular permeability, and barrier integrity; produced by astrocytes and ECs	Increased levels in response to hypoxic, inflammatory, or traumatic stimuli can disrupt BBB integrity
Albumin quotient (CSF/serum albumin ratio)	Ratio of albumin levels in cerebrospinal fluid (CSF) to serum; albumin normally has low levels in CSF under physiological conditions	An elevated ratio indicates increased BBB permeability and dysfunction
Matrix Metalloproteinases (MMP-2 & MMP-9)	Endopeptidases involved in tissue development, regeneration, and BBB remodeling; facilitate the degradation of tight junctions and basement membrane substrates	Overexpression leads to BBB barrier disruption, increased permeability, neuroinflammation, and further tissue damage

Soluble Intercellular Adhesion Molecule-I (sICAM-I), Soluble Vascular Cell Adhesion Molecule-I (sVCAM-I), Platelet Endothelial Cell Adhesion Molecule (PECAM-I)

Glial Fibrillary Acidic Protein (GFAP) GFAP is a cytoskeletal structural component, primarily expressed by astrocytes. It serves as a marker of astrocytic activation or injury. Elevated levels of GFAP in serum or CSF indicate astrocyte damage and are associated with BBB disruption. GFAP quantification is particularly useful for assessing the severity of BBB compromise and the extent of glial response (Kadry et al. 2020).

Vascular Endothelial Growth Factor (VEGF) It is a key signaling protein involved in angiogenesis promotion, vascular permeability, and barrier integrity. When produced by astrocytes, VEGF supports BBB maintenance and angiogenesis, while when secreted by ECs, it contributes to vascular permeability and repair (Profaci et al. 2020). VEGF levels increase in response to hypoxic, inflammatory, or traumatic stimuli, and its presence can disrupt the BBB integrity. Since its levels change rapidly in response to injury or disease progression, it provides a real-time reflection of BBB status (Zhao et al. 2021).

Albumin Quotient (CSF/Serum Albumin Ratio) Human serum albumin accounts for approximately 50% of all blood plasma proteins. While its primary role is to regulate blood oncotic pressure, it also binds to and transports various ligands. Under normal physiological conditions, albumin levels in the CSF are low. However, in cases of BBB dysfunction, increased permeability allows albumin to pass into the CNS. The CSF-to-serum albumin ratio is a well-established biomarker in clinical and research settings to assess BBB integrity, with elevated levels indicating increased BBB permeability (Wallin et al. 2018).

Matrix Metalloproteinases (MMPs) These endopeptidases play an important role in the physiological processes of tissue development and regeneration, including cell migration, angiogenesis, tissue formation, neuronal network formation, and BBB remodeling. They are also implicated in pathophysiological processes, like healing and inflammation. MMPs facilitate the degradation of TJ-related molecules and BM substrates. Under normal physiological conditions, MMP activity is tightly regulated; however, in pathological states, the overexpression of MMPs, specifically MMP-2 and MMP-9, can lead to barrier disruption (Kumar and Sandhir 2022). As a result, there is increased permeability that allows harmful molecules and immune cells to infiltrate the CNS, contributing to neuroinflammation and further tissue damage. MMPs have been recognized as both biomarkers and therapeutic targets in conditions where BBB dysfunction is a central feature (Hussain et al. 2021).

Dynamic Contrast-Enhanced Magnetic Resonance Imaging (DCE-MRI) While serum biomarkers can reliably indicate and quantify BBB leakage, they do not provide information on the specific location of the disruption, or whether the permeability is regional or global. Recently developed DCE-MRI, an MRI technique employing gadolinium-based contrast agents, allows for the detection, localization, and quantification of BBB leakage (Raja et al. 2018).

Gadolinium-based contrast agents enhance the visibility of blood vessels and permeable tissues during imaging. Under normal conditions, these agents do not cross an intact BBB due to their hydrophilic and polar properties, as well as their reliance on albumin transport. However, they may extravasate from the bloodstream into the brain tissue when the BBB is compromised, even partially (Löscher and Friedman 2020). DCE-MRI evaluates the BBB permeability by measuring the rate and extent of contrast agent penetration into brain tissues, providing valuable information about its distribution and concentration. This non-invasive imaging technique, increasingly utilized in research and clinical settings, offers real-time data on vascular dynamics and BBB changes over time, enabling the monitoring of disease progression and therapeutic response (Raja et al. 2018) (Fig. 1.2).

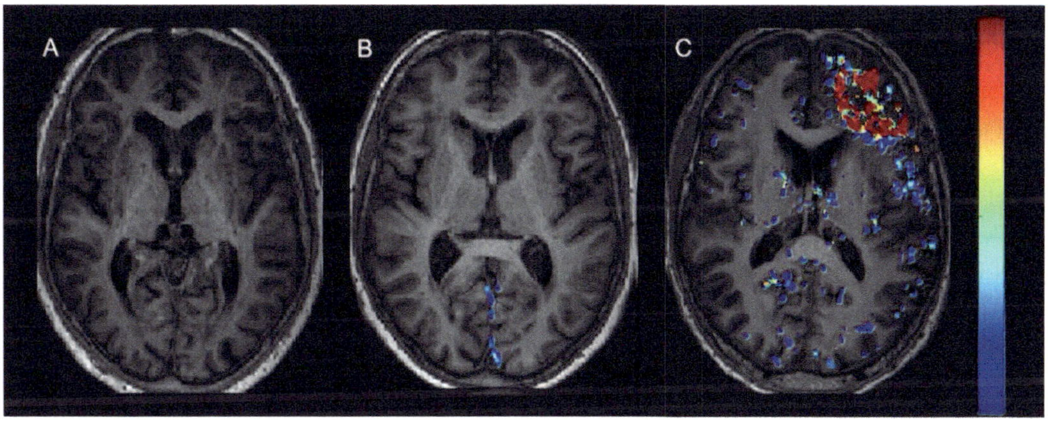

Fig. 1.2 These DCE-MRI images illustrate different patterns of BBB leakage. (**a**) MRI of a young individual with no evidence of BBB dysfunction. (**b**) Mild leakage was observed in the medial-posterior cortex of a healthy aging individual. (**c**) Significant disruption of the BBB in the left frontal region, seen in a patient with a high-grade glioma. (Data obtained from ongoing in-house studies conducted at Neurovascular Genetics Laboratory, Trinity College Dublin)

1.7 BBB under Physiological Conditions

1.7.1 BBB and Sex

Biological sex strongly influences the permeability, integrity, and response of the BBB to stress and injury. Sex hormones, particularly estrogen and testosterone, are key regulators of BBB structure and function. Estrogen enhances BBB integrity, while testosterone is necessary for its normal functioning. These hormones modulate the activity of TJ and BBB transport proteins, influence the inflammatory response profile (Erickson et al. 2018), and facilitate cell communication. Such differences might contribute to potential sex-specific susceptibilities to various neurological and psychiatric conditions, including the development and progression of neurodegenerative diseases (Moon et al. 2021).

In general, most research has shown that females tend to exhibit better BBB integrity compared to males. Estrogen exerts a protective role in maintaining CNS vasculature and BBB integrity. Its effects are mediated through membrane-bound receptors, ERα and ERβ, expressed in the brain endothelium. Estrogen enhances nitric oxide (NO) production, which dilates cerebral vasculature, improving blood flow in response to metabolic demands (Weber and Clyne 2021). At the BBB level, estrogen promotes EC preservation, upregulates TJ protein expression (Maggioli et al. 2016; Pellegrini et al. 2023), and reduces TJ disruption (Shin et al. 2016). An animal study found that female mice have higher mRNA expression of claudin-1, -5, and -12, occludin, JAM-A, and ZO-1 compared to males (Weber and Clyne 2021; Kamm 2019). Estrogen is also a potent astrocyte regulator. At baseline, it affects the growth of astrocyte processes (Garcia-Segura et al. 1989), and modulates intracellular calcium levels (Chaban et al. 2004), influencing cell-to-cell communication within the BBB. Estrogen's action on astrocytes is linked to several neuroprotective pathways, including reduced neuroinflammation and maintaining neuronal homeostasis (Moon et al. 2021).

Estrogen mitigates BBB inflammation by upregulating annexin A1 (ANXA1), an anti-inflammatory mediator, intercellular adhesion molecule 1 (ICAM-1), and VCAM-1. This promotes TJ protein expression and reduces lymphocyte migration across the barrier (Maggioli et al. 2016). Furthermore, estrogen attenuates the expression of MMPs, helping to preserve TJ and BM integrity and maintaining BBB stability (Na et al. 2015).

During menopause or in conditions like ovariectomy, estrogen-dependent NO production declines, potentially contributing to the deterioration of female cerebrovascular health and a 500-fold increase in BBB permeability (Wilson et al. 2008). In animal models, ovariectomized female mice display drastically higher levels of brain proinflammatory cytokines, suggesting that low estrogen levels may drive BBB breakdown and neuroinflammation (Dion-Albert et al. 2022). Although associated with increased cardiovascular morbidity and risk for stroke in some studies (Wilson et al. 1985), estrogen replacement therapy in postmenopausal females has shown to restore BBB integrity (Dion-Albert et al. 2022).

Estrogen levels fluctuate significantly with age and reproductive cycles, adding complexity to the research into its role in BBB regulation. The mechanisms by which fluctuations in estrogen levels affect BBB permeability over time, age, and dosage remain to be fully elucidated (Dion-Albert et al. 2022).

In contrast to estrogen, fewer studies have explored the impact of androgens on neurovascular function. Research on testosterone's role in vascular health has shown that it promotes endothelial and smooth muscle cell proliferation and migration (Wilson et al. 1985). Growing evidence suggests that lower testosterone levels in men exacerbate cardiovascular disease and increase stroke risk (Campelo et al. 2012).

Under physiological conditions, testosterone plays a protective role in BBB integrity. In animal models, chronic testosterone depletion leads to increased BBB permeability, reduced expression of claudin-5 and ZO-1, and TJ disorganization. This permeability is associated with neuroinflammation, evidenced by the upregulation of inflammatory markers (Muller et al. 2004). Under pathological conditions, such as stroke, testosterone has been shown to impair BBB integrity by downregulating claudin-5 and upregulating the proinflammatory cytokines (Atallah et al. 2017). However, testosterone supplementation in castrated male animals restored BBB integrity and reduced inflammation, further highlighting its protective role (Abi-Ghanem et al. 2020).

Excessive testosterone levels may negatively affect BBB integrity. For example, a study examining the effects of endurance training combined with testosterone supplementation in rats revealed that testosterone treatment reduced the expression of claudin-5, occludin, and VE-Cadherin compared to non-supplemented animals, suggesting potential adverse impacts on barrier function (Nierwińska et al. 2019).

1.7.2 BBB and Temperature

In humans, normal body temperature in humans oscillates between 36 and 37.5 °C (Rzechorzek et al. 2022), while brain temperature tends to be slightly higher, around 38.5 °C, due to the brain's high metabolic rate (Wang et al. 2016). Recent studies suggest that the brain temperature varies by region and age, following a circadian rhythm that reaches its lowest point at night (Rzechorzek et al. 2022).

Hypothermia is defined as a core body temperature below 35 °C (Cheshire Jr. 2016), and is commonly caused by exposure to cold weather, water immersion, drug-induced disruptions in thermoregulation, metabolic and endocrine disorders like hypothyroidism and adrenal insufficiency, and trauma, due to blood loss and shock (McCullough and Arora 2004). Hyperthermia, on the other hand, occurs when body temperature exceeds 40.5 °C (Cheshire Jr. 2016). It is often triggered by infections, environmental heat exposure, physical activity, adverse drug reactions to sympathomimetics, anticholinergics, and serotonergic agents, and medical conditions such as malignant hyperthermia and heat stroke (Cheshire Jr. 2016).

Temperature plays a critical role in regulating BBB permeability, altering water homeostasis, structure, and viability of CNS cells, which are sensitive to thermal damage. These effects are highly dependent on the duration of temperature exposure (Laptook 2014).

The impact of hypothermia on the neuronal tissue and BBB has been predominantly studied in the setting of brain injuries like TBI, ischemic encephalopathy, and stroke (Yenari and Hemmen 2010). Therapeutic hypothermia therapy (THT), achieved through systemic cooling or selective head cooling, has shown to protect BBB integrity. Studies suggest that reducing brain temperature between 32 and 34 °C for 12–24 h can provide neuroprotection by enhancing the BBB function, lowering cerebral metabolic rates, and reducing inflammation and apoptosis (Kuczynski et al. 2020; Bhatt et al. 2020).

THT started within one hour of a hypoxic-ischemic event and concomitant with reperfusion treatment, has shown to reduce the expression of MMP-2, MMP-9, and vascular endothelial growth factor (VEGF), while promoting the activity of tissue inhibitors of metalloproteinases (TIMPs). This supports the preservation of TJs, AJs, cell-BM integrity, and overall BBB structure (Laptook 2014; Li et al. 2017). Additionally, pericyte morphology and its umbrella-like coverage over the EC gaps are maintained, which promotes the maintenance of vascular perfusion to ischemic tissue (Laptook 2014).

Extreme hypothermia can cause BBB disruption and selective glial activation due to brain dehydration (Kiyatkin and Sharma 2009).

Hyperthermia generally has detrimental effects on the CNS, increasing cerebral metabolic rate, oxidative stress, and inflammation (Shepley et al. 1985). Studies in animals, where subjects are exposed to prolonged whole-body hyperthermia starting at 38 °C, showed BBB disruption, with increased paracellular transport in multiple brain regions, as well as microhemorrhages in the youngest subjects, but not in the adult ones (Zhao et al. 2023). Temperatures above 38.5 °C result in increased albumin and GFAP immunoreactivity, indicating astrocytic activation and BBB breakdown. Peak damage occurs at temperatures of

41–42 °C, marked by a substantial increase in brain water and electrolytes, as well as cellular abnormalities, indicative of acute edema and irreversible damage (Kiyatkin and Sharma 2009). In vitro studies have revealed that cell death occurs at temperatures above 42 °C, though viability varies depending on cell type and duration of heat exposure (Zhao et al. 2023).

Therapeutic hyperthermia involves controlled increases in temperature and is utilized for the treatment of various diseases. Systemic therapy typically raises body temperature by 1–1.5 °C, which increases blood flow and might benefit prevention and treatment of cardiovascular diseases, particularly heart failure (Ye et al. 2020). In neuro-oncology, local hyperthermia therapy, including hyperthermic chemotherapy and radiofrequency ablation, is employed to treat brain tumors by inducing cell death and increasing BBB permeability in the targeted regions (Skandalakis et al. 2020).

1.7.3 BBB and Stress

Stress is characterized by the disruption of homeodynamic balance in response to physical, psychological, intrinsic, or extrinsic stimuli perceived as challenging. The stress response system helps to restore balance by adjusting energy distribution to meet the organism's needs. While short-term stress promotes adaptive responses, prolonged or excessive stress can overwhelm the system's ability to regulate itself, leading to maladaptive effects and a persistently imbalanced state (Agorastos and Chrousos 2022).

Psychosocial stress affects all bodily systems, and the BBB is not immune to its effect. The hypothalamic-pituitary-adrenal axis and inflammatory signaling pathways mediate stress-induced BBB disruption. Elevated cortisol levels and pro-inflammatory cytokines prompt a series of NVU alterations that affect the BBB dynamics, increasing its permeability (Welcome and Mastorakis 2020). Notably, stress downregulates claudin-5 expression in ECs, directly compromising TJs' integrity. This effect shows a sex-dependent topographic pattern in animal models and post-mortem studies (Dion-Albert et al. 2022). In females, the stress-induced loss of claudin-5 occurs predominantly in the prefrontal cortex, responsible for executive function and decision-making, whereas in males, it primarily affects the nucleus accumbens, fundamental modulating stress responses, and mood disorders (Dion-Albert et al. 2023).

Stress also induces pericyte proliferation, which exhibits a dysfunctional reduction in the expression of key markers, like α1-integrin, essential for maintaining BBB stability. This leads to a hindered pericyte-endothelium interaction, further perpetuating neurovascular dysfunction (Bhowmick et al. 2019). Astrocytic regulation of neuronal excitability and synaptic transmission is also affected by stress, which induces a reduction of glutamate uptake, leading to excitotoxicity and neuronal damage (David et al. 2009). In addition, downregulation of potassium (K+) channel Kir4.1 increases extracellular K+, promoting neuronal depolarization, and hyperexcitability (Nwaobi et al. 2016). Stress also induces astrocytic overexpression of VEGF and cytokines, exacerbating barrier disruption and inflammatory signaling (Dion-Albert et al. 2023). Microglia activate and adopt a pro-inflammatory phenotype when facing stress, increasing the expression of inflammatory cytokines and oxidative stress markers. Additionally, microglia can phagocytose astrocytic end-feet, further compromising BBB function (Ronaldson and Davis 2020).

Chronic social stress and sub-chronic variable stress induce sex-specific BBB changes in the BBB, with females showcasing more pronounced BBB leakiness and associated depressive behaviors (Dion-Albert et al. 2022). Chronic stress and sustained BBB disruption can lead to cortical atrophy, particularly in the limbic system (Sathyanesan et al. 2017), impaired neurogenesis, and cognitive deficits (Lee et al. 2018). However, studies in animal models have shown that treatment with the antidepressant imipramine for 35 days restored BBB integrity despite chronic stress exposure (Dion-Albert et al. 2022).

1.7.4 BBB and Aging

Aging is a time-related physiological process, consisting of the deterioration of cellular functions necessary for survival and reproduction, ultimately resulting in organ failure and death. A poorly preserved cellular identity, the accumulation of errors in DNA replication and repair, and reduced metabolic clearance create a stressed and vulnerable cellular environment (Vijg 2000). In the CNS, aging causes significant changes in the vasculature. Capillary density decreases (Reeson et al. 2018), along with the neovascularization capacity (Rivard et al. 2000), and overall cerebral blood flow (Tarumi and Zhang 2018). In addition to the direct effects of aging on the NVU cellular components, the regulatory function of the BBB declines, increasing its permeability and disrupting the CNS homeostasis (Hyman et al. 2012).

Even in healthy aging, the BBB can exhibit degenerative features. The CNS endothelium becomes progressively dysfunctional, losing its secretory capacity and ability to supply growth and nutritional factors to the brain (Marques et al. 2013). There is a decline in the expression of occludin, claudin-5, and ZO-1, leading to TJ breakdown (Elahy et al. 2015). Simultaneously, TJ and basal lamina-degrading enzymes become overactive (Costea et al. 2019). Impaired paracellular and transcellular transport regulation allows the entrance of molecules such as albumin, fibrinogen, IgG, and autoantibodies into the brain (Montagne et al. 2017). Additionally, defective efflux transporters fail to clear these substances, leading to the build-up of toxic molecules (Xu et al. 2022). The BM thickens, and the expression of its components changes, like increased production of collagen and MMP-9 (Candiello et al. 2010). The pericyte population is reduced as a consequence of the suppression of cell viability and upregulation of apoptosis (Schultz et al. 2018). Pericytes also undergo structural changes, including enlarged mitochondria, vesicular inclusions, and foamy transformations (Hussain et al. 2021). These senile pericytes generate ROS and RNS, exacerbating oxidative stress and amplifying BBB dysregulation (Sweeney et al. 2016). Astrocytes' end-feet also display age-related changes. While their size increases, their vascular coverage diminishes, alongside reduced expression of AQP4. This is accompanied by elevated GFAP expression, indicating increased reactive astrogliosis (Middeldorp and Hol 2011).

Aging is further marked by a chronic, low-grade inflammatory state, which contributes to BBB dysfunction. Elevated levels of pro-inflammatory cytokines such as IFN-γ, IL-1ß, IL-6, and TNF-a, along with activated microglial and peripheral immune cells, have been identified in uninjured aged brains, resembling the inflammation state after a TBI (Kumagai et al. 2007). In this aged CNS, response to an insult is more severe, due to the amplification of a pro-inflammatory response and the attenuation of its anti-inflammatory counterpart (Delaney and Campbell 2017).

The hippocampus has been identified as one of the first brain regions to experience BBB breakdown in aging and cognitively normal populations, particularly in the CA1 and dentate gyrus areas (Montagne et al. 2015). This is thought to be related to a baseline hippocampal BBB vulnerability due to a dual blood supply from anterior and posterior cerebral circulation and greater neuronal sensitivity to aging and damage (Ivanidze et al. 2019). However, the extent of BBB leakage is small and variable among individuals (Knopp et al. 2023). Further studies have reported variable BBB leakage, with some findings showing increased permeability in the orbitofrontal and cingulate cortex (Verheggen et al. 2020), and occipital white matter and thalamus (Ha et al. 2021). It has also been suggested that BBB disruption increases during normal aging secondarily to white matter integrity loss and a decrease in cortical thickness (Wang et al. 2003).

In healthy individuals, these age-related BBB changes are generally mild and occur as part of a physiological process. While they do not cause pathology on their own, they may predispose the

CNS to chronic inflammation and long-term damage, increasing the risk of developing age-associated neuropsychiatric and degenerative disorders (Delaney and Campbell 2017).

1.8 BBB under Pathological Conditions

Alzheimer disease (AD), multiple sclerosis (MS), and epilepsy are among the neurological conditions with the highest incidence and prevalence worldwide. While the pathophysiology of these disorders continues to be elucidated, it is clear that BBB dysfunction plays a significant role, either by contributing to disease onset or as a consequence of its development.

1.8.1 BBB and Alzheimer Disease

Alzheimer disease (AD) is an irreversible neurodegenerative condition and the leading cause of dementia worldwide, accounting for approximately 60–70% of all cases. AD commonly manifests after age 60 and progresses over 8–12 years. Clinically, it is characterized by amnestic cognitive decline and progressively impaired daily functioning. The disease is defined by the accumulation of β-amyloid (Aβ) plaques and neurofibrillary tangles in the brain, which result in neuronal death, synaptic loss, and brain atrophy, corresponding to its clinical manifestations (Kamatham et al. 2024). AD development involves a complex interaction between genetic and environmental factors (Bagheri et al. 2024).

At the molecular level, two key pathological features of AD have been identified: the accumulation of β-amyloid plaques and the formation of tau neurofibrillary tangles. Aβ plaques are extracellular deposits primarily composed of Aβ, a peptide derived from the amyloid precursor protein (APP). Recent studies suggest that, in low concentrations, Aβ modulates synaptic plasticity and memory (Puzzo and Arancio 2013). Among Aβ1–40 and Aβ1–42, the most common peptide forms (Sehar et al. 2022), Aβ1–42 is especially important in AD pathogenesis, as it is the one that aggregates extracellularly, triggering a cascade of toxic events leading to synaptic dysfunction, neuronal death, and neurodegeneration (Sehar et al. 2022). The precise mechanisms leading to Aβ pathologic accumulation and AD pathology remain an area of active investigation.

While Aβ drives extracellular pathology, intracellular changes are dominated by tubulin-associated unit (Tau) protein dysfunction. Tau is a cytoskeleton protein with a primary role in maintaining the stability of microtubule neuronal axons. Tau is abundant in the CNS, predominantly in the cerebral cortex (Shin et al. 1991). In AD, Tau becomes hyperphosphorylated, losing its stabilizing ability and aggregating in intracellular neurofibrillary tangles, impairing synaptic function, and promoting neuronal death (Alonso and Cohen 2018).

Beyond these protein abnormalities, AD is also characterized by early disruptions in the BBB, which often precede clinical symptoms and neurodegeneration (Nation et al. 2019). Several mechanisms contribute to BBB breakdown in AD, including oxidative stress, inflammation, and alterations in key transporters and receptors, all of which favor the accumulation of Aβ and tau proteins. BBB dysfunction not only amplifies but also perpetuates this pathological cycle (Bell et al. 2012). One of the most significant genetic risk factors for both BBB breakdown and Aβ accumulation is the presence of the Apolipoprotein E-ε4 (ApoE4) allele (Perry et al. 2002).

The role of ApoE4 is closely tied to its interactions with other proteins, such as the low-density lipoprotein receptor-related protein 1 (LRP-1). ApoE is, physiologically, a lipid transport protein. The ApoE gene consists of three alleles, ε2, ε3, and ε4, translated into respective isoforms. ApoE3 is the most abundant isoform in humans (77.9%), while ApoE4 and ApoE2 distributions are 13.7 and 8.4%, respectively (Farrer et al. 1997). Astrocytes produce ApoE in the CNS, while the liver does it for the rest of the body. ApoE plays an essential role in maintaining BBB integrity by interacting with LRP-1 on pericytes to modulate MMP activity (Bell et al. 2012; Deane et al. 2004). In individuals carrying the ApoE4 allele, accelerated degeneration of pericytes, along with overexpression of Cyclophilin A and MMP-9, leads to BBB disruption. This has been demonstrated in animal models expressing human APOE4, where cerebral vasculature shows a thinner BM, reduced TJ expression, and a compromised BBB (Bell et al. 2012). Additionally, ApoE4 binds to Aβ, further hindering its clearance, and, together with LRP-1 downregulation, promotes Aβ accumulation (Deane et al. 2008).

LRP-1 is a primary receptor for Aβ clearance. It mediates uptake and internalization of Aβ for degradation within neurons, astrocytes, microglia, and vascular smooth muscle cells, and also acts as an Aβ efflux transporter on the ECs' surface. Reduced LRP1 levels have been observed in aged brains, in humans and animal models. This downregulation is thought to be driven by oxidative stress and inflammation, contributing to Aβ retention in the brain (Deane et al. 2004; Storck et al. 2016).

In addition to LRP-1, other transport systems such as P-glycoprotein (P-gp) play a significant role in regulating Aβ levels at the BBB. Highly expressed in the luminal aspect of brain ECs, P-gp, also known as ABCB1, is an efflux pump with a multitude of substrates, including Aβ. Its role extends beyond the BBB to neurons, where it mediates the export of Aβ peptides (van Assema et al. 2012). P-gp deficiency in mouse models reduces Aβ clearance by half. Conversely, Aβ40 triggers P-gp internalization, establishing a detrimental feedback loop that exacerbates pathology (Hartz et al. 2016). Studies have found that P-gp expression and activity are inversely correlated with aging, Aβ deposition, and AD progression (van Assema et al. 2012).

The receptor for advanced glycation end products (RAGE), another key player in Aβ transport, is a transmembrane protein present in ECs, among other tissues, like microglia and neurons, mediating the influx of Aβ from the bloodstream into the brain, working in opposition to LRP-1. RAGE expression is upregulated at the BBB of patients with AD as a consequence of its interaction with advanced glycation end products (AGE) and Aβ. This leads to further Aβ accumulation, tau hyperphosphorylation, neurofibrillary tangle formation, increased oxidative stress, and inflammation in ECs, exacerbating BBB breakdown and perpetuating neuronal degeneration (Deane et al. 2003).

As Aβ aggregates and plaque formation progresses, reactive inflammation follows, inducing perivascular aggregation of proteins, including fibrinogen, albumin, thrombin, and immunoglobulin (IgG), while compromising tight junctions and degenerating ECs and pericytes (Kinnecom et al. 2007; Nelson et al. 2016). Glucose transporter protein (GLUT-1), enriched in CNS endothelium, is known to be repressed in brain microvessels of AD patients, reducing glucose availability in the brain. Whether this reduction represents lower energy demands or primary BBB dysfunction remains unclear. Astrocytes in the AD brain also undergo morphological changes, with endfeet depolarization, reduced expression, and altered distribution of AQP4, also responsible for facilitating Aβ clearance, further contributing to Aβ buildup and neuroinflammation (Hoshi et al. 2012; Rasmussen et al. 2018).

Microglia play a complex role in AD, as they are implicated in both the progression and potential mitigation of the disease. They activate in response to Aβ plaques and neurofibrillary tangles, leading to the release of pro-inflammatory cytokines, leukocyte migration across the BBB, and further neuroinflammation. This chronic inflammation exacerbates neuronal damage and contributes to disease progression. On the other hand, microglia phagocytes Aβ and Tau, facilitating the clearance of these pathological molecules; however, in the later stages of AD, this mechanism fails (Merighi et al. 2022).

Emerging evidence suggests that BBB dysfunction in AD may begin in the hippocampus, further implicating this region in the early stages of the disease (Montagne et al. 2015; Miners et al. 2019). Some studies have proposed BBB dysfunction as an early biomarker of AD, given its presence before the onset of clinical symptoms and neurodegeneration (Montagne et al. 2015; Nation et al. 2019; van de Haar et al. 2016).

1.8.2 BBB and Multiple Sclerosis

Multiple sclerosis (MS) is a chronic, autoimmune, inflammatory, and neurodegenerative disease of the CNS, characterized by demyelination and axonal damage. It affects the brain, spinal cord, and optic nerves while sparing structures of the peripheral nervous system. MS primarily affects young adults, with a mean age of onset between 20 and 30 years, and it is more prevalent in women. This disease is driven by both inflammatory and neurodegenerative processes, resulting in intermittent neurological symptoms, followed by the gradual accumulation of disability. Common clinical features of MS include motor and sensory deficits, facial paralysis, nystagmus, speech and vision impairment, ataxia, tremor, fatigue, and cognitive decline (McGinley et al. 2021).

MS is clinically classified into three main forms: relapsing-remitting (RR), primary progressive (PP), and secondary progressive (SP). RR-MS, the most common form, is characterized by "attacks" of new or worsening neurologic symptoms, known as relapses, followed by periods of partial or complete recovery, termed remissions. During remissions, symptoms might either resolve or become permanent. Over time, RR-MS often transitions into the SP stage, where functional decline becomes progressive (McGinley et al. 2021).

Despite extensive research, the exact cause of MS remains unclear; however, advances in drug therapies have highlighted the critical role of immune cell migration across the BBB in the pathophysiology of the disease. As BBB dysfunction is central to both the onset and progression of MS, it has been a focus of many studies. The NVU plays a key role in regulating immune cell infiltration into the brain, promoting the development of a CNS-specific immune response response (Guttmann et al. 2016; Gaitán et al. 2011).

During an MS attack, activated T cells, B cells, leukocytes, and peripheral macrophages infiltrate the brain through a disrupted BBB, targeting the myelin sheath surrounding axons (Sweeney et al. 2019). Cytotoxic factors such as pro-inflammatory cytokines, proteases, ROS, and RNS accumulate and contribute to myelin destruction. This breakdown of the BBB, while transient, often recurs in different locations over time, leading to irregular lesion development involving further BBB leakage, immunologically mediated demyelination, and axonal damage (Ortiz et al. 2014).

Activated immune cells enter the CNS' perivascular space through postcapillary venules (Greter et al. 2005), where they express cell adhesion molecules (Carrithers et al. 2000), that activate the endothelium. Leukocytes then travel across the endothelial surfaces in search of intercellular gaps, where they alter the expression and organization of TJ, allowing for firm adhesion and trans-endothelial migration (Hernández-Pedro et al. 2013), later accessing the parenchyma by breaking down the BM through MMP action (Wu and Tsirka 2009). This process activates astrocytes and damages both cellular and non-cellular components of the BBB (Schreiner et al. 2022).

Once inside the CNS, T-cells recognizing CNS-specific autoantigens are reactivated, leading to their clonal expansion and the amplification of the immune response (Fletcher et al. 2010). These T-cells, in collaboration with antigen-presenting cells (mainly perivascular macrophages), release cytokines, chemokines, and other inflammatory mediators that cause structural changes in the BBB. This promotes the secondary influx of immune cells, activates resident microglia, and perpetuates a cascade of demyelination and axonal damage (Ricci et al. 2009).

Although the exact triggers of inflammation in MS are not fully understood, it is widely accepted that environmental factors play a significant role in genetically predisposed individuals, activating a T-cell-mediated immune response against the CNS (Ricci et al. 2009).

Neuroimaging studies using contrast enhancement have demonstrated that BBB disruption is an early feature of MS (Alvarez et al. 2011b). While gadolinium enhancement is typically associated with active MS lesions, MRI scans have revealed BBB disruption in seemingly normal white matter even before lesion enhancement and in non-enhancing areas (Vos et al. 2005). These observations suggest that BBB dysfunction may precede visible lesion formation and could play a critical role in the early stages of disease development (Sweeney et al. 2019).

Increasing interest in the BBB as a therapeutic target for MS has led to the development of treatments aimed at either restoring BBB integrity or reducing local inflammatory responses and limiting CNS-specific immune cell infiltration (Ortiz et al. 2014).

1.8.3 BBB and Epilepsy

Epilepsy is a chronic neurological disease characterized by a predisposition to recurrent seizures (Fisher et al. 2014). Its etiology is diverse and includes genetic, structural, metabolic, immune, infectious, and idiopathic causes (Thijs et al. 2019). Epilepsy can be focal, where seizures arise from networks limited to a specific cortical region, or generalized, involving both hemispheres from onset (Fisher et al. 2014).

Epilepsy affects individuals of all ages, sexes, and ethnicities, with a bimodal incidence peak in infancy and older age groups. Its impact on quality of life is substantial, stemming not only from the seizures but also from associated comorbidities, including cognitive deficits, mood disorders, and the side effects of treatment (Devinsky et al. 2018).

The relationship between epilepsy and BBB is multifaceted. The concept of BBB dysfunction as a hallmark of a seizing brain is the product of extensive research on the correlation between epilepsy and BBB disruption (Löscher and Friedman 2020). Whether BBB impairment is a cause or a consequence of epileptic seizures remains to be completely elucidated but is believed to be a bidirectional relationship (Sweeney et al. 2019). Seizures can impair BBB integrity, while BBB dysfunction can facilitate epileptogenesis and exacerbate seizure activity. This is evident in conditions like neuroinfections, inflammation, stroke, and TBI, which compromise BBB integrity and can trigger seizures and lead to epilepsy (Oby and Janigro 2006; van Vliet et al. 2007).

Acute epileptogenic insults, such as a traumatic injury or a hypoxic-ischemic lesion, lead to the release of glutamate, pro-inflammatory mediators, and free radicals, which in turn induce endothelial activation, TJ disruption, increased MMP activity, and pericyte dysfunction, ultimately causing BBB breakdown. This activates resident glial cells and attracts leukocyte migration into the brain parenchyma, thereby expanding cerebrovascular inflammation and disturbing local homeostasis. Consequently, the spreading depolarization threshold decreases, neurons become hyperexcitable, and seizures occur. Impaired glial function and neuronal excitability are further linked to synaptogenesis and neuroplasticity, processes that may contribute to epileptogenesis and cognitive deterioration (Löscher and Friedman 2020).

Prolonged seizure activity, like in the case of status epilepticus (SE) in its most extreme expression, is followed by spreading cortical depression (van Vliet et al. 2015), to which brain vasculature responds with altered vascular tone. This causes either transient hyperperfusion in otherwise healthy tissue or severe hypoperfusion in areas at risk of progressive damage (Winkler et al. 2012), leading to cellular injury and BBB disruption.

While the exact mechanisms of epileptogenesis remain incompletely understood, it is well-established that BBB dysfunction and brain inflammation lower the seizure threshold, creating a self-perpetuating cycle that drives epilepsy progression (van Vliet et al. 2015). However, BBB disruption alone is insufficient to induce seizure activity or cause epilepsy (van Vliet et al. 2015), as seen in other conditions like AD, MS, or stroke, where seizures can occur but are not a consistent feature. Epileptogenesis is believed to depend not only on the nature and extent of BBB permeability increase but also on the affected brain networks, among other multiple alterations (Löscher and Friedman 2020).

In established epilepsy, several studies have documented peri-ictal and interictal BBB dysfunction, often correlating temporally and anatomically with the region of seizure onset (Rüber et al. 2018; Reiter et al. 2024). In temporal lobe epilepsy (TLE)—the most common focal epileptic disorder in adults (Tatum 2012)—resected hippocampus revealed aberrant angiogenesis, increased microvascular density, loss of TJs, and increased permeability to IgGs (Sweeney et al. 2019). In animal models of TLE, BBB leakage is observed not only in the hippocampus but also in the entorhinal and piriform cortexes, thalamus, amygdala, septum, and substantia nigra, with a consistent metabolic activation during seizures (van Vliet et al. 2015).

Other studies on epilepsy have shown extravasation of serum proteins like albumin, increased pinocytosis, TJ abnormalities, BM thickening, pericyte-microglia clustering, astrocytic activation accompanied by altered expression of K+ channels, and AQP4 in epileptic brains (Löscher and Friedman 2020), and efflux transporters overexpression, which may limit the penetration of antiepileptic drugs, contributing to drug resistance (Hartz et al. 2019).

A deeper understanding of these processes holds the potential to drive the development of novel therapeutic strategies that are disease-modifying rather than merely seizure-suppressive.

1.9 Conclusion

The BBB is a highly specialized and dynamic system, fundamental to CNS homeostasis. Its structural and functional complexity, resulting from the interplay of cellular and molecular components within the NVU, ensures precise regulation of nutrient transport, waste clearance, and neuronal protection. ECs, pericytes, astrocytes, and BMs form interconnected networks supported by TJs, transport systems, and signaling molecules. Together, these elements maintain the barrier's selective permeability, protecting the brain from external insults while preserving a stable microenvironment, essential for neuronal function. This sophisticated orchestration highlights the BBB's evolutionary refinement to meet the unique metabolic demands of the brain, and its critical role as a gatekeeper for the CNS.

While this complexity enables the BBB to adapt to most physiological challenges, it is not immune to deterioration. Factors like aging, hormonal imbalances, stress, and temperature extremes can weaken its integrity, increasing permeability and making the CNS more susceptible to damage. In aging, for instance, a decline in the expression of TJ proteins, BM alterations, and chronic inflammation contributes to increased BBB permeability, making the CNS vulnerable to injury and disease.

BBB dysfunction is increasingly recognized as a key factor in the pathogenesis of numerous neurological disorders, emphasizing its critical role in neural health. In AD, the breakdown of the BBB contributes to the accumulation of β-amyloid plaques and neurofibrillary tangles, exacerbating neuronal degeneration. In MS, immune cell infiltration through a compromised BBB initiates demyelination and axonal damage, driving disease progression. Similarly, in epilepsy, BBB disruption promotes neuroinflammation and excitotoxicity, lowering the seizure threshold and perpetuating epileptogenesis. These examples illustrate both the pathological consequences of barrier failure and the potential as a therapeutic target in some of these neurological disorders.

As our understanding of the BBB continues to grow, it remains a keystone of neuroscience research, representing both a challenge and an opportunity in the pursuit of strategies to protect and repair the CNS.

Acknowledgments The Campbell lab is supported by grants from SFI (Eye-D-21/SPP/3732) to MC, the Irish Research Council (IRC), and a research grant from SFI under grant number 21/RC/10294_P2 and co-funded under the European Regional Development Fund by FutureNeuro industry partners to MC. The Campbell lab is also supported by a grant from the European Research Council (ERC—Retina-Rhythm). The lab is also supported by Fighting Blindness Ireland.

References

Abbott NJ, Rönnbäck L, Hansson E (2006) Astrocyte-endothelial interactions at the blood-brain barrier. Nat Rev Neurosci 7(1):41–53. https://doi.org/10.1038/nrn1824. PMID: 16371949

Abi-Ghanem C, Robison LS, Zuloaga KL (2020) Androgens' effects on cerebrovascular function in health and disease. Biol Sex Differ 11(1):35. https://doi.org/10.1186/s13293-020-00309-4. PMID: 32605602; PMCID: PMC7328272

Agorastos A, Chrousos GP (2022) The neuroendocrinology of stress: the stress-related continuum of chronic disease development. Mol Psychiatry 27(1):502–513. https://doi.org/10.1038/s41380-021-01224-9. Epub 2021 Jul 21. PMID: 34290370

Aird WC (2007) Phenotypic heterogeneity of the endothelium: I. Structure, function, and mechanisms. Circ Res 100(2):158–173. https://doi.org/10.1161/01.RES.0000255691.76142.4a. PMID: 17272818

Alahmari A (2021) Blood-brain barrier overview: structural and functional correlation. Neural Plast 2021: 6564585. https://doi.org/10.1155/2021/6564585. PMID: 34912450; PMCID: PMC8668349

Almutairi MM, Gong C, Xu YG, Chang Y, Shi H (2016) Factors controlling permeability of the blood-brain barrier. Cell Mol Life Sci. 73(1):57–77. https://doi.org/10.1007/s00018-015-2050-8. Epub 2015 Sep 24. PMID: 26403789; PMCID: PMC11108286

Al-Onaizi M, Al-Khalifah A, Qasem D, ElAli A (2020) Role of microglia in modulating adult neurogenesis in health and neurodegeneration. Int J Mol Sci 21(18):6875. https://doi.org/10.3390/ijms21186875. PMID: 32961703; PMCID: PMC7555074

Alonso AD, Cohen LS (2018) Our Tau tales from normal to pathological behavior. J Alzheimers Dis 64(s1):S507–S516. https://doi.org/10.3233/JAD--179906. PMID: 29614672

Alvarez JI, Dodelet-Devillers A, Kebir H, Ifergan I, Fabre PJ, Terouz S, Sabbagh M, Wosik K, Bourbonnière L, Bernard M, van Horssen J, de Vries HE, Charron F, Prat A (2011a) The hedgehog pathway promotes blood-brain barrier integrity and CNS immune quiescence. Science 334(6063):1727–1731. https://doi.org/10.1126/science.1206936. Epub 2011 Dec 1. PMID: 22144466

Alvarez JI, Cayrol R, Prat A (2011b) Disruption of central nervous system barriers in multiple sclerosis. Biochim Biophys Acta 1812(2):252–264. https://doi.org/10.1016/j.bbadis.2010.06.017. Epub 2010 Jul 7. PMID: 20619340

Alvarez JI, Katayama T, Prat A (2013) Glial influence on the blood brain barrier. Glia 61(12):1939–1958. https://doi.org/10.1002/glia.22575. Epub 2013 Oct 7. PMID: 24123158; PMCID: PMC4068281

Atallah A, Mhaouty-Kodja S, Grange-Messent V (2017) Chronic depletion of gonadal testosterone leads to blood-brain barrier dysfunction and inflammation in male mice. J Cereb Blood Flow Metab. 37(9):3161–3175. https://doi.org/10.1177/0271678X16683961. Epub 2016 Jan 1. PMID: 28256950; PMCID: PMC5584691

Aurrand-Lions M, Duncan L, Ballestrem C, Imhof BA (2001) JAM-2, a novel immunoglobulin superfamily molecule, expressed by endothelial and lymphatic cells. J Biol Chem 276(4):2733–2741. https://doi.org/10.1074/jbc.M005458200. Epub 2000 Oct 25. PMID: 11053409

Bagheri S, Saboury AA, Saso L (2024) Sequence of molecular events in the development of Alzheimer's disease: cascade interactions from Beta-amyloid to other involved proteins. Cells 13(15):1293. https://doi.org/10.3390/cells13151293. PMID: 39120323; PMCID: PMC11312137

Bell RD, Winkler EA, Sagare AP, Singh I, LaRue B, Deane R, Zlokovic BV (2010) Pericytes control key neurovascular functions and neuronal phenotype in the adult brain and during brain aging. Neuron 68(3):409–427. https://doi.org/10.1016/j.neuron.2010.09.043. Erratum in: Neuron. 2023;111(19):3131–3132. https://doi.org/10.1016/j.neuron.2023.09.007. PMID: 21040844; PMCID: PMC3056408

Bell RD, Winkler EA, Singh I, Sagare AP, Deane R, Wu Z, Holtzman DM, Betsholtz C, Armulik A, Sallstrom J, Berk BC, Zlokovic BV (2012) Apolipoprotein E controls cerebrovascular integrity via cyclophilin A. Nature 485(7399):512–516. https://doi.org/10.1038/nature11087. Erratum in: Nature 2023;617(7961):E12. https://doi.org/10.1038/s41586-023-06118-0. PMID: 22622580; PMCID: PMC4047116

Bernacki J, Dobrowolska A, Nierwińska K, Małecki A (2008) Physiology and pharmacological role of the blood-brain barrier. Pharmacol Rep 60(5):600–622. PMID: 19066407

Bertrand L, Cho HJ, Toborek M (2019) Blood-brain barrier pericytes as a target for HIV-1 infection. Brain 142(3):502–511. https://doi.org/10.1093/brain/awy339. PMID: 30668645; PMCID: PMC6391611

Bhatt DR, Reddy N, Ruiz R, Bustos DV, Peacock T, Dizon RA, Weerasinghe S, Braun DX, Ramanathan R (2020) Perinatal quality improvement bundle to decrease hypothermia in extremely low birthweight infants with birth weight less than 1000 g: single-center experience over 6 years. J Investig Med. 68(7):1256–1260. https://doi.org/10.1136/jim-2020-001334. Epub 2020 Jul 19. PMID: 32690596; PMCID: PMC7525784

Bhowmick S, D'Mello V, Caruso D, Wallerstein A, Abdul-Muneer PM (2019) Impairment of pericyte-endothelium crosstalk leads to blood-brain barrier dysfunction following traumatic brain injury. Exp Neurol 317:260–270. https://doi.org/10.1016/j.expneurol.2019.03.014. Epub 2019 Mar 26. PMID: 30926390

Bodor N, Buchwald P (2002) Barriers to remember: brain-targeting chemical delivery systems and Alzheimer's disease. Drug Discov Today 7(14):766–774. https://doi.org/10.1016/s1359-6446(02)02332-2. PMID: 12547033

Bouldin TW, Krigman MR (1975) Differential permeability of cerebral capillary and choroid plexus to lanthanum ion. Brain Res 99(2):444–448. https://doi.org/10.1016/0006-8993(75)90053-0. PMID: 1182566

Brunner N, Stein L, Cornelius V, Knittel R, Fallier-Becker P, Amasheh S (2020) Blood-Brain barrier protein claudin-5 expressed in paired Xenopus laevis oocytes mediates cell-cell interaction. Front Physiol 11:857.

https://doi.org/10.3389/fphys.2020.00857. PMID: 32848831; PMCID: PMC7396581

Burek M, König A, Lang M, Fiedler J, Oerter S, Roewer N, Bohnert M, Thal SC, Blecharz-Lang KG, Woitzik J, Thum T, Förster CY (2019) Hypoxia-induced MicroRNA-212/132 alter blood-brain barrier integrity through inhibition of tight junction-associated proteins in human and mouse brain microvascular endothelial cells. Transl Stroke Res 10(6):672–683. https://doi.org/10.1007/s12975-018-0683-2. Epub 2019 Jan 8. PMID: 30617994; PMCID: PMC6842347

Campelo AE, Cutini PH, Massheimer VL (2012) Testosterone modulates platelet aggregation and endothelial cell growth through nitric oxide pathway. J Endocrinol 213(1):77–87. https://doi.org/10.1530/JOE-11-0441. Epub 2012 Jan 26. PMID: 22281525

Campos-Bedolla P, Walter FR, Veszelka S, Deli MA (2014) Role of the blood-brain barrier in the nutrition of the central nervous system. Arch Med Res 45(8):610–638. https://doi.org/10.1016/j.arcmed.2014.11.018. Epub 2014 Dec 3. PMID: 25481827

Candiello J, Cole GJ, Halfter W (2010) Age-dependent changes in the structure, composition and biophysical properties of a human basement membrane. Matrix Biol 29(5):402–410. https://doi.org/10.1016/j.matbio.2010.03.004. Epub 2010 Mar 31. PMID: 20362054

Cardoso FL, Brites D, Brito MA (2010) Looking at the blood-brain barrier: molecular anatomy and possible investigation approaches. Brain Res Rev 64(2):328–363. https://doi.org/10.1016/j.brainresrev.2010.05.003. Epub 2010 May 26. PMID: 20685221

Carman CV, Martinelli R (2015) T lymphocyte-endothelial interactions: emerging understanding of trafficking and antigen-specific immunity. Front Immunol 6:603. https://doi.org/10.3389/fimmu.2015.00603. PMID: 26635815; PMCID: PMC4657048

Carrithers MD, Visintin I, Kang SJ, Janeway CA Jr (2000) Differential adhesion molecule requirements for immune surveillance and inflammatory recruitment. Brain 123(Pt 6):1092–1101. https://doi.org/10.1093/brain/123.6.1092. PMID: 10825349

Chaban VV, Lakhter AJ, Micevych P (2004) A membrane estrogen receptor mediates intracellular calcium release in astrocytes. Endocrinology 145(8):3788–3795. https://doi.org/10.1210/en.2004-0149. Epub 2004 May 6. PMID: 15131017

Chen Y, Liu L (2012) Modern methods for delivery of drugs across the blood-brain barrier. Adv Drug Deliv Rev 64(7):640–665. https://doi.org/10.1016/j.addr.2011.11.010. Epub 2011 Nov 28. PMID: 22154620

Cheshire WP Jr (2016) Thermoregulatory disorders and illness related to heat and cold stress. Auton Neurosci 196:91–104. https://doi.org/10.1016/j.autneu.2016.01.001. Epub 2016 Jan 6. PMID: 26794588

Costea L, Mészáros Á, Bauer H, Bauer HC, Traweger A, Wilhelm I, Farkas AE, Krizbai IA (2019) The blood-brain barrier and its intercellular junctions in age-related brain disorders. Int J Mol Sci 20(21):5472. https://doi.org/10.3390/ijms20215472. PMID: 31684130; PMCID: PMC6862160

D'Ambrosio A, Pontecorvo S, Colasanti T, Zamboni S, Francia A, Margutti P (2015) Peripheral blood biomarkers in multiple sclerosis. Autoimmun Rev 14(12):1097–1110. https://doi.org/10.1016/j.autrev.2015.07.014. Epub 2015 Jul 28. PMID: 26226413

Daneman R, Prat A (2015) The blood-brain barrier. Cold Spring Harb Perspect Biol 7(1):a020412. https://doi.org/10.1101/cshperspect.a020412. PMID: 25561720; PMCID: PMC4292164

David Y, Cacheaux LP, Ivens S, Lapilover E, Heinemann U, Kaufer D, Friedman A (2009) Astrocytic dysfunction in epileptogenesis: consequence of altered potassium and glutamate homeostasis? J Neurosci 29(34):10588–10599. https://doi.org/10.1523/JNEUROSCI.2323-09.2009. PMID: 19710312; PMCID: PMC2875068

Deane R, Du Yan S, Submamaryan RK, LaRue B, Jovanovic S, Hogg E, Welch D, Manness L, Lin C, Yu J, Zhu H, Ghiso J, Frangione B, Stern A, Schmidt AM, Armstrong DL, Arnold B, Liliensiek B, Nawroth P, Hofman F, Kindy M, Stern D, Zlokovic B (2003) RAGE mediates amyloid-beta peptide transport across the blood-brain barrier and accumulation in brain. Nat Med 9(7):907–913. https://doi.org/10.1038/nm890. PMID: 12808450

Deane R, Wu Z, Sagare A, Davis J, Du Yan S, Hamm K, Xu F, Parisi M, LaRue B, Hu HW, Spijkers P, Guo H, Fletcher X, Lenting PJ, Van Nostrand WE, Zlokovic BV (2004) LRP/amyloid beta-peptide interaction mediates differential brain efflux of Abeta isoforms. Neuron 43(3):333–344. https://doi.org/10.1016/j.neuron.2004.07.017. PMID: 15294142

Deane R, Sagare A, Zlokovic BV (2008) The role of the cell surface LRP and soluble LRP in blood-brain barrier Abeta clearance in Alzheimer's disease. Curr Pharm Des 14(16):1601–1605. https://doi.org/10.2174/138161208784705487. PMID: 18673201; PMCID: PMC2895311

Del Zoppo GJ, Milner R, Mabuchi T, Hung S, Wang X, Koziol JA (2006) Vascular matrix adhesion and the blood-brain barrier. Biochem Soc Trans 34(Pt 6):1261–1266. https://doi.org/10.1042/BST0341261. PMID: 17073798

Delaney C, Campbell M (2017) The blood brain barrier: Insights from development and ageing. Tissue Barriers 5(4):e1373897. https://doi.org/10.1080/21688370.2017.1373897. Epub 2017 Sep 28. PMID: 28956691; PMCID: PMC5788423

DellaValle B, Hasseldam H, Johansen FF, Iversen HK, Rungby J, Hempel C (2019) Multiple soluble components of the Glycocalyx are increased in patient plasma after ischemic stroke. Stroke 50(10):2948–2951. https://doi.org/10.1161/STROKEAHA.119.025953. Epub 2019 Aug 14. PMID: 31409270

Devinsky O, Vezzani A, O'Brien TJ, Jette N, Scheffer IE, de Curtis M, Perucca P (2018) Epilepsy Nat

Rev Dis Primers 4:18024. https://doi.org/10.1038/nrdp.2018.24. PMID: 29722352

Ding S, Khan AI, Cai X, Song Y, Lyu Z, Du D, Dutta P, Lin Y (2020) Overcoming blood-brain barrier transport: advances in nanoparticle-based drug delivery strategies. Mater Today (Kidlington) 37:112–125. https://doi.org/10.1016/j.mattod.2020.02.001. Epub 2020 Mar 4. PMID: 33093794; PMCID: PMC7575138

Dion-Albert L, Cadoret A, Doney E, Kaufmann FN, Dudek KA, Daigle B, Parise LF, Cathomas F, Samba N, Hudson N, Lebel M, Signature Consortium, Campbell M, Turecki G, Mechawar N, Menard C (2022) Vascular and blood-brain barrier-related changes underlie stress responses and resilience in female mice and depression in human tissue. Nat Commun 13:164. https://doi.org/10.1038/s41467-021-27604-x. PMID: 35013188; PMCID: PMC8748803

Dion-Albert L, Dudek KA, Russo SJ, Campbell M, Menard C (2023) Neurovascular adaptations modulating cognition, mood, and stress responses. Trends Neurosci 46(4):276–292. https://doi.org/10.1016/j.tins.2023.01.005. Epub 2023 Feb 17. PMID: 36805768

Dore-Duffy P (2008) Pericytes: pluripotent cells of the blood brain barrier. Curr Pharm Des 14(16):1581–1593. https://doi.org/10.2174/138161208784705469. PMID: 18673199

Dyrna F, Hanske S, Krueger M, Bechmann I (2013) The blood-brain barrier. J Neuroimmune Pharmacol 8(4):763–773. https://doi.org/10.1007/s11481-013-9473-5. Epub 2013 Jun 6. PMID: 23740386

Ehrlich P (1885) Sauerstoff-Bedürfniss des Organismus-Eine Farbenanalytische Studie

Elahy M, Jackaman C, Mamo JC, Lam V, Dhaliwal SS, Giles C, Nelson D, Takechi R (2015) Blood-brain barrier dysfunction developed during normal aging is associated with inflammation and loss of tight junctions but not with leukocyte recruitment. Immun Ageing 12:2. https://doi.org/10.1186/s12979-015-0029-9. PMID: 25784952; PMCID: PMC4362825

Engelhardt B, Liebner S (2014) Novel insights into the development and maintenance of the blood-brain barrier. Cell Tissue Res 355(3):687–699. https://doi.org/10.1007/s00441-014-1811-2. Epub 2014 Mar 4. PMID: 24590145; PMCID: PMC3972432

Engelhardt B, Ransohoff RM (2012) Capture, crawl, cross: the T cell code to breach the blood-brain barriers. Trends Immunol 33(12):579–589. https://doi.org/10.1016/j.it.2012.07.004. Epub 2012 Aug 25. PMID: 22926201

Erickson KK, Sundstrom JM, Antonetti DA (2007) Vascular permeability in ocular disease and the role of tight junctions. Angiogenesis 10(2):103–117. https://doi.org/10.1007/s10456-007-9067-z. Epub 2007 Mar 6. PMID: 17340211

Erickson MA, Liang WS, Fernandez EG, Bullock KM, Thysell JA, Banks WA (2018) Genetics and sex influence peripheral and central innate immune responses and blood-brain barrier integrity. PLoS One 13(10):e0205769. https://doi.org/10.1371/journal.pone.0205769. PMID: 30325961; PMCID: PMC6191122

Fanning AS, Jameson BJ, Jesaitis LA, Anderson JM (1998) The tight junction protein ZO-1 establishes a link between the transmembrane protein occludin and the actin cytoskeleton. J Biol Chem 273(45):29745–29753. https://doi.org/10.1074/jbc.273.45.29745. PMID: 9792688

Farrer LA, Cupples LA, Haines JL, Hyman B, Kukull WA, Mayeux R, Myers RH, Pericak-Vance MA, Risch N, van Duijn CM (1997) Effects of age, sex, and ethnicity on the association between apolipoprotein E genotype and Alzheimer disease. A meta-analysis. APOE and Alzheimer disease meta analysis consortium. JAMA 278(16):1349–1356. PMID: 9343467

Fernández-López D, Natarajan N, Ashwal S, Vexler ZS (2014) Mechanisms of perinatal arterial ischemic stroke. J Cereb Blood Flow Metab 34(6):921–932. https://doi.org/10.1038/jcbfm.2014.41. Epub 2014 Mar 26. PMID: 24667913; PMCID: PMC4050239

Fischer S, Wobben M, Marti HH, Renz D, Schaper W (2002) Hypoxia-induced hyperpermeability in brain microvessel endothelial cells involves VEGF-mediated changes in the expression of zonula occludens-1. Microvasc Res 63(1):70–80. https://doi.org/10.1006/mvre.2001.2367. PMID: 11749074

Fisher RS, Acevedo C, Arzimanoglou A, Bogacz A, Cross JH, Elger CE, Engel J Jr, Forsgren L, French JA, Glynn M, Hesdorffer DC, Lee BI, Mathern GW, Moshé SL, Perucca E, Scheffer IE, Tomson T, Watanabe M, Wiebe S (2014) ILAE official report: a practical clinical definition of epilepsy. Epilepsia 55(4):475–482. https://doi.org/10.1111/epi.12550. Epub 2014 Apr 14. PMID: 24730690

Fletcher JM, Lalor SJ, Sweeney CM, Tubridy N, Mills KH (2010) T cells in multiple sclerosis and experimental autoimmune encephalomyelitis. Clin Exp Immunol 162(1):1–11. https://doi.org/10.1111/j.1365--2249.2010.04143.x. PMID: 20682002; PMCID: PMC2990924

Furuse M, Hirase T, Itoh M, Nagafuchi A, Yonemura S, Tsukita S, Tsukita S (1993) Occludin: a novel integral membrane protein localizing at tight junctions. J Cell Biol 123(6 Pt 2):1777–1788. https://doi.org/10.1083/jcb.123.6.1777. PMID: 8276896; PMCID: PMC2290891

Gaitán MI, Shea CD, Evangelou IE, Stone RD, Fenton KM, Bielekova B, Massacesi L, Reich DS (2011) Evolution of the blood-brain barrier in newly forming multiple sclerosis lesions. Ann Neurol 70(1):22–29. https://doi.org/10.1002/ana.22472. Epub 2011 Jun 27. PMID: 21710622; PMCID: PMC3143223

Garcia-Segura LM, Torres-Aleman I, Naftolin F (1989) Astrocytic shape and glial fibrillary acidic protein immunoreactivity are modified by estradiol in primary rat hypothalamic cultures. Brain Res Dev Brain Res 47(2):298–302. https://doi.org/10.1016/0165--3806(89)90186-7. PMID: 2743562

Ghersi-Egea JF, Babikian A, Blondel S, Strazielle N (2015) Changes in the cerebrospinal fluid circulatory system of the developing rat: quantitative volumetric analysis and effect on blood-CSF permeability interpretation. Fluids Barriers CNS. 12:8. https://doi.org/10.1186/s12987-015-0001-2. PMID: 25793105; PMCID: PMC4365764

Ginhoux F, Prinz M (2015) Origin of microglia: current concepts and past controversies. Cold Spring Harb Perspect Biol 7(8):a020537. https://doi.org/10.1101/cshperspect.a020537. PMID: 26134003; PMCID: PMC4526747

Ginhoux F, Lim S, Hoeffel G, Low D, Huber T (2013) Origin and differentiation of microglia. Front Cell Neurosci 7:45. https://doi.org/10.3389/fncel.2013.00045. PMID: 23616747; PMCID: PMC3627983

Greene C, Hanley N, Campbell M (2019) Claudin-5: gatekeeper of neurological function. Fluids Barriers CNS. 16(1):3. https://doi.org/10.1186/s12987-019-0123-z. PMID: 30691500; PMCID: PMC6350359

Greenwood J, Wang Y, Calder VL (1995) Lymphocyte adhesion and transendothelial migration in the central nervous system: the role of LFA-1, ICAM-1, VLA-4 and VCAM-1. off. Immunology 86(3):408–415. PMID: 8550078; PMCID: PMC1383944

Greter M, Heppner FL, Lemos MP, Odermatt BM, Goebels N, Laufer T, Noelle RJ, Becher B (2005) Dendritic cells permit immune invasion of the CNS in an animal model of multiple sclerosis. Nat Med 11(3):328–334. https://doi.org/10.1038/nm1197. Epub 2005 Feb 27. PMID: 15735653

Gumbiner B, Lowenkopf T, Apatira D (1991) Identification of a 160-kDa polypeptide that binds to the tight junction protein ZO-1. Proc Natl Acad Sci USA 88(8):3460–3464. https://doi.org/10.1073/pnas.88.8.3460. PMID: 2014265; PMCID: PMC51467

Guo R, Sakamoto H, Sugiura S, Ogawa M (2007) Endothelial cell motility is compatible with junctional integrity. J Cell Physiol 211(2):327–335. https://doi.org/10.1002/jcp.20937. PMID: 17167782.

Guo S, Wang H, Yin Y (2022) Microglia polarization from M1 to M2 in neurodegenerative diseases. Front Aging Neurosci 16(14):815347. https://doi.org/10.3389/fnagi.2022.815347. PMID: 35250543; PMCID: PMC8888930

Guttmann CR, Rousset M, Roch JA, Hannoun S, Durand-Dubief F, Belaroussi B, Cavallari M, Rabilloud M, Sappey-Marinier D, Vukusic S, Cotton F (2016) Multiple sclerosis lesion formation and early evolution revisited: a weekly high-resolution magnetic resonance imaging study. Mult Scler 22(6):761–769. https://doi.org/10.1177/1352458515600247. Epub 2015 Sep 11. PMID: 26362901

Ha IH, Lim C, Kim Y, Moon Y, Han SH, Moon WJ (2021) Regional differences in blood-brain barrier permeability in cognitively Normal elderly subjects: a dynamic contrast-enhanced MRI-based study. Korean J Radiol 22(7):1152–1162. https://doi.org/10.3348/kjr.2020.0816. Epub 2021 Mar 9. PMID: 33739632; PMCID: PMC8236362

Harati R, Villégier AS, Banks WA, Mabondzo A (2012) Susceptibility of juvenile and adult blood-brain barrier to endothelin-1: regulation of P-glycoprotein and breast cancer resistance protein expression and transport activity. J Neuroinflammation 9:273. https://doi.org/10.1186/1742-2094-9-273. PMID: 23253775; PMCID: PMC3547749

Hartz AM, Zhong Y, Wolf A, LeVine H 3rd, Miller DS, Bauer B (2016) Aβ40 reduces P-glycoprotein at the Blood-Brain barrier through the ubiquitin-proteasome pathway. J Neurosci 36(6):1930–1941. https://doi.org/10.1523/JNEUROSCI.0350-15.2016. PMID: 26865616; PMCID: PMC4748076

Hartz AMS, Rempe RG, Soldner ELB, Pekcec A, Schlichtiger J, Kryscio R, Bauer B (2019) Cytosolic phospholipase A2 is a key regulator of blood-brain barrier function in epilepsy. FASEB J. 33(12):14281–14295. https://doi.org/10.1096/fj.201901369RR. Epub 2019 Oct 29. PMID: 31661303; PMCID: PMC6894050

Haseloff RF, Dithmer S, Winkler L, Wolburg H, Blasig IE (2015) Transmembrane proteins of the tight junctions at the blood-brain barrier: structural and functional aspects. Semin Cell Dev Biol 38:16–25. https://doi.org/10.1016/j.semcdb.2014.11.004. Epub 2014 Nov 26. PMID: 25433243

Hawkins BT, Davis TP (2005) The blood-brain barrier/neurovascular unit in health and disease. Pharmacol Rev 57(2):173–185. https://doi.org/10.1124/pr.57.2.4. PMID: 15914466

Hernández-Pedro NY, Espinosa-Ramirez G, de la Cruz VP, Pineda B, Sotelo J (2013) Initial immunopathogenesis of multiple sclerosis: innate immune response. Clin Dev Immunol. 2013:413465. https://doi.org/10.1155/2013/413465. Epub 2013 Sep 24. PMID: 24174969; PMCID: PMC3794540

Hindle SJ, Munji RN, Dolghih E, Gaskins G, Orng S, Ishimoto H, Soung A, DeSalvo M, Kitamoto T, Keiser MJ, Jacobson MP, Daneman R, Bainton RJ (2017) Evolutionarily conserved roles for Blood-Brain barrier xenobiotic transporters in endogenous steroid partitioning and behavior. Cell Rep 21(5):1304–1316. https://doi.org/10.1016/j.celrep.2017.10.026. PMID: 29091768; PMCID: PMC5774027

Hirase T, Staddon JM, Saitou M, Ando-Akatsuka Y, Itoh M, Furuse M, Fujimoto K, Tsukita S, Rubin LL (1997) Occludin as a possible determinant of tight junction permeability in endothelial cells. J Cell Sci 110(Pt 14):1603–1613. https://doi.org/10.1242/jcs.110.14.1603. PMID: 9247194

Hoshi A, Yamamoto T, Shimizu K, Ugawa Y, Nishizawa M, Takahashi H, Kakita A (2012) Characteristics of aquaporin expression surrounding senile plaques and cerebral amyloid angiopathy in Alzheimer disease. J Neuropathol Exp Neurol 71(8):750–759. https://doi.org/10.1097/NEN.0b013e3182632566. PMID: 22805778

Hrvatin S, Hochbaum DR, Nagy MA, Cicconet M, Robertson K, Cheadle L, Zilionis R, Ratner A, Borges-Monroy R, Klein AM, Sabatini BL, Greenberg ME

(2018) Single-cell analysis of experience-dependent transcriptomic states in the mouse visual cortex. Nat Neurosci. 21(1):120–129. https://doi.org/10.1038/s41593-017-0029-5. PMID: 29230054; PMCID: PMC5742025

Huang SH, Wang L, Chi F, Wu CH, Cao H, Zhang A, Jong A (2013) Circulating brain microvascular endothelial cells (cBMECs) as potential biomarkers of the blood-brain barrier disorders caused by microbial and non-microbial factors. PLoS One 8(4):e62164. https://doi.org/10.1371/journal.pone.0062164. PMID: 23637989; PMCID: PMC3637435

Huber JD, Witt KA, Hom S, Egleton RD, Mark KS, Davis TP (2001) Inflammatory pain alters blood-brain barrier permeability and tight junctional protein expression. Am J Physiol Heart Circ Physiol 280(3):H1241–H1248. https://doi.org/10.1152/ajpheart.2001.280.3.H1241. PMID: 11179069

Hussain B, Fang C, Chang J (2021) Blood-brain barrier breakdown: an emerging biomarker of cognitive impairment in normal aging and dementia. Front Neurosci 15:688090. https://doi.org/10.3389/fnins.2021.688090. PMID: 34489623; PMCID: PMC8418300

Hyman BT, Phelps CH, Beach TG, Bigio EH, Cairns NJ, Carrillo MC, Dickson DW, Duyckaerts C, Frosch MP, Masliah E, Mirra SS, Nelson PT, Schneider JA, Thal DR, Thies B, Trojanowski JQ, Vinters HV, Montine TJ (2012) National Institute on Aging-Alzheimer's Association guidelines for the neuropathologic assessment of Alzheimer's disease. Alzheimers Dement 8(1):1–13. https://doi.org/10.1016/j.jalz.2011.10.007. PMID: 22265587; PMCID: PMC3266529

Itoh M, Nagafuchi A, Yonemura S, Kitani-Yasuda T, Tsukita S, Tsukita S (1993) The 220-kD protein colocalizing with cadherins in non-epithelial cells is identical to ZO-1, a tight junction-associated protein in epithelial cells: cDNA cloning and immunoelectron microscopy. J Cell Biol 121(3):491–502. https://doi.org/10.1083/jcb.121.3.491. PMID: 8486731; PMCID: PMC2119563

Ivanidze J, Mackay M, Hoang A, Chi JM, Cheng K, Aranow C, Volpe B, Diamond B, Sanelli PC (2019) Dynamic contrast-enhanced MRI reveals unique Blood-Brain barrier permeability characteristics in the hippocampus in the Normal Brain. AJNR Am J Neuroradiol 40(3):408–411. https://doi.org/10.3174/ajnr.A5962. Epub 2019 Feb 7. PMID: 30733256; PMCID: PMC7028673

Johnson-Léger C, Imhof BA (2003) Forging the endothelium during inflammation: pushing at a half-open door? Cell Tissue Res 314(1):93–105. https://doi.org/10.1007/s00441-003-0775-4. Epub 2003 Sep 3. PMID: 12955495

Kadry H, Noorani B, Cucullo L (2020) A blood-brain barrier overview on structure, function, impairment, and biomarkers of integrity. Fluids Barriers CNS 17(1):69. https://doi.org/10.1186/s12987-020-00230-3. PMID: 33208141; PMCID: PMC7672931

Kamatham PT, Shukla R, Khatri DK, Vora LK (2024) Pathogenesis, diagnostics, and therapeutics for Alzheimer's disease: breaking the memory barrier. Ageing Res Rev 101:102481. https://doi.org/10.1016/j.arr.2024.102481. Epub ahead of print. PMID: 39236855

Kamm D (2019) The effects of diet and sex differences on cortical tight-junction protein expression in senescence-accelerated mouse-prone 8 (SAMP8) mice. Master's thesis. Southern Illinois University, Edwardsville

Keaney J, Campbell M (2015) The dynamic blood-brain barrier. FEBS J 282(21):4067–4079. https://doi.org/10.1111/febs.13412. Epub 2015 Sep 8. PMID: 26277326

Kinnecom C, Lev MH, Wendell L, Smith EE, Rosand J, Frosch MP, Greenberg SM (2007) Course of cerebral amyloid angiopathy-related inflammation. Neurology 68(17):1411–1416. https://doi.org/10.1212/01.wnl.0000260066.98681.2e. PMID: 17452586

Kisler K, Nelson AR, Montagne A, Zlokovic BV (2017) Cerebral blood flow regulation and neurovascular dysfunction in Alzheimer disease. Nat Rev Neurosci 18(7):419–434. https://doi.org/10.1038/nrn.2017.48. Epub 2017 May 18. PMID: 28515434; PMCID: PMC5759779

Kiyatkin EA, Sharma HS (2009) Permeability of the blood-brain barrier depends on brain temperature. Neuroscience 161(3):926–939. https://doi.org/10.1016/j.neuroscience.2009.04.004. Epub 2009 Apr 9. PMID: 19362131; PMCID: PMC2694729

Knopp RC, Erickson MA, Rhea EM, Reed MJ, Banks WA (2023) Cellular senescence and the blood-brain barrier: implications for aging and age-related diseases. Exp Biol Med (Maywood) 248(5):399–411. https://doi.org/10.1177/15353702231157917. Epub 2023 Apr 3. PMID: 37012666; PMCID: PMC10281623

Kuczynski AM, Ospel JM, Demchuk AM, Goyal M, Mitha AP, Almekhlafi MA (2020) Therapeutic hypothermia in patients with malignant ischemic stroke and hemicraniectomy—a systematic review and meta-analysis. World Neurosurg 141:e677–e685. https://doi.org/10.1016/j.wneu.2020.05.277. Epub 2020 Jun 19. PMID: 32569760

Kumagai N, Chiba Y, Hosono M, Fujii M, Kawamura N, Keino H, Yoshikawa K, Ishii S, Saitoh Y, Satoh M, Shimada A, Hosokawa M (2007) Involvement of pro-inflammatory cytokines and microglia in an age-associated neurodegeneration model, the SAMP10 mouse. Brain Res 1185:75–85. https://doi.org/10.1016/j.brainres.2007.09.021. Epub 2007 Sep 20. PMID: 17949696

Kumar M, Sandhir R (2022) Hydrogen sulfide attenuates hyperhomocysteinemia-induced blood-brain barrier permeability by inhibiting MMP-9. Int J Neurosci 132(11):1061–1071. https://doi.org/10.1080/00207454.2020.1860967. Epub 2021 Feb 1. PMID: 33287606

Kutuzov N, Flyvbjerg H, Lauritzen M (2018) Contributions of the glycocalyx, endothelium, and extravascular compartment to the blood-brain barrier. Proc Natl Acad Sci USA 115:E9429–E9438. https://doi.org/10.1073/pnas.1802155115

Kvichansky AA, Volobueva MN, Spivak YS, Tret'yakova LV, Gulyaeva NV, Bolshakov AP (2019) Expression of mRNAs for IL-1β, IL-6, IL-10, TNFα, CX3CL1, and TGFβ1 cytokines in the brain tissues: assessment of contribution of blood cells with and without perfusion. Biochemistry (Mosc) 84(8):905–910. https://doi.org/10.1134/S0006297919080066. PMID: 31522672

Lampron A, Elali A, Rivest S (2013) Innate immunity in the CNS: redefining the relationship between the CNS and its environment. Neuron 78(2):214–232. https://doi.org/10.1016/j.neuron.2013.04.005. PMID: 23622060

Laptook A (2014) The importance of temperature on the neurovascular unit. Early Hum Dev 90(10):713–717. https://doi.org/10.1016/j.earlhumdev.2014.06.008. Epub 2014 Jul 18. PMID: 25043136

Lee S, Kang BM, Kim JH, Min J, Kim HS, Ryu H, Park H, Bae S, Oh D, Choi M, Suh M (2018) Real-time in vivo two-photon imaging study reveals decreased cerebro-vascular volume and increased blood-brain barrier permeability in chronically stressed mice. Sci Rep 8(1):13064. https://doi.org/10.1038/s41598-018-30875-y. PMID: 30166586; PMCID: PMC6117335

Lewandowsky M (1900) Zur Lehre der Zerebrospinalflüssigkeit. Z. Klin. Med

Li J, Li C, Yuan W, Wu J, Li J, Li Z, Zhao Y (2017) Mild hypothermia alleviates brain oedema and blood-brain barrier disruption by attenuating tight junction and adherens junction breakdown in a swine model of cardiopulmonary resuscitation. PLoS One 12(3):e0174596. https://doi.org/10.1371/journal.pone.0174596. PMID: 28355299; PMCID: PMC5371345

Li W, Pan R, Qi Z, Liu KJ (2018) Current progress in searching for clinically useful biomarkers of blood-brain barrier damage following cerebral ischemia. Brain Circ 4(4):145–152. https://doi.org/10.4103/bc.bc_11_18. Epub 2018 Dec 31. PMID: 30693340; PMCID: PMC6329218

Liebner S, Dijkhuizen RM, Reiss Y, Plate KH, Agalliu D, Constantin G (2018) Functional morphology of the blood-brain barrier in health and disease. Acta Neuropathol 135(3):311–336. https://doi.org/10.1007/s00401-018-1815-1. Epub 2018 Feb 6. PMID: 29411111; PMCID: PMC6781630

Liu X, Tu M, Kelly RS, Chen C, Smith BJ (2004) Development of a computational approach to predict blood-brain barrier permeability. Drug Metab Dispos 32(1):132–139. https://doi.org/10.1124/dmd.32.1.132. PMID: 14709630

Löscher W, Friedman A (2020) Structural, molecular, and functional alterations of the blood-brain barrier during epileptogenesis and epilepsy: a cause, consequence, or both? Int J Mol Sci 21(2):591. https://doi.org/10.3390/ijms21020591. PMID: 31963328; PMCID: PMC7014122

Maggioli E, McArthur S, Mauro C, Kieswich J, Kusters DHM, Reutelingsperger CPM, Yaqoob M, Solito E (2016) Estrogen protects the blood-brain barrier from inflammation-induced disruption and increased lymphocyte trafficking. Brain Behav Immun 51:212–222. https://doi.org/10.1016/j.bbi.2015.08.020. Epub 2015 Aug 29. PMID: 26321046

Marques F, Sousa JC, Sousa N, Palha JA (2013) Blood-brain-barriers in aging and in Alzheimer's disease. Mol Neurodegener 8:38. https://doi.org/10.1186/1750--1326-8-38. PMID: 24148264; PMCID: PMC4015275

Martìn-Padura I, Lostaglio S, Schneemann M, Williams L, Romano M, Fruscella P, Panzeri C, Stoppacciaro A, Ruco L, Villa A, Simmons D, Dejana E (1998) Junctional adhesion molecule, a novel member of the immunoglobulin superfamily that distributes at intercellular junctions and modulates monocyte transmigration. J Cell Biol 142(1):117–127. https://doi.org/10.1083/jcb.142.1.117. PMID: 9660867; PMCID: PMC2133024

McCullough L, Arora S (2004) Diagnosis and treatment of hypothermia. Am Fam Physician 70(12):2325–2332. PMID: 15617296

McGinley MP, Goldschmidt CH, Rae-Grant AD (2021) Diagnosis and treatment of multiple sclerosis: a review. JAMA 325(8):765–779. https://doi.org/10.1001/jama.2020.26858. Erratum in: JAMA. 2021;325(21):2211. https://doi.org/10.1001/jama.2021.7928. PMID: 33620411

Merighi S, Nigro M, Travagli A, Gessi S (2022) Microglia and Alzheimer's disease. Int J Mol Sci 23(21):12990. https://doi.org/10.3390/ijms232112990. PMID: 36361780; PMCID: PMC9657945

Miao Y, Liao JK (2014) Potential serum biomarkers in the pathophysiological processes of stroke. Expert Rev Neurother. 14(2):173–185. https://doi.org/10.1586/14737175.2014.875471. Epub 2014 Jan 13. PMID: 24417214; PMCID: PMC3939679

Michalicova A, Majerova P, Kovac A (2020) Tau protein and its role in Blood-Brain barrier dysfunction. Front Mol Neurosci 13:570045. https://doi.org/10.3389/fnmol.2020.570045. PMID: 33100967; PMCID: PMC7554615

Michell-Robinson MA, Touil H, Healy LM, Owen DR, Durafourt BA, Bar-Or A, Antel JP, Moore CS (2015) Roles of microglia in brain development, tissue maintenance and repair. Brain 138(Pt 5):1138–1159. https://doi.org/10.1093/brain/awv066. Epub 2015 Mar 29. PMID: 25823474; PMCID: PMC5963417

Michinaga S, Koyama Y (2019) Dual roles of astrocyte-derived factors in regulation of Blood-Brain barrier function after Brain damage. Int J Mol Sci 20(3):571. https://doi.org/10.3390/ijms20030571. PMID: 30699952; PMCID: PMC6387062

Middeldorp J, Hol EM (2011) GFAP in health and disease. Prog Neurobiol 93(3):421–443. https://doi.org/10.1016/j.pneurobio.2011.01.005. Epub 2011 Jan 8. PMID: 21219963

Miller DS (2015) Regulation of ABC transporters blood-brain barrier: the good, the bad, and the ugly. Adv

Cancer Res 125:43–70. https://doi.org/10.1016/bs.acr.2014.10.002. Epub 2015 Jan 8. PMID: 25640266

Miners JS, Kehoe PG, Love S, Zetterberg H, Blennow K (2019) CSF evidence of pericyte damage in Alzheimer's disease is associated with markers of blood-brain barrier dysfunction and disease pathology. Alzheimers Res Ther 11(1):81. https://doi.org/10.1186/s13195-019-0534-8. PMID: 31521199; PMCID: PMC6745071

Montagne A, Barnes SR, Sweeney MD, Halliday MR, Sagare AP, Zhao Z, Toga AW, Jacobs RE, Liu CY, Amezcua L, Harrington MG, Chui HC, Law M, Zlokovic BV (2015) Blood-brain barrier breakdown in the aging human hippocampus. Neuron 85(2):296–302. https://doi.org/10.1016/j.neuron.2014.12.032. PMID: 25611508; PMCID: PMC4350773

Montagne A, Zhao Z, Zlokovic BV (2017) Alzheimer's disease: a matter of blood-brain barrier dysfunction? J Exp Med. 214(11):3151–3169. https://doi.org/10.1084/jem.20171406. Epub 2017 Oct 23. PMID: 29061693; PMCID: PMC5679168

Moon Y, Lim C, Kim Y, Moon WJ (2021) Sex-related differences in regional blood-brain barrier integrity in non-demented elderly subjects. Int J Mol Sci 22(6):2860. https://doi.org/10.3390/ijms22062860. PMID: 33799794; PMCID: PMC8001339

Muller M, van den Beld AW, Bots ML, Grobbee DE, Lamberts SW, van der Schouw YT (2004) Endogenous sex hormones and progression of carotid atherosclerosis in elderly men. Circulation 109(17):2074–2079. https://doi.org/10.1161/01.CIR.0000125854.51637.06. Epub 2004 Apr 19. PMID: 15096452

Na W, Lee JY, Kim WS, Yune TY, Ju BG (2015) 17β-estradiol ameliorates tight junction disruption via repression of MMP transcription. Mol Endocrinol 29(9):1347–1361. https://doi.org/10.1210/ME.2015-1124. Epub 2015 Jul 13. PMID: 26168035; PMCID: PMC5414681

Nagasawa K, Chiba H, Fujita H, Kojima T, Saito T, Endo T, Sawada N (2006) Possible involvement of gap junctions in the barrier function of tight junctions of brain and lung endothelial cells. J Cell Physiol 208(1):123–132. https://doi.org/10.1002/jcp.20647. PMID: 16547974

Nation DA, Sweeney MD, Montagne A, Sagare AP, D'Orazio LM, Pachicano M, Sepehrband F, Nelson AR, Buennagel DP, Harrington MG, Benzinger TLS, Fagan AM, Ringman JM, Schneider LS, Morris JC, Chui HC, Law M, Toga AW, Zlokovic BV (2019) Blood-brain barrier breakdown is an early biomarker of human cognitive dysfunction. Nat Med. 25(2):270–276. https://doi.org/10.1038/s41591-018-0297-y. Epub 2019 Jan 14. PMID: 30643288; PMCID: PMC6367058

Nelson AR, Sweeney MD, Sagare AP, Zlokovic BV (2016) Neurovascular dysfunction and neurodegeneration in dementia and Alzheimer's disease. Biochim Biophys Acta. 1862(5):887–900. https://doi.org/10.1016/j.bbadis.2015.12.016. Epub 2015 Dec 17. PMID: 26705676; PMCID: PMC4821735

Nierwińska K, Nowacka-Chmielewska M, Bernacki J, Jagsz S, Chalimoniuk M, Langfort J, Małecki A (2019) The effect of endurance training and testosterone supplementation on the expression of blood spinal cord barrier proteins in rats. PLoS One 14(2):e0211818. https://doi.org/10.1371/journal.pone.0211818. PMID: 30742658; PMCID: PMC6370194

Nishijima T, Piriz J, Duflot S, Fernandez AM, Gaitan G, Gomez-Pinedo U, Verdugo JM, Leroy F, Soya H, Nuñez A, Torres-Aleman I (2010) Neuronal activity drives localized blood-brain-barrier transport of serum insulin-like growth factor-I into the CNS. Neuron 67(5):834–846. https://doi.org/10.1016/j.neuron.2010.08.007. PMID: 20826314

Nitta T, Hata M, Gotoh S, Seo Y, Sasaki H, Hashimoto N, Furuse M, Tsukita S (2003) Size-selective loosening of the blood-brain barrier in claudin-5-deficient mice. J Cell Biol 161(3):653–660. https://doi.org/10.1083/jcb.200302070. PMID: 12743111; PMCID: PMC2172943

Nwaobi SE, Cuddapah VA, Patterson KC, Randolph AC, Olsen ML (2016) The role of glial-specific Kir4.1 in normal and pathological states of the CNS. Acta Neuropathol 132(1):1–21. https://doi.org/10.1007/s00401-016-1553-1. Epub 2016 Mar 9. PMID: 26961251; PMCID: PMC6774634

Obermeier B, Daneman R, Ransohoff RM (2013) Development, maintenance and disruption of the blood-brain barrier. Nat Med 19(12):1584–1596. https://doi.org/10.1038/nm.3407. Epub 2013 Dec 5. PMID: 24309662; PMCID: PMC4080800

Oby E, Janigro D (2006) The blood-brain barrier and epilepsy. Epilepsia 47(11):1761–1774. https://doi.org/10.1111/j.1528-1167.2006.00817.x. PMID: 17116015

Ortiz GG, Pacheco-Moisés FP, Macías-Islas MÁ, Flores-Alvarado LJ, Mireles-Ramírez MA, González-Renovato ED, Hernández-Navarro VE, Sánchez-López AL, Alatorre-Jiménez MA (2014) Role of the blood-brain barrier in multiple sclerosis. Arch Med Res 45(8):687–697. https://doi.org/10.1016/j.arcmed.2014.11.013. Epub 2014 Nov 26. PMID: 25431839

Pardridge WM (2005) The blood-brain barrier: bottleneck in brain drug development. NeuroRx 2(1):3–14. https://doi.org/10.1602/neurorx.2.1.3. PMID: 15717053; PMCID: PMC539316

Patching SG (2017) Glucose transporters at the blood-brain barrier: function, regulation and gateways for drug delivery. Mol Neurobiol 54(2):1046–1077. https://doi.org/10.1007/s12035-015-9672-6. Epub 2016 Jan 22. PMID: 26801191

Patel MM, Patel BM (2017) Crossing the blood-brain barrier: recent advances in drug delivery to the brain. CNS Drugs 31(2):109–133. https://doi.org/10.1007/s40263-016-0405-9. PMID: 28101766

Pellegrini E, Fernezelian D, Malleret C, Gueguen MM, Patche-Firmin J, Rastegar S, Meilhac O, Diotel

N (2023) Estrogenic regulation of claudin 5 and tight junction protein 1 gene expression in zebrafish: a role on blood-brain barrier? J Comp Neurol 531(17):1828–1845. https://doi.org/10.1002/cne.25543. Epub 2023 Oct 9. PMID: 37814509

Perry G, Cash AD, Smith MA (2002) Alzheimer disease and oxidative stress. J Biomed Biotechnol 2(3):120–123. https://doi.org/10.1155/S1110724302203010. PMID: 12488575; PMCID: PMC161361

Pillinger NL, Kam P (2017) Endothelial glycocalyx: basic science and clinical implications. Anaesth Intensive Care 45(3):295–307. https://doi.org/10.1177/0310057X1704500305. PMID: 28486888

Pimentel E, Sivalingam K, Doke M, Samikkannu T (2020) Effects of drugs of abuse on the blood-brain barrier: a brief overview. Front Neurosci 14:513. https://doi.org/10.3389/fnins.2020.00513. PMID: 32670001; PMCID: PMC7326150

Profaci CP, Munji RN, Pulido RS, Daneman R (2020) The blood-brain barrier in health and disease: important unanswered questions. J Exp Med 217(4):e20190062. https://doi.org/10.1084/jem.20190062. PMID: 32211826; PMCID: PMC7144528

Pun PB, Lu J, Moochhala S (2009) Involvement of ROS in BBB dysfunction. Free Radic Res 43(4):348–364. https://doi.org/10.1080/10715760902751902. Epub 2009 Feb 24. PMID: 19241241

Puzzo D, Arancio O (2013) Amyloid-β peptide: Dr. Jekyll or Mr. Hyde? J Alzheimers Dis 33(Suppl 1):S111–S120. https://doi.org/10.3233/JAD-2012-129033. PMID: 22735675; PMCID: PMC3696497

Raja R, Rosenberg GA, Caprihan A (2018) MRI measurements of blood-brain barrier function in dementia: a review of recent studies. Neuropharmacology. 134(Pt B):259–271. https://doi.org/10.1016/j.neuropharm.2017.10.034. Epub 2017 Oct 28. PMID: 29107626; PMCID: PMC6044415

Ransohoff RM, Engelhardt B (2012) The anatomical and cellular basis of immune surveillance in the central nervous system. Nat Rev Immunol 12(9):623–635. https://doi.org/10.1038/nri3265. Epub 2012 Aug 20. PMID: 22903150

Rasmussen MK, Mestre H, Nedergaard M (2018) The glymphatic pathway in neurological disorders. Lancet Neurol 17(11):1016–1024. https://doi.org/10.1016/S1474-4422(18)30318-1. PMID: 30353860; PMCID: PMC6261373

Rasmussen MK, Mestre H, Nedergaard M (2022) Fluid transport in the brain. Physiol Rev. 102(2):1025–1151. https://doi.org/10.1152/physrev.00031.2020. Epub 2021 May 5. PMID: 33949874; PMCID: PMC8897154

Reese TS, Karnovsky MJ (1967) Fine structural localization of a blood-brain barrier to exogenous peroxidase. J Cell Biol 34(1):207–217. https://doi.org/10.1083/jcb.34.1.207. PMID: 6033532; PMCID: PMC2107213

Reeson P, Choi K, Brown CE (2018) VEGF signaling regulates the fate of obstructed capillaries in mouse cortex. elife 26(7):e33670. https://doi.org/10.7554/eLife.33670. PMID: 29697373; PMCID: PMC5919759

Reiter JT, Schulte F, Bauer T, David B, Endler C, Isaak A, Schuch F, Bitzer F, Witt JA, Hattingen E, Deichmann R, Attenberger U, Becker AJ, Helmstaedter C, Radbruch A, Surges R, Friedman A, Rüber T (2024) Evidence for interictal blood-brain barrier dysfunction in people with epilepsy. Epilepsia 65(5):1462–1474. https://doi.org/10.1111/epi.17929. Epub 2024 Mar 4. PMID: 38436479

Rempe RG, Hartz AMS, Bauer B (2016) Matrix metalloproteinases in the brain and blood-brain barrier: Versatile breakers and makers. J Cereb Blood Flow Metab 36(9):1481–1507. https://doi.org/10.1177/0271678X16655551. Epub 2016 Jun 20. PMID: 27323783; PMCID: PMC5012524

Ricci G, Volpi L, Pasquali L, Petrozzi L, Siciliano G (2009) Astrocyte-neuron interactions in neurological disorders. J Biol Phys. 35(4):317–336. https://doi.org/10.1007/s10867-009-9157-9. Epub 2009 May 14. PMID: 19669420; PMCID: PMC2750745

Ridley H (1695) The anatomy of the Brain. Printers to the Royal Society

Rivard A, Principe N, Andrés V (2000) Age-dependent increase in c-fos activity and cyclin A expression in vascular smooth muscle cells. A potential link between aging, smooth muscle cell proliferation and atherosclerosis. Cardiovasc Res 45(4):1026–1034. https://doi.org/10.1016/s0008-6363(99)00385-5. PMID: 10728429

Ronaldson PT, Davis TP (2020) Regulation of blood-brain barrier integrity by microglia in health and disease: a therapeutic opportunity. J Cereb Blood Flow Metab. 40(1_suppl):S6–S24. https://doi.org/10.1177/0271678X20951995. Epub 2020 Sep 14. PMID: 32928017; PMCID: PMC7687032

Rüber T, David B, Lüchters G, Nass RD, Friedman A, Surges R, Stöcker T, Weber B, Deichmann R, Schlaug G, Hattingen E, Elger CE (2018) Evidence for peri-ictal blood-brain barrier dysfunction in patients with epilepsy. Brain 141(10):2952–2965. https://doi.org/10.1093/brain/awy242. PMID: 30239618

Rzechorzek NM, Thrippleton MJ, Chappell FM, Mair G, Ercole A, Cabeleira M, CENTER-TBI High Resolution ICU (HR ICU) Sub-Study Participants and Investigators; Rhodes J, Marshall I, O'Neill JS (2022) A daily temperature rhythm in the human brain predicts survival after brain injury. Brain 145(6):2031–2048. https://doi.org/10.1093/brain/awab466. PMID: 35691613; PMCID: PMC9336587

Sakakibara A, Furuse M, Saitou M, Ando-Akatsuka Y, Tsukita S (1997) Possible involvement of phosphorylation of occludin in tight junction formation. J Cell Biol 137(6):1393–1401. https://doi.org/10.1083/jcb.137.6.1393. PMID: 9182670; PMCID: PMC2132539

Sathyanesan M, Haiar JM, Watt MJ, Newton SS (2017) Restraint stress differentially regulates inflammation and glutamate receptor gene expression in the hippocampus of C57BL/6 and BALB/c mice. Stress 20(2):197–204. https://doi.org/10.1080/10253890.2

017.1298587. Epub 2017 Mar 8. PMID: 28274152; PMCID: PMC5724770

Schaeffer S, Iadecola C (2021) Revisiting the neurovascular unit. Nat Neurosci 24(9):1198–1209. https://doi.org/10.1038/s41593-021-00904-7. Epub 2021 Aug 5. PMID: 34354283; PMCID: PMC9462551

Schlageter KE, Molnar P, Lapin GD et al (1999) Microvessel organization and structure in experimental brain tumors: microvessel populations with distinctive structural and functional properties. Microvasc Res 58(3):312–328

Schreiner TG, Romanescu C, Popescu BO (2022) The blood-brain barrier—a key player in multiple sclerosis disease mechanisms. biomol ther 12(4):538. https://doi.org/10.3390/biom12040538. PMID: 35454127; PMCID: PMC9025898

Schultz N, Brännström K, Byman E, Moussaud S, Nielsen HM, Olofsson A, Wennström M, Netherlands Brain Bank (2018) Amyloid-beta 1-40 is associated with alterations in NG2+ pericyte population ex vivo and in vitro. Aging Cell 17(3):e12728. https://doi.org/10.1111/acel.12728. Epub 2018 Feb 17. PMID: 29453790; PMCID: PMC5946076

Sehar U, Rawat P, Reddy AP, Kopel J, Reddy PH (2022) Amyloid beta in aging and Alzheimer's disease. Int J Mol Sci 23(21):12924. https://doi.org/10.3390/ijms232112924. PMID: 36361714; PMCID: PMC9655207

Shen J, Xin W, Li Q, Gao Y, Yuan L, Zhang J (2019) Methylene blue reduces neuronal apoptosis and improves blood-brain barrier integrity after traumatic Brain injury. Front Neurol 10:1133. https://doi.org/10.3389/fneur.2019.01133. PMID: 31787917; PMCID: PMC6856146

Shepley BR, Ainslie PN, Hoiland RL, Donnelly J, Sekhon MS, Zetterberg H, Blennow K, Bain AR (1985) Negligible influence of moderate to severe hyperthermia on blood-brain barrier permeability and neuronal parenchymal integrity in healthy men. J Appl Physiol 130(3):792–800. https://doi.org/10.1152/japplphysiol.00645.2020. Epub 2021 Jan 14. PMID: 33444119

Shimizu F, Sano Y, Maeda T, Abe MA, Nakayama H, Takahashi R, Ueda M, Ohtsuki S, Terasaki T, Obinata M, Kanda T (2008) Peripheral nerve pericytes originating from the blood-nerve barrier expresses tight junctional molecules and transporters as barrier-forming cells. J Cell Physiol 217(2):388–399. https://doi.org/10.1002/jcp.21508. PMID: 18543246

Shin K, Margolis B (2006) ZOning out tight junctions. Cell 126(4):647–649. https://doi.org/10.1016/j.cell.2006.08.005. PMID: 16923383

Shin RW, Iwaki T, Kitamoto T, Tateishi J (1991) Hydrated autoclave pretreatment enhances tau immunoreactivity in formalin-fixed normal and Alzheimer's disease brain tissues. Lab Investig 64(5):693–702. PMID: 1903170

Shin JA, Yoon JC, Kim M, Park EM (2016) Activation of classical estrogen receptor subtypes reduces tight junction disruption of brain endothelial cells under ischemia/reperfusion injury. Free Radic Biol Med 92:78–89. https://doi.org/10.1016/j.freeradbiomed.2016.01.010. Epub 2016 Jan 16. PMID: 26784014

Sixt M, Engelhardt B, Pausch F, Hallmann R, Wendler O, Sorokin LM (2001) Endothelial cell laminin isoforms, laminins 8 and 10, play decisive roles in T cell recruitment across the blood-brain barrier in experimental autoimmune encephalomyelitis. J Cell Biol 153(5):933–946. https://doi.org/10.1083/jcb.153.5.933. PMID: 11381080; PMCID: PMC2174323

Skandalakis GP, Rivera DR, Rizea CD, Bouras A, Jesu Raj JG, Bozec D, Hadjipanayis CG (2020) Hyperthermia treatment advances for brain tumors. Int J Hyperth 37(2):3–19. https://doi.org/10.1080/02656736.2020.1772512. PMID: 32672123; PMCID: PMC7756245

Sorokin L (2010) The impact of the extracellular matrix on inflammation. Nat Rev Immunol 10(10):712–723. https://doi.org/10.1038/nri2852. PMID: 20865019

Stamatovic SM, Johnson AM, Keep RF, Andjelkovic AV (2016) Junctional proteins of the blood-brain barrier: new insights into function and dysfunction. Tissue Barriers 4(1):e1154641. https://doi.org/10.1080/21688370.2016.1154641. PMID: 27141427; PMCID: PMC4836471

Stevenson BR, Siliciano JD, Mooseker MS, Goodenough DA (1986) Identification of ZO-1: a high molecular weight polypeptide associated with the tight junction (zonula occludens) in a variety of epithelia. J Cell Biol 103(3):755–766. https://doi.org/10.1083/jcb.103.3.755. PMID: 3528172; PMCID: PMC2114282

Storck SE, Meister S, Nahrath J, Meißner JN, Schubert N, Di Spiezio A, Baches S, Vandenbroucke RE, Bouter Y, Prikulis I, Korth C, Weggen S, Heimann A, Schwaninger M, Bayer TA, Pietrzik CU (2016) Endothelial LRP1 transports amyloid-β(1–42) across the blood-brain barrier. J Clin Invest. 126(1):123–136. https://doi.org/10.1172/JCI81108. Epub 2015 Nov 30. PMID: 26619118; PMCID: PMC4701557

Streit WJ (2005) Microglia and neuroprotection: implications for Alzheimer's disease. Brain Res Brain Res Rev 48(2):234–239. https://doi.org/10.1016/j.brainresrev.2004.12.013. Epub 2005 Jan 15. PMID: 15850662

Sun P, Bu F, Min JW, Munshi Y, Howe MD, Liu L, Koellhoffer EC, Qi L, McCullough LD, Li J (2019) Inhibition of calcium/calmodulin-dependent protein kinase kinase (CaMKK) exacerbates impairment of endothelial cell and blood-brain barrier after stroke. Eur J Neurosci 49(1):27–39. https://doi.org/10.1111/ejn.14223. Epub 2018 Nov 29. PMID: 30422362; PMCID: PMC6590081

Sweeney MD, Ayyadurai S, Zlokovic BV (2016) Pericytes of the neurovascular unit: key functions and signaling pathways. Nat Neurosci 19(6):771–783. https://doi.org/10.1038/nn.4288. PMID: 27227366; PMCID: PMC5745011

Sweeney MD, Sagare AP, Zlokovic BV (2018) Blood-brain barrier breakdown in Alzheimer disease and other neurodegenerative disorders. Nat Rev Neurol 14(3):133–150. https://doi.org/10.1038/nrneurol.2017.188. Epub 2018 Jan 29. PMID: 29377008; PMCID: PMC5829048

Sweeney MD, Zhao Z, Montagne A, Nelson AR, Zlokovic BV (2019) Blood-Brain barrier: from physiology to disease and back. Physiol Rev 99(1):21–78. https://doi.org/10.1152/physrev.00050.2017. PMID: 30280653; PMCID: PMC6335099

Takeshita Y, Ransohoff RM (2012) Inflammatory cell trafficking across the blood-brain barrier: chemokine regulation and in vitro models. Immunol Rev 248(1):228–239. https://doi.org/10.1111/j.1600-065X.2012.01127.x. PMID: 22725965; PMCID: PMC3383666

Tarumi T, Zhang R (2018) Cerebral blood flow in normal aging adults: cardiovascular determinants, clinical implications, and aerobic fitness. J Neurochem 144(5):595–608. https://doi.org/10.1111/jnc.14234. Epub 2017 Nov 7. PMID: 28986925; PMCID: PMC5874160

Tatum WO 4th (2012) Mesial temporal lobe epilepsy. J Clin Neurophysiol 29(5):356–365. https://doi.org/10.1097/WNP.0b013e31826b3ab7. PMID: 23027091

Thijs RD, Surges R, O'Brien TJ, Sander JW (2019) Epilepsy in adults. Lancet 393(10172):689–701. https://doi.org/10.1016/S0140-6736(18)32596-0. Epub 2019 Jan 24. PMID: 30686584

Umeda K, Matsui T, Nakayama M, Furuse K, Sasaki H, Furuse M, Tsukita S (2004) Establishment and characterization of cultured epithelial cells lacking expression of ZO-1. J Biol Chem 279(43):44785–44794. https://doi.org/10.1074/jbc.M406563200. Epub 2004 Jul 30. PMID: 15292177

van Assema DM, Lubberink M, Bauer M, van der Flier WM, Schuit RC, Windhorst AD, Comans EF, Hoetjes NJ, Tolboom N, Langer O, Müller M, Scheltens P, Lammertsma AA, van Berckel BN (2012) Blood-brain barrier P-glycoprotein function in Alzheimer's disease. Brain 135(Pt 1):181–189. https://doi.org/10.1093/brain/awr298. Epub 2011 Nov 26. PMID: 22120145

van de Haar HJ, Jansen JFA, van Osch MJP, van Buchem MA, Muller M, Wong SM, Hofman PAM, Burgmans S, Verhey FRJ, Backes WH (2016) Neurovascular unit impairment in early Alzheimer's disease measured with magnetic resonance imaging. Neurobiol Aging 45:190–196. https://doi.org/10.1016/j.neurobiolaging.2016.06.006. Epub 2016 Jun 17. PMID: 27459939

van Vliet EA, da Costa AS, Redeker S, van Schaik R, Aronica E, Gorter JA (2007) Blood-brain barrier leakage may lead to progression of temporal lobe epilepsy. Brain 130(Pt 2):521–534. https://doi.org/10.1093/brain/awl318. Epub 2006 Nov 22. PMID: 17124188

van Vliet EA, Aronica E, Gorter JA (2015) Blood-brain barrier dysfunction, seizures and epilepsy. Semin Cell Dev Biol 38:26–34. https://doi.org/10.1016/j.semcdb.2014.10.003. Epub 2014 Nov 6. PMID: 25444846

Vasile F, Dossi E, Rouach N (2017) Human astrocytes: structure and functions in the healthy brain. Brain Struct Funct 222(5):2017–2029. https://doi.org/10.1007/s00429-017-1383-5. Epub 2017 Mar 9. PMID: 28280934; PMCID: PMC5504258

Verheggen ICM, de Jong JJA, van Boxtel MPJ, Gronenschild EHBM, Palm WM, Postma AA, Jansen JFA, Verhey FRJ, Backes WH (2020) Increase in blood-brain barrier leakage in healthy, older adults. Geroscience 42(4):1183–1193. https://doi.org/10.1007/s11357-020-00211-2. Epub 2020 Jun 29. PMID: 32601792; PMCID: PMC7394987

Verscheijden LFM, Koenderink JB, de Wildt SN, Russel FGM (2019) Development of a physiologically-based pharmacokinetic pediatric brain model for prediction of cerebrospinal fluid drug concentrations and the influence of meningitis. PLoS Comput Biol 15(6):e1007117. https://doi.org/10.1371/journal.pcbi.1007117. PMID: 31194730; PMCID: PMC6592555

Vijg J (2000) Somatic mutations and aging: a re-evaluation. Mutat Res 447(1):117–135. https://doi.org/10.1016/s0027-5107(99)00202-x. PMID: 10686308

Vos CM, Geurts JJ, Montagne L, van Haastert ES, Bö L, van der Valk P, Barkhof F, de Vries HE (2005) Blood-brain barrier alterations in both focal and diffuse abnormalities on postmortem MRI in multiple sclerosis. Neurobiol Dis 20(3):953–960. https://doi.org/10.1016/j.nbd.2005.06.012. Epub 2005 Jul 21. PMID: 16039866

Wallin A, Román GC, Esiri M, Kettunen P, Svensson J, Paraskevas GP, Kapaki E (2018) Update on vascular cognitive impairment associated with subcortical small-vessel disease. J Alzheimers Dis 62(3):1417–1441. https://doi.org/10.3233/JAD-170803. PMID: 29562536; PMCID: PMC5870030

Wang L, Swank JS, Glick IE, Gado MH, Miller MI, Morris JC, Csernansky JG (2003) Changes in hippocampal volume and shape across time distinguish dementia of the Alzheimer type from healthy aging. NeuroImage 20(2):667–682. https://doi.org/10.1016/S1053-8119(03)00361-6. PMID: 14568443

Wang H, Kim M, Normoyle KP, Llano D (2016) Thermal regulation of the brain—an anatomical and physiological review for clinical neuroscientists. Front Neurosci 9:528. https://doi.org/10.3389/fnins.2015.00528. PMID: 26834552; PMCID: PMC4720747

Weber CM, Clyne AM (2021) Sex differences in the blood-brain barrier and neurodegenerative diseases. APL Bioeng 5(1):011509. https://doi.org/10.1063/5.0035610. PMID: 33758788; PMCID: PMC7968933

Welcome MO, Mastorakis NE (2020) Stress-induced blood brain barrier disruption: Molecular mechanisms and signaling pathways. Pharmacol Res. 157:104769. https://doi.org/10.1016/j.phrs.2020.104769. Epub 2020 Apr 8

Wilson PW, Garrison RJ, Castelli WP (1985) Postmenopausal estrogen use, cigarette smoking, and cardiovascular morbidity in women over 50. The framingham study. N Engl J Med 313(17):1038–1043. https://doi.org/10.1056/NEJM198510243131702. PMID: 2995808

Wilson ME, Westberry JM, Prewitt AK (2008) Dynamic regulation of estrogen receptor-alpha gene expression in the brain: a role for promoter methylation? Front Neuroendocrinol 29(3):375–385. https://doi.org/10.1016/j.yfrne.2008.03.002. Epub 2008 Mar 13. PMID: 18439661; PMCID: PMC2460564

Winger RC, Koblinski JE, Kanda T, Ransohoff RM, Muller WA (2014) Rapid remodeling of tight junctions during paracellular diapedesis in a human model of the blood-brain barrier. J Immunol 193(5):2427–2437. https://doi.org/10.4049/jimmunol.1400700. Epub 2014 Jul 25. PMID: 25063869; PMCID: PMC4138548

Winkler EA, Bell RD, Zlokovic BV (2011) Central nervous system pericytes in health and disease. Nat Neurosci 14(11):1398–1405. https://doi.org/10.1038/nn.2946. PMID: 22030551; PMCID: PMC4020628

Winkler MK, Chassidim Y, Lublinsky S, Revankar GS, Major S, Kang EJ, Oliveira-Ferreira AI, Woitzik J, Sandow N, Scheel M, Friedman A, Dreier JP (2012) Impaired neurovascular coupling to ictal epileptic activity and spreading depolarization in a patient with subarachnoid hemorrhage: possible link to blood-brain barrier dysfunction. Epilepsia 53(Suppl 6):22–30. https://doi.org/10.1111/j.1528-1167.2012.03699.x. PMID: 23134492; PMCID: PMC3625730

Wu M, Tsirka SE (2009) Endothelial NOS-deficient mice reveal dual roles for nitric oxide during experimental autoimmune encephalomyelitis. Glia 57(11):1204–1215. https://doi.org/10.1002/glia.20842. PMID: 19170181; PMCID: PMC2706940

Xu L, Nirwane A, Yao Y (2018) Basement membrane and blood-brain barrier. Stroke Vasc Neurol 4(2):78–82. https://doi.org/10.1136/svn-2018-000198. PMID: 31338215; PMCID: PMC6613871

Xu W, Bai Q, Dong Q, Guo M, Cui M (2022) Blood-brain barrier dysfunction and the potential mechanisms in chronic cerebral hypoperfusion induced cognitive impairment. Front Cell Neurosci 16:870674. https://doi.org/10.3389/fncel.2022.870674. PMID: 35783093; PMCID: PMC9243657

Yang J, Ma K, Zhang C, Liu Y, Liang F, Hu W, Bian X, Yang S, Fu X (2020) Burns impair blood-brain barrier and mesenchymal stem cells can reverse the process in mice. Front Immunol 11:578879. https://doi.org/10.3389/fimmu.2020.578879. PMID: 33240266; PMCID: PMC7677525

Yao Y, Chen ZL, Norris EH, Strickland S (2014) Astrocytic laminin regulates pericyte differentiation and maintains blood brain barrier integrity. Nat Commun 5:3413. https://doi.org/10.1038/ncomms4413. PMID: 24583950; PMCID: PMC3992931

Yasuda T, Mochizuki H (2010) Use of growth factors for the treatment of Parkinson's disease. Expert Rev Neurother 10(6):915–924. https://doi.org/10.1586/ern.10.55. PMID: 20518608

Ye WN, Thipse M, Mahdi MB, Azad S, Davies R, Ruel M, Silver MA, Hakami L, Mesana T, Leenen F, Mussivand T (2020) Can heat therapy help patients with heart failure? Artif Organs 44(7):680–692. https://doi.org/10.1111/aor.13659. Epub 2020 Mar 8. PMID: 32017138

Yenari MA, Hemmen TM (2010) Therapeutic hypothermia for brain ischemia: where have we come and where do we go? Stroke 41(10 Suppl):S72–S74. https://doi.org/10.1161/STROKEAHA.110.595371. PMID: 20876510; PMCID: PMC2953728

Zhang W, Liu QY, Haqqani AS, Leclerc S, Liu Z, Fauteux F, Baumann E, Delaney CE, Ly D, Star AT, Brunette E, Sodja C, Hewitt M, Sandhu JK, Stanimirovic DB (2020) Differential expression of receptors mediating receptor-mediated transcytosis (RMT) in brain microvessels, brain parenchyma and peripheral tissues of the mouse and the human. Fluids Barriers CNS. 17(1):47. https://doi.org/10.1186/s12987-020-00209-0. PMID: 32698806; PMCID: PMC7376922

Zhao Z, Nelson AR, Betsholtz C, Zlokovic BV (2015) Establishment and dysfunction of the Blood-Brain barrier. Cell 163(5):1064–1078. https://doi.org/10.1016/j.cell.2015.10.067. PMID: 26590417; PMCID: PMC4655822

Zhao Y, Gan L, Ren L, Lin Y, Ma C, Lin X (2021) Factors influencing the blood-brain barrier permeability. Brain Res. 1788:147937. https://doi.org/10.1016/j.brainres.2022.147937. Epub 2022 May 11. PMID: 35568085

Zhao N, Chung TD, Guo Z, Jamieson JJ, Liang L, Linville RM, Pessell AF, Wang L, Searson PC (2023) The influence of physiological and pathological perturbations on blood-brain barrier function. Front Neurosci 17:1289894. https://doi.org/10.3389/fnins.2023.1289894. Erratum in: Front Neurosci. 2024;17:1328902. https://doi.org/10.3389/fnins.2023.1328902. PMID: 37937070; PMCID: PMC10626523

Zhu J, Li X, Yin J, Hu Y, Gu Y, Pan S (2018) Glycocalyx degradation leads to blood-brain barrier dysfunction and brain edema after asphyxia cardiac arrest in rats. J Cereb Blood Flow Metab 38(11):1979–1992. https://doi.org/10.1177/0271678X17726062. Epub 2017. PMID: 28825336; PMCID: PMC6259325

Zlokovic BV (2008) The blood-brain barrier in health and chronic neurodegenerative disorders. Neuron 57(2):178–201. https://doi.org/10.1016/j.neuron.2008.01.003. PMID: 18215617

Noradrenaline Regulation of Brain-Body Communication

2

Giselle Fernandes, Ruby Lam, Jiho Park, Gabrielle T. Drummond, and Mriganka Sur

Abstract

Norepinephrine (NE) and its methylated form, epinephrine (Epi), are catecholamines that serve as both neurotransmitters and hormones, regulating a wide array of physiological functions. In the central nervous system (CNS), NE acts as a neurotransmitter, primarily released by the locus coeruleus (LC), to regulate vigilance states and optimize behavioral performance. It is also the primary neurotransmitter of the peripheral, sympathetic nervous system, responsible for tonic and reflexive changes to cardiovascular tone. Conversely, Epi operates primarily as a hormone released by the adrenal medulla, to regulate metabolic homeostasis and orchestrate the body's response to acute stress. While there is no evidence for direct connections between the central and peripheral adrenergic systems, various feedforward and feedback circuits enable them to collaborate and synchronize brain-body responses to stimuli. A notable example of this interaction is evident in the body's response to stress, where multiple adrenergic systems work in concert to elicit a fight-or-flight response to perceived danger. Reciprocal communication between the LC, other adrenergic systems in the brain, and the NE-driven sympathetic nervous system induces a state of heightened arousal, enhances the processing of sensory stimuli, and facilitates adaptive behavioral responses to stressors.

In this chapter, we will explore adrenergic systems in the brain and body and share evidence for multiple channels of communication between them. We will trace the anatomical connections and physiological characteristics of NE-producing neurons in the brain and the peripheral nervous system. We will also describe adrenergic systems in the body, projections of the sympathetic post-ganglionic neurons, the vasal ganglia of the cranial nerve, and release of NE and Epi by the adrenal medulla into the bloodstream. Importantly, this chapter will elucidate the circuits and pathways of communication between NE in the central and peripheral nervous systems that regulate and synchronize cognitive processes and physiological responses. Finally, we will explore various neurological disorders, including chronic pain and

G. Fernandes · J. Park · G. T. Drummond · M. Sur (✉)
Picower Institute for Learning and Memory, Brain and Cognitive Sciences, Massachusetts Institute of Technology, Cambridge, MA, USA
e-mail: msur@mit.edu

R. Lam
Picower Institute for Learning and Memory, Brain and Cognitive Sciences, Massachusetts Institute of Technology, Cambridge, MA, USA

Boston Children's Hospital and Harvard Medical School teaching hospital, Boston, USA, 02115, 300 Longwood Avenue, MA

D. Tropea, E. Giacometti (eds.), *Brain-Body Connections*, Advances in Experimental Medicine and Biology 1477, https://doi.org/10.1007/978-3-031-89525-8_2

Alzheimer's disease (AD), that show evidence for deficits in adrenergic systems and disruption of the communication between the brain and body.

Keywords

Norepinephrine · Locus Coeruleus · Sympathetic ganglia · Arousal · Fight-or-flight response

Abbreviations

Ach	Acetyl choline
ACTH	Adrenocorticotropic hormone
AD	Alzheimer's disease
ADHD	Attention deficit hyperactivity disorder
ARAS	Ascending reticular activating system
ATP	Adenosine triphosphate
CeA	Central amygdala
CGRP	Calcitonin gene-related peptide
CNS	Central nervous system
CRH	Corticotropin releasing hormone
DRG	Dorsal root ganglia
DSP4	N-(2-chloroethyl)-N-ethyl-2-bromobenzylamine hydrochloride
EEG	Electroencephalogram
Epi	Epinephrine
GANE	Glutamate amplifies noradrenergic effects
GPCRs	G-protein coupled receptors
HPA	Hypothalamic-pituitary-adrenal axis
LC	Locus Coeruleus
LMTRS	Low threshold mechanoreceptors
LTP	Long-term potentiation
MERFISH	Multiplexed error-robust fluorescence in situ hybridization
NAc	Nucleus accumbens
NE	Norepinephrine
NGF	Nerve growth factor
NPY	Neuropeptide Y
NREM	Non-rapid eye movement
NTS	Nucleus tractus solitarius
PAG	Periaqueductal gray
PKA	Protein kinase A
PTSD	Post-traumatic stress disorder
PVN	Paraventricular nucleus of the hypothalamus
REM	Rapid eye movement
RPE	Reward prediction error
SAM	Sympathetic-adrenomedullary axis
TH	Tyrosine hydroxylase

Norepinephrine (NE, also known as noradrenaline) and epinephrine (Epi, also known as adrenaline) are monoamine neurotransmitters and hormones that are widely distributed in the brain and body. NE is widely released in the central nervous system (CNS), while epinephrine (Epi) is present in relatively low amounts in the mammalian brain. Outside the brain, NE is one of the primary neurotransmitters of the autonomic nervous system and also functions as a hormone released by the adrenal medulla. On the other hand, Epi functions primarily as a hormone, mainly secreted by the adrenal gland and by a small number of neurons in the medulla oblongata. NE and Epi are synthesized in the body from dietary amino acids through a series of enzymatic reactions and are the targets of numerous clinically significant drugs that modify the noradrenergic system (Fernstrom and Fernstrom 2007). Together, these molecules play a vital role in the maintenance of homeostasis through the autonomic nervous system, the modulation of memory through adaptive gain of neuronal circuits, and the regulation of behavior through changes in sensorimotor processing.

2.1 Noradrenaline in the Brain

NE-producing neurons in the CNS form a relatively small group of cells primarily residing in portions of the medulla oblongata, pontine, and dorsal vagal complex. Early studies of monoaminergic distribution and histochemical staining identified 7 groups of cells (named A1 - > A7) based on their location and projection patterns (Felten and Sladek 1983), with the A6 group or the Locus Coeruleus (LC) being the most widely studied. The LC is the largest of these nuclei, comprising almost 50% of NE neurons in the brain. Though small in number, noradrenergic neurons project widely throughout the brain and spinal cord, innervating almost every brain region and regulating multiple aspects of sensory processing, learning, and motor control. Groups of NE neurons in the ventral and dorsal tegmentum of the medulla oblongata and the pons (A1, A2, A3, A7) form the ventral ascending noradrenergic pathway while the LC forms the primary dorsal ascending pathway. Ascending NE fibers arising from the A1, A2, A5, and A7 nuclei project to the midbrain reticular formation, the entire hypothalamus, and parts of the limbic cortex. Descending fibers from these nuclei terminate in the ventral and intermediate nuclear columns of the spinal cord (Felten and Sladek 1983; Guyenet 1991).

The A1 group of neurons, located in the caudal ventrolateral medulla of the brainstem, is thought to be primarily involved in the regulation of cardiovascular control, including maintenance of blood pressure. A large fraction of A1 cells co-express neuropeptide Y (NPY) and a smaller fraction express galanin (Guyenet 1991; Everitt et al. 1984). Tracing studies indicate that the A1 nucleus projects to various regions in the hypothalamus, the median preoptic area, the bed nucleus of stria terminalis, and the nucleus tractus solitarius (NTS) (Guyenet 1991). The A2 group of cells lies within the hindbrain dorsal vagal complex and supports its role in vagal sensory-motor reflex flow. Extensive reciprocal connections to the majority of the NTS, various hypothalamic nuclei, and various limbic and cortical forebrain regions make the A2 nucleus an important crossroad of visceral motor outflow and interoceptive feedback to the brain (Rinaman 2011). Projections from the A2 cell group are thought to relay gastric and cardiovascular (i.e., visceral) signals to regions such as the amygdala and nucleus accumbens (NAc), modulating their activity and shaping learning of emotional stimuli (Rinaman 2011; Kerfoot et al. 2008). The A5 and A7 nuclei form 2 clusters of noradrenergic cells in the pontine tegmentum. While the vast majority of A5 neurons innervate the intermediolateral cell column of the spinal cord, tracing studies also demonstrate projections to the central amygdala, hypothalamus, periaqueductal gray (PAG), and NTS (Byrum and Guyenet 1987). Chemogenetic activation of A5 neurons drives sympathetic nerve activity to increase blood pressure (Souza et al. 2022a), suggesting a role for these cells in the modulation of blood flow. On the other hand, A7 neurons relay antinociceptive signals from the ventral PAG to the dorsal horn of the spinal cord (Bajic et al. 2001; Clark and Proudfit 1991).

While the various noradrenergic nuclei display territorial specialization in their downstream projections, they all play a part in both relaying visceral information to the brain and modulating bodily functions through the regulation of autonomic and endocrine functions. However, direct modulation of complex behavior is mediated almost exclusively by the LC (or the A6 group), which is the sole noradrenergic nucleus innervating the cerebral cortex.

2.1.1 The Locus Coeruleus and Its Connections in the Brain

The majority of NE-producing neurons are located in the LC, or "blue spot":a small nucleus located deep in the brainstem that supplies far-reaching noradrenergic innervations to the brain (Poe et al. 2020). Despite its small number of neurons, ~3000 neurons in the rodent brain (Sara 2009) and ~50,000 in the human brain (Sharma et al. 2010), the LC sends out dense projections to most cortical and subcortical areas (Schwarz and Luo 2015) (illustrated in Fig. 2.1). Importantly, the LC is the only source of NE for the hippocampus and neocortex, areas essential for higher cognitive and emotional processes (Poe et al. 2020; Aston-Jones and Cohen 2005; Breton-Provencher et al. 2021). Similarly, LC neurons are thought to receive excitatory inputs from a diverse set of brain regions including most brainstem nuclei and forebrain areas (Schwarz and Luo 2015; Breton-Provencher et al. 2021). Additionally, the dense core of NE-producing neurons that make up the LC receives several

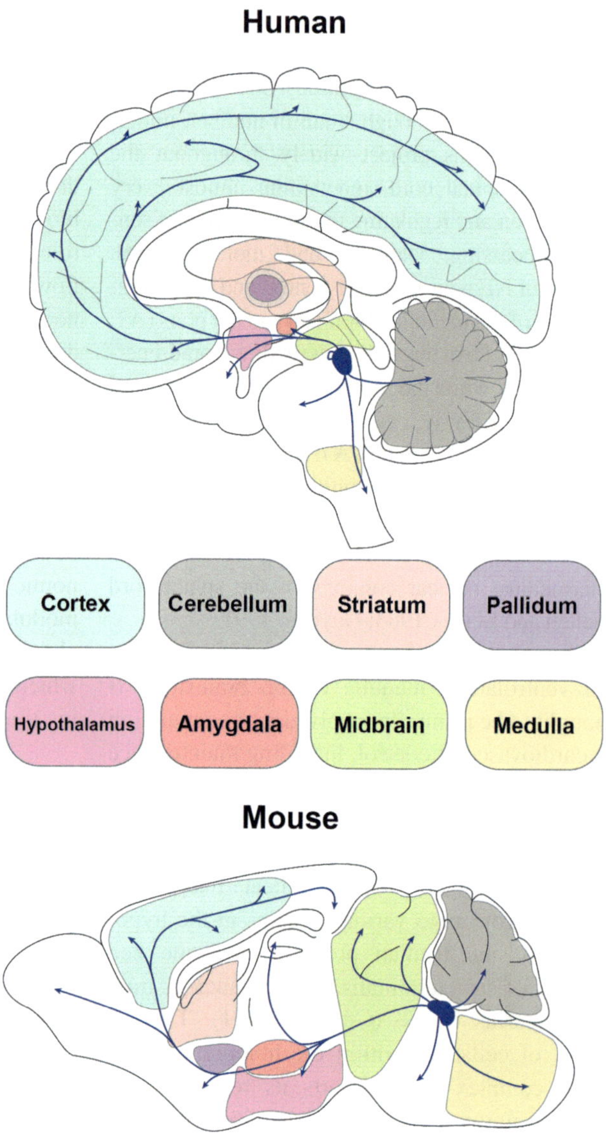

Fig. 2.1 NE connections in the brain. Projections of the Locus Coeruleus, the brain's primary noradrenergic nucleus, in the human *(top)* and mouse *(bottom)* brain. (Adapted with permission from (Breton-Provencher et al. 2021))

local (Breton-Provencher and Sur 2019; Jones 1991a) and long-range (Ennis and Aston-Jones 1989; Dimitrov et al. 2013) GABAergic connections as well as inputs from cholinergic and neuropeptide-S expressing neurons (Jones 1991a; Xu et al. 2004). The wide range of convergent inputs to LC-NE neurons suggests their recruitment in the detection of salient sensory stimuli as well as in the regulation of goal-directed behaviors. For example, the LC receives afferent projections from the gigantocellular reticular nucleus in the brainstem—which responds to tactile, visual, auditory, vestibular, and olfactory stimuli (Martin et al. 2010)—resulting in LC activation following salient sensory stimulation. On the other hand, projections from the central amygdala recruit LC-NE neurons to the body's stress response (Bouret et al. 2003; Curtis et al. 2002; Van Bockstaele et al. 1998), while LC inputs from distinct subregions of the medial prefrontal cortex suggest a role for LC-NE in the modulation of learning and memory (Sara and Hervé-Minvielle 1995; Jodo and Aston-Jones 1997; Jodo et al. 1998).

Historically, the LC was considered to be a homogenous nucleus, with no distinct organization of inputs along its rostro-caudal axis and a uniform release of NE into the forebrain, cerebellum, brainstem, and spinal cord (Aston-Jones and Waterhouse 2016). This suggests that the LC integrates information from and broadcasts signals to many brain regions, consistent with its key function in modulating global brain states. In an important study, Schwarz et al. (2015) used the TRIO (tracing the relationship between input and output) method to virally trace the input and output relationships of LC neurons. They reported that LC neurons have a largely indiscriminate input-output relationship, with LC neurons that project to diverse brain regions also receiving similar densities of inputs from the same regions. Despite this highly integrative architecture of LC-NE connectivity, they also found some heterogeneity of LC sub-circuits. LC-NE neurons that project to the medulla received disproportionately fewer inputs from the central amygdala than those projecting to other areas. This implies that medulla-projecting LC neurons are preferentially immune to stress-related activation by the amygdala, creating a modularity of networks within the LC circuit. Since then, multiple studies have shown spatial heterogeneity of LC-NE projections, with anatomical evidence for target-specific distributions of LC axons. MAPseq (Multiplexed Analysis of Projections by Sequencing) analysis using axon-specific RNA barcodes to distinguish between the projections of specific neurons revealed that individual LC neurons have preferred cortical targets (Kebschull et al. 2016). Multiple studies have demonstrated differential roles of LC projections to different hippocampal sub-regions (CA1, CA3, and dentate gyrus) in mediating various aspects of learning and memory (Wagatsuma et al. 2018; Chowdhury et al. 2022). Subpopulations of LC-NE neurons projecting to the medial prefrontal cortex, an area involved in cognition and executive function, and the motor cortex that mediates goal-directed movement, show discrete anatomical, biochemical, and electrophysiological properties (Breton-Provencher et al. 2022; Chandler et al. 2014). These distinct projection patterns allow LC-NE neurons to modulate the activity of different brain regions in a task-dependent manner. For example, specific NE release in the motor cortex facilitates the execution of movement, while a salient reinforcement signal is more broadly distributed among cortical regions to improve subsequent performance (Breton-Provencher et al. 2022). Similarly, LC neurons projecting to the amygdala facilitate aversive learning, while an independent population projecting to the medial prefrontal cortex extinguishes those aversive responses to enable flexible behaviors (Uematsu et al. 2017). Such modular organization of LC neurons, along with variations in NE receptor densities and subtypes across different brain regions, allow LC-NE neurons to modulate distinct brain regions in a flexible, task-specific manner.

2.1.2 Theories of the Role of LC-NE in Brain Function

Although NE is primarily recognized for its role in regulating arousal, it also plays a pivotal role in modulating learning and memory. NE profoundly shapes how the brain processes, adapts to, and integrates new information. Several theories propose distinct mechanisms by which LC-NE activity influences cognitive flexibility, decision-making, and behavioral adaptation, underscoring its critical involvement in learning-related processes (Poe et al. 2020; Breton-Provencher et al. 2021). One such theory is the *Network Reset Theory*, proposed by Bouret and Sara, which posits that the LC becomes highly active in response to surprising outcomes or contextual shifts that demand behavioral change (Bouret and Sara 2005). This manifests as a phasic activation of LC-NE, which precedes changes in neuronal activity across other cortical regions within single trials. Phasic LC activation triggers a widespread release of NE, particularly in the cortex, effectively "resetting" cortical networks. By disrupting ongoing activity patterns, this resetting mechanism primes the brain to adapt more flexibly to new contexts. Such an adaptive process is especially apparent in behavioral tasks where stimulus-reinforcement contingencies change abruptly. In these situations, NE serves as a catalyst for behavioral shifts, enabling the brain to disengage from existing routines or predictions and prioritize the processing of novel stimuli. Such a mechanism is crucial in learning environments that require rapid adjustment to new circumstances, allowing for greater behavioral flexibility. This theory also aligns with the broader concept of NE as a modulator of attention, facilitating the brain's ability to focus resources on the most relevant and salient stimuli while suppressing distractions from familiar, less pertinent information. This theory is complemented by the work of Yu and Dayan (2005), who suggest that NE globally signals "unexpected uncertainty" in the external environment (Yu and Dayan 2005). In their model, NE release is triggered when there is a mismatch between expected and actual outcomes. This release enhances bottom-up information processing and updates the brain's internal predictions, or "priors," to optimize future responses. Unlike the Network Reset Theory, which emphasizes rapid, trial-specific NE activation, their model describes a slower timescale of NE release that unfolds across multiple trials. Both theories highlight the critical role of outcome predictability and reliability in triggering the LC-NE response.

Another prominent theory for explaining the role of NE in learning and behavior is the *Adaptive Gain Theory*, developed by Aston-Jones and Cohen (2005). This theory proposes that NE's influence on cognitive performance and behavioral flexibility can be understood through two distinct modes of activity in the LC: phasic and tonic. In the phasic mode, the LC exhibits brief, rapid bursts of NE release in response to task-relevant stimuli, particularly in contexts where an elevated level of attention is required. Phasic NE release sharpens focus on important stimuli, enhancing sensory processing and facilitating task performance. This allows the brain to exploit known information and optimize behavior for efficient learning and execution in a goal-directed behavior. In contrast, the tonic mode is characterized by a sustained increase in baseline NE release that is not associated with specific task demands. This mode is linked to increased distractibility and reduced task performance. Unlike phasic NE activity, the tonic mode allows the brain to disengage from current tasks and promotes exploration of new contingencies in novel or uncertain environments. In this way, tonic NE activity is essential for enabling behavioral flexibility when the exploitation of existing strategies is no longer effective.

These two modes of NE activity are well illustrated by the contrasting behavioral performance observed in animals performing simple decision-making tasks (Aston-Jones et al. 1994). For instance, in a signal detection task where an animal is trained to associate one stimulus with a behavioral response to receive a reward, and to avoid responding to another stimulus to prevent a punishment, measurements of NE release can be linked to specific behaviors. Phasic NE activity is predominantly observed during periods of strong behavioral performance, particularly when the animal responds to reward-predicting stimuli or adjusts behavior after a reversal in stimulus-reward contingencies. On the other hand, tonic NE activity is associated with poor behavioral performance, during which the animal's ability to discriminate between stimuli and its response threshold to those stimuli decreases. The balance between phasic and tonic LC-NE activity is thought to mediate a key trade-off between exploitation and exploration. In situations that demand focus and precise action, phasic bursts of NE facilitate exploitation by enhancing attention and suppressing distractions. Conversely, during periods of uncertainty or low task engagement, increased tonic NE levels enable the brain to shift toward exploration and consider alternative strategies. This dynamic switch between modes allows the LC-NE system to regulate learning and decision-making in a flexible, context-dependent manner.

The *GANE (glutamate amplifies noradrenergic effects)* model, also called the glutamate amplification of learning model, offers a more mechanistic view of NE's role in learning. This model, proposed by Mather and colleagues, suggests that NE's influence on learning is not uniform across brain regions but is highly specific to areas where strong excitatory activity already exists (Poe et al. 2020; Mather et al. 2016). This model posits that when neurons release elevated levels of glutamate in response to a stimulus, nearby NE varicosities and astrocytes sense this increase and signal for the release of NE from the LC. This local NE release, in turn, amplifies the effects of glutamate, creating a feedback loop that enhances neural activity in circuits engaged in processing the most relevant or salient information. As a result, the combination of glutamate and NE potentiates the neural representation of important stimuli, prioritizing learning and memory formation in these regions.

The GANE model introduces a layer of specificity to the role of NE in learning by emphasizing how NE selectively amplifies activity in highly activated circuits rather than uniformly affecting the entire brain. This stands in contrast to earlier models that focus primarily on widespread NE release during states of heightened arousal. According to the GANE model, NE's action is more targeted and input-specific, affecting regions like the hippocampus, amygdala, and prefrontal cortex—areas heavily involved in spatial learning, emotional memory, and decision-making. In these regions, the NE-glutamate interaction promotes long-term potentiation (LTP), strengthening synaptic connections and enhancing the consolidation of memories. Moreover, the GANE model helps explain how NE can both enhance and suppress neural activity depending on the context. When glutamate levels are high, NE amplifies the excitatory signals, boosting attention and learning for the most critical stimuli. Conversely, in areas with lower glutamate activity, NE can suppress activity, helping to filter out irrelevant information and focus cognitive resources on more important tasks. This dual role of NE in enhancing relevant inputs while dampening distractions allows for highly efficient processing of information, optimizing learning in complex environments where multiple stimuli compete for attention. Additionally, unlike previous theories, the GANE model highlights the importance of secondary messengers like glutamate and astrocytes. It introduces an astrocyte-mediated mechanism of NE release, which modulates the effects of LC-NE on learning and behavior. This astrocyte involvement provides a more elaborate understanding of how NE modulates neural activity, making the GANE model particularly significant in explaining the localized and dynamic nature of NE's influence on learning and memory.

Recent findings have expanded our understanding of how NE contributes to learning and memory, further refining existing theories like the Network Reset and Adaptive Gain models. For instance, Brenton-Provencher et al. demonstrated that phasic LC-NE activity is closely tied to behavioral performance in reinforcement learning tasks, particularly during surprising outcomes (Breton-Provencher et al. 2022). Their findings suggest that LC-NE activity not only signals unexpected events but also encodes a reward prediction error (RPE), influencing subsequent behavioral adjustments. This dynamic role of NE in updating predictions based on surprising outcomes highlights its critical function in learning, suggesting that NE activity supports the optimization of behavior by refining internal models of the environment. Such insights underscore the importance of NE in shaping cognitive flexibility and learning through mechanisms that integrate prediction, uncertainty, and behavioral feedback.

2.2 NE and Epinephrine in the Body

A major source of NE in the body is the adrenal glands. The adrenal glands are small, pyramidal-shaped organs that sit on top of each kidney. The names noradrenaline and epinephrine are derived from this anatomical location, with "ad" meaning "near," epi meaning "upon," and "nephros" meaning "kidneys." The adrenal gland has an outer cortex and an inner medulla consisting of chromaffin cells. Cholinergic preganglionic neurons in the spinal cord project onto chromaffin cells on the adrenal medulla and activate nicotinic receptors on these cells. The chromaffin cells synthesize NE and store it in chromaffin granules where it can be released into the bloodstream or converted to Epi by phenylethanolamine-N-methyl transferase. This release of NE and Epi into the bloodstream can influence various tissues and organs that express adrenergic receptors including the heart, blood vessels, and skin (Axelrod 1962; Wurtman and Axelrod 1966).

In addition to the adrenal gland mediated release of NE and EPI into the bloodstream, NE also plays a significant role in transmission of information throughout the body via the autonomic nervous system. The autonomic nervous system consists of the sympathetic and parasympathetic nervous system. While the primary neurotransmitter of the parasympathetic system is acetylcholine (ACh), it works in concert with the sympathetic nervous system and other adrenergic systems to coordinate various interactions between the brain and body. Additionally, NE is a major driver of the sympathetic nervous system where postganglionic neurons in the spinal cord send out adrenergic projections to most major organs in the body (Boron and Boulpaep 2009).

2.2.1 The Sympathetic Nervous System

The sympathetic nervous system begins with neurons in the spinal cord (sympathetic preganglionic neurons) projecting to ganglia in the peripheral nervous system (sympathetic postganglionic/chromaffin cells). Postganglionic sympathetic neurons receive synapses from preganglionic neurons residing in the lateral horn (intermediolateral column, IML) of the spinal cord which send their axons out through the ventral roots. Sympathetic postganglionic neurons all share a common progenitor in the dorsal neural tube and thus have conserved roles (Wehrwein et al. 2016). The axons of these postganglionic neurons project extensively throughout the body and have dense-core vesicles that contain NE. These post-ganglionic neurons produce most of the NE used in the sympathetic nervous system. The abundance of NE-filled vesicles along each axon allows for many en passant synapses which can modulate the impact of affected cells (Smolen 1988). The following section details the effects of NE release from different postganglionic sympathetic neurons on various tissues (also illustrated in Fig. 2.2).

2.2.2 Sympathetic Postganglionic Chain Ganglia

The foremost ganglia within the chain of ganglia are the superior cervical ganglia, which projects many tissues in the head and face including the pupils, conjunctiva, salivary, and lacrimal glands. In smooth muscles that regulate the iris, NE can elicit a contraction without triggering an action potential. When NE is released in the salivary gland, salivation increases. The superior cervical ganglia also travel alongside the internal carotid and vertebral arteries as they enter the skull and branche in tandem with the brain's vasculature. NE release can induce contraction or dilation of vascular smooth muscle cells depending on the expression of distinct types of adrenergic receptors. In arteries and veins, a burst of action potentials in sympathetic ganglia releases NE along with NPY or adenosine triphosphate (ATP), resulting in a vasoconstrictor response. After an injury, blood vessels reinnervated by sympathetic neurons exhibit heightened vasoconstriction and appear more reactivatable (Boron and Boulpaep 2009; Klimaschewski et al. 1996).

Located inferior to the cervical ganglia are the middle cervical ganglion, stellate ganglion, and thoracic postganglionic sympathetic chain. These three ganglionic chains provide sympathetic innervation to the majority of tissues in the upper limbs and torso including sweat glands, hair follicles, lungs, skeletal muscle as well as cardiac tissues. In the sweat glands of the mouse paw, NE facilitates the development of the sweat response. Interestingly, in rodents, sympathetic neurons that innervate sweat glands switch from using NE to ACh during late postnatal development (Tafari

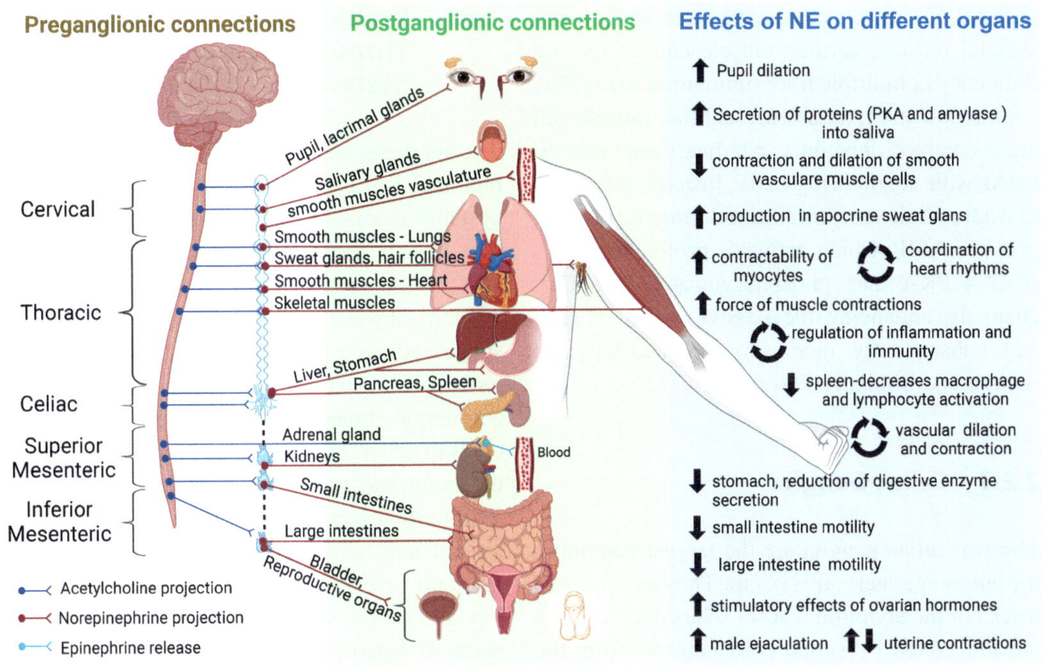

Fig. 2.2 NE connections in the body. Projections of the sympathetic nervous system to various organs in the body and the effect of NE release on their function. Preganglionic neurons in the spinal cord send cholinergic projections to postganglionic neurons *(left)*. These then innervate various organs throughout the body, releasing NE *(middle)* and modulating their function *(right)*

et al. 1997). In hair follicles, the arrector pili muscle anchors the sympathetic innervation from the chain ganglia. These sympathetic nerves form synapse-like connections with hair follicle stem cells, which communicate with NE and affect skin conditions, hair follicle immune system, hair pigmentation loss, and hair growth (Zhang et al. 2021). In lung tissue, NE stimulation relaxes airways and reduces constriction. Skeletal muscles receive sympathetic input from the chain ganglia which may influence the function of the neuromuscular junction, metabolism, and regeneration of muscle fibers. The effects of NE on muscle cells depend significantly on the organ system the muscle is a part of. In some tissues, adrenergic stimulation increases and in others it decreases activity. Experimentally, infusing muscles with NE results in repeated contractions measured as a tetanus or a single long twitch that lasts 10 times longer than non-NE mediated muscle contractions (Boron and Boulpaep 2009).

The heart receives intricate innervations by noradrenergic fibers from the sympathetic system and cholinergic nerve fibers from the parasympathetic system. The balance between these opposing innervations allows for normal heart rhythms as sympathetic neurons accelerate, while parasympathetic postganglionic neurons slow the heart rate. To do this, the sinoatrial node is densely innervated by both NE and ACh fibers to modulate the pacemaker cell's depolarization rate. In the atrioventricular node, NE stimulation increases conduction velocity to decrease the delay of electrical impulse to allow for proper filling of the ventricles. The bundle of His and the ventricular muscle mostly receive NE fibers. NE binding to β1 receptors on the ventricular myocytes increases extracellular calcium influx and results in enhancing the contractile force of the muscles. Furthermore, NE plays a crucial role in coordinating the contractile forces of each muscle in the heart, which is essential for adapting to changes in heart rate during exercise. Unlike skeletal muscle, cardiac muscle cannot rely on frequency or multiple fiber summation to regulate contraction strength, as each cardiac muscle cell must contract only once per heartbeat, coordinated with neighboring cells. Instead, NE binds to β adrenergic receptors, stimulating the production of cAMP, which activates protein kinase A (PKA). PKA then phosphorylates L-type Ca^{2+} channels, enhancing the passive influx of Ca^{2+} and subsequently increasing contractile force (Boron and Boulpaep 2009).

2.2.3 Celiac Ganglia

The two celiac ganglia are the largest ganglia in the autonomic nervous system. They sit on celiac trucks of the abdominal aorta. The celiac ganglia are innervated by spinal cord neurons from the lower thoracic and upper lumbar spinal levels (T12-L1) and innervate the upper abdominal cavities including much of the upper digestive and accessory structures including the stomach, gallbladder, pancreases, and liver. In the enteric system, including the stomach and upper gastrointestinal tract, NE release inhibits digestion. NE release into the liver regulates the production and release of glucose while relaxing smooth muscles in the gallbladder and pancreas. Overall, NE stimulation results in the inhibition of biliary and pancreatic digestive function. The celiac ganglia are also connected to several other sympathetic ganglia, including the thoracic-aorticorenal ganglia, which innervate the kidneys, the lumbar superior mesenteric ganglia which innervate the small intestines, and the inferior mesenteric ganglia/prevertebral ganglia which innervate the colon, anal sphincter, urinary bladder, and genitalia. NE in the colon slows down digestion and induces contractions of smooth muscles in the anal sphincter, urinary bladder, and induces ejaculation (Gutman and Weil-Malherbe 1967; Kimura et al. 2009).

2.3 Brain-Body Connections Through the Adrenergic System

The synthesis and release of NE and Epi by various nuclei in the central and peripheral nervous systems underscore their crucial role in transmitting signals throughout the brain and body. In addition, the release of NE and Epi into the blood, primarily by the adrenal glands, leads to widespread adrenergic signaling across organ systems. This extensive signaling network of adrenergic molecules makes them ideal candidates to integrate information processing between the brain and body. This adrenergic network is vital to the execution of a wide range of physiological and behavioral functions, most notably the fight-or-flight response to threats. In the following section, we describe the anatomical connections between the brain and body driven by adrenergic signaling and explore physiological processes that are coordinated through these connections.

2.3.1 Anatomical Adrenergic Connections Between the Brain and Body

NE influences the autonomic and peripheral nervous system which has direct connections to and from the brain. The sympathetic nervous system originates with neurons in the spinal cord (sympathetic preganglionic) that project to ganglia in the peripheral (sympathetic postganglionic/ chromaffin cells) nervous system. This effectively connects the brain to the body via two synapses, one predominantly using ACh and the other NE. Various NE sympathetic postganglionic contributions to several major functions are already outlined above. Another channel of communication between the central and peripheral nervous system is through the vagal nerve. The vagal nerve is one of the longest nerves in the body and is named for its "wandering" projections. The vagal nerve contains both motor and sensory neurons. While the cholinergic motor neurons are housed in the dorsal motor nucleus of the medulla oblongata in the brainstem, the sensory neurons reside distally through the jugular foramen in the jugular and nodose ganglia outside of the skull. This nerve then "wanders" and projects to most of the organs in the abdominal cavity, including the heart, lungs, liver, and gastrointestinal tract (Paintal 1973; Berthoud and Neuhuber 2000; Prescott and Liberles 2022). Thus, the vagal nerve is a direct conduit between the periphery and the brain, transmitting signals related to various bodily functions (Zhao et al. 2022), including heart rate (Coleridge and Coleridge 1994; Zeng et al. 2018), digestion (Powley et al. 2019), and immune response (Jin et al. 2024; Bin et al. 2023). Direct stimulation of the vagal nerve activates the NTS neurons, which project to the LC (Ter Horst et al. 1991) and induces NE release in the brain (Groves et al. 2005; Hulsey et al. 2017). Additionally, peripheral NE signaling, which influences arousal and stress responses, can interact with the vagus nerve to relay information about the body's physiological state. This information helps the brain integrate these signals, facilitating appropriate behavioral and physiological responses.

2.3.2 Influences on Major Sensory Systems

The LC-NE system plays a crucial role in shaping sensory perception across all modalities. Primary sensory neurons, located in the dorsal root ganglia and the cranial nerve V, can receive NE and Epi signals directly through the bloodstream. These pseudo unipolar neurons express adrenergic receptors in various levels and can be organized into functionally meaningful populations such as proprioceptors, cool sensing neurons, peptidergic Aβ rapidly adapting Aβ-RA-LTMRS (low threshold mechanoreceptors) (Le Pichon and Chesler 2014), and Aδ-LTMRs. These neurons innervate the majority of the periphery. NE can modulate the activity of neurons in various sensory pathways resulting in enhancing signal processing, filtering out noise, and optimizing perceptions for appropriate behavioral outcomes. These effects on sensory processing are dependent on factors such as arousal, attentional state, and specific task demands. The following is a brief summary of how NE affects each of the sensory modalities (Waterhouse and Navarra 2019).

Olfaction NE plays a critical role in the modality of olfaction as evidenced by the many NE inputs from the LC to the olfactory bulb and accessory olfactory bulb. For instance, about 30% of the LC neurons in rats project to the olfactory bulb, most of which project to the subglomerular layers, specifically the internal plexiform layer and the granule cell layer (Shipley et al. 1985). The projection to this area is nearly ten times greater than to any other part of the cerebral cortex (Basbaum 2008). Additionally, the MERFISH (multiplexed error-robust fluorescence in situ hybridization) spatial transcriptomic data from the Allen Brain Atlas shows the highest expression of α1 adrenergic receptors in the olfactory bulb of the mouse (Yao et al. 2023; Kunst et al. 2023; Zeng et al. 2023) (also see Box 2.1 and Fig. 2.2). This strong connection between the LC and the olfactory bulb aligns with theories that suggest a crucial role of LC neurons in enhanced sensory vigilance and regulation of growth and adaptability of neural connections. Previous studies have shown that presentation of olfactory cues in behaving animals elicits high LC activity and subsequent release of NE in the olfactory bulb. Such an olfactory bulb-specific NE increase was not seen after exposing the animal to a loud noise or an electrical stimulation of its forepaw. This NE increase to odorant stimuli persisted even when sympathetic cervical ganglionectomy was performed (Chase and Kopin 1968; Aston-Jones and Bloom 1981). The formation of olfactory memories to pheromones (Brennan et al. 1990) and the regulation of pregnancy and maternal behaviors (Kaba and Keverne 1988) require intact NE terminals in the olfactory bulb. Increases of NE turnover in the bulb are also seen in the immediate 4 hours after mating, but not to re-exposures to the same pheromones. This allows new males to block the impregnation by males who have previously copulated with the same female. Lesioning the olfactory bulb in the females blocks memory formation. NE inputs to the olfactory system have also been shown to influence the formation of cross modal associations. For example, in neonatal rats, NE release as a result of pleasant gentle tactile stimulation leads to a preference for an associatively paired odorant (Kaba et al. 1989). On the other hand, evidence for a specific role of NE in olfaction at the cellular and network level is somewhat inconsistent. In one study, stimulating the LC before performing paired pulse stimulations of the lateral olfactory tract resulted in immediate depression of the synapses. Two minutes later, paired pulse inhibition was significantly potentiated (Okutani et al. 1998). Others found that LC stimulation resulted in depression of granule cell field potentials (Perez et al. 1987). Iontophoretic injections of NE inhibit mitral cells in cats (McLennan 1971) and rabbits (Salmoiraghi et al. 1964), but activates mitral cells in turtle olfactory bulbs (Kaba et al. 1989; Jahr and Nicoll 1982).

Box 2.1: Adrenergic Receptors in the Brain and Body

The experimental infusion of NE in the brain commonly elicits inconsistent and mixed responses from postsynaptic neurons. One simple reason for this could be due to the diversity of adrenergic receptors that bind NE and the different intracellular signals they elicit. Adrenergic receptors are grouped into three subtypes which preferentially couple with a major subfamily of G-protein coupled receptors or GPCRS (Gs, Gq/11, and Gi). α1-adrenergic receptors (α1A, α1B, and α1D) preferentially couple with the Gq subfamily, α2-adrenergic receptors (α2A, α2B, and α2C) preferentially couple with Gi, and β-adrenergic receptors (β1, β2, and β3) preferentially couple with Gs (Motiejunaite et al. 2021). Recent advancements in single cell sequencing have provided a comprehensive view of adrenergic receptor expression levels in different tissues. The following section highlights the findings from several single cell sequencing datasets in the mouse central and peripheral nervous systems.

Figure 2.3 demonstrates the extensive distribution of adrenergic receptors *in the brain*, where both neuronal and non-neuronal cells across various brain regions express a diverse array of receptors (Yao et al. 2023; Zeng et al. 2023). Of the α1-adrenergic receptors, α1A is the most broadly and strongly expressed receptor in the brain and can be found in both neurons and astrocytes. It is most prominently expressed in GABAergic neurons of the cortical caudal ganglionic eminence and the olfactory bulb. α1B is highly expressed in GABAergic cerebral cortex caudal ganglionic eminence neurons, glutamatergic thalamic neurons, intratelencephalic-extratelencephalic neurons, near-projecting corticothalamic layer 6b neuron as well as ependymal astrocytes and oligodendrocytes. α1D receptors have the lowest overall expression of adrenergic receptors with sparse expression levels in glutamatergic dentate gyrus and intra telencephalic-extra telencephalic neurons. Of the α2-adrenergic receptors, α2A is the most broadly expressed. These receptors are most prevalent in ependymal astrocytes, vasculature, GABAergic neurons in cerebral nuclei lateral ganglionic eminence, and glutamatergic neurons in near-projecting corticothalamic layer 6b. α2B receptors have fairly low overall expression: they are expressed in glutamatergic thalamic neurons. The α2C receptors are mainly expressed in glutamatergic neurons of the cerebral nuclei lateral ganglionic eminence and dentate gyrus. Of the β-adrenergic receptors, β1 receptors are broadly expressed across cell types. They have the highest expression in glutamatergic intra telencephalic–extra telencephalic neurons, glutamatergic thalamic neurons, glutamatergic medial habenula–lateral habenula neurons, glutamatergic dentate gyrus immature neurons, ependymal astrocytes, as well as immune cells. β2 and β3 receptors have exceptionally low expression levels in the brain. β2 receptors are seen in immune, vascular, and GABAergic cerebellum cells while

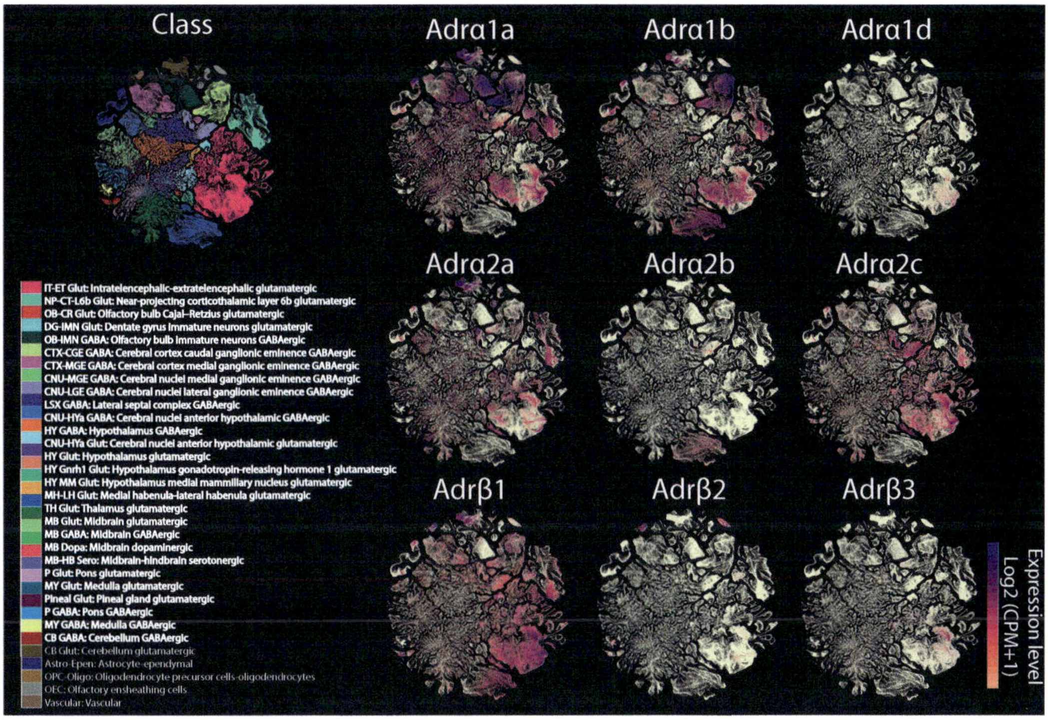

Fig. 2.3 Adrenergic receptors in the brain. Uniform Manifold Approximation and Projection (UMAP) representation of transcriptomic cell types of the whole mouse brain with cell classes differentiated by color *(left)*. Adrenergic receptor subtype expression profiles overlaid onto UMPAP colored in magenta *(right)*. (This figure was generated from the whole brain 10x scRNAseq dataset of the Allen Brain Cell Atlas (Yao et al. 2023; Zeng et al. 2023). https://portal.brain-map.org/atlases-and-data/bkp/abc-atlas)

β3 receptors are expressed by glutamatergic cerebral nuclei anterior hypothalamic and midbrain dopaminergic neurons.

In the periphery, sequencing studies reveal that adrenergic receptors are expressed on different cell types within the dorsal root ganglia (Sharma et al. 2020). These expression patterns allude to functionally specific roles NE could have in different somatosensory modalities. α1B receptors are expressed in proprioceptors, Aβ-RA-LTMRS (low threshold mechanoreceptors), and Aδ- LTMRs, while α1A and α1D have low to no overall expression in any sensory neurons. α2A receptors are expressed in calcitonin gene-related peptide (CGRP)-eta, CGRP-epsilon, and in early developmental somatostatin neurons. α2B receptors are selectively expressed in just proprioceptors. α2C receptors are broadly expressed in Aβ-RA-LTMRS, Aδ-LTMRs, proprioceptors, Aβ-field LTMRS, CGRP-eta, CGRP-zeta, CGRP-epsilon peptidergic nociceptors, and Trpm8 cooling neurons. β1receptors are present at exceptionally low levels throughout primary sensory neurons but are expressed in c-LTMRS, Aδ- LTMRs, and proprioceptors (Sharma et al. 2020). *In vagal neurons* of the jugular and nodose ganglia, neurons predominantly express α2-adrenergic receptors. α2A receptors are highly expressed in several classes of nodose ganglia (NG10, NG8) as well as in a population of jugular neurons (JG4). α2C receptors are expressed by nearly every class of jugular and nodose neurons with the highest expression seen in NG6, JG4, and NG1 (Kupari et al. 2019). *In the heart*, NE activates β-adrenergic receptors on cardiac pacemaker neurons, leading to tachycardia, and also stimulates β receptors on cardiac myocytes, increasing cardiac contractility.

Additionally, NE can induce vasoconstriction by activating α1 and α2 adrenergic receptors in the blood vessels (Myagmar et al. 2017).

The diversity and specific expression in different tissues of adrenergic receptors make them an exciting target for clinical interventions. For example, smooth muscles in the prostate predominantly express α1A-receptors; therefore, α1A-adrenergic antagonists could be used therapeutically to treat prostatic hypertrophy. β3-adrenergic receptors are found uniquely in brown adipose tissue and the gallbladder and could be targeted for obesity treatments. In the heart, β-1 and α-1B receptors are present in all myocytes. α1A is present in 60%, and β-2 and β-3 are detected in only 5% of myocytes (Myagmar et al. 2017). An α2-receptor agonist, dexmedetomidine, is routinely used in operating rooms to manage pain and sedation (Brown et al. 2018).

Somatosensation NE can affect the somatosensory system at multiple levels of the ascending somatosensory pathway beginning with sensory neurons in the periphery (trigeminal ganglia/cranial nerve 5 and dorsal root ganglia) and centrally in the spinal cord, brainstem, thalamus, and somatosensory cortex. Since LC neurons project so widely across the somatosensory pathway, they can have bidirectional effects on somatosensation. For example, in the spinal cord, they can both enhance and suppress pain signals depending on the receptors the tissues express. Furthermore, there is evidence that NE centrally can improve tactile discrimination. Iontophoretic NE application in the rat somatosensory cortex enhances excitatory and inhibitory transmission without directly altering neuron firing (Waterhouse et al. 1990a). Differential patterns of LC activation can differentially facilitate somatosensory signals from whisker-stimulation (Waterhouse et al. 1990a; Devilbiss et al. 2006; Devilbiss and Waterhouse 2004). Specifically, stimulating LC potentiates the responses of layer 5 barrel cortex neurons to whisker deflection and sharpens their receptive field. These support that idea that NE establishes a sensory processing state that selectively influences neuronal responsiveness to afferent signals during different phases of whisker movement.

Vision NE can affect the visual system at levels of the thalamic lateral geniculate nucleus and the visual cortex. Iontophoretic application of NE into the lateral geniculate nucleus enhanced neuronal responses to afferent optic stimulation, an effect abolished when all NE stores are pharmacologically depleted (Rogawski and Aghajanian 1980). Additionally, in vivo and in vitro experiments demonstrated that the effects of NE in the visual system relied on adrα1 receptors, which mediated the conversion of lateral geniculate neuron activity from burst firing to single spike firing and improves signal transmission (Holdefer and Jacobs 1994; McCormick et al. 1991). The switch from a state of rhythmic oscillation and low excitability to a state of increased excitability and responsiveness supports the role of NE in facilitating changes from the drowsiness seen in slow-wave sleep to periods of waking and attentiveness. NE also influences neuronal activity in the visual cortex. Specifically, NE enhances the signal-to-noise ratio for visual inputs, particularly through α2 receptor-mediated pathways (Kolta and Reader 1989). NE's influence increases the precision of visual processing, such as direction selectivity and velocity tuning, by modulating responses to threshold-level stimuli (Waterhouse et al. 1990a). Additionally, NE reuptake inhibitors, like atomoxetine, were found to improve cortical excitability and response accuracy in visuomotor tasks, emphasizing NE's impact on attention and sensory-motor integration within cortical circuits (Grefkes et al. 2010).

Auditory In the auditory system, noradrenergic effects are bidirectional and follow an inverted-U response profile. While minimal LC activity and low NE levels provide limited modulation, increasing LC discharge and NE release optimally enhance neuronal responses to synaptic input. However, this facilitation diminishes as LC activity and NE levels rise further, ultimately hindering neuronal responsiveness. Experimental local administration of NE improved the selectivity of auditory cortical neurons for pure-tone frequencies (Manunta and Edeline 1997) but did not boost signal-to-noise ratios or activate subthreshold synapses (Edeline et al. 2011). These varied responses to NE are likely the result of differences in receptor expression. Slice experiments demonstrate that NE modulates GABAergic signaling differently through receptor-specific actions. NE increases GABA release by acting on $\alpha 2$ and β receptors, enhancing inhibitory presynaptic input, while $\alpha 1$ receptor activation suppresses postsynaptic GABA responses, amplifying responses to excitatory stimuli (Salgado et al. 2012). This dual action is thought to enhance sensory focus on relevant auditory signals. Recent research further demonstrated that NE enhances behavioral auditory discrimination and perceptual salience in long-term tuning of auditory cortex responses to specific tones (Martins and Froemke 2015).

2.3.3 Noradrenergic Modulation of Arousal

Recent advances in neuroscience have uncovered numerous physiological functions influenced by the interplay of central and peripheral catecholamines. One such function is the control of arousal. While the concept of arousal has an intuitive appeal, a formal, unified characterization of the term remains elusive, with researchers instead describing various aspects of arousal such as an awakening from sleep or sedation, or a heightened sense of emotional reactivity and alertness. For the purposes of this chapter, we define general arousal as a continuum of increased sensitivity to environmental stimuli (Berridge 2008). The body is in a heightened state of relative responsiveness, preparing for action through the activation of various neurological (e.g., cortical and sympathetic neural activation) and bodily systems. The physiological signs of arousal are varied, encompassing elevated heart rate, blood pressure, respiration rate, muscle tension, perspiration, and metabolic rate. The importance of precise control over arousal levels is undeniable. Dampened arousal can lead to drowsiness, sleep, and even unconsciousness (Aston-Jones and Cohen 2005). On the other hand, heightened arousal, triggered by a salient event or emotionally strong memories can facilitate attention, motivation, and learning while driving behavioral responses, but can also lead to distractibility and anxiety (Aston-Jones and Cohen 2005; Breton-Provencher et al. 2021; Sara and Bouret 2012). In 1908, Yerkes and Dodson described a nonlinear relationship between arousal and behavioral performance (Yerkes and Dodson 1908). Their study examined how task difficulty and stress affect performance in a discrimination task. Since then, numerous studies across various animal models have investigated this connection, culminating in the "Yerkes-Dodson Law," which describes an inverted U-shaped relationship between arousal and behavioral performance (Aston-Jones and Cohen 2005; Breton-Provencher et al. 2021). Cognitive tasks requiring attention are optimally performed at intermediate arousal levels, with performance declining significantly if arousal is too high or too low.

Arousal involves components across the central, sympathetic, and peripheral nervous systems, as well as the circulatory, respiratory, and visceral motor systems. Various brain regions, parts of the autonomic nervous system, and the endocrine system work collectively to modulate the levels of brain-body arousal. In the brain, neural regulation of arousal involves activation of multiple systems across the brain, including alterations in hippocampal theta oscillations (dos Santos Lima et al. 2019; Jarosiewicz and Skaggs 2004), thalamic bursting (Bastos et al. 2021; Murata and Colonnese 2018; Miller et al. 1989; McCormick and Bal 1997), and cortical activation (Tauber et al. 2024; Eisen et al. 2024; Ährlund-Richter et al. 2024). In the late 1940s, the pivotal work of Moruzzi and Magoun first identified the role of the brainstem reticular formation in mediating arousal states (Moruzzi and Magoun 1949). Activation of these regions resulted in a shift in cortical activity from a high-voltage slow-wave electroencephalogram (EEG) signal to a low-voltage, fast activity pattern. As clinicians, Moruzzi and Magoun noted that this shift in cortical activity is similar to that seen during an arousal response to natural stimuli and is abolished during anesthesia. The brain regions involved in mediating this arousal response collectively came to be known as the ascending reticular activating system (ARAS). The central noradrenergic system, driven by the LC, is one of the major components of ARAS in the brain. In 1988, Adams and Foote developed a combined recording-infusion probe that uses electrophysiological recordings to enable precise placement of pharmacological agents in the vicinity of LC neurons (Adams and Foote 1988), enabling selective and verifiable activation of the LC. Using this technique, Berridge and Foote showed that LC activation reliably induced a shift in frontal cortical EEG activity from low-frequency, high-amplitude to high-frequency, low-amplitude, along with induction of an intense hippocampal theta rhythm (Adams and Foote 1988). Such shifts in forebrain activity, reflected in altered EEG signals, have long been established as a signature of changes in arousal states (Bonnet and Arand 2001; Makeig and Inlow 1993). Since then, numerous studies have demonstrated a causal relationship between LC-NE and arousal states. LC-NE neurons are silent during rapid eye movement or REM sleep, exhibit low activity levels during non-REM (NREM) sleep, and are most active during wakefulness and periods of high vigilance (Osorio-Forero et al. 2021; Liang et al. 2021; Swift et al. 2018; Eschenko et al. 2012; Hayat et al. 2020). Importantly, changes in LC neuronal activity precede transitions in arousal state (Aston-Jones and Cohen 2005; Breton-Provencher and Sur 2019; Foote et al. 1980), thereby implying that they play a causal role in driving these transitions (Berridge and Waterhouse 2003). Moreover, the relationship between LC-NE activation and optimal behavioral performance follows the typical Yerkes-Dodson inverted-U curve, consistent with the view that the LC mediates arousal (Aston-Jones and Cohen 2005; Breton-Provencher et al. 2021; Sara and Bouret 2012).

Control of arousal states is thought to arise through two basic aspects of LC function. First, extensive projections from the LC to various sensory regions of the cerebral cortex and thalamus enhance the accumulation of salient sensory information. Second, its numerous connections to other brain regions such as the prefrontal cortex, amygdala, and hippocampus, modulate the processing of external stimuli to drive behavioral responses. In addition, projections to other arousal centers in the brain serve to amplify LC control over vigilance states. The LC is also thought to promote wakefulness via activation of α and $\beta 1$ receptors in the general regions of the medial septal area and the medial preoptic area (Berridge 2008). In the somatosensory cortex, phasic LC activation or local NE infusions inhibit spontaneous activity (Waterhouse and Navarra 2019; Waterhouse et al. 1988) and reduce response latency and jitter (Lecas et al. 2004), while facilitating cortical representation of sensory stimuli (Vazey et al. 2018). Safaai and colleagues demonstrated a coupling between phasic LC activity and cortical fluctuations that can enhance cortical responses to weaker stimuli (Safaai et al. 2015). Similar effects of LC-NE activation have been observed in the auditory (Kiissp and Vater 1989; Manunta and Edeline 1999), visual (Waterhouse et al. 1990b; McLean and Waterhouse 1994) and olfactory (Hasselmo et al. 1997; Ciombor et al. 1999) cortices. Thus, LC control of arousal states could stem from a heightened mode of sensory processing. Similarly, NE modulation over critical regions involved in learning and memory as well as motor function demonstrates its role in facilitating behavioral execution (see reviews by (Aston-Jones and Cohen 2005; Breton-Provencher et al. 2021)). LC activity in the hippocampus modulates the encoding and integration (Chowdhury et al. 2022) of spatial memories, while reinforcement signals to the PFC improve performance accuracy in subsequent trials (Breton-Provencher et al. 2022). Noradrenergic signals also enhance the responsiveness of motoneurons in the spinal cord (Fung and Barnes 1987; Witts et al. 2023) as well as pyramidal neurons in the motor cortex

(Fung and Barnes 1987; Dymecki et al. 2022), thereby driving the initiation or execution of motor responses (Breton-Provencher et al. 2021; Breton-Provencher et al. 2022). Thus, the LC appears to improve the detection and processing of salient sensory stimuli leading to a switch in vigilance states to facilitate an optimal behavioral response.

In addition to the LC, other noradrenergic cell groups in the brain contribute to certain aspects of arousal. For example, A2 cells in the nucleus of the solitary tract receive sensory feedback from the cardiovascular, respiratory, and alimentary systems via inputs from the vagal nerve, the area postrema, and direct visceral afferents of the solitary tract (Kalia and Sullivan 1982; Appleyard et al. 2007; Cunningham et al. 1994). Consequently, these cells respond to a broad array of interoceptive signals. Medullary A2 neurons are thought to engage arousal circuits by relaying gastric and cardiovascular signals to the amygdala and neurons in the shell of the NAc (Kerfoot et al. 2008; Kirouac And and Ciriello 1997; Mehendale et al. 2004). Moreover, A2 cells have been shown to innervate LC-NA neurons, providing a direct feedback link between visceral signals and the central arousal circuit in the brain (Van Bockstaele et al. 1999).

Noradrenergic activation in the brain via either the LC or the adrenal medulla results in well-defined patterns of autonomic activation. The LC sends direct connections to both the sympathetic and parasympathetic nervous system through its descending projections in the brainstem and spinal cord. In general, NE release by LC neurons increases sympathetic activity via the activation of α_1-adrenoceptors on preganglionic sympathetic neurons (Lewis and Coote 1990) and reduces parasympathetic activity via the activation of α_2-adrenoceptors on preganglionic parasympathetic neurons (Yaci et al. 2002; Unnerstall et al. 1984; Smith et al. 1995). In turn, sympathetic nerves carry NE signals to smooth muscles of the heart and sweat glands, constricting local blood vessels to induce a moderate increase in heart rate and blood pressure (Boron and Boulpaep 2009). Barosensitive C1 adrenergic cells in the rostral ventrolateral medulla respond to these changes in blood pressure and help to tightly control heart rate and blood pressure during periods of high arousal (Lipski et al. 1995). A key aspect of the central adrenergic system's involvement in arousal regulation is demonstrated through its control of pupillary activity. In addition to changes in blood flow and respiration, fluctuations in pupil size have been intricately linked to levels of sedation and alertness and have been demonstrated as a reliable index of arousal (Hou et al. 2006; Bradley et al. 2008). Numerous studies have shown that fluctuations in pupil size track both LC activity (Breton-Provencher and Sur 2019; Reimer et al. 2016) and cortical state (Reimer et al. 2014). Descending projections from the LC to noradrenergic neurons of the intermediolateral column in the spinal cord activate sympathetic control of pupil size, activating iris sphincter muscles and dilating the pupil. In parallel, LC inhibition of cholinergic preganglionic motoneurons in the Edinger-Westphal nucleus inhibits parasympathetic drive and relaxes the sphincter muscles to permit dilation (Larsen and Waters 2018). Thus, the consequences of LC activation are observable as an increase in pupil diameter (Breton-Provencher and Sur 2019; Reimer et al. 2016), increases in heart rate and blood pressure (Lightman et al. 1984; Ward and Gunn 1976), suppression of the baroreflex response (Hwang et al. 1998), and a reduction in salivation (Samuels and Szabadi 2008).

2.3.4 Fight-or-Flight and Stress Response

A particularly important circumstance of precise control over arousal levels is demonstrated by the execution of the body's stress response. In 1915, Canon first coined the term "fight-or-flight" to describe the body's stress response which compiled various visceral changes occurring as a consequence of nociceptive events, hunger, cold, exercise, and strong emotions (Cannon 1915). The stress response is an acute physiological reaction to a real or perceived threat. Cognitive evaluation of sensory inputs and internal states elicits widespread bodily responses aimed at adaptation or escape from the stressful stimuli. The brain shifts to a state of hyperarousal and heightened attention, accompanied by surges of adrenaline and cortisol in the blood. This in turn increases heart rate and muscle tension, while dilating the pupils to enhance vision, increase sensory processing, and promote adaptive physiological reactions. The stress response to physical or psychological stressors involves the activation of two different networks or "axes" of cascading signals between various regions of the brain and body—sympathetic-adrenomedullary axis (SAM) and the hypothalamic-pituitary-adrenal axis (HPA). The diverse visceral and peripheral connections of adrenergic and noradrenergic neurons described in previous sections allow them to respond to a range of psychological, physical, and metabolic stressors. Multiple lines of evidence indicate that noradrenergic neurons in the brain are activated by both physical and psychological stressors. Studies of immediate early-gene expression following aversive electric shocks show activation of brainstem adrenergic and noradrenergic nuclei (Pezzone et al. 1993).

The SAM Axis The SAM axis triggers a quick, transient response to stress, thereby heightening arousal and vigilance. This allows for a rapid assessment of threats and formulation of strategic responses to tackle them. Perception of acute threats or stressors leads to the activation of the LC and other adrenergic and noradrenergic nuclei in the brainstem (Pezzone et al. 1993; Valentino and Van Bockstaele 2008). The LC is thought to receive information about stressful stimuli through multiple convergent pathways. In addition, components of the limbic circuits, mainly the prefrontal cortex, amygdala, and hippocampus are each activated by psychological stressors (McEwen et al. 2015; Roozendaal et al. 2009; Janak and Tye 2015). The interconnections between these regions and the LC create a feedback loop that regulates the stress response. The NTS in the brainstem serves as an important interface between visceral sensory signals and the brain, relaying information about physical perturbations such as pain, blood loss, inflammation, or respiratory distress (Ulrich-Lai and Herman 2009). For example, noradrenergic A2 neurons in the NTS are consistently activated by stimulation of afferent fibers of the solitary tract (Appleyard et al. 2007). Adrenergic C1 neurons in the rostral ventrolateral medulla drive an increase in sympathetic nerve activity to maintain blood pressure following hemorrhagic shock (Souza et al. 2022b). A2 neurons in the NTS trigger sympathetic responses following acute restraint stress, raising heart rate and corticosterone levels (Bundzikova-Osacka et al. 2015). NTS neurons also respond to interoceptive inputs from the gut and airways (Ran et al. 2022). Projections from the NTS relay information about these stressors to other regions of the brain, including the paraventricular nucleus (PVN) of the hypothalamus (Bailey et al. 2006) and the central amygdala (CeA) (He et al. 2022), triggering the stress response. The SAM response to these various stressors is mediated primarily by catecholamines, primarily Epi and NE from the adrenal medulla and NE from sympathetic nerves. Descending projections from the PVN, LC, and adrenal medulla through the spinal cord transmit stressor responses from the brain to various effector tissues. One such instance involves post-ganglionic activation of the chromaffin cells in the adrenal medulla, which promptly releases substantial amounts of both Epi and NE into the bloodstream. Epi and NE from the autonomic nerves and in the blood bind to adrenergic receptors in smooth muscles and various organs. Preganglionic neurons also activate the sympathetic response, elevating blood pressure, cardiac output, and metabolic activity, thereby preparing the body to either "fight" or "flee" from the perceived threat (Godoy et al. 2018). In addition, circulating Epi activates peripheral β-adrenergic receptors on vagal afferents (Roozendaal et al. 2009), sending stress signals back to the NTS and the brain, thus amplifying the stress response.

The HPA Axis The hypothalamus-pituitary-adrenal (HPA) axis recruits the endocrine system and hormonal messengers to deliver an amplified and more prolonged response to stress. Neurosecretory cells within the PVN secrete corticotropin releasing hormone (CRH, previously called corticotropin-releasing factor or CRF) into the hypophyseal portal system. CRH then stimulates the anterior pituitary to release adrenocorticotropic hormone (ACTH) into the bloodstream, stimulating the synthesis and release of glucocorticoids from the adrenal glands into the blood (cortisol in humans, corticosterone in rats and mice) (Herman et al. 1997). Glucocorticoids then bind to mineralocorticoid and glucocorticoid receptors in various stress-responsive regions of the brain to induce a behavioral response to the stressful stimulus (McEwen et al. 2015; Roozendaal et al. 2009). Furthermore, brain circuits that are responsible for regulating the stress response intersect with those involved in learning and memory, such as the LC, amygdala, hippocampus, and prefrontal cortex, thus providing a means to tailor the fight-or-flight response based on past experience and predicted outcomes (McEwen et al. 2015; Roozendaal et al. 2009; Janak and Tye 2015; McEwen 1999). While glucocorticoids are thought to be the main drivers of the HPA axis, several studies have demonstrated the influence and participation of central and peripheral catecholamines in the modulation of this stress response. For example, inhibition of postganglionic sympathetic nerves attenuated corticosterone release following chronic restraint stress (Lowrance et al. 2016). The LC and various adrenergic nuclei in the NTS express glucocorticoid receptors which allow them to be activated by the HPA response. Additionally, the amygdala, hippocampus, and medial prefrontal cortex all respond to emotionally significant stimuli, subsequently activating the PVN, LC, and other catecholaminergic cells in the ventrolateral medulla and NTS (Godoy et al. 2018; Myers 2017). Moreover, glucocorticoid receptors co-localize with adrenoreceptors, facilitating an interaction between the SAM and HPA axes (Godoy et al. 2018). The LC-NE network is also a major target of CRH. CRH immunoreactive fibers have been shown to innervate both the inner nucleus of the LC and the pericoerulear zones (Valentino et al. 1992), and are thought to boost tonic LC activity and induce anxiety-like behaviors (Curtis et al. 2012; Curtis et al. 1997; McCall et al. 2015). Interestingly, the release of CRH within the LC is regulated by glucocorticoids, (Valentino and Van Bockstaele 2008), allowing the HPA axis control over shifts in arousal and thereby modulating behavioral flexibility (Snyder et al. 2012). Thus, a hierarchical organization relays stress-related information from limbic structures to sympathetic and parasympathetic effectors, which is necessary to translate emotional stimuli into physiological responses.

The consequence of Epi release from the adrenal medulla is thought to be a largely acute and short-lived response, initiated by events in the SAM axis. However, prolonged stress and circulating glucocorticoids, released by the adrenal gland during signaling via the HPA axis, in turn controls Epi levels in the bloodstream. Numerous studies have demonstrated that PNMT activity, and hence Epi synthesis, is affected by glucocorticoids and the stress response (Wurtman and Axelrod 1966; Ross et al. 1990). It has long since been recognized that prolonged exposure to highly emotional or traumatic stressors contributes to the development of mood and anxiety disorders, including post-traumatic stress disorder (PTSD) and major depressive disorder (Roozendaal et al. 2009). A prolonged period of elevated adrenergic and noradrenergic signaling contributes to the development of these disorders. For instance, repeated stress-induced elevation of Epi induces sustained vasoconstriction and elevated blood pressure, increasing the risk for stress-induced hypertension (Ulrich-Lai and Herman 2009). *Chapter 11 of this book further details the diverse responses of the brain and body to various stressors and the maladaptation of these responses due to chronic stress.*

2.4 Disrupted Adrenergic Modulation in Disease

2.4.1 Pain

Central NE and Pain The integration of neural circuits that modulate the perception of pain includes LC-NE neurons that project axons into the spinal cord. In general, NE in the motoneurons of the spinal ventral horn facilitates nociceptive spinal reflex responses, while NE in the spinal dorsal sensory horn suppresses these nociceptive reflex responses. Increased rates of NE turnover in the dorsal spinal cord have been observed in rat models of arthritis pain, indicating that pain activates descending noradrenergic systems (Weil-Fugazza et al. 1986). The stimulation of these dorsal descending axons that release NE can exert an antinociceptive effect (Proudfit 1988; Jones 1991b). Experiments in animals and clinical studies in humans have shown that central administration of NE, or of synthetic alpha adrenergic receptor agonists into the spinal cord, produces an anti-nociceptive effect (Pertovaara 2007). Conversely, administration of α2-adrenoceptor antagonists or knocking out α2-adrenoreceptors reverses these analgesic effects and increases nociceptive responses. Selective depletion of spinal NE by intrathecal injections of DSP4 (N-(2-chloroethyl)-N-ethyl-2-bromobenzylamine hydrochloride) attenuates morphine-induced analgesia (Zhong et al. 1985). These experiments clearly demonstrate how important NE fibers are in descending spinal modulation.

Sympathetic Nervous System and Pain The sympathetic arm of the autonomic nervous system is critically involved in the generation of pain. In addition to its conventional role of transmitting signals from the CNS to peripheral target cells such as smooth muscle cells and secretory cells, the peripheral sympathetic noradrenergic neurons have several direct and indirect roles associated with protection of bodily tissues via pain.

While NE administration to healthy subjects does not alter mechanical sensitivity, it has several possible interactions under pathological conditions. Some of the most direct responses the sympathetic system has in reaction to noxious stimuli occur immediately after injury. When the sciatic nerve is injured by ligation or axotomy, sympathetic fibers sprout into the dorsal root ganglia (DRG), where the somata of the injured primary sensory neurons reside and wrap around these neurons with elaborate baskets of noradrenergic endings (McLachlan et al. 1993). Experimentally, nerve growth factor (NGF) can similarly induce sprouting of tyrosine hydroxylase (TH) immunoreactive fibers in the DRG (Chung et al. 1993; Jones et al. 1999). α2 adrenergic receptor expression is upregulated in both the DRG neurons and the NE axons. Some of these primary sensory DRG neurons, especially the nociceptive fibers and A-δ fibers, have been shown to depolarize directly from sympathetic NE stimulation (Birder and Perl 1999). Another direct role of NE is its role in the generation of sympathetic maintained pain after trauma. In many human clinical cases, stimulation of the sympathetic postganglionic neurons or increasing NE by intracutaneous injections can produce pain. Conversely, inhibition of these neurons via alpha-adrenergic blockers or pharmacological depletion of NE from vesicle stores with guanethidine can relieve pain (Jänig 2020).

Sympathetic neurons can also have indirect effects that contribute to pain after an injury. NE can activate the α adrenoreceptors on the blood vessels and result in vasoconstriction. NE can also activate β adrenoreceptors on macrophages and induce cytokine release. Experimental evidence has demonstrated that newly reinnervated blood vessels after lesions are hypersensitive to NE and exhibit enhanced vasoconstriction. Such NE-mediated effects can result in increasing the activity of afferent fibers, increasing inflammation, and generally modulating the experience of pain.

2.4.2 Aging and Alzheimer's Disease (AD)

Comprehensive studies on individuals aged 25 to 85 have revealed a gradual decline in noradrenaline levels across various brain regions as people age, ultimately resulting in a loss of more than half of the initial cell density in the LC (Mann 1983). This decline is especially exacerbated in neurodegenerative disorders that accelerate neuronal deterioration such as Alzheimer's disease. The LC is one of the earliest regions to be affected by tau pathology and progressively deteriorates with the progression of AD and dementia (Jacobs et al. 2021; Braak et al. 2011; Braak and Del Tredici 2011). LC neurons progressively degenerate during the course of the disease and studies show a significant correlation between cell loss in the LC and the duration and progression of AD (German et al. 1992). Neurofibrillary changes in the LC have been detected in patients with mild cognitive impairment (Grudzien et al. 2007). Further, the number of neurofibrillary tangles correlates with the severity of dementia observed in AD patients (Bondareff et al. 1987). Experimental lesions of the LC in animal models of AD suggest a role for increased inflammation in the progression of AD. DSP4-induced degradation of LC projections significantly elevates glial inflammation as well as augmented deposits of amyloid plaques (Heneka et al. 2006). Recruitment of microglia to Aβ sites, as well as phagocytosis of plaques is disrupted with degradation of the LC. Conversely, restoring NE function rescues these impairments (Heneka et al. 2010). Together these studies suggest a key role for the noradrenergic system in the progression of AD. The early susceptibility of LC neurons to protein aggregates suggests that the noradrenergic system could be a promising target for early detection markers of AD.

2.5 Conclusion

Neural systems facilitating brain-body communication form intricate networks that integrate sensory and proprioceptive information with executive motor commands. As a pivotal chemical messenger with extensive networks across brain regions and diverse receptor expression in multiple organs within the body, we propose that NE plays a crucial role in linking neural control to physiological functions. Salient sensory and visceral stimuli alter LC-NE activity in the brain, prompting the LC to modulate the gain of sensory processing and motor output. Meanwhile, NE release through sympathetic nerves directly modulates the function of various organs, such as the heart, lungs, digestive tract, and muscles to organize coordinated responses to external stimuli. This dual role equips the noradrenergic system to facilitate processes that require integration of signals across the brain and body, such as the fight-or-flight response to stress. In addition, NE signaling in the brain influences the memory of salient events, thus allowing for flexibility in the control of physiological systems and adaptation of behavioral responses to stimuli. These diverse roles of NE offer promising prospects for a range of clinical applications, from the treatment of attention disorders and PTSD to the management of chronic pain and anesthesia.

References

Adams LM, Foote SL (1988) Effects of locally infused pharmacological agents on spontaneous and sensory-evoked activity of locus coeruleus neurons. Brain Res Bull 21(3):395–400

Ährlund-Richter S, Osako Y, Jenks KR, Odom E, Huang H, Arnold DB, Sur M (2024) Prefrontal Cortex subregions provide distinct visual and behavioral feedback modulation to the Primary Visual Cortex. bioRxiv 2024.08.06.606894

Appleyard SM, Marks D, Kobayashi K, Okano H, Low MJ, Andresen MC (2007) Visceral afferents directly activate catecholamine neurons in the solitary tract nucleus. J Neurosci 27(48):13292–13302

Aston-Jones G, Bloom FE (1981) Norepinephrine-containing locus coeruleus neurons in behaving rats exhibit pronounced responses to non-noxious environmental stimuli. J Neurosci 1(8):887–900

Aston-Jones G, Cohen JD (2005) An integrative theory of locus coeruleus-norepinephrine function: adaptive gain and optimal performance. Annu Rev Neurosci 28:403–450

Aston-Jones G, Waterhouse B (2016) Locus coeruleus: from global projection system to adaptive regulation of behavior. Brain Res 1645:75–78

Aston-Jones G, Rajkowski J, Kubiak P, Alexinsky T (1994) Locus coeruleus neurons in monkey are selectively activated by attended cues in a vigilance task. J Neurosci 14(7):4467–4480

Axelrod J (1962) Purification and properties of Phenylethanolamine-N-methyl transferase. J Biol Chem 237:1657–1660

Bailey TW, Hermes SM, Andresen MC, Aicher SA (2006) Cranial visceral afferent pathways through the nucleus of the solitary tract to caudal ventrolateral medulla or paraventricular hypothalamus: target-specific synaptic reliability and convergence patterns. J Neurosci 26(46):11893–11902

Bajic D, Van Bockstaele EJ, Proudfit HK (2001) Ultrastructural analysis of ventrolateral periaqueductal gray projections to the A7 catecholamine cell group. Neuroscience 104(1):181–197

Basbaum AI (ed) (2008) The senses: a comprehensive reference, vol 4, 1st edn. Elsevier, Amsterdam

Bastos AM, Donoghue JA, Brincat SL, Mahnke M, Yanar J, Correa J et al (2021) Neural effects of propofol-induced unconsciousness and its reversal using thalamic stimulation. elife 10:e60824

Berridge CW (2008) Noradrenergic modulation of arousal. Brain Res Rev 58(1):1–17

Berridge CW, Waterhouse BD (2003) The locus coeruleus-noradrenergic system: modulation of behavioral state and state-dependent cognitive processes. Brain Res Rev (Elsevier) 42:33–84

Berthoud HR, Neuhuber WL (2000) Functional and chemical anatomy of the afferent vagal system. Auton Neurosci 85(1–3):1–17

Bin NR, Prescott SL, Horio N, Wang Y, Chiu IM, Liberles SD (2023) An airway-to-brain sensory pathway mediates influenza-induced sickness. Nature 615(7953):660–667

Birder LA, Perl ER (1999) Expression of α2-adrenergic receptors in rat primary afferent neurones after peripheral nerve injury or inflammation. J Physiol 515(2):533–542

Bondareff W, Mountjoy CQ, Roth M, Rossor MN, Iversen LL, Reynolds GP, Hauser DL (1987) Neuronal degeneration in locus ceruleus and cortical correlates of Alzheimer disease. Alzheimer Dis Assoc Disord 1(4):256–262

Bonnet MH, Arand DL (2001) Impact of activity and arousal upon spectral EEG parameters. Physiol Behav 74(3):291–298

Boron WF, Boulpaep EL (2009) Medical physiology, 2nd edn. Saunders/Elsevier, Philadelphia

Bouret S, Sara SJ (2005) Network reset: a simplified overarching theory of locus coeruleus noradrenaline function. Trends Neurosci 28(11):574–582

Bouret S, Duvel A, Onat S, Sara SJ (2003) Phasic activation of locus ceruleus neurons by the central nucleus of the amygdala. J Neurosci 23(8):3491–3497

Braak H, Del Tredici K (2011) The pathological process underlying Alzheimer's disease in individuals under thirty. Acta Neuropathol 121(2):171–181

Braak H, Thal DR, Ghebremedhin E, Del TK (2011) Stages of the pathologic process in Alzheimer disease: age categories from 1 to 100 years. J Neuropathol Exp Neurol 70(11):961–969

Bradley MM, Miccoli L, Escrig MA, Lang PJ (2008) The pupil as a measure of emotional arousal and autonomic activation. Psychophysiology 45(4):602–607

Brennan P, Kaba H, Keverne EB (1990) Olfactory recognition: a simple memory system. Science 250(4985):1223–1226

Breton-Provencher V, Sur M (2019) Active control of arousal by a locus coeruleus GABAergic circuit. Nat Neurosci 22(2):218–228

Breton-Provencher V, Drummond GT, Sur M (2021) Locus Coeruleus norepinephrine in learned behavior: anatomical modularity and spatiotemporal integration in targets. Front Neural Circuits 15:638007

Breton-Provencher V, Drummond GT, Feng J, Li Y, Sur M (2022) Spatiotemporal dynamics of noradrenaline during learned behavior. Nature 606(7915):732–738

Brown EN, Pavone KJ, Naranjo M (2018) Multimodal general anesthesia: theory and practice. Anesth Analg 127(5):1246–1258

Bundzikova-Osacka J, Ghosal S, Packard BA, Ulrich-Lai YM, Herman JP (2015) Role of nucleus of the solitary tract noradrenergic neurons in post-stress cardiovascular and hormonal control in male rats. Stress 18(2):221–232

Byrum CE, Guyenet PG (1987) Afferent and efferent connections of the A5 noradrenergic cell group in the rat. J Comp Neurol 261(4):529–542

Cannon WB (1915) Bodily changes in pain, hunger, fear and rage: an account of recent researches into the function of emotional excitement. D Appleton & Company, New York

Chandler DJ, Gao WJ, Waterhouse BD (2014) Heterogeneous organization of the locus coeruleus projections to prefrontal and motor cortices. Proc Natl Acad Sci USA 111(18):6816–6821

Chase T, Kopin I (1968) Stimulus-induced release of substances from Olfactory Bulb using the Push–Pull Cannula. Nature 217:466–467

Chowdhury A, Luchetti A, Fernandes G, Filho DA, Kastellakis G, Tzilivaki A, Ramirez EM, Tran MY, Poirazi P, Silva AJ (2022) A locus coeruleus-dorsal CA1 dopaminergic circuit modulates memory linking. Neuron 110(20):3374–3388.e8

Chung K, Kim HJ, Na HS, Park MJ, Chung JM (1993) Abnormalities of sympathetic innervation in the area of an injured peripheral nerve in a rat model of neuropathic pain. Neurosci Lett 162(1–2):85–88

Ciombor KJ, Ennis M, Shipley MT (1999) Norepinephrine increases rat mitral cell excitatory responses to weak olfactory nerve input via alpha-1 receptors in vitro. Neuroscience 90(2):595–606

Clark FM, Proudfit HK (1991) The projection of noradrenergic neurons in the A7 catecholamine cell group to the spinal cord in the rat demonstrated by anterograde tracing combined with immunocytochemistry. Brain Res 547(2):279–288

Coleridge HM, Coleridge JCG (1994) Pulmonary reflexes: neural mechanisms of pulmonary defense. Annu Rev Physiol 56(1):69–91

Cunningham ET, Miselis RR, Sawchenko PE (1994) The relationship of efferent projections from the area postrema to vagal motor and brain stem catecholamine-containing cell groups: an axonal transport and

immunohistochemical study in the rat. Neuroscience 58(3):635–648
Curtis AL, Lechner SM, Pavcovich LA, Valentino RJ (1997) Activation of the locus coeruleus noradrenergic system by intracoerulear microinfusion of corticotropin-releasing factor: effects on discharge rate, cortical norepinephrine levels and cortical electroencephalographic activity. J Pharmacol Exp Ther 281(1):163–172
Curtis AL, Bello NT, Connolly KR, Valentino RJ (2002) Corticotropin-releasing factor neurones of the central nucleus of the amygdala mediate locus coeruleus activation by cardiovascular stress. J Neuroendocrinol 14(8):667–682
Curtis AL, Leiser SC, Snyder K, Valentino RJ (2012) Predator stress engages corticotropin-releasing factor and opioid systems to alter the operating mode of locus coeruleus norepinephrine neurons. Neuropharmacology 62(4):1737–1745
Devilbiss DM, Waterhouse BD (2004) The effects of tonic locus ceruleus output on sensory-evoked responses of ventral posterior medial thalamic and barrel field cortical neurons in the awake rat. J Neurosci 24(48):10773–10785
Devilbiss DM, Page ME, Waterhouse BD (2006) Locus ceruleus regulates sensory encoding by neurons and networks in waking animals. J Neurosci 26(39):9860–9872
Dimitrov EL, Yanagawa Y, Usdin TB (2013) Forebrain GABAergic projections to locus coeruleus in mouse. J Comp Neurol 521(10):2373–2397
dos Santos Lima GZ, Lobao-Soares B, Corso G, Belchior H, Lopes SR, de Lima PT et al (2019) Hippocampal and cortical communication around micro-arousals in slow-wave sleep. Sci Rep 9(1):5876
Dymecki S, Bouret S, Wildner H, Waterhouse BD, Predale HK, Plummer NW et al (2022) Probing the structure and function of locus coeruleus projections to CNS motor centers. Front Neural Circuits 16:895481
Edeline JM, Manunta Y, Hennevin E (2011) Induction of selective plasticity in the frequency tuning of auditory cortex and auditory thalamus neurons by locus coeruleus stimulation. Hear Res 274(1–2):75–84
Eisen AJ, Kozachkov L, Bastos AM, Donoghue JA, Mahnke MK, Brincat SL et al (2024) Propofol anesthesia destabilizes neural dynamics across cortex. Neuron 112(16):2799–2813.e9
Ennis M, Aston-Jones G (1989) GABA-mediated inhibition of locus coeruleus from the dorsomedial rostral medulla. J Neurosci 9(8):2973–2981
Eschenko O, Magri C, Panzeri S, Sara SJ (2012) Noradrenergic neurons of the locus coeruleus are phase locked to cortical up-down states during sleep. Cerebral Cortex [Internet] 22:426–435
Everitt BJ, Hökfelt T, Terenius L, Tatemoto K, Mutt V, Goldstein M (1984) Differential co-existence of neuropeptide Y (NPY)-like immunoreactivity with catecholamines in the central nervous system of the rat. Neuroscience 11(2):443462
Felten DL, Sladek JR (1983) Monoamine distribution in primate brain V. Monoaminergic nuclei: anatomy, pathways and local organization. Brain Res Bull 10(2):171–284
Fernstrom JD, Fernstrom MH (2007) Tyrosine, phenylalanine, and catecholamine synthesis and function in the brain. J Nutr 137(6):1539S–1547S
Foote SL, Aston-Jones G, Bloom FE (1980) Impulse activity of locus coeruleus neurons in awake rats and monkeys is a function of sensory stimulation and arousal. Proc Natl Acad Sci USA 77(5):3033–3037
Fung SJ, Barnes CD (1987) Membrane excitability changes in hindlimb motoneurons induced by stimulation of the locus coeruleus in cats. Brain Res 402(2):230–242
German DC, Manaye KF, White CL 3rd, Woodward DJ, McIntire DD, Smith WK, Kalaria RN, Mann DM (1992) Disease-specific patterns of locus coeruleus cell loss. Ann Neurol 32(5):667–676
Godoy LD, Rossignoli MT, Delfino-Pereira P, Garcia-Cairasco N, de Lima Umeoka EH (2018) A comprehensive overview on stress neurobiology: basic concepts and clinical implications. Front Behav Neurosci (Frontiers Media S.A.) 12:127
Grefkes C, Wang LE, Eickhoff SB, Fink GR (2010) Noradrenergic modulation of cortical networks engaged in visuomotor processing. Cereb Cortex 20(4):783–797
Groves DA, Bowman EM, Brown VJ (2005) Recordings from the rat locus coeruleus during acute vagal nerve stimulation in the anaesthetised rat. Neurosci Lett 379(3):174–179
Grudzien A, Shaw P, Weintraub S, Bigio E, Mash DC, Mesulam MM (2007) Locus coeruleus neurofibrillary degeneration in aging, mild cognitive impairment and early Alzheimer's disease. Neurobiol Aging 28(3):327–335
Gutman Y, Weil-Malherbe H (1967) Subcellular distribution of norepinephrine in uteri of some species. Nature 214:108–109
Guyenet PG (1991) Central noradrenergic neurons: the autonomic connection. Prog Brain Res 88:365–380
Hasselmo ME, Linster C, Patil M, Ma D (1997) Noradrenergic suppression of synaptic transmission may influence cortical signal-to-noise ratio. J Neurophysiol 77(6):3326–3339
Hayat H, Regev N, Matosevich N, Sales A, Paredes-Rodriguez E, Krom AJ, Bergman L, Li Y, Lavigne M, Kremer EJ, Yizhar O, Pickering AE, Nir Y (2020) Locus coeruleus norepinephrine activity mediates sensory-evoked awakenings from sleep. Sci Adv 6(15):eaaz4232
He S, Huang X, Zheng J, Zhang Y, Ruan X (2022) An NTS-CeA projection modulates depression-like behaviors in a mouse model of chronic pain. Neurobiol Dis 174:105893
Heneka MT, Ramanathan M, Jacobs AH, Dumitrescu-Ozimek L, Bilkei-Gorzo A, Debeir T, Sastre M, Galldiks N, Zimmer A, Hoehn M, Heiss WD, Klockgether T, Staufenbiel M (2006) Locus ceru-

leus degeneration promotes Alzheimer pathogenesis in amyloid precursor protein 23 transgenic mice. J Neurosci 26(5):1343–1354

Heneka MT, Nadrigny F, Regen T, Martinez-Hernandez A, Dumitrescu-Ozimek L, Terwel D, Jardanhazi-Kurutz D, Walter J, Kirchhoff F, Hanisch UK, Kummer MP (2010) Locus ceruleus controls Alzheimer's disease pathology by modulating microglial functions through norepinephrine. Proc Natl Acad Sci USA 107(13):6058–6063

Herman JP, Cullinan WE, Herman JP, Cullinan WE (1997) Neurocircuitry of stress: central control of the hypothalamo-pituitary-adrenocortical axis. Trends Neurosci 20:78–84

Holdefer RN, Jacobs BL (1994) Phasic stimulation of the locus coeruleus: effects on activity in the lateral geniculate nucleus. Exp Brain Res 100(3):444–452

Hou RH, Scaife J, Freeman C, Langley RW, Szabadi E, Bradshaw CM (2006) Relationship between sedation and pupillary function: comparison of diazepam and diphenhydramine. Br J Clin Pharmacol 61(6):752–760

Hulsey DR, Riley JR, Loerwald KW, Rennaker RL, Kilgard MP, Hays SA (2017) Parametric characterization of neural activity in the locus coeruleus in response to vagus nerve stimulation. Exp Neurol 289:21–30

Hwang KR, Chan SHH, Chan JYH (1998) Noradrenergic neurotransmission at PVN in locus ceruleus-induced baroreflex suppression in rats. Am J Phys Heart Circ Phys 274(4):H1284–H1292

Jacobs HIL, Riphagen JM, Ramakers IHGB, Verhey FRJ (2021) Alzheimer's disease pathology: pathways between central norepinephrine activity, memory, and neuropsychiatric symptoms. Mol Psychiatry 26(3):897–906

Jahr CE, Nicoll RA (1982) Noradrenergic modulation of dendrodendritic inhibition in the olfactory bulb. Nature 297(5863):227–229

Janak PH, Tye KM (2015) From circuits to behaviour in the amygdala. Nature 517(7534):284–292

Jänig W (2020) Sympathetic nervous system and pain. In: The senses: a comprehensive reference: volume 1-7, Second Edition, vol 5, pp 349–378

Jarosiewicz B, Skaggs WE (2004) Level of arousal during the small irregular activity state in the rat hippocampal EEG. J Neurophysiol 91(6):2649–2657

Jin H, Li M, Jeong E, Castro-Martinez F, Zuker CS (2024) A body–brain circuit that regulates body inflammatory responses. Nature 630(8017):695–703

Jodo E, Aston-Jones G (1997) Activation of locus coeruleus by prefrontal cortex is mediated by excitatory amino acid inputs. Brain Res 768(1–2):327–332

Jodo E, Chiang C, Aston-Jones G (1998) Potent excitatory influence of prefrontal cortex activity on noradrenergic locus coeruleus neurons. Neuroscience 83(1):63–79

Jones BE (1991a) Noradrenergic locus coeruleus neurons: their distant connections and their relationship to neighboring (including cholinergic and GABAergic) neurons of the central gray and reticular formation. In: Barnes CD, Pompeiano O (eds) Progress in brain research. Elsevier, pp 15–30

Jones SL (1991b) Descending noradrenergic influences on pain. Prog Brain Res 88(C):381–394

Jones MG, Munson JB, Thompson SWN (1999) A role for nerve growth factor in sympathetic sprouting in rat dorsal root ganglia. Pain 79(1):21–29

Kaba H, Keverne EB (1988) The effect of microinfusions of drugs into the accessory olfactory bulb on the olfactory block to pregnancy. Neuroscience 25(3):1007–1011

Kaba H, Rosser A, Keverne B (1989) Neural basis of olfactory memory in the context of pregnancy block. Neuroscience 32(3):657–662

Kalia M, Sullivan JM (1982) Brainstem projections of sensory and motor components of the vagus nerve in the rat. J Comp Neurol 211(3):248–264

Kebschull JM, Garcia da Silva P, Reid AP, Peikon ID, Albeanu DF, Zador AM (2016) High-throughput mapping of single-neuron projections by sequencing of barcoded RNA. Neuron 91(5):975–987

Kerfoot EC, Chattillion EA, Williams CL (2008) Functional interactions between the nucleus tractus solitarius (NTS) and nucleus accumbens shell in modulating memory for arousing experiences. Neurobiol Learn Mem 89(1):47–60

Kiissp M, Vater M (1989) Noradrenaline enhances temporal auditory contrast and neuronal timing precision in the cochlear nucleus of the mustached bat. J Neurosci 9(12):4169–4178

Kimura Y, Tadano T, Sakai K, Miyamoto A, Kisara K (2009) Changes of ejaculation due to suppression of noradrenaline biosynthesis by diethyldithiocarbamate. Andrologia 12(1):85–91

Kirouac And GJ, Ciriello J (1997) Medullary inputs to nucleus accumbens neurons. Am J Phys Regul Integr Comp Phys 6(273):R2080–R2088

Klimaschewski L, Kummer W, Heym C (1996) Localization, regulation and functions of neurotransmitters and neuromodulators in cervical sympathetic ganglia. Microsc Res Tech 35(1):44–68

Kolta A, Reader TA (1989) Modulatory effects of catecholamines on neurons of the rat visual cortex: single cell iontophoretic studies. Can J Physiol Pharmacol 67(6):615–623

Kunst M, McMillen D, Close J, Campos J, Hupp M, Martin N, et al (2023) Whole Mouse Brain Transcriptomic Cell Type Atlas – MERSCOPE v1. Available from: https://doi.org/10.35077/g.610

Kupari J, Häring M, Agirre E, Castelo-Branco G, Ernfors P (2019) An atlas of vagal sensory neurons and their molecular specialization. Cell Rep 27(8):2508–2523.e4

Larsen RS, Waters J (2018) Neuromodulatory correlates of pupil dilation. Front Neural Circuits (Frontiers Media S.A.) 12:21

Le Pichon CE, Chesler AT (2014) The functional and anatomical dissection of somatosensory subpopulations using mouse genetics. Front Neuroanat 8:21

Lecas JC, Et P, Curie M, Bernard S (2004) Locus coeruleus activation shortens synaptic drive while decreasing spike latency and jitter in sensorimotor cortex.

Implications for neuronal integration. European Journal of Neuroscience 19:2519–2530
Lewis DI, Coote JH (1990) Excitation and inhibition of rat sympathetic preganglionic neurones by catecholamines. Brain Res 530(2):229–234
Liang Y, Shi W, Xiang A, Hu D, Wang L, Zhang L (2021) The NAergic locus coeruleus-ventrolateral preoptic area neural circuit mediates rapid arousal from sleep. Curr Biol 31(17):3729–3742.e5
Lightman SL, Todd K, Everitt BJ (1984) Ascending noradrenergic projections from the brainstem: evidence for a major role in the regulation of blood pressure and vasopressin secretion*. Exp Brain Res 55:145–151
Lipski J, Kanjhan R, Kruszewska B, Smith M (1995) Barosensitive neurons in the rostral ventrolateral medulla of the rat in vivo: morphological properties and relationship to C1 adrenergic neurons. Neuroscience 69(2):601–618
Lowrance SA, Ionadi A, McKay E, Douglas X, Johnson JD (2016) Sympathetic nervous system contributes to enhanced corticosterone levels following chronic stress. Psychoneuroendocrinology 68:163–170
Makeig S, Inlow M (1993) Lapse in alertness: coherence of fluctuations in performance and EEG spectrum. Electroencephalogr Clin Neurophysiol 86(1):23–35
Mann DMA (1983) The locus coeruleus and its possible role in ageing and degenerative disease of the human central nervous system. Mech Ageing Dev 23(1):73–94
Manunta Y, Edeline J (1997) Effects of noradrenaline on frequency tuning of rat auditory cortex neurons. Eur J Neurosci 9(4):833–847
Manunta Y, Edeline JM (1999) Effects of noradrenaline on frequency tuning of auditory cortex neurons during wakefulness and slow-wave sleep. Eur J Neurosci 11:2134–2150
Martin EM, Pavlides C, Pfaff D (2010) Multimodal sensory responses of nucleus reticularis gigantocellularis and the responses' relation to cortical and motor activation. J Neurophysiol 103(5):2326–2338
Martins ARO, Froemke RC (2015) Coordinated forms of noradrenergic plasticity in the locus coeruleus and primary auditory cortex. Nat Neurosci 18(10):1483–1492
Mather M, Clewett D, Sakaki M, Harley CW (2016) Norepinephrine ignites local hotspots of neuronal excitation: how arousal amplifies selectivity in perception and memory. Behav Brain Sci 39:e200
McCall JG, Al-Hasani R, Siuda ER, Hong DY, Norris AJ, Ford CP et al (2015) CRH engagement of the locus coeruleus noradrenergic system mediates stress-induced anxiety. Neuron 87(3):605–620
McCormick DA, Bal T (1997) Sleep and arousal: thalamocortical mechanisms. Annu Rev Neurosci 20(1):185–215
McCormick DA, Pape HC, Williamson A (1991) Actions of norepinephrine in the cerebral cortex and thalamus: implications for function of the central noradrenergic system. Prog Brain Res 88(C):293–305
McEwen BS (1999) Stress and hippocampal plasticity. Annu Rev Neurosci 22(1):105–122
McEwen BS, Bowles NP, Gray JD, Hill MN, Hunter RG, Karatsoreos IN et al (2015) Mechanisms of stress in the brain. Nat Neurosci 18(10):1353–1363
McLachlan EM, Wilfid J, Marshall D, Martin M (1993) Peripheral nerve injury triggers noradrenergic sprouting within dorsal root ganglia. Nature 363:543–546
McLean J, Waterhouse BD (1994) Noradrenergic modulation of cat area 17 neuronal responses to moving visual stimuli. Brain Res 667(1):83–97
McLennan H (1971) The pharmacology of inhibition of mitral cells in the olfactory bulb. Brain Res 29(2):177–184
Mehendale S, Xie JT, Aung HH, Guan XF, Yuan CS (2004) Nucleus accumbens receives gastric vagal inputs. Acta Pharmacol Sin 25(3):271–275
Miller JW, Hall CM, Holland KD, Ferrendelli JA (1989) Identification of a median thalamic system regulating seizures and arousal. Epilepsia 30(4):493–500
Moruzzi G, Magoun HW (1949) Brain stem reticular formation and activation of the EEG. Electroencephalogr Clin Neurophysiol 1(1–4):455–473
Motiejunaite J, Amar L, Vidal-Petiot E (2021) Adrenergic receptors and cardiovascular effects of catecholamines. Ann Endocrinol (Paris) 82(3–4):193–197
Murata Y, Colonnese MT (2018) Thalamus controls development and expression of arousal states in visual cortex. J Neurosci 38(41):8772–8786
Myagmar BE, Flynn JM, Cowley PM, Swigart PM, Montgomery MD, Thai K et al (2017) Adrenergic receptors in individual ventricular myocytes: the beta-1 and alpha-1B are in all cells, the alpha-1A is in a subpopulation, and the beta-2 and beta-3 are mostly absent. Circ Res 120(7):1103–1115
Myers B (2017) Corticolimbic regulation of cardiovascular responses to stress. Physiol Behav 172:49–59
Okutani F, Kaba H, Takahashi S, Seto K (1998) The biphasic effects of locus coeruleus noradrenergic activation on dendrodendritic inhibition in the rat olfactory bulb. Brain Res 783(2):272–279
Osorio-Forero A, Cardis R, Vantomme G, Guillaume-Gentil A, Katsioudi G, Devenoges C, Fernandez LMJ, Lüthi A (2021) Noradrenergic circuit control of non-REM sleep substates. Curr Biol 31(22):5009–5023.e7
Paintal AS (1973) Vagal sensory receptors and their reflex effects. Physiol Rev 53(1):159–227
Perez H, Hernandez A, Almli CR (1987) Locus coeruleus stimulation modulates olfactory bulb evoked potentials. Brain Res Bull 18(6):767–770
Pertovaara A (2007) Noradrenergic pathology and pain. In: Brain norepinephrine: neurobiology and therapeutics. Cambridge University Press, pp 385–407
Pezzone MA, Sen LW, Hoffman GE, Pezzone KM, Rabin BS (1993) Activation of brainstem catecholaminergic neurons by conditioned and unconditioned aversive stimuli as revealed by c-Fos immunoreactivity. Brain Res 608(2):310–318
Poe GR, Foote S, Eschenko O, Johansen JP, Bouret S, Aston-Jones G et al (2020) Locus coeruleus: a new look at the blue spot. Nat Rev Neurosci Nat Res 21:644–659

Powley TL, Jaffey DM, McAdams J, Baronowsky EA, Black D, Chesney L et al (2019) Vagal innervation of the stomach reassessed: brain–gut connectome uses smart terminals. Ann N Y Acad Sci 1454(1):14–30

Prescott SL, Liberles SD (2022) Internal senses of the vagus nerve. Neuron 110(4):579–599

Proudfit HK (1988) Chapter 27 Pharmacologic evidence for the modulation of nociception by noradrenergic neurons. Prog Brain Res 77(C):357–370

Ran C, Boettcher JC, Kaye JA, Gallori CE, Liberles SD (2022) A brainstem map for visceral sensations. Nature 609(7926):320–326

Reimer J, Froudarakis E, Cadwell CR, Yatsenko D, Denfield GH, Tolias AS (2014) Pupil fluctuations track fast switching of cortical states during quiet wakefulness. Neuron 84(2):355–362

Reimer J, McGinley MJ, Liu Y, Rodenkirch C, Wang Q, McCormick DA, Tolias AS (2016) Pupil fluctuations track rapid changes in adrenergic and cholinergic activity in cortex. Nat Commun 7:13289

Rinaman L (2011) Hindbrain noradrenergic A2 neurons: diverse roles in autonomic, endocrine, cognitive, and behavioral functions. Am J Physiol Regul Integr Comp Physiol 300(2):R222–R235

Rogawski MA, Aghajanian GK (1980) Modulation of lateral geniculate neurone excitability by noradrenaline microiontophoresis or locus coeruleus stimulation. Nature 283

Roozendaal B, McEwen BS, Chattarji S (2009) Stress, memory and the amygdala. Nat Rev Neurosci 10:423–433

Ross ME, Evinger MJ, Hyman SE, Carroll JM, Mucke L, Comb M, Reis DJ, Joh TH, Goodman HM (1990) Identification of a functional glucocorticoid response element in the phenylethanolamine N-methyltransferase promoter using fusion genes introduced into chromaffin cells in primary culture. J Neurosci 10(2):520–530

Safaai H, Neves R, Eschenko O, Logothetis NK, Panzeri S (2015) Modeling the effect of locus coeruleus firing on cortical state dynamics and single-trial sensory processing. Proc Natl Acad Sci USA 112(41):12834–12839

Salgado H, Garcia-Oscos F, Martinolich L, Hall S, Restom R, Tseng KY et al (2012) Pre- and postsynaptic effects of norepinephrine on γ-aminobutyric acid-mediated synaptic transmission in layer 2/3 of the rat auditory cortex. Synapse 66(1):20–28

Salmoiraghi GC, Bloom FE, Costa E (1964) Adrenergic mechanisms in rabbit olfactory bulb. Am J Physiol-Legacy Content 207(6):1417–1424

Samuels ER, Szabadi E (2008) Functional neuroanatomy of the noradrenergic locus coeruleus: its roles in the regulation of arousal and autonomic function part II: physiological and pharmacological manipulations and pathological alterations of locus coeruleus activity in humans. Curr Neuropharmacol 6:254–285

Sara SJ (2009) The locus coeruleus and noradrenergic modulation of cognition. Nat Rev Neurosci 10(3):211–223

Sara SJ, Bouret S (2012) Orienting and reorienting: the locus Coeruleus mediates cognition through arousal. Neuron 76(1):130–141

Sara SJ, Hervé-Minvielle A (1995) Inhibitory influence of frontal cortex on locus coeruleus neurons. Proc Natl Acad Sci USA 92(13):6032–6036

Schwarz LA, Luo L (2015) Organization of the locus coeruleus-norepinephrine system. Curr Biol (Cell Press) 25:R1051–R1056

Schwarz LA, Miyamichi K, Gao XJ, Beier KT, Weissbourd B, DeLoach KE, Ren J, Ibanes S, Malenka RC, Kremer EJ, Luo L (2015) Viral-genetic tracing of the input-output organization of a central noradrenaline circuit. Nature 524(7563):88–92

Sharma Y, Xu T, Graf WM, Fobbs A, Sherwood CC, Hof PR et al (2010) Comparative anatomy of the locus coeruleus in humans and nonhuman primates. J Comp Neurol 518(7):963–971

Sharma N, Flaherty K, Lezgiyeva K, Wagner DE, Klein AM, Ginty DD (2020) The emergence of transcriptional identity in somatosensory neurons. Nature 577(7790):392–398

Shipley MT, Halloran FJ, de la Torre J (1985) Surprisingly rich projection from locus coeruleus to the olfactory bulb in the rat. Brain Res 329(1–2):294–299

Smith MS, Schambra UB, Wilson KH, Page SO, Hulette C, Light AR et al (1995) α2-Adrenergic receptors in human spinal cord: specific localized expression of mRNA encoding α2-adrenergic receptor subtypes at four distinct levels. Mol Brain Res 34(1):109–117

Smolen AJ (1988) Morphology of synapses in the autonomic nervous system. J Electron Microsc Tech 10(2):187–204

Snyder K, Wang WW, Han R, McFadden K, Valentino RJ (2012) Corticotropin-releasing factor in the norepinephrine nucleus, locus coeruleus, facilitates behavioral flexibility. Neuropsychopharmacology 37(2):520–530

Souza GMPR, Stornetta DS, Vitali AJ, Wildner H, Zeilhofer HU, Campbell JN et al (2022a) Chemogenetic activation of noradrenergic A5 neurons increases blood pressure and visceral sympathetic activity in adult rats. Am J Physiol Regul Integr Comp Physiol 323(4):R512–R531

Souza GMPR, Stornetta RL, Stornetta DS, Guyenet PG, Abbott SBG (2022b) Adrenergic C1 neurons monitor arterial blood pressure and determine the sympathetic response to hemorrhage. Cell Rep 38(10)

Swift KM, Gross BA, Frazer MA, Bauer DS, Clark KJD, Vazey EM, Aston-Jones G, Li Y, Pickering AE, Sara SJ, Poe GR (2018) Abnormal locus coeruleus sleep activity alters sleep signatures of memory consolidation and impairs place cell stability and spatial memory. Curr Biol 28(22):3599–3609.e4

Tafari AT, Thomas SA, Palmiter RD (1997) Norepinephrine facilitates the development of the murine sweat response but is not essential. J Neurosci 17(11):4275–4281

Tauber JM, Brincat SL, Stephen EP, Donoghue JA, Kozachkov L, Brown EN et al (2024) Propofol-

mediated unconsciousness disrupts progression of sensory signals through the cortical hierarchy. J Cogn Neurosci 36(2):394–413
Ter Horst GJ, Toes GJ, van Willigen JD (1991) Locus coeruleus projections to the dorsal motor vagus nucleus in the rat. Neuroscience 45(1):153–160
Uematsu A, Tan BZ, Ycu EA, Cuevas JS, Koivumaa J, Junyent F et al (2017) Modular organization of the brainstem noradrenaline system coordinates opposing learning states. Nat Neurosci 20(2):1602
Ulrich-Lai YM, Herman JP (2009) Neural regulation of endocrine and autonomic stress responses. Nat Rev Neurosci 10:397–409
Unnerstall JR, Kopajtic TA, Kuhar MJ (1984) Distribution of α2 agonist binding sites in the rat and human central nervous system: analysis of some functional, anatomic correlates of the pharmacologic effects of clonidine and related adrenergic agents. Brain Res Rev 7(1):69–101
Valentino RJ, Van Bockstaele E (2008) Convergent regulation of locus coeruleus activity as an adaptive response to stress. Eur J Pharmacol 583(2–3):194–203
Valentino RJ, Page M, Van Bockstaele E, Aston-Jones G (1992) Corticotropin-releasing factor innervation of the locus coeruleus region: distribution of fibers and sources of input. Neuroscience 48(3):689–705
Van Bockstaele EJ, Colago EEO, Valentino RJ, Van Bockstaele E (1998) Amygdaloid corticotropin-releasing factor targets locus coeruleus dendrites: substrate for the co-ordination of emotional and cognitive limbs of the stress response. J Neuroendocrinol 10:743–757
Van Bockstaele EJ, Peoples J, Telegan P (1999) Efferent projections of the nucleus of the solitary tract to peri-locus coeruleus dendrites in rat brain: evidence for a monosynaptic pathway. J Comp Neurol 412(3):410–428
Vazey EM, Moorman DE, Aston-Jones G (2018) Phasic locus coeruleus activity regulates cortical encoding of salience information. Proc Natl Acad Sci USA 115(40):E9439–E9448
Wagatsuma A, Okuyama T, Sun C, Smith LM, Abe K, Tonegawa S (2018) Locus coeruleus input to hippocampal CA3 drives single-trial learning of a novel context. Proc Natl Acad Sci USA 115(2):E310–E316
Ward DG, Gunn CG (1976) Locus coeruleus complex: elicitation of a pressor response and a brain stem region necessary for its occurence. Brain Res 107(2):401–406
Waterhouse BD, Navarra RL (2019) The locus coeruleus-norepinephrine system and sensory signal processing: a historical review and current perspectives. Brain Res 1709:1–15
Waterhouse BD, Sessler FM, Jung-Tung C, Woodward DJ, Azizi SA, Moises HC (1988) New evidence for a gating action of norepinephrine in central neuronal circuits of mammalian brain. Brain Res Bull 21(3):425–432
Waterhouse BD, Ausim Azizi S, Burne RA, Woodward DJ (1990a) Modulation of rat cortical area 17 neuronal responses to moving visual stimuli during norepinephrine and serotonin microiontophoresis. Brain Res 514(2):276–292
Waterhouse BD, Ausim Azizi S, Burne RA, Woodward DJ (1990b) Modulation of rat cortical area 17 neuronal responses to moving visual stimuli during norepinephrine and serotonin micro iontophoresis. Brain Res 514(2):276–292
Wehrwein EA, Orer HS, Barman SM (2016) Overview of the anatomy, physiology, and pharmacology of the autonomic nervous system. Compr Physiol 6(3):1239–1278
Weil-Fugazza J, Godefroy F, Manceau V, Besson JM (1986) Increased norepinephrine and uric acid levels in the spinal cord of arthritic rats. Brain Res 374(1):190–194
Witts EC, Mathews MA, Murray AJ (2023) The locus coeruleus directs sensory-motor reflex amplitude across environmental contexts. Curr Biol 33(21):4679–4688.e3
Wurtman RJ, Axelrod J (1966) Control of enzymatic synthesis of adrenaline in the adrenal medulla by adrenal cortical steroids. J Biol Chem 241(10):2301–2305
Xu YL, Reinscheid RK, Huitron-Resendiz S, Clark SD, Wang Z, Lin SH et al (2004) Neuropeptide S: a neuropeptide promoting arousal and anxiolytic-like effects. Neuron 43(4):487–497
Yaci ED, Rampin O, Calas A, Jestin A, McKenna KE, Leclerc P et al (2002) α2a and α2c adrenoceptors on spinal neurons controlling penile erection. Neuroscience 114(4):945–960
Yao Z, van Velthoven CTJ, Kunst M, Zhang M, McMillen D, Lee C et al (2023) A high-resolution transcriptomic and spatial atlas of cell types in the whole mouse brain. Nature 624(7991):317–332
Yerkes RM, Dodson JD (1908) The relation of strength of stimulus to rapidity of habit-formation. J Comp Neurol Psychol 18:459–482
Yu AJ, Dayan P (2005) Uncertainty, neuromodulation, and attention. Neuron 46(4):681–692
Zeng H, Yao Z, van Velthoven C, Smith KA, Tasic B, Lee C, et al (2023) Whole Mouse Brain Transcriptomic Cell Type Atlas – 10x scRNAseq whole brain. [Dataset]. Available from https://assets.nemoarchive.org/dat-qg7n1b0
Zeng WZ, Marshall KL, Min S, Daou I, Chapleau MW, Abboud FM, Liberles SD, Patapoutian A (2018) PIEZOs mediate neuronal sensing of blood pressure and the baroreceptor reflex. Science 362(6413):464–467
Zhang J, Chen R, Wen L, Fan Z, Guo Y, Hu Z, Miao Y (2021) Recent progress in the understanding of the effect of sympathetic nerves on hair follicle growth. Front Cell Dev Biol 9:736738
Zhao Q, Yu CD, Wang R, Xu QJ, Dai Pra R, Zhang L et al (2022) A multidimensional coding architecture of the vagal interoceptive system. Nature 603(7903):878–884
Zhong F-X, Ji X-Q, Tsou K (1985) Intrathecal DSP4 selectively depletes spinal noradrenaline and attenuates morphine analgesia. Eur J Pharmacol 116(3):327–330

Brain-Body Communication in Glucose Metabolism

3

Astrid A. S. van Irsen, Andries Kalsbeek, and Susanne E. la Fleur

Abstract

Glucose is an essential fuel for the brain, and its concentration must be maintained within strict boundaries for optimal fitness. Maintaining glucose homeostasis involves a balance between glucose uptake and output, as well as the management of daily rhythms in glucose concentrations. This chapter explores the roles of various brain regions in glucose homeostasis and their connections through the sympathetic and parasympathetic nervous systems to peripheral organs such as the pancreas and liver. Key hypothalamic nuclei, including the arcuate nucleus and the ventromedial hypothalamus, are well established in their roles in glucose regulation. Additionally, cortico-limbic areas, such as the nucleus accumbens and amygdala, contribute to the modulation of glucose metabolism. These brain regions communicate with the pancreas and liver via autonomic pathways, influencing insulin secretion, hepatic glucose production, and overall metabolic balance. By examining the neural circuits and mechanisms involved, this chapter aims to provide a comprehensive understanding of how brain-body interactions maintain glucose homeostasis and their implications for metabolic health.

Keywords

Glucose homeostasis · Hypothalamic nuclei · Cortico-limbic areas · Autonomic nervous system

Abbreviations

GLUT	Glucose transporter
BBB	Blood-brain barrier
GE	Glucose-excited
GI	Glucose-inhibited
K_{ATP}	ATP-sensitive potassium
ATP	Adenosine triphosphate
ADP	Adenosine diphosphate
TRPC3	Transient receptor potential canonical type 3
VMH	Ventromedial hypothalamus
CRR	Counterregulatory responses
ARC	Arcuate nucleus
AgRP	agouti-related peptide
NPY	Neuropeptide Y
POMC	Pro-opiomelanocortin
CART	Cocaine-amphetamine related transcript
MC4R	Melanocortin 4 receptor
PVN	Paraventricular nucleus
IR	Insulin receptors

A. A. S. van Irsen · A. Kalsbeek · S. E. la Fleur (✉)
Department of Laboratory Medicine, Endocrine Laboratory, Amsterdam, The Netherlands
e-mail: s.e.lafleur@amsterdamumc.nl

D. Tropea, E. Giacometti (eds.), *Brain-Body Connections*, Advances in Experimental Medicine and Biology 1477, https://doi.org/10.1007/978-3-031-89525-8_3

CSF	Cerebrospinal fluid
LHA	Lateral hypothalamic area
MCH	Melanin-concentrating hormone
K_{2P}	tandem-pore potassium channels
NTS	Nucleus of the solitary tract
DMV	Dorsal motor nucleus of the vagus
AP	Area postrema
GABA	Gamma aminobutyric acid
CLS	Cortico-limbic system
PFC	Prefrontal cortex
PVT	Paraventricular thalamus
NAc	Nucleus accumbens
MSNs	Medium spiny neurons
Gck	Glucokinase
EGP	Endogenous glucose production
BOLD	Blood-oxygen-level-dependent
ACh	Acetylcholine
NA	Nucleus ambiguous
CRF	Corticotropin-releasing factor
ICV	Intracerebroventricular
SCN	Suprachiasmatic nucleus
IML	Intermediolateral nucleus
AMH	Dorsomedial hypothalamic nucleus
MPO	Median preoptic area
NE	Norepinephrine
CCK	Cholecystokinin
IGF1	Insulin-like growth factor one
DM	Diabetes mellitus
HFD	High-fat diet
2-DG	2-deoxy-D-glucose

3.1 Introduction

Despite making up only 2% of the body's weight, the brain utilizes more glucose than all other organs (Mergenthaler et al. 2013); however, it has only limited storage capacity. Considering the brain's limited storage capacity, it is not surprising that it plays an important role in maintaining glucose homeostasis with continuous communication between the brain and the body. In order to ensure adequate glucose availability, it senses glucose, initiates feeding behavior, and controls glucose production by the liver. To ensure the brain functions optimally, it is crucial to tightly regulate glucose levels—low levels impair brain function while excessive levels are toxic to brain cells. The brain is thus instrumental in keeping glucose levels within the normal range, which is referred to as euglycemia. Human brain euglycemia typically falls between 1 and 2.5 millimoles per liter (mM), varying based on the specific brain region and its level of activity (Ritter 2017). Plasma glucose levels are normally regulated between 4 and 6 mM (Donovan and Watts 2014), highlighting a clear discrepancy between brain and plasma levels. Different glucose challenges during the day ask for an adequate response to maintain glucose homeostasis. When glucose levels drop, for example during starvation, the liver needs to produce glucose to ensure enough energy for the brain. On the other hand, when glucose levels rise, for example, after food consumption, when experiencing stress, or because of medication, dehydration, and illness, tissues must absorb this excess glucose.

When glucose levels rise, beta cells in the pancreas respond by secreting the hormone insulin, which facilitates glucose uptake in skeletal muscle and adipose tissue. This process, known as facilitated diffusion, relies on specialized glucose transporters (GLUTs) across the plasma membrane. The entry of glucose into the brain relies on GLUTs located at the blood-brain barrier (BBB). Due to its hydrophilic nature, glucose cannot traverse the tight junctions of the BBB (Koepsell 2020). This transport process is meticulously regulated, accounting for fluctuations in energy demand associated with brain activity, as well as variations in blood glucose levels and blood pressure. The human genome encodes 14 types of GLUTs, but most of our knowledge is based on research in rodents on GLUT1-4. Different brain cells express different GLUTs, with particular emphasis on GLUT1, GLUT2, and GLUT3 for brain function (Koepsell 2020). Meanwhile, GLUT2 also plays a role in glucose uptake by the liver and pancreas, while GLUT4 is essential for glucose uptake in adipose tissue and striated muscle (Koepsell 2020). Within the brain, GLUTs are found on specific cell types. GLUT1 is not found in neurons, but in the astrocytes making up the BBB. Both GLUT3 and GLUT4 are found in neurons, while the latter is also found in endothelial cells of the microvessels. GLUT2, found in neurons, astrocytes, oligodendrocytes, ependymal cells, and tanycytes, transports glucose as well as fructose. While GLUT5 exclusively transports fructose and is abundantly found in microglial cells in the brain (Mizuno et al. 2021).

When blood glucose levels drop, the body activates counterregulatory hormones to restore the balance. One such hormone is glucagon, which is secreted by pancreatic alpha cells. Glucagon prompts the liver to convert stored glycogen into glucose, to synthesize new glucose from non-carbohydrate sources (gluconeogenesis), and to inhibit glucose breakdown (glycolysis) and glycogen formation (glycogenesis). The effect of these processes ensures healthy glucose levels during periods like nighttime and between meals. On the other hand, insulin suppresses glucagon secretion, while low insulin levels lead to glucagon release. The stress hormone cortisol also contributes; it opposes insulin by temporarily increasing glucose production and making cells more resistant to insulin. Cortisol is produced by the adrenal gland and rises early in the morning to aid waking up. Chronic stress can lead to elevated glucose levels and worsened insulin resistance. Similarly, adrenaline (the "fight or flight" hormone) is released during stress or physical activity from the adrenal glands. It signals the liver to produce glucose and increases insulin resistance in cells, leading to higher blood glucose levels. Exercise stimulates non-insulin-dependent glucose uptake by muscles and regular exercise actually reduces adrenaline and cortisol levels, which is beneficial for overall health. Additionally, during growth, the release of growth hormone by the pituitary gland becomes crucial. It acts on the liver, kidneys, and fat cells to boost glucose levels. Fat cells (adipocytes) release leptin, signaling the brain about the present energy reserves. In contrast, ghrelin, the "hunger" hormone, originates from the stomach and is released when the stomach is empty.

It is important to understand how glucose metabolism is normally regulated to gain insight into how brain dysfunction, as occurs with obesity, may complicate the treatment of diseases characterized by obesity and aberrant glucose control (including type 2 diabetes mellitus). In this chapter, we will discuss how the brain senses glucose and regulates it, how the brain communicates with the liver and pancreas to regulate glucose metabolism, and what the clinical implications are of a dysfunctional glucose regulation by the brain.

3.2 Brain Glucose "Sensing" and Regulation

Several neurons in the brain respond to extracellular glucose concentration. A specialized subset of these neurons is called glucose-sensing neurons and are defined as neurons that alter their activity directly upon sensing changes in extracellular glucose levels. Other neurons indirectly respond to glucose and are more broadly referred to as glucose-responsive neurons (Tu et al. 2022). There are two types of neurons that "sense" changing glucose levels: glucose-excited (GE) and glucose-inhibited (GI) neurons, becoming active or inactive upon increased glucose levels, respectively. GE neurons seem to signal rising glucose levels through glucokinase and ATP-sensitive potassium (K_{ATP}) channels. Glucose changes the ATP-to-ADP ratio inactivating the K_{ATP} channels, leading to dominating inward currents depolarizing the membrane, leading to cellular activity (Ashcroft and Rorsman 2013). Adenosine triphosphate (ATP) is the nucleotide that provides energy for metabolic processes, and upon use, it is converted to adenosine diphosphate (ADP). On the contrary, K_{ATP}-independent mechanisms have been found to play a role in high glucose sensing in the hypothalamus (Fioramonti et al. 2004), which is likely driven through the transient receptor potential canonical type 3 (TRPC3) channels (Chretien et al. 2017). GI neurons also express glucokinase, but they are active under low glucose levels which is caused by a decrease in intracellular ATP levels activating adenosine monophosphate-activated protein kinase (AMPK) leading to inhibition of chloride channels, in turn depolarizing the cell (Murphy et al. 2009). Glucose-responsive neurons do not all use these metabolism-dependent approaches to alter their activity; they, for example, respond to glucose-sensing neurons or circulating signals like leptin and insulin.

Initially, researchers identified the hypothalamus and brainstem as brain regions involved in glucose homeostasis (Boychuk et al. 2015; Lamy et al. 2014)—regions that are also known to be involved in feeding behavior and body weight regulation. Over time, additional brain regions have been recognized for their contributions to glucose metabolism including the nucleus accumbens (Diepenbroek et al. 2024; Papp et al. 2007), amygdala (Nakano et al. 1986; Zhou et al. 2010), hippocampus (Tingley et al. 2021), the medial prefrontal cortex (Nagy et al 2012), and dorsal raphe nucleus (Nakamura et al. 2024; Ouchi et al. 2009).

3.2.1 Hypothalamic Perspective on Homeostatic Glucose Metabolism

For decades, the hypothalamus and brainstem have been established to play a pivotal role as the neural networks that govern feeding, energy expenditure, and glucose metabolism.

The *ventromedial hypothalamus (VMH)* was the first subregion of the hypothalamus identified to contain glucose-sensing neurons. When hypothalamic glucose levels rise, VMH neurons show increased activity, while lateral hypothalamic neurons exhibit variable responses (Oomura et al. 1969). The hypothalamus contains distinct nuclei with their various neuropeptides and neurotransmitters, all crucial for maintaining energy homeostasis. Subsequent research revealed that the VMH contains both GE and GI neurons, which play essential roles in regulating pancreatic hormones—insulin and glucagon—in response to hyperglycemia (Ruud et al. 2017; Stanley et al. 2016). GE neurons restore blood glucose levels by modulating insulin release, while GI neurons initiate counterregulatory responses (CRR) during hypoglycemia, including glucagon and glucocorticoid release (W. P. Borg et al. 1995; Tong et al. 2007). The K_{ATP} channels are important for the initiation of CRR; therefore, it seems that both the GE and GI neurons are important for the CRR (Beall et al. 2012; Fioramonti et al. 2011). Within the VMH, many neurons express glucokinase, a key enzyme in glycolysis that phosphorylates glucose to glucose-6-phosphate. Activation of VMH glucokinase-containing neurons increases blood glucose in mice (Stanley et al. 2016), whereas deletion of glucokinase reduces glucagon secretion (Steinbusch et al. 2016). Intriguingly, mutations in the glucokinase gene lead to glucokinase-maturity-onset diabetes of the young (Meglasson and Matschinsky 1984), a disorder characterized by mild hyperglycemia typically not requiring treatment except during pregnancy.

The *arcuate nucleus (ARC)* of the hypothalamus, known as the infundibular nucleus in humans, lies adjacent to the third ventricle and the median eminence. This last structure does not contain a BBB, thereby allowing entry of peripheral signals and release of hypothalamic hormones into the portal system before reaching the general circulation (Korf and Moller 2021). Two counteracting neuronal populations in the ARC regulate appetite and homeostatic feeding—the orexigenic (appetite stimulating) agouti-related peptide/neuropeptide Y-expressing (AgRP/NPY) neurons and the anorexigenic pro--opiomelanocortin/cocaine-amphetamine-related transcript-expressing neurons (POMC/CART). These neurons exert their function through the postsynaptic melanocortin 4 receptor (MC4R); in particular, the MC4R-expressing neurons in the paraventricular nucleus (PVN) are important for energy balance (Balthasar et al. 2005; Kim et al. 2000). Mutations in MC4R are the most common monogenic cause of obesity (Yeo et al. 1998). The AgRP neurons exert glucose regulation by managing peripheral insulin sensitivity as acute activation causes impaired insulin sensitivity (Engstrom Ruud et al. 2020; Steculorum et al. 2016), leading to reduced peripheral glucose deposition and higher plasma glucose concentrations when there is no access to food. Insulin inhibits AgRP neurons through action on the insulin receptors (IR), leading to suppressed hepatic glucose production (Konner et al. 2007; Lin et al. 2010). The role of the POMC neurons in glucose regulation is more complex. The inhibition of POMC neurons decreased glucose levels in mice (Uner et al. 2019); therefore, some expected these neurons to be GE. However, a year later a paper appeared, describing that selective activation of POMC neurons innervating the liver led to increased glucose levels (Kwon et al. 2020). It turns out that the POMC neuron population is quite heterogeneous, further supported by the findings that POMC neurons are either inhibited or activated by insulin (Dodd et al. 2018; Qiu et al. 2014). Different classification techniques like electrophysiology, chemogenetics, Cre transgenic animal lines, and RNA-scope may give a biased view on the function of the whole population of POMC neurons, when in reality looking at different subtypes (Quarta et al. 2021). Future experiments might show whether these neurons are truly glucose-sensing. Neurons are not the only cells in the brain that play a role in glucose metabolism. Tanycytes are special ependymal cells lining the third and fourth ventricles, making contact to both the cerebrospinal fluid (CSF) and neuronal nuclei important for metabolism. Glucose stimulation of tanycyte cell bodies evokes ATP-mediated calcium responses in brain slices (Frayling et al. 2011). Tanycytes contain an insulin receptor that regulates insulin access to the ARC in mice, and lacking this receptor on tanycytes leads to systemic insulin resistance without affecting food intake or energy expenditure (Porniece Kumar et al. 2021). Tanycytes express GLUT1 and adjust their expression to ensure energy homeostasis, for example, GLUT1 expression is increased upon fasting (Rodriguez-Cortes et al. 2022).

The *lateral hypothalamic area (LHA)* is a heterogeneous region comprising multiple nuclei, namely tuberal region, perifornical part, peduncular part, and magnocellular nucleus, including GE and GI neurons. Within the LHA, there are different non-overlapping populations of neurons responding to glucose changes, namely NPY, orexin/hypocretin, and melanin-concentrating hormone (MCH)-containing neurons. The NPY population is activated upon fasting and insulin-induced hypoglycemia. Upon increased extracellular glucose levels, the LHA NPY neurons are hyperpolarized by activated potassium currents (Marston et al. 2011). This might be a different approach for glucose-sensing than the originally established method that reacts to decreases in intracellular ATP. The activated potassium currents point to a similar mechanism observed in orexin neurons in the LHA. In orexin neurons, inhibition by glucose is mediated by tandem-pore potassium channels (K_{2P}) (Burdakov et al. 2006) independent of ATP, which is referred to as metabolism-independent glucose sensing (Lopez-Gambero et al. 2019).

In mice brain slices, increased glucose levels inhibits *orexin neurons* (Burdakov et al. 2005). Moreover, it was found that low glucose levels led to activation of the LHA orexin neurons in cell recordings (Sheng et al. 2014). Activating orexin cells in LHA increases blood glucose levels via an increase in hepatic glucose production (Yi et al. 2009). Activity of orexin cells is influenced by metabolic status: leptin reduces, while ghrelin enhances activation under normal glucose levels (Sheng et al. 2014). Furthermore, leptin dampens while ghrelin enhances the effect of decreased glucose on orexin cells. Inhibition of orexin neurons by glucose is reduced by increased ghrelin and decreased leptin following overnight fasting (Sheng et al. 2014). Body weight loss in mice reduces glucose sensitivity of LHA orexin neurons, resulting in increased activation when glucose levels go down (Teegala et al. 2023). Glucose sensing by orexin cells thus depends on the energy status of the animal.

Most *melanin-concentrating hormone* (MCH) neurons show an increase in excitability upon increased glucose levels in mouse brain slices (Burdakov et al. 2005). Defective glucose sensing in these MCH neurons, through expression of mutant Kir6.2, led to impaired glucose tolerance in mice (Kong et al. 2010), whereas excitability of these neurons through UCP2 deletion lowered fasting glucose and enhanced glucose tolerance without altering plasma insulin levels (Kong et al. 2010). Together these data point toward a role for glucose-induced MCH neuronal activation in lowering glucose production and increasing glucose uptake by peripheral tissue. This seems in contrast to studies where all MCH neurons were targeted, as MCH knockdown in rats, wild-type mice, or leptin-deficient mice resulted in enhanced glucose tolerance and insulin sensitivity (Jeon et al. 2006; Mul et al. 2010). However, overall MCH deletion also led to lower body weight and/or enhanced locomotor activity which in turn could explain the enhanced glucose tolerance.

Not all MCH neurons are glucose sensing as a relative amount of MCH neurons did not respond to glucose and also did not express the specific K_{ATP} channel required for glucose sensing (Kong et al. 2010). It was shown that activating all MCH neurons did not affect glucose tolerance nor insulin sensitivity (Dilsiz et al. 2020), whereas acute inhibition of all MCH neurons improved both glucose tolerance and insulin sensitivity in mice (Dilsiz et al. 2020). The latter might be explained by secondary effects on locomotor activity, but it could also be that inhibition of the 25% non-sensing glucose neurons affects glucose metabolism via other mechanisms. It needs to be determined how these glucose sensing and non-glucose sensing MCH neurons differ and whether they have different roles in glucose regulation.

The *brainstem* connects the hypothalamus to the spinal cord and peripheral tissues and is involved through these connections in glucose regulation. However, the brainstem also harbors glucose-responsive neurons and thus is also involved in direct glucose control. Glucose-responsive neurons are found in the rostral ventrolateral region of the medulla and the dorsal vagal complex, the latter consisting of the nucleus of the solitary tract (NTS), the dorsal motor nucleus of the vagus (DMV), and the area postrema (AP). These regions surround the cerebrospinal fluid containing fourth ventricle and aqueduct, therefore, having close access to glucose entering the brain. At first it was thought that the AP must be involved in glucose homeostasis, because its ablation led to altered food intake (Adachi et al. 1995; Bird et al. 1983; Edwards and Ritter 1981; Kott et al. 1984). Only recently it was found that during pancreatic euglycemic clamps in rats, infusion of glucose into the AP, like in the NTS, suppressed glucose production (Li et al. 2023), while a high-fat diet disrupted this glucose-sensing. The NTS contains gamma aminobutyric acid (GABA)ergic GI neurons projecting to the DMV stimulating vagal activity in response to hypoglycemia (Boychuk et al. 2015) through inhibition of local tonic inhibitory DMV neurons (Babic et al. 2011), thereby increasing glucagon secretion (Lamy et al. 2014). The NTS also receives peripheral signals from the glucose sensors in the hepatic portal vein, showing that the this nucleus integrates signals from brain and body (Adachi et al. 1995). A feedback loop between NTS and PVN may optimally control glucose homeostasis : noradrenergic and adrenergic NTS neurons both express NPY and project to the PVN, in turn the PVN neurons project onto the NTS neurons. Next to the NTS lies the DMV, where preganglionic neurons of the vagus nerve are localized that activate the parasympathetic pathway to the pancreas (Mussa and Verberne 2013) (innervation of the pancreas will be discussed later in more detail).

3.2.2 Expanding Horizons: Glucose Metabolism Regulation beyond the Hypothalamus

Glucose regulation is not only controlled by the hypothalamus and brainstem, but is also regulated by brain sites beyond the hypothalamus. The *cortico-limbic system* (CLS) contains many sites that have been shown to express glucose sensing markers that also exhibit glucose-excited and glucose-inhibited properties (for review see Koekkoek et al. 2017). Areas within the CLS, including nuclei such as the amygdala, hippocampus, nucleus accumbens, septum, prefrontal cortex (PFC), and paraventricular thalamus (PVT), are involved in regulating emotions, feeding motivated behavior and memory by processing sensory stimuli, and as a circuit are connected to the endocrine and autonomic nervous systems (Blessing 1997). Thus, these structures within the cortico-limbic system are good candidates for influencing glucose metabolism.

An area instrumental for food-motivated behavior and reward is the *nucleus accumbens (NAc)*. This area receives information from many cortico-limbic areas resulting in adaptation of behavior. Within the NAc, it has been suggested that glucose-sensing neurons are present (Papp et al. 2007). The NAc comprises a central core surrounded by a shell. Notably, in the shell, most of the neurons that react to the micro-electrophoretic administration of glucose are GI, while the core predominantly features GE neurons (Papp et al. 2007). Importantly, the majority of neurons in both subregions remained unresponsive to glucose, with the administered glucose concentration being 500 mM, significantly higher than typical physiological concentrations. Thus, it remains to be determined, with more suitable techniques like patch-clamp electrophysiology or calcium imaging whether these cells are truly glucose-sensing. Interestingly, many areas such as the amygdala and the PVT contain proper glucose-sensing neurons (Labouebe et al. 2016; Nakano et al. 1986; Zhou et al. 2010), and these have direct contact with the medium spiny neurons (MSNs) in the shell of the NAc. For example, within the PVT, two different neuronal populations have been identified, of which one population expresses GLUT2 and the other glucokinase (Gck), two markers for glucose sensing (Kessler et al. 2021; Labouebe et al. 2016; Lamy et al. 2014). GLUT2-PVT neurons are responding to hypoglycemia, whereas GCK-PVT neurons respond to hyperglycemia. These population both project to the shell of the NAc and influence feeding behavior in an opposite manner (Kessler et al. 2021; Labouebe et al. 2016).

Recent findings have highlighted the role of the *NAc-shell* in glucose metabolism. Initially, we demonstrated that electrical stimulation of the shell (and not the core) increased plasma levels of glucose and glucagon and activated the LHA in rats (Diepenbroek et al. 2013). Electrical stimulation of the NAc has also been reported to alter serotonin and dopamine concentrations (Figee et al. 2014; Sesia et al. 2010). Subsequently, we found that increased blood glucose concentrations, accompanied by higher endogenous glucose production (EGP) by the liver, occurred when serotonin levels in the NAc-shell were elevated (Diepenbroek et al. 2017). More recently, we showed that increasing dopamine level in the NAcshell of rats decreases EGP and prevents glucose rising over time (Diepenbroek et al. 2024). Mechanistic studies on the underlying circuitry revealed that direct inhibitory projections from neurons within the NAc shell to the LHA and the vagal nerve innervating the liver play a role in controlling glucose metabolism (Diepenbroek et al. 2024). Additionally, we found that infusion of the μ-opioid receptor agonist into the NAc core of rats did not acutely affect blood glucose levels, but enhanced the glycemic response during both an insulin and glucose tolerance test (Koekkoek et al. 2021). In humans, altering dopamine signaling in the striatum (which includes the NAc) through methods such as deep brain stimulation has been shown to affect glucose metabolism, providing evidence that the NAc-liver connection is also involved in glucose regulation in humans (Ter Horst et al. 2018).

There is an interaction between neurotransmitters and glucose detection in the NAc shell. The link between neuromodulators and glucose *metabolism is also discussed in Chap 2 "Noradrenaline regulation of brain-body communication."* The authors explain that noradrenergic nuclei in the brain contribute to some aspects of arousal. For example, medullary A2 neurons are thought to engage arousal circuits by relaying gastric and cardiovascular signals to the amygdala and to neurons in NAc shell.

As mentioned earlier, the *amygdala*, which is crucial for processing emotions, also responds to glucose changes. Initially, it was shown in macaques that local glucose delivery decreased neuronal activity (Nakano et al. 1986). Later studies in rats gave an indication of the presence of glucose-excited (GE) and glucose-inhibited (GI) neurons in the amygdala, which express glucokinase (Zhou et al. 2010). Infusion of 2-deoxy-D-glucose (2-DG), a non-metabolizable glucose analog, directly into the medial amygdala during hyperinsulinemic hypoglycemia increased glucagon and epinephrine levels, thus stimulating a counterregulatory response (Zhou et al. 2010). This might point to physiological glucose sensing in the amygdala. Additionally, peripheral alterations in glucose concentrations were shown to affect amygdala activity. Intragastric glucose administration in rats increased the blood-oxygen-level-dependent (BOLD) response in the amygdala (Tsurugizawa et al. 2008), while subcutaneous 2-DG administration, mimicking hypoglycemia, decreased neural activity in the central amygdala (Dodd et al. 2010). These findings collectively indicate that the amygdala responds to changes in peripheral glucose concentrations and can modulate glucose metabolism. However, further studies are needed to validate these findings and understand the underlying mechanisms.

3.3 Unraveling the Hidden Wires of Liver and Pancreas

Before describing in detail the neural innervation of the liver and pancreas involved in glucose metabolism, it is essential to shortly review the organization of the peripheral autonomic nervous system. The sympathetic nervous system ("fight or flight" response) leads to overall activity and attention, blood pressure and heart rate increase, glycogenolysis follows, and gastrointestinal processes are halted. On the other hand, the parasympathetic nervous system promotes rest and digestion. In both systems, the preganglionic neurons utilize acetylcholine (ACh) as their main neurotransmitter, although they have their own projection route. The sympathetic pathways travel from the preganglionic neurons in the thoracolumbar division of the spinal cord to the celiac ganglion (for pancreas and liver) where they synapse on postganglionic neurons expressing norepinephrine or epinephrine (LeBouef et al. 2024). These postganglionic neurons project through the body to their respective target organ. The parasympathetic pathway starts with preganglionic neurons in the dorsal motor nucleus of the vagus (DMV) and the nucleus ambiguous (NA) in the brainstem that reach the pancreas and liver through the vagal nerve to release ACh (Borgmann and Fenselau 2024). Various challenges ask for appropriate physiological glucose levels; in the next section, we will explore how the precise coordination of pancreas and liver is orchestrated by the central nervous system and how they communicate with each other.

The pancreas plays an important role in glucose metabolism by producing insulin during hyperglycemia and glucagon during hypoglycemia. The pancreas is innervated by both the sympathetic and the parasympathetic nervous systems. Norepinephrine is the primary sympathetic neurotransmitter; additionally, neuropeptides including galanin and neuropeptide Y are expressed and released by pancreatic sympathetic nerves (Ahren 2000; Dunning and Taborsky Jr. 1989; Sheikh et al. 1988). All of them inhibit insulin and stimulate glucagon secretion, thereby increasing glucose availability (Dunning and Taborsky Jr. 1989; Hamilton et al. 2018; Imai et al. 2007; Opara et al. 1991; Schwetz et al. 2013; Sluga et al. 2022). Recently, it has been shown that chemogenetic activation of the pancreatic cholinergic innervation reduced basal blood glucose and improved glucose tolerance mediated by an increase in plasma insulin in 6 hour fasted male mice. Surprisingly, female mice did not show any of these changes (Jimenez-Gonzalez et al. 2022).

Several brain areas have been shown to directly or indirectly connect to the pancreas. From retrograde viral tracing studies, we know that the pre-autonomic neurons projecting onto the sympathetic preganglionic neurons are located in the PVN, LHA, locus coeruleus, noradrenergic cell group A5 and ventrolateral medulla (Buijs et al. 2001; Jansen et al. 1997). Higher up neurons that project to these pre-autonomic neurons (so-called third-order neurons) are found in DMV, NTS, VMH, ARC, and amygdala (Buijs et al. 2001). For some of these areas, it has been shown that they are involved in glucose metabolism by influencing pancreatic hormonal output.

Within the *VMH*, GABA(A) receptor activity plays a key role in regulating glucagon release by the pancreas during hypoglycemic events. A GABA agonist infused into the VMH reduced, whereas a GABA(A) antagonist increased glucagon levels during a hyperinsulinemic-hypoglycemic glucose clamp (Chan et al. 2006). Moreover, glucose infusions in the VMH inhibit glucagon, epinephrine, and norepinephrine release in rats, while infusion of non-metabolizable glucose (2DG) increased these concentrations (Borg et al. 1997). In addition, it has been shown that insulin and corticotropin-releasing factor (CRF) within the VMH can influence the glucagon response to a hypoglycemic event as well (Chen et al. 2007; Paranjape et al. 2010). The effect of CRF on glucagon response was specific to the VMH as injections in the PVN did not exert effects on glucagon release (Chen et al. 2007). Thus, the VMH indirectly controls pancreatic glucagon release, which could be through either the parasympathetic or sympathetic innervation or even both.

A study investigating mice lacking a protein for mitochondrial fusion in *ARC*-POMC neurons found glucose intolerance due to defective insulin secretion by the pancreas. This was due to an increased sympathetic tone as these effects were rescued by an alpha2-adrenergic antagonist but not a cholinergic agonist (Ramirez et al. 2017). Besides the arcuate POMC neurons, also the *PVN* seems to have a sympathetic connection to the pancreas. Increased activity of PVN oxytocin neurons suppressed insulin secretion and led to hyperglycemia in mice, while its inhibition led to hypoglycemia (Papazoglou et al. 2022). These data strongly suggest a PVN projection to sympathetic preganglionic neurons in regulating insulin secretion by β-cells. Others showed that sympathectomy in rats increases body weight and fat pads and leads to metabolic dysfunctions (Prates et al. 2018) characterized by glucose intolerance, due to lower insulin levels. These findings were explained by altered muscarinic and adrenergic assays to test for pancreas function, indicating a disrupted balance between both branches of the autonomic nervous system (Prates et al. 2018).

Within the *ARC*, a large population of cells express neuropeptide Y. These cells project throughout the hypothalamus, and receptors for NPY are localized at several sites in the brain. Chronic intracerebroventricular (ICV) infusion of NPY leads to hyperphagia, hyperinsulinemia, muscle insulin resistance, and white adipose tissue insulin hyperresponsiveness (Sainsbury et al. 1997; Zarjevski et al. 1993). Preventing hyperphagia by pair-feeding with vehicle-infused control rats still leads to hyperinsulinemia, insulin resistance, and adipose tissue hyperresponsiveness. Bilateral vagotomy (Sainsbury et al. 1997) or adrenalectomy (Sainsbury et al. 1997) before NPY infusion prevented the basal and glucose-induced hyperinsulinemia, indicating that NPY hyperinsulinemia is mediated by the parasympathetic nervous system innervating the pancreas and through changes in corticosterone levels via effects on the adrenals. There are however also indications that effects of central NPY infusion also run through autonomic nerve innervation of the liver (van den Hoek et al. 2008).

The liver, like the pancreas, is innervated by parasympathetic and sympathetic nerves to maintain optimal glucose levels. A key difference between the sympathetic liver innervation in humans and commonly studied animal models, such as rats and mice, is that these animals lack distinct parenchymal sympathetic innervation (Fukuda et al. 1996). In the human liver (and also in the guinea pig), sympathetic noradrenaline-immunoreactive fibers extend deep into the lobule, reaching the hepatocytes. Despite this, the information of nerve terminals in species without parenchymal sympathetic innervation can be transmitted electrically through cell-to-cell gap junctions, thereby compensating for the limited direct innervation of hepatocytes (Beckh et al. 1990; Hertzberg and Gilula 1979; Seseke et al. 1992).

Tracing studies have explored hepatic autonomic innervation to distinguish between para- and sympathetic innervation. When a trans-synaptic retrograde tracer was injected into the liver, labeling was observed in the DMV. This labeling persisted after sympathetic liver denervation, but not when the parasympathetic nerves were denervated (Kalsbeek et al. 2004; la Fleur et al. 2000). Without denervation of the sympathetic or parasympathetic route, first-order neurons were also observed in the intermediolateral nucleus in the spinal cord, second-order neurons were labeled in NTS, DMV, A5, locus coeruleus, LHA, PVN, and PFC, and third-order neurons were labeled in suprachiasmatic nucleus (SCN) and amygdala (la Fleur et al. 2000). Upon parasympathetic liver denervation, first-order neurons were found in the intermediolateral nucleus (IML), A5, locus coeruleus; second-order neurons were observed in PVN and LHA; and third-order neurons in NTS, dorsomedial hypothalamic nucleus (DMH), ARC, VMH, and SCN (Kalsbeek et al. 2004). However, sympathetic liver denervation showed first-order neurons in DMV; second-order neurons in brainstem, PVN, DMH, and median preoptic area (MPO); and third-order neurons in the SCN and VMH. These tracing studies indicate the extensive connection of the brain to the liver through both autonomic nervous routes, showing not only discriminate but also a lot of overlap in brain regions (see Fig. 3.1 for overview).

Experimental studies gave more insight into the function of these specific routes. Already in the 1960s, the role of the hypothalamus influencing the liver in glucose regulation was established in rabbits (Shimazu et al. 1966). They discovered that electrically stimulating the VMH triggers the sympathetic nervous system, leading to higher glucose levels in the blood by breaking down glycogen in the liver and inhibiting its production (Shimazu et al. 1966). This effect persisted even after adrenalectomy or pancreatectomy (Shimazu and Fukuda 1965). The sympathetic regulation of glycogenolysis mainly involves the activation of α-adrenergic receptors, especially Adra1b, as its deletion causes an abnormal buildup of glycogen (Burcelin et al. 2004). Stimulation of either VMH or the hepatic sympathetic nerve increased glucose both output and norepinephrine (NE) outflow from the liver (Takahashi et al. 1997). The sympathetic nervous system also influences liver glucose production by releasing neuropeptides. Galanin amplifies the liver's response to NE to increase blood glucose levels, while NPY promotes glucose uptake in the liver and inhibits glucose release triggered by glucagon and NE

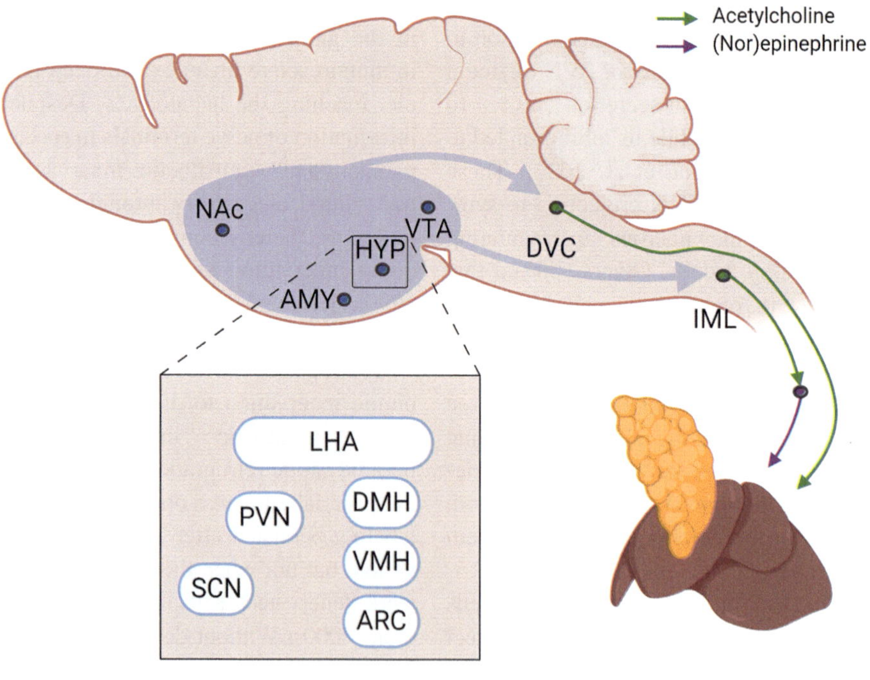

Fig. 3.1 Overview of brain regions crucial for regulating glucose metabolism through the sympathetic (via IML) and parasympathetic (via DVC) nervous system, which innervate the liver and pancreas. Figure created using BioRender.com. Abbreviations: *NAc* Nucleus accumbens, *AMY* amygdala, *HYP* hypothalamus, *VTA* ventral tegmental area, *DVC* dorsal vagal complex, *IML* intermediolateral nucleus, *LHA* lateral hypothalamic area, *PVN* paraventricular nucleus, *SCN* suprachiasmatic nucleus, *DMH* dorsomedial hypothalamus, *VMH* ventromedial hypothalamus, *ARC* arcuate nucleus

(Mundinger and Taborsky Jr. 2000). Overexpression of NPY in NE neurons led to increased white adipose tissue and liver triglyceride accumulation without hyperphagia or increased body weight. After 6 months, hyperinsulinemia and impaired glucose tolerance developed with increased circulating ghrelin levels compared to wild-type mice (Ruohonen et al. 2008). However, an NPY increase can also be favorable, for example, during fasting NPY release increases and stimulates the sympathetic nervous system to maintain triglyceride-rich very low-density lipoproteins in the liver at a post-prandial level (Bruinstroop et al. 2012). Intracerebroventricular infusion of NPY hinders the inhibition of endogenous glucose production by insulin in mice (van den Hoek et al. 2004), which is completely blocked by sympathetic, but not parasympathetic, liver denervation (van den Hoek et al. 2008). Moreover, hepatic insulin sensitivity is decreased upon administration of the glucocorticoid receptor agonist dexamethasone to the ARC in rats, mediated by neuropeptide Y through sympathetic fibers to the liver (Yi et al. 2012).

Glucose levels show a natural daily rhythm, which is driven by the SCN via the sympathetic route to the liver (Cailotto et al. 2005). Glucose levels rise just before the onset of the active period, likely triggered by the SCN reducing GABAergic input to sympathetic neurons in the PVN, which leads to increased hepatic glucose production (Kalsbeek et al. 2004). Interestingly, either a hepatic sympathectomy or parasympathectomy disrupts the daily plasma glucose rhythm, indicating that unilateral denervation of the liver has a more detrimental effect on maintaining rhythmic liver metabolism than the complete removal of both neural branches (Cailotto et al. 2008). Just like we have seen for the pancreas, also in the liver, a balance between para- and sympathetic divisions seems crucial for optimal glucose metabolism.

More recently it was shown that a subset of *ARC* POMC neurons innervate the liver via preganglionic parasympathetic neurons in the DMV. Stimulation of these neurons elevated blood glucose levels by inhibiting the parasympathetic cholinergic outflow to the liver (Kwon et al. 2020). These effects were abolished by a MC4R antagonist or knockdown of the MC4R gene in the DMV. Early work showed that stimulation of the LHA leads to activation of the parasympathetic nerve and increases glycogen synthesis in the liver of rabbits (Shimazu et al. 1966). This is contrary to the finding that disinhibiting the LHA, using an GABA antagonist, enhanced glucose production and glucose levels (Yi et al. 2009). This was mediated by the sympathetic innervation of the liver and prevented with an orexin antagonist (Yi et al. 2009). The targeting of Yi et al. was more precise and localized to the perifornical area of the LHA and this could explain the difference between the two studies. In line with the study of Yi et al. we recently showed that elevating dopamine within the NAc shell decreases hepatic glucose production through inhibition at the level of the LHA and mediated by parasympathetic innervation (Diepenbroek et al. 2024). Also this effect was probably mediated by orexin cells within the LHA (Diepenbroek et al. 2024). From the LHA, this signal must lead to increased activity of the vagal nerve. Mentioned earlier, chemogenetic activation of parasympathetic neurons in the DMV improved glucose tolerance, while inhibition impaired glucose tolerance (NamKoong et al. 2019). These neurons might be activated indirectly by the increased dopamine in the NAc shell (Diepenbroek et al. 2024).

The liver not only directly influences the available glucose levels but also indirectly modulates the insulin sensitivity of other organs. Hepatic parasympathetic denervation increases insulin resistance and decreases glucose uptake by peripheral tissues (Lautt et al. 2001). At the same time, insulin suppresses hepatic glucose production. Knockdown of NPY-expressing neurons in the DMH projecting to the DMV increased insulin's ability to suppress hepatic glucose production (Li et al. 2016). This effect was abolished by hepatic vagotomy, but not by vagal deafferentiation, meaning the efferent fibers (from brain to liver) are important for the inhibitory effects of central insulin on hepatic glucose production (Pocai et al. 2005). Later it was found that the cholecystokinin (CCK) signal from the gut enhances hepatic insulin sensitivity and inhibits glucose production, which relied on intact hepatic efferent nerve (Cheung et al. 2009). Intact hepatic vagal innervation is also important for the glucose-lowering effect of postprandial insulin (Fernandes et al. 2011) and for the daily rhythm of plasma glucose and insulin in rats (Inoue et al. 1995), although the latter is most-likely mediated through afferent hepatic fibers, indicating the importance of cross-talk between the brain and liver to regulate glucose metabolism. The liver is also the main producer of insulin-like growth factor one (IGF1). *As described in Chap. 4 "IGF1 as Brain-Body Connector,"* this factor can cross the BBB and modulate brain states connecting the information related to body metabolism to the brain.

In the *intestine*, microorganisms can metabolize primary bile acids into secondary bile acids and promote the activation of bile acid receptors, thereby regulating host glucose and lipid metabolism, immune response, and the proliferation of intestinal pathogens. *In Chap. 6 "Microbiome—A Key Regulator of Body-Brain Interactions,"* the author describes in depth enteroendocrine signaling and hormonal modulation as well as microbial metabolites and structural components. Gut bacteria, including *Lactobacillus* and *Bifidobacterium*, express bile salt hydrolase enzymes which can deconjugate bile acids, enhancing their survival in the gut. This microbial activity increases bile acid (BA) diversity in the host, affecting metabolism and weight regulation. BAs are crucial for maintaining the gut barrier and regulating immune responses, which are essential for gut and brain function and health.

3.4 Aberrant Glucose Metabolism: Clinical Consequences and Insights

Disruptions in glucose metabolism, particularly in obesity and diabetes mellitus (DM), can have profound effects on the brain health and function. Obesity and diabetes are closely linked through a common pathway of insulin resistance. In obesity, excess fat tissue leads to chronic inflammation and the release of adipokines, which interfere with insulin signaling (Guillemot-Legris and Muccioli 2017). Over time, this can lead to the development of type 2 diabetes, further exacerbating metabolic dysfunction in the brain.

Insulin resistance not only occurs in the periphery but also in the brain. Insulin activates the POMC cells in the ARC, while it inhibits the NPY neurons, thereby decreasing food intake and effecting insulin sensitivity (Benoit et al. 2002; Schwartz et al. 1992). The way insulin acts in the hypothalamus is similar to the well-known effect of leptin, indicating that these two hormones work collectively in the hypothalamus (Baskin et al. 1999). In addition, a diet high in fat (HFD) in rodents leads to impaired glucose sensing in POMC neurons (Parton et al. 2007). After only 2 days of a HFD in rats, systemic glucose-induced insulin secretion is elevated and sympathetic tone is changed (Cruciani-Guglielmacci et al. 2005). This is accompanied by changes in the VMH showing lower NE decrease after a glucose challenge and glucose intolerance worsened over time. In a diabetic mouse model, it has been shown that PVN neurons, that are known to connect to the liver, have increased excitability (Gao et al. 2017); this is also found upon HFD feeding in wild-type mice (Molinas et al. 2023). A high-fat high-sugar diet (HFHS) in rats decreased hypothalamic glucose sensing which affected insulin secretion (Desmoulins et al. 2019). Taken together, it is clear that when animals become obese, insulin resistance leads to aberrant glucose sensing and a dysfunctional central control over systemic glucose metabolism.

Within humans, insulin resistance in the brain has also been described using different neuroimaging techniques. It has been shown that brain insulin resistance goes along with peripheral glucose disturbances. The effects of brain insulin resistance in humans have been extensively described in a recent review by Martin Heni (Heni 2024). In addition to brain insulin sensitivity, it has also been shown that the dopamine system (i.e., striatal dopamine D2/3 receptor availability) within the human brain is altered in individuals with obesity, which has been linked to peripheral insulin resistance (de Weijer et al. 2011; Wang et al. 2001). D2/D3 receptor availability was increased again 3 years (on average) after bariatric surgery; however, it was still lower than in lean individuals (van der Zwaal et al. 2016). It remains to be determined which underlying mechanism is involved in the effects of striatal dopamine on dysregulated glucose metabolism.

In conclusion, disruptions in glucose metabolism, particularly in the context of obesity and diabetes mellitus, have profound and clear effects on brain health and function. The intricate link between obesity and diabetes through disrupted glucose metabolism underscores the critical need for preventive measures. Impaired insulin signaling in obesity not only exacerbates metabolic dysfunction but also contributes to the development of type 2 diabetes, further compromising brain function. The evidence of altered neuronal activity and glucose metabolism in various brain regions, such as the hypothalamus and nucleus accumbens, highlights the detrimental impact of high-fat diets. Altered glucose sensing upon high-fat diets drives the further metabolic deterioration that transforms into obesity. These findings emphasize the importance of adopting healthy dietary habits and lifestyle changes to prevent obesity and its associated metabolic disorders.

References

Adachi A, Kobashi M, Funahashi M (1995) Glucose-responsive neurons in the brainstem. Obes Res 3(Suppl 5):735S–740S. https://doi.org/10.1002/j.1550-8528.1995.tb00493.x

Ahren B (2000) Autonomic regulation of islet hormone secretion—implications for health and disease. Diabetologia 43(4):393–410. https://doi.org/10.1007/s001250051322

Ashcroft FM, Rorsman P (2013) K(ATP) channels and islet hormone secretion: new insights and controversies. Nat Rev Endocrinol 9(11):660–669. https://doi.org/10.1038/nrendo.2013.166

Babic T, Browning KN, Travagli RA (2011) Differential organization of excitatory and inhibitory synapses within the rat dorsal vagal complex. American Journal of Physiology-Gastrointestinal and Liver Physiology 300(1):G21–G32. https://doi.org/10.1152/ajpgi.00363.2010

Balthasar N, Dalgaard LT, Lee CE, Yu J, Funahashi H, Williams T et al (2005) Divergence of melanocortin pathways in the control of food intake and energy expenditure. Cell 123(3):493–505. https://doi.org/10.1016/j.cell.2005.08.035

Baskin DG, Figlewicz Lattemann D, Seeley RJ, Woods SC, Porte D Jr, Schwartz MW (1999) Insulin and leptin: dual adiposity signals to the brain for the regulation of food intake and body weight. Brain Res 848(1–2):114–123. https://doi.org/10.1016/s0006-8993(99)01974-5

Beall C, Ashford ML & McCrimmon RJ (2012) The physiology and pathophysiology of the neural control of the counterregulatory response. Am J Physiol Regul Integr Comp Physiol 302(2):R215–223. https://doi.org/10.1152/ajpregu.00531.2011

Beckh K, Fuchs E, Balle C, Jungermann K (1990) Activation of glycogenolysis by stimulation of the hepatic nerves in perfused livers of Guinea pig and tree shrew as compared to rat: differences in the mode of action. Biol Chem Hoppe Seyler 371(2):153–158. https://doi.org/10.1515/bchm3.1990.371.1.153

Benoit SC, Air EL, Coolen LM, Strauss R, Jackman A, Clegg DJ et al (2002) The catabolic action of insulin in the brain is mediated by melanocortins. J Neurosci 22(20):9048–9052. https://doi.org/10.1523/JNEUROSCI.22-20-09048.2002

Bird E, Cardone CC, Contreras RJ (1983) Area postrema lesions disrupt food intake induced by cerebroventricular infusions of 5-thioglucose in the rat. Brain Res 270(1):193–196. https://doi.org/10.1016/0006-8993(83)90815-6

Blessing WW (1997) Inadequate frameworks for understanding bodily homeostasis. Trends Neurosci 20(6):235–239. https://doi.org/10.1016/s0166-2236(96)01029-6

Borg WP, Sherwin RS, During MJ, Borg MA, Shulman GI (1995) Local ventromedial hypothalamus glucopenia triggers counterregulatory hormone release. Diabetes 44(2):180–184. https://doi.org/10.2337/diab.44.2.180

Borg MA, Sherwin RS, Borg WP, Tamborlane WV, Shulman GI (1997) Local ventromedial hypothalamus glucose perfusion blocks counterregulation during systemic hypoglycemia in awake rats. J Clin Invest 99(2):361–365. https://doi.org/10.1172/JCI119165

Borgmann D, Fenselau H (2024) Vagal pathways for systemic regulation of glucose metabolism. Semin Cell Dev Biol 156:244–252. https://doi.org/10.1016/j.semcdb.2023.07.010

Boychuk CR, Gyarmati P, Xu H, Smith BN (2015) Glucose sensing by GABAergic neurons in the mouse nucleus tractus solitarii. J Neurophysiol 114(2):999–1007. https://doi.org/10.1152/jn.00310.2015

Bruinstroop E, Pei L, Ackermans MT, Foppen E, Borgers AJ, Kwakkel J et al (2012) Hypothalamic neuropeptide Y (NPY) controls hepatic VLDL-triglyceride secretion in rats via the sympathetic nervous system.

Diabetes 61(5):1043–1050. https://doi.org/10.2337/db11-1142
Buijs RM, Chun SJ, Niijima A, Romijn HJ, Nagai K (2001) Parasympathetic and sympathetic control of the pancreas: a role for the suprachiasmatic nucleus and other hypothalamic centers that are involved in the regulation of food intake. J Comp Neurol 431(4):405–423. https://doi.org/10.1002/1096-9861(20010319)431:4<405::aid-cne1079>3.0.co;2-d
Burcelin R, Uldry M, Foretz M, Perrin C, Dacosta A, Nenniger-Tosato M et al (2004) Impaired glucose homeostasis in mice lacking the alpha1b-adrenergic receptor subtype. J Biol Chem 279(2):1108–1115. https://doi.org/10.1074/jbc.M307788200
Burdakov D, Gerasimenko O, Verkhratsky A (2005) Physiological changes in glucose differentially modulate the excitability of hypothalamic melanin-concentrating hormone and orexin neurons in situ. J Neurosci 25(9):2429–2433. https://doi.org/10.1523/JNEUROSCI.4925-04.2005
Burdakov D, Jensen LT, Alexopoulos H, Williams RH, Fearon IM, O'Kelly I et al (2006) Tandem-pore K+ channels mediate inhibition of orexin neurons by glucose. Neuron 50(5):711–722. https://doi.org/10.1016/j.neuron.2006.04.032
Cailotto C, La Fleur SE, Van Heijningen C, Wortel J, Kalsbeek A, Feenstra M et al (2005) The suprachiasmatic nucleus controls the daily variation of plasma glucose via the autonomic output to the liver: are the clock genes involved? Eur J Neurosci 22(10):2531–2540. https://doi.org/10.1111/j.1460-9568.2005.04439.x
Cailotto C, van Heijningen C, van der Vliet J, van der Plasse G, Habold C, Kalsbeek A et al (2008) Daily rhythms in metabolic liver enzymes and plasma glucose require a balance in the autonomic output to the liver. Endocrinology 149(4):1914–1925. https://doi.org/10.1210/en.2007-0816
Chan O, Zhu W, Ding Y, McCrimmon RJ, Sherwin RS (2006) Blockade of GABA(A) receptors in the ventromedial hypothalamus further stimulates glucagon and sympathoadrenal but not the hypothalamo-pituitary-adrenal response to hypoglycemia. Diabetes 55(4):1080–1087. https://doi.org/10.2337/diabetes.55.04.06.db05-0958
Chen SS, Zhang Y, Santomango TS, Williams PE, Lacy DB, McGuinness OP (2007) Glucagon chronically impairs hepatic and muscle glucose disposal. Am J Physiol Endocrinol Metab 292(3):E928–E935. https://doi.org/10.1152/ajpendo.00063.2006
Cheung GW, Kokorovic A, Lam CK, Chari M, Lam TK (2009) Intestinal cholecystokinin controls glucose production through a neuronal network. Cell Metab 10(2):99–109. https://doi.org/10.1016/j.cmet.2009.07.005
Chretien C, Fenech C, Lienard F, Grall S, Chevalier C, Chaudy S, . . . Fioramonti X (2017) Transient receptor potential canonical 3 (TRPC3) channels are required for hypothalamic glucose detection and energy homeostasis. Diabetes 66(2):314–324. https://doi.org/10.2337/db16-1114
Cruciani-Guglielmacci C, Vincent-Lamon M, Rouch C, Orosco M, Ktorza A, Magnan C (2005) Early changes in insulin secretion and action induced by high-fat diet are related to a decreased sympathetic tone. Am J Physiol Endocrinol Metab 288(1):E148–E154. https://doi.org/10.1152/ajpendo.00225.2004
de Weijer BA, van de Giessen E, van Amelsvoort TA, Boot E, Braak B, Janssen IM et al (2011) Lower striatal dopamine D2/3 receptor availability in obese compared with non-obese subjects. EJNMMI Res 1(1):37. https://doi.org/10.1186/2191-219X-1-37
Desmoulins L, Chretien C, Paccoud R, Collins S, Cruciani-Guglielmacci C, Galinier A, . . . Leloup C (2019) Mitochondrial dynamin-related protein 1 (DRP1) translocation in response to cerebral glucose is impaired in a rat model of early alteration in hypothalamic glucose sensing. Mol Metab 20:166–177. https://doi.org/10.1016/j.molmet.2018.11.007
Diepenbroek C, van der Plasse G, Eggels L, Rijnsburger M, Feenstra MG, Kalsbeek A et al (2013) Alterations in blood glucose and plasma glucagon concentrations during deep brain stimulation in the shell region of the nucleus accumbens in rats. Front Neurosci 7:226. https://doi.org/10.3389/fnins.2013.00226
Diepenbroek C, Rijnsburger M, Eggels L, van Megen KM, Ackermans MT, Fliers E et al (2017) Infusion of fluoxetine, a serotonin reuptake inhibitor, in the shell region of the nucleus accumbens increases blood glucose concentrations in rats. Neurosci Lett 637:85–90. https://doi.org/10.1016/j.neulet.2016.11.045
Diepenbroek C, Rijnsburger M, van Irsen AAS, Eggels L, Kisner A, Foppen E et al (2024) Dopamine in the nucleus accumbens shell controls systemic glucose metabolism via the lateral hypothalamus and hepatic vagal innervation in rodents. Metabolism 150:155696. https://doi.org/10.1016/j.metabol.2023.155696
Dilsiz P, Aklan I, Sayar Atasoy N, Yavuz Y, Filiz G, Koksalar F et al (2020) MCH neuron activity is sufficient for reward and reinforces feeding. Neuroendocrinology 110(3–4):258–270. https://doi.org/10.1159/000501234
Dodd GT, Williams SR, Luckman SM (2010) Functional magnetic resonance imaging and c-Fos mapping in rats following a glucoprivic dose of 2-deoxy-D-glucose. J Neurochem 113(5):1123–1132. https://doi.org/10.1111/j.1471-4159.2010.06671.x
Dodd GT, Michael NJ, Lee-Young RS, Mangiafico SP, Pryor JT, Munder AC et al (2018) Insulin regulates POMC neuronal plasticity to control glucose metabolism. elife 7. https://doi.org/10.7554/eLife.38704
Donovan CM, Watts AG (2014) Peripheral and central glucose sensing in hypoglycemic detection. Physiology (Bethesda) 29(5):314–324. https://doi.org/10.1152/physiol.00069.2013
Dunning BE, Taborsky GJ Jr (1989) Galanin release during pancreatic nerve stimulation is sufficient to influence islet function. Am J Phys 256(1 Pt 1):E191–E198. https://doi.org/10.1152/ajpendo.1989.256.1.E191
Edwards GL, Ritter RC (1981) Ablation of the area postrema causes exaggerated consumption of preferred

foods in the rat. Brain Res 216(2):265–276. https://doi.org/10.1016/0006-8993(81)90129-3
Engstrom Ruud L, Pereira MMA, de Solis AJ, Fenselau H, Bruning JC (2020) NPY mediates the rapid feeding and glucose metabolism regulatory functions of AgRP neurons. Nat Commun 11(1):442. https://doi.org/10.1038/s41467-020-14291-3
Fernandes AB, Patarrao RS, Videira PA, Macedo MP (2011) Understanding postprandial glucose clearance by peripheral organs: the role of the hepatic parasympathetic system. J Neuroendocrinol 23(12):1288–1295. https://doi.org/10.1111/j.1365-2826.2011.02226.x
Figee M, de Koning P, Klaassen S, Vulink N, Mantione M, van den Munckhof P et al (2014) Deep brain stimulation induces striatal dopamine release in obsessive-compulsive disorder. Biol Psychiatry 75(8):647–652. https://doi.org/10.1016/j.biopsych.2013.06.021
Fioramonti X, Lorsignol A, Taupignon A, Penicaud L (2004) A new ATP-sensitive K+ channel-independent mechanism is involved in glucose-excited neurons of mouse arcuate nucleus. Diabetes 53(11):2767–2775. https://doi.org/10.2337/diabetes.53.11.2767
Fioramonti X, Song Z, Vazirani RP, Beuve A, Routh VH (2011) Hypothalamic nitric oxide in hypoglycemia detection and counterregulation: a two-edged sword. Antioxid Redox Signal 14(3):505–517. https://doi.org/10.1089/ars.2010.3331
Frayling C, Britton R, Dale N (2011) ATP-mediated glucosensing by hypothalamic tanycytes. J Physiol 589(Pt 9):2275–2286. https://doi.org/10.1113/jphysiol.2010.202051
Fukuda Y, Imoto M, Koyama Y, Miyazawa Y, Hayakawa T (1996) Demonstration of noradrenaline-immunoreactive nerve fibres in the liver. J Int Med Res 24(6):466–472. https://doi.org/10.1177/030006059602400603
Gao H, Molinas AJR, Miyata K, Qiao X, Zsombok A (2017) Overactivity of liver-related neurons in the paraventricular nucleus of the hypothalamus: electrophysiological findings in db/db mice. J Neurosci 37(46):11140–11150. https://doi.org/10.1523/JNEUROSCI.1706-17.2017
Guillemot-Legris O, Muccioli GG (2017) Obesity-induced neuroinflammation: beyond the hypothalamus. Trends Neurosci 40(4):237–253. https://doi.org/10.1016/j.tins.2017.02.005
Hamilton A, Zhang Q, Salehi A, Willems M, Knudsen JG, Ringgaard AK et al (2018) Adrenaline stimulates glucagon secretion by Tpc2-dependent Ca(2+) mobilization from acidic stores in pancreatic alpha-cells. Diabetes 67(6):1128–1139. https://doi.org/10.2337/db17-1102
Heni M (2024) The insulin resistant brain: impact on whole-body metabolism and body fat distribution. Diabetologia 67(7):1181–1191. https://doi.org/10.1007/s00125-024-06104-9
Hertzberg EL, Gilula NB (1979) Isolation and characterization of gap junctions from rat liver. J Biol Chem 254(6):2138–2147. Retrieved from https://www.ncbi.nlm.nih.gov/pubmed/422573
Imai Y, Patel HR, Hawkins EJ, Doliba NM, Matschinsky FM, Ahima RS (2007) Insulin secretion is increased in pancreatic islets of neuropeptide Y-deficient mice. Endocrinology 148(12):5716–5723. https://doi.org/10.1210/en.2007-0404
Inoue S, Satoh S, Saito M, Naitoh M, Suzuki H, Egawa M (1995) Effects of selective vagotomy on circadian rhythms of plasma glucose, insulin and food intake in control and ventromedial hypothalamic (VMH) lesioned rats. Obes Res 3(Suppl 5):747S–752S. https://doi.org/10.1002/j.1550-8528.1995.tb00495.x
Jansen AS, Hoffman JL, Loewy AD (1997) CNS sites involved in sympathetic and parasympathetic control of the pancreas: a viral tracing study. Brain Res 766(1–2):29–38. https://doi.org/10.1016/s0006-8993(97)00532-5
Jeon JY, Bradley RL, Kokkotou EG, Marino FE, Wang X, Pissios P, Maratos-Flier E (2006) MCH−/− mice are resistant to aging-associated increases in body weight and insulin resistance. Diabetes 55(2):428–434. https://doi.org/10.2337/diabetes.55.02.06.db05-0203
Jimenez-Gonzalez M, Li R, Pomeranz LE, Alvarsson A, Marongiu R, Hampton RF et al (2022) Mapping and targeted viral activation of pancreatic nerves in mice reveal their roles in the regulation of glucose metabolism. Nat Biomed Eng 6(11):1298–1316. https://doi.org/10.1038/s41551-022-00909-y
Kalsbeek A, La Fleur S, Van Heijningen C, Buijs RM (2004) Suprachiasmatic GABAergic inputs to the paraventricular nucleus control plasma glucose concentrations in the rat via sympathetic innervation of the liver. J Neurosci 24(35):7604–7613. https://doi.org/10.1523/JNEUROSCI.5328-03.2004
Kessler S, Labouebe G, Croizier S, Gaspari S, Tarussio D, Thorens B (2021) Glucokinase neurons of the paraventricular nucleus of the thalamus sense glucose and decrease food consumption. iScience 24(10):103122. https://doi.org/10.1016/j.isci.2021.103122
Kim MS, Rossi M, Abusnana S, Sunter D, Morgan DG, Small CJ et al (2000) Hypothalamic localization of the feeding effect of agouti-related peptide and alpha-melanocyte-stimulating hormone. Diabetes 49(2):177–182. https://doi.org/10.2337/diabetes.49.2.177
Koekkoek LL, Mul JD, la Fleur SE (2017) Glucose-sensing in the reward system. Front Neurosci 11:716. https://doi.org/10.3389/fnins.2017.00716
Koekkoek LL, Kool T, Eggels L, van der Gun LL, Lamuadni K, Slomp M et al (2021) Activation of nucleus accumbens mu-opioid receptors enhances the response to a glycaemic challenge. J Neuroendocrinol 33(10):e13036. https://doi.org/10.1111/jne.13036
Koepsell H (2020) Glucose transporters in brain in health and disease. Pflugers Arch 472(9):1299–1343. https://doi.org/10.1007/s00424-020-02441-x
Kong D, Vong L, Parton LE, Ye C, Tong Q, Hu X et al (2010) Glucose stimulation of hypothalamic MCH neurons involves K(ATP) channels, is modulated by UCP2, and regulates peripheral glucose homeostasis. Cell Metab 12(5):545–552. https://doi.org/10.1016/j.cmet.2010.09.013

Konner AC, Janoschek R, Plum L, Jordan SD, Rother E, Ma X et al (2007) Insulin action in AgRP-expressing neurons is required for suppression of hepatic glucose production. Cell Metab 5(6):438–449. https://doi.org/10.1016/j.cmet.2007.05.004

Korf HW, Moller M (2021) Arcuate nucleus, median eminence, and hypophysial pars tuberalis. Handb Clin Neurol 180:227–251. https://doi.org/10.1016/B978-0-12-820107-7.00015-X

Kott JN, Ganfield CL, Kenney NJ (1984) Area postrema/nucleus of the solitary tract ablations: analysis of the effects of hypophagia. Physiol Behav 32(3):429–435. https://doi.org/10.1016/0031-9384(84)90258-0

Kwon E, Joung HY, Liu SM, Chua SC Jr, Schwartz GJ, Jo YH (2020) Optogenetic stimulation of the liver-projecting melanocortinergic pathway promotes hepatic glucose production. Nat Commun 11(1):6295. https://doi.org/10.1038/s41467-020-20160-w

la Fleur SE, Kalsbeek A, Wortel J, Buijs RM (2000) Polysynaptic neural pathways between the hypothalamus, including the suprachiasmatic nucleus, and the liver. Brain Res 871(1):50–56. https://doi.org/10.1016/s0006-8993(00)02423-9

Labouebe G, Boutrel B, Tarussio D, Thorens B (2016) Glucose-responsive neurons of the paraventricular thalamus control sucrose-seeking behavior. Nat Neurosci 19(8):999–1002. https://doi.org/10.1038/nn.4331

Lamy CM, Sanno H, Labouebe G, Picard A, Magnan C, Chatton JY, Thorens B (2014) Hypoglycemia-activated GLUT2 neurons of the nucleus tractus solitarius stimulate vagal activity and glucagon secretion. Cell Metab 19(3):527–538. https://doi.org/10.1016/j.cmet.2014.02.003

Lautt WW, Macedo MP, Sadri P, Takayama S, Duarte Ramos F, Legare DJ (2001) Hepatic parasympathetic (HISS) control of insulin sensitivity determined by feeding and fasting. Am J Physiol Gastrointest Liver Physiol 281(1):G29–G36. https://doi.org/10.1152/ajpgi.2001.281.1.G29

LeBouef T, Yaker Z, Whited L (2024) Physiology, autonomic nervous system. StatPearls, Treasure Island

Li L, de La Serre CB, Zhang N, Yang L, Li H, Bi S (2016) Knockdown of neuropeptide Y in the dorsomedial hypothalamus promotes hepatic insulin sensitivity in male rats. Endocrinology 157(12):4842–4852. https://doi.org/10.1210/en.2016-1662

Li RJW, Chiu JFM, Bruce K, Zhang SY, Barros DR, Yue JTY, Lam TKT (2023) A glucose-sensing mechanism with glucose transporter 1 and pyruvate kinase in the area postrema regulates hepatic glucose production in rats. J Biol Chem 299(5):104633. https://doi.org/10.1016/j.jbc.2023.104633

Lin HV, Plum L, Ono H, Gutierrez-Juarez R, Shanabrough M, Borok E et al (2010) Divergent regulation of energy expenditure and hepatic glucose production by insulin receptor in agouti-related protein and POMC neurons. Diabetes 59(2):337–346. https://doi.org/10.2337/db09-1303

Lopez-Gambero AJ, Martinez F, Salazar K, Cifuentes M, Nualart F (2019) Brain glucose-sensing mechanism and energy homeostasis. Mol Neurobiol 56(2):769–796. https://doi.org/10.1007/s12035-018-1099-4

Marston OJ, Hurst P, Evans ML, Burdakov DI, Heisler LK (2011) Neuropeptide Y cells represent a distinct glucose-sensing population in the lateral hypothalamus. Endocrinology 152(11):4046–4052. https://doi.org/10.1210/en.2011-1307

Meglasson MD, Matschinsky FM (1984) New perspectives on pancreatic islet glucokinase. Am J Phys 246(1 Pt 1):E1–E13. https://doi.org/10.1152/ajpendo.1984.246.1.E1

Mergenthaler P, Lindauer U, Dienel GA, Meisel A (2013) Sugar for the brain: the role of glucose in physiological and pathological brain function. Trends Neurosci 36(10):587–597. https://doi.org/10.1016/j.tins.2013.07.001

Mizuno TM, Lew PS, Jhanji G (2021) Regulation of the fructose transporter gene Slc2a5 expression by glucose in cultured microglial cells. Int J Mol Sci, 22(23). https://doi.org/10.3390/ijms222312668

Molinas AJR, Desmoulins LD, Davis RK, Gao H, Satou R, Derbenev AV, Zsombok A (2023) High-fat diet modulates the excitability of neurons within the brain-liver pathway. Cells 12(8). https://doi.org/10.3390/cells12081194

Mul JD, Yi CX, van den Berg SA, Ruiter M, Toonen PW, van der Elst MC et al (2010) Pmch expression during early development is critical for normal energy homeostasis. Am J Physiol Endocrinol Metab 298(3):E477–E488. https://doi.org/10.1152/ajpendo.00154.2009

Mundinger TO, Taborsky GJ Jr (2000) Differential action of hepatic sympathetic neuropeptides: metabolic action of galanin, vascular action of NPY. Am J Physiol Endocrinol Metab 278(3):E390–E397. https://doi.org/10.1152/ajpendo.2000.278.3.E390

Murphy BA, Fakira KA, Song Z, Beuve A, Routh VH (2009) AMP-activated protein kinase and nitric oxide regulate the glucose sensitivity of ventromedial hypothalamic glucose-inhibited neurons. Am J Physiol Cell Physiol 297(3):C750–C758. https://doi.org/10.1152/ajpcell.00127.2009

Mussa BM, Verberne AJ (2013) The dorsal motor nucleus of the vagus and regulation of pancreatic secretory function. Exp Physiol 98(1):25–37. https://doi.org/10.1113/expphysiol.2012.066472

Nagy B, Szabo I, Papp S, Takacs G, Szalay C, Karadi Z (2012) Glucose-monitoring neurons in the mediodorsal prefrontal cortex. Brain Res 1444:38–44. https://doi.org/10.1016/j.brainres.2012.01.025

Nakano Y, Oomura Y, Lenard L, Nishino H, Aou S, Yamamoto T, Aoyagi K (1986) Feeding-related activity of glucose- and morphine-sensitive neurons in the monkey amygdala. Brain Res 399(1):167–172. https://doi.org/10.1016/0006-8993(86)90613-x

Nakamura S, Sasaki T, Uenoyama Y, Inoue N, Nakanishi M, Yamada K, ...Tsukamura H (2024) Raphe glucose-sensing serotonergic neurons stimulate KNDy neurons to enhance LH pulses via 5HT2CR: rat and goat studies. Sci Rep 14(1):10190. https://doi.org/10.1038/s41598-024-58470-4

NamKoong C, Song WJ, Kim CY, Chun DH, Shin S, Sohn JW, Choi HJ (2019) Chemogenetic manipulation of parasympathetic neurons (DMV) regulates feeding behavior and energy metabolism. Neurosci Lett 712:134356. https://doi.org/10.1016/j.neulet.2019.134356

Oomura Y, Ono T, Ooyama H, Wayner MJ (1969) Glucose and osmosensitive neurones of the rat hypothalamus. Nature 222(5190):282–284. https://doi.org/10.1038/222282a0

Opara EC, Burch WM, Taylor IL, Akwari OE (1991) Pancreatic hormone response to neuropeptide Y (NPY) perifusion in vitro. Regul Pept 34(3):225–233. https://doi.org/10.1016/0167-0115(91)90181-f

Ouchi Y, Yoshikawa E, Futatsubashi M, Yagi S, Ueki T, Nakamura K (2009) Altered brain serotonin transporter and associated glucose metabolism in Alzheimer disease. J Nucl Med 50(8):1260–1266. https://doi.org/10.2967/jnumed.109.063008

Papazoglou I, Lee JH, Cui Z, Li C, Fulgenzi G, Bahn YJ et al (2022) A distinct hypothalamus-to-beta cell circuit modulates insulin secretion. Cell Metab 34(2):285–298 e287. https://doi.org/10.1016/j.cmet.2021.12.020

Papp S, Lukats B, Takacs G, Szalay C, Karadi Z (2007) Glucose-monitoring neurons in the nucleus accumbens. Neuroreport 18(15):1561–1565. https://doi.org/10.1097/WNR.0b013e3281667eca

Paranjape SA, Chan O, Zhu W, Horblitt AM, McNay EC, Cresswell JA et al (2010) Influence of insulin in the ventromedial hypothalamus on pancreatic glucagon secretion in vivo. Diabetes 59(6):1521–1527. https://doi.org/10.2337/db10-0014

Parton LE, Ye CP, Coppari R, Enriori PJ, Choi B, Zhang CY et al (2007) Glucose sensing by POMC neurons regulates glucose homeostasis and is impaired in obesity. Nature 449(7159):228–232. https://doi.org/10.1038/nature06098

Pocai A, Lam TK, Gutierrez-Juarez R, Obici S, Schwartz GJ, Bryan J et al (2005) Hypothalamic K(ATP) channels control hepatic glucose production. Nature 434(7036):1026–1031. https://doi.org/10.1038/nature03439

Porniece Kumar M, Cremer AL, Klemm P, Steuernagel L, Sundaram S, Jais A et al (2021) Insulin signalling in tanycytes gates hypothalamic insulin uptake and regulation of AgRP neuron activity. Nat Metab 3(12):1662–1679. https://doi.org/10.1038/s42255-021-00499-0

Prates KV, de Oliveira JC, Malta A, Matiusso CCI, Miranda RA, Ribeiro TA et al (2018) Sympathetic innervation is essential for metabolic homeostasis and pancreatic beta cell function in adult rats. Mol Cell Endocrinol 462(Pt B):119–126. https://doi.org/10.1016/j.mce.2017.09.031

Qiu J, Zhang C, Borgquist A, Nestor CC, Smith AW, Bosch MA et al (2014) Insulin excites anorexigenic proopiomelanocortin neurons via activation of canonical transient receptor potential channels. Cell Metab 19(4):682–693. https://doi.org/10.1016/j.cmet.2014.03.004

Quarta C, Claret M, Zeltser LM, Williams KW, Yeo GSH, Tschop MH et al (2021) POMC neuronal heterogeneity in energy balance and beyond: an integrated view. Nat Metab 3(3):299–308. https://doi.org/10.1038/s42255-021-00345-3

Ramirez S, Gomez-Valades AG, Schneeberger M, Varela L, Haddad-Tovolli R, Altirriba J et al (2017) Mitochondrial dynamics mediated by mitofusin 1 is required for POMC neuron glucose-sensing and insulin release control. Cell Metab 25(6):1390–1399 e1396. https://doi.org/10.1016/j.cmet.2017.05.010

Ritter S (2017) Monitoring and maintenance of brain glucose supply: importance of hindbrain catecholamine neurons in this multifaceted task. In: Harris RBS (ed) Appetite and food intake: central control, 2nd edn, Boca Raton, pp 177–204

Rodriguez-Cortes B, Hurtado-Alvarado G, Martinez-Gomez R, Leon-Mercado LA, Prager-Khoutorsky M, Buijs RM (2022) Suprachiasmatic nucleus-mediated glucose entry into the arcuate nucleus determines the daily rhythm in blood glycemia. Curr Biol 32(4):796–805 e794. https://doi.org/10.1016/j.cub.2021.12.039

Ruohonen ST, Pesonen U, Moritz N, Kaipio K, Roytta M, Koulu M, Savontaus E (2008) Transgenic mice overexpressing neuropeptide Y in noradrenergic neurons: a novel model of increased adiposity and impaired glucose tolerance. Diabetes 57(6):1517–1525. https://doi.org/10.2337/db07-0722

Ruud J, Steculorum SM, Bruning JC (2017) Neuronal control of peripheral insulin sensitivity and glucose metabolism. Nat Commun 8:15259. https://doi.org/10.1038/ncomms15259

Sainsbury A, Cusin I, Rohner-Jeanrenaud F, Jeanrenaud B (1997) Adrenalectomy prevents the obesity syndrome produced by chronic central neuropeptide Y infusion in normal rats. Diabetes 46(2):209–214. https://doi.org/10.2337/diab.46.2.209

Sainsbury A, Rohner-Jeanrenaud F, Cusin I, Zakrzewska KE, Halban PA, Gaillard RC, Jeanrenaud B (1997) Chronic central neuropeptide Y infusion in normal rats: status of the hypothalamo-pituitary-adrenal axis, and vagal mediation of hyperinsulinaemia. Diabetologia 40(11):1269–1277. https://doi.org/10.1007/s001250050820

Schwartz MW, Sipols AJ, Marks JL, Sanacora G, White JD, Scheurink A et al (1992) Inhibition of hypothalamic neuropeptide Y gene expression by insulin. Endocrinology 130(6):3608–3616. https://doi.org/10.1210/endo.130.6.1597158

Schwetz TA, Ustione A, Piston DW (2013) Neuropeptide Y and somatostatin inhibit insulin secretion through different mechanisms. Am J Physiol Endocrinol Metab 304(2):E211–E221. https://doi.org/10.1152/ajpendo.00374.2012

Seseke FG, Gardemann A, Jungermann K (1992) Signal propagation via gap junctions, a key step in the regulation of liver metabolism by the sympathetic hepatic nerves. FEBS Lett 301(3):265–270. https://doi.org/10.1016/0014-5793(92)80254-e

Sesia T, Bulthuis V, Tan S, Lim LW, Vlamings R, Blokland A et al (2010) Deep brain stimulation of the nucleus accumbens shell increases impulsive behavior and tissue levels of dopamine and serotonin. Exp Neurol 225(2):302–309. https://doi.org/10.1016/j.expneurol.2010.06.022

Sheikh SP, Holst JJ, Skak-Nielsen T, Knigge U, Warberg J, Theodorsson-Norheim E et al (1988) Release of NPY in pig pancreas: dual parasympathetic and sympathetic regulation. Am J Phys 255(1 Pt 1):G46–G54. https://doi.org/10.1152/ajpgi.1988.255.1.G46

Sheng Z, Santiago AM, Thomas MP, Routh VH (2014) Metabolic regulation of lateral hypothalamic glucose-inhibited orexin neurons may influence midbrain reward neurocircuitry. Mol Cell Neurosci 62:30–41. https://doi.org/10.1016/j.mcn.2014.08.001

Shimazu T, Fukuda A (1965) Increased activities of glycogenolytic enzymes in liver after splanchnic-nerve stimulation. Science 150(3703):1607–1608. https://doi.org/10.1126/science.150.3703.1607

Shimazu T, Fukuda A, Ban T (1966) Reciprocal influences of the ventromedial and lateral hypothalamic nuclei on blood glucose level and liver glycogen content. Nature 210(5041):1178–1179. https://doi.org/10.1038/2101178a0

Sluga N, Krizancic Bombek L, Kercmar J, Sarikas S, Postic S, Pfabe J et al (2022) Physiological levels of adrenaline fail to stop pancreatic beta cell activity at unphysiologically high glucose levels. Front Endocrinol (Lausanne) 13:1013697. https://doi.org/10.3389/fendo.2022.1013697

Stanley SA, Kelly L, Latcha KN, Schmidt SF, Yu X, Nectow AR et al (2016) Bidirectional electromagnetic control of the hypothalamus regulates feeding and metabolism. Nature 531(7596):647–650. https://doi.org/10.1038/nature17183

Steculorum SM, Ruud J, Karakasilioti I, Backes H, Engstrom Ruud L, Timper K et al (2016) AgRP neurons control systemic insulin sensitivity via myostatin expression in Brown adipose tissue. Cell 165(1):125–138. https://doi.org/10.1016/j.cell.2016.02.044

Steinbusch LK, Picard A, Bonnet MS, Basco D, Labouebe G, Thorens B (2016) Sex-specific control of fat mass and counterregulation by hypothalamic Glucokinase. Diabetes 65(10):2920–2931. https://doi.org/10.2337/db15-1514

Takahashi A, Ishimaru H, Ikarashi Y, Kishi E, Maruyama Y (1997) Effects of ventromedial hypothalamus stimulation on glycogenolysis in rat liver using in vivo microdialysis. Metabolism 46(8):897–901. https://doi.org/10.1016/s0026-0495(97)90076-7

Teegala SB, Sarkar P, Siegel DM, Sheng Z, Hao L, Bello NT et al (2023) Lateral hypothalamus hypocretin/orexin glucose-inhibited neurons promote food seeking after calorie restriction. Mol Metab 76:101788. https://doi.org/10.1016/j.molmet.2023.101788

Ter Horst KW, Lammers NM, Trinko R, Opland DM, Figee M, Ackermans MT et al (2018) Striatal dopamine regulates systemic glucose metabolism in humans and mice. Sci Transl Med 10(442). https://doi.org/10.1126/scitranslmed.aar3752

Tingley D, McClain K, Kaya E, Carpenter J, Buzsaki G (2021) A metabolic function of the hippocampal sharp wave-ripple. Nature 597(7874):82–86. https://doi.org/10.1038/s41586-021-03811-w

Tong Q, Ye C, McCrimmon RJ, Dhillon H, Choi B, Kramer MD et al (2007) Synaptic glutamate release by ventromedial hypothalamic neurons is part of the neurocircuitry that prevents hypoglycemia. Cell Metab 5(5):383–393. https://doi.org/10.1016/j.cmet.2007.04.001

Tsurugizawa T, Kondoh T, Torii K (2008) Forebrain activation induced by postoral nutritive substances in rats. Neuroreport 19(11):1111–1115. https://doi.org/10.1097/WNR.0b013e328307c414

Tu L, Fukuda M, Tong Q, Xu Y (2022) The ventromedial hypothalamic nucleus: watchdog of whole-body glucose homeostasis. Cell Biosci 12(1):71. https://doi.org/10.1186/s13578-022-00799-2

Uner AG, Kecik O, Quaresma PGF, De Araujo TM, Lee H, Li W et al (2019) Role of POMC and AgRP neuronal activities on glycaemia in mice. Sci Rep 9(1):13068. https://doi.org/10.1038/s41598-019-49295-7

van den Hoek AM, Voshol PJ, Karnekamp BN, Buijs RM, Romijn JA, Havekes LM, Pijl H (2004) Intracerebroventricular neuropeptide Y infusion precludes inhibition of glucose and VLDL production by insulin. Diabetes 53(10):2529–2534. https://doi.org/10.2337/diabetes.53.10.2529

van den Hoek AM, van Heijningen C, der Elst JPSV, Ouwens DM, Havekes LM, Romijn JA et al (2008) Intracerebroventricular administration of neuropeptide Y induces hepatic insulin resistance via sympathetic innervation. Diabetes 57(9):2304–2310. https://doi.org/10.2337/db07-1658

van der Zwaal EM, de Weijer BA, van de Giessen EM, Janssen I, Berends FJ, van de Laar A et al (2016) Striatal dopamine D2/3 receptor availability increases after long-term bariatric surgery-induced weight loss. Eur Neuropsychopharmacol 26(7):1190–1200. https://doi.org/10.1016/j.euroneuro.2016.04.009

Wang GJ, Volkow ND, Logan J, Pappas NR, Wong CT, Zhu W et al (2001) Brain dopamine and obesity. Lancet 357(9253):354–357. https://doi.org/10.1016/s0140-6736(00)03643-6

Yeo GS, Farooqi IS, Aminian S, Halsall DJ, Stanhope RG, O'Rahilly S (1998) A frameshift mutation in MC4R associated with dominantly inherited human obesity. Nat Genet 20(2):111–112. https://doi.org/10.1038/2404

Yi CX, Serlie MJ, Ackermans MT, Foppen E, Buijs RM, Sauerwein HP et al (2009) A major role for perifornical orexin neurons in the control of glucose metabolism in rats. Diabetes 58(9):1998–2005. https://doi.org/10.2337/db09-0385

Yi CX, Foppen E, Abplanalp W, Gao Y, Alkemade A, la Fleur SE et al (2012) Glucocorticoid signaling in the arcuate nucleus modulates hepatic insulin sensitivity. Diabetes 61(2):339–345. https://doi.org/10.2337/db11-1239

Zarjevski N, Cusin I, Vettor R, Rohner-Jeanrenaud F, Jeanrenaud B (1993) Chronic intracerebroventricular neuropeptide-Y administration to normal rats mimics hormonal and metabolic changes of obesity. Endocrinology 133(4):1753–1758. https://doi.org/10.1210/endo.133.4.8404618

Zhou L, Podolsky N, Sang Z, Ding Y, Fan X, Tong Q et al (2010) The medial amygdalar nucleus: a novel glucose-sensing region that modulates the counterregulatory response to hypoglycemia. Diabetes 59(10):2646–2652. https://doi.org/10.2337/db09-0995

4 Insulin- Like Growth Factor-1 as Brain-Body Connector

Seraphina Ho, Snaiha Iyer Narayan, and Daniela Tropea

Abstract

Insulin-like growth factor 1 (IGF1) plays a vital role in the brain and the body, acting to promote growth and repair, as well as aiding neuronal connections and brain states. Its action is modulated by IGF binding proteins (IGFBPs), which regulate its transport and access to its receptor, the type 1 IGF1 receptor (IGF1R). The activation of IGF1R regulates several downstream pathways, predominantly mitogen- activated protein kinases (MAPK) and Phosphoinositide 3-kinase (PI3K). In the body, IGF1 is primarily secreted by hepatocytes and transported in the bloodstream to act as a driver of growth, protein anabolism, and feedback for the action of growth hormone (GH). In the brain, IGF1 is produced by several cell types, and it is a regulator of several systems, exerting its action on neurogenesis, modulation of neuronal excitability and plasticity, and regulation of brain states.

IGF1 concentration needs to be finely regulated, since its deficits or overexpression can have detrimental and potentially lethal effects. In addition, IGF1 plays an important role in pathologies including cancer, metabolic diseases, stroke, and brain disorders.

IGF1 is one of the few molecules whose transport between the brain and body is regulated by a specialized mechanism present in the blood-brain barrier (BBB). Considering that IGF1 is produced in both brain and body, it is not clear why IGF1 needs to be transported in the brain from the body. We propose that IGF1 acts as a connector to improve brain function in response to the body's metabolic states and other factors. Therefore, IGF1 transport from body to brain represents a strategy to coordinate the whole-body well-being.

Keywords

IGF1 · GH · Neurodevelopment · Neurodegeneration

S. Ho · S. I. Narayan
Trinity College Dublin Medical School, Dublin, Ireland

D. Tropea (✉)
Trinity Translational Medicine Institute, St James Hospital, Dublin, Ireland

FutureNeuro Research Ireland Centre, Trinity College Dublin, Dublin, Ireland

Trinity College Institute for Neuroscience, Dublin, Ireland
e-mail: tropead@tcd.ie

D. Tropea, E. Giacometti (eds.), *Brain-Body Connections*, Advances in Experimental Medicine and Biology 1477, https://doi.org/10.1007/978-3-031-89525-8_4

Abbreviations

(1-3) IGF1	IGF1 active tripeptide
ADAMS	The Anxiety Depression and Mood Scale
ALS	Amyotrophic Lateral Sclerosis
AN	Anorexia nervosa
BBB	Blood-brain barrier
CBP S436A	CBP phosphorylation site
CBP	CREB binding protein
CD1	Cyclin D1
CKD	Chronic kidney disease
des-IGF1	des-N-(1-3) IGF1
GH	Growth hormone
GluR1	Glutamate receptor 1
GnRH	Gonadotropin hormone releasing hormone
GPE	Glycine-proline-glutamate
Grb2	Growth factor receptor bound protein 2
HPA	Hypothalamic-pituitary axis
IGF1	Insulin like growth factor 1
IGF1R	IGF1 receptor
IGFALS	Insulin like growth factor binding protein, acid labile subunit
IGFBPs	IGF binding proteins
IGFs	Insulin-like growth Factors
IGHD	Idiopathic growth hormone deficiency
InsR	Insulin receptor
IRS-1	Insulin receptor substrate 1
IS	Infantile spasms
LS	Laron syndrome
MAD	Multiple ascending dose
MAP	Mitogen-activated protein
MAPK	Mitogen-activated protein kinase
MCUc complex	Mitochondrial calcium uniporter
MetS	Metabolic syndrome
mRNA	Messenger RNA
OLE	Open label IGF-1 treatment
PI3-K	Phosphatidylinositol-3 kinase
rhIGF-1	Recombinant human IGF-1
ROS	Reactive Oxygen Species
RTKs	Receptor tyrosine kinases
RTT	Rett syndrome
SDS	Standard deviation score
SSPE	Subacute sclerosing panencephalitis
TE	Transdermal estradiol

4.1 Introduction

Insulin-like growth factors (IGFs) are a family of insulin-related peptides consisting of 70 amino acids with a molecular weight of 7649 Da. They play a significant role in homeostatic functioning in several tissues. The IGF system is made up of two ligands—IGF1 and IGF2—that regulate many physiological processes such as metabolism and growth, acting through the Type 1 IGF Receptor (IGF1R) (Forbes et al. 2020). Other receptors also involved in IGF signaling are the insulin receptor and mannose 6-phosphate receptor. IGF1, the primary focus of this chapter, binds with high affinity to IGF1R and has different roles in the brain and body.

In the brain, IGF1 modulates a variety of functions—from hippocampal neurogenesis and apoptosis, to angiogenesis; these inherently are neuroprotective and hold reparative capacity (Cao et al. 2023). In the rest of the body, IGF1 is mainly secreted by the liver, and it performs many other functions, influencing the cardiovascular system, maintaining glucose metabolism, and contributing to the immune response and wound healing. IGF1 is the second messenger of growth hormone (GH) and contributes to bone and tissue growth and development (Dixit et al. 2021). In the skeletomuscular system, IGF is secreted by cartilaginous cells.

IGF1 dysregulation also plays a critical role in pathologies such as cancer and neurodegenerative disorders associated with aging. IGF1 dysfunction can also lead to developmental abnormalities (Réthelyi et al. 2023).

The transport and binding of IGF1 onto its receptors is a tightly regulated process; IGF Binding Proteins (IGFBPs), which extend the bioavailability of IGF1 in the bloodstream, allow its transport, and—usually—homeostatically mediate its potential interaction with the Insulin receptor (InsR) (Allard and Duan 2018). IGF1R is a transmembrane receptor tyrosine kinase. The binding of IGF1 to IGF1R leads to receptor dimerization and activation of its intracellular tyrosine kinase domain, triggering downstream signaling pathways.

This chapter will overview the role of IGF1 in the brain and body, its transport, and homeostatic mechanisms, alongside its therapeutic potential in pathologies as either a drug target or a clinical adjunct itself.

4.2 The Synthesis of IGF1

The gene encoding insulin-like growth factor 1 (*IGF1*) is located on chromosome 12q23–23 and comprises six exons. Conversely, the IGF1R protein is encoded by a single-copy gene with high similarity to the InsR, spanning more than 100 kilobases of genomic DNA situated at the terminal end of chromosome 15q25–26.

Upon *IGF1* gene activation, transcription to messenger RNA (mRNA) occurs, and within the cytoplasm, ribosomes engage in the translation process, synthesizing a linear sequence of amino acids to construct a nascent polypeptide chain termed pre-pro-IGF1. After its synthesis, pre-pro-IGF1 undergoes proteolytic cleavage and a series of post-translational modifications within the endoplasmic reticulum and Golgi apparatus, yielding pro-IGF1, which is further processed to generate mature IGF1. In the brain, mRNA staining for IGF1 has shown that expression occurs in neurons, astrocytes, and microglia in several regions (Suh et al. 2013). It acts at a local level, and its action is mediated in space and time through different IGFBPs. In the body, IGF1 synthesis primarily occurs in the hepatocytes. Following the above-described synthesis cascade, mature IGF1 is liberated into the systemic circulation, where it is bound by IGFBP3 and exerts pleiotropic effects on target tissues.

During embryogenesis, the *IGF1R* gene is expressed by nearly all tissue types. In the liver, the organ with the highest IGF1 ligand expression, IGF1R mRNA is almost undetectable, possibly because of the "downregulation" of the receptor by the local production of IGF1 (Adams et al. 2000).

IGF1R is a heterotetrametric transmembrane glycoprotein, and its internal portion contains a tyrosine kinase domain, which is vital for IGF1-related downstream signaling (Forbes et al. 2020). IGF1 binding triggers receptor autophosphorylation (Kavran et al. 2014), which in turn allows for the further addition of phosphate groups to proteins such as insulin receptor substrate 1 (IRS-1) (Li et al. 2019). The autophosphorylation of IGF1R also triggers a cascade of intracellular events for further cellular signal transduction via the pathways outlined in Table 4.1 below (Werner 2023).

Overall, this cascade of enzyme activation triggered by the IGF1R plays a critical role in transmitting the signal from the cell surface to the nucleus, where it ultimately results in a temporal transcriptional response leading to a wide range of biological processes, as shown in Fig. 4.1 (Xiu et al. 2021). Some studies have shown that the truncated versions of IGF1R, even those with shorter (carboxyl) C-terminal tails, retain their catalytic activity (Adams et al. 2000).

4.3 IGF1 Binding Proteins and Receptors

The transport of IGF1 between the body and the brain is a tightly regulated process, which is crucial for maintaining brain function and overall physiological homeostasis. IGF1, primarily produced in the liver in response to GH stimulation, circulates in the bloodstream bound to insulin-like growth factor binding proteins (IGFBPs). These binding proteins not only regulate the bioavailability and stability of IGF1, but also serve as carriers, facilitating its transport across the blood-brain barrier (BBB).

In blood plasma, 80% of circulating IGF1 is bound to a complex consisting of one molecule of IGFBP3, and one molecule of an 88 kDa acid-labile subunit-based protein, which has been known to be synthesized primarily in the fetal liver. The remaining 20% is complexed with an array of binding proteins that homeostatically modulate its tissue availability. So far, there are six identified IGF1-binding proteins (Clemmons et al. 1993). IGFBP3 is regulated mainly by GH

Table 4.1 The downstream pathways associated with IGF1

Phosphatidylinositol-3 kinase (PI3-K):	PI3-K is activated by the phosphorylated IGF1-R and generates molecules called phosphoinositides, which regulate cell growth, survival, and metabolism via action on glucose transport and glycogen synthesis proteins, play a role in the downstream generation of PIP3, and assist in cytokine/hormone signaling.
Grb2 (growth factor receptor bound protein 2):	Grb2 is a protein that binds with high affinity to the autophosphorylated IGF1R, and then recruits another protein called Sos, which in turn activates the Ras protein.
Syp (a phosphotyrosine phosphatase):	Syp is an enzyme that counteracts intracellular phosphorylation, primarily performed by kinases, as it removes phosphate groups from proteins in accordance with homeostatic needs.
Nck (an oncogenic protein):	Nck is a protein that organizes downstream signaling complexes, though its role in the IGF1 signaling pathway is not fully understood yet.
Shc (src homology domain protein):	Shc is an adapter protein that binds to the autophosphorylated IGF1R and activates the Raf protein.
Raf:	Raf is a kinase, activated by Shc, which initiates a phosphorylation cascade, including the activation of mitogen-activated protein (MAP) kinase.
MAP kinase (MAPK):	MAPK regulates gene expression, cellular growth, differentiation, and survival via the phosphorylation of various target proteins.

but also to some degree by IGF1 and other proteins. It is primarily synthesized and released by the liver, and once bound to IGF1, or even IGF2, its complexes are internalized into the cell either via the IGF1 Receptor (IGF1R) or via endocytic pathways, after which IGFBP3 is recycled in the nucleus. The association of IGF1 with IGFBPs increases ligand stability, allowing prolonged signaling (Bastian et al. 1997). However, the role of IGFBPs in IGF1 signaling is complicated by the fact that they associate with the ligands with nearly equal affinity as the receptor and thus compete with the receptor for binding, potentially decreasing receptor activation. It is also important to note that because free IGF1 is not enzymatically stable, it is cleaved at its N-terminal to release Glycine-proline-glutamate (GPE) (Guan and Gluckman 2009) and form des-N-(1-3) IGF1 (des-IGF1) in plasma and brain tissue (Guan et al. 2023). GPE is then transformed into cGP, which binds the IGFBPs competing with the IGF1 binding site. Thus, IGFBPs have a modulating action in protecting IGF1 from degradation and preventing its binding to the receptor. Overall, IGFBPs regulate the local bioavailability of IGF1 and are specific according to tissue and developmental stage (Guan and Gluckman 2009).

4.4 IGF1 in the Brain and Body

Despite IGF1 binding to the same receptor (IGF1R) in the brain and the body, its main functions differ in each area. In the brain, IGF1 production is promoted by 17-β estradiol, and it is vital for brain function mainly through its neurotrophic effects and regulation of BBB permeability (Lu et al. 2020). This hormone and the related synthesis and action of IGF1 will be further explored in its role between males and females in the Sect. 4.7. IGF1 binds to IGF1R located on microvascular endothelial cells that are part of the BBB and enters the brain parenchyma. Subsequently, it can also be taken up by astrocytes or neurons themselves. This versatility allows IGF1 to exert its effects within the brain through endocrine, paracrine, or autocrine pathways. The mechanisms governing these transport processes include receptor-mediated transcytosis, adsorptive-mediated transcytosis, and carrier-mediated transport (Nieto-Estévez et al. 2016). IGF1 is not the only molecule transported through

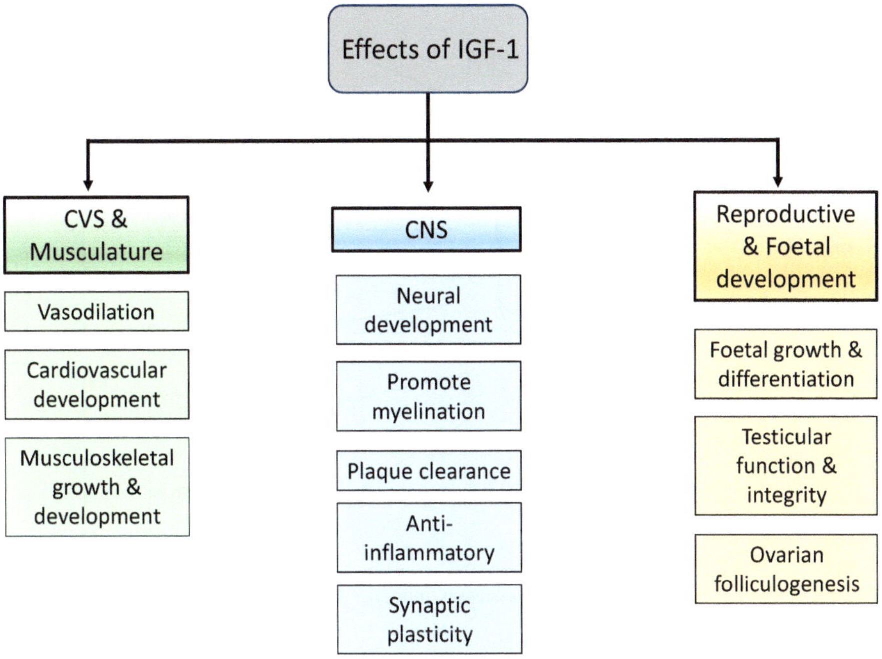

Fig. 4.1 List of the biological processes influenced by IGF1 in different systems

the blood-brain barrier. *Other examples are described in the Chap. 1*.

In early development, IGF1 can be found to modulate progenitor cells responsible for producing new neurons in the dentate gyrus and in the subventricular zone of the hippocampus. This neurogenic ability was shown in mice overexpressing IGF1, which display augmented brain size characterized by an increase in neuronal number. Furthermore, dysfunction in the *IGF1* or *IGF1R* gene leads to brain growth retardation in both humans and animals (Anderson et al. 2002). In vitro culture studies have demonstrated IGF1's pleiotropic functions in all neural cells via increasing cell number and facilitating myelination and maturation. IGF1's neurotrophic actions are further displayed by its effect on neural stem cells that results in higher neural progenitor cell proliferation and longer maintenance in cell culture (Wrigley et al. 2017). The combination of its neuroprotective and neurogenic functions in the brain allows IGF1 to be vital for cognitive function, shown by its significant link with better performance in cognitive and perceptual-motor speed tests. This result is significant in the context of aging, where the GH/IGF1 axis decreases significantly with aging and is potentially involved in the related decline in cognitive function (Aleman et al. 1999). IGF1 can also cross the BBB via the IGF1R located on the cells of microvessels. This passage is crucial in facilitating higher IGF1 levels in the brain (which typically has lower IGF1 levels than the body) to allow it to serve its neuroprotective functions, for example, in degenerative processes (Pan and Kastin 2000).

In these ways, IGF1 has a positive action in neuronal survival, growth, and—as recently found—even synaptic plasticity, thus contributing to cognitive function and neuroprotection. These effects of IGF1 are displayed in Fig. 4.2. In animal cerebellar deafferentation models, the administration of IGF1 led to the restoration and reinnervation of olivo-cerebellar pathways in both juvenile and mature rats, effectively minimizing observable ataxic deficits. Furthermore, IGF1 administration normalizes the levels of several proteins crucial for synaptic function, such as calbindin, glutamate receptor 1 (GluR1), GABA, and glutamic acid, which are typically

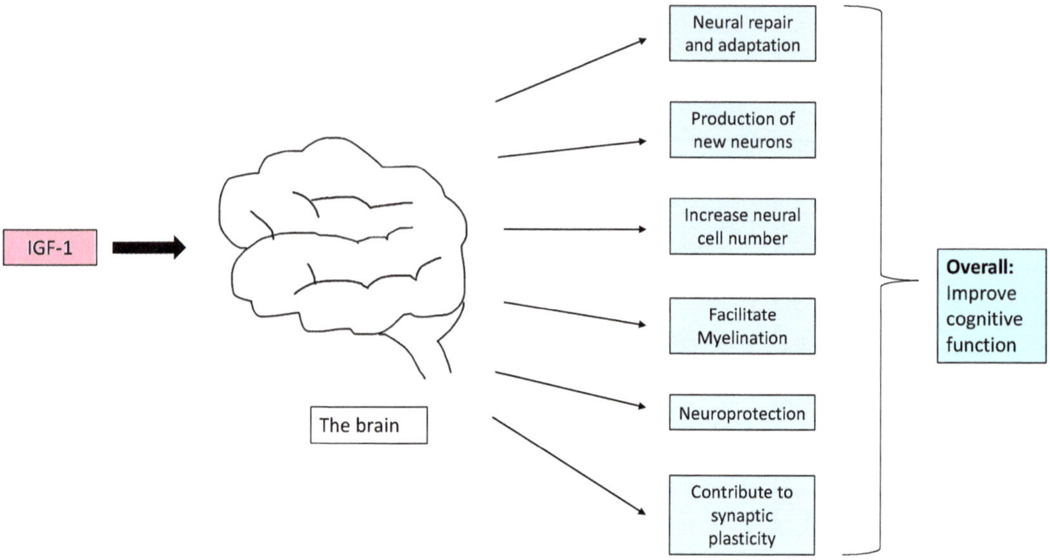

Fig. 4.2 Effects of IGF-1 on the brain. List of processes influenced by IGF1 in the brain. The overall effect is to improve cognition

affected by deafferentation (Dyer et al. 2016). These findings underscore the significant role of IGF1 in cellular neuroplasticity. IGF1 has also demonstrated its importance in reparative plasticity in models of kanamycin-induced deafness and excitotoxic lesions in the central gyrus. In both models, IGF1 promotes the thickening of postsynaptic densities, enhances dendritic arborization, and supports functional and cellular neuroplasticity in the CNS. These results highlight the multifaceted effects of IGF1 in promoting neural repair and adaptation to various pathological conditions affecting the CNS. Furthermore, dysregulation of IGF1 transport and signaling in the brain has been implicated in various neurological disorders, including Alzheimer's disease, Parkinson's disease, and stroke (Dyer et al. 2016). While it is unclear whether IGF1's passage across the BBB is bidirectional, there is strong evidence to suggest that its uptake and presence in the CNS is also crucial in influencing the peripheral circulation. For example, in conditions such as metabolic syndrome (MetS), local and systemic deficiencies of IGF1 lead to a disturbed BBB and altered cerebrovascular microenvironment. This transport then affects the body via MetS-related symptoms such as hypertension and hyperglycemia. IGF1 levels in the brain are thus crucial for ensuring adequate neuroprotection and regulation of certain peripheral systems, allowing its breadth of function to be widespread and multi-functional (Arjunan et al. 2023).

In the body, IGF1 is mainly produced in the liver following GH stimulation, but it can also be secreted by multiple tissues for paracrine or autocrine functions (Puche and Castilla-Cortázar 2012). Binding of IGF1 to IGF1R in the body results in two main signaling cascades being activated. The first is the PI3K pathway that results in IGF-induced DNA synthesis, cell differentiation, and cell survival in different cell types (Hakuno and Takahashi 2018). In this pathway, the activation of Akt suppresses TSC, which functions as a GTPase-activating protein for Rheb. mTORC1 enhances protein synthesis, metabolic processes, and cellular growth by regulating S6K1 and 4EBP1. The pathway is exemplified in Fig. 4.3 below, as adapted from Paplomata and O'Regan (2014). This pathway mainly mediates the metabolic function of IGF1 (Forbes et al. 2020).

The second pathway is the MAPK pathway that is also required for the IGF-induced DNA synthesis and cell survival (Hakuno and

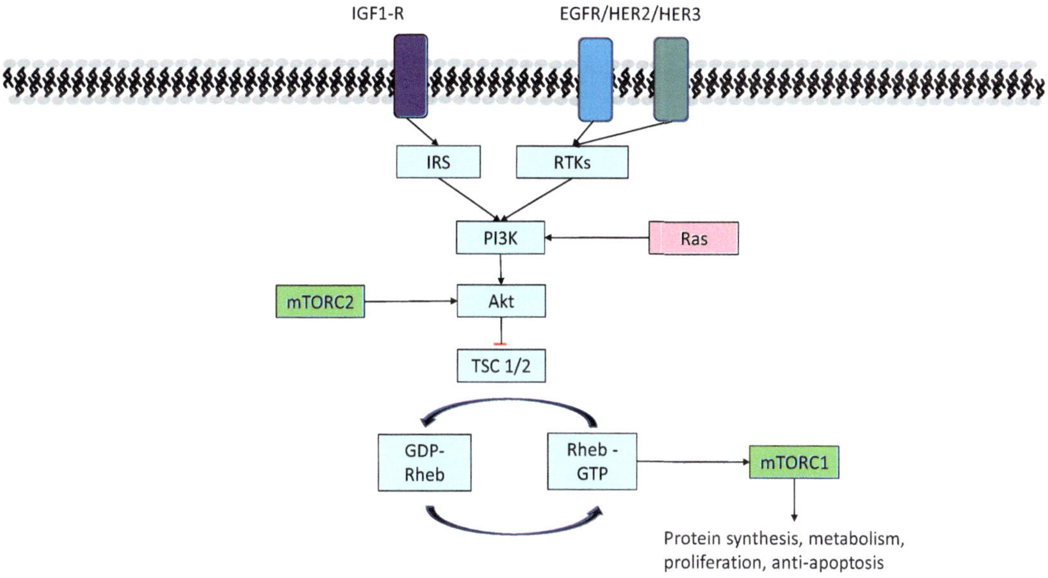

Fig. 4.3 PI3K Pathway. One of the main pathways activated by IGF1 is the PI3K pathway, which is represented here

Takahashi 2018). Firstly, the p38 kinase can be activated by cytokine stimulation through the TGFβ pathway, leading to TP53 activation under normal conditions. TGFβ-induced p38 activation occurs independently of canonical SMAD signaling but relies on the TAB/TAK1 complex and MKK3/6 kinases. The canonical MAPK pathway starts with growth factor binding to receptor tyrosine kinases (RTKs), triggering downstream activation of RAS, RAF, and MEK, ultimately activating ERK1/2, as seen in Fig. 4.4 below. The PI3K/AKT/mTOR pathway can also be activated by RTKs and RAS, mainly influencing metabolic signaling and protein synthesis to support cell growth (Braicu et al. 2019). Furthermore, IGF1R activation of the Shc and IRS adaptor proteins causes Shc to stimulate Raf and several other proteins via phosphorylation to cause cell proliferation (Józefiak et al. 2021).

Overall, IGF1 has multiple functions in all areas of the body. Yet, its main functions differ largely between the brain and the body. In the brain, it is crucial in neuroprotection via its downstream pathways, like the activation of PI3K, MAPK, and Raf1, which increase during cerebral ischemia; reactive oxygen species also act as feedback mechanisms for the activation of these downstream pathways (Zhou et al. 2015). Moreover, IGF1 acts as a vital neurotrophic factor. In the body, it is essential for overall cell growth and proliferation.

4.5 IGF1 Brain-Body Interactions: Feedback and Homeostasis

One of the main mechanisms regulating IGF1 homeostasis is the Hypothalamic-Pituitary Axis (HPA). In this book, *the HPA axis has been discussed in other chapters, such as Chap. 6,* where the authors describe its role in stress and psychiatric disorders and metabolic disturbances. *Also, in the Chap. 7, it is presented as one of the key mediators of stress.*

The axis' feedback loop, where GH is secreted by the anterior pituitary after hypothalamic stimulation, also influences IGF1R and GH levels, therefore impacting IGF1 concentration (as seen in Fig. 4.5). Elevated IGF1 levels inhibit GH secretion via negative feedback, ensuring precise control over GH levels and preventing excessive growth (Bianchi et al. 2017). Conversely, GH controls levels of circulating IGF1 through rapidly activating *IGF1* gene transcription, while IGF1 controls GH secretion from the pituitary

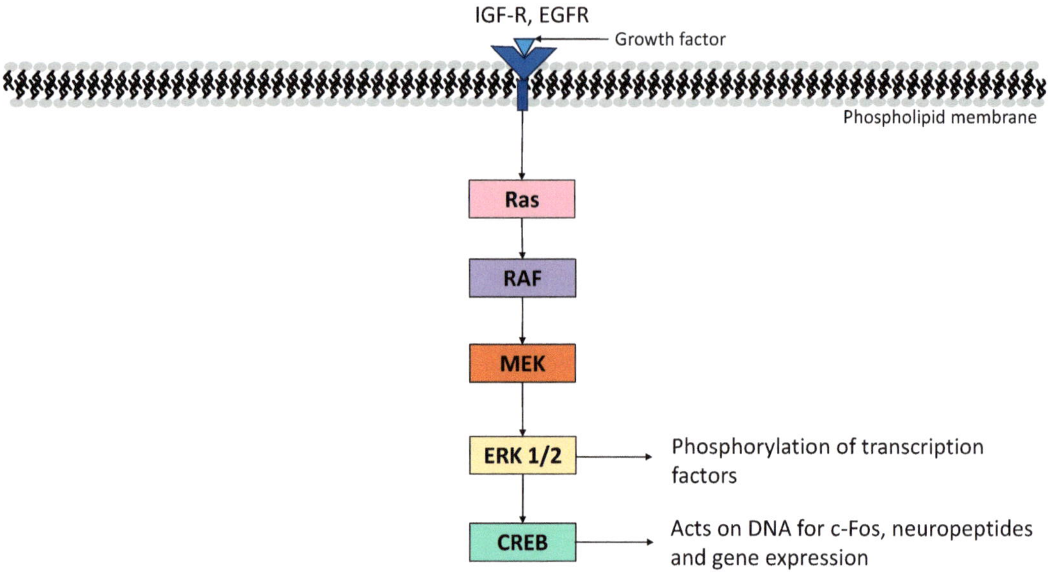

Fig. 4.4 MAPK pathway. Another main pathway activated by IGF1 is the MAPK pathway, which is represented here

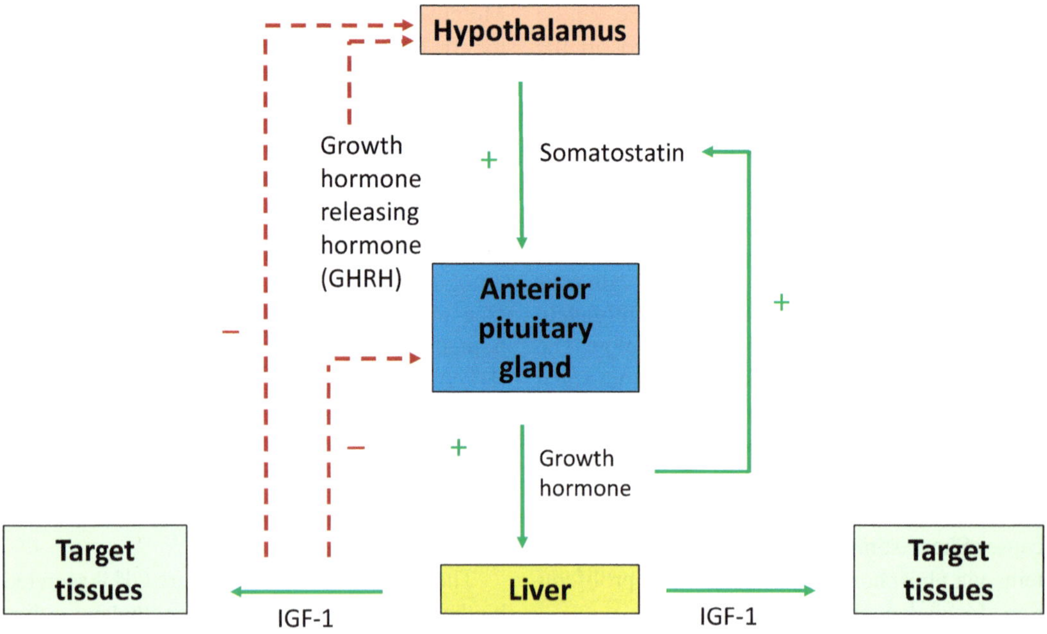

Fig. 4.5 The IGF-1-GH Axis. IGF1 is the second messenger of growth hormone. GH induces IGF1 expression in the liver. IGF1 then is distributed to several target tissues through the blood

gland in a negative feedback loop. This mechanism is shown in Fig. 4.6.

IGF1, via this mechanism, also mediates somatic growth and development. The mentioned axis has a plethora of actions on many areas of the body. In bone, it facilitates bone growth, exemplified by it being necessary for the development of the dento-alveolar complex and bone repair/regeneration (Chesnokova 2022). Moreover, this axis aids intracellular signaling in

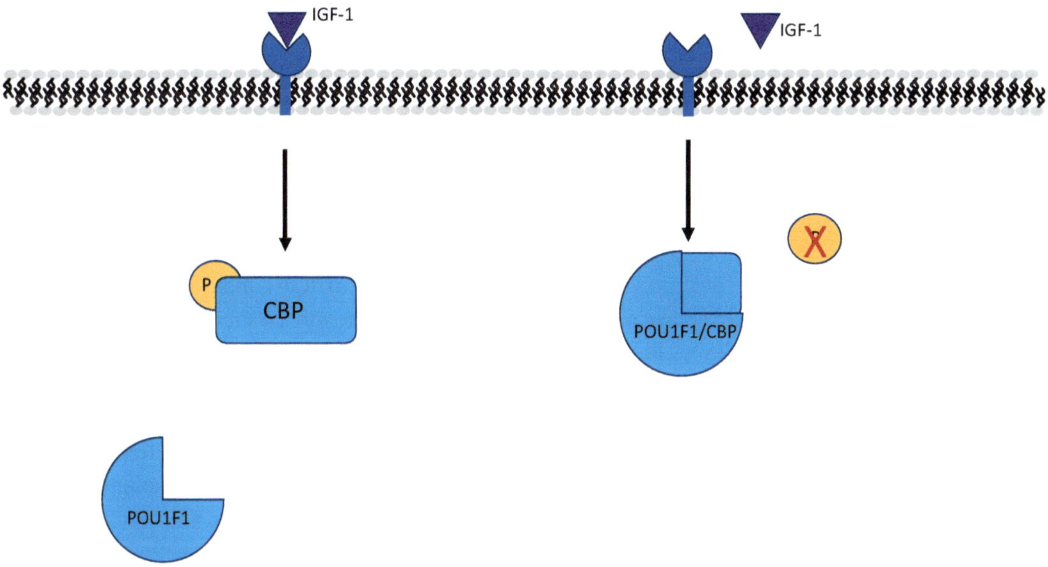

Fig. 4.6 The POU1F1-IGF-1 interaction. IGF1 affect gene transcription through the CREB BINDING PROTEIN (CBP), which is activated by the IGF1 signaling

lung development, growth, and repair in both cellular and animal models via Growth Hormone Releasing Hormone control (Chesnokova 2022). However, defects in its proliferative functions can promote oncogenic properties and cause diseases, such as acromegaly (Kasuki et al. 2022).

Furthermore, the IGF1R also acts as a negative feedback regulator of GH. Several studies indicate that IGF1 can decrease *GH* gene expression and hormone release (Romero et al. 2012). IGF1 signaling phosphorylates, and thus interferes with, a protein known as CREB binding protein (CBP) that is necessary for *GH* gene expression; this occurs via the disruption of the formation of the POU1F1/CBP complex, reducing their capacity to associate with the *GH* promoter, resulting in the downregulation of *GH* gene expression. Further, a mouse model harboring a mutation in a CBP phosphorylation site (CBP S436A) showed a lack of IGF1 influence on it, while increased serum GH levels were observed, indicating reduced feedback to the somatotropic cells, as shown in Fig. 4.6. Interestingly, however, there was no increase in growth. Mice with the CBP gene mutation described before had decreased fat mass; this shows the vital role of CBP in somatotropic cell regulation.

In the brain, it has been shown that IGF1 is responsible for stable firing homeostasis in neuronal circuits, which is important to protect the brain against neurodegeneration and cognitive decline. This result was found by inactivation of the IGF1R and studying firing rates in response to activity level changes; the mechanism was seen to be linked to alterations in mitochondrial calcium signaling mediated by the mitochondrial calcium uniporter complex (MCUc) (Katsenelson et al. 2022).

In both the body and the brain, IGF1 acts to enhance insulin sensitivity indirectly by suppressing GH, which counteracts insulin; this effect is particularly important in skeletal muscle and the kidneys, where IGF1 plays a significant role in regulating glucose metabolism. Moreover, IGF1 binds to InsRs, mimicking insulin's actions and further amplifying insulin sensitivity (Clemmons 2006).

It is vital to maintain IGF1 homeostasis, which is essential for development, survival, metabolism, maintaining glucose homeostasis, and regulating neuronal activity. These findings suggest potential therapeutic implications for developmental anomalies and age-related cognitive neurodegenerative disorders (Nieto-Estévez et al. 2016).

4.6 The Role of IGF-1 in Lifespan and Aging

The IGF1 signaling pathway also plays a key role in the process of aging (Wrigley et al. 2017). It has been established that reduced exposure of tissues to endocrine or autocrine/paracrine IGF1 is associated with a markedly extended lifespan (Yang et al. 2005). While the studies on IGF1 signaling regulation of lifespan report different and sometimes contradictory results, it has been postulated that longevity is associated with a tightly regulated HPA. The impact of brain IGF1R on lifespan has been addressed by specific inactivation of the IGF1R in embryonic mice brains. Partial abrogation of the receptor inhibited the GH and IGF1 pathways, causing metabolic and developmental anomalies in postnatal life, including a reduction in adult body size (Yang et al. 2005).

IGF1 levels have been shown to fluctuate in both the brain and body throughout life. In the body, IGF1 acts synergistically with GH to promote protein anabolism, contributes to skeletal development and linear growth, and increases bone mineral density (Al-Samerria and Radovick 2021). In childhood, serum levels of IGF1 are generally low. However, around adolescence, the concentrations of IGF1 increase as it is vital for somatic cell growth (Ashpole et al. 2015), and these peak at puberty. This surge is followed by a sharp decline in the third decade and a more gradual decrease from the third to the eighth decade (Janssen 2019). Circulating levels of IGF1 may even be negligible in patients over 60. However, transcription of the IGF1R can increase with age as a potential compensation mechanism for IGF1 decrease (Ashpole et al. 2015). The changing levels of IGF1 expression during development are illustrated in Fig. 4.7.

As displayed in Fig. 4.7, IGF1 levels reach a peak during puberty in both males and females, with an additional peak in males at about 14 years old. However, female levels generally remain higher than males throughout life until about the fourth decade. During puberty, this can be

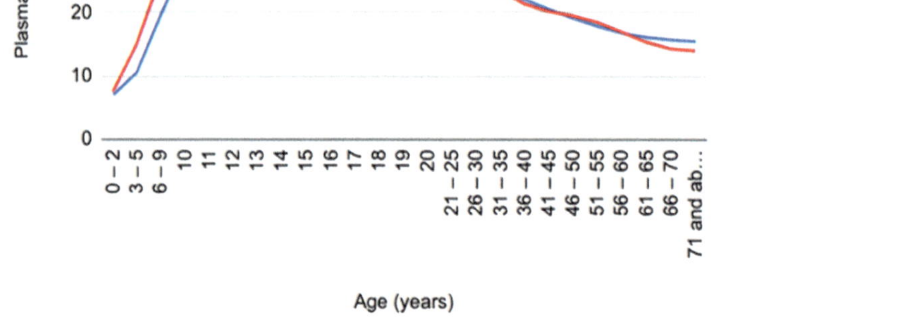

Fig. 4.7 Graph depicting average IGF-1 levels in the blood in males and females as per values from Gloucestershire Hospitals NHS Foundation Trust (2023)

explained by the beginning of menstruation in females, which correlates to higher levels of GH and thus IGF1 (Span et al. 2000).

The levels of IGF1 in the blood can be compared to the levels of IGF binding proteins that help to regulate circulating IGF-1 (Bach 2018). Similar to IGF1 levels, IGFBP3—the most abundant IGFBP in the blood—peaks in puberty (Alberti et al. 2011; Kim 2013). This trend can be attributed to the modulation of IGF1 by IGFBP3, which transports IGF to its target cells. IGFBP3 participates in IGF1-mediated activation of downstream signaling pathways (Varma Shrivastav et al. 2020), facilitating the fast-paced growth an individual would typically undergo during puberty. Although IGFBP3 levels seem to decrease after puberty, further studies will need to be performed to assess its levels in the third decade and onwards, to provide better comparison with the concentration in the previous years.

IGF1 has further been studied for its effects on CNS preparation and plasticity, alongside neurogenesis in the adult brain, using injury and sensory deprivation models. These studies have indicated that a reduction in IGF1 levels as aging progresses may indirectly lead to reduced reparation, remodeling, and resistance to stress (Wrigley et al. 2017).

In the brain, IGF1 decreases with age. Age-related reductions in IGF1 have been shown to cause cognitive deficits; thus, its use could be potentially beneficial to maintain brain function in aging (Junnila et al. 2013). However, due to its numerous functions, its role in the aging brain remains controversial and complex (Gubbi et al. 2018). This role is often rooted in how the brain is inherently a complex mosaic (Vitale et al. 2019). Additionally, clinical studies regarding IGF1 are hard to conduct, not only given its fluctuations throughout one's lifespan but also between sexes (as further explored in this chapter under Sect. 4.7 below). It is also important to note the many controversial results about IGF1 in a range of cases, such as aging, metabolism, and neurodegeneration, are due to the difference in IGF1 and IGF1R expression (Delafontaine et al. 2003).

Decreases in IGF1 levels over time are consistent with many age-related pathologies, neurodegenerative disorders, and an overall higher risk of illness as IGF1 is essential for tissue function and normal development (Ashpole et al. 2015).

4.7 Sex-Dependence of IGF1 Expression and Function

The sex dependence of IGF1 can be further explored in the context of the brain and the body. As mentioned earlier, IGF1 production is induced by 17-β estradiol in the brain, which is the most potent form of estrogen in nonpregnant females (Gillies and McArthur 2010). One difference in the IGF1 system between males and females could be the levels of IGF1R expression in the developing cerebellum. IGF1R levels are elevated in males during the first postnatal week and fall afterwards. In females, IGF1R expression reaches a peak at 2 weeks postnatally. The different activation of IGF1 signaling in males and females could explain why males have more Purkinje neurons than females. Additionally, IGF1R expression reaches its peak at different times in the hippocampus between males and females, indicating different influences in the development and function of this brain area (Sohrabji 2015). These differences in IGF1 levels between males and females are probably dependent on higher levels of estrogen in females.

Conversely, estrogens effect on the body causes premenopausal women to have lower IGF1 levels than men. This is rooted in estrogen's ability to desensitize growth hormone receptors in the liver, while testosterone acts in an opposite manner and amplifies GH-stimulated IGF-1 production (Giraldi et al. 2022).

Different mechanisms can explain differences in IGF1 levels between males and females. First, estrogen has been shown to be a vital protective factor in females. For example, mitochondrial estrogen receptors are involved in cyto-protection from apoptosis and oxidative stress, and a deficiency in this mechanism has been shown to be causally linked to Parkinson's disease pathogenesis (Gillies and McArthur 2010). Second, the difference in estrogen levels could depend on the release of luteinizing hormone, which is dynamic in females while it is tonic in males. Typically (90%) estradiol has negative feedback on gonadotropin hormone-releasing hormone (GnRH) in females, but it turns into positive feedback prior to ovulation, leading to a massive, coordinated release of GnRH. In males, instead, there is a steady release of testosterone, which results in constant negative feedback on GnRH release.

In addition, the difference in estrogen between the brain and the body may be due to estrogen and IGF1's similar neuroprotective properties. Both estrogen and IGF1 can increase neuronal differentiation, neuroendocrine regulation, synaptic plasticity, and reproductive behavior. Additionally, they are both able to work in a concerted manner to activate Akt and stimulate proliferation, although the mechanism of action is unknown (Mendez et al. 2003). Proliferation is further stimulated by 17-β estradiol being able to transiently activate the IGF1R via tyrosine phosphorylation, not only resulting in an interaction between estrogen receptor alpha and IGF1R, but also in an enhanced interaction with aspects of the PI3K pathway (Sohrabji 2015).

4.8 Pathologies Related to IGF1

There are several pathologies in the brain and body that are related to IGF1 deficiencies and arise due to a multitude of reasons, including genetic bases and the functionality of both ligand and receptor (Puche and Castilla-Cortázar 2012). These pathologies are summarized in Table 4.2, alongside their neurological and more widespread effects (Puche and Castilla-Cortázar 2012):

Table 4.2 Pathologies related to IGF-1 deficiency

Pathology	Description
Laron syndrome (LS)	LS, also known as Laron-type dwarfism, is a rare autosomal recessive disorder characterized by severe short stature and facial dysmorphisms due to mutations in the GH receptor gene. Interestingly, LS leads to cancer resistance despite high GH levels, due to recently proven gene downregulation for growth and upregulation for detoxification. Lifelong low IGF1 exposure boosts protective responses against cancer, including apoptosis and oxidative stress.
IGF1 and/or IGF1R gene mutations	*IGF1* and *IGF1R* gene mutations can cause either the formation of deficient/ defective IGF1 or a lack of elicited response due to receptor dysfunction, symptomatically manifesting as severe intrauterine abnormalities, prenatal and postnatal growth retardation, developmental delays, microcephaly, and sensorineural deafness.
IGFALS (ALS) gene mutations	Mutations in the *IGFALS* gene can cause ALS, resulting in reduced circulating levels of IGF1 and IGFBP3, also leading to short stature, microcephaly, and dysmorphic features.
IGF1 resistance due to IGF1 antibodies, IGF1R/ GH receptor mutations, and/ or GH resistance due to post-receptor defects	These pathologies are all delineated by short stature, microcephaly, and dysmorphic features, alongside occasional developmental delays typical of IGF1 and/or GH deficiency. These mimic LS symptomatically.

Furthermore, IGF1's role has been delineated in several disorders via either the mechanisms outlined above or idiopathically; some novel findings from clinical studies investigating the role of IGF1 in neurological pathologies are shown below in Table 4.3:

Lastly, it has been found that IGF1 dysregulation also plays a crucial role in the pathogenesis of neurodegenerative conditions like Alzheimer's disease, Parkinson's disease, ALS, and prion disorders (Bianchi et al. 2017) via mechanisms like alterations in IGF1 production, receptor expression, or downstream signaling cascades, eventually causing impaired neuronal viability, synaptic dysfunction, and neuroinflammation (Arjunan et al. 2023).

Understanding the role of IGF1 in the disease pathogenesis can provide insights into disease mechanisms and provide potential therapeutic targets for intervention.

Table 4.3 IGF1-related pathologies

Disease	Role of IGF1 in disease (Riikonen 2017)
Neonatal Asphyxia	IGF1 reduces neuroinflammation and is vital for brain growth and development. Reduced IGF1 levels are indicative of poor neural development observed in preterm infants.
Infantile Spasms (IS)	A reduced concentration of IGF1 in the CSF is correlated with IS that is heavily symptomatic, cortical damage, therapeutic resistance, and mental retardation. Normalization of the IGF1 axis may contribute to normal development.
PEHO Syndrome	PEHO syndrome, like IS, is characterized by low IGF1 levels in the CSF; it impacts cerebellar neurons and causes encephalopathy, hypsarrhythmia, and optic atrophy.
Autism Spectrum Disorder	Low IGF1 concentrations in the CSF have been found in autistic children; this may suggest disruption of normal neural development—reducing Purkinje cell survival and leading to abnormal growth patterns within the brain.
Rett Syndrome	Despite normal IGF1 levels in the CSF in Rett syndrome, other neurotrophic factors like NGF are reduced and cause language and motor deficits.
Subacute Sclerosing Panencephalitis (SSPE)	Increased IGF1 levels in the CSF are found in the advanced stages of SSPE, which is induced by the measles virus, although further research is needed to elucidate the relationship between IGF-1, measles IgG titers, and tau protein in SSPE.

4.9 The Role of IGF-1 in Cancer

Several studies report—directly or indirectly—a positive relationship between high levels of IGF1 signaling and various types of cancer. IGF1 is needed for cell cycle entry, and the assembly between Cyclin D1 (CD1) and CDK4 is an essential factor to the G1 phase progression (Grimberg 2003). IGF1 further contributes to tumor and continued cell growth as a pro-angiogenesis agent. In fact, IGF1 facilitates the activation of angiogenesis growth factors with the activation of the kinase/Akt signaling pathway (Lin et al. 2017).

Additional data show that IGF1 contributes to the occurrence of cancer; this hormone can boost androgen receptor signaling by increasing the production of androgens in both the adrenal glands and gonads; additionally, it enhances the activation of androgen receptors by producing exclusion of a protein called FoxO1 from the cell nucleus. Since FoxO1 decreases mitochondrial respiration and ROS formation, increased IGF1 signaling coupled with downregulated FoxO levels creates an imbalanced elimination of ROS. This imbalance increases the risk of cancer development, creating a favorable environment for angiogenesis (Melnik et al. 2011), as illustrated in Fig. 4.8. Altogether, these data show that activation of IGF1 signaling can contribute to cancer progression.

4.10 IGF1 Role in Metabolic Pathologies

Diet indirectly upregulates the expression of IGF1 signaling; high glycemic foods increase free IGF1 bioavailability in the plasma. As stated by Melnik et al. (2011), individuals consuming cow milk shift the somatotropic axis to higher levels and increase the serum levels of GH and IGF1.

This hormone is also associated with metabolic pathologies; this is commonly known as Metabolic Syndrome. Briefly, a decrease in InsR signaling creates a sensitive environment for insulin resistance and diabetes, and IGF1 deficiency enhances the risks of cardiovascular disease and atherosclerosis. According to Reaven (1988), clinical features of several syndromes and conditions are associated with insulin resistance (Arai et al. 2009).

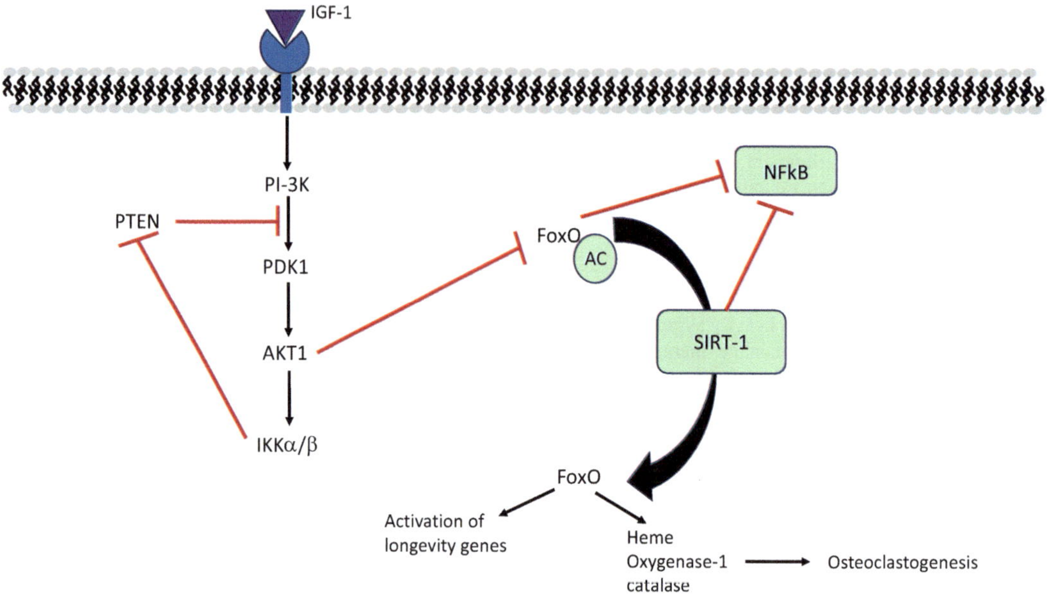

Fig. 4.8 IGF-1 Signaling and FoxO. Activation of IGF1 signaling paired with a downregulated production of FoxO increase the risk of cancer and favor angiogenesis

4.11 IGF as a Diagnostic and Therapeutic Agent

IGF1, due to its multifaceted responses, can be used as a diagnostic, prognostic, and therapeutic agent. Serum IGF1 test acts as an indicator of the GH-HPA feedback loop and, therefore, can aid in the diagnosis of a pituitary tumor. Other tests, including X-rays to measure bone health and thyroid functionality, are used as adjuncts when making diagnoses, since IGF1 levels vary depending on factors like age, sex, insulin resistance, and history. Moreover, IGF1 and GH levels are directly correlated with disorders such as dwarfism and acromegaly; thus, the serum IGF1 test is highly clinically informative.

IGF1 and its derivatives are also used as treatments. Besides IGF1 for the treatment of dwarfism, IGF1 active tripeptide ((1-3) IGF1) and cGP play a role in neuroprotection and neurological conditions (Guan and Gluckman 2009; Guan et al. 2023). The mechanisms of action of these molecules are only being recently investigated, and the predominant hypothesis is that they enhance the expression of synaptic markers (Corvin et al. 2012).

IGF1 has also been tested as a treatment for neurodevelopmental disorders such as Rett Syndrome (RTT), Fragile X, and Phelan-McDermid Syndrome. In RTT—a severe neurodevelopmental disorder resulting in developmental regression—the first results came from preclinical studies (Tropea et al. 2009; Castro et al. 2014), and follow-up clinical trials proved its safety and efficacy in patients. In 2023, Trofnetide—a synthetic analogue of (1-3) IGF1—has been approved as the first treatment for RTT (Kennedy et al. 2024; Tropea 2024). These findings underline the use of IGF1 and related molecules as therapeutic agents and as tools for understanding RTT pathology and other disorders.

During insulin-based treatment regimens, individuals may paradoxically experience hypoglycemia due to mismatched glucose responses and compensatory hyperinsulinemia, due to the subjective sensitivity of insulin resistance arising from genetic variance (McDonald et al. 2007). Therefore, the recombinant human IGF1 therapy, alone or with IGFBP3, offers a promising alternative to disorders stemming from insulin resistance. IGF1 also acts on the InsR due to its structural similarity with IGF1R, as previously discussed, and improves glucose disposal and potentially aids in the preservation of beta-cell function (McDonald et al. 2007).

However, the clinical limitations of IGF1 must be addressed. The reliability of IGF1 as a biomarker for assessing treatment efficacy and dictating dose adjustments in pediatric patients undergoing GH therapy has shown inconsistencies. Data from a study in a cohort of prepubertal children receiving GH treatment in Korea revealed discrepancies between IGF1 levels and growth response. Despite there being a positive correlation between IGF1 levels and weight standard deviation score (SDS) in selected groups, no significant association was found between pretreatment IGF1 levels and growth outcomes (Kim et al. 2021). Notably, in patients with idiopathic GH deficiency (IGHD), a growth response only occurs when the change in IGF1 SDS surpasses a threshold value of 1. Consequently, the limitations of using IGF1 as a sole determinant of GH-based treatment efficacy have been underscored, advocating instead for their role in monitoring treatment compliance in pediatric patients with short stature undergoing GH therapy (Kim et al. 2021).

Many other trials tested the efficacy of IGF1 treatment in several conditions with different outcomes. Only a few trials are listed here: Duchenne Muscular Dystrophy (NCT01207908), hemodialysis Patients (NCT01209403), Amyotrophic Lateral Sclerosis-ALS (NCT00035815), patients with cognitive impairment (NCT02139410), patients with heart attacks (NCT01438086), insulin resistance (NCT00004419), individuals with short stature (NCT02636270), Phelan-McDermid Syndrome (NCT01525901), and advanced tumors expressing IGF1R (NCT02045368). It is noteworthy that IGF1 is also used to treat a spectrum of clinical disorders, confirming that IGF1 action is widespread to several systems in the brain and body.

Table 4.4 below outlines a list of clinical trials utilizing an IGF-1 based therapeutic regimen, alongside the results yielded thus far.

The important role of IGF1 signaling is confirmed from an evolutionary standpoint, since expression of IGF1 and the molecules of its signaling pathway are conserved from flies to humans. This continuity across species highlights the role of IGF1 and IGF2 in activating receptor tyrosine kinases, specifically the IGF1 receptor and the InsR, and mentions the regulatory function of IGF binding proteins. Their overexpression in cancer has thus far been associated with tumor development, acquired resistance to therapies, and metastasis. Moreover, insulin resistance is characterized in the brain by reduced signaling through InsRs and IGF1Rs, and these mechanisms may contribute to neurodegenerative disease development (Rosenzweig 2020).

Despite a myriad of research being conducted on the physiology of IGF1 signaling, there are yet several complexities as part of the pathways that need to be investigated to understand the true clinical potential of IGF1. Studies on levels of IGF binding proteins and receptors during aging would yield an improved analysis of IGF-1 functionality (Zhang et al. 2020). Firstly, a comparison between binding proteins and IGF levels will help assess their extent of control over IGF concentrations since binding proteins modulate IGF1 signaling by blocking its potential binding to its receptor. Additionally, a comparison of IGF1R expression with IGF1 levels would help us evaluate whether receptor upregulation is an effective compensation mechanism to decrease IGF levels with age (Alberti et al. 2011). It is likely that these comparisons will be dependent on the area studies and the developmental stage.

4.12 The Role of IGF1 as Brain-Body Connector

So far, we have examined the role of IGF1 in the body and brain, highlighting the main functions and applications of this molecule. In the brain, IGF1 synthesis occurs in many cell types, and the local synthesis is crucial for neuronal growth, survival, and synaptic plasticity, directly impacting brain function and development. In addition, IGF1 can also function as a supplement in neuronal degeneration and some nervous system dis-

Table 4.4 Clinical trials using IGF1 as therapy

Trial	Aim of study	Outcome
Treatment of Rett Syndrome with recombinant human IGF-1 (rhIGF-1)	Exploring the safeness of IGF-1 therapy for girls with Rett Syndrome, if IGF-1 has any effect on the unstable vital signs, see its effect on their behavior and mood, their cortical function, and changes occurring within the symptoms they experience particularly flushing of the cheeks, cold hands/feet and abnormal respiration.	There was a reduction found in the apnea index compared from pre-MAD to post-OLE. This determines an improved outcome.
Research conducted by Walter E. Kaufmann, MD and sponsored by Boston Children's hospital	There are 2 groups, the open label IGF-1 treatment (OLE) and the Multiple ascending dose (MAD)	The Anxiety Depression and Mood Scale (ADAMS) demonstrated a negative value; hence illustrating an improved outcome.
Publication: Pubmed		All participants experienced side effects.
IGF1 in Amyotrophic lateral sclerosis (ALS)	Discovering if IGF1 can influence ALS patients by slowing down the progressive weakness. The participants are followed for 2 years, half will receive placebo whilst the other half will receive active IGF1.	The IGF-1 group has a higher MMT score than the placebo group, however both groups still have low MMT scores. Similarly, the IGF-1 group responded better to the ALS functional rating scale than the placebo group.
Research conducted by Eric J. Sorenson, MD and sponsored by Mayo clinic		10.78% experienced serious side effects, whereas 61.68% experienced other side effects.
Publication: Pubmed		
Effect of Anorexia Nervosa (AN) on Peak Bone Mass	The study will test the influence of giving rhIGF-1 with estrogen and compare it to giving estrogen alone on bone metabolism in adolescent girls with AN.	When assessed over 12 months via the lumbar spine BMD score, rhIGF-1 given with transdermal estradiol (TE) has a lower score than placebo with TE indicating that TE monotherapy has better outcome, however both demonstrate an unfavorable outcome.
Research conducted by Madhusmita Misra, MD, MPH and sponsored by Massachusetts General Hospital	The aim is to see if the co-administration increases bone mineral density, the bone mass to reach the level found in healthy girls, and is able to imitate regular puberty to enhance bone formation through IGF-1 mediated anabolic effect.	Similarly, TE monotherapy had a better outcome than co-administration of rhIGF-1 with TE, although both displayed an unfavorable outcome.
		18.42% on co-administration and 32.43% on TE monotherapy experienced serious side effects.
		21.05% on co-administration and 51.35% on TE monotherapy experienced other side effects.
Optimize GH Dosing in children with chronic kidney disease (CKD) by measuring IGF-1 levels in the blood	The aim is to measure the quantity of IGF1 produced by the body in response to giving 2 different GH doses in children for 7 days; this results with researchers being able to dose GH based on the amount of IGF1 produced by the body regardless of the child's weight.	The average change in the total amount of IGF1 produced within day 1 and day 8 was found higher in those who got a standard dose of GH rather than those who received a low dose of GH.
Research conducted by Dr. Amira Al-Uzri and sponsored by Oregon Health and Science University	Researchers aspire to find the favorable level of IGF1 produced for decent growth after 7 days of GH in relation to CKD and shorten the overall duration of GH therapy.	The change in height increased in the participants when compared from week 1 to week 56.
Publications: Pubmed	The study will also evaluate the effects of inflammatory cytokine Il-6 and TNF-α on GH insensitivity and therefore IGF-1 generation test in the group of children.	None experienced side effects.

eases due to its neuroprotective functions (Ioachimescu et al. 2019).

Considering that IGF1 is produced in both the brain and body, it is not clear why IGF1 needs to be transported into the brain. One possibility is that IGF1 acts as a connector to coordinate brain function in response to the body's metabolic state. This view was put forward by Dr. Torres Aleman (2010) and fosters the role of IGF1 as a regulator of energy allocation. Since neuronal activity requires energy, it is not surprising that the flow of IGF1 inside the brain is higher in areas that are more active neuronally—and hence metabolically (Nishijima et al. 2010). The action of IGF1 in the brain could be related to the increase of glutamate metabolism, to the regulation of mitochondrial function, and to the activation of mechanisms of cell repair linked to high metabolism (for example, the activation of autophagy—to decrease the risks coming from the generation of ROS).

We propose that there are two reasons that may drive IGF1 transport into the brain: (1) The metabolism in the body is high; hence, the body moves more IGF1 into the brain. (2) The brain activity is high, so IGF1 is requested from the body to stimulate metabolic processes that sustain brain activity. The first option is supported by experiments run in mice, where a rise in IGF1 serum levels induces an increase in the levels of IGF1R. This increase happens in physiological conditions and under pharmacological stimulation (Trueba-Saiz et al. 2017) but it is not always linked to an increase of IGF1 in the brain. Indeed, changes in peripheral IGF1 do not always influence brain levels of IGF1 (Cioana et al. 2020). In the second scenario, the imbalance between IGF1 production and the number of IGF1R expressed in the brain may represent a strategy to gauge the need of brain IGF1 and to modulate IGF1 transport in the brain.

Alternatively, transport mechanisms may exist due to the lack of widespread production of IGF1 in the brain. Although IGF1Rs are widely distributed in the brain, sites of IGF1 synthesis are constricted to areas such as the olfactory bulb, cerebellum, hippocampus, or spinal cord (Torres-Aleman 2019). Thus, there must be measures in place to allow higher levels of IGF1 to serve its vital protective functions. This hypothesis is further supported by the observation that these areas are critically linked to cognition, further strengthening the idea that IGF1 deficiencies may contribute to cognitive impairment.

Another possibility is that IGF1—besides its role in metabolism—is correlated to the brain's different states. Indeed, IGF1 acts similarly to a neuromodulator to control different brain states (fear arousal and others) by interacting with neurotransmitters, hormones, growth factors, and systems that modulate the homeostasis of neuronal excitability and transmission (Trueba Saiz et al. 2013). This complex role fits very well with the role of IGF1 as a brain-body modulator, and it suggests that the transport and synthesis of IGF1 rely not only on the BBB transcytosis but also on hormones, cytokines, microbiota, and other factors. There is important evidence that this is the case (Szalecki et al. 2018; Jensen et al. 2020; Zhao et al. 2019), but there are not many studies reporting how the brain concentration of IGF1 changes its levels in the blood: an increase in the IGF1 peripheral levels was observed when the IGF1R activity was genetically reduced in neurons (De Magalhaes Filho et al. 2017), but the changes of IGF1 levels in the brain do not always induce changes at the peripheral level.

Overall, the view of IGF1 as a brain-body connector allows us to re-evaluate IGF1's function and better interpret the role of IGF1 in pathologies. For example, considering that the transport in the brain is dependent on the transcytosis of IGF1 through the BBB, it is possible that conditions that affect the integrity of the BBB (such as stress and viral infections) could mediate the inefficacy of this transport and lead to IGF1-related pathologies (Fig. 4.9).

Also, the delicate interplay of IGF1 levels in the body and brain may need to be considered in future treatment options. The transport of IGF1 across the BBB suggests that when IGF1 is administered peripherally, a certain amount will also enter the brain. As mentioned earlier, IGF1 is used as a treatment option in diseases such as Donohue's and Rabson-Mendenhall syndromes to milder forms of type A insulin resistance or in

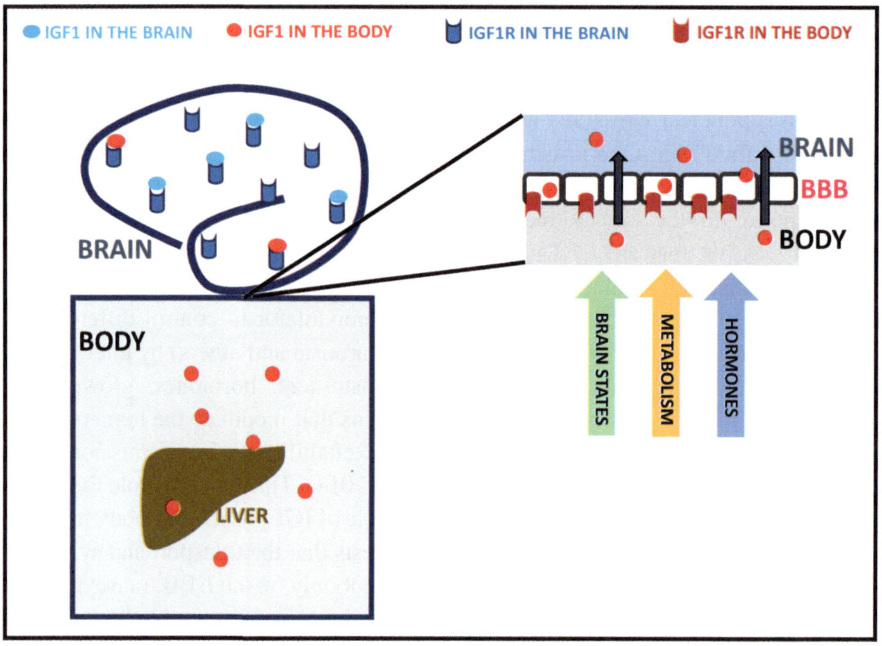

Fig. 4.9 IGF1 as brain-body connector
Schematic representation of the production and transport of IGF1 in body and in brain, including the transcytosis through blood-brain barrier, and the factors influencing the processes

diseases such as Laron's syndrome (Puche and Castilla-Cortázar 2012). Vice versa, treatments that boost the IGF1 levels in the brain for the treatment of neurodevelopmental and neurodegenerative disorders (Kolevzon et al. 2022) may also increase the IGF1 levels in the body. While IGF1 can elicit positive effects in the brain, influencing neurogenesis, synaptic plasticity, and cognitive function, the synthesis of IGF1 within the brain could provide a more immediate and localized response to the brain's needs, while an excess of IGF1 may cause uncontrolled proliferation in the form of neoplasms or alteration in the glycemia levels.

However, it is important to note that IGF1 signaling is strongly regulated, and it is harder to predict the effects of IGF1 administration. In fact, the hypothesis of IGF1 as a homeostatic regulator (Nunez et al. 2023) implies that its effects will be limited in baseline conditions, and considering the importance of IGF1 signaling in the body and brain, it is comprehensible that many feedback systems and checkpoints work simultaneously to assure that the system works properly. In addition, the activation of IGF1 signaling relies on the action of IGFBPs, which are also involved in IGF1's local and trans-compartment transport. Therefore, a thorough understanding of the IGF1 mode of action will require deepening our understanding of IGFBP's function.

Ultimately, IGF's role expands to include many vital aspects of human health, from maintaining glucose metabolism to assisting with the functionality and integrity of the nervous system and regulating various processes in the body, such as growth, homeostasis, bone remodeling, and keeping the cardiovascular and skeletal systems functional. As research progresses, understanding the interplay between circulating and locally synthesized IGF1 as well as its role as a brain-body connector will be key to unlocking new therapeutic strategies for conditions such as Parkinson's disease, cognitive decline, and other neurological disorders, potentially revolutionizing treatment paradigms in neurobiology and beyond.

Acknowledgments This chapter is dedicated to Snaiha, and her great love for learning. This publication has emanated from research supported in part by a research grant from Research Ireland under Grant Number 21/RC/10294_P2 and co-funded under the European Regional Development Fund and by FutureNeuro industry partners.

References

(2023) NHS choices. Available at: https://www.gloshospitals.nhs.uk/about-us/ (Accessed: 15 April 2024)

Adams TE et al (2000) Structure and function of the type 1 insulin-like growth factor receptor. Cell Mol Life Sci 57(7):1050–1093. https://doi.org/10.1007/pl00000744

Alberti C et al (2011) Serum concentrations of insulin-like growth factor (IGF)-1 and IGF binding protein-3 (IGFBP-3), IGF-1/IGFBP-3 ratio, and markers of bone turnover: reference values for French children and adolescents and Z-score comparability with other references. Clin Chem 57(10):1424–1435. https://doi.org/10.1373/clinchem.2011.169466

Aleman A et al (1999) Insulin-like growth factor-I and cognitive function in healthy older men. J Clin Endocrinol Metabol 84(2):471–475. https://doi.org/10.1210/jcem.84.2.5455

Allard JB, Duan C (2018) IGF-binding proteins: why do they exist and why are there so many? Front Endocrinol 9. https://doi.org/10.3389/fendo.2018.00117

Al-Samerria S, Radovick S (2021) The role of insulin-like growth factor-1 (IGF-1) in the control of neuroendocrine regulation of growth. Cells 10(10):2664. https://doi.org/10.3390/cells10102664

Anderson MF et al (2002) Insulin-like growth factor-I and neurogenesis in the adult mammalian brain. Dev Brain Res 134(1–2):115–122. https://doi.org/10.1016/s0165-3806(02)00277-8

Arai Y et al (2009) The metabolic syndrome, IGF-1, and insulin action. Mol Cell Endocrinol 299(1):124–128. https://doi.org/10.1016/j.mce.2008.07.002

Arjunan A, Sah DK, Woo M, Song J (2023) Identification of the molecular mechanism of insulin-like growth factor-1 (IGF-1): a promising therapeutic target for neurodegenerative diseases associated with metabolic syndrome – cell & bioscience. BioMed Central. https://cellandbioscience.biomedcentral.com/articles/10.1186/s13578-023-00966-z

Ashpole NM et al (2015) Growth hormone, insulin-like growth factor-1 and the aging brain. Exp Gerontol 68:76–81. https://doi.org/10.1016/j.exger.2014.10.002

Bach LA (2018) 40 years of IGF1: IGF-binding proteins. J Mol Endocrinol 61(1). https://doi.org/10.1530/jme-17-0254

Bastian SE, Walton PE, Belford DA (1997) Paracellular transport of insulin-like growth factor-I (IGF-I) across human umbilical vein endothelial cell monolayers. J Cell Physiol 170(3):290–298

Bianchi V, Locatelli V, Rizzi L (2017) Neurotrophic and neuroregenerative effects of GH/IGF1. Int J Mol Sci 18(11):2441. https://doi.org/10.3390/ijms18112441

Braicu et al (2019) A comprehensive review on MAPK: a promising therapeutic target in cancer. Cancers 11(10):1618. https://doi.org/10.3390/cancers11101618

Cao Z et al (2023) Circulating insulin-like growth factor-1 and brain health: evidence from 369,711 participants in the UK biobank. Alzheimers Res Ther 15(1). https://doi.org/10.1186/s13195-023-01288-5

Castro J et al (2014) Functional recovery with recombinant human IGF1 treatment in a mouse model of Rett syndrome. Proc Natl Acad Sci 111(27):9941–9946. https://doi.org/10.1073/pnas.1311685111

Chesnokova V (2022) The multiple faces of the GH/igf axis. Cells 11(2):217. https://doi.org/10.3390/cells11020217

Cioana M, Michalski B, Fanhestock M (2020) Insulin-like growth factor and insulin-like growth factor receptor expression in human idiopathic autism fusiform gyrus tissue. Autism Res 13(6):897–907. https://doi.org/10.1002/aur.2291

Clemmons D (2006) Involvement of insulin-like growth factor-i in the control of glucose homeostasis. Curr Opin Pharmacol 6(6):620–625

Clemmons DR, Jones JI, Busby WH, Wright G (1993) Role of insulin-like growth factor binding proteins in modifying IGF actions. Ann N Y Acad Sci. https://pubmed.ncbi.nlm.nih.gov/7692784/

Corvin AP, Molinos I, Little G, Donohue G, Gill M, Morris DW, Tropea D (2012) Insulin-like growth factor 1 (IGF1) and its active peptide (1–3)IGF1 enhance the expression of synaptic markers in neuronal circuits through different cellular mechanisms. Neurosci Lett 520(1):51–56. https://doi.org/10.1016/j.neulet.2012.05.029

De Magalhaes Filho CD, Kappeler L, Dupont J, Solinc J, Villapol S, Denis C, Nosten-Bertrand M, Jean-Marie Billard JM, Blaise A, Tronche F, Giros B, Charriaut-Marlangue C, Saba Aïd S, Le Bouc Y, Holzenberger M (2017) Deleting IGF-1 receptor from forebrain neurons confers neuroprotection during stroke and upregulates endocrine somatotropin. J Cereb Blood Flow Metab 37(2):396–412. https://doi.org/10.1177/0271678X15626718

Delafontaine P, Song Y-H, Li Y (2003) Expression, regulation, and function of IGF-1, IGF-1R, and IGF-1 binding proteins in blood vessels. Arterioscler Thromb Vasc Biol. https://www.ahajournals.org/doi/full/10.1161/01.ATV.0000105902.89459.09

Dixit M et al (2021) Effects of GH/Igf Axis on bone and cartilage. Mol Cell Endocrinol 519:111052. https://doi.org/10.1016/j.mce.2020.111052

Dyer AH et al (2016) The role of insulin-like growth factor 1 (IGF-1) in brain development, maturation and neuroplasticity. Neuroscience 325:89–99. https://doi.org/10.1016/j.neuroscience.2016.03.056

Forbes BE, Blyth AJ, Wit JM (2020) Disorders of igfs and IGF-1R signaling pathways. Mol Cell

Endocrinol 518:111035. https://doi.org/10.1016/j.mce.2020.111035

Gillies GE, McArthur S (2010) Estrogen actions in the brain and the basis for differential action in men and women: a case for sex-specific medicines. Pharmacol Rev. https://www.ncbi.nlm.nih.gov/pmc/articles/PMC2879914/

Giraldi E, Allen JW, loachimescu AG (2022) Pituitary incidentalomas: best practices and looking ahead. Endocr Pract. https://www.sciencedirect.com/science/article/pii/S1530891X22006450. Epub 2022 Oct 18

Grimberg A (2003) Mechanisms by which IGF-I may promote cancer. Cancer Biol Ther 2(6):628–633. https://doi.org/10.4161/cbt.2.6.678

Guan J, Gluckman PD (2009) IGF-1 derived small neuropeptides and analogues: a novel strategy for the development of pharmaceuticals for neurological conditions. Br J Pharmacol. https://www.ncbi.nlm.nih.gov/pmc/articles/PMC2737647/

Guan J, Li F, Kang D, Anderson T, Pitcher T, Dalrymple-Alford J, Shorten P, Singh-Mallah G (2023) Cyclic glycine-proline (CGP) normalises insulin-like growth factor-1 (IGF-1) function: clinical significance in the ageing brain and in age-related neurological conditions. Molecules (Basel, Switzerland). https://www.ncbi.nlm.nih.gov/pmc/articles/PMC9919809/

Gubbi S, Quipildor GF, Barzilai N, Huffman DM, Milman S (2018) 40 years of IGF1: IGF1: the Jekyll and Hyde of the aging brain. jme. https://jme.bioscientifica.com/view/journals/jme/61/1/JME-18-0093.xml#:~:text=Abstract-,The%20insulin%2Dlike%20growth%20factor%201%20(IGF1)%20signaling%20pathway,brain%20remains%20complex%20and%20controversial

Hakuno F, Takahashi S-I (2018) 40 years of IGF1: IGF1 receptor signaling pathways. J Mol Endocrinol 61(1). https://doi.org/10.1530/jme-17-0311

Handbook of Biologically Active Peptides (n.d.) ScienceDirect. Available at: https://www.sciencedirect.com/book/9780123850959/handbook-of-biologically-active-peptides#book-info (Accessed: 13 April 2024)

Insulin-like growth factor (n.d.) University of Rochester Medical Center. Available at: https://www.urmc.rochester.edu/encyclopedia/content.aspx?ContentTypeID=167&ContentID=insulin_like_growth_factor#:~:text=The%20test%20is%20done%20with,in%20your%20arm%20or%20hand (Accessed: 15 April 2024)

Insulin-like growth factor binding proteins (n.d.) Insulin-like growth factor binding proteins – an overview | ScienceDirect Topics. Available at: https://www.sciencedirect.com/topics/agricultural-and-biological-sciences/insulin-like-growth-factor-binding-proteins#:~:text=It%20is%20well%20established%20that,Dattani%20and%20Gevers%2C%202016 (Accessed: 12 April 2024)

Insulin-like growth factor-1 (IGF-1) (2021). Testing.com. Available at: https://www.testing.com/tests/insulin-growth-factor-1-igf-1/ (Accessed: 12 April 2024)

Ioachimescu AG, Handa T, Goswami N, Pappy AL, Veledar E, Oyesiku NM (2019) Gender differences and temporal trends over two decades in acromegaly: a single center study in 112 patients – endocrine. SpringerLink. https://link.springer.com/article/10.1007/s12020-019-02123-4

Janssen JAMJL (2019) IGF-I and the endocrinology of aging. Curr Opin Endocr Metab Res 5:1–6. https://doi.org/10.1016/j.coemr.2018.11.001

Jensen EA, Young JA, Samuel C, Mathes SC, List EO, Carroll RK, Kuhn J, Onusko M, Kopchick JJ, Murphy ER, Berryman DE (2020) Crosstalk between the growth hormone/insulin-like growth factor-1 axis and the gut microbiome: a new frontier for microbial endocrinology. Growth Horm IGF Res 53-54:101333. https://doi.org/10.1016/j.ghir.2020.101333

Józefiak A et al (2021) The IGF-1 signaling pathway in viral infections. Viruses 13(8):1488. https://doi.org/10.3390/v13081488

Junnila RK et al (2013) The GH/IGF-1 axis in ageing and longevity. Nat Rev Endocrinol 9(6):366–376. https://doi.org/10.1038/nrendo.2013.67

Kasuki L, Maia B, Gadelha MR (2022) Acromegaly and colorectal neoplasm: an update. Front Endocrinol 13. https://doi.org/10.3389/fendo.2022.924952

Katsenelson M et al (2022) IGF-1 receptor regulates upward firing rate homeostasis via the mitochondrial calcium uniporter. Proc Natl Acad Sci 119(33). https://doi.org/10.1073/pnas.2121040119

Kavran JM et al (2014) How IGF-1 activates its receptor. elife 3. https://doi.org/10.7554/elife.03772

Kennedy M, Glass L, Glaze DG, Kaminsky S, Percy AK, Neul JL, Jones NE, Tropea D, Horrigan JP, Nues P, Bishop KM, Youakim JM (2024) Development of trofinetide for the treatment of Rett Syndrome: from bench to bedside. Front Pharmacol. https://pubmed.ncbi.nlm.nih.gov/38318312/

Kolevzon A, Breen MS, Siper PM, Halpern D, Frank Y, Rieger H, Weismann J, Trelles MP, Lerman B, Rapaport R, Buxbaum JD (2022) Clinical trial of insulin-like growth factor-1 in Phelan-McDermid syndrome. Mol Autism 13(1):17

Kim H-S (2013) Role of insulin-like growth factor binding protein-3 in glucose and lipid metabolism. Ann Pediatr Endocrinol Metab 18(1):9. https://doi.org/10.6065/apem.2013.18.1.9

Kim JH et al (2021) Factors affecting IGF-I level and correlation with growth response during growth hormone treatment in LG Growth Study. PLoS One 16(7). https://doi.org/10.1371/journal.pone.0252283

Li J et al (2019) Structural basis of the activation of type 1 insulin-like growth factor receptor. Nat Commun 10(1). https://doi.org/10.1038/s41467-019-12564-0

Lin S, Zhang Q, Shao X, Zhang T, Xue C, Shi S, Zhao D, Lin Y (2017) IGF-1 promotes angiogenesis in endothelial cells/adipose-derived stem cells co-culture system with activation of PI3K/AKT Signal pathway. Cell Prolif. Available at: https://pubmed.ncbi.nlm.nih.gov/28960620/ (Accessed: 24 March 2024)

Lu Y et al (2020) Neuron-derived estrogen is critical for astrocyte activation and neuroprotection of the ischemic brain. J Neurosci 40(38):7355–7374. https://doi.org/10.1523/jneurosci.0115-20.2020

McDonald A et al (2007) IGF-I treatment of insulin resistance. Eur J Endocrinol 157(suppl_1). https://doi.org/10.1530/eje-07-0271

Melnik BC, John SM, Schmitz G (2011) Over-stimulation of insulin/IGF-1 signalling by western diet may promote diseases of civilization: lessons learnt from laron syndrome. Nutr Metab (Lond) 8:41. https://doi.org/10.1186/1743-7075-8-41

Mendez P, Azcoitia I, Garcia-Segura LM (2003) Estrogen receptor alpha forms estrogen-dependent multimolecular complexes with insulin-like growth factor receptor and phosphatidylinositol 3-kinase in the adult Rat Brain. Mol Brain Res. https://www.sciencedirect.com/science/article/pii/S0169328X03000883?via=ihub

Nieto-Estévez V, Defterali Ç, Vicario-Abejón C (2016) IGF-I: a key growth factor that regulates neurogenesis and synaptogenesis from embryonic to adult stages of the brain. Front Neurosci. Available at: https://www.ncbi.nlm.nih.gov/pmc/articles/PMC4763060/ (Accessed: 18 April 2024)

Nishijima T, Piriz J, Duflot S, Fernandez AM, Gaitan G, Gomez-Pinedo U, Garcia Verdugo JM, Leroy F, Soya H, Nunez A, Torres-Aleman I (2010) Neuronal activity drives localized blood-brain-barrier transport of serum insulin-like growth factor-I into the CNS. Neuron. https://www.sciencedirect.com/science/article/pii/S0896627310006173

Nuñez A, Zegarra-Valdivia J, Fernandez de Sevilla D, Pignatelli J, Torres Aleman I (2023) The neurobiology of insulin-like growth factor I: from neuroprotection to modulation of brain states. Mol Psychiatry 28(8):3220–3230

Pan W, Kastin AJ (2000) Interactions of IGF-1 with the blood-brain barrier in vivo and in situ. Neuroendocrinology 72(3):171–178. https://doi.org/10.1159/000054584

Paplomata E, O'Regan R (2014) The PI3K/AKT/mtor pathway in breast cancer: targets, trials and biomarkers. Ther Adv Med Oncol 6(4):154–166. https://doi.org/10.1177/1758834014530023

Puche JE, Castilla-Cortázar I (2012) Human conditions of insulin-like growth factor-i (IGF-I) deficiency. J Transl Med 10(1). https://doi.org/10.1186/1479-5876-10-224

Réthelyi JM et al (2023) The role of insulin/IGF1 signalling in neurodevelopmental and neuropsychiatric disorders – evidence from human neuronal cell models. Neurosci Biobehav Rev 153:105330. https://doi.org/10.1016/j.neubiorev.2023.105330

Riikonen R (2017) Insulin-like growth factors in the pathogenesis of neurological diseases in children. Int J Mol Sci 18(10):2056. https://doi.org/10.3390/ijms18102056

Romero CJ et al (2012) Insulin-like growth factor 1 mediates negative feedback to somatotroph GH expression via POU1F1/CREB binding protein interactions. Mol Cell Biol 32(21):4258–4269. https://doi.org/10.1128/mcb.00171-12

Rosenzweig SA (2020) The continuing evolution of insulin-like growth factor signaling. F1000Research 9:205. https://doi.org/10.12688/f1000research.22198.1

Sohrabji F (2015) Estrogen-IGF-1 interactions in neuroprotection: ischemic stroke as a case study. Front Neuroendocrinol. https://www.ncbi.nlm.nih.gov/pmc/articles/PMC4247812/

Span JP et al (2000) Gender difference in insulin-like growth factor I response to growth hormone (GH) treatment in GH-deficient adults: role of sex hormone replacement. J Clin Endocrinol Metabol 85(3):1121–1125. https://doi.org/10.1210/jcem.85.3.6463

Suh H-S, Zhao M-L, Derico L, Choi N, Lee SC (2013) Insulin-like growth factor 1 and 2 (IGF1, IGF2) expression in human microglia: differential regulation by inflammatory mediators. J Neuroinflammation. BioMed Central. https://jneuroinflammation.biomedcentral.com/articles/10.1186/1742-2094-10-37

Szalecki M, Malinowska A, Prokop-Piotrkowska M, Janas R (2018) Interactions between the growth hormone and cytokines – a review. Adv Med Sci 63(2):285–289. https://doi.org/10.1016/j.advms.2018.03.001

Torres-Aleman A (2010) Toward a comprehensive neurobiology of IGF1. Dev Neurobiol 7:384–396. https://doi.org/10.1002/dneu.20778

Torres-Aleman I (2019) Serum growth factors and neuroprotective surveillance. Mol Neurobiol. SpringerLink. https://link.springer.com/article/10.1385/MN:21:3:153

Tropea D (2024) Trofinetide treatment for Rett syndrome: lessons to learn. Med 5(10):1194–1196. https://doi.org/10.1016/j.medj.2024.07.028

Tropea D, Giacometti E, Wilson N, Beard C, McCurry C, Fu DD, Flannery R, Jaenisch R, Sur M (2009) Partial reversal of rett syndrome-like symptoms in MeCP2 mutant mice. PNAS 106 (6):2029–2034

Trueba-Saiz A, Fernandez AM, Nishijima T, Mecha M, Santi AA, Munive V (2017) Circulating insulin-like growth factor I regulates its receptor in the brain of male mice. Endocrinology 158(2):349–355. https://doi.org/10.1210/en.2016-1468

Trueba-Sáiz A, Cavada C, Fernandez AM, Leon T, González DA, Fortea Ormaechea J, Lleó A, Del Ser T, Nuñez A, Torres-Aleman I (2013) Loss of serum IGF-I input to the brain as an early biomarker of disease onset in Alzheimer mice. Transl Psychiatry 3(12):e330

Varma Shrivastav S et al (2020) Insulin-like growth factor binding protein-3 (IGFBP-3): unraveling the role in mediating IGF-independent effects within the cell. Front Cell Dev Biol 8. https://doi.org/10.3389/fcell.2020.00286

Vitale G, Pellegrino G, Vollery M, Hofland LJ (2019) Role of IGF-1 system in the modulation of longevity: controversies and new insights from a centenarians' perspective. Front Endocrinol. https://www.frontiersin.org/journals/endocrinology/articles/10.3389/fendo.2019.00027/full

Werner H (2023) The IGF1 signaling pathway: from basic concepts to therapeutic opportunities. Int J Mol Sci 24(19):14882. https://doi.org/10.3390/ijms241914882

Wrigley S, Arafa D, Tropea D (2017) Insulin-like growth factor 1: at the crossroads of brain development and aging. Front Cell Neurosci 11. https://doi.org/10.3389/fncel.2017.00014

Xiu M, Huan X, Ou Y, Ying S, Wang J (2021) The basic route of nuclear-targeted transport of IGF-1/IGF-1R and potential biological functions in intestinal epithelial cells. Cell Prolif 54(6):e13030

Yang D, Caraty A, Dupont J (2005) Molecular mechanisms involved in LH release by the ovine pituitary cells. Domest Anim Endocrinol 29(3):488–507

Zhang WB et al (2020) Insulin-like growth factor-1 and IGF binding proteins predict all-cause mortality and morbidity in older adults. Cells 9(6):1368. https://doi.org/10.3390/cells9061368

Zhao YL, Wu J, Zhang TP, Cheng QY, Wang WP, Gu MM, Pan HF, Ye DQ (2019) Circulating insulin-like growth factor-1 levels in patients with rheumatoid arthritis: a meta-analysis. Curr Pharm Des 25(10):1091–1098. https://doi.org/10.2174/138161282566619031912400 9

Zhou J et al (2015) Crosstalk between MAPK/ERK and PI3K/AKT signal pathways during brain ischemia/reperfusion. ASN Neuro 7(5):175909141560246. https://doi.org/10.1177/1759091415602463

Neuronal Synaptic Communication and Mitochondrial Energetics in Human Health and Disease

5

Jill R. Glausier

Abstract

The human brain is an energetically costly organ, consuming 20–25% of all biochemical energy produced in the body, despite comprising only 2–3% of total body mass. Most energy in the brain is consumed to support synaptic neurotransmission, which is the primary means of neuron-to-neuron communication between. This energy is in the form of adenosine triphosphate (ATP), and within neural cells, nearly all ATP is produced by mitochondria via the process of oxidative phosphorylation (OXPHOS). To ensure that ATP is readily available, mitochondria are trafficked to areas of greater energy use, such as neuronal synapses. The balance between energetic supply and synaptic communication is essential for proper brain functioning. This chapter begins with a brief introduction to key features of neuronal synapses and mitochondrial energy production in the cerebral cortex. Next, the tight and bidirectional coupling of neuronal synaptic activity and mitochondrial OXPHOS is examined from functional, ultrastructural, and molecular perspectives. The effects of brain and non-brain organ system perturbations on synaptic-mitochondrial coupling are then examined within the context of (1) primary brain disorders, such as schizophrenia and bipolar disorder, and (2) primary peripheral disorders, such as diabetes and obesity. Finally, a discussion of potential intervention strategies that may restore neural communication and mitochondrial bioenergetics, within the framework of the brain-body connection, is provided.

Keywords

Prefrontal cortex · Electron transport chain · Brain bioenergetics · Schizophrenia · Bipolar disorder · Glucose regulation

Abbreviations

ATP	Adenosine triphosphate
ANLS	Astrocyte-neuron lactate shuttle
COX	Cytochrome c oxidase
DLPFC	Dorsolateral prefrontal cortex
EEG	Electroencephalogram
EM	Electron microscopy
ETC	Electron transport chain
fMRI	Functional magnetic resonance imaging
GABA	Gamma-aminobutyric acid
GWAS	Genome-wide association studies
GLUT1	Glucose transporter 1

J. R. Glausier (✉)
Department of Psychiatry, University of Pittsburgh School of Medicine, Pittsburgh, PA, USA
e-mail: glausierjr@upmc.edu

D. Tropea, E. Giacometti (eds.), *Brain-Body Connections*, Advances in Experimental Medicine and Biology 1477, https://doi.org/10.1007/978-3-031-89525-8_5

GLUT3	Glucose transporter 3
MRS	Magnetic resonance spectroscopy
mtDNA	Mitochondrial DNA
NALS	Neuron-astrocyte lactate shuttle
nDNA	Nuclear DNA
OXPHOS	Oxidative phosphorylation
PV	Parvalbumin
Phe	Phenylalanine
PKU	Phenylketonuria
PET	Positron emission tomography
PSD	Postsynaptic density
PFC	Prefrontal cortex
ROS	Reactive oxygen species
Redox	Reduction/oxidation
WGCNA	Weighted gene co-expression analysis

5.1 Introduction

The adult human brain contains approximately 86 billion neurons (Herculano-Houzel 2009; von Bartheld et al. 2016; Andrade-Moraes et al. 2013; Goriely 2024), and the coordinated communication between these neurons underlies the emergent abilities of the brain such as consciousness, perception, learning, and memory. Communication between neurons primarily occurs via chemical synaptic neurotransmission at discrete morphological sites termed synapses (Shepherd and Erulkar 1997). In its most elemental form, a synapse is comprised of three components (Palay 1964; Peters et al. 1991): (1) presynaptic axonal bouton which contains synaptic vesicles filled with neurotransmitter molecules such as glutamate or GABA, (2) postsynaptic structure of the recipient neuron, such as a dendritic spine, which contains neurotransmitter receptors and signal transduction proteins, and (3) synaptic cleft, which is the small (~20 nm) space separating the pre- and postsynaptic structures that are directly apposed to each other (Fig. 5.1). In the cerebral cortex, which represents roughly 80% of total brain mass in the adult human (Herculano-Houzel 2009), each of the approximate 16 billion cortical neurons receives thousands to tens of thousands of synapses (Wildenberg et al. 2021, 2023). Approximately 75–80% of cortical neurons use the excitatory neurotransmitter glutamate, and the remaining 20–25% are interneurons that use the inhibitory neurotransmitter gamma-aminobutyric acid (GABA) (Hendry et al. 1987; Jones 1993). Across cortical regions, approximately 80–90% of synapses are glutamatergic, 10–20% are GABAergic, and <10% of synapses represent other neurotransmitter systems such as dopamine, serotonin, or neurepinephrine (DeFelipe et al. 1999, 2002; Peters et al. 2008). A complex choreography of neuronal organelles, ion channels, neurotransmitter receptors, effector proteins, and cytoskeletal elements, as well as molecular reuptake and shuttling functions by surrounding astrocytic processes, is required at synapses for effective communication between neurons. Thus, it comes as no surprise that synaptic signaling represents the major energy-consuming process in the brain (Attwell and Iadecola 2002; Attwell and Laughlin 2001; Harris et al. 2012; Howarth et al. 2012; Lezmy et al. 2021; Sibson et al. 1998a, b; Hyder et al. 2006, 2013a). Indeed, despite comprising only 2–3% of total body mass, brain work uses 20–25% of all biochemical energy produced in the body (Aiello and Wheeler 1995).

Synaptic energy is provided in the form of adenosine triphosphate (ATP) molecules, and nearly all synaptic ATP is generated within mitochondria (Attwell and Laughlin 2001; Harris et al. 2012; Howarth et al. 2012; Rangaraju et al. 2014; Ivannikov et al. 2013; Hall et al. 2012; Ma et al. 2009; Verstreken et al. 2005; Li et al. 2020; Seager et al. 2020). Though truly multi-functional organelles (references herein and see other chapters), the most vital function of mitochondria is to aerobically produce the ATP that fuels nearly all neuronal functions. It is this ability that bestows the ubiquitous "powerhouse of the cell" moniker (Siekevitz 1957). Mitochondrial production of ATP represents a critical 'brain-body' interaction since aerobic respiration requires oxygen and nutrients (e.g., glucose) to be delivered via blood flow through the brain's capillary network (Clarke and Sokoloff 1999). Indeed, as exemplified by hypoglycemic states, a perfectly healthy brain can quickly begin to malfunction if there is

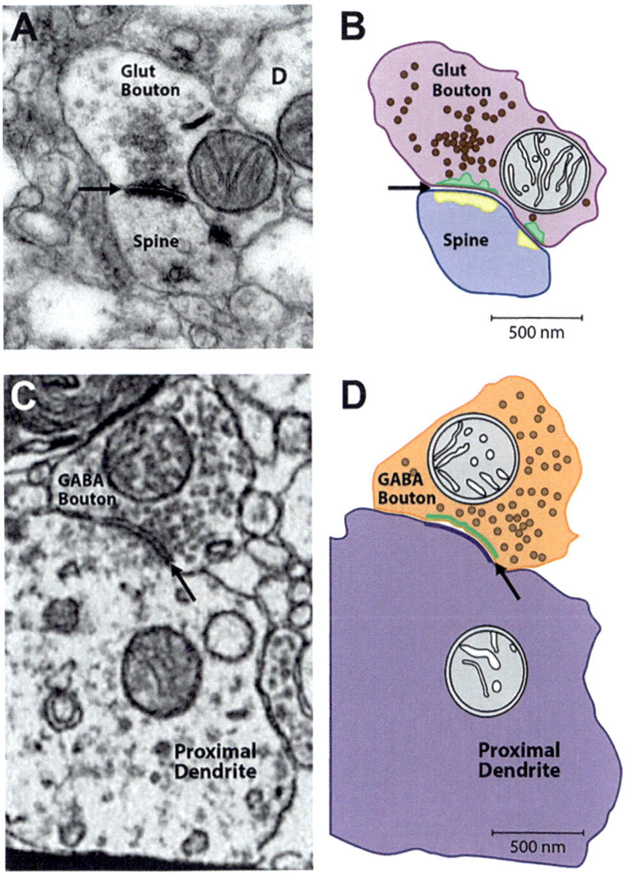

Fig. 5.1 Synaptic structures. (**a**) Electron micrograph illustrating the core features of a glutamatergic synapse in the brain. (**b**) Schematic drawing of the glutamatergic synaptic structure shown in panel A. The glutamatergic axonal bouton (pink) containing synaptic vesicles (brown), a mitochondrion (grey) and an active zone (green) forms a Type 1, "asymmetric" synapse onto a dendritic spine (blue) containing a perforated PSD. The synaptic cleft is visible (black arrow). (**c**) Electron micrograph illustrating the core features of a GABAergic synapse in the brain. (**d**) Schematic drawing of the GABAergic synaptic structure shown in C. The GABAergic axonal bouton (orange) containing synaptic vesicles (light brown), a mitochondrion (grey) and a thin active zone (green) forms a Type 2, "symmetric" synapse onto a proximal dendritic segment (purple). No pronounced PSD is present; only a slight thickening of the postsynaptic membrane. The synaptic cleft is visible (black arrow), and the postsynaptic dendrite contains a mitochondrion and other organelles in the cytoplasm

an insufficient supply of the glucose energy substrate within blood (Cryer 2007).

The primary source of glucose in the body is from food and drink; complex carbohydrates are converted by several organ systems into the simple sugar glucose which is released into the bloodstream. Glucose molecules are transported across the blood-brain barrier into the interstitial space, where they are available for transport into neurons primarily via glucose transporter 3 (GLUT3), and to a lesser extent, GLUT1 (Shah et al. 2012; Sullivan et al. 2018a; Maher et al. 1991). The process of aerobic respiration ultimately converts a single glucose molecule into 30–32 molecules of ATP. These ATP molecules are produced during the final step of aerobic respiration: oxidative phosphorylation (OXPHOS). The process of OXPHOS occurs across five protein complexes, collectively known as the electron transport chain (ETC), that are embedded within the inner mitochondrial membrane folds termed "cristae" (Fig. 5.2). Thus, OXPHOS

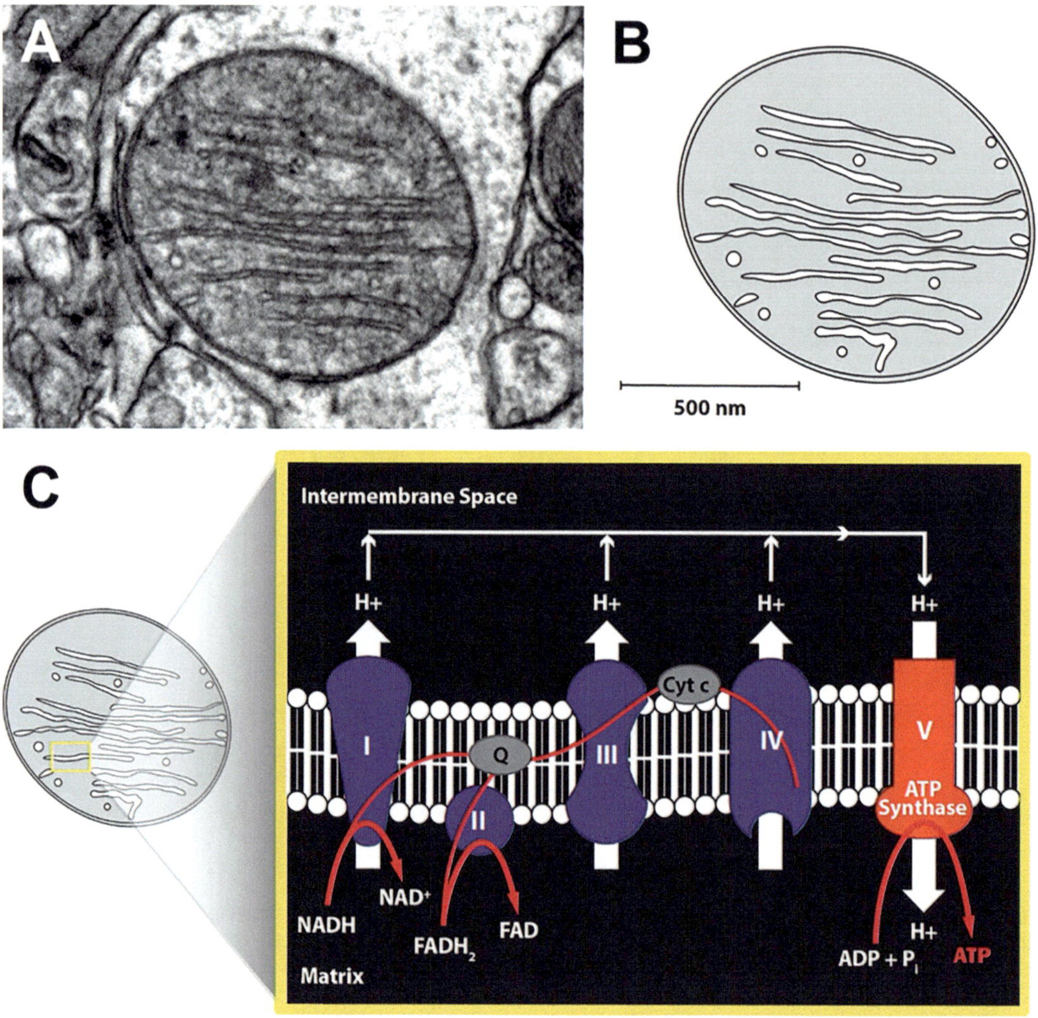

Fig. 5.2 OXPHOS in mitochondrial cristae. (**a**) Single mitochondrion in a neuronal dendritic segment in the human brain. Note the mitochondrial-unique double membrane, organized cristae and homogeneous, relatively electron-dense, matrix. (**b**) Schematic drawing of the mitochondrion in A. (**c**) Schematic drawing of the relative positioning of Complexes I–V of the electron transport chain (ETC), and the order of processes, to produce ATP via oxidative phosphorylation. NADH (nicotinamide adenine dinucleotide) and FADH2 (flavin adenine dinucleotide) are produced by the Krebs cycle, which exclusively occurs within in the mitochondrial matrix. NADH and FADH2 are the substrates for Complexes I and II, respectively. Coenzyme Q (Q) carries the electrons harvested from these substrates to Complex III, which catalyzes the transfer of electrons to cytochrome c (cyt c). Complex IV is the final enzyme that transfers electrons, oxidizing the reduced cyt c created by Complex III. At this point in the process, sufficient proton (H+) motive force has accumulated to drive Complex V to synthesize ATP

exclusively occurs within mitochondria (Erecinska and Wilson 1982; Wainio 1970). Reduction/oxidation (redox) reactions occur sequentially across complexes I–IV, creating an electrochemical gradient across the inner mitochondrial membrane (i.e., from the intermembrane space to the matrix). This gradient ultimately directs protons back into the matrix through complex V, driving ATP synthesis. ATP molecules are exported from the matrix into the mitochondrial intermembrane space via the ADP/ATP carrier protein, where ATP can then diffuse out into the cellular cytosol. Once in the cytosol, ATP is available to fuel energy-dependent processes. In neurons, most ATP is consumed to support synaptic communication sub-processes such as synaptic vesicle loading, neurotransmitter reuptake, and presynaptic and postsynaptic ion pumping across neuronal membranes (Attwell and Iadecola 2002; Attwell and Laughlin 2001; Harris et al. 2012; Howarth et al. 2012; Lezmy et al. 2021; Sibson et al. 1998a, b; Hyder et al. 2006, 2013a).

5.2 Synaptic Communication and Mitochondrial OXPHOS Are Tightly Coupled

The highly differentiated and complex morphology of neurons, some with axonal and dendritic arbors that can extend distances greater than 600 mm, necessitates that mitochondria be efficiently mobilized to areas of energy consumption. Further, the abundance of synapses that an individual neuron makes and receives generates a substantial need for localized ATP production throughout neuronal dendritic and axonal arbors (Misgeld and Schwarz 2017). In order to properly power synaptic communication and related processes, mitochondria are preferentially trafficked to these areas of greater energy consumption (MacAskill et al. 2010; López-Doménech and Kittler 2023). Thus, a dynamic relationship that is coordinated in a spatiotemporal manner exists between neuronal synaptic activity and mitochondrial energetics. Findings from functional, ultrastructural, and molecular studies provide convergent support identifying this relationship as a core feature of neural communication that is conserved across mammalian species.

5.2.1 Functional Studies

Studies evaluating the relationship between neuronal communication and mitochondrial OXPHOS span multiple levels of anatomical resolution, from functional in vivo imaging of whole brain to electrophysiological measures of individual synapses. Oxidative ATP production has a strong linear relationship with measures indexing neurotransmitter release and neuronal firing, and this relationship is evidenced across multiple levels of anatomical resolution, species and brain regions, in vitro systems, and as measured by different techniques (Sibson et al. 1998a, b; Hyder et al. 2006, 2013a; Rothman et al. 2019; Kida and Hyder 2006). The functional relationship between neuroenergetics and synaptic signaling has been an active area of research for more than half a century and is comprehensively reviewed in (Harris et al. 2012; Hyder et al. 2006; Rothman et al. 2011, 2019; Shulman et al. 2004). Below, I provide a brief overview of the literature and recent developments.

5.2.1.1 In Vivo Whole Brain Imaging

A linear relationship between the levels of cortical glutamate/GABA neuron synaptic neurotransmission and oxidative ATP production was initially identified in rat brain using ^{13}C magnetic resonance spectroscopy (MRS); these studies also estimated that ~80% of all neuronal ATP production fuels processes associated with synaptic neurotransmission (Sibson et al. 1997, 1998a, b). Subsequent studies utilizing different 1) techniques, such as functional magnetic resonance imaging (fMRI) and positron emission tomography (PET); 2) species, such as the tree shew and humans; and 3) functional states, such as anesthetized or with sensory stimulation, provide convergent support of this finding (reviewed in Rothman et al. 2011). Computational studies incorporating in vivo, in vitro, and ex vivo data have further informed the relationship between synaptic activity and oxidative ATP production (see Sect. 5.2.1.3). However, data interpretations remain consistent with initial studies: measures indexing synaptic neurotransmission and neuronal firing and OXPHOS have a strong relationship (Sibson et al. 1998a, b; Hyder et al. 2013a) such that persistent reductions in neuronal firing and synaptic signaling lower mitochondrial OXPHOS and vice versa (Harris et al. 2012; Hall et al. 2012; Wong-Riley 2012).

5.2.1.2 In Vitro Energetic Measures by Functional Assays and Light Microscopy

Studies in brain slice preparations from experimental animal models permit precise manipulation of multiple factors that influence the functional relationship between mitochondrial OXPHOS and synaptic neurotransmission. The dependence of synaptic signaling on mitochondrial OXPHOS has been demonstrated in a multitude of biological systems and technical approaches (for examples, see (Harris et al. 2012; Howarth et al. 2012; Rangaraju et al. 2014; Hall et al. 2012; Pathak et al. 2015; Sobieski et al. 2017; Yellen 2018; Sun et al. 2013; Díaz-García and Yellen 2019)). In a seminal study, Hall et al. demonstrated in rat hippocampal slices that synaptic signaling is principally powered by OXPHOS (Hall et al. 2012), though glycolysis certainly has a role in synaptic signaling (Díaz-García and Yellen 2019; Li et al. 2023). Studies of high frequency oscillations, particularly those in the gamma (30-90 Hz) range, further support the tight coupling of synaptic signaling and mitochondria OXPHOS (Kann 2011, 2016; Kann et al. 2011, 2014, 2016; Kann and Kovacs 2007). In order for neural circuits to maintain these oscillations (which are critical for proper cognitive functioning (Miller et al. 2018)), mitochondrial OXPHOS is required to meet the substantial ATP consumption.

5.2.1.3 Mathematical Modeling Incorporating In Vivo and In Vitro Findings

Studies which incorporate data from multiple modalities and species have modeled and estimated neuronal energy usage (Hyder et al. 2006, 2013a, b; Lennie 2003; Levy and Calvert 2021; Yu et al. 2018; Herculano-Houzel and Rothman 2022). Generally, these studies find similar relative cerebral energetic values in rats and human and non-human primates. For example, in both rats and humans, a linear relationship is present between neural firing and oxidative ATP synthesis (Hyder et al. 2013a), and a calculated 70–80% of that ATP is used for synaptic neurotransmission. A recent study incorporated existing histological cytoarchitectonic maps and estimates of synaptic densities along with PET imaging data to interrogate the relationship between oxidative energy production and synaptic neurotransmission across different cortical and sub-cortical regions in humans (Yu et al. 2023). When incorporating data from these multiple modalities and levels of anatomical resolution, the authors calculate that ~70% of oxidative energy synthesis in the human cortex is devoted to neural signaling, with some cortical regions that act as "information hubs" devoting even more to synaptic signaling. Finally, calculations that parse neural communication (e.g., passage of information at a single synapse) and neural computation (e.g., integration and transformation of that information by a neuron) find that communication utilizes 35 times more ATP than does computation in the human cortex (Levy and Calvert 2021; Balasubramanian 2021).

5.2.2 Ultrastructural Studies

As described in the Introduction section, the morphological substrate of neural communication is the synapse, and the morphological substrate of OXPHOS is the ETC, which is exclusively located in mitochondrial cristae. Direct analysis of individual synapses and mitochondria can only be achieved with electron microscopy (EM) approaches, which provide an ultrastructural level of resolution (~1–20 nm). A rich literature exists demonstrating the interdependent nature of ultrastructure and function for synapses and mitochondria (Li et al. 2020; Ovtscharoff et al. 2008; Matsuzaki et al. 2004; Sun et al. 2021; Nusser et al. 1997, 1998; Chen et al. 2020; Schikorski and Stevens 2001, 1997; Yuste and Bonhoeffer 2001; Glebov et al. 2016, 2017; Arellano et al. 2007; Meyer et al. 2014; Holler et al. 2021; Rollenhagen et al. 2018; Bosch et al. 2014; Murthy et al. 2001; Darcy et al. 2006; Turner et al. 2022; Becker et al. 2008; Bourne et al. 2013; Bourne and Harris 2011, 2012; Holtmaat and Svoboda 2009; Propst and Ko 1987; Clarke et al. 2012; Ackermann et al. 2015; Sudhof 2004; Bartol et al. 2015; Glancy et al.

2020; Devine and Kittler 2018; Justs et al. 2022; Li and Sheng 2021; Rossi and Pekkurnaz 2019; Mendelsohn et al. 2022; Hackenbrock 1981; Scalettar et al. 1991; Cserép et al. 2018; Kubota and Kawaguchi 2000; Kubota et al. 2015; Glausier et al. 2024; Hintze et al. 2024). For example, in glutamatergic synapses (e.g., Fig. 5.1a, b), postsynaptic density (PSD) size and excitatory postsynaptic potential amplitude are strongly correlated across individual cortical synapses (Holler et al. 2021), as is the presynaptic glutamate release probability and PSD size (Holler et al. 2021; Harris and Weinberg 2012). Similarly, in GABA synapses, synaptic size and amount of GABA receptors are strongly correlated (Nusser et al. 1997), and axonal bouton size is greatest in GABA neurons that are highly active (Kubota and Kawaguchi 2000; Kubota et al. 2015). In mitochondria, a greater relative level of ATP production is associated with a globular mitochondrial shape and larger cristae:matrix volume ratio, whereas lower levels of ATP production are associated with an elongated mitochondrial shape and smaller cristae:matrix volume (Mendelsohn et al. 2022; Cserép et al. 2018; Perkins et al. 1997, 2003; Hyder et al. 2013a; Matsuzaki et al. 2004). Indeed, mitochondrial cristae are highly dynamic, expanding and contracting with increasing and decreasing ATP production (Ovtscharoff et al. 2008). When energy utilization is high, mitochondria may also exhibit a "condensed" morphology, whereas when energy utilization is low, mitochondria switch to an "orthodox" conformation (Fig. 5.3) (Hackenbrock 1972a, b, 1981). Indeed, mitochondria in a condensed conformation can produce up to 15 times more ATP than orthodox mitochondria (Sun et al. 2021).

The interdependent nature of synaptic activity and oxidative energy production is also reflected in ultrastructural features across neuronal populations and individual synapses. In neurons, mitochondria are trafficked to presynaptic boutons and postsynaptic dendritic shafts in order to provide targeted ATP production to support synaptic signaling (Herculano-Houzel and Rothman 2022; Yu et al. 2023; Balasubramanian 2021). Highly active neurons, such as glutamatergic pyramidal neurons and parvalbumin (PV)-containing GABA neurons, have more synaptic mitochondria than less active neuronal populations (Ivannikov et al. 2013). Accordingly, the presence of synaptic mitochondria predict larger bouton size and greater abundance of presynaptic vesicles, as well as a correspondingly larger PSD (Justs et al. 2022; Glausier et al. 2024; Hintze et al. 2024). Synapses with few, or no, mitochondria have smaller boutons, fewer vesicles, and correspondingly smaller PSDs (Cserép et al. 2018). Consistent with these findings, experimental manipulations that induce long-term blockage of action potentials and synaptic signaling reduce mitochondrial size, number and/or amount of ETC complexes within mitochondria, as shown by EM (Nie and Wong-Riley 1995, 1996a, b; Rodriguez-Moreno et al. 2018; Wong-Riley and Carroll 1984a, b; Wong-Riley 1989; Wong-Riley et al. 1978, 1989, 1994; Wong-Riley and Norton 1988; Wong-Riley and Welt 1980). Together, these findings support that synaptic and mitochondrial ultrastructures are coordinately regulated at the level of individual synapses, and their ultrastructural features reflect relative levels of ATP production and synaptic activation and plasticity.

5.2.3 Molecular Studies

OXPHOS and synaptic neurotransmission are also coupled at the transcriptional (mRNA) and translational (protein) levels. Cells, including neurons, normally make coordinated adjustments in the expression of OXPHOS-related molecules, especially the subunits that comprise complexes I-V of the ETC, to meet changes in energy usage (Wong-Riley 2012; van Waveren and Moraes 2008; Devin and Rigoulet 2007; Williams et al. 1987; Barshad et al. 2018). Indeed, expression of ETC genes is highly correlated within a variety of cell types and tissues (van Waveren and Moraes 2008; Barshad et al. 2018; Garbian et al. 2010). This relationship reflects the fact that the coordinated expression and cooperative activity of ETC components is required for ATP synthesis via OXPHOS within mitochondria (Harris et al. 2012; Hall et al. 2012; Whittaker et al. 2011).

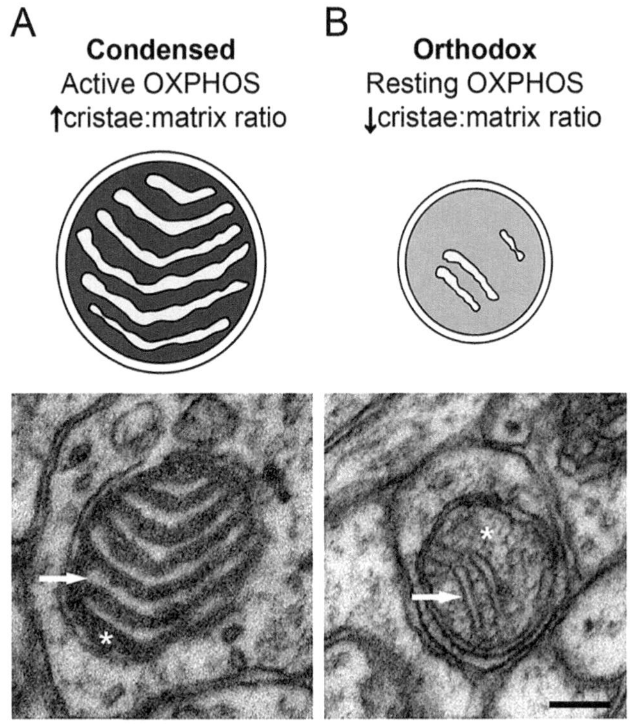

Fig. 5.3 Mitochondrial conformations reflect relative ATP production via OXPHOS. Schematic drawings and electron micrographs of mitochondria in monkey cortex. (**a**) During high ATP production, mitochondria are in a condensed conformation, characterized by an electron-dense, compacted matrix (asterisk), and larger, open cristae (arrow). (**b**) During lower ATP production, mitochondria convert to an orthodox conformation, characterized by a less electron-dense and larger matrix (asterisk), and smaller, fewer cristae (arrow). Scale bar is 100 nm

The coordinated relationship between ETC transcription/translation and neural firing is exemplified by ETC complex IV, also called cytochrome C oxidase (COX). COX, the terminal and rate-limiting OXPHOS enzyme, is a 13-subunit enzyme complex (Miller et al. 2018). COX subunits 1–3 form the catalytic core and are encoded by mitochondrial DNA (mtDNA). The remaining 10 structural subunits are encoded by nuclear DNA (nDNA) and are required for proper COX functioning (Lennie 2003; Levy and Calvert 2021). Expression of COX subunits and mitochondrial morphology are all regulated by neuronal signaling (Hyder et al. 2013b). Quantifying COX enzyme activity or subunit expression is an established and reliable method to detect relative changes or cell type-specific differences within neuronal activity (Hyder et al. 2013b; Yu et al. 2018). Indeed, the transcription and translation of all 13 COX subunits are tightly controlled by neuronal firing (Hyder et al. 2013b). For example, chronically depolarizing primary cultured neurons cause a significant increase in all COX subunit mRNAs (Aiello and Wheeler 1995). Chronic reductions in cortical excitation by tetrodotoxin or silencing thalamic afferents cause a significant decrease in all COX mRNAs as well as COX enzyme activity (Aiello and Wheeler 1995; Ivannikov et al. 2013; Hall et al. 2012; Ma et al. 2009; Verstreken et al. 2005; Li et al. 2020; Seager et al. 2020; Siekevitz 1957).

5.3 The Role of Astrocytes and Lactate in Regulating Neural Communication and Bioenergetics

Although this chapter focuses on the relationship between neuronal oxidative bioenergetics and synaptic signaling, non-neuronal cells can also influence these processes. For example, *Chap. 1 of this book* describes the importance of blood-brain barrier cells (endothelial cells, pericytes, astrocytes, and basement membrane cells) to brain functioning and bioenergetics. One of these cells, astrocytes, also have a notable role in maintaining proper synaptic and energetic operations. Astrocytes are a specialized type of glial cell whose extensive processes provide structural, molecular, and nutritive support to neurons, particularly at synaptic sites (Chung et al. 2015). A substantial literature exists regarding the influence of astrocytes on synaptic signaling and mitochondrial OXPHOS in human health and disease (e.g., see reviews (Lee et al. 2022; Mulica et al. 2021)). However, a topic that remains hotly debated is the existence and function of lactate shuttling between neurons and astrocytes to support neuronal energetic needs during synaptic signaling. Support exists for (1) an astrocyte-neuron lactate shuttle (Pellerin and Magistretti 1994) (ANLS), (2) a neuron-astrocyte lactate shuttle (Mangia et al. 2009) (NALS), (3) the directionality of lactate shuttling depending upon neural signaling dynamics (Patsatzis et al. 2019), and (4) lactate contributions as energetic fuel being modest irrespective of the directionality (DiNuzzo et al. 2010). Detailed discussion is outside the scope of this review, but the mechanics and utility of a lactate shuttle is defended or questioned in many review articles that provide comprehensive and rigorous assessments (e.g (Mangia et al. 2009; 2011; Dienel 2012; Kann 2024; Tang 2018)).

5.4 Synaptic Communication and Neuroenergetics in Brain Disorders

5.4.1 Schizophrenia

Schizophrenia is a complex psychiatric disorder whose clinical features can be divided into three domains: positive, negative, and cognitive. Positive symptoms are experiences that occur in addition to our shared reality, such as hallucinations (i.e., sensory experiences in the absence of a stimulus), delusions (i.e., fixed and false personal beliefs that do not change even when presented evidence to the contrary), and thought disorder (i.e., disorganized thought processes that cause abnormal and confused communication) (American Psychiatric Association 2013). Negative symptoms reflect a loss of feelings and emotions common to most people, such as diminished emotional expression (i.e., flattened affect), inability to derive pleasure from activities that were previously enjoyable (i.e., anhedonia), poverty of speech (i.e., alogia), and inability to initiate and persist in goal-directed activities (i.e., avolition) (American Psychiatric Association 2013). Finally, cognitive symptoms reflect impairments in core cognitive processes such as cognitive control (i.e., ability to adjust thoughts or behaviors to achieve goals) and working memory (i.e., ability to maintain information over a short period of time to guide future thoughts and behaviors) (American Psychiatric Association 2013; Green and Harvey 2014). Approximately 0.3–1.0% of adults in the world are afflicted with schizophrenia, and this is associated with a substantial economic, societal, and individual burden (American Psychiatric Association 2013; Solmi et al. 2023; Jin and Mosweu 2017; Chong et al. 2016).

Although positive and negative symptoms may be more salient at the initial presentation of the disorder, the severity of cognitive symptoms is the best predictor of important functional outcomes such as relapse frequency, social integration, and employment (Green and Harvey 2014; Green 2016; McCutcheon et al. 2023). However, current pharmacotherapies address positive, and to a lesser degree negative (typically a combination of antipsychotic and antidepressant medications), symptoms (Puranen et al. 2023; Schneider-Thoma et al. 2022); there are no currently available therapeutics that address cognitive dysfunction in schizophrenia. Consequently, it is critical to identify the pathophysiology underlying cognitive dysfunction in individuals with schizophrenia in order to identify novel therapeutic targets.

The etiology of schizophrenia is unknown (Tandon et al. 2024), but findings from large-scale genome-wide association studies (GWAS) clearly demonstrate that schizophrenia is a disease originating in the brain. These studies also show that risk-conferring loci converge on fundamental biological processes related to neuronal firing and synaptic signaling, particularly of the glutamate and GABA systems (Fromer et al. 2014, 2016; Ripke et al. 2013, 2020; Trubetskoy et al. 2022; Goncalves et al. 2018; Schizophrenia Working Group of the Psychiatric Genomics, C 2014; Jaffe et al. 2018; Smeland et al. 2020; Kirov et al. 2012; Hall et al. 2015; Genovese et al. 2016; Skene et al. 2018; Singh et al. 2020). Consistent with genetic findings, in vivo brain imaging (Onwordi et al. 2020, 2021; Radhakrishnan et al. 2021; Merritt et al. 2016, 2021; Sydnor and Roalf 2020; Nakahara et al. 2021; Adams et al. 2022; Jeon et al. 2021; Howells et al. 2022) and molecular, cellular, and circuity studies of postmortem brain tissue (Fromer et al. 2014, 2016; Jaffe et al. 2018; Gandal et al. 2018a, b; Girdhar et al. 2022; MacDonald et al. 2017, 2020; Pierri et al. 1999; Kim et al. 2021; Osimo et al. 2019; Hu et al. 2021; Chung et al. 2016; Hoftman et al. 2018; Arion et al. 2015, 2017; Glausier et al. 2010; Kimoto et al. 2016; Datta et al. 2017; Enwright Iii et al. 2018; Glantz and Lewis 2000; Kolluri et al. 2005; Dienel et al. 2019; Banovac et al. 2020; Konopaske et al. 2014, 2018; Somenarain and Jones 2010; Beneyto and Meador-Woodruff 2006, 2008) identify alterations that reflect dysfunctional cortical glutamate and/or GABA synaptic neurotransmission (also see (Howes and Onwordi 2023; Glausier and Lewis 2013; Schmidt and Mirnics 2015) for reviews).

A brain region that has emerged as a locus of synaptic abnormalities, and plays a critical role in cognitive control and working memory, is the dorsolateral prefrontal cortex (DLPFC) (Miller and Cohen 2001; Lewis et al. 2012; Arnsten and Jin 2014). A substantial literature (reviewed in (Barch and Sheffield 2014; Glausier and Lewis 2018)), confirmed by meta-analyses (Callicot et al. 1998; Glahn et al. 2005, 2008; Minzenberg et al. 2009), demonstrates that DLPFC dysfunction is associated with cognitive impairments in schizophrenia. For example, individuals diagnosed with schizophrenia show reduced prefrontal cortex (PFC) activation as measured by functional magnetic resonance imaging (fMRI) while performing working memory tasks (Glahn et al. 2005; Yao et al. 2024; Shimodera et al. 2012; Minzenberg et al. 2009; Davidson and Heinrichs 2003). These fMRI findings are corroborated by electroencephalogram (EEG) and positron emission tomography (PET) studies, which show lower power of high-frequency neural oscillations in the gamma range and decreased energy metabolism in the PFC of schizophrenia subjects during working memory performance (Davidson and Heinrichs 2003; Park et al. 2006; Cho et al. 2007; Minzenberg et al. 2010; Dreher et al. 2012). This reduced PFC activity appears to be an integral part of the disease pathology, and not simply a reflection of poor performance (Dreher et al. 2012). A substantial literature also demonstrates molecular (Glausier and Lewis 2018; Hall and Bray 2022; Aryal et al. 2023; Uno and Coyle 2019), structural (Glausier and Lewis 2018; Prasad et al. 2022), and physiological (Krystal et al. 2017; Kraguljac and Lahti 2021) markers that reflect dysfunctional glutamate and/or GABA cortical synaptic neurotransmission in the DLPFC in schizophrenia.

Given the tight coupling of synaptic communication and mitochondrial energetics, lower measures of mitochondrial OXPHOS are predicted to accompany synaptic impairment. Consistent with studies supporting the presence of less cortical glutamate synaptic signaling in SZ (Fromer et al. 2014, 2016; Jaffe et al. 2018; Onwordi et al. 2020; Radhakrishnan et al. 2021; Onwordi et al. 2021; Merritt et al. 2016, 2021; Sydnor and Roalf 2020; Nakahara et al. 2021; Adams et al. 2022; Jeon et al. 2021; Howells et al. 2022; Gandal et al. 2018a, b; MacDonald et al. 2017, 2020; Pierri et al. 1999; Kim et al. 2021; Osimo et al. 2019; Hu et al. 2021; Chung et al. 2016; Hoftman et al. 2018), accumulating in vivo and postmortem studies of energetic measures indicate lower OXPHOS in DLPFC in SZ (Arion et al. 2015, 2017; Minzenberg et al. 2009; Park et al. 2006; Hazlett et al. 2019; Schoonover et al. 2024; Glausier et al. 2020; Enwright III and Lewis 2021; Bralet et al. 2016; Sukumar et al. 2020; Fujimoto et al. 2007; Mitelman et al.

2018). For example, transcriptomic analyses in DLPFC total gray matter (Gandal et al. 2018a; Glausier et al. 2020; Enwright III and Lewis 2021), glutamate pyramidal neurons (Enwright Iii et al. 2018; Glausier et al. 2020), and GABAergic interneurons (Enwright et al. 2018) showed the stereotypic coordinated reduction in OXPHOS gene expression that occurs in response to chronic reductions in neuronal firing and synaptic neurotransmission (Attwell and Laughlin 2001; Harris et al. 2012; Hall et al. 2012; Wong-Riley 2012). Furthermore, targeted analysis of a gene set that that indexed the diversity of mitochondrial functions, including OXPHOS, Ca2+ buffering and reactive species generation, showed a specific and selective reduction in functional pathways related to OXPHOS in SZ. Application of weighted gene co-expression analysis (WGCNA) further revealed that OXPHOS-related network structure was completely preserved, as would be predicted if neurons were making coordinated adjustments to respond to lower ATP consumption (Glausier et al. 2020). In further support of this interpretation, a recent cell type-specific study of individual glutamatergic cells in the DLPFC showed that lower markers of glutamate synaptic signaling are correlated with lower markers of complex IV/COX (Schoonover et al. 2024). EM studies of cortical brain regions in schizophrenia also show that mitochondria exhibit ultrastructural features consistent with lower levels of mitochondrial OXPHOS due to lower ATP demand (reviewed in (Roberts 2021)). Thus, many studies of brain from individuals with "idiopathic" schizophrenia converge on the interpretation that mitochondrial OPXHOS is not dysfunctional or defective per se, but that lower measures of OXPHOS reflect the expected consequence of long-term reductions in synaptic signaling.

General consensus in the field supports the interpretation that less ATP is produced via mitochondrial OXPHOS in the brains of individuals with schizophrenia. However, whether these findings reflect a consequence of synaptic dysfunction that causes neurons to fire less, or reflects impairments to mitochondrial OXPHOS that results in insufficient energy to properly maintain synaptic function, is an outstanding question. These, and other possible interpretations, are discussed below.

Cognitive dysfunction and schizophrenia-like symptoms are often present in traditional mitochondrial diseases (Seeger et al. 2010; Clay et al. 2011; Park and Park 2012; Debray et al. 2014), so much so that some mitochondrial disorders can be initially misdiagnosed as schizophrenia (Clay et al. 2011; Fattal et al. 2006), suggesting a possible shared pathophysiology. Some genetic analyses of idiopathic schizophrenia suggest a role for mtDNA or nDNA OXPHOS genes and that an epistatic relationship may exist between synapse- and OXPHOS-related genes (Schulmann et al. 2019; Whitehurst and Howes 2022; Hjelm et al. 2015). Frank genetic disorders in which large sections of DNA are deleted, such as the 22q11.2 deletion syndrome (Li et al. 2021) and 3q29 deletion syndrome (Purcell et al. 2023), increase the risk for developing schizophrenia up to 40-fold. The deleted regions are known to include genes critical for metabolism in general, and specifically mitochondrial function, demonstrating that mitochondrial dysfunction can be an etiological component of some schizophrenias (Gokhale et al. 2019, 2021). Indeed, development of schizophrenia in the presence of these deletions is not considered idiopathic, but genetic (Bassett and Chow 2008; Bassett et al. 2008). Experimentally restricting mitochondrial OXPHOS capacity significantly reduces gamma oscillation power in model systems (Kann 2011, 2016; Kann et al. 2011, 2014, 2016; Kann and Kovacs 2007), and as described above, lower gamma power is characteristic of schizophrenia. These model system studies also showed a particularly important role for ETC complex I in maintaining sufficient OXPHOS, and many studies show lower levels of complex I transcripts, proteins, and activity in brain or derived cells from individuals with schizophrenia (reviewed in (Ben-Shachar 2017; Bergman and Ben-Shachar 2016; Holper et al. 2019)).

Other interpretations regarding brain energetics and synaptic function in schizophrenia emphasize other aspects of these systems that may result in dysfunction, such as oxidative stress (Cuenod et al. 2022), bioenergetic uncoupling (Henkel et al. 2022; Sullivan et al. 2018b), neuroinflammation (Murphy et al. 2021), affected lactate shuttling (Sullivan et al. 2018b), and cumulative effects of environmental insults (such as trauma) (Sturmberg et al. 2017). Given the equifinal nature of schizophrenia (i.e., schizophrenia syndrome can emerge from multiple etiological origins) (Green and Glausier 2015), the likely reality is that any given individual with "idiopathic" schizophrenia may have some combination of the above states and risk factors. Future efforts toward personalized medicine, and biology-based stratification of individuals experiencing psychiatric illness, hold great promise to advance the search for novel, research-based therapeutic targets.

5.4.2 Bipolar Disorder

Bipolar disorder is characterized by the cycling between mood states of mania and depression, with periods of intervening euthymia (i.e., balanced mood and emotions) (American Psychiatric Association 2013). Mania can be characterized by abnormally high levels of physical activity, racing thoughts, feelings of grandiosity, and participation in risky behaviors. Depressed states in individuals with bipolar disorder are often indistinguishable from unipolar depression, including feelings of hopelessness and sadness, fatigue, and anhedonia. While cognitive deficits have been identified in individuals with bipolar disorder, their prevalence and severity are much less than that seen in people with schizophrenia (Kahn 2019). Approximately 0.5% of the global population is afflicted with bipolar disorder, and it represents a significant burden, especially for young people (Grande et al. 2016; World Health Organization 2024; Mental Disorders Collaborators G.M.D 2022; Alonso et al. 2019). As with most psychiatric disorders, the etiology of bipolar disorder is unknown. GWAS studies point to similar biological themes as for schizophrenia: risk alleles preferential for brain-enriched genes, and, of those, enrichment of genes broadly involved with neural communication (Mullins et al. 2021). However, additional genetic studies have identified links with mitochondrial-related genes (reviewed in (Clay et al. 2011; Machado et al. 2016; Manji et al. 2012; Kato 2017; Lam et al. 2023)), including mtDNA-encoded genes identified by deep exome sequencing (Nishioka et al. 2023). As described below, bipolar disorder is increasingly being conceptualized not only as a synaptic disorder but also as a disease of mitochondrial, metabolic, and/or inflammation defects (Freyberg et al. 2024).

In vivo structural and functional imaging (Xia et al. 2019; Phillips 2023; Yu et al. 2010) and molecular (Gandal et al. 2018a; Arbabi et al. 2024; Dean et al. 2016; Zandi et al. 2022; Focking et al. 2016) and microscopy (Konopaske et al. 2014) studies of brain from individuals with bipolar disorder suggest synaptic perturbations in the illness. However, those alterations may be more nuanced than what is seen in schizophrenia. Similarly, mitochondrial perturbations are suggested by in vivo imaging (Kraguljac et al. 2012; Dogan et al. 2018; Yuksel et al. 2015; Stork and Renshaw 2005) and molecular (Gandal et al. 2018a; Arbabi et al. 2024; Sun et al. 2006; Andreazza et al. 2010) and microscopy (Cataldo et al. 2010) studies of brain in individuals with bipolar disorder. OXPHOS-specific mitochondrial alterations are sometimes identified in bipolar disorder. For example, a recent study used a fractionation approach to obtain DLPFC samples enriched for type 1 synapses and evaluated the disease effect of bipolar disorder on the synaptic proteome (Aryal et al. 2023). Functional pathways related to synaptic signaling and mitochondrial functions, including OXPHOS, were identified as downregulated. However, many studies find different effects of bipolar disorder on synaptic and mitochondrial OXPHOS measures (Arion et al. 2017; Datta et al. 2017; Glausier et al. 2020; Enwright III and Lewis 2021; MacDonald et al. 2006; Eastwood and Harrison 2010; Tobe et al. 2017; Sibille et al. 2011; Okuda et al. 2024; Dienel et al. 2023; Lanz et al. 2019; Guan et al. 2019; Ramaker et al. 2017; Ellis et al. 2016; Du et al. 2018; Duong et al. 2016; Bodenstein et al. 2019; Sahay et al.

2024). For example, transcriptomic profiling studies that showed significant schizophrenia disease effects on mitochondrial OXPHOS gene expression in cortical gray matter and glutamatergic neurons found no significant effect of bipolar disorder on these measures (Arion et al. 2017; Glausier et al. 2020; Enwright III and Lewis 2021). Thus, unlike schizophrenia, the multifunctional aspects of mitochondria may be more engaged with the bipolar disease process.

Consistent with the above, mitochondrial-related alterations to Ca^{2+} regulation, apoptosis, inflammation, and reactive oxygen species (ROS) generation are identified in brains of individuals with bipolar disorder (reviewed in (Manji et al. 2012; Kato 2017; Brown et al. 2014; Giménez-Palomo et al. 2021; Fraunberger et al. 2016; Kato and Kato 2000; Scaini et al. 2021)). Interestingly, recent perspectives of bioenergetic alterations in bipolar disorder focus on disruptions to the Krebs cycle that may ultimately lead to downstream consequences which induce oxidative stress and cellular dysfunction (Lam et al. 2023). Furthermore, mitochondria are proposed to be a lynchpin in mediating the effects of dysregulated metabolism and inflammatory/immune systems in the bipolar disorder disease process (Bernard et al. 2024). As "master regulators of inflammatory responses (Marchi et al. 2023)," mitochondria are in a key position to exacerbate any existing disordered inflammatory processes and may initiate a vicious cycle that impairs multiple aspects of mitochondrial functioning (Bernard et al. 2024). For example, a recent study that integrated findings from genomic and transcriptomic datasets identified glial cell dysfunction, including microglia which are the brain's resident immune cell, as a potential pathogenic mechanism in the brain in bipolar disorder (Wei et al. 2024).

Ongoing research is focused on biology-informed stratification of bipolar disorder subjects using novel peripheral biomarkers and functional and molecular analyses of individual patient-derived induced pluripotent stem cells. These studies may identify individuals more amenable to specific therapeutic interventions that affect mitochondrial bioenergetics/Krebs cycle (e.g., ketogenic diet (Freyberg et al. 2024)), inflammatory processes (e.g., nonsteroidal anti-inflammatory drugs (Fitton et al. 2022)), or oxidative stress (e.g., nutraceuticals (Giménez-Palomo et al. 2021)).

5.5 Effects of Physical Illness on Brain Bioenergetics and Synaptic Function

The phenomenon of high comorbidity between psychiatric and physical illnesses is well documented (Na et al. 2024; De Hert et al. 2011a; Launders et al. 2021; Correll et al. 2017; Tian et al. 2023). For example, individuals with psychiatric disorders have significantly higher rates of cardiovascular disease (Correll et al. 2017; Osborn et al. 2007), type 2 diabetes (Chien et al. 2009; Wändell et al. 2014; Charles et al. 2016), and obesity (Vancampfort et al. 2015; Annamalai et al. 2017), as well as chronic infectious illnesses such as hepatitis B and C (Gutiérrez-Rojas et al. 2023; Chong et al. 2018; Hughes et al. 2016). Psychotropic medication side-effects (Correll et al. 2015; Bou Khalil 2012), substance use (licit and illicit) (Hartz et al. 2014), and insufficient medical care (De Hert et al. 2011b; Scott et al. 2012) certainly contribute to the high comorbidity of "brain and body disorders." For example, systemic glucose dysregulation, including hyperglycemia, insulin resistance, and type 2 diabetes can be induced or exacerbated by antipsychotic drug treatment (reviewed in (Lieberman 3rd 2004)). However, a substantial literature demonstrates that the elevated comorbidity between psychiatric and physical illnesses cannot be exclusively attributed to these factors, at least in some individuals (reviewed in (Agarwal et al. 2020a; Miller et al. 2013)). Using the example of glucose dysregulation and insulin resistance in schizophrenia, the below discussion illustrates the functional relationship between physical and psychiatric disorders and their effects on brain bioenergetics and synaptic function.

5.5.1 Glucose Dysregulation and Schizophrenia: Possible Common Origins

Hyperglycemia occurs when blood glucose levels are not properly regulated by insulin, typically due to impaired insulin sensitivity of peripheral cells (e.g., fat and liver cells). This phenomenon is termed insulin resistance and often leads to type 2 diabetes. A relationship between psychosis and hyperglycemia has been noted since the early twentieth century (Kohen 2004), approximately 30 years before the introduction of antipsychotic drugs. Subsequent studies have supported these early observations. For example, a study of first episode, antipsychotic drug-naïve subjects with schizophrenia showed that ~25% exhibited hyperglycemia and insulin resistance, whereas none of the demographically matched comparison subjects showed this profile (Chen et al. 2016). This pattern of findings was supported by meta-analysis (Pillinger et al. 2017; Perry et al. 2016; Greenhalgh et al. 2017) and a recent study of N > 700 first episode, antipsychotic drug-naive subjects with schizophrenia (Wu et al. 2024). Together, these and other findings suggest that the pathogenic processes of schizophrenia and glucose dysregulation/insulin resistance may have common origins, at least in some individuals (reviewed in (Agarwal et al. 2020a, b; Zhuo et al. 2024; Lin and Shuldiner 2010; Holt and Mitchell 2015)).

Three explanatory models have been proposed regarding this relationship (Lin and Shuldiner 2010), and while each model could be mutually exclusive within-subject, evidence supportive of each has been identified across individuals with schizophrenia (Fig. 5.4). As discussed below, existing epidemiological, clinical, and genetic studies provide support for the notion that Model 1, 2, or 3 could be operative in any given individual with schizophrenia.

In support of Model 1, some studies find statistically significant relationships between genetic predictors of schizophrenia and glucose dysregulation/insulin resistance (Tomasik et al. 2019; Hackinger et al. 2018; Maj et al. 2020; Cao et al. 2017; So et al. 2019; Li et al. 2018), suggesting there may be pleiotropic effects. Twin and family studies also find that unaffected and unmedicated siblings or parents of the schizophrenia proband tend to also have measures reflective of peripheral and/or brain glucose dysregulation (Chouinard et al. 2019; Sullivan et al. 2003; Almgren et al. 2011; Willemsen et al. 2015). Along with convergent environmental risk factors, like fetal malnutrition, these gene x environment interactions may increase susceptibility to both schizophrenia and glucose dysregulation/insulin resistance (Lin and Shuldiner 2010). A recent study combining clinical and preclinical transcriptomic data found similar gene expression profiles in peripheral samples of antipsychotic drug-naive schizophrenia subjects and brain gene expression profiles of rodent models with an induced central insulin perturbation, suggesting that the schizophrenia disease process elicits a central insulin pathology (Lee et al. 2024). However, not all genetic studies find significant positive relationships between schizophrenia and measures of glucose dysregulation, including well-powered studies using modern analytical approaches such as Mendelian randomization (Li et al. 2018; Habtewold et al. 2020; Polimanti et al. 2018; Perry et al. 2022; Aoki et al. 2022; Shamsutdinova et al. 2023).

Model 2 proposes that glucose dysregulation/insulin resistance is caused by factors commonly associated with schizophrenia such as (1) elevated psychosocial stress, (2) "diagnostic overshadowing" that can occur when complaints are ascribed to an existing psychiatric illness by physicians that many result in inadequate medical care, or (3) sedentary lifestyles that results from negative symptoms. Epidemiological, clinical, and preclinical studies provide support for this model. For example, the recent large N study by Wu and colleagues (Wu et al. 2024) demonstrate a strong positive correlation between indices of glucose dysregulation (including glycemic and insulin resistance measures) and negative symptoms. Post-hoc analysis revealed that schizophrenia subjects with predominately negative symptoms had significantly greater glucose dysregulation than those with predominately positive symptoms. However, the negative-symptom

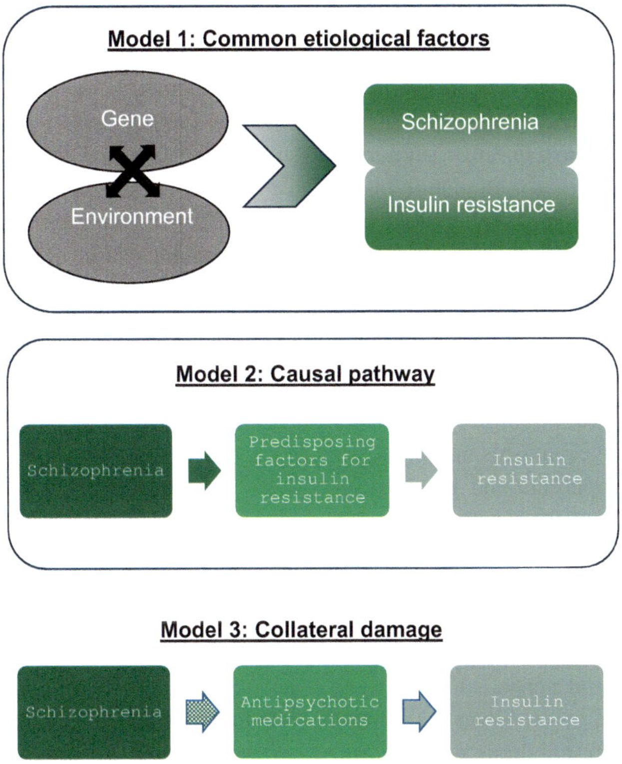

Fig. 5.4 Explanatory models for comorbidity of schizophrenia and insulin resistance. These models, proposed by Lin and Shuldiner, describe three possibilities of multifactorial interactions that could underlie the high comorbidity of schizophrenia with glucose dysregulation/insulin resistance. The hatched arrow in Model 3 reflects that treatment with antipsychotic medications links schizophrenia and glucose dysregulation/insulin resistance. Current epidemiological, clinical and preclinical findings suggest that any one of these models may accurately capture the mechanism of comorbidity in any given individual with schizophrenia. (Modified from Lin and Shuldiner (2010). With permission from XXX)

group also had nearly twice the average duration of illness than did the positive-symptom group. Similarly, a recent longitudinal study showed increased insulin resistance over a 12-month period in individuals with clinical high risk for schizophrenia that converted and in individuals experiencing their first psychotic episode at intake, independent of treatment with antipsychotic medications (Armio et al. 2024). Together, these findings suggest that glucose dysregulation may index consequences of being ill with schizophrenia.

Model 3 proposes that glucose dysregulation/insulin resistance is almost exclusively due to the side effects of antipsychotic medications. Both clinical and preclinical studies support the feasibility of this model to be operative in some individuals with schizophrenia (reviewed in (Lieberman 3rd 2004; Au et al. 2024; Freyberg et al. 2017)). For example, a study of 15 female cynomolgus monkeys showed that long-term administration of the atypical antipsychotic olanzapine induced insulin resistance (Cunningham et al. 2019). On the other hand, a brain transcriptomics study found that monkeys chronically administered typical or atypical antipsychotic drugs did not have any effects on mitochondrial gene sets, such as OXPHOS, and had a transcriptomic signature completely unlike individuals with schizophrenia (Glausier et al. 2020). Thus, in the context of all available data, Model 3 appears to be the least common explanatory rationale for glucose dysregulation/insulin resistance across individuals with schizophrenia.

In sum, existing epidemiological, clinical, and genetic studies do provide support for the notion that Model 1, 2, or 3 could be operative in any given individual with schizophrenia. When considered in aggregate, most findings support an interpretation that glucose dysregulation/insulin resistance in schizophrenia likely results from a multi-factorial process that reflects gene x environment interactions.

5.5.1.1 Glucose Dysregulation, Schizophrenia, and Synaptic Mitochondria

Although the precise mechanisms underlying the comorbidity of schizophrenia and glucose dysregulation/insulin resistance remain unknown, perturbations to these systems can affect brain bioenergetics and synaptic function. As described in detail in the previous sections, the primary substrate of aerobic mitochondrial respiration in the brain is glucose, and persistent hyperglycemia (i.e., high glucose in the blood) can impair mitochondrial functioning in brain cells, causing oxidative stress and damage to neuronal compartments (reviewed in (González et al. 2023; Wardelmann et al. 2019)). Additionally, insulin signaling in the brain regulates glucose production by the liver (Agarwal et al. 2020a), and a "brain-centered glucoregulatory system" can regulate blood glucose levels via a central insulin-dependent manner (Schwartz et al. 2013). Insulin signaling in the brain is associated with synaptic plasticity, synaptic signaling, and neural remodeling (Wardelmann et al. 2019; Edwards et al. 2010; Huang et al. 2010); altered brain insulin signaling can impair memory performance in preclinical models (reviewed in (Ottomana et al. 2023)), and much indirect evidence in humans also supports a key role of insulin signaling and cognitive abilities (reviewed in (Gupta et al. 2023; Zhang et al. 2023)). Thus, these two systems, neuronal synaptic signaling and bioenergetics in brain and peripheral glucose regulation, may appear disparate upon casual observation but are highly intertwined in both the healthy and disease states.

Indeed in Chap. 3 "Brain-Body Communication in Glucose Metabolism," the authors explore the roles of various brain regions in glucose homeostasis and their connections through the sympathetic and parasympathetic nervous systems to peripheral organs, such as the pancreas and liver, to provide a comprehensive description of how brain-body interactions maintain glucose homeostasis and their implications for metabolic health.

5.6 "Brain-Body" Intervention Strategies to Improve Brain Bioenergetics and Synaptic Signaling in Psychiatric Illness: Challenges and Strategies

As described above, a substantial literature indicates that direct, and bidirectional, relationships between psychiatric and physical illnesses contribute to high comorbidity. The evidence is overwhelmingly clear: health outcomes for individuals with psychiatric illnesses are worse when medical providers do not take an integrated treatment approach and instead maintain a rigid separation of "brain and body" (Tian et al. 2023). However, the multi-factorial, and mostly unknown, etiological and pathogenetic processes underlying psychiatric and physical illnesses make difficult the widespread implementation of this approach.

One approach is to enact interventions that have little to no likelihood of deleterious side effects, such as treating substance use disorders (Manseau and Bogenschutz 2016), completing regular physical exercise (Ashton et al. 2020), and nutraceutical (Puri et al. 2022) (i.e., foods and dietary supplements intended to improve health) interventions (Giménez-Palomo et al. 2021). Indeed, the role of diet in mental and physical health has been recognized since Hippocrates (460–375 BCE) (Cardenas 2013), and reports from the Ming Dynasty (1368–1644) in China indicate that the most important feature of imperial cuisine was to maintain good health (上官丰, and 章挺权 1998). Dietary restrictions remain a core treatment approach for some genetic disorders like phenylketonuria (PKU). In PKU, phenylalanine (Phe) metabolism is completely prevented, and Phe accumulation in the body and brain can cause depression and anxiety and ultimately leads to intellectual disability (Ashe et al. 2019). By restricting dietary sources of Phe, an amino acid found in all sources of protein, the genetic disorder can be completely managed.

For psychiatric disorders, a potential therapeutic role of the ketogenic diet is emerging (Freyberg et al. 2024; Sethi and Ford 2022). The ketogenic diet is defined by a severe restriction in carbohydrate intake so that the body and brain enter ketosis, a metabolic state where the primary substrate for ATP production is fat and not glucose. How are synaptic signaling and mitochondrial bioenergetics maintained in the brain when adhering to a ketogenic diet, since glucose is the primary substrate for aerobic respiration? Ketone bodies, molecules that are produced when non-carbohydrate sources of energy (i.e., fat) are metabolized, replace glucose as the source for acetyl-CoA to initiate the Krebs cycle (Kolb et al. 2021). The ketogenic diet has been a part of medical practice for over a 100 years. Indeed, it was a cornerstone of epilepsy treatment, particularly for children, in the early twentieth century before anti-epileptic medications were widely available (Wheless 2004).

A recent pilot study (Sethi et al. 2024) showed significant improvements in psychiatric measures, such as mood and anxiety, and in physical measures, such as insulin resistance and glycemic control, in bipolar disorder individuals that adhered to a ketogenic diet. A convincing argument has been made that the benefits of the ketogenic diet on bipolar disorder clinical outcomes occur, at least in part, through modulating levels of excitatory (glutamate) and inhibitory (GABA) neurotransmitters via actions at the Krebs cycle, which occurs in the mitochondrial matrix (Freyberg et al. 2024). Thus, application of the ketogenic diet as an adjunctive treatment for psychiatric illness has traction, since it affects (1) multiple brain systems perturbed in psychiatric illnesses (i.e., synaptic neurotransmission and mitochondrial bioenergetics) and (2) bioenergetics and metabolism in peripheral systems, reflecting a true "brain-body" therapeutic approach.

A second, and not mutually exclusive, approach is "precision psychiatry." The unique "brain-body" relationships between glucose regulation, insulin signaling, and risk for schizophrenia and bipolar disorder provide a powerful opportunity to investigate and develop strategies for identifying peripheral biomarkers predictive of disease risk and optimize treatment strategies. This approach—grouping people with a common disorder (e.g., schizophrenia or bipolar disorder) into disease sub-populations that are defined by quantitative factors like genetic markers (e.g., variants in risk loci), molecular metabolic profiles (e.g., heightened markers of inflammation or hyperglycemia), and/or comorbidities (type 2 diabetes, hepatitis C)—is termed "stratification." The subject inclusion/exclusion criteria applied in the ketogenic diet pilot study described above (Sethi et al. 2024) could be considered a type of stratification. Included subjects (1) were already taking psychotropic medications, (2) were overweight and had gained weight while taking psychotropic medications, and (3) exhibited metabolic abnormalities on a comprehensive blood metabolic panel. This experimental design, which restricted subjects to those with multiple features indicative of metabolic disturbance, is likely a key factor contributing to the pilot study's success.

Emerging studies continue to push the precision psychiatry/stratification approach forward, for example, by utilizing artificial intelligence or machine learning to incorporate multiple sources of information to generate precision stratifications (Grzenda and Widge 2024; Bzdok and Meyer-Lindenberg 2018; Delgadillo and Lutz 2020). Given the incredible biological complexity of psychiatric illnesses, and the substantial person-to-person heterogeneity at nearly all levels of investigation (genetic, molecular, metabolic, anatomical, and behavioral), these approaches hold great promise to identify rational stratification parameters that will improve the prevention, treatment, and outcomes of individuals with psychiatric disorders. Thus, we may find that highly curated clinical and biological features will be a tractable entre into precision psychiatry.

Acknowledgments The authors gratefully acknowledge the digital graphics expertise of Mary Brady. Funding support was provided by NIH grants R01MH132544 and S10MH133575.

References

Ackermann F, Waites CL, Garner CC (2015) Presynaptic active zones in invertebrates and vertebrates. EMBO Rep 16:923–938. https://doi.org/10.15252/embr.201540434

Adams RA, Pinotsis D, Tsirlis K, Unruh L, Mahajan A, Horas AM, Convertino L, Summerfelt A, Sampath H, Du XM et al (2022) Computational modeling of electroencephalography and functional magnetic resonance imaging paradigms indicates a consistent loss of pyramidal cell synaptic gain in Schizophrenia. Biol Psychiatry 91:202–215. https://doi.org/10.1016/j.biopsych.2021.07.024

Agarwal SM, Caravaggio F, Costa-Dookhan KA, Castellani L, Kowalchuk C, Asgariroozbehani R, Graff-Guerrero A, Hahn M (2020a) Brain insulin action in schizophrenia: something borrowed and something new. Neuropharmacology 163:107633. https://doi.org/10.1016/j.neuropharm.2019.05.010

Agarwal SM, Kowalchuk C, Castellani L, Costa-Dookhan KA, Caravaggio F, Asgariroozbehani R, Chintoh A, Graff-Guerrero A, Hahn M (2020b) Brain insulin action: implications for the treatment of schizophrenia. Neuropharmacology 168:107655. https://doi.org/10.1016/j.neuropharm.2019.05.032

Aiello LC, Wheeler P (1995) The expensive-tissue hypothesis: the brain and the digestive system in human and primate evolution. Curr Anthropol 36:199–221

Almgren P, Lehtovirta M, Isomaa B, Sarelin L, Taskinen M-R, Lyssenko V, Tuomi T, Groop L, Group, B.S (2011) Heritability and familiality of type 2 diabetes and related quantitative traits in the Botnia Study. Diabetologia 54:2811–2819

Alonso J, Vilagut G, Mortier P, Auerbach RP, Bruffaerts R, Cuijpers P, Demyttenaere K, Ebert DD, Ennis E, Gutiérrez-García RA et al (2019) The role impairment associated with mental disorder risk profiles in the WHO World Mental Health International College student initiative. Int J Methods Psychiatr Res 28:e1750. https://doi.org/10.1002/mpr.1750

American Psychiatric Association (2013) DSM-V: diagnostic and statistical manual of mental disorders, 5th edn. American Psychiatric Publishing. https://doi.org/10.1176/appi.books.9780890425596.744053

Andrade-Moraes CH, Oliveira-Pinto AV, Castro-Fonseca E, da Silva CG, Guimarães DM, Szczupak D, Parente-Bruno DR, Carvalho LR, Polichiso L, Gomes BV et al (2013) Cell number changes in Alzheimer's disease relate to dementia, not to plaques and tangles. Brain 136:3738–3752. https://doi.org/10.1093/brain/awt273

Andreazza AC, Shao L, Wang JF, Young LT (2010) Mitochondrial complex I activity and oxidative damage to mitochondrial proteins in the prefrontal cortex of patients with bipolar disorder. Arch Gen Psychiatry 67:360–368. https://doi.org/10.1001/archgenpsychiatry.2010.22

Annamalai A, Kosir U, Tek C (2017) Prevalence of obesity and diabetes in patients with schizophrenia. World J Diabetes 8:390–396. https://doi.org/10.4239/wjd.v8.i8.390

Aoki R, Saito T, Ninomiya K, Shimasaki A, Ashizawa T, Ito K, Ikeda M, Iwata N (2022) Shared genetic components between metabolic syndrome and schizophrenia: genetic correlation using multipopulation data sets. Psychiatry Clin Neurosci 76:361–366

Arbabi K, Newton DF, Oh H, Davie MC, Lewis DA, Wainberg M, Tripathy SJ, Sibille E (2024) Transcriptomic pathology of neocortical microcircuit cell types across psychiatric disorders. Mol Psychiatry 30:1057. https://doi.org/10.1038/s41380-024-02707-1

Arellano JI, Benavides-Piccione R, DeFelipe J, Yuste R (2007) Ultrastructure of dendritic spines: correlation between synaptic and spine morphologies. Front Neurosci 1:131–143

Arion D, Corradi JP, Tang S, Datta D, Boothe F, He A, Cacace AM, Zaczek R, Albright CF, Tseng G, Lewis DA (2015) Distinctive transcriptome alterations of prefrontal pyramidal neurons in schizophrenia and schizoaffective disorder. Mol Psychiatry 20:1397–1405. https://doi.org/10.1038/mp.2014.171

Arion D, Huo Z, Enwright JF, Corradi JP, Tseng G, Lewis DA (2017) Transcriptome alterations in prefrontal pyramidal cells distinguish schizophrenia from bipolar and major depressive disorders. Biol Psychiatry 82:594–600. https://doi.org/10.1016/j.biopsych.2017.03.018

Armio R-L, Laurikainen H, Ilonen T, Walta M, Sormunen E, Tolvanen A, Salokangas RKR, Koutsouleris N, Tuominen L, Hietala J (2024) Longitudinal study on hippocampal subfields and glucose metabolism in early psychosis. Schizophrenia 10:66. https://doi.org/10.1038/s41537-024-00475-z

Arnsten AF, Jin LE (2014) Molecular influences on working memory circuits in dorsolateral prefrontal cortex. Prog Mol Biol Transl Sci 122:211–231. https://doi.org/10.1016/b978-0-12-420170-5.00008-8

Aryal S, Bonanno K, Song B, Mani DR, Keshishian H, Carr SA, Sheng M, Dejanovic B (2023) Deep proteomics identifies shared molecular pathway alterations in synapses of patients with schizophrenia and bipolar disorder and mouse model. Cell Rep 42:112497. https://doi.org/10.1016/j.celrep.2023.112497

Ashe K, Kelso W, Farrand S, Panetta J, Fazio T, De Jong G, Walterfang M (2019) Psychiatric and cognitive aspects of Phenylketonuria: the limitations of diet and promise of new treatments. Front Psych 10:561. https://doi.org/10.3389/fpsyt.2019.00561

Ashton MM, Mohebbi M, Turner A, Marx W, Berk M, Malhi GS, Ng CH, Cotton SM, Dodd S, Sarris J (2020) Physical activity as a predictor of clinical trial outcomes in bipolar depression: a subanalysis of a mitochondrial-enhancing nutraceutical randomized controlled trial. Can J Psychiatry 65:306–318

Attwell D, Iadecola C (2002) The neural basis of functional brain imaging signals. Trends Neurosci 25:621–625

Attwell D, Laughlin SB (2001) An energy budget for signaling in the grey matter of the brain. J Cereb Blood Flow Metab 21:1133–1145. https://doi.org/10.1097/00004647-200110000-00001

Au E, Panganiban KJ, Wu S, Sun K, Humber B, Remington G, Agarwal SM, Giacca A, Pereira S, Hahn M (2024) Antipsychotic-induced dysregulation of glucose metabolism through the central nervous system: a scoping review of animal models. Biol Psychiatry Cogn Neurosci Neuroimaging 10:244. https://doi.org/10.1016/j.bpsc.2024.10.001

Balasubramanian V (2021) Brain power. Proc Natl Acad Sci USA 118. https://doi.org/10.1073/pnas.2107022118

Banovac I, Sedmak D, Rojnić Kuzman M, Hladnik A, Petanjek Z (2020) Axon morphology of rapid Golgi-stained pyramidal neurons in the prefrontal cortex in schizophrenia. Croat Med J 61:354–365. https://doi.org/10.3325/cmj.2020.61.354

Barch DM, Sheffield JM (2014) Cognitive impairments in psychotic disorders: common mechanisms and measurement. World Psychiatry 13:224–232. https://doi.org/10.1002/wps.20145

Barshad G, Blumberg A, Cohen T, Mishmar D (2018) Human primitive brain displays negative mitochondrial-nuclear expression correlation of respiratory genes. Genome Res 28:952–967. https://doi.org/10.1101/gr.226324.117

Bartol TM Jr, Bromer C, Kinney J, Chirillo MA, Bourne JN, Harris KM, Sejnowski TJ (2015) Nanoconnectomic upper bound on the variability of synaptic plasticity. eLife 4:e10778. https://doi.org/10.7554/eLife.10778

Bassett AS, Chow EW (2008) Schizophrenia and 22q11.2 deletion syndrome. Curr Psychiatry Rep 10:148–157. https://doi.org/10.1007/s11920-008-0026-1

Bassett A, Chow E, Hodgkinson K (2008) Psychiatric genetics: applications in clinical practice. American Psychiatric Publishing, Inc

Becker N, Wierenga CJ, Fonseca R, Bonhoeffer T, Nägerl UV (2008) LTD induction causes morphological changes of presynaptic boutons and reduces their contacts with spines. Neuron 60:590–597. https://doi.org/10.1016/j.neuron.2008.09.018

Beneyto M, Meador-Woodruff JH (2006) Lamina-specific abnormalities of AMPA receptor trafficking and signaling molecule transcripts in the prefrontal cortex in schizophrenia. Synapse 60:585–598. https://doi.org/10.1002/syn.20329

Beneyto M, Meador-Woodruff JH (2008) Lamina-specific abnormalities of NMDA receptor-associated postsynaptic protein transcripts in the prefrontal cortex in schizophrenia and bipolar disorder. Neuropsychopharmacology 33:2175–2186

Ben-Shachar D (2017) Mitochondrial multifaceted dysfunction in schizophrenia; complex I as a possible pathological target. Schizophr Res 187:3–10. https://doi.org/10.1016/j.schres.2016.10.022

Bergman O, Ben-Shachar D (2016) Mitochondrial oxidative phosphorylation system (OXPHOS) deficits in schizophrenia: possible interactions with cellular processes. Can J Psychiatr 61:457–469. https://doi.org/10.1177/0706743716648290

Bernard J, Tamouza R, Godin O, Berk M, Andreazza AC, Leboyer M (2024) Mitochondria at the crossroad of dysregulated inflammatory and metabolic processes in bipolar disorders. Brain Behav Immun 123:456–465. https://doi.org/10.1016/j.bbi.2024.10.008

Bodenstein DF, Kim HK, Brown NC, Navaid B, Young LT, Andreazza AC (2019) Mitochondrial DNA content and oxidation in bipolar disorder and its role across brain regions. NPJ Schizophr 5:21. https://doi.org/10.1038/s41537-019-0089-5

Bosch M, Castro J, Saneyoshi T, Matsuno H, Sur M, Hayashi Y (2014) Structural and molecular remodeling of dendritic spine substructures during long-term potentiation. Neuron 82:444–459. https://doi.org/10.1016/j.neuron.2014.03.021

Bou Khalil R (2012) Atypical antipsychotic drugs, schizophrenia, and metabolic syndrome in non-Euro-American societies. Clin Neuropharmacol 35:141–147. https://doi.org/10.1097/WNF.0b013e31824d5288

Bourne JN, Harris KM (2011) Coordination of size and number of excitatory and inhibitory synapses results in a balanced structural plasticity along mature hippocampal CA1 dendrites during LTP. Hippocampus 21:354–373. https://doi.org/10.1002/hipo.20768

Bourne JN, Harris KM (2012) Nanoscale analysis of structural synaptic plasticity. Curr Opin Neurobiol 22:372–382. https://doi.org/10.1016/j.conb.2011.10.019

Bourne JN, Chirillo MA, Harris KM (2013) Presynaptic ultrastructural plasticity along CA3→CA1 axons during long-term potentiation in mature hippocampus. J Comp Neurol 521:3898–3912. https://doi.org/10.1002/cne.23384

Bralet MC, Buchsbaum MS, DeCastro A, Shihabuddin L, Mitelman SA (2016) FDG-PET scans in patients with Kraepelinian and non-Kraepelinian schizophrenia. Eur Arch Psychiatry Clin Neurosci 266:481–494. https://doi.org/10.1007/s00406-015-0633-x

Brown NC, Andreazza AC, Young LT (2014) An updated meta-analysis of oxidative stress markers in bipolar disorder. Psychiatry Res 218:61–68. https://doi.org/10.1016/j.psychres.2014.04.005

Bzdok D, Meyer-Lindenberg A (2018) Machine learning for precision psychiatry: opportunities and challenges. Biol Psychiatry Cogn Neurosci Neuroimaging 3:223–230. https://doi.org/10.1016/j.bpsc.2017.11.007

Callicot JH, Ramsey NF, Tallent K, Bertolino A, Knable MB, Coppola R, Goldberg T, van Gelderen P, Mattay

VS, Frank JA et al (1998) Functional magnetic resonance imaging brain mapping in psychiatry: methodological issues illustrated in a study of working memory in schizophrenia. Neuropsychopharmacology 18:186–196
Cao H, Chen J, Meyer-Lindenberg A, Schwarz E (2017) A polygenic score for schizophrenia predicts glycemic control. Transl Psychiatry 7:1295
Cardenas D (2013) Let not thy food be confused with thy medicine: the hippocratic misquotation. e-SPEN J 8:e260–e262. https://doi.org/10.1016/j.clnme.2013.10.002
Cataldo AM, McPhie DL, Lange NT, Punzell S, Elmiligy S, Ye NZ, Froimowitz MP, Hassinger LC, Menesale EB, Sargent LW et al (2010) Abnormalities in mitochondrial structure in cells from patients with bipolar disorder. Am J Pathol 177:575–585. https://doi.org/10.2353/ajpath.2010.081068
Charles EF, Lambert CG, Kerner B (2016) Bipolar disorder and diabetes mellitus: evidence for disease-modifying effects and treatment implications. Int J Bipolar Disord 4:13. https://doi.org/10.1186/s40345-016-0054-4
Chen DC, Du XD, Yin GZ, Yang KB, Nie Y, Wang N, Li YL, Xiu MH, He SC, Yang FD et al (2016) Impaired glucose tolerance in first-episode drug-naïve patients with schizophrenia: relationships with clinical phenotypes and cognitive deficits. Psychol Med 46:3219–3230. https://doi.org/10.1017/S0033291716001902
Chen F, Danladi J, Wegener G, Madsen TM, Nyengaard JR (2020) Sustained ultrastructural changes in rat hippocampal formation after repeated electroconvulsive seizures. Int J Neuropsychopharmacol 23:446–458. https://doi.org/10.1093/ijnp/pyaa021
Chien IC, Hsu JH, Lin CH, Bih SH, Chou YJ, Chou P (2009) Prevalence of diabetes in patients with schizophrenia in Taiwan: a population-based National Health Insurance study. Schizophr Res 111:17–22
Cho RY, Konecky RO, Carter CS (2007) Impaired cognitive control and frontal cortical gamma-band synchrony in schizophrenia. Proc Natl Acad Sci 103:19878–19883
Chong HY, Teoh SL, Wu DB, Kotirum S, Chiou CF, Chaiyakunapruk N (2016) Global economic burden of schizophrenia: a systematic review. Neuropsychiatr Dis Treat 12:357–373. https://doi.org/10.2147/ndt.S96649
Chong LW, Hsu CC, Lee CY, Chou RH, Lin CL, Chang KH, Hsu YC (2018) Association of viral hepatitis and bipolar disorder: a nationwide population-based study. J Transl Med 16:173. https://doi.org/10.1186/s12967-018-1542-3
Chouinard VA, Henderson DC, Dalla Man C, Valeri L, Gray BE, Ryan KP, Cypess AM, Cobelli C, Cohen BM, Öngür D (2019) Impaired insulin signaling in unaffected siblings and patients with first-episode psychosis. Mol Psychiatry 24:1513–1522. https://doi.org/10.1038/s41380-018-0045-1
Chung WS, Allen NJ, Eroglu C (2015) Astrocytes control synapse formation, function, and elimination. Cold Spring Harb Perspect Biol 7:a020370. https://doi.org/10.1101/cshperspect.a020370
Chung DW, Fish KN, Lewis DA (2016) Pathological basis for deficient excitatory drive to cortical parvalbumin interneurons in Schizophrenia. Am J Psychiatry 173:1131–1139. https://doi.org/10.1176/appi.ajp.2016.16010025
Clarke DD, Sokoloff L (1999) Circulation and energy metabolism. In: Siegel GJ, Agranoff BW, Albers RW, Fisher SK, Uhler MD (eds) Basic neurochemistry: molecular, cellular, and medical aspects. Lippincott-Raven, pp 637–669
Clarke G, Chen J, Nishimune H (2012) Presynaptic active zone density during development and synaptic plasticity. Front Mol Neurosci 5. https://doi.org/10.3389/fnmol.2012.00012
Clay HB, Sillivan S, Konradi C (2011) Mitochondrial dysfunction and pathology in bipolar disorder and schizophrenia. Int J Dev Neurosci 29:311–324
Correll CU, Detraux J, De Lepeleire J, De Hert M (2015) Effects of antipsychotics, antidepressants and mood stabilizers on risk for physical diseases in people with schizophrenia, depression and bipolar disorder. World Psychiatry 14:119–136. https://doi.org/10.1002/wps.20204
Correll CU, Solmi M, Veronese N, Bortolato B, Rosson S, Santonastaso P, Thapa-Chhetri N, Fornaro M, Gallicchio D, Collantoni E et al (2017) Prevalence, incidence and mortality from cardiovascular disease in patients with pooled and specific severe mental illness: a large-scale meta-analysis of 3,211,768 patients and 113,383,368 controls. World Psychiatry 16:163–180. https://doi.org/10.1002/wps.20420
Cryer PE (2007) Hypoglycemia, functional brain failure, and brain death. J Clin Invest 117:868–870. https://doi.org/10.1172/jci31669
Cserép C, Pósfai B, Schwarcz AD, Dénes Á (2018) Mitochondrial ultrastructure is coupled to synaptic performance at axonal release sites. eneuro 5, ENEURO.0390-0317.2018. https://doi.org/10.1523/eneuro.0390-17.2018
Cuenod M, Steullet P, Cabungcal JH, Dwir D, Khadimallah I, Klauser P, Conus P, Do KQ (2022) Caught in vicious circles: a perspective on dynamic feed-forward loops driving oxidative stress in schizophrenia. Mol Psychiatry 27:1886–1897. https://doi.org/10.1038/s41380-021-01374-w
Cunningham JI, Eyerman DJ, Todtenkopf MS, Dean RL, Deaver DR, Sanchez C, Namchuk M (2019) Samidorphan mitigates olanzapine-induced weight gain and metabolic dysfunction in rats and non-human primates. J Psychopharmacol 33:1303–1316. https://doi.org/10.1177/0269881119856850
Darcy KJ, Staras K, Collinson LM, Goda Y (2006) Constitutive sharing of recycling synaptic vesicles between presynaptic boutons. Nat Neurosci 9:315–321. https://doi.org/10.1038/nn1640
Datta D, Arion D, Roman KM, Volk DW, Lewis DA (2017) Altered expression of ARP2/3 complex signaling pathway genes in Prefrontal Layer 3 pyramidal

cells in Schizophrenia. Am J Psychiatry 174:163–171. https://doi.org/10.1176/appi.ajp.2016.16020204
Davidson LL, Heinrichs RW (2003) Quantification of frontal and temporal lobe brain-imaging findings in schizophrenia: a meta-analysis. Psychiatry Res 122:69–87
De Hert M, Correll CU, Bobes J, Cetkovich-Bakmas M, Cohen D, Asai I, Detraux J, Gautam S, Möller HJ, Ndetei DM et al (2011a) Physical illness in patients with severe mental disorders. I. Prevalence, impact of medications and disparities in health care. World Psychiatry 10:52–77. https://doi.org/10.1002/j.2051-5545.2011.tb00014.x
De Hert M, Cohen D, Bobes J, Cetkovich-Bakmas M, Leucht S, Ndetei DM, Newcomer JW, Uwakwe R, Asai I, Möller HJ et al (2011b) Physical illness in patients with severe mental disorders. II. Barriers to care, monitoring and treatment guidelines, plus recommendations at the system and individual level. World Psychiatry 10:138–151. https://doi.org/10.1002/j.2051-5545.2011.tb00036.x
Dean B, Gibbons AS, Boer S, Uezato A, Meador-Woodruff J, Scarr E, McCullumsmith RE (2016) Changes in cortical N-methyl-d-aspartate receptors and post-synaptic density protein 95 in schizophrenia, mood disorders and suicide. Aust N Z J Psychiatry 50:275–283. https://doi.org/10.1177/0004867415586601
Debray FG, Seneca S, Gonce M, Vancampenhaut K, Bianchi E, Boemer F, Weekers L, Smet J, Van Coster R (2014) Mitochondrial encephalomyopathy with cytochrome c oxidase deficiency caused by a novel mutation in the MTCO1 gene. Mitochondrion 17:101–105
DeFelipe J, Marco P, Busturia I, Merchán-Pérez A (1999) Estimation of the number of synapses in the cerebral cortex: methodological considerations. Cereb Cortex 9:722–732. https://doi.org/10.1093/cercor/9.7.722
DeFelipe J, Alonso-Nanclares L, Arellano JI (2002) Microstructure of the neocortex: comparative aspects. J Neurocytol 31:299–316
Delgadillo J, Lutz W (2020) A development pathway towards precision mental health care. JAMA Psychiatry 77:889–890. https://doi.org/10.1001/jamapsychiatry.2020.1048
Devin A, Rigoulet M (2007) Mechanisms of mitochondrial response to variations in energy demand in eukaryotic cells. Am J Physiol Cell Physiol 292:C52–C58. https://doi.org/10.1152/ajpcell.00208.2006
Devine MJ, Kittler JT (2018) Mitochondria at the neuronal presynapse in health and disease. Nat Rev Neurosci 19:63–80. https://doi.org/10.1038/nrn.2017.170
Díaz-García CM, Yellen G (2019) Neurons rely on glucose rather than astrocytic lactate during stimulation. J Neurosci Res 97:883–889. https://doi.org/10.1002/jnr.24374
Dienel GA (2012) Brain lactate metabolism: the discoveries and the controversies. J Cereb Blood Flow Metab 32:1107–1138. https://doi.org/10.1038/jcbfm.2011.175
Dienel SJ, Enwright JF 3rd, Hoftman GD, Lewis DA (2019) Markers of glutamate and GABA neurotransmission in the prefrontal cortex of schizophrenia subjects: Disease effects differ across anatomical levels of resolution. Schizophr Res 217:86. https://doi.org/10.1016/j.schres.2019.06.003
Dienel SJ, Dowling KF, Barile Z, Bazmi HH, Liu A, Vespoli JC, Fish KN, Lewis DA (2023) Diagnostic specificity and association with cognition of molecular alterations in prefrontal somatostatin neurons in schizophrenia. JAMA Psychiatry 80:1235–1245
DiNuzzo M, Mangia S, Maraviglia B, Giove F (2010) Changes in glucose uptake rather than lactate shuttle take center stage in subserving neuroenergetics: evidence from mathematical modeling. J Cereb Blood Flow Metab 30:586–602. https://doi.org/10.1038/jcbfm.2009.232
Dogan AE, Yuksel C, Du F, Chouinard VA, Ongur D (2018) Brain lactate and pH in schizophrenia and bipolar disorder: a systematic review of findings from magnetic resonance studies. Neuropsychopharmacology 43:1681–1690. https://doi.org/10.1038/s41386-018-0041-9
Dreher JC, Koch P, Kohn P, Apud J, Weinberger DR, Berman KF (2012) Common and differential pathophysiological features accompany comparable cognitive impairments in medication-free patients with schizophrenia and in healthy aging subjects. Biol Psychiatry 71:890
Du F, Yuksel C, Chouinard VA, Huynh P, Ryan K, Cohen BM, Ongur D (2018) Abnormalities in high-energy phosphate metabolism in first-episode bipolar disorder measured using (31)P-magnetic resonance spectroscopy. Biol Psychiatry 84:797–802. https://doi.org/10.1016/j.biopsych.2017.03.025
Duong A, Che Y, Ceylan D, Pinguelo A, Andreazza AC, Trevor Young L, Berk M (2016) Regulators of mitochondrial complex I activity: a review of literature and evaluation in postmortem prefrontal cortex from patients with bipolar disorder. Psychiatry Res 236:148–157. https://doi.org/10.1016/j.psychres.2015.12.015
Eastwood SL, Harrison PJ (2010) Markers of glutamate synaptic transmission and plasticity are increased in the anterior cingulate cortex in bipolar disorder. Biol Psychiatry 67:1010–1016. https://doi.org/10.1016/j.biopsych.2009.12.004
Edwards JL, Quattrini A, Lentz SI, Figueroa-Romero C, Cerri F, Backus C, Hong Y, Feldman EL (2010) Diabetes regulates mitochondrial biogenesis and fission in mouse neurons. Diabetologia 53:160–169. https://doi.org/10.1007/s00125-009-1553-y
Ellis SE, Panitch R, West AB, Arking DE (2016) Transcriptome analysis of cortical tissue reveals shared sets of downregulated genes in autism and schizophrenia. Transl Psychiatry 6:e817. https://doi.org/10.1038/tp.2016.87
Enwright JF III, Lewis DA (2021) Similarities in cortical transcriptome alterations between schizophrenia and bipolar disorder are related to the presence of psychosis. Schizophr Bull 47:1442–1451. https://doi.org/10.1093/schbul/sbaa195

Enwright Iii JF, Huo Z, Arion D, Corradi JP, Tseng G, Lewis DA (2018) Transcriptome alterations of prefrontal cortical parvalbumin neurons in schizophrenia. Mol Psychiatry 23:1606–1613. https://doi.org/10.1038/mp.2017.216

Enwright JFI, Huo Z, Arion D, Corradi JP, Tseng G, Lewis DA (2018) Transcriptome alterations of prefrontal cortical parvalbumin neurons in schizophrenia. Mol Psychiatry 23:1606–1613

Erecinska M, Wilson DF (1982) Regulation of cellular energy metabolism. J Membr Biol 70:1–14

Fattal O, Budur K, Vaughan AJ, Franco K (2006) Review of the literature on major mental disorders in adult patients with mitochondrial diseases. Psychosomatics 47:1–7

Fitton R, Sweetman J, Heseltine-Carp W, van der Feltz-Cornelis C (2022) Anti-inflammatory medications for the treatment of mental disorders: a scoping review. Brain Behav Immun Health 26:100518. https://doi.org/10.1016/j.bbih.2022.100518

Focking M, Dicker P, Lopez LM, Hryniewiecka M, Wynne K, English JA, Cagney G, Cotter DR (2016) Proteomic analysis of the postsynaptic density implicates synaptic function and energy pathways in bipolar disorder. Transl Psychiatry 6:e959. https://doi.org/10.1038/tp.2016.224

Fraunberger EA, Scola G, Laliberté VL, Duong A, Andreazza AC (2016) Redox modulations, antioxidants, and neuropsychiatric disorders. Oxidative Med Cell Longev 2016:4729192. https://doi.org/10.1155/2016/4729192

Freyberg Z, Aslanoglou D, Shah R, Ballon JS (2017) Intrinsic and antipsychotic drug-induced metabolic dysfunction in schizophrenia. Front Neurosci 11:432. https://doi.org/10.3389/fnins.2017.00432

Freyberg Z, Andreazza AC, McClung CA, Phillips ML (2024) Linking mitochondrial dysfunction, neurotransmitter, and neural network abnormalities and mania: elucidating neurobiological mechanisms of the therapeutic effect of the ketogenic diet in bipolar disorder. Biol Psychiatry Cogn Neurosci Neuroimaging 10:267. https://doi.org/10.1016/j.bpsc.2024.07.011

Fromer M, Pocklington AJ, Kavanagh DH, Williams HJ, Dwyer S, Gormley P, Georgieva L, Rees E, Palta P, Ruderfer DM et al (2014) De novo mutations in schizophrenia implicate synaptic networks. Nature 506:179–184. https://doi.org/10.1038/nature12929

Fromer M, Roussos P, Sieberts SK, Johnson JS, Kavanagh DH, Perumal TM, Ruderfer DM, Oh EC, Topol A, Shah HR et al (2016) Gene expression elucidates functional impact of polygenic risk for schizophrenia. Nat Neurosci 19:1442–1453. https://doi.org/10.1038/nn.4399

Fujimoto T, Takeuch K, Matsumoto T, Kamimura K, Hamada R, Nakamura K, Kato N (2007) Abnormal glucose metabolism in the anterior cingulate cortex in patients with schizophrenia. Psychiatry Res 154:49–58. https://doi.org/10.1016/j.pscychresns.2006.04.002

Gandal MJ, Haney JR, Parikshak NN, Leppa V, Ramaswami G, Hartl C, Schork AJ, Appadurai V, Buil A, Werge TM et al (2018a) Shared molecular neuropathology across major psychiatric disorders parallels polygenic overlap. Science 359:693–697. https://doi.org/10.1126/science.aad6469

Gandal MJ, Zhang P, Hadjimichael E, Walker RL, Chen C, Liu S, Won H, van Bakel H, Varghese M, Wang Y et al (2018b) Transcriptome-wide isoform-level dysregulation in ASD, schizophrenia, and bipolar disorder. Science 362. https://doi.org/10.1126/science.aat8127

Garbian Y, Ovadia O, Dadon S, Mishmar D (2010) Gene expression patterns of oxidative phosphorylation complex I subunits are organized in clusters. PLoS One 5:e9985. https://doi.org/10.1371/journal.pone.0009985

Genovese G, Fromer M, Stahl EA, Ruderfer DM, Chambert K, Landén M, Moran JL, Purcell SM, Sklar P, Sullivan PF et al (2016) Increased burden of ultra-rare protein-altering variants among 4,877 individuals with schizophrenia. Nat Neurosci 19:1433–1441. https://doi.org/10.1038/nn.4402

Giménez-Palomo A, Dodd S, Anmella G, Carvalho AF, Scaini G, Quevedo J, Pacchiarotti I, Vieta E, Berk M (2021) The role of mitochondria in mood disorders: from physiology to pathophysiology and to treatment. Front Psych 12. https://doi.org/10.3389/fpsyt.2021.546801

Girdhar K, Hoffman GE, Bendl J, Rahman S, Dong P, Liao W, Hauberg ME, Sloofman L, Brown L, Devillers O et al (2022) Chromatin domain alterations linked to 3D genome organization in a large cohort of schizophrenia and bipolar disorder brains. Nat Neurosci 25:474–483. https://doi.org/10.1038/s41593-022-01032-6

Glahn DC, Ragland JD, Abramoff A, Barrett J, Laird AR, Bearden CE, Velligan DI (2005) Beyond hypofrontality: a quantitative meta-analysis of functional neuroimaging studies of working memory in schizophrenia. Hum Brain Mapp 25:60–69

Glahn DC, Laird AR, Ellison-Wright I, Thelen SM, Robinson JL, Lancaster JL, Bullmore E, Fox PT (2008) Meta-analysis of gray matter anomalies in schizophrenia: application of anatomic likelihood estimation and network analysis. Biol Psychiatry 64:774–781

Glancy B, Kim Y, Katti P, Willingham TB (2020) The functional impact of mitochondrial structure across subcellular scales. Front Physiol 11:541040. https://doi.org/10.3389/fphys.2020.541040

Glantz LA, Lewis DA (2000) Decreased dendritic spine density on prefrontal cortical pyramidal neurons in schizophrenia. Arch Gen Psychiatry 57:65–73. https://doi.org/10.1001/archpsyc.57.1.65

Glausier JR, Lewis DA (2013) Dendritic spine pathology in schizophrenia. Neuroscience 251:90–107. https://doi.org/10.1016/j.neuroscience.2012.04.044

Glausier JR, Lewis DA (2018) Mapping pathologic circuitry in schizophrenia. Handb Clin Neurol 150:389–417. https://doi.org/10.1016/b978-0-444-63639-3.00025-6

Glausier JR, Bazmi HH, Lewis DA (2010) GABA A alpha 1 subunit mRNA expression in pyramidal cells and

interneurons in the dorsolateral prefrontal cortex of schizophrenia subjects. Biol Psychiatry 67:74S–74S
Glausier JR, Enwright JF 3rd, Lewis DA (2020) Diagnosis- and cell type-specific mitochondrial functional pathway signatures in schizophrenia and bipolar disorder. Am J Psychiatry 177:1140–1150. https://doi.org/10.1176/appi.ajp.2020.19111210
Glausier JR, Bouchet-Marquis C, Maier M, Banks-Tibbs T, Wu K, Ning J, Melchitzky D, Lewis DA, Freyberg Z (2024) Volume electron microscopy reveals 3D synaptic nanoarchitecture in postmortem human prefrontal cortex. bioRxiv. https://doi.org/10.1101/2024.02.26.582174
Glebov OO, Cox S, Humphreys L, Burrone J (2016) Neuronal activity controls transsynaptic geometry. Sci Rep 6:22703. https://doi.org/10.1038/srep22703
Glebov OO, Jackson RE, Winterflood CM, Owen DM, Barker EA, Doherty P, Ewers H, Burrone J (2017) Nanoscale structural plasticity of the active zone matrix modulates presynaptic function. Cell Rep 18:2715–2728. https://doi.org/10.1016/j.celrep.2017.02.064
Gokhale A, Hartwig C, Freeman AAH, Bassell JL, Zlatic SA, Sapp Savas C, Vadlamudi T, Abudulai F, Pham TT, Crocker A et al (2019) Systems analysis of the 22q11.2 microdeletion syndrome converges on a mitochondrial interactome necessary for synapse function and behavior. J Neurosci 39:3561–3581. https://doi.org/10.1523/jneurosci.1983-18.2019
Gokhale A, Lee CE, Zlatic SA, Freeman AAH, Shearing N, Hartwig C, Ogunbona O, Bassell JL, Wynne ME, Werner E et al (2021) Mitochondrial proteostasis requires genes encoded in a neurodevelopmental syndrome locus. J Neurosci 41:6596–6616. https://doi.org/10.1523/jneurosci.2197-20.2021
Goncalves VF, Cappi C, Hagen CM, Sequeira A, Vawter MP, Derkach A, Zai CC, Hedley PL, Bybjerg-Grauholm J, Pouget JG et al (2018) A comprehensive analysis of nuclear-encoded mitochondrial genes in Schizophrenia. Biol Psychiatry 83:780–789. https://doi.org/10.1016/j.biopsych.2018.02.1175
González P, Lozano P, Ros G, Solano F (2023) Hyperglycemia and oxidative stress: an integral, updated and critical overview of their metabolic interconnections. Int J Mol Sci 24. https://doi.org/10.3390/ijms24119352
Goriely A (2024) Eighty-six billion and counting: do we know the number of neurons in the human brain? Brain 148:689. https://doi.org/10.1093/brain/awae390
Grande I, Berk M, Birmaher B, Vieta E (2016) Bipolar disorder. Lancet 387:1561–1572. https://doi.org/10.1016/S0140-6736(15)00241-X
Green MF (2016) Impact of cognitive and social cognitive impairment on functional outcomes in patients with schizophrenia. J Clin Psychiatry 77(Suppl 2):8–11. https://doi.org/10.4088/JCP.14074su1c.02
Green IW, Glausier JR (2015) Different paths to core pathology: the equifinal model of the schizophrenia syndrome. Schizophr Bull 42:542. https://doi.org/10.1093/schbul/sbv136
Green MF, Harvey PD (2014) Cognition in schizophrenia: past, present, and future. Schizophr Res Cogn 1:e1–e9. https://doi.org/10.1016/j.scog.2014.02.001
Greenhalgh AM, Gonzalez-Blanco L, Garcia-Rizo C, Fernandez-Egea E, Miller B, Arroyo MB, Kirkpatrick B (2017) Meta-analysis of glucose tolerance, insulin, and insulin resistance in antipsychotic-naïve patients with nonaffective psychosis. Schizophr Res 179:57–63. https://doi.org/10.1016/j.schres.2016.09.026
Grzenda A, Widge AS (2024) Electronic health records and stratified psychiatry: bridge to precision treatment? Neuropsychopharmacology 49:285–290. https://doi.org/10.1038/s41386-023-01724-y
Guan J, Cai JJ, Ji G, Sham PC (2019) Commonality in dysregulated expression of gene sets in cortical brains of individuals with autism, schizophrenia, and bipolar disorder. Transl Psychiatry 9:152. https://doi.org/10.1038/s41398-019-0488-4
Gupta M, Pandey S, Rumman M, Singh B, Mahdi AA (2023) Molecular mechanisms underlying hyperglycemia associated cognitive decline. IBRO Neurosci Rep 14:57–63. https://doi.org/10.1016/j.ibneur.2022.12.006
Gutiérrez-Rojas L, de la Gándara Martín JJ, García Buey L, Uriz Otano JI, Mena Á, Roncero C (2023) Patients with severe mental illness and hepatitis C virus infection benefit from new pangenotypic direct-acting antivirals: Results of a literature review. Gastroenterol Hepatol 46:382–396. https://doi.org/10.1016/j.gastrohep.2022.06.001
Habtewold TD, Islam MA, Liemburg EJ, Bartels-Velthuis AA, van Beveren NJ, Cahn W, de Haan L, Delespaul P, Meijer CJ, Myin-Germeys I (2020) Polygenic risk score for schizophrenia was not associated with glycemic level (HbA1c) in patients with non-affective psychosis: genetic risk and outcome of psychosis (GROUP) cohort study. J Psychosom Res 132:109968
Hackenbrock CR (1972a) States of activity and structure in mitochondrial membranes. Ann NY Acad Sci 195:492–505
Hackenbrock CR (1972b) Energy-linked ultrastructural transformations in isolated liver mitochondria and mitoplasts. Preservation of configurations by freeze-cleaving compared to chemical fixation. J Cell Biol 53:450–465
Hackenbrock CR (1981) Energy-linked condensed-orthodox ultrastructural transformations in mitochondria. Chemotherapy 27(Suppl 2):21–26
Hackinger S, Prins B, Mamakou V, Zengini E, Marouli E, Brčić L, Serafetinidis I, Lamnissou K, Kontaxakis V, Dedoussis G (2018) Evidence for genetic contribution to the increased risk of type 2 diabetes in schizophrenia. Transl Psychiatry 8:252
Hall J, Bray NJ (2022) Schizophrenia genomics: convergence on synaptic development, adult synaptic plasticity, or both? Biol Psychiatry 91:709–717. https://doi.org/10.1016/j.biopsych.2021.10.018
Hall CN, Klein-Flugge MC, Howarth C, Attwell D (2012) Oxidative phosphorylation, not glycolysis, powers presynaptic and postsynaptic mechanisms

underlying brain information processing. J Neurosci 32:8940–8951
Hall J, Trent S, Thomas KL, O'Donovan MC, Owen MJ (2015) Genetic risk for schizophrenia: convergence on synaptic pathways involved in plasticity. Biol Psychiatry 77:52–58. https://doi.org/10.1016/j.biopsych.2014.07.011
Harris KM, Weinberg RJ (2012) Ultrastructure of synapses in the mammalian brain. Cold Spring Harb Perspect Biol 4. https://doi.org/10.1101/cshperspect.a005587
Harris JJ, Jolivet R, Attwell D (2012) Synaptic energy use and supply. Neuron 75:762–777. https://doi.org/10.1016/j.neuron.2012.08.019
Hartz SM, Pato CN, Medeiros H, Cavazos-Rehg P, Sobell JL, Knowles JA, Bierut LJ, Pato MT, Consortium, f.t.G.P.C (2014) Comorbidity of severe psychotic disorders with measures of substance use. JAMA Psychiatry 71:248–254. https://doi.org/10.1001/jamapsychiatry.2013.3726
Hazlett EA, Vaccaro DH, Haznedar MM, Goldstein KE (2019) (F-18) Fluorodeoxyglucose positron emission tomography studies of the schizophrenia spectrum: the legacy of Monte S. Buchsbaum, M.D. Psychiatry Res 271:535–540. https://doi.org/10.1016/j.psychres.2018.12.030
Hendry SHC, Schwark HD, Jones EG, Yan J (1987) Numbers and proportions of GABA-immunoreactive neurons in different areas of monkey cerebral cortex. J Neurosci 7:1503–1519
Henkel ND, Wu X, O'Donovan SM, Devine EA, Jiron JM, Rowland LM, Sarnyai Z, Ramsey AJ, Wen Z, Hahn MK, McCullumsmith RE (2022) Schizophrenia: a disorder of broken brain bioenergetics. Mol Psychiatry 27:2393. https://doi.org/10.1038/s41380-022-01494-x
Herculano-Houzel S (2009) The human brain in numbers: a linearly scaled-up primate brain. Front Hum Neurosci 3:31. https://doi.org/10.3389/neuro.09.031.2009
Herculano-Houzel S, Rothman DL (2022) From a demand-based to a supply-limited framework of brain metabolism. Front Integr Neurosci 16:818685. https://doi.org/10.3389/fnint.2022.818685
Hintze A, Lange F, Steyer AM, Anstatt J, Möbius W, Jakobs S, Wichmann C (2024) Developmental changes of the mitochondria in the murine anteroventral cochlear nucleus. iScience 27:108700. https://doi.org/10.1016/j.isci.2023.108700
Hjelm BE, Rollins B, Mamdani F, Lauterborn JC, Kirov G, Lynch G, Gall CM, Sequeira A, Vawter MP (2015) Evidence of mitochondrial dysfunction within the complex genetic etiology of schizophrenia. Mol Neuropsychiatry 1:201–219. https://doi.org/10.1159/000441252
Hoftman GD, Dienel SJ, Bazmi HH, Zhang Y, Chen K, Lewis DA (2018) Altered gradients of glutamate and gamma-aminobutyric acid transcripts in the cortical visuospatial working memory network in Schizophrenia. Biol Psychiatry 83:670–679. https://doi.org/10.1016/j.biopsych.2017.11.029
Holler S, Köstinger G, Martin KAC, Schuhknecht GFP, Stratford KJ (2021) Structure and function of a neocortical synapse. Nature 591:111–116. https://doi.org/10.1038/s41586-020-03134-2
Holper L, Ben-Shachar D, Mann JJ (2019) Multivariate meta-analyses of mitochondrial complex I and IV in major depressive disorder, bipolar disorder, schizophrenia, Alzheimer disease, and Parkinson disease. Neuropsychopharmacology 44:837–849. https://doi.org/10.1038/s41386-018-0090-0
Holt RIG, Mitchell AJ (2015) Diabetes mellitus and severe mental illness: mechanisms and clinical implications. Nat Rev Endocrinol 11:79–89. https://doi.org/10.1038/nrendo.2014.203
Holtmaat A, Svoboda K (2009) Experience-dependent structural synaptic plasticity in the mammalian brain. Nat Rev Neurosci 10:647–658. https://doi.org/10.1038/nrn2699
Howarth C, Gleeson P, Attwell D (2012) Updated energy budgets for neural computation in the neocortex and cerebellum. J Cereb Blood Flow Metab 32:1222–1232. https://doi.org/10.1038/jcbfm.2012.35
Howells FM, Hsieh JH, Temmingh HS, Baldwin DS, Stein DJ (2022) Capacity for cortical excitation is reduced in psychotic disorders: an investigation of the TMS-EMG cortical silent period. Schizophr Res 240:73–77. https://doi.org/10.1016/j.schres.2021.12.012
Howes OD, Onwordi EC (2023) The synaptic hypothesis of schizophrenia version III: a master mechanism. Mol Psychiatry 28:1843. https://doi.org/10.1038/s41380-023-02043-w
Hu B, Won H, Mah W, Park RB, Kassim B, Spiess K, Kozlenkov A, Crowley CA, Pochareddy S, Li Y et al (2021) Neuronal and glial 3D chromatin architecture informs the cellular etiology of brain disorders. Nat Commun 12:3968. https://doi.org/10.1038/s41467-021-24243-0
Huang C-C, Lee C-C, Hsu K-S (2010) The role of insulin receptor signaling in synaptic plasticity and cognitive function. Chang Gung Med J 33:115–125
Hughes E, Bassi S, Gilbody S, Bland M, Martin F (2016) Prevalence of HIV, hepatitis B, and hepatitis C in people with severe mental illness: a systematic review and meta-analysis. Lancet Psychiatry 3:40–48. https://doi.org/10.1016/S2215-0366(15)00357-0
Hyder F, Patel AB, Gjedde A, Rothman DL, Behar KL, Shulman RG (2006) Neuronal-glial glucose oxidation and glutamatergic-GABAergic function. J Cereb Blood Flow Metab 26:865–877. https://doi.org/10.1038/sj.jcbfm.9600263
Hyder F, Rothman DL, Bennett MR (2013a) Cortical energy demands of signaling and nonsignaling components in brain are conserved across mammalian species and activity levels. Proc Natl Acad Sci USA 110:3549–3554. https://doi.org/10.1073/pnas.1214912110
Hyder F, Fulbright RK, Shulman RG, Rothman DL (2013b) Glutamatergic function in the resting awake human brain is supported by uniformly high oxidative

energy. J Cereb Blood Flow Metab 33:339–347. https://doi.org/10.1038/jcbfm.2012.207

Ivannikov MV, Sugimori M, Llinás RR (2013) Synaptic vesicle exocytosis in hippocampal synaptosomes correlates directly with total mitochondrial volume. J Mol Neurosci 49:223–230. https://doi.org/10.1007/s12031-012-9848-8

Jaffe AE, Straub RE, Shin JH, Tao R, Gao Y, Collado-Torres L, Kam-Thong T, Xi HS, Quan J, Chen Q et al (2018) Developmental and genetic regulation of the human cortex transcriptome illuminate schizophrenia pathogenesis. Nat Neurosci 21:1117–1125. https://doi.org/10.1038/s41593-018-0197-y

Jeon P, Limongi R, Ford SD, Mackinley M, Dempster K, Théberge J, Palaniyappan L (2021) Progressive changes in glutamate concentration in early stages of schizophrenia: a longitudinal 7-Tesla MRS study. Schizophr Bull Open 2:sgaa072. https://doi.org/10.1093/schizbullopen/sgaa072

Jin H, Mosweu I (2017) The societal cost of Schizophrenia: a systematic review. PharmacoEconomics 35:25–42. https://doi.org/10.1007/s40273-016-0444-6

Jones EG (1993) GABAergic neurons and their role in cortical plasticity in primates. Cereb Cortex 3:361–372

Justs KA, Lu Z, Chouhan AK, Borycz JA, Lu Z, Meinertzhagen IA, Macleod GT (2022) Presynaptic mitochondrial volume and packing density scale with presynaptic power demand. J Neurosci 42:954–967. https://doi.org/10.1523/jneurosci.1236-21.2021

Kahn RS (2019) On the specificity of continuous cognitive decline in schizophrenia. Am J Psychiatry 176:774–776. https://doi.org/10.1176/appi.ajp.2019.19080794

Kann O (2011) The energy demand of fast neuronal network oscillations: insights from brain slice preparations. Front Pharmacol 2:90

Kann O (2016) The interneuron energy hypothesis: implications for brain disease. Neurobiol Dis 90:75–85. https://doi.org/10.1016/j.nbd.2015.08.005

Kann O (2024) Lactate as a supplemental fuel for synaptic transmission and neuronal network oscillations: potentials and limitations. J Neurochem 168:608–631. https://doi.org/10.1111/jnc.15867

Kann O, Kovacs R (2007) Mitochondria and neuronal activity. Am J Physiol Cell Physiol 292:C641–C657

Kann O, Huchzermeyer C, Kovacs R, Wirtz S, Schuelke M (2011) Gamma oscillations in the hippocampus require high complex I gene expression and strong functional performance of mitochondria. Brain 134:345–358

Kann O, Papageorgiou IE, Draguhn A (2014) Highly energized inhibitory interneurons are a central element for information processing in cortical networks. J Cereb Blood Flow Metab 34:1270–1282. https://doi.org/10.1038/jcbfm.2014.104

Kann O, Hollnagel JO, Elzoheiry S, Schneider J (2016) Energy and potassium ion homeostasis during gamma oscillations. Front Mol Neurosci 9:47. https://doi.org/10.3389/fnmol.2016.00047

Kato T (2017) Neurobiological basis of bipolar disorder: Mitochondrial dysfunction hypothesis and beyond. Schizophr Res 187:62–66. https://doi.org/10.1016/j.schres.2016.10.037

Kato T, Kato N (2000) Mitochondrial dysfunction in bipolar disorder. Bipolar Disord 2:180–190. https://doi.org/10.1034/j.1399-5618.2000.020305.x

Kida I, Hyder F (2006) Physiology of functional magnetic resonance imaging: energetics and function. Methods Mol Med 124:175–195. https://doi.org/10.1385/1-59745-010-3:175

Kim M, Haney JR, Zhang P, Hernandez LM, Wang LK, Perez-Cano L, Loohuis LMO, de la Torre-Ubieta L, Gandal MJ (2021) Brain gene co-expression networks link complement signaling with convergent synaptic pathology in schizophrenia. Nat Neurosci 24:799–809. https://doi.org/10.1038/s41593-021-00847-z

Kimoto S, Glausier JR, Fish KN, Volk DW, Bazmi HH, Arion D, Datta D, Lewis DA (2016) Reciprocal alterations in regulator of G protein signaling 4 and microRNA16 in Schizophrenia. Schizophr Bull 42:396–405. https://doi.org/10.1093/schbul/sbv139

Kirov G, Pocklington AJ, Holmans P, Ivanov D, Ikeda M, Ruderfer D, Moran J, Chambert K, Toncheva D, Georgieva L et al (2012) De novo CNV analysis implicates specific abnormalities of postsynaptic signalling complexes in the pathogenesis of schizophrenia. Mol Psychiatry 17:142–153. https://doi.org/10.1038/mp.2011.154

Kohen D (2004) Diabetes mellitus and schizophrenia: historical perspective. Br J Psychiatry Suppl 47:S64–S66

Kolb H, Kempf K, Röhling M, Lenzen-Schulte M, Schloot NC, Martin S (2021) Ketone bodies: from enemy to friend and guardian angel. BMC Med 19:313. https://doi.org/10.1186/s12916-021-02185-0

Kolluri N, Sun Z, Sampson AR, Lewis DA (2005) Lamina-specific reductions in dendritic spine density in the prefrontal cortex of subjects with schizophrenia. Am J Psychiatry 162:1200–1202. https://doi.org/10.1176/appi.ajp.162.6.1200

Konopaske GT, Lange N, Coyle JT, Benes FM (2014) Prefrontal cortical dendritic spine pathology in schizophrenia and bipolar disorder. JAMA Psychiatry 71:1323–1331. https://doi.org/10.1001/jamapsychiatry.2014.1582

Konopaske GT, Balu DT, Presti KT, Chan G, Benes FM, Coyle JT (2018) Dysbindin-1 contributes to prefrontal cortical dendritic arbor pathology in schizophrenia. Schizophr Res 201:270–277

Kraguljac NV, Lahti AC (2021) Neuroimaging as a window into the pathophysiological mechanisms of schizophrenia. Front Psych 12:613764. https://doi.org/10.3389/fpsyt.2021.613764

Kraguljac NV, Reid M, White D, Jones R, den Hollander J, Lowman D, Lahti AC (2012) Neurometabolites in schizophrenia and bipolar disorder – a systematic review and meta-analysis. Psychiatry Res 203:111–125. https://doi.org/10.1016/j.pscychresns.2012.02.003

Krystal JH, Anticevic A, Yang GJ, Dragoi G, Driesen NR, Wang XJ, Murray JD (2017) Impaired tuning of neural ensembles and the pathophysiology of schizophrenia: a translational and computational neuroscience perspective. Biol Psychiatry 81:874–885

Kubota Y, Kawaguchi Y (2000) Dependence of GABAergic synaptic areas on the interneuron type and target size. J Neurosci 20:375–386

Kubota Y, Kondo S, Nomura M, Hatada S, Yamaguchi N, Mohamed AA, Karube F, Lubke J, Kawaguchi Y (2015) Functional effects of distinct innervation styles of pyramidal cells by fast spiking cortical interneurons. elife 4. https://doi.org/10.7554/eLife.07919

Lam X-J, Xu B, Yeo P-L, Cheah P-S, Ling K-H (2023) Mitochondria dysfunction and bipolar disorder: from pathology to therapy. IBRO Neurosci Rep 14:407–418. https://doi.org/10.1016/j.ibneur.2023.04.002

Lanz TA, Reinhart V, Sheehan MJ, Rizzo SJS, Bove SE, James LC, Volfson D, Lewis DA, Kleiman RJ (2019) Postmortem transcriptional profiling reveals widespread increase in inflammation in schizophrenia: a comparison of prefrontal cortex, striatum, and hippocampus among matched tetrads of controls with subjects diagnosed with schizophrenia, bipolar or major depressive disorder. Transl Psychiatry 9:151. https://doi.org/10.1038/s41398-019-0492-8

Launders N, Hayes JF, Price G, Osborn DP (2021) Clustering of physical health multimorbidity in 68,392 people with severe mental illness and matched comparators: a lifetime prevalence analysis of United Kingdom primary care data. medRxiv, 2021.2004.2030.21256296. https://doi.org/10.1101/2021.04.30.21256296

Lee HG, Wheeler MA, Quintana FJ (2022) Function and therapeutic value of astrocytes in neurological diseases. Nat Rev Drug Discov 21:339–358. https://doi.org/10.1038/s41573-022-00390-x

Lee J, Xue X, Au E, McIntyre WB, Asgariroozbehani R, Tseng GC, Papoulias M, Panganiban K, Agarwal SM, McCullumsmith R et al (2024) Central insulin dysregulation in antipsychotic-naïve first-episode psychosis: In silico exploration of gene expression signatures. Psychiatry Res 331:115636. https://doi.org/10.1016/j.psychres.2023.115636

Lennie P (2003) The cost of cortical computation. Curr Biol 13:493–497. https://doi.org/10.1016/s0960-9822(03)00135-0

Levy WB, Calvert VG (2021) Communication consumes 35 times more energy than computation in the human cortex, but both costs are needed to predict synapse number. Proc Natl Acad Sci USA 118. https://doi.org/10.1073/pnas.2008173118

Lewis DA, Curley AA, Glausier JR, Volk DW (2012) Cortical parvalbumin interneurons and cognitive dysfunction in schizophrenia. Trends Neurosci 35:57–67. https://doi.org/10.1016/j.tins.2011.10.004

Lezmy J, Harris JJ, Attwell D (2021) Optimising the energetic cost of the glutamatergic synapse. Neuropharmacology 197:108727. https://doi.org/10.1016/j.neuropharm.2021.108727

Li S, Sheng ZH (2021) Energy matters: presynaptic metabolism and the maintenance of synaptic transmission. Nat Rev Neurosci 23:4. https://doi.org/10.1038/s41583-021-00535-8

Li Z, Chen P, Chen J, Xu Y, Wang Q, Li X, Li C, He L, Shi Y (2018) Glucose and insulin-related traits, type 2 diabetes and risk of schizophrenia: a Mendelian randomization study. EBioMedicine 34:182–188

Li S, Xiong GJ, Huang N, Sheng ZH (2020) The crosstalk of energy sensing and mitochondrial anchoring sustains synaptic efficacy by maintaining presynaptic metabolism. Nat Metab 2:1077–1095. https://doi.org/10.1038/s42255-020-00289-0

Li J, Tran OT, Crowley TB, Moore TM, Zackai EH, Emanuel BS, McDonald-McGinn DM, Gur RE, Wallace DC, Anderson SA (2021) Association of mitochondrial biogenesis with variable penetrance of schizophrenia. JAMA Psychiatry 78:911–921. https://doi.org/10.1001/jamapsychiatry.2021.0762

Li H, Guglielmetti C, Sei YJ, Zilberter M, Le Page LM, Shields L, Yang J, Nguyen K, Tiret B, Gao X et al (2023) Neurons require glucose uptake and glycolysis in vivo. Cell Rep 42. https://doi.org/10.1016/j.celrep.2023.112335

Lieberman JA 3rd (2004) Metabolic changes associated with antipsychotic use. Prim Care Companion J Clin Psychiatry 6:8–13

Lin PI, Shuldiner AR (2010) Rethinking the genetic basis for comorbidity of schizophrenia and type 2 diabetes. Schizophr Res 123:234–243. https://doi.org/10.1016/j.schres.2010.08.022

López-Doménech G, Kittler JT (2023) Mitochondrial regulation of local supply of energy in neurons. Curr Opin Neurobiol 81:102747. https://doi.org/10.1016/j.conb.2023.102747

Ma H, Cai Q, Lu W, Sheng ZH, Mochida S (2009) KIF5B motor adaptor syntabulin maintains synaptic transmission in sympathetic neurons. J Neurosci 29:13019–13029. https://doi.org/10.1523/jneurosci.2517-09.2009

MacAskill AF, Atkin TA, Kittler JT (2010) Mitochondrial trafficking and the provision of energy and calcium buffering at excitatory synapses. Eur J Neurosci 32:231–240

MacDonald ML, Naydenov A, Chu M, Matzilevich D, Konradi C (2006) Decrease in creatine kinase messenger RNA expression in the hippocampus and dorsolateral prefrontal cortex in bipolar disorder. Bipolar Disord 8:255–264. https://doi.org/10.1111/j.1399-5618.2006.00302.x

MacDonald ML, Alhassan J, Newman JT, Richard M, Gu H, Kelly RM, Sampson AR, Fish KN, Penzes P, Wills ZP et al (2017) Selective loss of smaller spines in Schizophrenia. Am J Psychiatry 174:586–594. https://doi.org/10.1176/appi.ajp.2017.16070814

MacDonald ML, Garver M, Newman J, Sun Z, Kannarkat J, Salisbury R, Glausier J, Ding Y, Lewis DA, Yates N, Sweet RA (2020) Synaptic proteome alterations in the primary auditory cortex of individuals with Schizophrenia. JAMA Psychiatry 77:86–95. https://doi.org/10.1001/jamapsychiatry.2019.2974

Machado AK, Pan AY, da Silva TM, Duong A, Andreazza AC (2016) Upstream pathways controlling mitochondrial function in major psychosis: a focus on bipolar disorder. Can J Psychiatr 61:446–456. https://doi.org/10.1177/0706743716648297

Maher F, Davies-Hill TM, Lysko PG, Henneberry RC, Simpson IA (1991) Expression of two glucose transporters, GLUT1 and GLUT3, in cultured cerebellar neurons: evidence for neuron-specific expression of GLUT3. Mol Cell Neurosci 2:351–360. https://doi.org/10.1016/1044-7431(91)90066-w

Maj C, Tosato S, Zanardini R, Lasalvia A, Favaro A, Leuci E, De Girolamo G, Ruggeri M, Gennarelli M, Bocchio-Chiavetto L (2020) Correlations between immune and metabolic serum markers and schizophrenia/bipolar disorder polygenic risk score in first-episode psychosis. Early Interv Psychiatry 14:507–511. https://doi.org/10.1111/eip.12906

Mangia S, Simpson IA, Vannucci SJ, Carruthers A (2009) The in vivo neuron-to-astrocyte lactate shuttle in human brain: evidence from modeling of measured lactate levels during visual stimulation. J Neurochem 109(Suppl 1):55–62. https://doi.org/10.1111/j.1471-4159.2009.06003.x

Mangia S, DiNuzzo M, Giove F, Carruthers A, Simpson IA, Vannucci SJ (2011) Response to 'comment on recent modeling studies of astrocyte-neuron metabolic interactions': much ado about nothing. J Cereb Blood Flow Metab 31:1346–1353. https://doi.org/10.1038/jcbfm.2011.29

Manji H, Kato T, Di Prospero NA, Ness S, Beal MF, Krams M, Chen G (2012) Impaired mitochondrial function in psychiatric disorders. Nat Rev Neurosci 13:293–307

Manseau M, Bogenschutz M (2016) Substance use disorders and schizophrenia. Focus (Am Psychiatr Publ) 14:333–342. https://doi.org/10.1176/appi.focus.20160008

Marchi S, Guilbaud E, Tait SWG, Yamazaki T, Galluzzi L (2023) Mitochondrial control of inflammation. Nat Rev Immunol 23:159–173. https://doi.org/10.1038/s41577-022-00760-x

Matsuzaki M, Honkura N, Ellis-Davies GC, Kasai H (2004) Structural basis of long-term potentiation in single dendritic spines. Nature 429:761–766

McCutcheon RA, Keefe RSE, McGuire PK (2023) Cognitive impairment in schizophrenia: aetiology, pathophysiology, and treatment. Mol Psychiatry 28:1902–1918. https://doi.org/10.1038/s41380-023-01949-9

Mendelsohn R, Garcia GC, Bartol TM, Lee CT, Khandelwal P, Liu E, Spencer DJ, Husar A, Bushong EA, Phan S et al (2022) Morphological principles of neuronal mitochondria. J Comp Neurol 530:886–902. https://doi.org/10.1002/cne.25254

Mental Disorders Collaborators G.M.D (2022) Global, regional, and national burden of 12 mental disorders in 204 countries and territories, 1990-2019: a systematic analysis for the Global Burden of Disease Study 2019. Lancet Psychiatry 9:137–150. https://doi.org/10.1016/s2215-0366(21)00395-3

Merritt K, Egerton A, Kempton MJ, Taylor MJ, McGuire PK (2016) Nature of glutamate alterations in schizophrenia: a meta-analysis of proton magnetic resonance spectroscopy studies. JAMA Psychiatry 73:665–674. https://doi.org/10.1001/jamapsychiatry.2016.0442

Merritt K, McGuire PK, Egerton A, Aleman A, Block W, Bloemen OJN, Borgan F, Bustillo JR, Capizzano AA, Coughlin JM et al (2021) Association of age, antipsychotic medication, and symptom severity in schizophrenia with proton magnetic resonance spectroscopy brain glutamate level: a mega-analysis of individual participant-level data. JAMA Psychiatry 78:667–681. https://doi.org/10.1001/jamapsychiatry.2021.0380

Meyer D, Bonhoeffer T, Scheuss V (2014) Balance and stability of synaptic structures during synaptic plasticity. Neuron 82:430–443. https://doi.org/10.1016/j.neuron.2014.02.031

Miller EK, Cohen JD (2001) An integrative theory of prefrontal cortex function. Annu Rev Neurosci 24:167–202. https://doi.org/10.1146/annurev.neuro.24.1.167

Miller AH, Haroon E, Raison CL, Felger JC (2013) Cytokine targets in the brain: impact on neurotransmitters and neurocircuits. Depress Anxiety 30:297–306. https://doi.org/10.1002/da.22084

Miller EK, Lundqvist M, Bastos AM (2018) Working Memory 2.0. Neuron 100:463–475. https://doi.org/10.1016/j.neuron.2018.09.023

Minzenberg MJ, Laird AR, Thelen S, Carter CS, Glahn DC (2009) Meta-analysis of 41 functional neuroimaging studies of executive function in schizophrenia. Arch Gen Psychiatry 66:811–822. https://doi.org/10.1001/archgenpsychiatry.2009.91

Minzenberg MJ, Firl AJ, Yoon JH, Gomes GC, Reinking C, Carter CS (2010) Gamma oscillatory power is impaired during cognitive control independent of medication status in first-episode schizophrenia. Neuropsychopharmacology 35:2590–2599

Misgeld T, Schwarz TL (2017) Mitostasis in neurons: maintaining mitochondria in an extended cellular architecture. Neuron 96:651–666. https://doi.org/10.1016/j.neuron.2017.09.055

Mitelman SA, Bralet MC, Mehmet Haznedar M, Hollander E, Shihabuddin L, Hazlett EA, Buchsbaum MS (2018) Positron emission tomography assessment of cerebral glucose metabolic rates in autism spectrum disorder and schizophrenia. Brain Imaging Behav 12:532–546. https://doi.org/10.1007/s11682-017-9721-z

Mulica P, Grünewald A, Pereira SL (2021) Astrocyte-neuron metabolic crosstalk in neurodegeneration: a mitochondrial perspective. Front Endocrinol (Lausanne) 12:668517. https://doi.org/10.3389/fendo.2021.668517

Mullins N, Forstner AJ, O'Connell KS, Coombes B, Coleman JRI, Qiao Z, Als TD, Bigdeli TB, Børte S, Bryois J et al (2021) Genome-wide association study of more than 40,000 bipolar disorder cases provides new insights into the underlying biology. Nat Genet 53:817–829. https://doi.org/10.1038/s41588-021-00857-4

Murphy CE, Walker AK, Weickert CS (2021) Neuroinflammation in schizophrenia: the role of nuclear factor kappa B. Transl Psychiatry 11:528. https://doi.org/10.1038/s41398-021-01607-0

Murthy VN, Schikorski T, Stevens CF, Zhu Y (2001) Inactivity produces increases in neurotransmitter release and synapse size. Neuron 32:673–682. https://doi.org/10.1016/s0896-6273(01)00500-1

Na EJ, Lee J, Sohn JH, Yang M, Park Y, Sim HB, Lee H (2024) Characteristics of comorbid physical disease in patients with severe mental illness in South Korea: a nationwide population-based study (2014–2019). Psychiatry Investig 21:361–370. https://doi.org/10.30773/pi.2023.0224

Nakahara T, Tsugawa S, Noda Y, Ueno F, Honda S, Kinjo M, Segawa H, Hondo N, Mori Y, Watanabe H et al (2021) Glutamatergic and GABAergic metabolite levels in schizophrenia-spectrum disorders: a meta-analysis of 1H-magnetic resonance spectroscopy studies. Mol Psychiatry 27:744. https://doi.org/10.1038/s41380-021-01297-6

Nie F, Wong-Riley MT (1995) Double labeling of GABA and cytochrome oxidase in the macaque visual cortex: quantitative EM analysis. J Comp Neurol 356:115–131

Nie F, Wong-Riley MT (1996a) Mitochondrial- and nuclear-encoded subunits of cytochrome oxidase in neurons: differences in compartmental distribution, correlation with enzyme activity, and regulation by neuronal activity. J Comp Neurol 373:139–155

Nie F, Wong-Riley MT (1996b) Differential glutamatergic innervation in cytochrome oxidase-rich and -poor regions of the macaque striate cortex: quantitative EM analysis of neurons and neuropil. J Comp Neurol 369:571–590

Nishioka M, Takayama J, Sakai N, Kazuno A-A, Ishiwata M, Ueda J, Hayama T, Fujii K, Someya T, Kuriyama S et al (2023) Deep exome sequencing identifies enrichment of deleterious mosaic variants in neurodevelopmental disorder genes and mitochondrial tRNA regions in bipolar disorder. Mol Psychiatry 28:4294–4306. https://doi.org/10.1038/s41380-023-02096-x

Nusser Z, Cull-Candy S, Farrant M (1997) Differences in synaptic GABA(A) receptor number underlie variation in GABA mini amplitude. Neuron 19:697–709

Nusser Z, Lujan R, Laube G, Roberts JD, Molnar E, Somogyi P (1998) Cell type and pathway dependence of synaptic AMPA receptor number and variability in the hippocampus. Neuron 21:545–559

Okuda T, Kimoto S, Kawabata R, Bian Y, Tsubomoto M, Okamura K, Enwright JF, Kikuchi M, Lewis DA, Hashimoto T (2024) Alterations in inhibitory neuron subtype-selective transcripts in the prefrontal cortex: comparisons across schizophrenia and mood disorders. Psychol Med 54:1–10. https://doi.org/10.1017/s0033291724002344

Onwordi EC, Halff EF, Whitehurst T, Mansur A, Cotel MC, Wells L, Creeney H, Bonsall D, Rogdaki M, Shatalina E et al (2020) Synaptic density marker SV2A is reduced in schizophrenia patients and unaffected by antipsychotics in rats. Nat Commun 11:246. https://doi.org/10.1038/s41467-019-14122-0

Onwordi EC, Whitehurst T, Mansur A, Statton B, Berry A, Quinlan M, O'Regan DP, Rogdaki M, Marques TR, Rabiner EA et al (2021) The relationship between synaptic density marker SV2A, glutamate and N-acetyl aspartate levels in healthy volunteers and schizophrenia: a multimodal PET and magnetic resonance spectroscopy brain imaging study. Transl Psychiatry 11:393. https://doi.org/10.1038/s41398-021-01515-3

Osborn DP, Levy G, Nazareth I, Petersen I, Islam A, King MB (2007) Relative risk of cardiovascular and cancer mortality in people with severe mental illness from the United Kingdom's general practice research database. Arch Gen Psychiatry 64:242–249

Osimo EF, Beck K, Reis Marques T, Howes OD (2019) Synaptic loss in schizophrenia: a meta-analysis and systematic review of synaptic protein and mRNA measures. Mol Psychiatry 24:549–561. https://doi.org/10.1038/s41380-018-0041-5

Ottomana AM, Presta M, O'Leary A, Sullivan M, Pisa E, Laviola G, Glennon JC, Zoratto F, Slattery DA, Macrì S (2023) A systematic review of preclinical studies exploring the role of insulin signalling in executive function and memory. Neurosci Biobehav Rev 155:105435. https://doi.org/10.1016/j.neubiorev.2023.105435

Ovtscharoff W Jr, Segal M, Goldin M, Helmeke C, Kreher U, Greenberger V, Herzog A, Michaelis B, Braun K (2008) Electron microscopic 3D-reconstruction of dendritic spines in cultured hippocampal neurons undergoing synaptic plasticity. Dev Neurobiol 68:870–876. https://doi.org/10.1002/dneu.20627

Palay SL (1964) The structural basis for neural action. In: Brazier MAB (ed) Brain function: RNA and brain function, memory and learning. University of California Press, pp 69–100

Park C, Park SK (2012) Molecular links between mitochondrial dysfunctions and schizophrenia. Mol Cells 33:105–110

Park HJ, Lee JD, Chun JW, Seok JH, Yun M, Oh MK, Kim JJ (2006) Cortical surface-based analysis of 18F-FDG PET: measured metabolic abnormalities in schizophrenia are affected by cortical structural abnormalities. NeuroImage 31:1434–1444

Pathak D, Shields LY, Mendelsohn BA, Haddad D, Lin W, Gerencser AA, Kim H, Brand MD, Edwards RH, Nakamura K (2015) The role of mitochondrially derived ATP in synaptic vesicle recycling. J Biol Chem 290:22325–22336. https://doi.org/10.1074/jbc.M115.656405

Patsatzis DG, Tingas EA, Goussis DA, Sarathy SM (2019) Computational singular perturbation analysis of brain lactate metabolism. PLoS One 14:e0226094. https://doi.org/10.1371/journal.pone.0226094

Pellerin L, Magistretti PJ (1994) Glutamate uptake into astrocytes stimulates aerobic glycolysis: a mechanism coupling neuronal activity to glucose utilization. Proc

Natl Acad Sci USA 91:10625–10629. https://doi.org/10.1073/pnas.91.22.10625
Perkins G, Renken C, Martone ME, Young SJ, Ellisman M, Frey T (1997) Electron tomography of neuronal mitochondria: three-dimensional structure and organization of cristae and membrane contacts. J Struct Biol 119:260–272. https://doi.org/10.1006/jsbi.1997.3885
Perkins GA, Ellisman MH, Fox DA (2003) Three-dimensional analysis of mouse rod and cone mitochondrial cristae architecture: bioenergetic and functional implications. Mol Vis 9:60–73
Perry BI, McIntosh G, Weich S, Singh S, Rees K (2016) The association between first-episode psychosis and abnormal glycaemic control: systematic review and meta-analysis. Lancet Psychiatry 3:1049–1058. https://doi.org/10.1016/s2215-0366(16)30262-0
Perry BI, Bowker N, Burgess S, Wareham NJ, Upthegrove R, Jones PB, Langenberg C, Khandaker GM (2022) Evidence for shared genetic aetiology between schizophrenia, cardiometabolic, and inflammation-related traits: genetic correlation and colocalization analyses. Schizophr Bull Open 3:sgac001
Peters A, Palay SL, Webster DF (1991) The fine structure of the nervous system. Oxford University Press
Peters A, Sethares C, Luebke JI (2008) Synapses are lost during aging in the primate prefrontal cortex. Neuroscience 152:970–981. https://doi.org/10.1016/j.neuroscience.2007.07.014
Phillips ML (2023) Building a neurobiological framework to elucidate neural mechanisms, aid early risk detection, and develop new treatments for bipolar disorder. Psychiatry Res 329:115521. https://doi.org/10.1016/j.psychres.2023.115521
Pierri JN, Edgar CL, Lewis DA (1999) Somal size of pyramidal neurons in deep layer 3 of prefrontal cortex of subjects with schizophrenia. Schizophr Res 36:84
Pillinger T, Beck K, Gobjila C, Donocik JG, Jauhar S, Howes OD (2017) Impaired glucose homeostasis in first-episode schizophrenia: a systematic review and meta-analysis. JAMA Psychiatry 74:261–269. https://doi.org/10.1001/jamapsychiatry.2016.3803
Polimanti R, Gelernter J, Stein DJ (2018) Genetically determined schizophrenia is not associated with impaired glucose homeostasis. Schizophr Res 195:286–289
Prasad K, Rubin J, Mitra A, Lewis M, Theis N, Muldoon B, Iyengar S, Cape J (2022) Structural covariance networks in schizophrenia: a systematic review Part II. Schizophr Res 239:176–191. https://doi.org/10.1016/j.schres.2021.11.036
Propst J, Ko C (1987) Correlations between active zone ultrastructure and synaptic function studied with freeze-fracture of physiologically identified neuromuscular junctions. J Neurosci 7:3654–3664. https://doi.org/10.1523/jneurosci.07-11-03654.1987
Puranen A, Koponen M, Lähteenvuo M, Tanskanen A, Tiihonen J, Taipale H (2023) Real-world effectiveness of antidepressant use in persons with schizophrenia: within-individual study of 61,889 subjects. Schizophrenia 9:34. https://doi.org/10.1038/s41537-023-00364-x
Purcell RH, Sefik E, Werner E, King AT, Mosley TJ, Merritt-Garza ME, Chopra P, McEachin ZT, Karne S, Raj N et al (2023) Cross-species analysis identifies mitochondrial dysregulation as a functional consequence of the schizophrenia-associated 3q29 deletion. Sci Adv 9:eadh0558. https://doi.org/10.1126/sciadv.adh0558
Puri V, Nagpal M, Singh I, Singh M, Dhingra GA, Huanbutta K, Dheer D, Sharma A, Sangnim T (2022) A comprehensive review on nutraceuticals: therapy support and formulation challenges. Nutrients 14. https://doi.org/10.3390/nu14214637
Radhakrishnan R, Skosnik PD, Ranganathan M, Naganawa M, Toyonaga T, Finnema S, Hillmer AT, Esterlis I, Huang Y, Nabulsi N et al (2021) In vivo evidence of lower synaptic vesicle density in schizophrenia. Mol Psychiatry 26:7690. https://doi.org/10.1038/s41380-021-01184-0
Ramaker RC, Bowling KM, Lasseigne BN, Hagenauer MH, Hardigan AA, Davis NS, Gertz J, Cartagena PM, Walsh DM, Vawter MP et al (2017) Postmortem molecular profiling of three psychiatric disorders. Genome Med 9:72. https://doi.org/10.1186/s13073-017-0458-5
Rangaraju V, Calloway N, Ryan TA (2014) Activity-driven local ATP synthesis is required for synaptic function. Cell 156:825–835. https://doi.org/10.1016/j.cell.2013.12.042
Ripke S, O'Dushlaine C, Chambert K, Moran JL, Kahler AK, Akterin S, Bergen SE, Collins AL, Crowley JJ, Fromer M et al (2013) Genome-wide association analysis identifies 13 new risk loci for schizophrenia. Nat Genet 45:1150–1159
Ripke S, Walters JT, O'Donovan MC (2020) Mapping genomic loci prioritises genes and implicates synaptic biology in schizophrenia. med Rxiv:2020.2009.2012.20192922. https://doi.org/10.1101/2020.09.12.20192922
Roberts RC (2021) Mitochondrial dysfunction in schizophrenia: with a focus on postmortem studies. Mitochondrion 56:91–101. https://doi.org/10.1016/j.mito.2020.11.009
Rodriguez-Moreno J, Rollenhagen A, Arlandis J, Santuy A, Merchan-Pérez A, DeFelipe J, Lübke JHR, Clasca F (2018) Quantitative 3D ultrastructure of thalamocortical synapses from the "lemniscal" ventral posteromedial nucleus in mouse barrel cortex. Cereb Cortex 28:3159–3175. https://doi.org/10.1093/cercor/bhx187
Rollenhagen A, Ohana O, Sätzler K, Hilgetag CC, Kuhl D, Lübke JHR (2018) Structural properties of synaptic transmission and temporal dynamics at excitatory layer 5b synapses in the adult rat somatosensory cortex. Front Synaptic Neurosci 10. https://doi.org/10.3389/fnsyn.2018.00024

Rossi MJ, Pekkurnaz G (2019) Powerhouse of the mind: mitochondrial plasticity at the synapse. Curr Opin Neurobiol 57:149–155. https://doi.org/10.1016/j.conb.2019.02.001
Rothman DL, De Feyter HM, de Graaf RA, Mason GF, Behar KL (2011) 13C MRS studies of neuroenergetics and neurotransmitter cycling in humans. NMR Biomed 24:943–957. https://doi.org/10.1002/nbm.1772
Rothman DL, de Graaf RA, Hyder F, Mason GF, Behar KL, De Feyter HM (2019) In vivo (13) C and (1) H-[(13) C] MRS studies of neuroenergetics and neurotransmitter cycling, applications to neurological and psychiatric disease and brain cancer. NMR Biomed 32:e4172. https://doi.org/10.1002/nbm.4172
Sahay S, Pulvender P, Rami Reddy M, McCullumsmith RE, O'Donovan SM (2024) Metabolic insights into neuropsychiatric illnesses and ketogenic therapies: a transcriptomic view. Int J Mol Sci 25. https://doi.org/10.3390/ijms25158266
Scaini G, Andrews T, Lima CNC, Benevenuto D, Streck EL, Quevedo J (2021) Mitochondrial dysfunction as a critical event in the pathophysiology of bipolar disorder. Mitochondrion 57:23–36. https://doi.org/10.1016/j.mito.2020.12.002
Scalettar BA, Abney JR, Hackenbrock CR (1991) Dynamics, structure, and function are coupled in the mitochondrial matrix. Proc Natl Acad Sci USA 88:8057–8061
Schikorski T, Stevens CF (1997) Quantitative ultrastructural analysis of hippocampal excitatory synapses. J Neurosci 17:5858–5867. https://doi.org/10.1523/jneurosci.17-15-05858.1997
Schikorski T, Stevens CF (2001) Morphological correlates of functionally defined synaptic vesicle populations. Nat Neurosci 4:391–395. https://doi.org/10.1038/86042
Schizophrenia Working Group of the Psychiatric Genomics, C (2014) Biological insights from 108 schizophrenia-associated genetic loci. Nature 511:421–427. https://doi.org/10.1038/nature13595
Schmidt MJ, Mirnics K (2015) Neurodevelopment, GABA system dysfunction, and schizophrenia. Neuropsychopharmacology 40:190–206. https://doi.org/10.1038/npp.2014.95
Schneider-Thoma J, Chalkou K, Dörries C, Bighelli I, Ceraso A, Huhn M, Siafis S, Davis JM, Cipriani A, Furukawa TA et al (2022) Comparative efficacy and tolerability of 32 oral and long-acting injectable antipsychotics for the maintenance treatment of adults with schizophrenia: a systematic review and network meta-analysis. Lancet 399:824–836. https://doi.org/10.1016/s0140-6736(21)01997-8
Schoonover KE, Miller NE, Fish KN, Lewis DA (2024) Scaling of smaller pyramidal neuron size and lower energy production in schizophrenia. Neurobiol Dis 191:106394. https://doi.org/10.1016/j.nbd.2023.106394
Schulmann A, Ryu E, Goncalves V, Rollins B, Christiansen M, Frye MA, Biernacka J, Vawter MP (2019) novel complex interactions between mitochondrial and nuclear DNA in schizophrenia and bipolar disorder. Mol Neuropsychiatry 5:13–27. https://doi.org/10.1159/000495658
Schwartz MW, Seeley RJ, Tschöp MH, Woods SC, Morton GJ, Myers MG, D'Alessio D (2013) Cooperation between brain and islet in glucose homeostasis and diabetes. Nature 503:59–66. https://doi.org/10.1038/nature12709
Scott D, Platania-Phung C, Happell B (2012) Quality of care for cardiovascular disease and diabetes amongst individuals with serious mental illness and those using antipsychotic medications. J Healthc Qual 34:15–21. https://doi.org/10.1111/j.1945-1474.2011.00155.x
Seager R, Lee L, Henley JM, Wilkinson KA (2020) Mechanisms and roles of mitochondrial localisation and dynamics in neuronal function. Neuronal Signals 4:Ns20200008. https://doi.org/10.1042/ns20200008
Seeger J, Schrank B, Pyle A, Stucka R, Lorcher U, Muller-Ziermann S, Abicht A, Czermin B, Holinski-Feder E, Lochmuller H, Horvath R (2010) Clinical and neuropathological findings in patients with TACO1 mutations. Neuromuscul Disord 20:720–724
Sethi S, Ford JM (2022) The role of ketogenic metabolic therapy on the brain in serious mental illness: a review. J Psychiatr Brain Sci 7:10.20900/jpbs.20220009
Sethi S, Wakeham D, Ketter T, Hooshmand F, Bjornstad J, Richards B, Westman E, Krauss RM, Saslow L (2024) Ketogenic diet intervention on metabolic and psychiatric health in bipolar and schizophrenia: a pilot trial. Psychiatry Res 335:115866. https://doi.org/10.1016/j.psychres.2024.115866
Shah K, Desilva S, Abbruscato T (2012) The role of glucose transporters in brain disease: diabetes and Alzheimer's disease. Int J Mol Sci 13:12629–12655. https://doi.org/10.3390/ijms131012629
Shamsutdinova D, Ajnakina O, Roberts A, Stahl D (2023) Schizophrenia polygenic risk score and type 2 diabetes onset in older adults with no schizophrenia diagnosis. Psychiatry Genet:33
Shepherd GM, Erulkar SD (1997) Centenary of the synapse: from Sherrington to the molecular biology of the synapse and beyond. Trends Neurosci 20:385–392. https://doi.org/10.1016/S0166-2236(97)01059-X
Shimodera S, Imai Y, Kamimura N, Morokuma I, Fujita H, Inoue S, Furukawa TA (2012) Mapping hypofrontality during letter fluency task in schizophrenia: a multi-channel near-infrared spectroscopy study. Schizophr Res 136:63–69
Shulman RG, Rothman DL, Behar KL, Hyder F (2004) Energetic basis of brain activity: implications for neuroimaging. Trends Neurosci 27:489–495. https://doi.org/10.1016/j.tins.2004.06.005
Sibille E, Morris HM, Kota RS, Lewis DA (2011) GABA-related transcripts in the dorsolateral prefrontal cortex in mood disorders. Int J Neuropsychopharmacol 14:721–734. https://doi.org/10.1017/S1461145710001616
Sibson NR, Dhankhar A, Mason GF, Behar KL, Rothman DL, Shulman RG (1997) In vivo 13C NMR measurements of cerebral glutamine synthesis as evidence

for glutamate-glutamine cycling. Proc Natl Acad Sci USA 94:2699–2704. https://doi.org/10.1073/pnas.94.6.2699

Sibson NR, Dhankhar A, Mason GF, Rothman DL, Behar KL, Shulman RG (1998a) Stoichiometric coupling of brain glucose metabolism and glutamatergic neuronal activity. Proc Natl Acad Sci USA 95:316–321. https://doi.org/10.1073/pnas.95.1.316

Sibson NR, Shen J, Mason GF, Rothman DL, Behar KL, Shulman RG (1998b) Functional energy metabolism: in vivo 13C-NMR spectroscopy evidence for coupling of cerebral glucose consumption and glutamatergic neuronalactivity. Dev Neurosci 20:321–330. https://doi.org/10.1159/000017327

Siekevitz P (1957) Powerhouse of the Cell. Sci Am 197:131

Singh T, Neale BM, Daly MJ, Consortium o.b.o.t.S.E.M.-A (2020) Exome sequencing identifies rare coding variants in 10 genes which confer substantial risk for schizophrenia. medRxiv, 2020.2009.2018.20192815. https://doi.org/10.1101/2020.09.18.20192815

Skene NG, Bryois J, Bakken TE, Breen G, Crowley JJ, Gaspar HA, Giusti-Rodriguez P, Hodge RD, Miller JA, Muñoz-Manchado AB et al (2018) Genetic identification of brain cell types underlying schizophrenia. Nat Genet 50:825–833. https://doi.org/10.1038/s41588-018-0129-5

Smeland OB, Bahrami S, Frei O, Shadrin A, O'Connell K, Savage J, Watanabe K, Krull F, Bettella F, Steen NE et al (2020) Genome-wide analysis reveals extensive genetic overlap between schizophrenia, bipolar disorder, and intelligence. Mol Psychiatry 25:844–853. https://doi.org/10.1038/s41380-018-0332-x

So HC, Chau KL, Ao FK, Mo CH, Sham PC (2019) Exploring shared genetic bases and causal relationships of schizophrenia and bipolar disorder with 28 cardiovascular and metabolic traits. Psychol Med 49:1286–1298. https://doi.org/10.1017/s0033291718001812

Sobieski C, Fitzpatrick MJ, Mennerick SJ (2017) Differential presynaptic ATP supply for basal and high-demand transmission. J Neurosci 37:1888–1899. https://doi.org/10.1523/jneurosci.2712-16.2017

Solmi M, Seitidis G, Mavridis D, Correll CU, Dragioti E, Guimond S, Tuominen L, Dargél A, Carvalho AF, Fornaro M et al (2023) Incidence, prevalence, and global burden of schizophrenia – data, with critical appraisal, from the Global Burden of Disease (GBD) 2019. Mol Psychiatry 28:5319–5327. https://doi.org/10.1038/s41380-023-02138-4

Somenarain L, Jones LB (2010) A comparative study of MAP 2 immunostaining in areas 9 and 17 in schizophrenia and Huntington chorea. J Psychiatr Res 44:694–699. https://doi.org/10.1016/j.jpsychires.2009.12.006

Stork C, Renshaw PF (2005) Mitochondrial dysfunction in bipolar disorder: evidence from magnetic resonance spectroscopy research. Mol Psychiatry 10:900–919. https://doi.org/10.1038/sj.mp.4001711

Sturmberg JP, Bennett JM, Martin CM, Picard M (2017) 'Multimorbidity' as the manifestation of network disturbances. J Eval Clin Pract 23:199–208. https://doi.org/10.1111/jep.12587

Sudhof TC (2004) The synaptic vesicle cycle. Annu Rev Neurosci 27:509–547. https://doi.org/10.1146/annurev.neuro.26.041002.131412

Sukumar N, Sabesan P, Anazodo U, Palaniyappan L (2020) Neurovascular uncoupling in schizophrenia: a bimodal meta-analysis of brain perfusion and glucose metabolism. Front Psych 11:754. https://doi.org/10.3389/fpsyt.2020.00754

Sullivan PF, Kendler KS, Neale MC (2003) Schizophrenia as a complex trait: evidence from a meta-analysis of twin studies. Arch Gen Psychiatry 60:1187–1192

Sullivan CR, Koene RH, Hasselfeld K, O'Donovan SM, Ramsey A, McCullumsmith RE (2018a) Neuron-specific deficits of bioenergetic processes in the dorsolateral prefrontal cortex in schizophrenia. Mol Psychiatry 24:1319. https://doi.org/10.1038/s41380-018-0035-3

Sullivan CR, O'Donovan SM, McCullumsmith RE, Ramsey A (2018b) Defects in bioenergetic coupling in Schizophrenia. Biol Psychiatry 83:739–750. https://doi.org/10.1016/j.biopsych.2017.10.014

Sun X, Wang JF, Tseng M, Young LT (2006) Downregulation in components of the mitochondrial electron transport chain in the postmortem frontal cortex of subjects with bipolar disorder. J Psychiatry Neurosci 31:189–196

Sun T, Qiao H, Pan PY, Chen Y, Sheng ZH (2013) Motile axonal mitochondria contribute to the variability of presynaptic strength. Cell Rep 4:413–419. https://doi.org/10.1016/j.celrep.2013.06.040

Sun Y, Smirnov M, Kamasawa N, Yasuda R (2021) Rapid ultrastructural changes in the PSD and surrounding membrane after induction of structural LTP in single dendritic spines. J Neurosci 41:7003–7014. https://doi.org/10.1523/jneurosci.1964-20.2021

Sydnor VJ, Roalf DR (2020) A meta-analysis of ultra-high field glutamate, glutamine, GABA and glutathione 1HMRS in psychosis: implications for studies of psychosis risk. Schizophr Res 226:61–69. https://doi.org/10.1016/j.schres.2020.06.028

Tandon R, Nasrallah H, Akbarian S, Carpenter WT Jr, DeLisi LE, Gaebel W, Green MF, Gur RE, Heckers S, Kane JM et al (2024) The schizophrenia syndrome, circa 2024: what we know and how that informs its nature. Schizophr Res 264:1–28. https://doi.org/10.1016/j.schres.2023.11.015

Tang BL (2018) Brain activity-induced neuronal glucose uptake/glycolysis: is the lactate shuttle not required? Brain Res Bull 137:225–228. https://doi.org/10.1016/j.brainresbull.2017.12.010

Tian YE, Di Biase MA, Mosley PE, Lupton MK, Xia Y, Fripp J, Breakspear M, Cropley V, Zalesky A (2023) Evaluation of brain-body health in individuals with common neuropsychiatric disorders. JAMA Psychiatry 80:567–576. https://doi.org/10.1001/jamapsychiatry.2023.0791

Tobe BTD, Crain AM, Winquist AM, Calabrese B, Makihara H, Zhao W-N, Lalonde J, Nakamura H, Konopaske G, Sidor M et al (2017) Probing the lithium-response pathway in hiPSCs implicates the phosphoregulatory set-point for a cytoskeletal modulator in bipolar pathogenesis. Proc Natl Acad Sci 114:E4462–E4471. https://doi.org/10.1073/pnas.1700111114

Tomasik J, Lago SG, Vázquez-Bourgon J, Papiol S, Suárez-Pinilla P, Crespo-Facorro B, Bahn S (2019) Association of insulin resistance with schizophrenia polygenic risk score and response to antipsychotic treatment. JAMA Psychiatry 76:864–867. https://doi.org/10.1001/jamapsychiatry.2019.0304

Trubetskoy V, Pardiñas AF, Qi T, Panagiotaropoulou G, Awasthi S, Bigdeli TB, Bryois J, Chen CY, Dennison CA, Hall LS et al (2022) Mapping genomic loci implicates genes and synaptic biology in schizophrenia. Nature 604:502–508. https://doi.org/10.1038/s41586-022-04434-5

Turner NL, Macrina T, Bae JA, Yang R, Wilson AM, Schneider-Mizell C, Lee K, Lu R, Wu J, Bodor AL et al (2022) Reconstruction of neocortex: organelles, compartments, cells, circuits, and activity. Cell 185:1082–1100.e1024. https://doi.org/10.1016/j.cell.2022.01.023

Uno Y, Coyle JT (2019) Glutamate hypothesis in schizophrenia. Psychiatry Clin Neurosci 73:204–215. https://doi.org/10.1111/pcn.12823

van Waveren C, Moraes CT (2008) Transcriptional co-expression and co-regulation of genes coding for components of the oxidative phosphorylation system. BMC Genomics 9:18

Vancampfort D, Stubbs B, Mitchell AJ, De Hert M, Wampers M, Ward PB, Rosenbaum S, Correll CU (2015) Risk of metabolic syndrome and its components in people with schizophrenia and related psychotic disorders, bipolar disorder and major depressive disorder: a systematic review and meta-analysis. World Psychiatry 14:339–347. https://doi.org/10.1002/wps.20252

Verstreken P, Ly CV, Venken KJ, Koh TW, Zhou Y, Bellen HJ (2005) Synaptic mitochondria are critical for mobilization of reserve pool vesicles at Drosophila neuromuscular junctions. Neuron 47:365–378. https://doi.org/10.1016/j.neuron.2005.06.018

von Bartheld CS, Bahney J, Herculano-Houzel S (2016) The search for true numbers of neurons and glial cells in the human brain: a review of 150 years of cell counting. J Comp Neurol 524:3865–3895. https://doi.org/10.1002/cne.24040

Wainio W (1970) The mammalian mitochondrial respiratory chain. Elsevier Science

Wändell P, Ljunggren G, Wahlström L, Carlsson AC (2014) Diabetes and psychiatric illness in the total population of Stockholm. J Psychosom Res 77:169–173. https://doi.org/10.1016/j.jpsychores.2014.06.012

Wardelmann K, Blümel S, Rath M, Alfine E, Chudoba C, Schell M, Cai W, Hauffe R, Warnke K, Flore T et al (2019) Insulin action in the brain regulates mitochondrial stress responses and reduces diet-induced weight gain. Mol Metab 21:68–81. https://doi.org/10.1016/j.molmet.2019.01.001

Wei W, Cheng B, Yang X, Chu X, He D, Qin X, Zhang N, Zhao Y, Shi S, Cai Q et al (2024) Single-cell multiomics analysis reveals cell/tissue-specific associations in bipolar disorder. Transl Psychiatry 14:323. https://doi.org/10.1038/s41398-024-03044-1

Wheless JW (2004) History and origin of the ketogenic diet. In: Stafstrom CE, Rho JM (eds) Epilepsy and the ketogenic diet. Humana Press, pp 31–50. https://doi.org/10.1007/978-1-59259-808-3_2

Whitehurst T, Howes O (2022) The role of mitochondria in the pathophysiology of schizophrenia: a critical review of the evidence focusing on mitochondrial complex one. Neurosci Biobehav Rev 132:449–464. https://doi.org/10.1016/j.neubiorev.2021.11.047

Whittaker RG, Turnbull DM, Whittington MA, Cunningham MO (2011) Impaired mitochondrial function abolishes gamma oscillations in the hippocampus through an effect on fast-spiking interneurons. Brain 134:e180

Wildenberg GA, Rosen MR, Lundell J, Paukner D, Freedman DJ, Kasthuri N (2021) Primate neuronal connections are sparse in cortex as compared to mouse. Cell Rep 36:109709. https://doi.org/10.1016/j.celrep.2021.109709

Wildenberg G, Li H, Sampathkumar V, Sorokina A, Kasthuri N (2023) Isochronic development of cortical synapses in primates and mice. Nat Commun 14:8018. https://doi.org/10.1038/s41467-023-43088-3

Willemsen G, Ward KJ, Bell CG, Christensen K, Bowden J, Dalgård C, Harris JR, Kaprio J, Lyle R, Magnusson PK (2015) The concordance and heritability of type 2 diabetes in 34,166 twin pairs from international twin registers: the discordant twin (DISCOTWIN) consortium. Twin Res Hum Genet 18:762–771

Williams RS, Garcia-Moll M, Mellor J, Salmons S, Harlan W (1987) Adaptation of skeletal muscle to increased contractile activity. Expression nuclear genes encoding mitochondrial proteins. J Biol Chem 262:2764–2767

Wong-Riley MT (1989) Cytochrome oxidase: an endogenous metabolic marker for neuronal activity. Trends Neurosci 12:94–101

Wong-Riley MT (2012) Bigenomic regulation of cytochrome c oxidase in neurons and the tight coupling between neuronal activity and energy metabolism. Adv Exp Med Biol 748:283–304. https://doi.org/10.1007/978-1-4614-3573-0_12

Wong-Riley M, Carroll EW (1984a) Effect of impulse blockage on cytochrome oxidase activity in monkey visual system. Nature 307:262–264

Wong-Riley MT, Carroll EW (1984b) Quantitative light and electron microscopic analysis of cytochrome oxidase-rich zones in V II prestriate cortex of the squirrel monkey. J Comp Neurol 222:18–37. https://doi.org/10.1002/cne.902220103

Wong-Riley MT, Norton TT (1988) Histochemical localization of cytochrome oxidase activity in the visual

system of the tree shrew: normal patterns and the effect of retinal impulse blockage. J Comp Neurol 272:562–578
Wong-Riley MT, Welt C (1980) Histochemical changes in cytochrome oxidase of cortical barrels after vibrissal removal in neonatal and adult mice. Proc Natl Acad Sci USA 77:2333–2337. https://doi.org/10.1073/pnas.77.4.2333
Wong-Riley MTT, Merzenich MM, Leake PA (1978) Changes in endogenous enzymatic reactivity to DAB induced by neuronal inactivity. Brain Res 141:185–192. https://doi.org/10.1016/0006-8993(78)90629-7
Wong-Riley MT, Trusk TC, Tripathi SC, Hoppe DA (1989) Effect of retinal impulse blockage on cytochrome oxidase-rich zones in the macaque striate cortex: II. Quantitative electron-microscopic (EM) analysis of neuropil. Vis Neurosci 2:499–514
Wong-Riley MT, Trusk TC, Kaboord W, Huang Z (1994) Effect of retinal impulse blockage on cytochrome oxidase-poor interpuffs in the macaque striate cortex: quantitative EM analysis of neurons. J Neurocytol 23:533–553
World Health Organization (2024). Bipolar disorder. WHO Fact Sheets. Published online 8 July 2024
Wu Q, Long Y, Peng X, Song C, Xiao J, Wang X, Liu F, Xie P, Yang J, Shi Z et al (2024) Prefrontal cortical dopamine deficit may cause impaired glucose metabolism in schizophrenia. Transl Psychiatry 14:79. https://doi.org/10.1038/s41398-024-02800-7
Xia M, Womer FY, Chang M, Zhu Y, Zhou Q, Edmiston EK, Jiang X, Wei S, Duan J, Xu K et al (2019) Shared and distinct functional architectures of brain networks across psychiatric disorders. Schizophr Bull 45:450–463. https://doi.org/10.1093/schbul/sby046
Yao Y, Zhang S, Wang B, Lin X, Zhao G, Deng H, Chen Y (2024) Neural dysfunction underlying working memory processing at different stages of the illness course in schizophrenia: a comparative meta-analysis. Cereb Cortex 34. https://doi.org/10.1093/cercor/bhae267
Yellen G (2018) Fueling thought: management of glycolysis and oxidative phosphorylation in neuronal metabolism. J Cell Biol 217:2235–2246. https://doi.org/10.1083/jcb.201803152
Yu K, Cheung C, Leung M, Li Q, Chua S, McAlonan G (2010) Are bipolar disorder and schizophrenia neuroanatomically distinct? An anatomical likelihood meta-analysis. Front Hum Neurosci 4:189. https://doi.org/10.3389/fnhum.2010.00189
Yu Y, Herman P, Rothman DL, Agarwal D, Hyder F (2018) Evaluating the gray and white matter energy budgets of human brain function. J Cereb Blood Flow Metab 38:1339–1353. https://doi.org/10.1177/0271678x17708691
Yu Y, Akif A, Herman P, Cao M, Rothman DL, Carson RE, Agarwal D, Evans AC, Hyder F (2023) A 3D atlas of functional human brain energetic connectome based on neuropil distribution. Cereb Cortex 33:3996–4012. https://doi.org/10.1093/cercor/bhac322
Yuksel C, Du F, Ravichandran C, Goldbach JR, Thida T, Lin P, Dora B, Gelda J, O'Connor L, Sehovic S et al (2015) Abnormal high-energy phosphate molecule metabolism during regional brain activation in patients with bipolar disorder. Mol Psychiatry 20:1079–1084. https://doi.org/10.1038/mp.2015.13
Yuste R, Bonhoeffer T (2001) Morphological changes in dendritic spines associated with long-term synaptic plasticity. Annu Rev Neurosci 24:1071–1089
Zandi PP, Jaffe AE, Goes FS, Burke EE, Collado-Torres L, Huuki-Myers L, Seyedian A, Lin Y, Seifuddin F, Pirooznia M et al (2022) Amygdala and anterior cingulate transcriptomes from individuals with bipolar disorder reveal downregulated neuroimmune and synaptic pathways. Nat Neurosci 25:381–389. https://doi.org/10.1038/s41593-022-01024-6
Zhang S, Zhang Y, Wen Z, Yang Y, Bu T, Bu X, Ni Q (2023) Cognitive dysfunction in diabetes: abnormal glucose metabolic regulation in the brain. Front Endocrinol (Lausanne) 14:1192602. https://doi.org/10.3389/fendo.2023.1192602
Zhuo C, Zhang Q, Wang L, Ma X, Li R, Ping J, Zhu J, Tian H, Jiang D (2024) Insulin resistance/diabetes and schizophrenia: potential shared genetic factors and implications for better management of patients with schizophrenia. CNS Drugs 38:33–44. https://doi.org/10.1007/s40263-023-01057-w
上官丰, and 章挺权 (1998) Chinese imperial cuisines and eating secrets, 1st edn. Panda Books; Distributed by China International Book Trading Corp

Microbiome: A Key Regulator of Body-Brain Interactions

6

Kenneth J. O'Riordan, Maria R. Aburto, Jatin Nagpal, Gerard Clarke, and John F. Cryan

Abstract

The gut microbiota, consisting of trillions of microorganisms, plays a critical role in regulating host physiology, including metabolism, immune responses, and brain function. This chapter examines the microbiota-gut-brain axis, a multifaceted bidirectional communication system connecting gut microbial activity with central nervous system processes through immune pathways, metabolic byproducts, and neural circuits like the vagus nerve. The evolution of the gut microbiota throughout an individual's life—from early developmental influences like birth mode and antibiotic use to changes associated with aging and neurodegenerative conditions—highlights its dynamic nature. The chapter reviews experimental approaches and microbiome-based interventions to demonstrate the influence of gut microbiota on neurological conditions such as autism spectrum disorder, anxiety, and Alzheimer's disease. Finally, it emphasises the importance of advancing microbiome-targeted therapies, integrating emerging technologies, and clinical trials to develop personalised strategies for enhancing brain health through gut microbiome modulation.

Keywords

Microbiota-Gut-Brain Axis · Gut Microbiota · Neuroinflammation · Short-Chain Fatty Acids (SCFAs) · Vagus Nerve · Blood-Brain Barrier (BBB) · Psychobiotics · Neurodevelopment

K. J. O'Riordan
APC Microbiome Ireland, University College Cork, Cork, Ireland
e-mail: ken.oriordan@ucc.ie

M. R. Aburto · J. F. Cryan (✉)
APC Microbiome Ireland, University College Cork, Cork, Ireland

Department of Anatomy and Neuroscience, University College Cork, Cork, Ireland
e-mail: maria.rodriguezaburto@ucc.ie; j.cryan@ucc.ie

J. Nagpal
APC Microbiome Ireland, University College Cork, Cork, Ireland

Department of Pharmacology and Therapeutics, School of Medicine, and School of Pharmacy, University College Cork, Cork, Ireland
e-mail: jatin.nagpal@ucc.ie

G. Clarke
APC Microbiome Ireland, University College Cork, Cork, Ireland

Department of Psychiatry and Neurobehavioural Science, University College Cork, Cork, Ireland
e-mail: g.clarke@ucc.ie

D. Tropea, E. Giacometti (eds.), *Brain-Body Connections*, Advances in Experimental Medicine and Biology 1477, https://doi.org/10.1007/978-3-031-89525-8_6

Abbreviations

Aβ	β-amyloid
ACTH	Adrenocorticotropic hormone
AD	Alzheimer's disease
ADHD	Attention deficit hyperactivity disorder
ASD	Autism spectrum disorder
ANS	Autonomic nervous system
BACHD	Bacterial artificial chromosome Huntingtin's disease
BAs	Bile acids
BBB	Blood-brain barrier
BMI	Body-mass index
CNS	Central nervous system
CRF	Corticotropin-releasing factor
DSM-5	Diagnostic and statistical manual of mental disorders
DSS	Dextran sodium sulphate
EAE	Experimental autoimmune encephalomyelitis
ENS	Enteric nervous system
EEC	Enteroendocrine cell
FFAR	Free fatty acid receptor
FMT	Faecal microbiota transplant
FXR	Farnesoid X receptor
FOS	Fructooligosaccharides
FGID	Functional gastrointestinal disorder
FODMAP	Fermentable oligosaccharides, disaccharides, monosaccharides and polyols
GABA	Gamma-aminobutyric acid
GLP-1	Glucagon-like peptide-1
GM-CSF	Granulocyte-macrophage colony-stimulating factor
GOS	Galactooligosaccharides
HDAC	Histone deacetylase
HFD	High-fat diet
HPA	Hypothalamic-pituitary-adrenal
hs-CRP	High-sensitivity C-reactive protein
IL	Interleukin
IBD	Inflammatory bowel disease
IBS	Irritable bowel syndrome
LPS	Lipopolysaccharides
MAMP	Microbial-associated molecular pattern
MD	Mediterranean diet
MDD	Major depressive disorder
miRNA	MicroRNA
mPFC	Medial prefrontal cortex
MS	Multiple sclerosis
NODs	Nucleotide-binding oligomerisation domain-containing proteins
NTS	Nucleus tractus solitarius
OCD	Obsessive-compulsive disorders
PD	Parkinson's disease
PepT1	Proton-coupled oligopeptide transporter 1
PRR	Pattern recognition receptor
PTSD	Post-traumatic stress disorder
PYY	Peptide YY
SCFA	Short-chain fatty acid
SSRI	Selective serotonin reuptake inhibitor
TGR5	G-protein-coupled bile acid receptor
TMAO	Trimethylamine-N-oxide
TLR	Toll-like receptor

6.1 Introduction

The human gastrointestinal tract hosts trillions of microorganisms, collectively known as the gut microbiota, which have co-evolved with humans and profoundly influence various aspects of host physiology, including metabolism, immunity, and brain function (Cryan et al. 2019). This chapter examines the intricate dynamics of the microbiota-gut-brain axis, a bidirectional communication network linking the gut microbiota with brain function through direct and indirect pathways including immune modulation, tryptophan metabolism, vagus nerve activity, the enteric nervous system, and microbial byproducts. The chapter underscores the role of the gut microbiota as a reservoir of bioactive compounds that significantly affect host physiology and behaviour. Microbe-host interactions are in a symbiotic relationship that evolves throughout life, from early developmental stages influenced by birth mode and antibiotic use to age-related changes in microbial diversity linked to neurodegenerative diseases. Understanding these changes will be crucial for developing targeted microbiota-based interventions.

Studying the microbiome-gut-brain axis involves diverse methodologies and experimental approaches from germ-free and antibiotic-treated preclinical models to interventions using prebiotics, probiotics, synbiotics, and postbiotics (Cryan et al. 2020 see Fig. 6.1). These studies have shed light on the microbiome's role in conditions such as autism spectrum disorder, attention-deficit hyperactivity disorder, anxiety, depression, obesity, and neurodegenerative diseases like Parkinson's and Alzheimer's disease. Factors such as diet and lifestyle significantly modulate the microbiome-gut-brain axis, influencing susceptibility to neurological disorders (Ross et al. 2024b; Schneider et al. 2024a). These findings open avenues for microbiome-targeted therapies aimed at restoring balance and improving brain health.

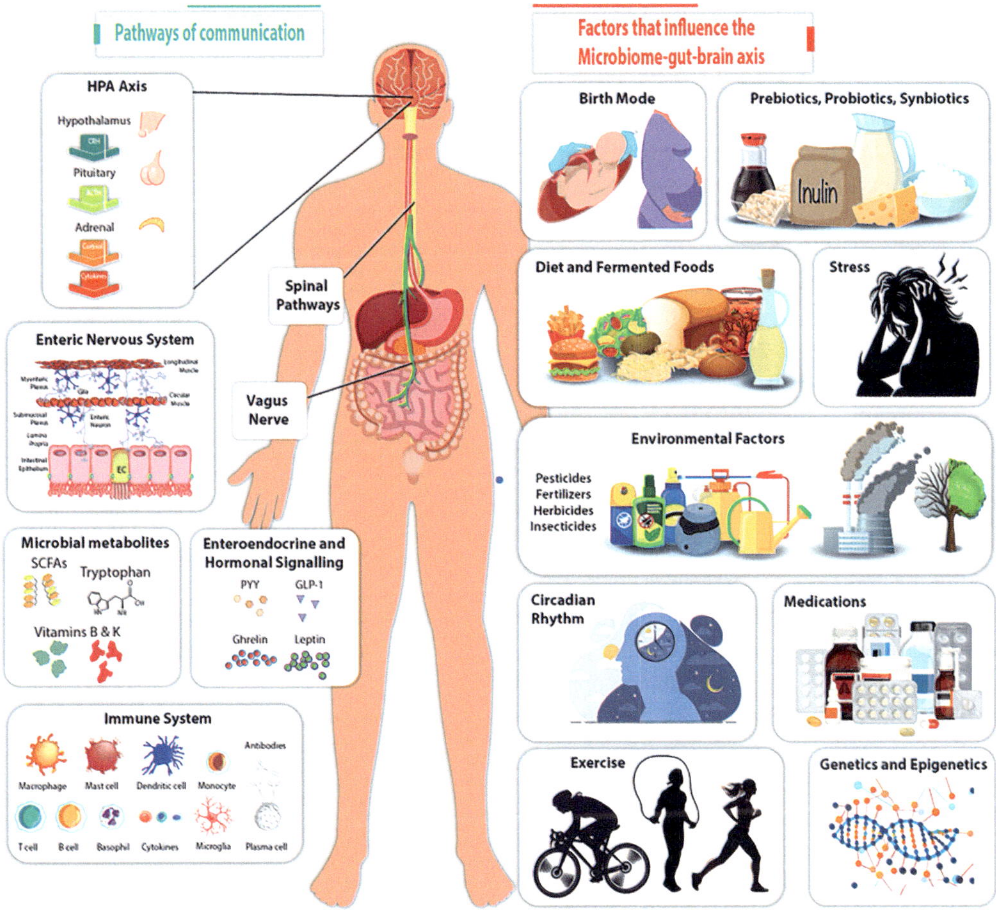

Fig. 6.1 The microbiota-gut-brain axis. Here we illustrate the complex interactions between the gut microbiota and the brain, highlighting the bidirectional communication that occurs via multiple pathways, including the hypothalamic-pituitary-adrenal (HPA) axis, the enteric nervous system, spinal pathways, and the vagus nerve. Key modulators of this axis include microbial metabolites (such as short-chain fatty acids and tryptophan), immune system mediators, and enteroendocrine hormones (like GLP-1, PYY, ghrelin, and leptin). The right side of the figure presents factors that influence the gut microbiota, including birth mode (vaginal vs. Caesarean), the intake of prebiotics, probiotics, and synbiotics, diet (particularly fermented foods), and exposure to stress, environmental chemicals (pesticides, fertilizers, herbicides, insecticides), circadian rhythm disturbances, physical activity (exercise), medications, and genetic and epigenetic factors. These elements collectively contribute to the regulation of gut-brain communication, potentially impacting health and disease

Future research should focus on elucidating the mechanisms underlying microbiome-brain interactions and developing microbial-based therapies for neuropsychiatric disorders. Integrating systems biology with longitudinal and randomised controlled trials will be essential for translating these findings into clinical practice, ultimately leading to precision therapeutic strategies that enhance brain health through gut microbiome modulation.

6.1.1 The Gut-Brain Connection

Over two millennia ago, Hippocrates, often hailed as the father of modern medicine, reportedly proclaimed, "All disease begins in the gut." Although the authenticity of the quote remains debated, the underlying wisdom continues to significantly influence the field of medicine, including the emergence of new fields like nutritional psychiatry (Cryan et al. 2020; Dinan et al. 2021; Marx et al. 2022). This chapter embarks on a journey through history and science to explore the connections between the gut, digestion, and overall health, with a particular focus on the microbiome-gut-brain axis.

The journey into the mysteries of digestion began in the early nineteenth century with a fortuitous event involving the Canadian fur trader Alexis St. Martin. A gunshot wound left St. Martin with a fistula, creating a unique opportunity for Dr. William Beaumont, a U.S. Army surgeon, to study human digestion in an unprecedented way (Beaumont 1833). Beaumont's meticulous observations and experiments, which included inserting food directly into St. Martin's stomach through the fistula and analysing gastric secretions, provided groundbreaking insights into the digestive process. He discovered that emotional states, such as anger and irritation, could significantly impact digestion, hinting at the complex interplay between the brain and the gut. This early work laid the foundation for the concept of the brain-gut axis.

Following Beaumont's pioneering work, other scientists, including Charles Darwin and Claude Bernard, advanced the understanding of physiology (Ayala 2009; Bernard 1878, 1949 (Originally 1865); Darwin 1859). Bernard's experiments with rabbits elucidated the process of fat digestion and the role of the pancreas. He introduced the concept of the "milieu intérieur," emphasising the necessity of a stable internal environment for healthy functioning, the cornerstone of homeostasis. Inspired by Beaumont, Ivan Pavlov developed the Pavlov pouch (LeDoux 2003; Pavlov 1897), a method to study digestive processes in canines. Pavlov's work on classical conditioning illustrated how external stimuli could affect internal physiological processes, reinforcing the concept of the brain-gut connection.

Such historical work laid the foundation for understanding the brain-gut connections' pivotal role in maintaining health and addressing disease. The advent of brain imaging technology in the 1980s revealed the bidirectional nature of a brain-gut axis, showing that gut distension activates crucial brain pathways, which are often heightened in conditions like irritable bowel syndrome (IBS) due to disruptions in the microbiota-gut-brain axis (Bonaz and Bernstein 2013; Mayer 2000). Additionally, the microbiota-gut-brain axis is recognised as a central component in mammalian interoception (Craig 2009).

6.1.2 The Microbial Ecosystem: Allies in Health and Disease

More recently the critical role of the trillions of microbes residing on our bodies and in our gut, collectively known as the human microbiota, has begun to emerge (Cryan et al. 2019). The human gut microbiota refers to the community of microorganisms that reside in the gut, whereas the gut microbiome refers to the collective genomes of all the microorganisms which includes bacteria, viruses, fungi, and other microbes. In humans, the gut microbiota is one of the most well-studied communities, and it plays a key role in digestion, immune function, and even brain health through the microbiota-gut-brain axis.

Advances in sequencing technology and bioinformatic pipelines have enabled scientists to sequence these microbes and understand their vast contribution to our physiology (Bastiaanssen et al. 2019, 2021, 2022; Coelho et al. 2022; Human Microbiome Project 2012; Sonnenburg and Sonnenburg 2019). As a result, and although the precise composition of healthy gut microbiota is still debated (Joos et al. 2024), the gut microbiome is now recognised as a key player in regulating our immune system, metabolism, and even mental health.

Studies utilising germ-free animals, rodents that have grown up without exposure to any microbes, have shown that the absence of the gut microbiota affects many elements of health, including brain development and behaviour (Castillo-Ruiz et al. 2023; Desbonnet et al. 2015; Jaggar et al. 2020; Sens et al. 2023). Experiments introducing specific bacterial strains to animals and observing resultant behavioural changes have demonstrated the significant influence these microbes have on the brain (Bravo et al. 2011; McKernan et al. 2010; Savignac et al. 2014). Human studies have mirrored these findings, highlighting the microbiota's potential to impact mental health and cognitive functions (Kullberg et al. 2024; Tian et al. 2023). Indeed, large collaborative initiatives like the Human Microbiome Project, MetaHIT, the American Gut Project, and the Dutch microbiome project, along with significant gut microbiota cohort analyses, have been crucial in mapping the gut microbiome at a population scale, providing foundational data that has informed subsequent studies focused on its role in mental health conditions such as depression, anxiety, and cognitive decline (Cotillard et al. 2022; Group et al. 2009; Lane et al. 2024; Li et al. 2014; Lopera-Maya et al. 2022; McDonald et al. 2018).

6.1.3 Microbiota Evolution: A Co-dependent Partnership in the Holobiont

To truly appreciate the role of the microbiota in host health, it is essential to consider its evolutionary context. Host organisms have never existed without their microbial friends, co-evolving and becoming mutually dependent for survival (Nagpal and Cryan 2021a). There has never been a time when mammals, including humans, existed without their microbial inhabitants (Bordenstein and Theis 2015). Moreover, the brain has always received signals from the gut (Stilling et al. 2014). This co-evolutionary relationship is best described by the concept of the holobiont, which views the host and its symbiotic microbiota as a single ecological unit. This perspective has led to the hologenome theory of evolution, proposing that the holobiont and its combined genetic material act as a unit of evolutionary selection (Theis et al. 2016; Zilber-Rosenberg and Rosenberg 2008). The hologenome theory highlights that genetic variation within the holobiont is driven by both the host genome and its associated microbial genome. This variation is further enhanced by the vertical and horizontal transmission of different microbial populations, including from parent to offspring and through social interactions, allowing optimal adaptation to environmental changes, such as diet, stress, and temperature (Walter and Hornef 2021). For instance, initial vertical transmission of maternal microbiota during birth and subsequent breastfeeding ensures that offspring acquire and establish critical microbial populations for immune and metabolic development (Sonnenburg and Backhed 2016). Horizontal transmission, on the other hand, occurs through interactions within social groups, such as grooming or shared food sources and environments, contributing to microbiota diversity and resilience (Moeller et al. 2016). This theory may also explain complex biological phenomena, including behaviours that promote social interactions among holobionts, thereby facilitating greater transmission of microbiota and enhancing genetic diversity (Archie and Tung 2015). Such transmission and assembly mechanisms reinforce the dynamic interplay between microbiota and host evolution, highlighting their collective role in shaping biological and behavioural traits. Understanding these evolutionary links between microbiota and the brain is crucial for studying human biology and the broader animal kingdom.

With this in mind, it is easier to understand how our microbial symbionts have significantly influenced brain development, physiology, and behaviour throughout evolution.

The Bi-directional Microbiota-Gut-Brain Axis: A Complex Communication Network With this information, researchers now refer to the bidirectional communication network connecting gut microbes to our brain, both directly and indirectly, as the microbiota-gut-brain axis, which integrates the enteric and central nervous systems. This connection explains how the gut can influence brain function and vice versa. Where early research focused on digestive functions and satiety (Konturek et al. 2004; Rinaman 2010), more recent work has explored the axis's impact on higher-order cognitive and psychological processes and behavioural mechanisms (Davidson et al. 2024; Hung and Margolis 2024; Ritz et al. 2024a). Disruptions in the microbiota-gut-brain axis have been linked to various disorders, including IBS (Mayer et al. 2023; Wilmes et al. 2021), inflammatory bowel disease (IBD; (Michaudel et al. 2023; Schneider et al. 2023), and mental health conditions like anxiety and depression (Audet 2021; Dalile et al. 2024; Diaz Heijtz 2024; McGuinness et al. 2024; Ross et al. 2024a). The concept of psychobiotics—including probiotics that can influence mental health—has emerged from these studies, offering new avenues for therapeutic interventions (Dinan et al. 2021). For example, there is evidence that alterations in the gut microbiota composition under cold temperatures can regulate norepinephrine release in the intestine and brown adipose tissue in Brandt's voles, thereby contributing to the modulation of energy intake and thermal homeostasis during winter (Ge et al. 2018).

Chapter 2 in this book provides an in-depth exploration of the effects of Noradrenaline and highlights its role as a key integrator of information processing between the brain and body. This adrenergic network is vital to the execution of a wide range of physiological and behavioural functions, most notably the fight-or-flight response to threats.

6.1.4 From Gut to Brain and Back: Decoding the Communication Pathways

The intricate relationship between the gut microbiota and the brain is facilitated by various pathways of communication, ranging from neural networks to small molecule signalling systems. Understanding these pathways is crucial to comprehending how the microorganisms in our gut can significantly influence our brain and behaviour.

6.1.4.1 The Autonomic Nervous System: Orchestrating Gut-Brain Symmetry

The ANS is a complex network of neurons that control bodily functions such as breathing, heartbeat, and digestion without conscious effort (Jänig 2006). The ANS consists of the sympathetic and parasympathetic branches and works closely with the enteric nervous system (ENS) and the central nervous system (CNS) to maintain physiological homeostasis (Wehrwein et al. 2016). The ANS communicates bidirectionally with the gut microbiota, influencing gut motility, permeability, secretion, and immune responses (Goehler et al. 2005; Sharkey and Mawe 2023). These communications are essential for processing visceral information and responding to various stimuli, ensuring rapid changes in gut physiology. The sympathetic branch of the ANS typically prepares the body for stressful or emergency situations by initiating the "fight or flight" response, which can influence gut motility and secretion through the release of neurotransmitters like norepinephrine (Kaelberer et al. 2020). Conversely, the parasympathetic branch, which includes the vagus nerve (see below), supports the "rest and digest" functions, promoting digestion and reducing stress responses.

The ANS also interacts with the hypothalamic-pituitary-adrenal (HPA) axis (see below), creating a comprehensive network for integrating signals from the gut microbiota and maintaining overall physiological balance (Cussotto et al. 2018). This intricate interplay between the ANS, HPA axis, and gut microbiota ensures that the body's response to internal and external stimuli is well-coordinated and adaptive.

6.1.4.2 The Vagus Nerve: A Superhighway for Gut-Brain Signals

The vagus nerve (tenth cranial nerve) is the primary conduit for communication between the gut and the brain (Fulling et al. 2019). Comprising mostly afferent fibres, it collects information from visceral organs and sends it to the brain directly and quickly. The vagus nerve plays an important role in regulating gut function, including motility and secretion (Grundy et al. 2006). It also modulates inflammatory responses through the vago-vagal anti-inflammatory reflex loop, influencing conditions like IBD and functional gastrointestinal disorders (FGIDs) (Pavlov and Tracey 2005). The vagus nerve's extensive innervation allows it to detect changes in the gut environment, such as the presence of bacterial metabolites, stretch, or chemical signals. It transmits these signals to the nucleus tractus solitarius (NTS) in the brainstem, which then relays information to other brain regions involved in autonomic control and emotional regulation, such as the hypothalamus, amygdala, and prefrontal cortex. Furthermore, the vagus nerve's efferent fibres carry signals from the brain back to the gut, modulating functions like gastric acid secretion, enzyme release, and gut motility (Bhattarai et al. 2018). This bidirectional communication is essential for maintaining gut-brain homeostasis and responding to changes in the gut environment.

6.1.4.3 The Enteric Nervous System: The Brain in Your Gut

The ENS, often referred to as the "second brain," is a network of neurons located in the walls of the gastrointestinal tract, specifically within the myenteric and submucosal plexuses (Benarroch 2007). The ENS serves as a critical interface between the gut microbiota and the CNS, responding to microbial signals through direct interactions with microbial metabolites and indirect pathways involving hormones, immune mediators, and the vagus nerve (Hyland and Cryan 2016). Indeed, recent advancements in high-throughput "omics" technologies, particularly single-cell RNA sequencing, have enabled the identification of specific expression signatures and potential subtypes of enteric neurons and glial cells in the ENS vulnerable to microbial signals (Dharshika and Gulbransen 2023). The ENS influences gut functions such as motility and fluid movement and is involved in the development and maintenance of gut health. The ENS comprises the myenteric plexus (primarily controls gut motility) and the submucosal plexus (regulates secretion and blood flow). The ENS operates independently of the CNS but can also interact with it through autonomic pathways, allowing for coordinated responses to changes in the gut environment. Studies involving germ-free animals have shown that the absence of a gut microbiota leads to significant abnormalities in the ENS, highlighting the microbiota's role in ENS development and function (De Vadder et al. 2018; Obata and Pachnis 2016; Wu and Wu 2012). During development, the gut microbiota plays a critical role in shaping the ENS by influencing the proliferation, differentiation, and maturation of enteric neurons (Obata and Pachnis 2016). Microbial metabolites such as short-chain fatty acids (SCFAs) and bile acids can modulate ENS activity, affecting gut motility and barrier function (Aburto and Cryan 2024; O'Riordan et al. 2022).

6.1.4.4 Spinal Pathways: Conveying Gut Distress to the Brain

Neural signalling from the gut to the brain involves both the vagus nerve and spinal pathways. Painful and non-painful stimuli are conveyed through different routes, with non-painful signals predominantly mediated by vagal innervation and painful stimuli by spinal splanchnic innervation (Rea et al. 2021; Vermeulen et al. 2014). These pathways ensure the brain receives comprehensive information about the gut's status, allowing for appropriate physiological responses. Pain and discomfort from the gut are transmitted through spinal afferent pathways to the brain, where they are processed and integrated with other sensory information (Vermeulen et al. 2014). These pathways involve the dorsal root ganglia and the spinal cord, which relay signals to brain regions such as the thalamus, hypothalamus, and cerebral cortex. Understanding these spinal mechanisms is crucial for comprehending how gut disturbances, such as inflammation or injury, can influence brain function and behaviour.

Bacterial infections can amplify pain signalling and induce long-term plastic changes in spinal nerve signalling by directly activating sensory afferents and modulating inflammation (Chiu et al. 2013). This heightened pain response, known as central sensitisation, is similar to long-term potentiation in the hippocampus, where neurotransmitters like glutamate and peptides such as substance P enhance synaptic transmission, leading to abnormal pain processing (Yu et al. 2016). Notably, *Lactobacillus reuteri* (DSM 17938) has been found to reduce spinal nerve firing in response to gut stimuli (Perez-Burgos et al. 2015).

6.1.4.5 Immune Interplay: Linking Gut and Brain Health

The gut is also home to a dense concentration of immune cells that continuously interact with the local microbes (Ratsika et al. 2023). The immune system, through the release of cytokines and chemokines, plays a significant role in gut-brain communication (Macpherson et al. 2023). These immune cells detect and respond to microbial-associated molecular patterns (MAMPs), such as lipopolysaccharides (LPS) from Gram-negative bacterial outer membranes and peptidoglycan (a polysaccharide, consisting of sugar and amino acids, which is a component of bacterial cell walls) (Gonzalez-Santana and Diaz Heijtz 2020). The recognition of these microbial patterns is mediated by pattern recognition receptors (PRRs), such as Toll-like receptors (TLRs) and nucleotide-binding oligomerisation domain-containing proteins (NODs) (Li and Wu 2021). For example, TLR4 specifically recognises LPS, leading to downstream immune responses, while NOD1 and NOD2 are involved in detecting peptidoglycan-derived molecules. This coordinated detection of microbial components is crucial for maintaining the balance between the host and the gut microbiota, preventing harmful pathogens from breaching the gut barrier. Recent research has shown that a diverse gut microbiota is essential for the proper functioning of microglia, the brain's resident immune cells, which are involved in neuroinflammatory responses and synaptic remodelling (Zhang et al. 2024a).

The adaptive immune system comprises B and T lymphocytes, which provide a more specific and long-lasting immune response. Microbial metabolites such as SCFAs can modulate immune cell function, promoting anti-inflammatory responses and maintaining gut barrier integrity. For example, microbial metabolites can stimulate the production of regulatory T cells, which help suppress excessive immune responses and prevent inflammation (Caffrey et al. 2024). This modulation of the immune system by the gut microbiota is essential for preventing chronic inflammation and maintaining overall health (Fung 2020). Indeed, various types of gut-innervating neurons influence intestinal mucosal immunity, dictating immune cell responses to neural signals during host defence and inflammatory processes providing a potential therapeutic target to address intestinal diseases (Yang et al. 2023).

One study investigated the microbiota-gut-brain axis using the dextran sodium sulphate (DSS)-induced colitis in mice, demonstrating that gut inflammation leads to pro-inflammatory immune responses that can negatively impact brain myelination, oxidative stress, and oxygenation (Boles et al. 2024). This suggests that gut inflammation not only disrupts intestinal health but also contributes to neuroinflammatory processes, potentially affecting cognitive and neurological function. A key player in this interaction is microglia that play a major role in regulating the integrity of the blood-brain barrier (BBB). The BBB is crucial for maintaining CNS function and homeostasis but becomes compromised with systemic inflammation, where gut-derived cytokines and other inflammatory mediators can disrupt BBB integrity, leading to neuroinflammation. This can further be exacerbated with age, contributing to various age-related neurodegenerative disorders, such as Alzheimer's disease (Aburto and Cryan 2024).

Different dietary interventions (vegan/ketogenic) can also modulate both gut microbiota composition and immune pathways, further influencing brain health (Boscaini et al. 2022). For instance, the ketogenic diet has been reported to enhance adaptive immune pathways, which may exert neuroprotective effects by reducing inflammation and promoting regulatory immune cell populations. In contrast, the vegan diet primarily promotes innate immune responses, particularly in antiviral immunity, highlighting the potential for tailored nutritional interventions to significantly influence host immunity and microbiota composition (Link et al. 2024). These findings underscore the potential for dietary strategies to modulate the microbiota-gut-brain axis, improve neuroinflammation, and mitigate the cognitive consequences of gut inflammation.

The brain can modulate peripheral inflammation by communicating with distinct populations of vagal neurons, manipulation of which can suppress pro-inflammatory responses while enhancing anti-inflammatory states (Jin et al. 2024). GABAergic signalling in the brain was shown to be crucial for regulating neuroinflammation-induced synaptic dysfunction, where inhibitory postsynaptic currents in glutamatergic neuronal activity were increased alongside excessive microglial activation (Jiang et al. 2022). Such neuroimmune function can be modifiable by diet. Prebiotic-enriched diets can reverse age-related immune priming, reduce monocyte infiltration into the brain, and decrease the activation of microglia, offering a potential route for the management of neuroinflammatory conditions (Boehme et al. 2020). A recent probiotic study in a diet-induced obese mouse model using *Bacteroides uniformis* CECT 7771 revealed that *B. uniformis* improved glucose tolerance and insulin secretion through adaptive immunity mechanisms that do not require the bacteria to be viable, while its viability was essential for preventing weight gain, indicating distinct pathways through which a probiotic can influence energy homeostasis and immunity (Romani-Perez et al. 2024b).

6.1.4.6 Stress Pathways: The HPA Axis and Microbiome Interactions

Upon encountering a stressor, the hypothalamus initiates the stress response by releasing corticotropin-releasing factor (CRF), which stimulates the pituitary gland to secrete adrenocorticotropic hormone (ACTH), which then prompts the adrenal glands to release glucocorticoids, such as cortisol, into the bloodstream (Wu et al. 2021). These hormones help the body adapt to stress by mobilising energy reserves and modulating immune responses. The HPA axis is a key neuroendocrine system that coordinates the body's response to stress. The gut microbiota can influence the HPA axis, as evidenced by studies showing exaggerated stress responses in germ-free animals (Clarke et al. 2013; Lyte et al. 2020; Sudo et al. 2004). The HPA axis interacts with other pathways of communication, including the vagus nerve and the immune system, forming a complex network that regulates the body's stress response and maintains homeostasis.

The HPA axis is tightly regulated by feedback mechanisms that involve glucocorticoid receptors in the brain and other tissues. Dysregulation of the HPA axis, as seen in chronic stress or psychiatric disorders, can lead to altered immune function, metabolic disturbances, and changes in gut microbiota composition (Audet 2019; Sheng et al. 2020). A recent pilot study investigated the association between gut microbiota composition and HPA axis reactivity in 34 healthy human infants by measuring salivary cortisol before and after a heel stick and conducting 16S rRNA amplicon sequencing to identify bacterial taxa (Rosin et al. 2021). It revealed a moderate positive correlation between α-diversity and cortisol reactivity at 1 month of age. Additionally, specific genera such as *Staphylococcus*, *Prevotella*, and *Lachnospiraceae* may be linked to cortisol responses, suggesting the microbiome's role as a mediator in the microbiome-gut-brain axis during infancy and highlighting the potential for manipulating gut microbes to enhance stress-response systems. Understanding the interplay between the HPA axis and the gut microbiota is essential for developing interventions to restore balance in stress-related disorders.

6.1.4.7 Hormonal Gatekeepers: Enteroendocrine Cells and Microbiota Influence

Enteroendocrine cells (EECs) are dispersed throughout the gut epithelium and can detect changes in the gut environment through their apical surface, which is exposed to the gut lumen. Upon detecting nutrients or microbial metabolites, EECs release hormones and neuropeptides into the bloodstream, or locally within the gut, to regulate various functions (Sharkey and Mawe 2023). For example, enteroendocrine L cells release glucagon-like peptide-1 (GLP-1) and peptide YY (PYY) in response to nutrient intake, which are involved in the regulation of appetite and glucose homeostasis (Zheng et al. 2024). GLP-1 and PYY act on hypothalamic and vagal afferents to reduce appetite and promote satiety, helping to regulate body weight and energy balance. They also influence insulin secretion and glucose homeostasis, playing a crucial role in metabolic health. GLP-1 and PYY are also influenced by bacterial metabolites like SCFAs, which activate G-protein-coupled receptors such as FFAR2 (GPR43) and FFAR3 (GPR41) on enteroendocrine L cells, triggering hormone secretion (Larraufie et al. 2018). SCFAs can also enhance L-cell activity by modulating cellular energy metabolism, promoting the release of GLP-1 and PYY into the bloodstream (O'Riordan et al. 2022). The gut microbiota can significantly affect EEC cell activity, thereby impacting host metabolism and behaviour. Recent work has explored the role of specific proteins secreted by *Akkermansia muciniphila* in improving metabolic health (Yoon et al. 2021) *A. muciniphila*-derived P9 improved metabolic health by enhancing GLP-1 secretion and reducing insulin resistance in mice. Further, pasteurised *A. muciniphila* can benefit metabolic health, suggesting live bacteria may not be essential for therapeutic effects (Plovier and Cani 2017).

Enterochromaffin cells produce the majority of the body's serotonin, which is involved in regulating gut motility, secretion, and pain perception (Sharkey and Mawe 2023). Enterochromaffin cells not only transduce noxious stimuli and induce visceral hypersensitivity following activation but also contribute to persistent pain and anxiety-like behaviours, with significant sex differences in their engagement (Morales-Soto et al. 2023; Bayrer et al. 2023). Not only do these gut cells produce and respond to serotonin, but so do bacteria, along with gamma-aminobutyric acid (GABA) and catecholamines (Wiley et al. 2021). These neuroactive molecules play critical roles in host mood, behaviour, and cognition. For instance, certain gut bacteria can modulate serotonin production, influencing the expression of tryptophan hydroxylase, the rate-limiting enzyme in serotonin synthesis (Hata et al. 2017). GABA is an inhibitory neurotransmitter that is produced by some gut bacteria and can affect gut motility and signalling to the brain (Mazzoli and Pessione 2016). Bacterial production of GABA can influence host behaviour, as evidenced by studies showing that GABA-producing probiotics can reduce anxiety-like and improve depression-like behaviours in animal models (Yu et al. 2020). Indeed, one study explored the impact of a GABA-producing *Bifidobacterium* strain on gut microbiota composition and metabolic activity in mice (Tames et al. 2024). Administering this GABA-producing strain leads to a distinct modulation of the gut microbiota compared to non-GABA-producing strains.

GABA also plays an important role in glucose metabolism and regulation. Within the (ventromedial) hypothalamus, GABA(A) receptor activity plays a key role in regulating glucagon release by the pancreas during hypoglycaemic events. *Chapter 3 in this book* explores the roles of different brain regions in glucose homeostasis and their connections to the pancreas and liver through the sympathetic and parasympathetic nervous systems.

Catecholamines like norepinephrine and dopamine are also produced by certain gut bacteria and can interact with host cells to influence various physiological processes (Lachmansingh et al. 2023). Gut microbiota can influence resilience and susceptibility to stress-induced psychopathologies, where mice susceptible to stress exhibited higher baseline and stress-induced levels of epinephrine and norepinephrine, alongside significant shifts in gut microbial composition, suggesting that microbial profiles may play a critical role in modulating the sympathoadrenal stress response (Tanelian et al. 2022). These studies highlight the complex communication network within the microbiota-gut-brain axis.

6.1.4.8 Metabolic Mediators: How Microbial Molecules Shape Brain Function

SCFAs, such as acetate, propionate, and butyrate, are primarily produced by microbial fermentation of non-digestible dietary fibres and have been implicated in various host processes, including gut health, immune function, and brain physiology (O'Riordan et al. 2022). These metabolites can be absorbed by the gut epithelium and enter the bloodstream, where they can exert systemic effects on the host. SCFAs can influence brain function indirectly by modulating the immune system and directly by crossing the blood-brain barrier (BBB) and affecting neuronal activity (Dalile et al. 2019). In the brain, SCFAs can affect neuronal function by modulating neurotransmitter synthesis and release, influencing BBB integrity and brain energy metabolism, and acting as histone deacetylase inhibitors to regulate gene expression (Knox et al. 2022b, 2023; Stachulski et al. 2023). Studies have shown that SCFAs can ameliorate stress-induced alterations in the brain and behaviour, highlighting their potential role in microbiota-gut-brain communication (Dalile et al. 2024, 2020). In the gut, SCFAs help maintain barrier integrity by promoting tight junction assembly and mucus production (Ma et al. 2022). They also modulate immune responses by influencing the activity of various immune cells, including regulatory T cells and macrophages, promoting an anti-inflammatory environment (Liu et al. 2023). Therapeutically, engineered bacterial biosensors are capable of sensing SCFA concentrations produced by gut microbes (Serebrinsky-Duek et al. 2023). By utilising genetic circuits that respond inversely to SCFA levels, the researchers created systems that not only quantify SCFA absence but also deliver therapeutic cytokines, such as granulocyte-macrophage colony-stimulating factor (GM-CSF), highlighting their potential as diagnostic and therapeutic tools for gut-related disorders.

Recently, a cranberry extract rich in (poly) phenols and oligosaccharides was given to 28 human subjects, revealing a significant bifidogenic effect and an increase in butyrate-producing bacteria after just 4 days of supplementation (Lessard-Lord et al. 2024). In particular, participants with a Prevotella-rich microbiota experienced an increase in butyrate-producing *Faecalibacterium* following cranberry extract supplementation. In fact, recent data from two large population-based cohorts in the Netherlands and Finland revealed that individuals with a higher relative abundance of butyrate-producing bacteria had a significantly reduced risk of hospitalisation or mortality due to infections, indicating that gut microbiota composition may play a protective role against severe infections in the general population (Kullberg et al. 2024). Indeed, reductions in butyrate-producing bacteria and butyrate production were linked to early changes in the microbiome-gut-brain axis that preceded Alzheimer's disease-related neuropathology and cognitive deficits in a mouse model (Chilton et al. 2024). Moreover, oral administration of tributyrin, a butyrate prodrug, mitigated these declines and delayed the onset of AD symptoms, suggesting that enhancing butyrate synthesis may be a promising therapeutic approach for AD prevention and treatment.

Bile Acids (BAs) are essential for absorbing dietary lipids and fat-soluble vitamins from the gut (Guzior and Quinn 2021). Synthesised in the liver from cholesterol, primary BAs such as cholic acid and chenodeoxycholic acid aid lipid digestion and are recycled back to the liver. Beyond their digestive roles, BAs are potent signalling molecules, influencing lipid, cholesterol, and glucose metabolism, as well as energy balance and immune homeostasis through receptors like farnesoid X receptor (FXR) and G-protein-coupled bile acid receptor (TGR5) (de Aguiar Vallim et al. 2013). Studies have revealed bidirectional communication between the host bile system and gut bacteria, impacting gut-brain signalling (Lavelle and Sokol 2020). BAs limit bacterial overgrowth in the small intestine through antimicrobial effects and activate FXR signalling to induce antimicrobial genes, protecting the gut epithelium from microbial invasion and inflammation (Long et al. 2017). Gut bacteria, including *Lactobacillus* and *Bifidobacterium*, express bile salt hydrolase enzymes which can deconjugate bile acids, enhancing their survival in the gut (Degirolamo et al. 2014; Long et al. 2017). This microbial activity increases BA diversity in the host, affecting metabolism and weight regulation.

BAs are crucial for maintaining the gut barrier and regulating immune responses, which are essential for gut and brain function and health. TGR5 activation by BAs has anti-inflammatory effects and helps maintain epithelial integrity by modulating the gastrointestinal immune system (Sarathy et al. 2017). Alterations in gut microbiota composition that change BA profiles can impact microbiota-gut-brain axis signalling, particularly through inflammatory pathways. Deficient BA deconjugation in a mouse model of autism spectrum disorder (ASD) was linked to impaired intestinal barrier function and behavioural changes, highlighting the importance of bile-metabolising capacities of specific probiotic strains such as *Bifidobacterium* and *Lactobacillus* (Golubeva et al. 2017). The probiotic VSL#3 enhances BA deconjugation and upregulates hepatic BA synthesis via the FXR/FGF15 pathway (Degirolamo et al. 2014). Further, FXR/FGF signalling in the ileum can suppress hypothalamic neurons involved in glucose regulation, improving glucose tolerance in obese mice (Liu et al. 2018b). Interestingly, BAs function as natural pheromones in mice, with the olfactory system distinguishing between male and female faeces based on BA composition, suggesting a direct role in gut-brain communication (Doyle et al. 2016). In certain liver diseases, individuals often experience cognitive impairment, also seen in a bile duct ligation rodent model. Here, bile duct ligation mice develop cognitive symptoms early, before the onset of cirrhosis or liver failure, including memory loss, BBB disruptions, and hippocampal network destabilisation. This was reversible with BA (obeticholic acid) treatment, restoring memory, preventing neuronal damage, and reversing cellular senescence in these mice (Gee et al. 2023). A pilot feasibility study investigating bile acid metabolites during the perinatal period revealed significant temporal variations and associations with mental health (Kimmel et al. 2022). Notably, changes in secondary bile acid levels, particularly ursodeoxycholic acid, correlate with microbial diversity and are linked to psychiatric history, suggesting that bile acid dynamics may play a crucial role in perinatal mental health adaptations.

Indoles are a group of tryptophan-derived metabolites produced by gut microbes (Lee et al. 2015; Turner et al. 2006). They play a critical role in gut-brain interactions, impacting host physiology through various mechanisms (Roager and Licht 2018). Indole also impairs locomotion and induces anxiety-like behaviour in mice via dose-dependent activation of the locus coeruleus, highlighting its role in gut-brain communication (Mir et al. 2024). Further, chronic indole production by the gut microbiota amplifies emotional responses to chronic mild stress in male mice and enhances catecholamine biosynthesis in the adrenal medulla, highlighting its role in stress susceptibility (Mir et al. 2020). Indole, along with indoleacetic acid and indolepropionic acid, influences the gut's immune and neuroendocrine systems by acting on the aryl hydrocarbon receptor (AhR) (Zelante et al. 2013). The AhR is a ligand-activated transcription factor that is responsive to a broad range of physiological ligands, including microbial metabolites such as indole and other tryptophan derivatives (Barroso et al. 2021). By engaging AHR, indoles modulate immune functions and contribute to maintaining mucosal homeostasis, which may protect against inflammation (Cervantes-Barragan et al. 2017).

In addition to their peripheral effects, indoles have been shown to influence brain health through gut-brain communication. Indolepropionic acid, for instance, has been associated with neuroprotective effects, particularly in reducing oxidative stress and inflammation, processes implicated in neurodegenerative diseases such as Alzheimer's disease (Sun et al. 2022). Furthermore, indoles may indirectly modulate neurotransmitter systems by supporting the intestinal barrier and reducing systemic inflammation, thereby preventing inflammation-induced disruptions in brain function (Loh et al. 2024). Recent work found that oral tryptophan supplementation in healthy individuals enhances AhR activation and increases microbial indole and host kynurenine pathway metabolites in the small intestine, suggesting potential for therapeutic application in managing small intestine inflammation (Rueda et al. 2024).

Indoles also support intestinal barrier integrity through pregnane X receptor (PXR) signalling, potentially reducing intestinal permeability (Bansal et al. 2010). This enhanced barrier function may have downstream effects on the brain, as gut barrier dysfunction is increasingly linked to neuroinflammation and behavioural changes. Moreover, indole stimulates the release of gut hormones like GLP-1, hinting at their influence on metabolic health (Chimerel et al. 2014). Further, gut epithelial Gpr35 can regulate depressive-like behaviour in mice, via alterations in microbial tryptophan metabolism, specifically decreasing indole-3-carboxaldehyde (IAld) and increasing indole-3-lactate (ILA), which have opposing effects on neuroplasticity in the nucleus accumbens (Cheng et al. 2024). Supplementing IAld alleviated depressive symptoms in mice and was found to be depleted in individuals with major depressive disorder (MDD), highlighting its potential as a biomarker and therapeutic target for depression. By connecting microbial metabolites to both local and systemic processes, indoles underscore the integral role of gut-derived molecules in maintaining gut and brain health.

Peptidoglycans are recognised by the host's immune system and can influence brain function (Gonzalez-Santana and Diaz Heijtz 2020). These molecules can cross the BBB and affect neurons directly, suggesting a direct link between gut bacteria and brain function (Arentsen et al. 2017). Research has shown that peptidoglycan recognition proteins (such as pattern recognition receptors including Toll-like receptors and nucleotide-binding oligomerisation domain-containing proteins (NODs)), are involved in both immune responses and behaviour, highlighting the significance of microbial components in gut-brain communication (Medzhitov 2007; Takeda and Akira 2004). These receptors initiate immune responses upon recognising microbial components, leading to the release of cytokines and other immune signalling molecules. Studies have shown that peptidoglycans can modulate neuronal function and behaviour by interacting with receptors in the brain (Xu et al. 2023b). For instance, NOD1 and NOD2 receptors, which recognise specific peptidoglycan motifs, are expressed in various brain regions and can influence neuronal activity and synaptic plasticity (Clarke et al. 2010; Gabanyi et al. 2022; Kawai and Akira 2009; Schneider and Thaiss 2022). Further, peptidoglycans were recently shown to play a key role in the interactions between neonatal gut microbiota, systemic inflammation, and brain development (Morin et al. 2025). Elevated circulating peptidoglycan levels were observed, suggesting that systemic inflammation combined with disrupted gut microbiota (due to Cesarean section delivery) facilitates the translocation of peptidoglycans into the bloodstream, potentially influencing microglial activation, brain connectivity, and myelination, ultimately contributing to neurodevelopmental deficits (Morin et al. 2025).

Understanding the role of peptidoglycans in gut-brain communication could provide crucial insight into how microbial components contribute to neurological and psychiatric disorders. It also highlights the potential for targeting these pathways in therapeutic strategies for brain-related conditions. The maternal intake of a multispecies probiotic revealed the role of bacterial peptidoglycan in neurodevelopment (Siegler Lathrop et al. 2024). Specifically, the expression of the proton-coupled oligopeptide transporter 1 (PepT1), which facilitates the transport of peptidoglycan motifs, was significantly increased in the prefrontal cortex of male offspring exposed to the probiotic.

Microbial extracellular vesicles are nanoscale particles released by gut bacteria that play a pivotal role in the microbiota-gut-brain axis (Mandelbaum et al. 2023). These vesicles can traverse biological barriers, including the blood-brain barrier, facilitating direct communication between gut microbiota and the central nervous system. Microbial extracellular vesicles carry various bioactive molecules, such as proteins, lipids, and nucleic acids, which can modulate immune responses and influence brain function (Kang et al. 2023). For instance, they can interact with immune receptors on glial cells, leading to the production of cytokines and inflammatory mediators, thereby impacting neuroinflammation and potentially contributing to neurological disorders.

Beyond the role of diet in shaping the gut microbiota and influencing metabolic and neuropsychiatric health, dietary components also interact with key appetite-regulatory peptides that link metabolism to behaviour. Ghrelin and GLP-1, two critical peptides involved in energy balance and feeding behaviour, not only regulate food intake but also play a role in reward processing and addictive behaviours. *This is covered in more detail in the Chap. 8* where we explore the involvement of these peptides in alcohol use disorder, shedding light on how metabolic signals influence substance-related behaviours and revealing potential therapeutic targets for addiction.

6.2 Microbiota-Gut-Brain Axis: Lifespan Dynamics and Clinical Implications

The microbiota-gut-brain axis is a dynamic system that evolves throughout our lives (Lynch et al. 2023). Small daily variations in microbiota composition are normal, but significant changes occur at different life stages. These transformations can be viewed as sensitive periods where the microbiota is highly influential on overall health and responsive to external factors, potentially offering therapeutic opportunities (Ratsika et al. 2023).

In early life, the *in utero* environment is understood to be sterile see: (Kennedy et al. 2023). Initial colonisation happens at birth, significantly shaping the microbiota's development (Roager et al. 2023). Early-life gut microbial succession follows conserved patterns globally, with a decline in *Bifidobacterium* and an increase in *Faecalibacterium prausnitzii* and other taxa as infants transition from breastfeeding to solid foods (Fahur Bottino et al. 2025). The gut microbiota then evolves rapidly in infancy, dominated by certain bacterial families, and reaches adult-like diversity within 1–3 years, influenced by diet and weaning (Palmer et al. 2007; Yatsunenko et al. 2012). Early life factors, including birth mode (Korpela and de Vos 2022), diet (McEvoy et al. 2024), and antibiotic use (Desbonnet et al. 2015), impact long-term health, with emerging research suggesting that early microbiota manipulations can alter developmental and health trajectories. Early-life infections, antibiotic use, stress, and various biomarkers—including infant microbiota and cord serum metabolites—have been linked to the development of neurodevelopmental disorders like intellectual disability and autism, purporting the possibility to establish a foundation for early-life prediction and intervention strategies (Ahrens et al. 2024).

The neonatal gut is enriched with neurotransmitters like serotonin, which are derived from both host cells and specific gut bacteria. Certain gut bacteria contribute to serotonin production that can inhibit monoamine oxidase A, reducing serotonin breakdown and supporting local signalling pathways (Sanidad et al. 2024). This modulation of serotonin levels can contribute to shaping the neonatal immune system, promoting long-term immune tolerance toward dietary antigens and commensal bacteria. This process is critical for preventing excessive inflammatory responses and establishing immune homeostasis in neonatal mice (Sanidad et al. 2024). These findings highlight the critical role of gut-derived serotonin in shaping T-cell responses and promoting immune tolerance during early development. Focus is also shifting to the early-life gut virome, where temperate bacteriophages may influence immune responses independently of the gut's bacterial population, particularly in the development of asthma ("Beyond bacteria: early-life gut virome link with childhood asthma development" 2024; Leal Rodriguez et al. 2024).

During adolescence, significant hormonal and brain changes coincide with shifts in social behaviour, diet, and exposure to new experiences, making this a vulnerable period for mental health issues

(Carbia et al. 2023; Ratsika et al. 2023). Although few studies have directly compared adolescent and adult microbiota, early research indicates gradual transitions rather than distinct adolescent profiles (Agans et al. 2011; Hopkins et al. 2001). Positive lifestyle factors—such as parental care?, school engagement, healthy nutrition, and physical activity—can counteract genetic risks associated with accelerated brain aging in adolescence (Petrican et al. 2024). Longitudinal analysis revealed that genetic vulnerability and lifestyle factors uniquely influence functional neurodevelopment, particularly in brain regions associated with emotional regulation, cognitive control, and stress responses. However, distinct microbiota trajectories between sexes emerge at puberty, driven by hormonal changes such as increased levels of sex hormones like oestrogen and testosterone (Korpela et al. 2021). These hormonal shifts can differentially shape the gut microbiota, influencing microbial composition, diversity, and function, which may contribute to sex-specific vulnerabilities to mental health disorders during adolescence. Diet and metabolic outcomes in adolescents correlate with microbiota composition (Kemp et al. 2024), suggesting further research is needed. Preclinical studies show that gut microbiota interventions during adolescence can influence long-term cognitive and behavioural outcomes, supporting the idea of adolescence as a sensitive period for microbiota-gut-brain interactions (Lynch et al. 2023).

Ageing brings a decline in various bodily functions and an increased prevalence of diseases, with notable changes in gut physiology affecting microbiota composition (Cruz-Pereira et al. 2023; Zhou et al. 2022). The stability of the adult gut microbiota deteriorates with age, influenced by diet and physical activity. Although studies show conflicting results regarding specific microbial taxa associated with ageing, decreased microbiota diversity is generally linked to adverse outcomes and unhealthy ageing (Ghosh et al. 2022). Some taxa, like *Bacteroidetes* and *Faecalibacterium prausnitzii*, are associated with reduced frailty and better health in the elderly (Ghosh et al. 2022; Lim and Nam 2023).

The BBB is disrupted with ageing, likely involving the altered function of microglia in regulating BBB integrity, leading to accelerated decay in ageing and neurodegeneration (Knox et al. 2022a). This disruption is further influenced by gut microbiota-derived metabolites, such as SCFAs and tryptophan metabolites, which can modulate microglial function and BBB permeability. Age-related changes in microbiota composition may exacerbate BBB dysfunction, linking gut health to brain ageing and the progression of neurodegenerative disorders (Zhang et al. 2024a).

The ageing gut microbiota contributes to "inflammaging," a state of chronic low-grade inflammation that accelerates aging and disease development (Fransen et al. 2017). Studies indicate that the gut microbiome plays a role in regulating neuroimmunity, with implications for therapeutic interventions targeting age-related cognitive decline and neurodegeneration (Boehme et al. 2021). Dietary interventions and caloric restriction have shown potential in modulating the gut microbiota to promote healthy aging, with some studies even suggesting that transferring microbiota from young to older individuals can extend lifespan (Boscaini et al. 2022). Intriguingly, dietary restriction in aging mice at risk for cognitive impairment following anaesthesia and surgery may alleviate cognitive dysfunction induced by anaesthesia/surgery by modulating gut microbiota (Ren et al. 2024). Reduced neuroinflammation in hippocampal microglia and enhanced neuronal repair and regeneration were noted. These insights suggest that dietary strategies could serve as innovative therapeutic approaches for managing the ageing brain.

Overall, the microbiota-gut-brain axis is crucial at various life stages, influencing health and disease outcomes. Understanding these dynamics offers opportunities for targeted interventions to improve health across the lifespan.

6.3 Shaping the Axis: Genetic, Environmental, and Lifestyle Factors

The interplay between the gut microbiota and the brain is influenced by various factors including genetics (Otaru et al. 2024), diet (including pre-, pro-, and synbiotics) (Schneider et al. 2024a), environment (Matsuzaki et al. 2023), exercise (Ni Lochlainn et al. 2024), medications (Cussotto et al. 2021a), and birth delivery mode (Dinan et al. 2022; Selma-Royo et al. 2024). Each of these elements can significantly impact both brain function and the composition of the gut microbiota, leading to a range of physiological and behavioural outcomes. Next, we will examine these factors, exploring their known effects and potential areas for future research.

6.3.1 How Genes and Epigenetics Drive Gut-Brain Communication

The influence of genetics on the microbiota-gut-brain axis is an emerging field of study (Goodrich et al. 2014). Host genetics play a crucial role in shaping the gut microbiota, with evidence suggesting a significant correlation between genetic makeup and microbial composition (Zhernakova et al. 2024). Studies on monozygotic twins have shown a higher similarity in gut microbiota compared to dizygotic twins, indicating a genetic component in microbiota composition (McEvoy et al. 2024). For instance, the family *Christensenellaceae* is highly heritable and associated with a lower body mass index, highlighting the gut microbiota's role in mediating genetic influences on phenotypes (Waters and Ley 2019).

Epigenetics, involving changes in gene expression without altering the DNA sequence, also plays a part in this axis. The gut microbiota can influence gene expression through mechanisms such as histone deacetylase (HDAC) activity. Inhibition of HDAC has been shown to reverse stress-induced behaviours in animal models, suggesting that microbial modulation of gene expression could impact brain function and behaviour (Liu et al. 2018a). Additionally, non-coding RNAs, particularly microRNAs (miRNAs), are crucial mediators in this interaction by regulating post-transcriptional gene expression. miRNAs modulate a wide range of biological processes, including neural development, synaptic plasticity, and immune responses, all of which are influenced by the gut microbiota (Moloney et al. 2019). miRNAs regulate gene expression and have been linked to neuropsychiatric disorders (Chen et al. 2017; Moloney et al. 2018). Research in germ-free mice has demonstrated that the gut microbiota affects miRNA expression in brain regions like the amygdala and prefrontal cortex, further underscoring the microbiota's role in gene regulation and brain health (Chen et al. 2017).

MicroRNAs can also modulate gut-brain connections. *As discussed in the Chap. 12* ongoing research is actively investigating the role of miRNAs originating in the gut and brain in contributing to various disease states.

6.3.2 How Birth Shapes the Microbiome: Vaginal Delivery Versus C-Section

The mode of delivery at birth is a significant determinant of an infant's initial microbiota (Zachariassen et al. 2021). Vaginal delivery allows for the transfer of maternal vaginal microbiota to the newborn, whereas C-section bypasses this exposure, leading to initial colonisation by skin and environmental bacteria. C-section delivery delays the transmission of beneficial maternal microbiota to infants, resulting in altered colonisation patterns, reduced early acquisition of key species like *Bifidobacterium longum*, and increased susceptibility to long-term health issues (Selma-Royo et al. 2024). However, breastfeeding plays a critical role in mitigating these effects by promoting the growth of beneficial bacteria and supporting gut and immune development, highlighting the importance of breastfeeding support and potential microbiota restoration strategies for C-section infants (Korpela and de Vos 2022; Selma-Royo et al. 2024). Vaginally delivered infants typically have a microbiota similar to their mothers' vaginal microbiota, rich in important commensal bacteria such as *Bifidobacterium* and *Lactobacillus*. In contrast, C-section-delivered infants often have a less diverse microbiota, with higher levels of *Staphylococcus* spp. and lower levels of important commensals like *Bacteroides* (Lynch et al. 2023; Shao et al. 2019).

These early differences can have long-term health implications. C-section delivery has been associated with an increased risk of developing conditions like obesity, type 1 diabetes, asthma, and allergies (Cabre et al. 2022). Additionally, C-section-born infants may have a higher risk of neonatal infections (Miller et al. 2020). Although these microbiota differences tend to diminish over time, the initial colonisation can have lasting effects on immune development and disease susceptibility (Dinan et al. 2022; Morais et al. 2020). C-section-born individuals can display an increased psychological vulnerability to stress and lower positive affect, along with altered immune profiles, suggesting a potentially lasting impact of delivery mode on psychological responses (Dinan et al. 2022).

In mice, the impact of early-life gut microbiota alterations due to C-section delivery reveals that C-section delivery leads to an excessively diverse microbiota too early in life, which triggers an inflammatory immune response, disrupts gut epithelium structure, and alters mucus production (Barone et al. 2023). Importantly, transferring the microbiota from mice born by C-section into germ-free mice confirmed the causal relationship between the altered microbiota and increased inflammatory susceptibility. Further, supplementation with *Lactobacilli* restored resilience against inflammation, further highlighting the critical role of early-life microbiota-host interactions in shaping long-term health outcomes. Such disruptions have been associated with various neurological and psychiatric conditions, suggesting potential implications for brain health too.

In rats, offspring born via C-section to obese mothers exhibited greater weight gain and caloric intake, alongside altered expression of appetite-regulating neuropeptides, while those from lean mothers born by C-section had higher stress hormone levels and reduced expression of genes associated with fat metabolism (Cho et al. 2021). The findings indicate that C-section delivery affects obesity risk differently depending on maternal obesity status, and such alterations in the gut microbiota due to C-section delivery could have systemic effects influencing brain function and behaviour through the microbiota-gut-brain axis, although more work is needed.

Strategies to mitigate the impact of C-section on the microbiota include vaginal seeding, where vaginal microbiota are transferred to the newborn, maternal faecal microbiota transplantation (FMT), and supplementation with probiotics and prebiotics (Korpela and de Vos 2022). A recent study investigated the impact of vaginal microbiota transfer (vaginal seeding) on neurodevelopment and gut microbiota composition in C-section-delivered infants (Zhou et al. 2023). Results indicated that vaginal seeding significantly improved neurodevelopmental outcomes at 6 months of age; this also accelerated gut microbiota maturation and positively influenced various metabolic functions within 42 days post-birth. While these preliminary results highlight the potential of vaginal seeding to support neurodevelopment and microbiota composition in C-section-delivered infants, the authors note that further research is needed to establish the long-term safety and efficacy of this intervention. These findings do not currently support routine clinical implementation but underscore the need for more studies to assess the approach's feasibility and broader health implications.

6.3.3 Dietary Drivers of Microbiota-Gut-Brain Health

Diet is a critical factor influencing the microbiota-gut-brain axis (Berding et al. 2023; Ross et al. 2024a; Schellekens et al. 2023). Different dietary patterns can significantly alter gut microbiota composition, affecting brain function and behaviour. The Western diet, high in sugar, salt, fat, and ultra-processed foods, is linked to negative health outcomes such as metabolic disorders and altered gut microbiota (McGuinness et al. 2024). This diet tends to decrease pro-health bacteria like *Bacteroidetes* and increase negative gut microbes like Proteobacteria (Ross et al. 2024b). Such shifts can lead to systemic inflammation and negatively impact brain health. Studies in mice have shown that a high-fat diet can drastically alter gut microbiota within 48 hours, indicating the rapid and profound effects of dietary changes (Fulling et al. 2020). Furthermore, HFD is associated with neuroinflammation, impaired blood-brain barrier integrity, and altered hippocampal plasticity, contributing to cognitive deficits and mood disturbances (Bruce-Keller et al. 2015). These effects are thought to be mediated through changes in gut microbiota composition, leading to systemic inflammation and increased permeability of the gut barrier and subsequent effects on the microbiota-gut-brain axis (de Paula et al. 2021).

The Mediterranean diet, rich in fruits, vegetables, whole grains, and healthy fats, is associated with numerous health benefits, including reduced risk of neurodegenerative diseases and improved mood (Bizzozero-Peroni et al. 2024; Conti et al. 2024; Park et al. 2024). Human studies have demonstrated that adhering to a Mediterranean diet can lower the incidence of depression and enhance cognitive function (Ross et al. 2024a). A first-of-its-kind randomised controlled trial provided evidence that a Mediterranean-style diet can significantly reduce depressive symptoms in individuals with moderate to severe depression (Jacka et al. 2017). Participants who improved their diet quality, with guidance from a dietitian, experienced notable mental health benefits, independent of weight changes. These findings highlight the potential of dietary improvement as an adjunctive strategy for managing depression in clinical settings.

The ketogenic diet, high in fat and low in carbohydrates, has been shown to benefit neurological conditions such as epilepsy and autism (Allan et al. 2024; Smolensky et al. 2023). This diet increases ketone bodies, which have neuroprotective effects, and alters gut microbiota composition to include important commensal bacteria like *Akkermansia* and *Parabacteroides* (Santangelo et al. 2023). These changes in gut microbiota composition and function are thought to contribute to the diet's therapeutic effects on brain health (Boscaini et al. 2022). One study examined the effects of low-carbohydrate diets, focusing on sex differences in metabolic responses and adipose tissue dynamics in mice (Smolensky et al. 2023). Although the ketogenic diet resulted in weight loss and reduced subcutaneous fat in males, it also revealed decreased insulin levels and increased ghrelin and Fibroblast growth factor-21 levels in both sexes, with female mice exhibiting unfavourable metabolic effects from the non-ketogenic moderate low-carbohydrate diet, suggesting a complex and sex-dependent response to these dietary interventions. Emerging evidence highlights the potential of the ketogenic diet as a therapeutic intervention for schizophrenia (Sarnyai et al. 2019). By providing ketone bodies as an alternative energy source, the diet compensates for glucose metabolism and mitochondrial dysfunction commonly observed in schizophrenia. Recent preclinical studies and case reports suggest that the ketogenic diet can restore synaptic communication, improve psychiatric symptoms, and mitigate metabolic side effects associated with antipsychotic treatment (Kraeuter et al. 2019). However, more rigorous randomised (Kennedy et al. 2007; Paoli et al. 2013) controlled trials are needed to validate these findings and explore their long-term safety and efficacy.

Carbohydrates, both digestible and indigestible, play a role in shaping gut microbiota and influencing body-brain interactions. Digestible carbohydrates, such as sugars, can increase certain commensal bacteria but may also contribute to metabolic disorders if consumed in excess (Lippert and Bruning 2022). Indigestible carbohydrates, like dietary fibre, are fermented by gut bacteria into a variety of compounds, including SCFAs, secondary bile acids, aromatic amino acids, neurotransmitters, and B vitamins. These metabolites support gut health and modulate brain function via anti-inflammatory properties and microbiota-gut-brain signalling pathways (Dalile et al. 2019; Grant et al. 2025). Indeed, SCFAs such as butyrate are known to cross the blood-brain barrier and influence neuroinflammation, neurotransmitter production, and cognitive processes (Silva et al. 2020). Individuals with insulin resistance have increased levels of faecal carbohydrates, particularly host-accessible monosaccharides, which correlate with microbial carbohydrate metabolism and inflammatory cytokines (Takeuchi et al. 2023). This systemic inflammation associated with insulin resistance can also affect neuroinflammation and brain health, contributing to conditions such as cognitive decline and Alzheimer's disease (Arnold et al. 2018). By identifying gut bacteria linked to insulin resistance and demonstrating their potential to improve host insulin sensitivity in a mouse model, microbial carbohydrate metabolism could be a promising therapeutic target for addressing insulin resistance, such as with type 2 diabetes. Such interventions may not only address insulin resistance and type 2 diabetes but also alleviate the associated cognitive and neurological complications through microbiota-gut-brain and immune pathways linked to brain health.

Protein intake, whether from animal or plant sources, affects gut microbiota diversity and composition. Animal-based proteins tend to increase bacteria associated with protein fermentation, while plant-based proteins promote a more diverse microbiota (Bartlett and Kleiner 2022). Indeed, omnivorous diet-associated microbes, influenced by meat consumption, were linked to inflammation and poorer cardiometabolic health, while vegan-associated microbes, enriched by plant-based diets, correlated with positive gut functions and improved cardiometabolic markers (Fackelmann et al. 2025). Vegetarian microbiomes displayed intermediate characteristics, with dairy consumption leaving a distinct microbial signature (Fackelmann et al. 2025). High-protein diets can enhance the production of SCFAs and branched-chain amino acids, which have various health benefits including anti-inflammatory effects and modulation of neurotransmitter synthesis (Awan et al. 2024; Mueller et al. 2021b). However, excessive protein intake, especially from animal sources, may increase the risk of gut inflammation and metabolic disorders, which are linked to systemic inflammation and impaired brain function (Armet et al. 2022; Zhu et al. 2015). Indeed, dietary proteins from different sources are digested and utilised in the gut differently by the host and the gut microbiota (Awan et al. 2024). Some proteins were preferentially utilised by the microbiota, such as the alpha subunit of β-conglycinin found in soy, lectin and legumin J present in peas, ovalbumin in egg white, and cell wall proteins from yeast. These variations in protein utilisation can influence gut microbial metabolites, which in turn impact brain health through the microbiota-gut-brain axis. For example, specific microbial metabolites derived from protein fermentation, such as aromatic amino acids, are precursors for neurotransmitters like serotonin and dopamine, which are crucial for mood regulation and cognitive function (Strandwitz 2018). This work highlights the significant differences in the fate of dietary proteins, suggesting that these variations may contribute to the differential health impacts associated with different protein sources.

The type of fat consumed can influence gut microbiota, overall health, and brain function. Saturated and trans fats, common in a Western diet, are linked to negative changes in the gut microbiota composition and increased risk of cardiovascular diseases, increased inflammation, and heightened risk of neurodegenerative diseases (Martinez et al. 2017; Zhang and Gerard 2022). In contrast, polyunsaturated fatty acids, such as omega-3 and omega-6, support the growth of commensal bacteria, have anti-inflammatory effects, and positively influence brain health (Leyrolle et al. 2022; Muth and Park 2021). Omega-3 fatty acids, in particular, have been shown to improve cognitive function and reduce the risk of depression and other neuropsychiatric conditions. In particular docosahexaenoic acid and eicosapentaenoic acid are critical for brain structure and function, as docosahexaenoic acid is a key component of neuronal membranes and synapses, contributing to neuroplasticity, neurotransmission, and cognitive function. For example, a recent study investigated the effects of various doses of eicosapentaenoic acid (an Omega-3 fatty acid) on inflammatory markers and depressive symptoms in a cohort of sixty-one unmedicated adults diagnosed with MDD and elevated high-sensitivity C-reactive protein (hs-CRP) levels (Mischoulon et al. 2022). A higher dose of daily EPA led to significant improvements in response rates and a correlation between reductions in hs-CRP and decreases in depressive symptoms. These findings highlight the role of omega-3 fatty acids in modulating both inflammation and brain health, emphasising their therapeutic potential for mood disorders and cognitive decline.

A recent study noted, after identifying five dietary patterns, that global dietary patterns are more strongly associated with variations in the gut microbiome than individual dietary components, emphasising the importance of a holistic approach to dietary evaluation for understanding gut health and informing personalised dietary strategies (Cotillard et al. 2022).

6.3.4 Fueling the Axis: Impact of Prebiotics and Fermented Foods

Prebiotics and fermented foods also play a crucial role in this field (Hutkins et al. 2025; Schneider et al. 2024a). Prebiotics are substrates selectively utilised by host microorganisms, conferring health benefits (Salminen et al. 2021). While many prebiotics are dietary fibres, such as resistant starch, inulin, fructooligosaccharides (FOS), and galactooligosaccharides (GOS), not all fibres meet the criteria to be classified as prebiotics (Liu et al. 2017b; Zhang et al. 2021b). Prebiotics can also include non-fibre compounds, such as certain polyphenols and human milk oligosaccharides, which have been shown to positively influence gut microbiota composition and function, thereby modulating host health (Donoso et al. 2020; Ferrier et al. 2022; Zhu et al. 2024a). These diverse prebiotics are found in sources like fruits, vegetables, grains, and human milk, each demonstrating unique effects on gut microbiota composition and metabolic activity.

The Western diet has led to a drop in prebiotic intake, along with a rise in inflammatory diseases and lifestyle disorders (Zhang et al. 2021b). Supplementing with prebiotics has been shown to reduce stress, anxiety, and depressive-like behaviour, positively influencing brain physiology and behaviour (Berding et al. 2023). The mechanisms of action for prebiotics involve altering gut microbiota composition, leading to increased production of health-promoting metabolites like SCFAs (Zhang et al. 2021b). These changes can enhance hippocampal synaptic efficacy, general hypothalamic neuronal activity, and cognitive functions (Church et al. 2023; Rusch et al. 2023). The dietary fibre polydextrose showed significant improvements in cognitive flexibility and sustained attention in the healthy female population (Berding et al. 2021). Although polydextrose supplementation did not alter microbial diversity, it did increase the abundance of *Ruminiclostridium* 5 and reduced inflammation, suggesting a potential mechanism for improved gut-to-brain communication and behavioural modulation (Berding et al. 2021).

Fermented foods are products rich in microbes and bioactive metabolites that interact with the gut microbiota, influencing the microbiota-gut-brain axis and promoting mental and physical health (Balasubramanian et al. 2024; Marco et al. 2021). They have emerged as a promising avenue for enhancing mental health through their ability to modulate the gut microbiota (Schneider et al. 2024a). Rich in microbes and bioactive metabolites, these foods interact with the microbiota-gut-brain axis, influencing the various pathways mentioned previously and impacting brain function and emotional well-being (Schneider et al. 2024b). By harnessing the power of fermented foods, we can tap into their potential to shape the gut microbiome, bolster immune responses, and promote cognitive health. For example, a recent study with critically ill patients received escalating doses of kefir (fermented dairy product), with no reports of serious side effects and a high adherence rate of 91% to the prescribed doses (Gupta et al. 2024). Although no significant changes in gut microbial diversity were noted, a marked improvement in the Gut Microbiome Wellness Index was observed, suggesting that kefir could positively influence gut health in this vulnerable population. A similar study using kefir noted improvements in relational memory tasks combined with a significant increase in Lactobacillus levels (Cannavale et al. 2023).

6.3.5 From Probiotics to Psychobiotics: Microbial Allies in Mental Health

Probiotics are live bacteria that confer health benefits to the host when ingested in adequate amounts and interact with host microbiota and the intestinal epithelium (Gupta et al. 2024; Hill et al. 2014; Mazzone et al. 2024; Samara et al. 2022). They have been shown to improve metabolism, immunity, and even brain function. Indeed, recent work has identified the Lab4 multistrain probiotic as reducing IBS symptom severity, anxiety, and depression, along with improved stool consistency in a cohort of individuals with IBS (Mullish et al. 2024). Further, children with ASD provided with the probiotic *L. reuteri* demonstrated noteworthy improvements in social functioning across various assessments, although it did not significantly affect overall autism severity or microbiota composition (Mazzone et al. 2024). A probiotic formulation featuring four *Bifidobacterium* strains and one *Lacticaseibacillus* strain was shown to facilitate the rapid transition toward a stable, term-like microbiome in extremely premature infants (Samara et al. 2022). Such work indicates that *Bifidobacterium* strains significantly influence microbiome maturation from an early age, contributing to a more interconnected microbial community and promoting an anti-inflammatory immune environment, thereby underscoring their role as key contributors to gut ecosystem development in this vulnerable population.

Research has documented probiotics' ability to modulate neurotransmission, neurogenesis, and neuroinflammation, impacting various disorders, from depression to autism (Audet 2021; Guzzetta et al. 2022; Mohle et al. 2016). For example, the probiotics *Lactobacillus plantarum* A7 and *Bifidobacterium animalis* delayed disease onset in an experimental autoimmune encephalomyelitis (EAE) animal model of Multiple Sclerosis (MS) (Salehipour et al. 2017). A marked inhibition of inflammatory cytokines and enhanced levels of regulatory T cells were noted, effectively promoting anti-inflammatory responses by favoring Treg differentiation.

A prebiotic or probiotic that, when ingested in adequate amounts, produces health benefits by positively influencing mental health and brain function through mechanisms such as modulation of the gut microbiota and its communication with the brain is called a psychobiotic (Slykerman et al. 2025). Probiotics like *Faecalibacterium prausnitzii* have demonstrated potential as psychobiotics, reducing anxiety and depression-like symptoms in animal models (Hao et al. 2019; Zhernakova et al. 2024). The ability of *F. prausnitzii* to produce butyrate under conditions of adequate dietary fibre highlights its therapeutic potential. However, its effects depend on the availability of specific substrates like dietary fibres or prebiotics and a conducive gut environment. Similarly, other psychobiotics may require specific conditions to produce bioactive metabolites, such as neurotransmitter precursors or SCFAs. Therefore, while the promise of probiotics for enhancing health is significant, it is essential to consider how factors such as dietary composition, availability of microbial substrates, and the complexity of the host's microbiota can influence their metabolic outputs, efficacy, and overall outcomes.

6.3.5.1 Beyond Pre- and Probiotics: The Rise of Synbiotics and Postbiotics

Synbiotics are "a mixture comprising live microorganisms and substrate(s) selectively utilised by host microorganisms that confers a health benefit on the host" (Swanson et al. 2020). The synergistic action of synbiotics aims to improve the balance of gut microbiota, support digestive health, and potentially provide various health benefits beyond those offered by probiotics or prebiotics alone. Synbiotics have been the focus of some important recent research. A randomised controlled trial involving 153 infants born via elective C-section was given a synbiotic or prebiotic-supplemented formula which showed significant differences in their gut microbiome development compared to vaginally born infants (Lay et al. 2021). Breastfed infants exhibited a more favourable hypoxic and acidic gut environment, preventing the establishment of an unwanted microbiome. One study investigated the impact of a synbiotic supplement on metabolic syndrome components in patients receiving antipsychotic treatment (Basafa-Roodi et al. 2024). They concluded that the synbiotic significantly reduced waist circumference and HbA1c levels, while also preventing BMI increases compared to the placebo group, suggesting a potential role for the synbiotic in managing metabolic health in this population. These findings have implications for the microbiota-gut-brain axis, as improved metabolic health and reduced inflammation are linked to better brain function and mental health outcomes in schizophrenia. By modulating gut microbiota and lowering systemic inflammation, the synbiotic may indirectly support cognitive function and mood regulation, addressing both physical and mental health challenges associated with this condition.

Postbiotics are the "preparation of inanimate microorganisms and/or their components that confers a health benefit on the host" (Salminen et al. 2021). These compounds can include SCFAs, peptides, and other metabolites that contribute to health benefits, such as supporting immune function, enhancing gut barrier integrity, and reducing inflammation (Braga et al. 2024; Salminen et al. 2021; Siciliano et al. 2021; Wang et al. 2024). Unlike probiotics, which are live microorganisms, postbiotics are not alive but still exert positive effects on the host's health. A recent study examining the postbiotic *Lactobacillus* LB, derived from heat-treated *Limosilactobacillus fermentum* and *Lactobacillus delbrueckii*, has been shown to enhance the growth of commensal bifidobacteria in both human faecal models and pure cultures, leading to improved metabolic outcomes such as increased acid production (Warda et al. 2021). Importantly, postbiotics may also play a role in brain health via the microbiota-gut-brain axis. For example, SCFAs, a class of microbial metabolites that may be delivered as part of a postbiotic preparation, have been shown to modulate microglial activation, reduce neuroinflammation, and influence brain plasticity, linking postbiotic activity to improved cognitive and emotional regulation (Dalile et al. 2019). Furthermore, GABA, a neurotransmitter critical for mood regulation and stress resilience, may act as part of postbiotic preparation since its levels in both circulation and the brain are influenced by gut microbiota composition. Recent findings suggest that both circulating and brain GABA levels are influenced by gut microbiota composition, positioning GABA as a potential mediator of the microbiota-gut-brain axis. Exploring GABA-producing microorganisms in the gut and dietary factors that enhance GABA production could lead to new therapeutic approaches for treating neurological and psychological disorders (Braga et al. 2024).

6.3.5.2 Environmental Influencers: Shaping Microbial Ecosystems

The environment we inhabit exerts a profound influence on our health outcomes and the composition of our gut microbiota (Gacesa et al. 2022; Matsuzaki et al. 2023; Sanmarco et al. 2022). Cross-cultural differences in health can often be traced back to environmental factors that shape our microbiome (Morais et al. 2024). Various members of the gut microbiota are involved in metabolising numerous environmental chemicals through enzymes such as azoreductase, nitroreductase, β-glucuronidases, sulfatases, and β-lyases (Claus et al. 2016; Walsh et al. 2020). Conversely, these chemicals can also modify the composition of the gut microbiota, impacting brain function and behaviour via the microbiota-gut-brain axis.

A notable example is bisphenol A, an endocrine-disrupting chemical used extensively in manufacturing plastic containers. Exposure to bisphenol A through diet has been shown to reduce the diversity of gut microbiota in rodents while increasing the abundance of *Proteobacteria* and *Helicobacteraceae* and decreasing *Clostridia* (Lai et al. 2016). Such alterations in microbiota composition could influence brain function and behaviour. These changes in microbiota composition have been linked to altered microbiota-gut-brain axis communication. For example, increased *Proteobacteria* are associated with heightened intestinal inflammation, which can promote neuroinflammation through immune signalling pathways (Heinzel et al. 2024). Additionally, reductions in bacteria such as *Clostridia* through environmental factors may impair the production of SCFAs, which are critical for maintaining blood-brain barrier integrity and regulating brain neurochemistry (Chiu et al. 2020).

Heavy metals such as cadmium, mercury, arsenic, and lead are highly toxic and can disrupt immune function, damage DNA, and cause oxidative stress (Jomova et al. 2025). Exposure to these metals also alters gut microbiota composition in both animals and humans (Lu et al. 2014; Thompson et al. 2022; Zhang et al. 2015). Studies have shown that heavy metal administration to animals perturbs their gut microbiota and metabolomic profile, increasing levels of *Lactobacillaceae* and *Erysipelotrichaceae* while reducing *Lachnospiraceae* (Breton et al. 2013; Zhu et al. 2024b), a genera of bacteria associated with depression in humans (Radjabzadeh et al. 2022). This suggests a potential microbiota-gut-brain pathway for the mental health impact of environmental pollutants.

Another environmental factor that can affect the gut microbiota is pesticides, which are widely used in modern agriculture (Matsuzaki et al. 2023). Preclinical research has demonstrated that pesticides can negatively affect gut microbiota composition, leading to changes in host physiology and behaviour, including increased anxiety-like and depressive-like behaviours (Curtis et al. 2023; Yue et al. 2024). As a result, it is becoming ever more critical to investigate whether pesticide-induced alterations in gut microbiota are responsible for these behavioural impairments. In humans, occupational pesticide exposure has been associated with altered gut microbiome composition, including decreased microbial diversity and changes in specific bacterial taxa (Gois et al. 2023). Chronic, low-level dietary exposure to highly toxic pesticides like Paraquat and Chlorpyrifos has been shown to negatively affect the human gut microbiota, potentially increasing systemic inflammation, compromising gut barrier integrity, and contributing to metabolic disorders, a concern that has not been fully addressed by regulatory agencies (Gama et al. 2022). The PELAGIE study investigated the relationship between prenatal and early childhood exposure to pyrethroid insecticides and neurobehavioural outcomes in children (Viel et al. 2017). They found that prenatal exposure (e.g. via diet) to certain pyrethroid metabolites was associated with internalising difficulties, while childhood exposure correlated with externalising behaviour problems and impaired prosocial skills. These findings highlight the potential influence of long-term pesticide exposure on the human gut microbiome and its possible health consequences, including neurological impacts. Better monitoring and regulation of pesticide residues in food is desperately needed, particularly in developing countries, where exposure could lead to prolonged and harmful effects on gut and brain health.

Recently, researchers identified unusual phospholipids, termed MmDEACLs, produced by *Morganella morganii*, which incorporate the environmental micropollutant diethanolamine (DEA) (Bang et al. 2025). DEA is an industrial chemical produced on a massive scale and widely used in agriculture, consumer products, and manufacturing. DEA is a non-natural micropollutant that enters the environment in significant quantities and is incorporated into these microbial metabolites via a promiscuous cardiolipin synthase enzyme in *M. morganii*. These DEA-modified phospholipids activate TLR2/TLR1 receptors, inducing the release of pro-inflammatory cytokines, particularly IL-6, known to be associated with MDD and other inflammatory conditions. This mechanism highlights the role of gut microbial metabolites, chronic inflammation, and environmental pollutants in mental health disorders.

In the Chap. 11 the authors present evidence that early immune system disruptions may significantly contribute to neurodevelopmental disorders. Specifically, altered immune responses during prenatal development—such as maternal immune activation (MIA) triggered by infections—have been linked to an increased risk of these disorders in offspring. Immune dysregulation, particularly elevated pro-inflammatory cytokines, can interfere with brain development and synaptic connectivity, resulting in lasting effects on brain function and behaviour.

6.3.5.3 Exercise and the Axis: Enhancing Gut-Brain Health Through Movement

Exercise is well known for its benefits to overall health, including brain health (Nota et al. 2023). It influences brain structures and functions, such as neurogenesis in rodents. Exercise can counteract age-related declines in hippocampal neurogenesis, potentially reversing aging effects within brain structures (Nicolas et al. 2024; Wang et al. 2023). Despite extensive research on exercise and brain health, the impact of exercise on gut health and the microbiota-gut-brain axis is less understood. The PROMOTe trial, a randomised controlled study investigating the effects of prebiotic supplementation on muscle strength, function, and cognition in older adults, demonstrated that prebiotic supplementation improved cognitive function in older adults, particularly in memory tasks linked to Alzheimer's risk, but had no significant impact on muscle strength or function over 12 weeks (Ni Lochlainn et al. 2024). The prebiotic (inulin and fructo-oligosaccharides) effectively modulated the gut microbiome, increasing *Bifidobacterium* abundance. These findings highlight the potential of prebiotics for addressing cognitive frailty, though further research is needed to explore their role in muscle health.

Moderate exercise can positively affect gut microbiota diversity and composition (Noetel et al. 2024; Sanchez et al. 2017). It reduces stress, enhances immunity, and improves energy homeostasis by modulating the hypothalamic-pituitary-adrenal (HPA) axis, leading to lower cortisol levels, a critical marker of stress (Hillman et al. 2008; Ni Lochlainn et al. 2024; Noetel et al. 2024; Rodrigues et al. 2025). Additionally, exercise enhances immunity by increasing the circulation of immune cells, such as natural killer cells, and modulating the production of anti-inflammatory cytokines (Campbell and Turner 2018). Exercise also improves energy homeostasis by influencing mitochondrial biogenesis and metabolic pathways that regulate glucose and lipid metabolism (Warren et al. 2020)

Studies in rodents have shown that voluntary exercise can significantly alter the balance of major bacterial phyla, such as Bacteroidetes and Firmicutes, preventing diet-induced obesity and normalising glucose tolerance (McNamara et al. 2021). Exercise increases commensal bacteria like *Lactobacillus* and *Bifidobacterium*, enhancing gut health and potentially influencing brain and behaviour.

Human studies have also highlighted the benefits of exercise on gut microbiota. For instance, the American Gut Project found that individuals who exercised more frequently had greater microbiota diversity, particularly in the Firmicutes phylum (Lane et al. 2024). Professional athletes have shown higher microbiota diversity compared to sedentary individuals, with an increase in the genus *Akkermansia*, associated with a healthier metabolic profile (Jie et al. 2021).

Regular physical activity positively correlates with increased microbiota diversity and specific bacteria. Studies involving various exercise regimens, including aerobic and resistance training, have shown modest but significant alterations in gut microbiota composition, suggesting that both exercise and diet drive gut biodiversity (Komulainen et al. 2021; Nota et al. 2023).

6.3.5.4 Pharmaceutical Impacts: How Medications Alter Microbiota

Antibiotics, although useful in the preclinical setting for studying the gut microbiota, are also widely used in agriculture and over-prescribed in human medicine, having a significant impact on natural ecosystems and the health of organisms, highlighting the need for a broader perspective on planetary health and antibiotic use (Llor and Bjerrum 2014; Manyi-Loh et al. 2018; Prestinaci et al. 2015). Antibiotics directly target the gut microbiota, but non-antibiotic drugs can also influence microbial composition and behaviour. Further, laxatives, antidepressants, and antihistamines also significantly alter the gut microbiota (Cussotto et al. 2021a, b; Rogers and Aronoff 2016). These changes can impact drug pharmacokinetics and host health.

Polypharmacy, the concurrent use of multiple medications, has been associated with reduced microbial diversity (Ticinesi et al. 2017). Certain drug classes, such as proton pump inhibitors, antidepressants, and antipsychotics, have strong associations with specific gut bacteria (Cussotto et al. 2019a, b; Imhann et al. 2016). Studies have shown that these medications can alter microbial gene expression related to drug metabolism, highlighting the need for careful consideration of polypharmacy and its impact on the microbiome (Bretler et al. 2019; Imhann et al. 2016; Xu et al. 2023a). Preclinical exposure to SSRIs combined with maternal immune activation leads to changes in immune and serotonin signalling pathways (Zengeler et al. 2023). These disruptions not only affect placental physiology but also influence the maternal and foetal microbiota, resulting in sex-specific neurodevelopmental and behavioural alterations in offspring through the microbiota-gut-brain axis.

6.3.5.5 Stress and the Axis: The Role of Microbiota in Resilience

Stress disrupts the normal homeostasis of an organism, triggering the HPA axis and resulting in cortisol release. Indeed, both traumatic and chronic stress can induce maladaptive changes that may increase vulnerability to gastrointestinal disorders and alter gut-brain communication, impacting neural and behavioural responses (Leigh et al. 2023). For example, chronic stress leads to HPA axis dysregulation and increased risk of various health issues, including anxiety and depression (Audet 2019; Clarke et al. 2013; Lyte et al. 2019, 2020). However, the effects of acute stress are less understood, but stress has been shown to impair gut barrier function, disrupt enteric nervous system circuitry, and modify the gut microbiota (Chen et al. 2024; Gheorghe et al. 2024). Stress has been shown to impair gut barrier function, disrupt enteric nervous system circuitry, and modify the gut microbiota (Chen et al. 2024; Gheorghe et al. 2024). Stress-related disorders have significant economic impacts, highlighting the importance of understanding the microbiota-gut-brain axis in stress regulation. Numerous studies have demonstrated that stress-induced alterations in gut microbiota can influence brain function and behaviour (Berding et al. 2023; Garcia-Cabrerizo and Cryan 2024; Ritz et al. 2024b). For instance, shifts in microbial composition, such as decreased *lactobacilli* and increased Helicobacter pylori levels in mice, may contribute to changes in neurotransmitter production and inflammatory pathways, which directly affect gut-brain communication (Carabotti et al. 2015; Guo et al. 2009; Palomar et al. 2014). Psychological stressors, such as maternal separation and overcrowding, have also been shown to change gut microbiota composition and disrupt microbial diversity, leading to altered brain signalling and behavioural outcomes (Chen et al. 2023; Nicolas et al. 2022b). Prenatal exposure to maternal stress alters the gut microbiota composition and stress-response pathways in offspring, with these effects persisting across generations (Chen et al. 2023). Further, maternal HFD exposure can alter the offspring's gut microbiota and metabolite profiles, contributing to metabolic and neurodevelopmental changes (Dzamko 2023; Nicolas et al. 2022a). These findings underscore the intricate interplay between stress, the gut microbiota, and brain health via the microbiota-gut-brain axis.

Maternal stress during pregnancy can alter the microbiota composition of offspring, affecting their stress response in adulthood (Block et al. 2022; Potter et al. 2023). Studies have shown that the transfer of maternal vaginal microbiota from stressed dams to non-stressed pups can alter their response to stress, suggesting that early-life microbial exposure plays a crucial role in stress regulation (Jasarevic et al. 2021; Mueller et al. 2021a; Wei et al. 2022; Zhou et al. 2023). In utilising a probiotic *Lactocaseibacillus rhamnosus* HN001 for the treatment of postpartum mood and caregiving behaviours after stress during pregnancy in a rodent model, researchers have discovered a potential therapy for pregnant and postpartum women (Lonstein et al. 2024). This work noted a reduction in postpartum anxiety-related behaviours and altered neurotransmitter levels in the prefrontal cortex. It also impacted maternal behaviours and gut microbiota diversity.

6.3.5.6 Circadian Clocks and Microbial Rhythms: Timing the Axis

The circadian rhythm, a 24-hour cycle regulating bodily functions, interacts with the microbiota and influences health outcomes (Teichman et al. 2020). Modern lifestyles often disrupt circadian rhythms, leading to metabolic and psychiatric disorders. Indeed, the gut microbiota has been shown to regulate stress responsivity via the circadian system (Tofani et al. 2025). The interplay between circadian rhythm and the microbiota is being further explored in the context of obesity, cardiovascular disease, diabetes, psychiatric disorders, and neurodegenerative diseases.

The mammalian circadian clock operates through a feedback loop involving transcription factors such as CLOCK and BMAL1 (Altaha et al. 2022; Heddes et al. 2022). Disruptions in circadian rhythm can alter gut microbiota composition, affecting metabolic and brain health (Heddes et al. 2022; Minnebo et al. 2023; Zhang et al. 2023). Studies have shown that changes in light, diet, or microbiota can influence peripheral clocks, leading to dysregulation of the microbiome and host homeostasis. Indeed, the microbiome itself may help stabilise circadian rhythms in the host gut, buffering against rapid environmental fluctuations (Zhang et al. 2023).

The rhythmicity of gut microbiota develops early on in life and is directly impacted by diet (Heppner et al. 2024). Different eating patterns, including meal frequency and timing, also influence the gut microbiome across various regions and niches of the colon (Minnebo et al. 2023). Although eating patterns moderately affect microbiome composition, time-restricted eating has an even more significant effect, increasing the abundance of commensal bacteria like *Bacteroides* and *Faecalibacterium*, highlighting the importance of eating patterns in dietary intervention studies (Minnebo et al. 2023). Formula supplemented with the prebiotic GOS in a recent study examining the impact of diet on gut microbiota in infants showed that the colonisation of the commensal *bifidobacteria* performed similarly to breastfed infants (Heppner et al. 2024). Intriguingly, circadian rhythmicity in gut bacterial communities and functional pathways became more pronounced with age, especially in both breastfed and GOS-supplemented infants.

Research in transgenic mice with disrupted circadian clock genes has demonstrated significant changes in microbiota composition, suggesting that circadian rhythm regulates microbiota function (Deaver et al. 2018; Voigt et al. 2016). Human studies have shown that disruptions such as jet lag and shift work can lead to arrhythmic gut microbiota, promoting metabolic imbalances (Dollish et al. 2024; Zhang et al. 2023). Using germ-free and antibiotic-treated mice, recent work revealed that tryptophan metabolism is highly responsive to a short acute stress event, with specific bacterial species involved in this metabolic process (Gheorghe et al. 2024). The gut's response to stress varies depending on the time of day and the makeup of the microbiota, resulting in significant changes to gut barrier function and tryptophan metabolism, particularly in the ileum and colon. These findings highlight the importance of understanding the relationship between circadian rhythms and the gut microbiota for potential therapeutic interventions.

The study of the microbiota-gut-brain axis is still in its early stages, but the insights gained so far underscore the profound impact of gut microbiota on overall health. As research progresses, the potential for developing targeted therapies to address neurological and psychiatric disorders through microbiota modulation continues to expand, promising new avenues for improving mental health and well-being.

6.4 The Gut Microbiome in Health and Disease: Bridging Body-Brain Communication

As research into the microbiota-gut-brain axis has advanced, it has become increasingly evident that the microbiota is involved in a variety of psychological and neurological disorders. This includes conditions influenced by stress exposure, dysregulation of the HPA axis, and individual resilience to stress, all of which are impacted by the gut microbiota. The depth of evidence varies across disorders, from early-stage studies showing correlations between gut microbiota composition and clinical populations to more robust findings supporting a causal role of the microbiota in certain disease processes. Now, we will consider the involvement of the gut microbiota in disease.

6.4.1 Autism and the Microbiome: Bridging Gut and Brain Development

ASD encompasses a range of neurodevelopmental conditions characterised by significant social deficits, repetitive behaviours, anxiety, and cognitive disturbances (American Psychiatric Association 2013; Hung and Margolis 2024). ASD affects approximately 1 in 68 children globally, with a higher prevalence in males. The behavioural symptoms are linked to changes in neuroinflammation, neurogenesis, neurotransmission, and the production of hormones like oxytocin and vasopressin. Although traditionally considered a neurological disorder, there is growing evidence of gastrointestinal involvement, with about 70% of children with ASD experiencing gastrointestinal issues such as bloating, constipation, and diarrhoea (Madra et al. 2020).

Studies have noted alterations in the gut microbiota of children with ASD, including a reduction in commensal *Bifidobacterium* and an increase in potentially negative *Desulfovibrio* and *Clostridia* (Morton et al. 2023; Zuffa et al. 2023). These microbial imbalances are linked to disruptions in the microbiota-gut-brain axis, with potential impacts on neuroinflammation and neurotransmitter synthesis. One study found that the *Clostridium perfringens* β2 toxin gene (*cpb2*), which encodes for the clostridial toxin b, is more expressed in *Clostridium perfringens* from ASD children, possibly explaining some gastrointestinal symptoms and their downstream effects on behaviour via vagus nerve signalling (Gora et al. 2018). Interestingly, *Clostridium perfringens* has also been reported increased in children born by C-section (Mueller et al. 2021a). Although interventions targeting the microbe, such as probiotics or dietary changes, have shown promise, more rigorous clinical trials are necessary (Cull et al. 2022; Wang et al. 2021).

Preclinical studies have provided further insight into the microbiota-gut-brain axis in ASD. For example, Shank-3 knockout mice, which exhibit autistic-like behaviours, also show microbiota alterations and respond differently to probiotic treatment based on sex (Osman et al. 2023; Rein et al. 2024). The BTBR mouse model, another tool for studying ASD, demonstrates altered gastrointestinal physiology and microbiota composition (O'Connor et al. 2021; Sen et al. 2022; van de Wouw et al. 2021). Environmental factors, such as *in utero* exposure to Poly I or valproic acid, have been shown to induce autistic-like behaviours and microbiota changes in animal models (Adiguzel et al. 2022; Avolio et al. 2022; Prince et al. 2024).

Probiotic interventions in preclinical models have shown some success. For instance, *L. reuteri* has been found to improve social behaviour and increase oxytocin production in male Shank-3 knockout mice, as well as significant improvements in social functioning in children with ASD (Mazzone et al. 2024; Tabouy et al. 2018). Similarly, *B. fragilis* has been shown to alleviate certain ASD-related behaviours in a maternal immune activation model (Buzun et al. 2024; Hsiao et al. 2013).

6.4.2 Microbial Influences in ADHD: Diet, Probiotics, and Beyond

ADHD is a neurodevelopmental condition marked by symptoms of inattention, hyperactivity, and impulsivity, either individually or in combination. Although research on the microbiome's role in ADHD is still in its infancy, the diet has long been recognised as influential in managing ADHD symptoms (American Psychiatric Association 2013; Bostock et al. 2017; Hontelez et al. 2021). A Western-style diet, characterised by high consumption of ultra-processed foods and sugars, has been suggested as a potential risk factor for ADHD, though the evidence remains debated (Howard et al. 2011). Similarly, elimination diets, especially those excluding artificial food colouring, have shown positive effects on symptom management in some studies, but findings are not universally agreed upon and require further investigation (Gkougka et al. 2022).

As diet profoundly influences gut microbiota, this suggests that the microbiome could play a role in ADHD symptomatology (Wang et al. 2020). A study involving a small clinical cohort of adults and adolescents with ADHD found an increased presence of the genus *Bifidobacterium* when compared to healthy controls, which might affect dopamine precursors in the gut (Partty et al. 2015). Furthermore, a perinatal probiotic intervention has been shown to reduce the risk of an ADHD diagnosis in later childhood, indicating the potential for microbiota-based treatments for ADHD (Allahyari et al. 2024). A few-foods diet can substantially reduce ADHD symptoms in children, although the mechanisms behind this effect are not yet well understood (Hontelez et al. 2021). Looking at the effects of supplementing a probiotic in children with ADHD for 8 weeks, those receiving probiotics showed significant reductions in total ADHD symptom scores and anxiety levels, suggesting that probiotics may offer beneficial effects on certain psychological symptoms in children with ADHD (Ahmadvand et al. 2021).

6.4.3 Depression and Microbiota: An Axis of Mood Regulation

MDD is marked by a persistently low mood, diminished interest or pleasure in activities, recurrent thoughts of death, and a range of cognitive and physical symptoms that can significantly impact the quality of life. MDD affects at least 4.4% of the global population and is also characterised by increased pro-inflammatory cytokines and HPA axis hyperactivation, along with a specific clinical diagnosis defined by established criteria in the DSM-5 (American Psychiatric Association 2013; Marx et al. 2023). Effective treatments for depression can reduce these inflammatory responses and HPA axis activity (Gheorghe et al. 2022). The microbiota-gut-brain axis has been implicated in MDD, with preclinical studies showing that certain probiotic and prebiotic interventions can reduce depressive-like behaviours and inflammation (Audet 2021; Bastiaanssen et al. 2020; Horne and Foster 2018).

Human studies have identified differences in the faecal microbiota of MDD patients compared to healthy controls, though findings remain inconsistent (Green et al. 2023; Jiang et al. 2015). Enema-delivered FMT was found to be feasible, acceptable, well-tolerated, and safe in a cohort of participants with MDD, showing high satisfaction and potential improvements in gastrointestinal symptoms and quality of life (Green et al. 2023). Some studies have reported reduced *Bifidobacterium* and *Lactobacillus*, while others have noted increased bacterial diversity or different bacterial populations (Brushett et al. 2023). One notable study involving over 1000 individuals found that depression was associated with a lower abundance of *Faecalibacterium* and *Coprococcus*, bacteria involved in producing the SCFA butyrate, which is linked to a higher quality of life (Nikolova et al. 2021; Val-Laillet et al. 2018; Valles-Colomer et al. 2019). Probiotic interventions have shown promise in improving mood and depressive symptoms in some human studies. For example, *Lacticaseibacillus casei* has improved mood in elderly individuals, while a combination of probiotics has shown benefits in MDD patients (Akkasheh et al. 2016). However, not all studies have found significant effects, highlighting the need for more research.

6.4.4 Microbiota, Depression, and Anxiety: Pathways to Calm

Anxiety often co-occurs with depression, and much of the evidence for the microbiota-gut-brain axis in anxiety comes from preclinical studies (Gyles et al. 2024). Germ-free mice, for example, show reduced anxiety-like behaviour and altered HPA axis responses (Lynch et al. 2024). One study highlighted a connection between gut microbiota composition and personality traits, including anxiety-related characteristics, where individuals with higher neuroticism, often linked to anxiety, exhibited reduced gut microbiota diversity and lower levels of commensal genera like *Faecalibacterium* and *Coprococcus* (Salia et al. 2025). Another study examining the role of gut microbiota in anxiety-like behaviours found that alterations in gut microbial homeostasis induced by chronic stress correlated with elevated anxiety levels in animal models (Chen et al. 2024). Such findings emphasise the importance of the microbiota-gut-brain axis, suggesting that microbial interventions may have therapeutic potential for anxiety disorders.

Interventional probiotic treatments have yielded mixed results in both animal models and human studies (Berding and Cryan 2022; Cannavale et al. 2023; Lonstein et al. 2024; Mullish et al. 2024; Mysonhimer et al. 2024). One study that performed maternal administration of *Lactocaseibacillus rhamnosus HN001* from early pregnancy through postpartum found that the probiotic reduced postpartum anxiety-related behaviours in rodents (Lonstein et al. 2024). The effects were associated with changes in prefrontal cortex dopamine and serotonin turnover and the prevention of postpartum reductions in gut microbiota diversity, highlighting the potential of probiotics in mitigating peripartum mood disorders through the microbiota-gut-brain axis. Further, prenatal intake of a multispecies probiotic formulation significantly influenced offspring neurodevelopment, with enduring sex-dependent effects; male offspring exhibited reduced anxiety-like behaviours, enhanced gut microbiota diversity, and an upregulation of anti-inflammatory and tight junction-related genes (Siegler Lathrop et al. 2024). Further, some probiotic strains have been shown to reduce anxiety in healthy individuals and those with chronic fatigue syndrome, such as *Lactobacillus plantarum* DR7, which has been identified as a potential psychobiotic capable of alleviating anxiety and stress-related symptoms in controlled clinical trials (Chong et al. 2019; Rudzki et al. 2019; Song et al. 2022; Wang et al. 2021). However, more translational research is needed to confirm these findings.

Dietary interventions for the treatment of depression and anxiety are also under investigation. A recent systematic review and meta-analysis of five randomised controlled trials explored the impact of a Mediterranean diet on alleviating depressive symptoms in adults with depression (Bizzozero-Peroni et al. 2024). Their findings concluded that adherence to the diet significantly reduced depressive symptoms compared to control conditions, although the overall risk of bias was moderate to high, and the certainty of the evidence was low, thus indicating that more rigorous, large-scale, and long-term trials are necessary to confirm the diets potential benefits for depression. One fascinating study investigated the connection between gut microbiota and cognitive function and depression in older adults with cognitive impairment (Kolobaric et al. 2024). Their work found that greater microbial diversity was associated with lower cognitive performance and higher depression levels, while certain bacterial abundances and metabolic activities were linked to future cognitive decline and increased depressive symptoms. They suggest that the gut microbiome could serve as a non-invasive biomarker and prognostic tool for cognitive and psychiatric health.

6.4.5 Post-traumatic Stress Disorder (PTSD) and Microbiota: Stress Recovery Through Gut-Brain Modulation

PTSD is a severe anxiety disorder resulting from exposure to traumatic events that has been gaining attention as a potential target for microbiome-based strategies (American Psychiatric Association 2013; Tanelian 2022 #7712). Although direct investigations into the microbiota's role in clinical PTSD are limited, preclinical and clinical evidence suggests that modulating the microbiota could influence stress responses

(Khan et al. 2025). By comparing the bacterial composition of trauma survivors with probable PTSD to trauma-exposed controls without PTSD, one study found no significant differences in overall microbial community structure, but the probable PTSD group exhibited notable changes in specific bacterial taxa (Geier et al. 2024). *Prevotella* and *Ruminococcaceae* negatively correlated with negative mood and positively associated with hyperarousal, suggesting that the microbiome may mediate the diverse psychological outcomes following trauma and could inform personalised psychobiotic interventions based on symptom profiles (Geier et al. 2024). Further preclinical research utilising a mouse model of PTSD-like symptoms induced by a single prolonged stress found that affected animals exhibited increased anxiety-like and depressive-like behaviours, a disruption in gut microbiota composition, and altered intestinal health (Khan et al. 2025). A probiotic treatment improved commensal microbiota and reduced anxiety and depression, suggesting a potential therapeutic approach for PTSD symptoms linked to gut health. Further research into the human gut microbiome profile under chronic stress conditions is necessary to understand its potential impact on PTSD.

6.4.6 Gut-Brain Connections in Bipolar Disorder: A New Frontier

Bipolar disorder, another serious neuropsychiatric condition, has also been linked to microbiota alterations (American Psychiatric Association 2013; McGuinness et al. 2024). Patients with bipolar mania are more likely to have been recently treated with antibiotics, which can disrupt gut microbial composition and potentially influence immune function and neuroinflammation (Yolken et al. 2016). Faecal microbiome analysis has shown decreased *Firmicutes*, particularly *Faecalibacterium*, a butyrate-producing genus, in bipolar patients (Nikolova et al. 2021). A compromised gut barrier, often referred to as "leaky gut," may allow microbial metabolites and inflammatory molecules, such as LPS, to enter the bloodstream, triggering systemic inflammation and potentially exacerbating mood instability via neuroimmune pathways.

A recent systematic review revealed that higher consumption of ultra-processed food is strongly associated with a range of adverse health outcomes, including bipolar disorder, although the quality of evidence varied from very low to moderate (Lane et al. 2024). This dietary pattern may contribute to chronic low-grade inflammation, oxidative stress, and metabolic disturbances, which are all increasingly recognised as key factors in mood disorders. Interestingly, a probiotic trial suggested that probiotic therapy could reduce rehospitalisation rates for patients recently discharged after mania (Dickerson et al. 2018). Such findings highlight the need for further research into microbiota-targeted interventions as an adjunctive strategy for bipolar disorder management, integrating dietary and probiotic therapies with conventional psychiatric treatments.

6.4.7 Microbial Profiles in Schizophrenia: Emerging Mechanisms

Schizophrenia is a severe psychiatric disorder marked by psychosis, cognitive dysfunction, and social withdrawal (American Psychiatric Association 2013; Tian et al. 2023). There is evidence suggesting a role for the gastrointestinal tract and immune response in its pathogenesis (Tian et al. 2023; Vasileva et al. 2024). Research into the gut microbiota of individuals with psychosis, including schizophrenia, reveals significant differences in microbiota composition compared to controls, with potential functional consequences (Murray et al. 2023). Although α-diversity did not consistently differ, β-diversity showed a marked contrast, indicating distinctive microbial profiles in those with psychosis (Nguyen et al. 2019). Notably, genera like *Bifidobacterium*, *Lactobacillus*, *Megaspheara*, and *Veillonella* are more prevalent in psychosis groups and are associated with metabolic pathways that influence inflammation and neurotransmitter function (Nguyen et al. 2021). This finding is surprising, as *Bifidobacterium* and *Lactobacillus* are typically associated with beneficial effects on gut and mental health, including anti-inflammatory properties and support for neurotransmitter function, making their increased prevalence in psychosis groups unexpected in terms of microbiota composition. These microbial patterns are also linked to behavioural and cognitive outcomes in both human and animal models, underscoring the microbiome's potential role in the pathophysiology of psychosis and schizophrenia (Basafa-Roodi et al. 2024; Misiak et al. 2024; Vasileva et al. 2024; Wei et al. 2024). Future research focusing on the functional impact of these microbial changes may offer new avenues for microbiome-targeted treatments for psychiatric conditions.

Further studies have found significant differences in the oropharyngeal microbiota of individuals with schizophrenia compared to healthy controls, including reduced microbial diversity and altered neurochemistry in germ-free mice receiving FMT from schizophrenic patients (Castro-Nallar et al. 2015; Wei et al. 2024). Although probiotic interventions have not yet demonstrated significant improvements in schizophrenia symptoms, some studies have reported alleviation of gastrointestinal issues (Lee et al. 2023). Further research is needed to explore the microbiota's role in schizophrenia and potential therapeutic interventions.

6.4.8 Gut Disturbance in Anorexia Nervosa: Targets for Recovery

Anorexia nervosa is one of the psychiatric illnesses with the highest mortality rates and has a profound impact on quality of life. It is characterised by a distorted body image leading to severe food intake restriction, significant weight loss, hyperactivity, and hypothermia (American Psychiatric Association 2013). The exact aetiology of anorexia nervosa remains unclear, and there are no effective medications currently available. Recent evidence, however, suggests a potential involvement of gut microbiota, identifying it as a possible target for intervention (Andreani et al. 2024; Fan et al. 2023). One longitudinal analysis involving adolescent inpatients diagnosed with anorexia nervosa revealed that microbiota composition changed significantly during acute starvation and weight recovery, suggesting that specific microbial profiles at admission may predict treatment outcomes and potential hospital readmission (Andreani et al. 2024). Faecal samples collected before and after clinical treatment from patients with eating disorders and non-eating disorder controls were characterised, revealing that while weight improvements were observed, the gut microbiota in anorexia nervosa patients post-treatment remained significantly different from non-eating disorder controls (Fouladi et al. 2022). Notably, pre-treatment anorexia nervosa microbiota signatures showed increased fermentation pathways and reduced carbohydrate degradation capabilities compared to controls, underscoring the intricate microbial communities in AN patients that warrant further investigation.

6.4.9 Addiction and the Microbiota: New Avenues for Intervention

Addiction is classified as a chronic relapsing disorder characterised by compulsive engagement in harmful behaviours despite adverse consequences (American Psychiatric Association 2013). Studies have mainly focused on the impact of substance abuse on the gut microbiome, particularly alcohol, which has been shown to alter gut microbiota, increase intestinal permeability, and promote hepatic inflammation (Garcia-Cabrerizo et al. 2021). Probiotics like *L. rhamnosus* GG have been found to mitigate some of these effects (Tette et al. 2022). Other substances, including tobacco, opioids, and methamphetamine, also alter gut microbiota, affecting intestinal permeability, inflammation, and metabolic processes (Garcia-Cabrerizo and Cryan 2024). Recent preclinical work has shown that the depletion of gut microbiota can significantly alter reward responses, diminishing the effects of cocaine while enhancing the reward derived from social interactions (Garcia-Cabrerizo et al. 2023). These findings suggest that targeting the gut microbiota in conjunction with promoting social engagement may offer an effective strategy for reducing the negative impacts of drug use, offering a potential therapeutic avenue, but requires further exploration.

6.4.10 Obsessive-Compulsive Disorders (OCD) and the Gut Microbiota: Compulsions in Context

Speculation about the involvement of the gut microbiota in OCD is growing, although current research is limited. Stress and antibiotic exposure, both of which affect microbiota composition, are proposed mechanisms potentially coinciding with the onset of OCD symptoms (American Psychiatric Association 2013). Although the literature on this connection remains limited, emerging evidence suggests that changes in the gut microbiota and high rates of bowel syndromes may be prevalent among OCD patients, particularly featuring an increase in pro-inflammatory bacterial species and a decrease in butyrate producers (Domènech et al. 2022; Turna et al. 2020). One study revealed significant reductions in α-diversity in OCD alongside notable differences in β-diversity across various taxonomic levels, with specific genera such as *Pseudomonas* and *Streptococcus* being less prevalent in OCD patients (Kang et al. 2023). Additionally, the microbial composition differed between early-onset and late-onset OCD, particularly regarding *Corynebacterium* and *Pelomonas*, highlighting the potential role of the microbiota-gut-brain interaction in the disorder's pathophysiology and underscoring the need for further longitudinal studies.

Transplanting faecal microbiota from first-episode, drug-naive OCD patients and healthy controls into antibiotic-treated specific pathogen-free mice revealed that colonisation with OCD-associated microbiota induced significant behavioural deficits, including anxiety-like and compulsive-like behaviours, along with altered neuronal morphology and function in the medial prefrontal cortex (mPFC) and inflammation in both the mPFC and colon (Zhang et al. 2024b). Metabolomic analyses identified elevated levels of succinic acid in OCD-colonised mice, which negatively impacted neuronal activity and contributed to inflammation, suggesting that gut microbiota may influence brain function through succinic acid signalling. Additionally, manipulation of dimethyl malonate showed neuroprotective effects, indicating potential therapeutic avenues for addressing OCD through gut-brain interactions.

Pre- and probiotics have been shown to influence compulsive behaviours, suggesting potential therapeutic benefits in future OCD treatments. One study investigated the effects of a multi-strain probiotic formulation on OCD-like behaviours in rats induced by quinpirole, revealing that the probiotic treatment effectively reduced compulsive behaviours and anxiety-like phenotypes (Ghuge et al. 2023). Probiotic administration also normalised inflammatory markers and neurotransmitter levels in the frontal cortex while promoting commensal gut microbiota changes, including an increased abundance of *Allobaculum* and specific *Bifidobacterium* species.

Longitudinal human studies are needed to clarify the gut microbiota's role in OCD symptomatology. Future evaluations of OCD should include comprehensive gastrointestinal assessments, addressing factors like physical comorbidities, nutritional habits, bowel movements, fluid intake, and exercise patterns, all of which influence gut health. Treatment implications indicate that understanding gastrointestinal issues can guide medication choices, such as using SSRIs or tricyclic antidepressants based on the presence of IBS symptoms.

6.4.11 Microbial Drivers of Obesity: Targets for Intervention

Obesity is a major public health issue that is often fuelled by diets high in fat and sugar that disrupt the balance between energy intake and expenditure and is associated with a range of serious health problems and centrally regulated disorders like depression and anxiety (González-Muniesa et al. 2017; Romani-Perez et al. 2024a). This dietary pattern shapes the gut microbiota, affecting neuroendocrine and immune functions within the intestine, which in turn influences gut-brain pathways that regulate appetite and metabolism (Boscaini et al. 2022). These interactions can elevate inflammation and disrupt metabolic processes, further promoting disease. Obesity has been associated with an altered gut microbiota composition, often characterised by an increased ratio of *Firmicutes* to *Bacteroidetes* (Ley et al. 2005). At a more specific taxonomic level, studies have reported elevated abundances of genera such as *Lactobacillus* and *Clostridium* (within Firmicutes) and reduced levels of *Bacteroides* and *Prevotella* (within Bacteroidetes) (Clarke et al. 2012c). These imbalances have been shown to impact metabolic pathways, such as those involved in energy harvest and short-chain fatty acid production, and can be modulated through dietary interventions (Tremmel et al. 2017; Turnbaugh et al. 2008, 2009). The gut microbiome plays a crucial role in the central regulation of food intake, and its composition is notably influenced by diet (Marungruang et al. 2020; Medawar et al. 2024). Moreover, the microbiota-gut-brain axis is emerging as an essential factor in obesity, as changes in gut microbiota can affect metabolic functions and energy homeostasis signalling through vagal afferent neurons (Daly et al. 2011).

Bariatric surgery is a notable intervention for morbid obesity that significantly alters the composition (diversity and abundance) of host gut microbiota, enhancing hormonal and inflammatory profiles, reducing adiposity, and improving insulin sensitivity (Aron-Wisnewsky and Clement 2014; Ribeiro et al. 2021). Research indicates that this surgical approach can lead to increased microbial diversity and altered composition, particularly with higher levels of *Gammaproteobacteria* and *Verrucomicrobia*, alongside a reduction in *Firmicutes* (Ulker and Yildiran 2019). Further findings have suggested that an FMT from post-bariatric surgery mice to germ-free animals can result in decreased fat mass and body weight (Liou et al. 2013). Additionally, gut-derived peptides such as ghrelin and oxyntomodulin, which are influenced by the gut microbiome, show potential as therapeutic targets for obesity treatment (Tschop et al. 2001; Wallace Fitzsimons et al. 2019). Another study in humans looked at the impact of bariatric surgery on cortical morphometry in obese individuals, examining changes in brain grey and white matter associated with food intake and cognitive-emotional regulation (Liu et al. 2019). Results indicated significant reductions in fasting plasma ghrelin, insulin, and leptin levels post-surgery, along with structural changes such as decreased cortical thickness in the praecuneus and increased thickness in various frontal and temporal regions, which were linked to weight loss. These findings suggest that bariatric surgery may promote structural recovery in brain areas involved in executive control and self-referential processing, supporting the idea that weight loss can lead to beneficial changes in brain morphology.

Probiotics and prebiotics are also being explored as potential interventions to restore normal microbiome-mediated gut-brain signalling and treat obesity (Schellekens et al. 2021; Torres-Fuentes et al. 2017). Indeed, one study investigated the anti-obesity potential of *Bifidobacterium longum* APC1472, demonstrating that supplementation led to decreased body weight and fat accumulation in high-fat diet-induced obese mice (Schellekens et al. 2021). Although the human trial did not show significant changes in body mass index or waist-to-hip ratio among healthy overweight/obese participants, it revealed a beneficial effect on fasting blood glucose levels, suggesting the potential of *B. longum* APC1472 as a valuable supplement for addressing specific obesity-related markers. Advances in understanding these diet-microbe-host interactions highlight potential microbiome-based interventions, like FMT, probiotics, and other biotherapeutics, which may offer promising avenues to address obesity, although significant knowledge gaps remain.

6.4.12 Irritable Bowel Syndrome and the Axis: Decoding Gut-Brain Dysfunction

IBS is a prevalent disorder of gut brain interaction, previously regarded as a functional gastrointestinal disorder, significantly impacting patients' quality of life and often presents with comorbid behavioural and psychological factors, including altered bowel habits, abdominal pain, bloating, and excessive flatulence (Enck et al. 2016). It often coexists with various psychosocial and environmental factors, making microbiota-gut-brain interactions a focal point in IBS research (Rodino-Janeiro et al. 2018). IBS is officially classified as a disorder of the microbiota-gut-brain axis, suggesting that diet-microbiota interactions may play a role in its core symptoms (Collins 2014; Quigley 2018; Wilmes et al. 2021). As a result, the microbiota-gut-brain axis is a prime target in IBS (Clarke et al. 2012a). A crucial mechanism in IBS is the serotonin signalling pathway, which regulates gastrointestinal functions, including secretion and motility; alterations in serotonergic function and tryptophan metabolism have been noted in IBS patients (Clarke et al. 2012b, 2009). Mono-colonisation of germ-free mice with *Ruminococcus gnavus* resulted in enhanced gastrointestinal transit and colonic secretion by stimulating serotonin production through the catabolism of dietary phenylalanine and tryptophan into metabolites like phenethylamine and tryptamine (Zhai et al. 2023). Further, altered levels of microbial metabolites like SCFAs have been reported in patients with diarrhoea-predominant IBS (Tian et al. 2020). Significantly higher levels of propionate and butyrate were found in the serum of these IBS patients, suggesting that elevated serum levels of propionic and butyric acids may be linked to the onset and progression of diarrhoea-predominant IBS. Indeed, the partial mechanistic function of the FMT may significantly increase the levels of butyrate, improving IBS symptoms (El-Salhy et al. 2021).

Emerging microbial and diet-based treatments for IBS show promise, with some studies highlighting the efficacy of probiotics like *L. acidophilus* and *B. longum* in alleviating stress-related gastrointestinal symptoms (Ait-Belgnaoui et al. 2018; Diop et al. 2008; Simren et al. 2010). Other research indicates that *B. longum* may improve mood in IBS patients, although inconsistent results have been reported with different probiotic strains (Crouzet et al. 2013; Dapoigny et al. 2012; De Palma et al. 2017; Pinto-Sanchez et al. 2017). The proposed mechanism for these benefits includes anti-inflammatory effects, which may also explain the positive outcomes associated with dietary interventions such as the low FODMAP (fermentable oligosaccharides, disaccharides, monosaccharides, and polyols) diet, known to reduce gastrointestinal distress by limiting certain fermentable carbohydrates (Whelan et al. 2018). Recent work investigating the efficacy of a multi-strain probiotic on symptom severity in female patients with IBS showed significant reductions in symptom severity, anxiety, and depression, along with improvements in bowel habits (Mullish et al. 2024). Although the overall diversity of the faecal microbiota remained largely unchanged, specific microbial differences were noted, suggesting that daily supplementation with this multi-strain probiotic could be a beneficial option for managing IBS symptoms in women. Intriguingly, a recent GWAS study identified six genetic loci associated with IBS that are linked to mood and anxiety disorders and are expressed in the nervous system (Eijsbouts et al. 2021). A strong genome-wide correlation was observed between IBS risk and traits such as anxiety, neuroticism, and depression, suggesting that shared pathogenic pathways underlie these conditions, rather than anxiety simply causing abdominal symptoms, highlighting the need for further investigation into the mechanisms of altered brain-gut interactions in IBS. Additionally, FMT from non-IBS individuals to those with IBS has gained popularity as a treatment option, although the overall efficacy remains a subject of ongoing investigation (El-Salhy et al. 2021; Johnsen et al. 2018; Verstockt et al. 2018).

6.4.13 Pain Disorders and the Microbiota: A Gut-Brain Perspective

Pain can be categorised as nociceptive, inflammatory, or pathological, with mechanisms and treatment targets varying across types (Colloca et al. 2017; Millan 2002). Nociceptive pain typically arises from tissue damage and is well understood, while the mechanisms behind visceral pain, neuropathic pain, and chronic pain remain complex and challenging to address therapeutically (Felice et al. 2015; Lagomarsino et al. 2021). Conditions like fibromyalgia, migraines, and gastrointestinal pain often involve intricate interactions between the nervous system and underlying pathophysiological changes, complicating the development of effective treatments (Drewes et al. 2020; Woolf et al. 1998). Gastrointestinal pain is prevalent in conditions like IBS, Crohn's disease, and pancreatitis; it presents a diagnostic challenge due to overlapping autonomic and somatic symptoms. Complex mechanisms, including peripheral and central sensitisation and autonomic nervous system involvement, contribute to gastrointestinal pain. Treatment varies based on pain type and underlying disorder, potentially involving pharmacological, behavioural, or surgical approaches.

Research into the role of gastrointestinal microbiota in pain response has predominantly focused on nociceptive and inflammatory disorders, particularly those affecting the gut, such as inflammatory bowel disease (IBD) and FGIDs (Holzer et al. 2017; Rea et al. 2017). Chronic visceral pain is a common symptom in these conditions, indicating a shared pathophysiological pathway regulated by both central and peripheral mechanisms, often referred to as disorders of the microbiota-gut-brain axis (Mayer et al. 2023; Pusceddu and Gareau 2018). Studies have identified altered gut microbiota profiles in patients with chronic visceral pain, revealing decreased levels of commensal genera like *Bifidobacterium* and *Lactobacillus* and shifts in the Firmicutes-to-Bacteroidetes ratio (often associated with metabolic health) although the causative relationships between microbiota changes and pain experiences remain unclear (Clarke et al. 2012b; Manor et al. 2020).

Research has indicated that modifying gut bacteria through diet, probiotics, or antibiotics could reduce symptoms in some chronic pain patients, although much remains to be clarified about the microbiome's impact on pain mechanisms (Baldi et al. 2022; Simren and Tack 2018; Tuck et al. 2022). Prebiotic, probiotic and FMT therapies have shown promise in treating gastrointestinal pain and associated symptoms, although outcomes can vary significantly among individuals (Basso et al. 2018). In fact, the prebiotic dietary inulin fibre was shown to influence the composition of murine gut microbiota, altering levels of microbiota-derived metabolites such as bile acids, leading to type 2 inflammation in the intestines and lungs (Arifuzzaman et al. 2022). This work highlights the role Inulin may play in promoting allergic inflammation and signalling host defence mechanisms.

Although FMT has been effective in treating some patients with *C. difficile* infections, the results for IBD treatment have been mixed, with reports of clinical remission rates varying widely (Green et al. 2020; Wei et al. 2022). Similarly, the efficacy of probiotics in managing pain related to arthritis and other inflammatory conditions is still being investigated, with current evidence inconclusive (Kragsnaes et al. 2018; Lin et al. 2023; Lowe et al. 2018). As the field progresses, further clinical trials are essential to clarify the effectiveness of microbiota-based interventions for visceral pain and to explore the mechanisms through which the gut microbiota may influence pain pathways, potentially offering new therapeutic avenues for managing chronic pain syndromes.

6.4.14 Parkinson's Disease (PD) and Microbial Contributions to Neurodegeneration

PD, a progressive neurodegenerative disorder, is characterised by cognitive and motor impairments and gut discomfort, which often precedes motor symptoms (Poewe et al. 2017). Disturbances in the gut microbiome are believed to play a role in neurodegeneration, particularly since constipation—a common non-motor symptom—can precede a PD diagnosis for many years (Abbott et al. 2001). Key features of PD include not only dopaminergic neuron loss and α-synuclein accumulation but also neuroinflammation (Dzamko 2023). Much research has demonstrated that α-synuclein, associated with PD pathology in the brain, is also found in the gut's nerve fibres and ganglia (Hilton et al. 2014). Preclinical models have indicated that gut microbiota may facilitate α-synuclein aggregation and impede the clearance of protein aggregates (Sampson et al. 2016), while evidence suggests that α-synuclein might travel from the gut to the brain via the vagus nerve, a crucial pathway for gut-brain communication (Holmqvist et al. 2014). Studies have shown that truncal vagotomy could lower PD risk and prevent the gut-to-brain spread of α-synucleinopathy in animal models (Kim et al. 2019; Liu et al. 2017a). Human studies have identified distinct microbial profiles in PD patients, indicating potential biomarkers for early diagnosis and new therapeutic targets (Stagaman et al. 2024). However, there remains no consensus on a specific microbial pattern linked to PD, highlighting the need for caution due to small study sizes and a lack of longitudinal data.

A Taiwanese study identified significant differences in microbiota composition between PD and non-PD patients, particularly a reduction in the commensal genus *Prevotella* across diverse populations (Lin et al. 2019). Increased levels of *Akkermansia muciniphila* have also been observed in PD patients, though its role remains ambiguous and may be associated with heightened constipation (Romano et al. 2021). Collectively, reduced SCFA-producing bacteria and elevated *A. muciniphila* may contribute to PD pathology, emphasising the microbiome's involvement in disease progression. FMT from PD patients to germ-free mice has induced PD-like symptoms, and antibiotic treatment has shown symptom improvement in preclinical models (Koutzoumis et al. 2020; Sampson et al. 2016). Future research should focus on microbiota composition across various stages of PD, including the prodromal phase, to better understand how specific bacterial changes can alter SCFA profiles and influence neuroinflammation in the brain.

6.4.15 Alzheimer's Disease (AD) and the Microbiota: Insights into Cognitive Decline

AD, the leading cause of dementia, is linked to β-amyloid plaques and tau protein abnormalities in the brain (Knopman et al. 2021). There is growing evidence for a microbial origin of AD, with bacteria and viruses potentially playing roles in its pathogenesis. Alterations in gut microbiota have been observed in AD patients, and preclinical studies have shown that probiotics can mitigate cognitive deficits and inflammation. Emerging evidence indicates a potential microbial origin for AD, suggesting that negative microbes may play a role in its development and progression (Grabrucker et al. 2023; Itzhaki et al. 2016; Lathe et al. 2023). Studies have found a correlation between increased *Bacteroides* and higher levels of β-amyloid and tau in AD mouse models, with gut alterations characterised by reduced microbial diversityin AD patients (Cattaneo et al. 2017; Jia et al. 2024; Vogt et al. 2017; Zhuang et al. 2018). Further, alterations in gut microbiota following antibiotic use can modulate neuroinflammation and influence Aβ deposition in mouse models of AD (Minter et al. 2017). One recent study identified a gut microbiome-brain axis in AD, showing that an imbalance of *Clostridium* and *Bacteroides* correlated with β-amyloid burden (Zha et al. 2025). Supplementation with *Bacteroides ovatus* or its metabolite lysophosphatidylcholine reduced Aβ pathology, improved cognitive function, and inhibited ferroptosis via activation of the orphan receptor GPR119 in AD mouse models. These findings suggest that gut microbiota-derived metabolites (e.g. lysophosphatidylcholine) could offer novel therapeutic avenues for AD by modulating neuroinflammation, lipid peroxidation, and neurodegeneration.

FMT from healthy individuals has shown promise in modulating the gut microbiome and reducing Aβ aggregates and tau pathology in AD mouse models (Kim et al. 2020). Although some clinical trials report cognitive improvements with probiotic treatment, results have been inconsistent, particularly in severe cases of AD (Agahi et al. 2018; Akbari et al. 2016; Tamtaji et al. 2017; Zha et al. 2025). Nonetheless, psychobiotic therapy presents a potential adjunctive treatment for AD, and ongoing research aims to elucidate the underlying mechanisms and identify biomarkers for early detection. The exploration of microbiota-based interventions continues to be a compelling area of study with the potential to transform the understanding and management of AD.

6.4.16 Stroke, Brain Injury, and the Gut: The Role of Microbiota in Recovery

Inadequate blood flow to the brain, leading to cell death, is a defining characteristic of stroke and a common consequence of brain injury (Campbell et al. 2019). Research has indicated that peripheral and systemic factors, particularly an enhanced inflammatory response, can exacerbate outcomes following a stroke (Brea et al. 2021). A recent study involving Chinese patients revealed a significant depletion of commensal gut microbes, such as *Bacteroides*, *Prevotella*, and *Faecalibacterium*, alongside an increase in opportunistic pathogens like *Enterobacter* and *Megasphaera* in individuals with stroke or transient ischemic attacks (Yin et al. 2015). Notably, patients with severe strokes exhibited a higher abundance of Proteobacteria compared to those with milder strokes. Further, levels of trimethylamine-N-oxide (TMAO), a low molecular weight microbial metabolite primarily generated from the metabolism of dietary substrates, strongly correlated with the development of numerous diseases, including stroke, other cardiovascular conditions, diabetes, certain cancers, and neurological disorders (Gatarek and Kaluzna-Czaplinska 2021). Using FMT from low- and high-TMAO-producing human subjects into germ-free mice, research demonstrated that both TMAO generation and stroke severity were transmissible traits (Zhu et al. 2021). This FMT conveyed functional deficits in pro-inflammatory T-cell trafficking and increased ischemic lesion volume.

Comparing germ-free mice to recolonised and conventional specific pathogen-free mice, another study showed that bacterial colonisation significantly reduced stroke volumes and enhanced cerebral expression of cytokines along with increased microglia and macrophage cell counts, contributing to improved stroke outcomes (Singh et al. 2018). Notably, this microbiome-mediated neuroprotection is lost in lymphocyte-deficient mice, indicating that lymphocyte-driven protective neuroinflammation, regulated by the microbiome, plays a crucial role in the brain's response to stroke. The use of broad-spectrum antibiotics before ischemic injury resulted in poorer outcomes, correlating with decreased trafficking of IL-17+ γδ T cells and reduced neuroinflammation at the ischemic site (Benakis et al. 2016).

Conversely, probiotic supplementation, such as with *C. butyricum*, has shown potential neuroprotective effects in traumatic brain injury models by enhancing intestinal barrier integrity and promoting the secretion of intestinal GLP-1, suggesting that pre- and probiotic treatments could mitigate some of the complications associated with stroke and cerebral ischemia (Li et al. 2018; Sun et al. 2016). Although some findings indicate promise for the use of probiotics and prebiotics, the current research is limited, and more work is needed.

6.4.17 Microbial Modulation in Multiple Sclerosis: Hope for Therapies

MS is an autoimmune disorder of the central nervous system characterised by motor deficits, blurred vision, and sensory changes that occur suddenly, often with few prodromal signals (Dendrou et al. 2015; Filippi et al. 2018). Given the immune basis of MS, there is growing interest in therapeutic interventions targeting the gut microbiota (Cantoni et al. 2022; Schumacher et al. 2024). Notably, FMT from MS patients into mice resulted in the development of autoimmune encephalomyelitis, a hallmark symptom of MS (Berer et al. 2017; Cekanaviciute et al. 2017), while germ-free mice showed resistance to this condition, indicating the influence of the gut microbiome on disease expression. In particular, the presence of gram-positive segmented filamentous bacteria and an increase in *Akkermansia* correlated with symptom development, while reduced levels of the anti-inflammatory species *Parabacteroides* distasonis were observed in MS patients (Cekanaviciute et al. 2017)

Looking at the complex interplay between the gut microbiome, immune system, metabolism, and diet in MS, a six-month longitudinal multi-omics analysis involving 49 participants found that MS patients exhibited significant alterations in gut microbiota composition (Cantoni et al. 2022). Specific microbes (including *B. thetaiotaomicron*) correlated with disease severity, alongside perturbations in blood metabolome pathways related to fatty acid metabolism and inflammation. Such findings suggest that the gut microbiota significantly contributes to MS pathophysiology and may be responsive to dietary interventions, warranting further research into targeted microbiota therapies

6.4.18 Huntington's Disease: Gut Microbiota as a Therapeutic Target

Huntington's disease, caused by a mutation in the Huntingtin gene, leads to progressive motor, cognitive, and psychiatric decline (McColgan and Tabrizi 2018). Altered gut microbiota-derived metabolites have been observed in early Huntington's disease, indicating potential microbiota-based interventions (Kong et al. 2020). Research in preclinical models suggests that microbiota may influence disease progression and symptomatology. Ultrastructural analysis of myelin plasticity and oligodendroglial dynamics in the BACHD mouse model of Huntington's disease revealed that germ-free BACHD mice exhibited alterations in myelin thickness and reduced levels of myelin-related proteins compared to specific-pathogen-free mice (Radulescu et al. 2019). They also reported a decrease in mature oligodendrocytes in the prefrontal cortex, indicating that the absence of gut microbiota can affect myelin characteristics, which highlights the adaptive properties of myelination in response to gut microbial changes.

In summary, the microbiota-gut-brain axis is a burgeoning field of research with implications for a variety of diseases, as well as psychological and neurological disorders. Although there is substantial evidence from preclinical studies, more rigorous and large-scale clinical trials are needed to establish effective microbiota-targeted therapies for these conditions.

6.5 Towards the Future: Translating Microbiome Science into Therapy

The exploration of the gut microbiota as a key regulator of body-brain interactions has unveiled a transformative understanding of how gut microbes influence both physical and mental health. This chapter has demonstrated that the microbiota-gut-brain axis represents a dynamic, bidirectional communication network that affects a variety of physiological and neurological processes. Through historical perspectives, cutting-edge research, and advanced methodologies, we now appreciate the profound roles that gut microbes play in immune function, metabolism, and brain health.

Evidence from both preclinical and clinical studies underscores the significant impact of gut microbes on conditions ranging from gastrointestinal disorders to neuropsychiatric and neurodegenerative diseases. The identification of microbial pathways and metabolites, such as SCFAs and neurotransmitter-like molecules, provides insight into how the gut microbiome communicates with the brain. These microbial signals traverse multiple pathways—including the autonomic nervous system, immune system, and endocrine signalling—to influence cognition, mood, and behaviour.

As we move forward, it is clear that future research must focus on unravelling the complexities of these communication pathways and the intricate relationships between the host and its microbial symbionts leveraging the range of preclinical model systems (Nagpal and Cryan 2021b). Advances in sequencing technologies and bioinformatics, along with the development of germ-free animal models and FMT, have equipped researchers with powerful tools to explore the causal mechanisms underpinning microbiota-brain interactions. This knowledge lays the groundwork for new therapeutic strategies targeting the microbiome to treat a wide range of disorders.

In conclusion, the microbiota-gut-brain axis offers tremendous potential for understanding the interplay between the brain and body. The ability to modulate this axis through diet, probiotics, prebiotics, and other microbial interventions opens the door to personalised medicine that enhances health across the lifespan. The convergence of microbiome science with neurobiology, immunology, and psychiatry promises to revolutionise how we approach disease prevention and treatment, positioning the microbiome at the forefront of modern medicine's quest to improve brain and body health.

Acknowledgements APC Microbiome Ireland (a research centre), JFC, GC, and KJOR are all funded by Science Foundation Ireland, through the Irish Government's National Development Plan (grant No. SFI/12/RC/2273_P2). JN was supported by a Government of Ireland Irish Research Council Postdoctoral fellowship (GOIPD/2019/714) and is currently funded by Science Foundation Ireland—Irish Research Council (now amalgamated and called Research Ireland) Pathway Programme (22/PATH-S/10876). MRA is supported by ERC_StG_RADIOGUT (project 101040951) and by SFI-IRC Pathway Program SFI21/PATH-S/9424.

Conflicts of Interest JFC has spoken at conferences organized by Mead Johnson, Ordesa, and Yakult and received research funding from Reckitt, Nutricia, Dupont/IFF, and Nestle. GC received honoraria from Janssen, Probi, Apsen and Boehringer Ingelheim; research funding from Pharmavite, Fonterra, Tate and Lyle, Nestle and Reckitt; and is or has been a paid consultant for Yakult, Zentiva and Bayer Healthcare. KJOR *received honoraria from Sanofi Genzyme and Danone.* This support did not influence or constrain the contents of this manuscript. JN and MRA report no conflicts.

References

Abbott RD, Petrovitch H, White LR, Masaki KH, Tanner CM, Curb JD, Grandinetti A, Blanchette PL, Popper JS, Ross GW (2001) Frequency of bowel movements and the future risk of Parkinson's disease. Neurology 57(3):456–462. https://doi.org/10.1212/wnl.57.3.456

Aburto MR, Cryan JF (2024) Gastrointestinal and brain barriers: unlocking gates of communication across the microbiota-gut-brain axis. Nat Rev Gastroenterol Hepatol 21(4):222–247. https://doi.org/10.1038/s41575-023-00890-0

Adiguzel E, Cicek B, Unal G, Aydin MF, Barlak-Keti D (2022) Probiotics and prebiotics alleviate behavioral deficits, inflammatory response, and gut dysbiosis in prenatal VPA-induced rodent model of autism. Physiol Behav 256:113961. https://doi.org/10.1016/j.physbeh.2022.113961

Agahi A, Hamidi GA, Daneshvar R, Hamdieh M, Soheili M, Alinaghipour A, Esmaeili Taba SM, Salami M (2018) Does severity of Alzheimer's disease contribute to its responsiveness to modifying gut microbiota? a double blind clinical trial. Front Neurol 9:662. https://doi.org/10.3389/fneur.2018.00662

Agans R, Rigsbee L, Kenche H, Michail S, Khamis HJ, Paliy O (2011) Distal gut microbiota of adolescent children is different from that of adults. FEMS Microbiol Ecol 77(2):404–412. https://doi.org/10.1111/j.1574-6941.2011.01120.x

de Aguiar Vallim TQ, Tarling EJ, Edwards PA (2013) Pleiotropic roles of bile acids in metabolism. Cell Metab 17(5):657–669. https://doi.org/10.1016/j.cmet.2013.03.013

Ahmadvand A, Sepehrmanesh Z, Shahzeidi A, Mansournia M, Ghaderi A (2021) Clinical and metabolic reaction to probiotic supplement in children suffering attention-deficit hyperactivity disorder: a randomized, double-blind, placebo-controlled experiment. Int Arch Health Sci 8(2):90–96. https://doi.org/10.4103/iahs.iahs_112_20

Ahrens AP, Hyotylainen T, Petrone JR, Igelstrom K, George CD, Garrett TJ, Oresic M, Triplett EW, Ludvigsson J (2024) Infant microbes and metabolites point to childhood neurodevelopmental disorders. Cell 187(8):1853–1873 e1815. https://doi.org/10.1016/j.cell.2024.02.035

Ait-Belgnaoui A, Payard I, Rolland C, Harkat C, Braniste V, Theodorou V, Tompkins TA (2018) Bifidobacterium longum and Lactobacillus helveticus synergistically suppress stress-related visceral hypersensitivity through hypothalamic-pituitary-adrenal axis modulation. J Neurogastroenterol Motil 24(1):138–146. https://doi.org/10.5056/jnm16167

Akbari E, Asemi Z, Daneshvar Kakhaki R, Bahmani F, Kouchaki E, Tamtaji OR, Hamidi GA, Salami M (2016) Effect of probiotic supplementation on cognitive function and metabolic status in Alzheimer's disease: a randomized, double-blind and controlled trial.

Front Aging Neurosci 8:256. https://doi.org/10.3389/fnagi.2016.00256

Akkasheh G, Kashani-Poor Z, Tajabadi-Ebrahimi M, Jafari P, Akbari H, Taghizadeh M, Memarzadeh MR, Asemi Z, Esmaillzadeh A (2016) Clinical and metabolic response to probiotic administration in patients with major depressive disorder: a randomized, double-blind, placebo-controlled trial. Nutrition 32(3):315–320. https://doi.org/10.1016/j.nut.2015.09.003

Allahyari P, Abbas Torki S, Aminnezhad Kavkani B, Mahmoudi Z, Mousavi Hoseini MS, Moradi M, Alami F, Keshavarz Mohammadian M, Bahoo Sele Bani S, Abbasi Mobarakeh K, Shafaei H, Khoshdooz S, Hajipour A, Doaei S, Gholamalizadeh M (2024) A systematic review of the beneficial effects of prebiotics, probiotics, and synbiotics on ADHD. Neuropsychopharmacol Rep 44(2):300–307. https://doi.org/10.1002/npr2.12437

Allan NP, Yamamoto BY, Kunihiro BP, Nunokawa CKL, Rubas NC, Wells RK, Umeda L, Phankitnirundorn K, Torres A, Peres R, Takahashi E, Maunakea AK (2024) Ketogenic diet induced shifts in the gut microbiome associate with changes to inflammatory cytokines and brain-related miRNAs in children with autism spectrum disorder. Nutrients 16(10). https://doi.org/10.3390/nu16101401

Altaha B, Heddes M, Pilorz V, Niu Y, Gorbunova E, Gigl M, Kleigrewe K, Oster H, Haller D, Kiessling S (2022) Genetic and environmental circadian disruption induce weight gain through changes in the gut microbiome. Mol Metab 66:101628. https://doi.org/10.1016/j.molmet.2022.101628

American Psychiatric Association (2013) Diagnostic and statistical manual of mental disorders, 5th edn. American Psychiatric Association

Andreani NA, Sharma A, Dahmen B, Specht HE, Mannig N, Ruan V, Keller L, Baines JF, Herpertz-Dahlmann B, Dempfle A, Seitz J (2024) Longitudinal analysis of the gut microbiome in adolescent patients with anorexia nervosa: microbiome-related factors associated with clinical outcome. Gut Microbes 16(1):2304158. https://doi.org/10.1080/19490976.2024.2304158

Archie EA, Tung J (2015) Social behavior and the microbiome. Curr Opin Behav Sci 6:28–34. https://doi.org/10.1016/j.cobeha.2015.07.008

Arentsen T, Qian Y, Gkotzis S, Femenia T, Wang T, Udekwu K, Forssberg H, Diaz Heijtz R (2017) The bacterial peptidoglycan-sensing molecule Pglyrp2 modulates brain development and behavior. Mol Psychiatry 22(2):257–266. https://doi.org/10.1038/mp.2016.182

Arifuzzaman M, Won TH, Li TT, Yano H, Digumarthi S, Heras AF, Zhang W, Parkhurst CN, Kashyap S, Jin WB, Putzel GG, Tsou AM, Chu C, Wei Q, Grier A, Consortium JILCB, Worgall S, Guo CJ, Schroeder FC, Artis D (2022) Inulin fibre promotes microbiota-derived bile acids and type 2 inflammation. Nature 611(7936):578–584. https://doi.org/10.1038/s41586-022-05380-y

Armet AM, Deehan EC, O'Sullivan AF, Mota JF, Field CJ, Prado CM, Lucey AJ, Walter J (2022) Rethinking healthy eating in light of the gut microbiome. Cell Host Microbe 30(6):764–785. https://doi.org/10.1016/j.chom.2022.04.016

Arnold SE, Arvanitakis Z, Macauley-Rambach SL, Koenig AM, Wang HY, Ahima RS, Craft S, Gandy S, Buettner C, Stoeckel LE, Holtzman DM, Nathan DM (2018) Brain insulin resistance in type 2 diabetes and Alzheimer disease: concepts and conundrums. Nat Rev Neurol 14(3):168–181. https://doi.org/10.1038/nrneurol.2017.185

Aron-Wisnewsky J, Clement K (2014) The effects of gastrointestinal surgery on gut microbiota: potential contribution to improved insulin sensitivity. Curr Atheroscler Rep 16(11):454. https://doi.org/10.1007/s11883-014-0454-9

Audet MC (2019) Stress-induced disturbances along the gut microbiota-immune-brain axis and implications for mental health: does sex matter? Front Neuroendocrinol 54:100772. https://doi.org/10.1016/j.yfrne.2019.100772

Audet MC (2021) Beyond the neuro-immune interplay in depression: could gut microbes be the missing link? Brain Behave Immune Health 16:100308. https://doi.org/10.1016/j.bbih.2021.100308

Avolio E, Olivito I, Rosina E, Romano L, Angelone T, De Bartolo A, Scimeca M, Bellizzi D, D'Aquila P, Passarino G, Alo R, Facciolo RM, Bagni C, De Lorenzo A, Canonaco M (2022) Modifications of behavior and inflammation in mice following transplant with fecal microbiota from children with autism. Neuroscience 498:174–189. https://doi.org/10.1016/j.neuroscience.2022.06.038

Awan A, Bartlett A, Blakeley-Ruiz JA, Richie T, Theriot CM, Kleiner M (2024) Dietary protein from different sources escapes host digestion and is differentially modified by the microbiota. bioRxiv. https://doi.org/10.1101/2024.06.26.600830

Ayala FJ (2009) Darwin and the scientific method. Proc Natl Acad Sci USA 106 Suppl 1(Suppl 1):10033–10039. https://doi.org/10.1073/pnas.0901404106

Balasubramanian R, Schneider E, Gunnigle E, Cotter PD, Cryan JF (2024) Fermented foods: harnessing their potential to modulate the microbiota-gut-brain axis for mental health. Neurosci Biobehav Rev 158:105562. https://doi.org/10.1016/j.neubiorev.2024.105562

Baldi S, Pagliai G, Dinu M, Di Gloria L, Nannini G, Curini L, Pallecchi M, Russo E, Niccolai E, Danza G, Benedettelli S, Ballerini G, Colombini B, Bartolucci G, Ramazzotti M, Sofi F, Amedei A (2022) Effect of ancient Khorasan wheat on gut microbiota, inflammation, and short-chain fatty acid production in patients with fibromyalgia. World J Gastroenterol 28(18):1965–1980. https://doi.org/10.3748/wjg.v28.i18.1965

Bang S, Shin YH, Park SM, Deng L, Williamson RT, Graham DB, Xavier RJ, Clardy J (2025) Unusual phospholipids from Morganella morganii linked to

depression. J Am Chem Soc. https://doi.org/10.1021/jacs.4c15158
Bansal T, Alaniz RC, Wood TK, Jayaraman A (2010) The bacterial signal indole increases epithelial-cell tight-junction resistance and attenuates indicators of inflammation. Proc Natl Acad Sci USA 107(1):228–233. https://doi.org/10.1073/pnas.0906112107
Barone M, Ramayo-Caldas Y, Estelle J, Tambosco K, Chadi S, Maillard F, Gallopin M, Planchais J, Chain F, Kropp C, Rios-Covian D, Sokol H, Brigidi P, Langella P, Martin R (2023) Gut barrier-microbiota imbalances in early life lead to higher sensitivity to inflammation in a murine model of C-section delivery. Microbiome 11(1):140. https://doi.org/10.1186/s40168-023-01584-0
Barroso A, Mahler JV, Fonseca-Castro PH, Quintana FJ (2021) The aryl hydrocarbon receptor and the gut-brain axis. Cell Mol Immunol 18(2):259–268. https://doi.org/10.1038/s41423-020-00585-5
Bartlett A, Kleiner M (2022) Dietary protein and the intestinal microbiota: an understudied relationship. iScience 25(11):105313. https://doi.org/10.1016/j.isci.2022.105313
Basafa-Roodi P, Jazayeri S, Hadi F, Paghaleh SJ, Khosravi-Darani K, Malakouti SK (2024) Effects of synbiotic supplementation on the components of metabolic syndrome in patients with schizophrenia: a randomized, double-blind, placebo-controlled trial. BMC Psychiatry 24(1):669. https://doi.org/10.1186/s12888-024-06061-y
Basso PJ, Camara NOS, Sales-Campos H (2018) Microbial-based therapies in the treatment of inflammatory bowel disease - an overview of human studies. Front Pharmacol 9:1571. https://doi.org/10.3389/fphar.2018.01571
Bastiaanssen TFS, Cowan CSM, Claesson MJ, Dinan TG, Cryan JF (2019) Making sense of ... the microbiome in psychiatry. Int J Neuropsychopharmacol 22(1):37–52. https://doi.org/10.1093/ijnp/pyy067
Bastiaanssen TFS, Cussotto S, Claesson MJ, Clarke G, Dinan TG, Cryan JF (2020) Gutted! unraveling the role of the microbiome in major depressive disorder. Harv Rev Psychiatry 28(1):26–39. https://doi.org/10.1097/HRP.0000000000000243
Bastiaanssen TFS, Gururajan A, van de Wouw M, Moloney GM, Ritz NL, Long-Smith CM, Wiley NC, Murphy AB, Lyte JM, Fouhy F, Stanton C, Claesson MJ, Dinan TG, Cryan JF (2021) Volatility as a concept to understand the impact of stress on the microbiome. Psychoneuroendocrinology 124:105047. https://doi.org/10.1016/j.psyneuen.2020.105047
Bastiaanssen TFS, Quinn TP, Loughman A (2022) Treating bugs as features: a compositional guide to the statistical analysis of the microbiome-gut-brain axis. Arxiv
Bayrer JR, Castro J, Venkataraman A, Touhara KK, Rossen ND, Morrie RD, Maddern J, Hendry A, Braverman KN, Garcia-Caraballo S, Schober G, Brizuela M, Castro Navarro FM, Bueno-Silva C, Ingraham HA, Brierley SM, Julius D (2023) Gut enterochromaffin cells drive visceral pain and anxiety. Nature 616(7955):137–142. https://doi.org/10.1038/s41586-023-05829-8
Beaumont W (1833) Experiments and observations on the gastric juice and the physiology of digestion. Maclachlan and Stewart, Edinburgh
Benakis C, Brea D, Caballero S, Faraco G, Moore J, Murphy M, Sita G, Racchumi G, Ling L, Pamer EG, Iadecola C, Anrather J (2016) Commensal microbiota affects ischemic stroke outcome by regulating intestinal gammadelta T cells. Nat Med 22(5):516–523. https://doi.org/10.1038/nm.4068
Benarroch EE (2007) Enteric nervous system: functional organization and neurologic implications. Neurology 69(20):1953–1957. https://doi.org/10.1212/01.wnl.0000281999.56102.b5
Berding K, Cryan JF (2022) Microbiota-targeted interventions for mental health. Curr Opin Psychiatry 35(1):3–9. https://doi.org/10.1097/YCO.0000000000000758
Berding K, Long-Smith CM, Carbia C, Bastiaanssen TFS, van de Wouw M, Wiley N, Strain CR, Fouhy F, Stanton C, Cryan JF, Dinan TG (2021) A specific dietary fibre supplementation improves cognitive performance-an exploratory randomised, placebo-controlled, crossover study. Psychopharmacology 238(1):149–163. https://doi.org/10.1007/s00213-020-05665-y
Berding K, Bastiaanssen TFS, Moloney GM, Boscaini S, Strain CR, Anesi A, Long-Smith C, Mattivi F, Stanton C, Clarke G, Dinan TG, Cryan JF (2023) Feed your microbes to deal with stress: a psychobiotic diet impacts microbial stability and perceived stress in a healthy adult population. Mol Psychiatry 28(2):601–610. https://doi.org/10.1038/s41380-022-01817-y
Berer K, Gerdes LA, Cekanaviciute E, Jia X, Xiao L, Xia Z, Liu C, Klotz L, Stauffer U, Baranzini SE, Kumpfel T, Hohlfeld R, Krishnamoorthy G, Wekerle H (2017) Gut microbiota from multiple sclerosis patients enables spontaneous autoimmune encephalomyelitis in mice. Proc Natl Acad Sci USA 114(40):10719–10724. https://doi.org/10.1073/pnas.1711233114
Bernard C (1878) Lectures on phenomena of life common to animals and plants. JB Balliere and Son, Paris
Bernard, C. (1949 (Originally 1865)) An introduction to the study of experimental medicine. Henry Schuman, Inc, Paris
Beyond bacteria: early-life gut virome link with childhood asthma development (2024) Nat Med 30(1):43–44. https://doi.org/10.1038/s41591-023-02747-0
Bhattarai Y, Williams BB, Battaglioli EJ, Whitaker WR, Till L, Grover M, Linden DR, Akiba Y, Kandimalla KK, Zachos NC, Kaunitz JD, Sonnenburg JL, Fischbach MA, Farrugia G, Kashyap PC (2018) Gut microbiota-produced tryptamine activates an Epithelial G-Protein-Coupled receptor to increase colonic secretion. Cell Host Microbe 23(6):775–785 e775. https://doi.org/10.1016/j.chom.2018.05.004

Bizzozero-Peroni B, Martinez-Vizcaino V, Fernandez-Rodriguez R, Jimenez-Lopez E, Nunez de Arenas-Arroyo S, Saz-Lara A, Diaz-Goni V, Mesas AE (2024) The impact of the Mediterranean diet on alleviating depressive symptoms in adults: a systematic review and meta-analysis of randomized controlled trials. Nutr Rev. https://doi.org/10.1093/nutrit/nuad176

Block CL, Eroglu O, Mague SD, Smith CJ, Ceasrine AM, Sriworarat C, Blount C, Beben KA, Malacon KE, Ndubuizu N, Talbot A, Gallagher NM, Chan Jo Y, Nyangacha T, Carlson DE, Dzirasa K, Eroglu C, Bilbo SD (2022) Prenatal environmental stressors impair postnatal microglia function and adult behavior in males. Cell Rep 40(5):111161. https://doi.org/10.1016/j.celrep.2022.111161

Boehme M, van de Wouw M, Bastiaanssen TFS, Olavarria-Ramirez L, Lyons K, Fouhy F, Golubeva AV, Moloney GM, Minuto C, Sandhu KV, Scott KA, Clarke G, Stanton C, Dinan TG, Schellekens H, Cryan JF (2020) Mid-life microbiota crises: middle age is associated with pervasive neuroimmune alterations that are reversed by targeting the gut microbiome. Mol Psychiatry 25(10):2567–2583. https://doi.org/10.1038/s41380-019-0425-1

Boehme M, Guzzetta KE, Bastiaanssen TFS, van de Wouw M, Moloney GM, Gual-Grau A, Spichak S, Olavarria-Ramirez L, Fitzgerald P, Morillas E, Ritz NL, Jaggar M, Cowan CSM, Crispie F, Donoso F, Halitzki E, Neto MC, Sichetti M, Golubeva AV et al (2021) Microbiota from young mice counteracts selective age-associated behavioral deficits. Nat Aging 1(8):666–676. https://doi.org/10.1038/s43587-021-00093-9

Boles JS, Krueger ME, Jernigan JE, Cole CL, Neighbarger NK, Uriarte Huarte O, Tansey MG (2024) A leaky gut dysregulates gene networks in the brain associated with immune activation, oxidative stress, and myelination in a mouse model of colitis. Brain Behav Immun 117:473–492. https://doi.org/10.1016/j.bbi.2024.02.007

Bonaz BL, Bernstein CN (2013) Brain-gut interactions in inflammatory bowel disease. Gastroenterology 144(1):36–49. https://doi.org/10.1053/j.gastro.2012.10.003

Bordenstein SR, Theis KR (2015) Host biology in light of the microbiome: ten principles of holobionts and hologenomes. PLoS Biol 13(8):e1002226. https://doi.org/10.1371/journal.pbio.1002226

Boscaini S, Leigh SJ, Lavelle A, Garcia-Cabrerizo R, Lipuma T, Clarke G, Schellekens H, Cryan JF (2022) Microbiota and body weight control: weight watchers within? Mol Metab 57:101427. https://doi.org/10.1016/j.molmet.2021.101427

Bostock EC, Kirkby KC, Taylor BV (2017) The current status of the ketogenic diet in psychiatry. Front Psych 8:43. https://doi.org/10.3389/fpsyt.2017.00043

Braga JD, Thongngam M, Kumrungsee T (2024) Gamma-aminobutyric acid as a potential postbiotic mediator in the gut-brain axis. NPJ Sci Food 8(1):16. https://doi.org/10.1038/s41538-024-00253-2

Bravo JA, Forsythe P, Chew MV, Escaravage E, Savignac HM, Dinan TG, Bienenstock J, Cryan JF (2011) Ingestion of Lactobacillus strain regulates emotional behavior and central GABA receptor expression in a mouse via the vagus nerve. Proc Natl Acad Sci USA 108(38):16050–16055. https://doi.org/10.1073/pnas.1102999108

Brea D, Poon C, Benakis C, Lubitz G, Murphy M, Iadecola C, Anrather J (2021) Stroke affects intestinal immune cell trafficking to the central nervous system. Brain Behav Immun 96:295–302. https://doi.org/10.1016/j.bbi.2021.05.008

Bretler T, Weisberg H, Koren O, Neuman H (2019) The effects of antipsychotic medications on microbiome and weight gain in children and adolescents. BMC Med 17(1):112. https://doi.org/10.1186/s12916-019-1346-1

Breton J, Massart S, Vandamme P, De Brandt E, Pot B, Foligne B (2013) Ecotoxicology inside the gut: impact of heavy metals on the mouse microbiome. BMC Pharmacol Toxicol 14:62. https://doi.org/10.1186/2050-6511-14-62

Bruce-Keller AJ, Salbaum JM, Luo M, Blanchard ET, Taylor CM, Welsh DA, Berthoud HR (2015) Obese-type gut microbiota induce neurobehavioral changes in the absence of obesity. Biol Psychiatry 77(7):607–615. https://doi.org/10.1016/j.biopsych.2014.07.012

Brushett S, Gacesa R, Vich Vila A, Brandao Gois MF, Andreu-Sanchez S, Swarte JC, Klaassen MAY, Collij V, Sinha T, Bolte LA, Wu J, Swertz M, de Kroon MLA, Reijneveld SA, Wijmenga C, Weersma RK, Fu J, van Loo HM, Kurilshikov A, Zhernakova A (2023) Gut feelings: the relations between depression, anxiety, psychotropic drugs and the gut microbiome. Gut Microbes 15(2):2281360. https://doi.org/10.1080/19490976.2023.2281360

Buzun E, Hsu CY, Sejane K, Oles RE, Vasquez Ayala A, Loomis LR, Zhao J, Rossitto LA, McGrosso DM, Gonzalez DJ, Bode L, Chu H (2024) A bacterial sialidase mediates early-life colonization by a pioneering gut commensal. Cell Host Microbe 32(2):181–190 e189. https://doi.org/10.1016/j.chom.2023.12.014

Cabre S, Ratsika A, Rea K, Stanton C, Cryan JF (2022) Animal models for assessing impact of C-section delivery on biological systems. Neurosci Biobehav Rev 135:104555. https://doi.org/10.1016/j.neubiorev.2022.104555

Caffrey EB, Sonnenburg JL, Devkota S (2024) Our extended microbiome: the human-relevant metabolites and biology of fermented foods. Cell Metab 36(4):684–701. https://doi.org/10.1016/j.cmet.2024.03.007

Campbell JP, Turner JE (2018) Debunking the myth of exercise-induced immune suppression: redefining the impact of exercise on immunological health across the lifespan [Review]. Front Immunol 9:648. https://doi.org/10.3389/fimmu.2018.00648

Campbell BCV, De Silva DA, Macleod MR, Coutts SB, Schwamm LH, Davis SM, Donnan GA (2019)

Ischaemic stroke. Nat Rev Dis Primers 5(1):70. https://doi.org/10.1038/s41572-019-0118-8

Cannavale CN, Mysonhimer AR, Bailey MA, Cohen NJ, Holscher HD, Khan NA (2023) Consumption of a fermented dairy beverage improves hippocampal-dependent relational memory in a randomized, controlled cross-over trial. Nutr Neurosci 26(3):265–274. https://doi.org/10.1080/1028415X.2022.2046963

Cantoni C, Lin Q, Dorsett Y, Ghezzi L, Liu Z, Pan Y, Chen K, Han Y, Li Z, Xiao H, Gormley M, Liu Y, Bokoliya S, Panier H, Suther C, Evans E, Deng L, Locca A, Mikesell R et al (2022) Alterations of host-gut microbiome interactions in multiple sclerosis. EBioMedicine 76:103798. https://doi.org/10.1016/j.ebiom.2021.103798

Carabotti M, Scirocco A, Maselli MA, Severi C (2015) The gut-brain axis: interactions between enteric microbiota, central and enteric nervous systems. Ann Gastroenterol 28(2):203–209

Carbia C, Bastiaanssen TFS, Iannone LF, Garcia-Cabrerizo R, Boscaini S, Berding K, Strain CR, Clarke G, Stanton C, Dinan TG, Cryan JF (2023) The microbiome-gut-brain axis regulates social cognition & craving in young binge drinkers. EBioMedicine 89:104442. https://doi.org/10.1016/j.ebiom.2023.104442

Castillo-Ruiz A, Gars A, Sturgeon H, Ronczkowski NM, Pyaram DN, Dauriat CJG, Chassaing B, Forger NG (2023) Brain effects of gestating germ-free persist in mouse neonates despite acquisition of a microbiota at birth. Front Neurosci 17:1130347. https://doi.org/10.3389/fnins.2023.1130347

Castro-Nallar E, Bendall ML, Perez-Losada M, Sabuncyan S, Severance EG, Dickerson FB, Schroeder JR, Yolken RH, Crandall KA (2015) Composition, taxonomy and functional diversity of the oropharynx microbiome in individuals with schizophrenia and controls. PeerJ 3:e1140. https://doi.org/10.7717/peerj.1140

Cattaneo A, Cattane N, Galluzzi S, Provasi S, Lopizzo N, Festari C, Ferrari C, Guerra UP, Paghera B, Muscio C, Bianchetti A, Volta GD, Turla M, Cotelli MS, Gennuso M, Prelle A, Zanetti O, Lussignoli G, Mirabile D et al (2017) Association of brain amyloidosis with pro-inflammatory gut bacterial taxa and peripheral inflammation markers in cognitively impaired elderly. Neurobiol Aging 49:60–68. https://doi.org/10.1016/j.neurobiolaging.2016.08.019

Cekanaviciute E, Yoo BB, Runia TF, Debelius JW, Singh S, Nelson CA, Kanner R, Bencosme Y, Lee YK, Hauser SL, Crabtree-Hartman E, Sand IK, Gacias M, Zhu Y, Casaccia P, Cree BAC, Knight R, Mazmanian SK, Baranzini SE (2017) Gut bacteria from multiple sclerosis patients modulate human T cells and exacerbate symptoms in mouse models. Proc Natl Acad Sci USA 114(40):10713–10718. https://doi.org/10.1073/pnas.1711235114

Cervantes-Barragan L, Chai JN, Tianero MD, Di Luccia B, Ahern PP, Merriman J, Cortez VS, Caparon MG, Donia MS, Gilfillan S, Cella M, Gordon JI, Hsieh CS, Colonna M (2017) Lactobacillus reuteri induces gut intraepithelial CD4(+)CD8alphaalpha(+) T cells. Science 357(6353):806–810. https://doi.org/10.1126/science.aah5825

Chen JJ, Zeng BH, Li WW, Zhou CJ, Fan SH, Cheng K, Zeng L, Zheng P, Fang L, Wei H, Xie P (2017) Effects of gut microbiota on the microRNA and mRNA expression in the hippocampus of mice. Behav Brain Res 322(Pt A):34–41. https://doi.org/10.1016/j.bbr.2017.01.021

Chen X, Hu C, Yan C, Tao E, Zhu Z, Shu X, Guo R, Jiang M (2023) Maternal separation leads to dynamic changes of visceral hypersensitivity and fecal metabolomics from childhood to adulthood. Sci Rep 13(1):7670. https://doi.org/10.1038/s41598-023-34792-7

Chen D, Lou Q, Song XJ, Kang F, Liu A, Zheng C, Li Y, Wang D, Qun S, Zhang Z, Cao P, Jin Y (2024) Microglia govern the extinction of acute stress-induced anxiety-like behaviors in male mice. Nat Commun 15(1):449. https://doi.org/10.1038/s41467-024-44704-6

Cheng L, Wu H, Cai X, Zhang Y, Yu S, Hou Y, Yin Z, Yan Q, Wang Q, Sun T, Wang G, Yuan Y, Zhang X, Hao H, Zheng X (2024) A Gpr35-tuned gut microbe-brain metabolic axis regulates depressive-like behavior. Cell Host Microbe 32(2):227–243 e226. https://doi.org/10.1016/j.chom.2023.12.009

Chilton PM, Ghare SS, Charpentier BT, Myers SA, Rao AV, Petrosino JF, Hoffman KL, Greenwell JC, Tyagi N, Behera J, Wang Y, Sloan LJ, Zhang J, Shields CB, Cooper GE, Gobejishvili L, Whittemore SR, McClain CJ, Barve SS (2024) Age-associated temporal decline in butyrate-producing bacteria plays a key pathogenic role in the onset and progression of neuropathology and memory deficits in 3xTg-AD mice. Gut Microbes 16(1):2389319. https://doi.org/10.1080/19490976.2024.2389319

Chimerel C, Emery E, Summers DK, Keyser U, Gribble FM, Reimann F (2014) Bacterial metabolite indole modulates incretin secretion from intestinal enteroendocrine L cells. Cell Rep 9(4):1202–1208. https://doi.org/10.1016/j.celrep.2014.10.032

Chiu IM, Heesters BA, Ghasemlou N, Von Hehn CA, Zhao F, Tran J, Wainger B, Strominger A, Muralidharan S, Horswill AR, Bubeck Wardenburg J, Hwang SW, Carroll MC, Woolf CJ (2013) Bacteria activate sensory neurons that modulate pain and inflammation. Nature 501(7465):52–57. https://doi.org/10.1038/nature12479

Chiu K, Warner G, Nowak RA, Flaws JA, Mei W (2020) The impact of environmental chemicals on the gut microbiome. Toxicol Sci 176(2):253–284. https://doi.org/10.1093/toxsci/kfaa065

Cho NA, Sales KM, Sampsell K, Wang W, Noye Tuplin EW, Lowry DE, Reimer RA (2021) C-section birth increases offspring obesity risk dependent on maternal diet and obesity status in rats. Obesity (Silver Spring) 29(10):1664–1675. https://doi.org/10.1002/oby.23258

Chong HX, Yusoff NAA, Hor YY, Lew LC, Jaafar MH, Choi SB, Yusoff MSB, Wahid N, Abdullah M, Zakaria N, Ong KL, Park YH, Liong MT (2019) Lactobacillus plantarum DR7 alleviates stress and anxiety in adults:

a randomised, double-blind, placebo-controlled study. Benef Microbes 10(4):355–373. https://doi.org/10.3920/BM2018.0135

Church JS, Bannish JAM, Adrian LA, Rojas Martinez K, Henshaw A, Schwartzer JJ (2023) Serum short chain fatty acids mediate hippocampal BDNF and correlate with decreasing neuroinflammation following high pectin fiber diet in mice. Front Neurosci 17:1134080. https://doi.org/10.3389/fnins.2023.1134080

Clarke G, Quigley EM, Cryan JF, Dinan TG (2009) Irritable bowel syndrome: towards biomarker identification. Trends Mol Med 15(10):478–489. https://doi.org/10.1016/j.molmed.2009.08.001

Clarke TB, Davis KM, Lysenko ES, Zhou AY, Yu Y, Weiser JN (2010) Recognition of peptidoglycan from the microbiota by Nod1 enhances systemic innate immunity. Nat Med 16(2):228–231. https://doi.org/10.1038/nm.2087

Clarke G, Cryan JF, Dinan TG, Quigley EM (2012a) Review article: probiotics for the treatment of irritable bowel syndrome--focus on lactic acid bacteria. Aliment Pharmacol Ther 35(4):403–413. https://doi.org/10.1111/j.1365-2036.2011.04965.x

Clarke G, McKernan DP, Gaszner G, Quigley EM, Cryan JF, Dinan TG (2012b) A distinct profile of tryptophan metabolism along the Kynurenine Pathway Downstream of toll-like receptor activation in irritable Bowel Syndrome. Front Pharmacol 3:90. https://doi.org/10.3389/fphar.2012.00090

Clarke SF, Murphy EF, Nilaweera K, Ross PR, Shanahan F, O'Toole PW, Cotter PD (2012c) The gut microbiota and its relationship to diet and obesity: new insights. Gut Microbes 3(3):186–202. https://doi.org/10.4161/gmic.20168

Clarke G, Grenham S, Scully P, Fitzgerald P, Moloney RD, Shanahan F, Dinan TG, Cryan JF (2013) The microbiome-gut-brain axis during early life regulates the hippocampal serotonergic system in a sex-dependent manner [Article]. Mol Psychiatry 18(6):666–673. https://doi.org/10.1038/mp.2012.77

Claus SP, Guillou H, Ellero-Simatos S (2016) The gut microbiota: a major player in the toxicity of environmental pollutants? NPJ Biofilms Microbiomes 2:16003. https://doi.org/10.1038/npjbiofilms.2016.3

Coelho LP, Alves R, Del Rio AR, Myers PN, Cantalapiedra CP, Giner-Lamia J, Schmidt TS, Mende DR, Orakov A, Letunic I, Hildebrand F, Van Rossum T, Forslund SK, Khedkar S, Maistrenko OM, Pan S, Jia L, Ferretti P, Sunagawa S et al (2022) Towards the biogeography of prokaryotic genes. Nature 601(7892):252–256. https://doi.org/10.1038/s41586-021-04233-4

Collins SM (2014) A role for the gut microbiota in IBS. Nat Rev Gastroenterol Hepatol 11(8):497–505. https://doi.org/10.1038/nrgastro.2014.40

Colloca L, Ludman T, Bouhassira D, Baron R, Dickenson AH, Yarnitsky D, Freeman R, Truini A, Attal N, Finnerup NB, Eccleston C, Kalso E, Bennett DL, Dworkin RH, Raja SN (2017) Neuropathic pain. Nat Rev Dis Primers 3(1):17002. https://doi.org/10.1038/nrdp.2017.2

Conti S, Perdixi E, Bernini S, Jesuthasan N, Severgnini M, Prinelli F, NutBrain Study, G (2024) Adherence to Mediterranean diet is inversely associated with depressive symptoms in older women: findings from the NutBrain Study. Br J Nutr 131(11):1892–1901. https://doi.org/10.1017/S0007114524000461

Cotillard A, Cartier-Meheust A, Litwin NS, Chaumont S, Saccareau M, Lejzerowicz F, Tap J, Koutnikova H, Lopez DG, McDonald D, Song SJ, Knight R, Derrien M, Veiga P (2022) A posteriori dietary patterns better explain variations of the gut microbiome than individual markers in the American Gut Project. Am J Clin Nutr 115(2):432–443. https://doi.org/10.1093/ajcn/nqab332

Craig AD (2009) How do you feel--now? The anterior insula and human awareness. Nat Rev Neurosci 10(1):59–70. https://doi.org/10.1038/nrn2555

Crouzet L, Gaultier E, Del'Homme C, Cartier C, Delmas E, Dapoigny M, Fioramonti J, Bernalier-Donadille A (2013) The hypersensitivity to colonic distension of IBS patients can be transferred to rats through their fecal microbiota. Neurogastroenterol Motil 25(4):e272–e282. https://doi.org/10.1111/nmo.12103

Cruz-Pereira JS, Moloney GM, Bastiaanssen TFS, Boscaini S, Fitzgerald P, Clarke G, Cryan JF (2023) Age-associated deficits in social behaviour are microbiota-dependent. Brain Behav Immun 110:119–124. https://doi.org/10.1016/j.bbi.2023.02.008

Cryan JF, O'Riordan KJ, Cowan CSM, Sandhu KV, Bastiaanssen TFS, Boehme M, Codagnone MG, Cussotto S, Fulling C, Golubeva AV, Guzzetta KE, Jaggar M, Long-Smith CM, Lyte JM, Martin JA, Molinero-Perez A, Moloney G, Morelli E, Morillas E et al (2019) The microbiota-gut-brain axis. Physiol Rev 99(4):1877–2013. https://doi.org/10.1152/physrev.00018.2018

Cryan JF, O'Riordan KJ, Sandhu K, Peterson V, Dinan TG (2020) The gut microbiome in neurological disorders. Lancet Neurol 19(2):179–194. https://doi.org/10.1016/S1474-4422(19)30356-4

Cull C, Singu VK, Cull BJ, Lechtenberg KF, Amachawadi RG, Schutz JS, Bryan KA (2022) Efficacy of two probiotic products fed daily to reduce clostridium perfringens-based adverse health and performance effects in dairy calves. Antibiotics (Basel) 11(11). https://doi.org/10.3390/antibiotics11111513

Curtis MA, Dhamsania RK, Branco RC, Guo JD, Creeden J, Neifer KL, Black CA, Winokur EJ, Andari E, Dias BG, Liu RC, Gourley SL, Miller GW, Burkett JP (2023) Developmental pyrethroid exposure causes a neurodevelopmental disorder phenotype in mice. PNAS Nexus 2(4):pgad085. https://doi.org/10.1093/pnasnexus/pgad085

Cussotto S, Sandhu KV, Dinan TG, Cryan JF (2018) The neuroendocrinology of the microbiota-gut-brain axis: a behavioural perspective. Front Neuroendocrinol 51:80–101. https://doi.org/10.1016/j.yfrne.2018.04.002

Cussotto S, Clarke G, Dinan TG, Cryan JF (2019a) Psychotropics and the microbiome: a chamber of secrets. Psychopharmacology 236(5):1411–1432. https://doi.org/10.1007/s00213-019-5185-8
Cussotto S, Strain CR, Fouhy F, Strain RG, Peterson VL, Clarke G, Stanton C, Dinan TG, Cryan JF (2019b) Differential effects of psychotropic drugs on microbiome composition and gastrointestinal function. Psychopharmacology 236(5):1671–1685. https://doi.org/10.1007/s00213-018-5006-5
Cussotto S, Clarke G, Dinan TG, Cryan JF (2021a) Psychotropic drugs and the microbiome. Mod Trends Psychiatry 32:113–133. https://doi.org/10.1159/000510423
Cussotto S, Walsh J, Golubeva AV, Zhdanov AV, Strain CR, Fouhy F, Stanton C, Dinan TG, Hyland NP, Clarke G, Cryan JF, Griffin BT (2021b) The gut microbiome influences the bioavailability of olanzapine in rats. EBioMedicine 66:103307. https://doi.org/10.1016/j.ebiom.2021.103307
Dalile B, Van Oudenhove L, Vervliet B, Verbeke K (2019) The role of short-chain fatty acids in microbiota-gut-brain communication. Nat Rev Gastroenterol Hepatol 16(8):461–478. https://doi.org/10.1038/s41575-019-0157-3
Dalile B, Vervliet B, Bergonzelli G, Verbeke K, Van Oudenhove L (2020) Colon-delivered short-chain fatty acids attenuate the cortisol response to psychosocial stress in healthy men: a randomized, placebo-controlled trial. Neuropsychopharmacology 45(13):2257–2266. https://doi.org/10.1038/s41386-020-0732-x
Dalile B, Fuchs A, La Torre D, Vervliet B, Van Oudenhove L, Verbeke K (2024) Colonic butyrate administration modulates fear memory but not the acute stress response in men: a randomized, triple-blind, placebo-controlled trial. Prog Neuro-Psychopharmacol Biol Psychiatry 131:110939. https://doi.org/10.1016/j.pnpbp.2024.110939
Daly DM, Park SJ, Valinsky WC, Beyak MJ (2011) Impaired intestinal afferent nerve satiety signalling and vagal afferent excitability in diet induced obesity in the mouse. J Physiol 589(Pt 11):2857–2870. https://doi.org/10.1113/jphysiol.2010.204594
Dapoigny M, Piche T, Ducrotte P, Lunaud B, Cardot JM, Bernalier-Donadille A (2012) Efficacy and safety profile of LCR35 complete freeze-dried culture in irritable bowel syndrome: a randomized, double-blind study. World J Gastroenterol 18(17):2067–2075. https://doi.org/10.3748/wjg.v18.i17.2067
Darwin C (1859) On the origin of species by means of natural selection. J Murray Page view. http://hdllocgov/locrbc/General174731
Davidson GL, Cienfuegos IA, Dalesman S (2024) Antibiotic-altered gut microbiota explain host memory plasticity and disrupt pace-of-life covariation for an aquatic snail. ISME J 18(1). https://doi.org/10.1093/ismejo/wrae078
De Palma G, Lynch MD, Lu J, Dang VT, Deng Y, Jury J, Umeh G, Miranda PM, Pigrau Pastor M, Sidani S, Pinto-Sanchez MI, Philip V, McLean PG, Hagelsieb MG, Surette MG, Bergonzelli GE, Verdu EF, Britz-McKibbin P, Neufeld JD et al (2017) Transplantation of fecal microbiota from patients with irritable bowel syndrome alters gut function and behavior in recipient mice. Sci Transl Med 9(379). https://doi.org/10.1126/scitranslmed.aaf6397
De Vadder F, Grasset E, Manneras Holm L, Karsenty G, Macpherson AJ, Olofsson LE, Backhed F (2018) Gut microbiota regulates maturation of the adult enteric nervous system via enteric serotonin networks. Proc Natl Acad Sci USA 115(25):6458–6463. https://doi.org/10.1073/pnas.1720017115
Deaver JA, Eum SY, Toborek M (2018) Circadian disruption changes gut microbiome taxa and functional gene composition. Front Microbiol 9:737. https://doi.org/10.3389/fmicb.2018.00737
Degirolamo C, Rainaldi S, Bovenga F, Murzilli S, Moschetta A (2014) Microbiota modification with probiotics induces hepatic bile acid synthesis via downregulation of the Fxr-Fgf15 axis in mice. Cell Rep 7(1):12–18. https://doi.org/10.1016/j.celrep.2014.02.032
Dendrou CA, Fugger L, Friese MA (2015) Immunopathology of multiple sclerosis. Nat Rev Immunol 15(9):545–558. https://doi.org/10.1038/nri3871
Desbonnet L, Clarke G, Traplin A, O'Sullivan O, Crispie F, Moloney RD, Cotter PD, Dinan TG, Cryan JF (2015) Gut microbiota depletion from early adolescence in mice: implications for brain and behaviour. Brain Behav Immun 48:165–173. https://doi.org/10.1016/j.bbi.2015.04.004
Dharshika C, Gulbransen BD (2023) Enteric neuromics: how high-throughput "Omics" deepens our understanding of enteric nervous system genetic architecture. Cell Mol Gastroenterol Hepatol 15(2):487–504. https://doi.org/10.1016/j.jcmgh.2022.10.019
Diaz Heijtz R (2024) Gut microbiota: a key susceptibility factor in social anxiety disorder. Proc Natl Acad Sci USA 121(5):e2320425121. https://doi.org/10.1073/pnas.2320425121
Dickerson F, Adamos M, Katsafanas E, Khushalani S, Origoni A, Savage C, Schweinfurth L, Stallings C, Sweeney K, Goga J, Yolken RH (2018) Adjunctive probiotic microorganisms to prevent rehospitalization in patients with acute mania: a randomized controlled trial. Bipolar Disord 20(7):614–621. https://doi.org/10.1111/bdi.12652
Dinan TG, Butler MI, Cryan JF (2021) Psychobiotics: evolution of novel antidepressants. Mod Trends Psychiatry 32:134–143. https://doi.org/10.1159/000510424
Dinan TG, Kennedy PJ, Morais LH, Murphy A, Long-Smith CM, Moloney GM, Bastiaanssen TFS, Allen AP, Collery A, Mullins D, Cusack AM, Berding K, O'Toole PW, Clarke G, Stanton C, Cryan JF (2022) Altered stress responses in adults born by Caesarean section. Neurobiol Stress 16:100425. https://doi.org/10.1016/j.ynstr.2021.100425

Diop L, Guillou S, Durand H (2008) Probiotic food supplement reduces stress-induced gastrointestinal symptoms in volunteers: a double-blind, placebo-controlled, randomized trial. Nutr Res 28(1):1–5. https://doi.org/10.1016/j.nutres.2007.10.001

Dollish HK, Tsyglakova M, McClung CA (2024) Circadian rhythms and mood disorders: time to see the light. Neuron 112(1):25–40. https://doi.org/10.1016/j.neuron.2023.09.023

Domènech L, Willis J, Alemany-Navarro M, Morell M, Real E, Escaramís G, Bertolín S, Sánchez Chinchilla D, Balcells S, Segalàs C, Estivill X, Menchón JM, Gabaldón T, Alonso P, Rabionet R (2022) Changes in the stool and oropharyngeal microbiome in obsessive-compulsive disorder. Sci Rep 12(1):1448. https://doi.org/10.1038/s41598-022-05480-9

Donoso F, Egerton S, Bastiaanssen TFS, Fitzgerald P, Gite S, Fouhy F, Ross RP, Stanton C, Dinan TG, Cryan JF (2020) Polyphenols selectively reverse early-life stress-induced behavioural, neurochemical and microbiota changes in the rat. Psychoneuroendocrinology 116:104673. https://doi.org/10.1016/j.psyneuen.2020.104673

Doyle WI, Dinser JA, Cansler HL, Zhang X, Dinh DD, Browder NS, Riddington IM, Meeks JP (2016) Faecal bile acids are natural ligands of the mouse accessory olfactory system [Article]. Nat Commun 7:11936. https://doi.org/10.1038/ncomms11936

Drewes AM, Olesen AE, Farmer AD, Szigethy E, Rebours V, Olesen SS (2020) Gastrointestinal pain. Nat Rev Dis Primers 6(1):1. https://doi.org/10.1038/s41572-019-0135-7

Dzamko N (2023) Cytokine activity in Parkinson's disease. Neuronal Signals 7(4):NS20220063. https://doi.org/10.1042/NS20220063

Eijsbouts C, Zheng T, Kennedy NA, Bonfiglio F, Anderson CA, Moutsianas L, Holliday J, Shi J, Shringarpure S, andMe Research, T, Voda AI, Bellygenes I, Farrugia G, Franke A, Hubenthal M, Abecasis G, Zawistowski M, Skogholt AH, Ness-Jensen E et al (2021) Genome-wide analysis of 53,400 people with irritable bowel syndrome highlights shared genetic pathways with mood and anxiety disorders. Nat Genet 53(11):1543–1552. https://doi.org/10.1038/s41588-021-00950-8

El-Salhy M, Valeur J, Hausken T, Gunnar Hatlebakk J (2021) Changes in fecal short-chain fatty acids following fecal microbiota transplantation in patients with irritable bowel syndrome. Neurogastroenterol Motil 33(2):e13983. https://doi.org/10.1111/nmo.13983

Enck P, Aziz Q, Barbara G, Farmer AD, Fukudo S, Mayer EA, Niesler B, Quigley EM, Rajilic-Stojanovic M, Schemann M, Schwille-Kiuntke J, Simren M, Zipfel S, Spiller RC (2016) Irritable bowel syndrome. Nat Rev Dis Primers 2:16014. https://doi.org/10.1038/nrdp.2016.14

Fackelmann G, Manghi P, Carlino N, Heidrich V, Piccinno G, Ricci L, Piperni E, Arre A, Bakker E, Creedon AC, Francis L, Capdevila Pujol J, Davies R, Wolf J, Bermingham KM, Berry SE, Spector TD, Asnicar F, Segata N (2025) Gut microbiome signatures of vegan, vegetarian and omnivore diets and associated health outcomes across 21,561 individuals. Nat Microbiol 10(1):41–52. https://doi.org/10.1038/s41564-024-01870-z

Fahur Bottino G, Bonham KS, Patel F, McCann S, Zieff M, Naspolini N, Ho D, Portlock T, Joos R, Midani FS, Schuroff P, Das A, Shennon I, Wilson BC, O'Sullivan JM, Britton RA, Murray DM, Kiely ME, Taddei CR et al (2025) Early life microbial succession in the gut follows common patterns in humans across the globe. Nat Commun 16(1):660. https://doi.org/10.1038/s41467-025-56072-w

Fan Y, Stoving RK, Berreira Ibraim S, Hyotylainen T, Thirion F, Arora T, Lyu L, Stankevic E, Hansen TH, Dechelotte P, Sinioja T, Ragnarsdottir O, Pons N, Galleron N, Quinquis B, Levenez F, Roume H, Falony G, Vieira-Silva S et al (2023) The gut microbiota contributes to the pathogenesis of anorexia nervosa in humans and mice. Nat Microbiol 8(5):787–802. https://doi.org/10.1038/s41564-023-01355-5

Felice VD, Moloney RD, Cryan JF, Dinan TG, O'Mahony SM (2015) Visceral pain and psychiatric disorders. Mod Trends Pharmacopsychiatry 30:103–119. https://doi.org/10.1159/000435936

Ferrier L, Eutamene H, Siegwald L, Marquard AM, Tondereau V, Chevalier J, Jacot GE, Favre L, Theodorou V, Vicario M, Rytz A, Bergonzelli G, Garcia-Rodenas CL (2022) Human milk oligosaccharides alleviate stress-induced visceral hypersensitivity and associated microbiota dysbiosis. J Nutr Biochem 99:108865. https://doi.org/10.1016/j.jnutbio.2021.108865

Filippi M, Bar-Or A, Piehl F, Preziosa P, Solari A, Vukusic S, Rocca MA (2018) Multiple sclerosis. Nat Rev Dis Prime 4(1):43. https://doi.org/10.1038/s41572-018-0041-4

Fouladi F, Bulik-Sullivan EC, Glenny EM, Thornton LM, Reed KK, Thomas S, Kleiman S, Watters A, Oakes J, Huh EY, Tang Q, Liu J, Djukic Z, Harper L, Trillo-Ordonez Y, Sun S, Blakely I, Mehler PS, Fodor AA et al (2022) Reproducible changes in the anorexia nervosa gut microbiota following inpatient therapy remain distinct from non-eating disorder controls. Gut Microbes 14(1):2143217. https://doi.org/10.1080/19490976.2022.2143217

Fransen F, van Beek AA, Borghuis T, Aidy SE, Hugenholtz F, van der Gaast-de Jongh C, Savelkoul HFJ, De Jonge MI, Boekschoten MV, Smidt H, Faas MM, de Vos P (2017) Aged gut microbiota contributes to systemical inflammaging after transfer to germ-free mice [Original Research]. Front Immunol 8(1385):1385. https://doi.org/10.3389/fimmu.2017.01385

Fulling C, Dinan TG, Cryan JF (2019) Gut microbe to brain signaling: what happens in Vagus. Neuron 101(6):998–1002. https://doi.org/10.1016/j.neuron.2019.02.008

Fulling C, Lach G, Bastiaanssen TFS, Fouhy F, O'Donovan AN, Ventura-Silva AP, Stanton C, Dinan TG, Cryan JF (2020) Adolescent dietary manipulations differentially affect gut microbiota composition and amygdala neuroimmune gene expression in male mice in adulthood. Brain Behav Immun 87:666–678. https://doi.org/10.1016/j.bbi.2020.02.013

Fung TC (2020) The microbiota-immune axis as a central mediator of gut-brain communication. Neurobiol Dis 136:104714. https://doi.org/10.1016/j.nbd.2019.104714

Gabanyi I, Lepousez G, Wheeler R, Vieites-Prado A, Nissant A, Chevalier G, Wagner S, Moigneu C, Dulauroy S, Hicham S, Polomack B, Verny F, Rosenstiel P, Renier N, Boneca IG, Eberl G, Lledo PM (2022) Bacterial sensing via neuronal Nod2 regulates appetite and body temperature. Science 376(6590):eabj3986. https://doi.org/10.1126/science.abj3986

Gacesa R, Kurilshikov A, Vich Vila A, Sinha T, Klaassen MAY, Bolte LA, Andreu-Sanchez S, Chen L, Collij V, Hu S, Dekens JAM, Lenters VC, Bjork JR, Swarte JC, Swertz MA, Jansen BH, Gelderloos-Arends J, Jankipersadsing S, Hofker M et al (2022) Environmental factors shaping the gut microbiome in a Dutch population. Nature 604(7907):732–739. https://doi.org/10.1038/s41586-022-04567-7

Gama J, Neves B, Pereira A (2022) Chronic effects of dietary pesticides on the gut microbiome and neurodevelopment. Front Microbiol 13:931440. https://doi.org/10.3389/fmicb.2022.931440

Garcia-Cabrerizo R, Cryan JF (2024) A gut (microbiome) feeling about addiction: interactions with stress and social systems. Neurobiol Stress 30:100629. https://doi.org/10.1016/j.ynstr.2024.100629

Garcia-Cabrerizo R, Carbia C, KJ OR, Schellekens H, Cryan JF (2021) Microbiota-gut-brain axis as a regulator of reward processes. J Neurochem 157(5):1495–1524. https://doi.org/10.1111/jnc.15284

Garcia-Cabrerizo R, Barros-Santos T, Campos D, Cryan JF (2023) The gut microbiota alone and in combination with a social stimulus regulates cocaine reward in the mouse. Brain Behav Immun 107:286–291. https://doi.org/10.1016/j.bbi.2022.10.020

Gatarek P, Kaluzna-Czaplinska J (2021) Trimethylamine N-oxide (TMAO) in human health. EXCLI J 20:301–319. https://doi.org/10.17179/excli2020-3239

Ge X, Pan J, Liu Y, Wang H, Zhou W, Wang X (2018) Intestinal crosstalk between microbiota and serotonin and its impact on gut motility. Curr Pharm Biotechnol 19(3):190–195. https://doi.org/10.2174/1389201019666180528094202

Gee LMV, Barron-Millar B, Leslie J, Richardson C, Zaki MYW, Luli S, Burgoyne RA, Cameron RIT, Smith GR, Brain JG, Innes B, Jopson L, Dyson JK, McKay KRC, Pechlivanis A, Holmes E, Berlinguer-Palmini R, Victorelli S, Mells GF et al (2023) Anti-Cholestatic therapy with obeticholic acid improves short-term memory in bile duct-ligated mice. Am J Pathol 193(1):11–26. https://doi.org/10.1016/j.ajpath.2022.09.005

Geier TJ, Atkinson SN, Pan AY, Mantz-Wichman M, Jazinski-Chambers K, Hillard CJ, de Roon-Cassini TA (2024) Differences in intestinal bacteria in traumatic injury survivors with and without probable posttraumatic stress disorder. J Affect Disord 361:528–535. https://doi.org/10.1016/j.jad.2024.06.075

Gheorghe CE, Cryan JF, Clarke G (2022) Debugging the gut-brain axis in depression. Cell Host Microbe 30(3):281–283. https://doi.org/10.1016/j.chom.2022.02.007

Gheorghe CE, Leigh SJ, Tofani GSS, Bastiaanssen TFS, Lyte JM, Gardellin E, Govindan A, Strain C, Martinez-Herrero S, Goodson MS, Kelley-Loughnane N, Cryan JF, Clarke G (2024) The microbiota drives diurnal rhythms in tryptophan metabolism in the stressed gut. Cell Rep 43(4):114079. https://doi.org/10.1016/j.celrep.2024.114079

Ghosh TS, Shanahan F, O'Toole PW (2022) The gut microbiome as a modulator of healthy ageing. Nat Rev Gastroenterol Hepatol 19(9):565–584. https://doi.org/10.1038/s41575-022-00605-x

Ghuge S, Rahman Z, Bhale NA, Dikundwar AG, Dandekar MP (2023) Multistrain probiotic rescinds quinpirole-induced obsessive-compulsive disorder phenotypes by reshaping of microbiota gut-brain axis in rats. Pharmacol Biochem Behav 232:173652. https://doi.org/10.1016/j.pbb.2023.173652

Gkougka D, Mitropoulos K, Tzanakaki G, Panagouli E, Psaltopoulou T, Thomaidis L, Tsolia M, Sergentanis TN, Tsitsika A (2022) Gut microbiome and attention deficit/hyperactivity disorder: a systematic review. Pediatr Res 92(6):1507–1519. https://doi.org/10.1038/s41390-022-02027-6

Goehler LE, Gaykema RP, Opitz N, Reddaway R, Badr N, Lyte M (2005) Activation in vagal afferents and central autonomic pathways: early responses to intestinal infection with Campylobacter jejuni. Brain Behav Immun 19(4):334–344. https://doi.org/10.1016/j.bbi.2004.09.002

Gois MFB, Fernandez-Pato A, Huss A, Gacesa R, Wijmenga C, Weersma RK, Fu J, Vermeulen RCH, Zhernakova A, Lenters VC, Kurilshikov A (2023) Impact of occupational pesticide exposure on the human gut microbiome. Front Microbiol 14:1223120. https://doi.org/10.3389/fmicb.2023.1223120

Golubeva AV, Joyce SA, Moloney G, Burokas A, Sherwin E, Arboleya S, Flynn I, Khochanskiy D, Moya-Perez A, Peterson V, Rea K, Murphy K, Makarova O, Buravkov S, Hyland NP, Stanton C, Clarke G, Gahan CGM, Dinan TG, Cryan JF (2017) Microbiota-related changes in bile acid & tryptophan metabolism are associated with gastrointestinal dysfunction in a mouse model of autism. EBioMedicine 24:166–178. https://doi.org/10.1016/j.ebiom.2017.09.020

González-Muniesa P, Mártinez-González M-A, Hu FB, Després J-P, Matsuzawa Y, Loos RJF, Moreno LA, Bray GA, Martinez JA (2017) Obesity. Nat Rev Dis Prime 3(1):17034. https://doi.org/10.1038/nrdp.2017.34

Gonzalez-Santana A, Diaz Heijtz R (2020) Bacterial peptidoglycans from microbiota in neurodevelopment and behavior. Trends Mol Med 26(8):729–743. https://doi.org/10.1016/j.molmed.2020.05.003

Goodrich JK, Waters JL, Poole AC, Sutter JL, Koren O, Blekhman R, Beaumont M, Van Treuren W,

Knight R, Bell JT, Spector TD, Clark AG, Ley RE (2014) Human genetics shape the gut microbiome. Cell 159(4):789–799. https://doi.org/10.1016/j.cell.2014.09.053

Gora B, Gofron Z, Grosiak M, Aptekorz M, Kazek B, Kocelak P, Radosz-Komoniewska H, Chudek J, Martirosian G (2018) Toxin profile of fecal Clostridium perfringens strains isolated from children with autism spectrum disorders. Anaerobe 51:73–77. https://doi.org/10.1016/j.anaerobe.2018.03.005

Grabrucker S, Marizzoni M, Silajdzic E, Lopizzo N, Mombelli E, Nicolas S, Dohm-Hansen S, Scassellati C, Moretti DV, Rosa M, Hoffmann K, Cryan JF, O'Leary OF, English JA, Lavelle A, O'Neill C, Thuret S, Cattaneo A, Nolan YM (2023) Microbiota from Alzheimer's patients induce deficits in cognition and hippocampal neurogenesis. Brain 146(12):4916–4934. https://doi.org/10.1093/brain/awad303

Grant ET, De Franco H, Desai MS (2025) Non-SCFA microbial metabolites associated with fiber fermentation and host health. Trends Endocrinol Metab 36(1):70–82. https://doi.org/10.1016/j.tem.2024.06.009

Green JE, Davis JA, Berk M, Hair C, Loughman A, Castle D, Athan E, Nierenberg AA, Cryan JF, Jacka F, Marx W (2020) Efficacy and safety of fecal microbiota transplantation for the treatment of diseases other than Clostridium difficile infection: a systematic review and meta-analysis. Gut Microbes 12(1):1–25. https://doi.org/10.1080/19490976.2020.1854640

Green JE, Berk M, Mohebbi M, Loughman A, McGuinness AJ, Castle D, Chatterton ML, Perez J, Strandwitz P, Athan E, Hair C, Nierenberg AA, Cryan JF, Jacka F (2023) Feasibility, acceptability, and safety of faecal microbiota transplantation in the treatment of major depressive disorder: a pilot randomized controlled trial. Can J Psychiatr 68(5):315–326. https://doi.org/10.1177/07067437221150508

Group, N. H. W, Peterson J, Garges S, Giovanni M, McInnes P, Wang L, Schloss JA, Bonazzi V, McEwen JE, Wetterstrand KA, Deal C, Baker CC, Di Francesco V, Howcroft TK, Karp RW, Lunsford RD, Wellington CR, Belachew T, Wright M et al (2009) The NIH human microbiome project. Genome Res 19(12):2317–2323. https://doi.org/10.1101/gr.096651.109

Grundy D, Al-Chaer ED, Aziz Q, Collins SM, Ke M, Tache Y, Wood JD (2006) Fundamentals of neurogastroenterology: basic science. Gastroenterology 130(5):1391–1411. https://doi.org/10.1053/j.gastro.2005.11.060

Guo G, Jia KR, Shi Y, Liu XF, Liu KY, Qi W, Guo Y, Zhang WJ, Wang T, Xiao B, Zou QM (2009) Psychological stress enhances the colonization of the stomach by Helicobacter pylori in the BALB/c mouse. Stress 12(6):478–485. https://doi.org/10.3109/10253890802642188

Gupta VK, Rajendraprasad S, Ozkan M, Ramachandran D, Ahmad S, Bakken JS, Laudanski K, Gajic O, Bauer B, Zec S, Freeman DW, Khanna S, Shah A, Skalski JH, Sung J, Karnatovskaia LV (2024) Safety, feasibility, and impact on the gut microbiome of kefir administration in critically ill adults. BMC Med 22(1):80. https://doi.org/10.1186/s12916-024-03299-x

Guzior DV, Quinn RA (2021) Review: microbial transformations of human bile acids. Microbiome 9(1):140. https://doi.org/10.1186/s40168-021-01101-1

Guzzetta KE, Cryan JF, O'Leary OF (2022) Microbiota-Gut-Brain axis regulation of adult Hippocampal Neurogenesis. Brain Plast 8(1):97–119. https://doi.org/10.3233/BPL-220141

Gyles TM, Nestler EJ, Parise EM (2024) Advancing preclinical chronic stress models to promote therapeutic discovery for human stress disorders. Neuropsychopharmacology 49(1):215–226. https://doi.org/10.1038/s41386-023-01625-0

Hao Z, Wang W, Guo R, Liu H (2019) Faecalibacterium prausnitzii (ATCC 27766) has preventive and therapeutic effects on chronic unpredictable mild stress-induced depression-like and anxiety-like behavior in rats. Psychoneuroendocrinology 104:132–142. https://doi.org/10.1016/j.psyneuen.2019.02.025

Hata T, Asano Y, Yoshihara K, Kimura-Todani T, Miyata N, Zhang XT, Takakura S, Aiba Y, Koga Y, Sudo N (2017) Regulation of gut luminal serotonin by commensal microbiota in mice. PLoS One 12(7):e0180745. https://doi.org/10.1371/journal.pone.0180745

Heddes M, Altaha B, Niu Y, Reitmeier S, Kleigrewe K, Haller D, Kiessling S (2022) The intestinal clock drives the microbiome to maintain gastrointestinal homeostasis. Nat Commun 13(1):6068. https://doi.org/10.1038/s41467-022-33609-x

Heinzel S, Jureczek J, Kainulainen V, Nieminen AI, Suenkel U, von Thaler AK, Kaleta C, Eschweiler GW, Brockmann K, Aho VTE, Auvinen P, Maetzler W, Berg D, Scheperjans F (2024) Elevated fecal calprotectin is associated with gut microbial dysbiosis, altered serum markers and clinical outcomes in older individuals. Sci Rep 14(1):13513. https://doi.org/10.1038/s41598-024-63893-0

Heppner N, Reitmeier S, Heddes M, Merino MV, Schwartz L, Dietrich A, List M, Gigl M, Meng C, van der Veen DR, Schirmer M, Kleigrewe K, Omer H, Kiessling S, Haller D (2024) Diurnal rhythmicity of infant fecal microbiota and metabolites: a randomized controlled interventional trial with infant formula. Cell Host Microbe 32(4):573–587 e575. https://doi.org/10.1016/j.chom.2024.02.015

Hill C, Guarner F, Reid G, Gibson GR, Merenstein DJ, Pot B, Morelli L, Canani RB, Flint HJ, Salminen S, Calder PC, Sanders ME (2014) Expert consensus document. The International Scientific Association for probiotics and prebiotics consensus statement on the scope and appropriate use of the term probiotic. Nat Rev Gastroenterol Hepatol 11(8):506–514. https://doi.org/10.1038/nrgastro.2014.66

Hillman CH, Erickson KI, Kramer AF (2008) Be smart, exercise your heart: exercise effects on brain and cognition. Nat Rev Neurosci 9(1):58–65. https://doi.org/10.1038/nrn2298

Hilton D, Stephens M, Kirk L, Edwards P, Potter R, Zajicek J, Broughton E, Hagan H, Carroll C (2014) Accumulation of alpha-synuclein in the bowel of

patients in the pre-clinical phase of Parkinson's disease. Acta Neuropathol 127(2):235–241. https://doi.org/10.1007/s00401-013-1214-6
Holmqvist S, Chutna O, Bousset L, Aldrin-Kirk P, Li W, Bjorklund T, Wang ZY, Roybon L, Melki R, Li JY (2014) Direct evidence of Parkinson pathology spread from the gastrointestinal tract to the brain in rats. Acta Neuropathol 128(6):805–820. https://doi.org/10.1007/s00401-014-1343-6
Holzer P, Farzi A, Hassan AM, Zenz G, Jacan A, Reichmann F (2017) Visceral inflammation and immune activation stress the brain. Front Immunol 8:1613. https://doi.org/10.3389/fimmu.2017.01613
Hontelez S, Stobernack T, Pelsser LM, van Baarlen P, Frankena K, Groefsema MM, Kleerebezem M, Rodrigues Pereira R, Postma EM, Smeets PAM, Stopyra MA, Zwiers MP, Aarts E (2021) Correlation between brain function and ADHD symptom changes in children with ADHD following a few-foods diet: an open-label intervention trial. Sci Rep 11(1):22205. https://doi.org/10.1038/s41598-021-01684-7
Hopkins MJ, Sharp R, Macfarlane GT (2001) Age and disease related changes in intestinal bacterial populations assessed by cell culture, 16S rRNA abundance, and community cellular fatty acid profiles. Gut 48(2):198–205. https://doi.org/10.1136/gut.48.2.198
Horne R, Foster JA (2018) Metabolic and microbiota measures as peripheral biomarkers in major depressive disorder. Front Psych 9:513. https://doi.org/10.3389/fpsyt.2018.00513
Howard AL, Robinson M, Smith GJ, Ambrosini GL, Piek JP, Oddy WH (2011) ADHD is associated with a "Western" dietary pattern in adolescents. J Atten Disord 15(5):403–411. https://doi.org/10.1177/1087054710365990
Hsiao EY, McBride SW, Hsien S, Sharon G, Hyde ER, McCue T, Codelli JA, Chow J, Reisman SE, Petrosino JF, Patterson PH, Mazmanian SK (2013) Microbiota modulate behavioral and physiological abnormalities associated with neurodevelopmental disorders. Cell 155(7):1451–1463. https://doi.org/10.1016/j.cell.2013.11.024
Human Microbiome Project, C (2012) Structure, function and diversity of the healthy human microbiome. Nature 486(7402):207–214. https://doi.org/10.1038/nature11234
Hung LY, Margolis KG (2024) Autism spectrum disorders and the gastrointestinal tract: insights into mechanisms and clinical relevance. Nat Rev Gastroenterol Hepatol 21(3):142–163. https://doi.org/10.1038/s41575-023-00857-1
Hutkins R, Walter J, Gibson GR, Bedu-Ferrari C, Scott K, Tancredi DJ, Wijeyesekera A, Sanders ME (2025) Classifying compounds as prebiotics — scientific perspectives and recommendations. Nat Rev Gastroenterol Hepatol 22(1):54–70. https://doi.org/10.1038/s41575-024-00981-6
Hyland NP, Cryan JF (2016) Microbe-host interactions: influence of the gut microbiota on the enteric nervous system. Dev Biol 417(2):182–187. https://doi.org/10.1016/j.ydbio.2016.06.027
Imhann F, Bonder MJ, Vich Vila A, Fu J, Mujagic Z, Vork L, Tigchelaar EF, Jankipersadsing SA, Cenit MC, Harmsen HJ, Dijkstra G, Franke L, Xavier RJ, Jonkers D, Wijmenga C, Weersma RK, Zhernakova A (2016) Proton pump inhibitors affect the gut microbiome. Gut 65(5):740–748. https://doi.org/10.1136/gutjnl-2015-310376
Itzhaki RF, Lathe R, Balin BJ, Ball MJ, Bearer EL, Braak H, Bullido MJ, Carter C, Clerici M, Cosby SL, Del Tredici K, Field H, Fulop T, Grassi C, Griffin WS, Haas J, Hudson AP, Kamer AR, Kell DB et al (2016) Microbes and Alzheimer's disease. J Alzheimers Dis 51(4):979–984. https://doi.org/10.3233/JAD-160152
Jacka FN, O'Neil A, Opie R, Itsiopoulos C, Cotton S, Mohebbi M, Castle D, Dash S, Mihalopoulos C, Chatterton ML, Brazionis L, Dean OM, Hodge AM, Berk M (2017) A randomised controlled trial of dietary improvement for adults with major depression (the 'SMILES' trial). BMC Med 15(1):23. https://doi.org/10.1186/s12916-017-0791-y
Jaggar M, Rea K, Spichak S, Dinan TG, Cryan JF (2020) You've got male: sex and the microbiota-gut-brain axis across the lifespan. Front Neuroendocrinol 56:100815. https://doi.org/10.1016/j.yfrne.2019.100815
Jänig W (2006) The integrative action of the autonomic nervous system neurobiology of homeostasis introduction. In: Integrative action of the autonomic nervous system: neurobiology of homeostasis. Cambridge University Press, pp 1–9
Jasarevic E, Hill EM, Kane PJ, Rutt L, Gyles T, Folts L, Rock KD, Howard CD, Morrison KE, Ravel J, Bale TL (2021) The composition of human vaginal microbiota transferred at birth affects offspring health in a mouse model. Nat Commun 12(1):6289. https://doi.org/10.1038/s41467-021-26634-9
Jia J, Ning Y, Chen M, Wang S, Yang H, Li F, Ding J, Li Y, Zhao B, Lyu J, Yang S, Yan X, Wang Y, Qin W, Wang Q, Li Y, Zhang J, Liang F, Liao Z, Wang S (2024) Biomarker changes during 20 Years preceding Alzheimer's disease. N Engl J Med 390(8):712–722. https://doi.org/10.1056/NEJMoa2310168
Jiang H, Ling Z, Zhang Y, Mao H, Ma Z, Yin Y, Wang W, Tang W, Tan Z, Shi J, Li L, Ruan B (2015) Altered fecal microbiota composition in patients with major depressive disorder. Brain Behav Immun 48:186–194. https://doi.org/10.1016/j.bbi.2015.03.016
Jiang J, Tang B, Wang L, Huo Q, Tan S, Misrani A, Han Y, Li H, Hu H, Wang J, Cheng T, Tabassum S, Chen M, Xie W, Long C, Yang L (2022) Systemic LPS-induced microglial activation results in increased GABAergic tone: a mechanism of protection against neuroinflammation in the medial prefrontal cortex in mice. Brain Behav Immun 99:53–69. https://doi.org/10.1016/j.bbi.2021.09.017
Jie Z, Yu X, Liu Y, Sun L, Chen P, Ding Q, Gao Y, Zhang X, Yu M, Liu Y, Zhang Y, Kristiansen K, Jia

H, Brix S, Cai K (2021) The baseline gut microbiota directs dieting-induced weight loss trajectories. Gastroenterology 160(6):2029–2042 e2016. https://doi.org/10.1053/j.gastro.2021.01.029

Jin H, Li M, Jeong E, Castro-Martinez F, Zuker CS (2024) A body-brain circuit that regulates body inflammatory responses. Nature 630(8017):695–703. https://doi.org/10.1038/s41586-024-07469-y

Johnsen PH, Hilpusch F, Cavanagh JP, Leikanger IS, Kolstad C, Valle PC, Goll R (2018) Faecal microbiota transplantation versus placebo for moderate-to-severe irritable bowel syndrome: a double-blind, randomised, placebo-controlled, parallel-group, single-centre trial. Lancet Gastroenterol Hepatol 3(1):17–24. https://doi.org/10.1016/S2468-1253(17)30338-2

Jomova K, Alomar SY, Nepovimova E, Kuca K, Valko M (2025) Heavy metals: toxicity and human health effects. Arch Toxicol 99(1):153–209. https://doi.org/10.1007/s00204-024-03903-2

Joos R, Boucher K, Lavelle A, Arumugam M, Blaser MJ, Claesson MJ, Clarke G, Cotter PD, De Sordi L, Dominguez-Bello MG, Dutilh BE, Ehrlich SD, Ghosh TS, Hill C, Junot C, Lahti L, Lawley TD, Licht TR, Maguin E et al (2024) Examining the healthy human microbiome concept. Nat Rev Microbiol. https://doi.org/10.1038/s41579-024-01107-0

Kaelberer MM, Rupprecht LE, Liu WW, Weng P, Bohorquez DV (2020) Neuropod cells: the emerging biology of gut-brain sensory transduction. Annu Rev Neurosci 43:337–353. https://doi.org/10.1146/annurev-neuro-091619-022657

Kang JI, Seo JH, Park CI, Kim ST, Kim YK, Jang JK, Kwon CO, Jeon S, Kim HW, Kim SJ (2023) Microbiome analysis of circulating bacterial extracellular vesicles in obsessive-compulsive disorder. Psychiatry Clin Neurosci 77(12):646–652. https://doi.org/10.1111/pcn.13593

Kawai T, Akira S (2009) The roles of TLRs, RLRs and NLRs in pathogen recognition. Int Immunol 21(4):317–337. https://doi.org/10.1093/intimm/dxp017

Kemp KM, Orihuela CA, Morrow CD, Judd SE, Evans RR, Mrug S (2024) Associations between dietary habits, socio-demographics and gut microbial composition in adolescents. Br J Nutr 131(5):809–820. https://doi.org/10.1017/S0007114523002271

Kennedy AR, Pissios P, Otu H, Roberson R, Xue B, Asakura K, Furukawa N, Marino FE, Liu FF, Kahn BB, Libermann TA, Maratos-Flier E (2007) A high-fat, ketogenic diet induces a unique metabolic state in mice. Am J Physiol Endocrinol Metab 292(6):E1724–E1739. https://doi.org/10.1152/ajpendo.00717.2006

Kennedy KM, de Goffau MC, Perez-Munoz ME, Arrieta MC, Backhed F, Bork P, Braun T, Bushman FD, Dore J, de Vos WM, Earl AM, Eisen JA, Elovitz MA, Ganal-Vonarburg SC, Ganzle MG, Garrett WS, Hall LJ, Hornef MW, Huttenhower C et al (2023) Questioning the fetal microbiome illustrates pitfalls of low-biomass microbial studies. Nature 613(7945):639–649. https://doi.org/10.1038/s41586-022-05546-8

Khan MF, Khodve G, Yadav S, Mallick K, Banerjee S (2025) Probiotic treatment improves post-traumatic stress disorder outcomes in mice. Behav Brain Res 476:115246. https://doi.org/10.1016/j.bbr.2024.115246

Kim S, Kwon SH, Kam TI, Panicker N, Karuppagounder SS, Lee S, Lee JH, Kim WR, Kook M, Foss CA, Shen C, Lee H, Kulkarni S, Pasricha PJ, Lee G, Pomper MG, Dawson VL, Dawson TM, Ko HS (2019) Transneuronal propagation of pathologic alpha-Synuclein from the gut to the brain models Parkinson's disease. Neuron 103(4):627–641 e627. https://doi.org/10.1016/j.neuron.2019.05.035

Kim MS, Kim Y, Choi H, Kim W, Park S, Lee D, Kim DK, Kim HJ, Choi H, Hyun DW, Lee JY, Choi EY, Lee DS, Bae JW, Mook-Jung I (2020) Transfer of a healthy microbiota reduces amyloid and tau pathology in an Alzheimer's disease animal model. Gut 69(2):283–294. https://doi.org/10.1136/gutjnl-2018-317431

Kimmel M, Jin W, Xia K, Lun K, Azcarate-Peril A, Plantinga A, Wu M, Ataei S, Rackers H, Carroll I, Meltzer-Brody S, Fransson E, Knickmeyer R (2022) Metabolite trajectories across the perinatal period and mental health: a preliminary study of tryptophan-related metabolites, bile acids and microbial composition. Behav Brain Res 418:113635. https://doi.org/10.1016/j.bbr.2021.113635

Knopman DS, Amieva H, Petersen RC, Chetelat G, Holtzman DM, Hyman BT, Nixon RA, Jones DT (2021) Alzheimer disease. Nat Rev Dis Primers 7(1):33. https://doi.org/10.1038/s41572-021-00269-y

Knox EG, Aburto MR, Clarke G, Cryan JF, O'Driscoll CM (2022a) The blood-brain barrier in aging and neurodegeneration. Mol Psychiatry 27(6):2659–2673. https://doi.org/10.1038/s41380-022-01511-z

Knox EG, Aburto MR, Tessier C, Nagpal J, Clarke G, O'Driscoll CM, Cryan JF (2022b) Microbial-derived metabolites induce actin cytoskeletal rearrangement and protect blood-brain barrier function. iScience 25(12):105648. https://doi.org/10.1016/j.isci.2022.105648

Knox EG, Lynch CMK, Lee YS, O'Driscoll CM, Clarke G, Cryan JF, Aburto MR (2023) The gut microbiota is important for the maintenance of blood-cerebrospinal fluid barrier integrity. Eur J Neurosci 57(2):233–241. https://doi.org/10.1111/ejn.15878

Kolobaric A, Andreescu C, Jasarevic E, Hong CH, Roh HW, Cheong JY, Kim YK, Shin TS, Kang CS, Kwon CO, Yoon SY, Hong SW, Aizenstein HJ, Karim HT, Son SJ (2024) Gut microbiome predicts cognitive function and depressive symptoms in late life. Mol Psychiatry 29(10):3064–3075. https://doi.org/10.1038/s41380-024-02551-3

Komulainen P, Tuomilehto J, Savonen K, Mannikko R, Hassinen M, Lakka TA, Hanninen T, Kiviniemi V, Jacobs DR, Kivipelto M, Rauramaa R (2021) Exercise, diet, and cognition in a 4-year randomized controlled trial: Dose-Responses to Exercise Training (DR's EXTRA). Am J Clin Nutr 113(6):1428–1439. https://doi.org/10.1093/ajcn/nqab018

Kong G, Cao KL, Judd LM, Li S, Renoir T, Hannan AJ (2020) Microbiome profiling reveals gut dysbiosis in a transgenic mouse model of Huntington's disease. Neurobiol Dis 135:104268. https://doi.org/10.1016/j.nbd.2018.09.001

Konturek SJ, Konturek JW, Pawlik T, Brzozowski T (2004) Brain-gut axis and its role in the control of food intake. J Physiol Pharmacol 55(1 Pt 2):137–154. https://www.ncbi.nlm.nih.gov/pubmed/15082874

Korpela K, de Vos WM (2022) Infant gut microbiota restoration: state of the art. Gut Microbes 14(1):2118811. https://doi.org/10.1080/19490976.2022.2118811

Korpela K, Kallio S, Salonen A, Hero M, Kukkonen AK, Miettinen PJ, Savilahti E, Kohva E, Kariola L, Suutela M, Tarkkanen A, de Vos WM, Raivio T, Kuitunen M (2021) Gut microbiota develop towards an adult profile in a sex-specific manner during puberty. Sci Rep 11(1):23297. https://doi.org/10.1038/s41598-021-02375-z

Koutzoumis DN, Vergara M, Pino J, Buddendorff J, Khoshbouei H, Mandel RJ, Torres GE (2020) Alterations of the gut microbiota with antibiotics protects dopamine neuron loss and improve motor deficits in a pharmacological rodent model of Parkinson's disease. Exp Neurol 325:113159. https://doi.org/10.1016/j.expneurol.2019.113159

Kraeuter AK, van den Buuse M, Sarnyai Z (2019) Ketogenic diet prevents impaired prepulse inhibition of startle in an acute NMDA receptor hypofunction model of schizophrenia. Schizophr Res 206:244–250. https://doi.org/10.1016/j.schres.2018.11.011

Kragsnaes MS, Kjeldsen J, Horn HC, Munk HL, Pedersen FM, Holt HM, Pedersen JK, Holm DK, Glerup H, Andersen V, Fredberg U, Kristiansen K, Christensen R, Ellingsen T (2018) Efficacy and safety of faecal microbiota transplantation in patients with psoriatic arthritis: protocol for a 6-month, double-blind, randomised, placebo-controlled trial. BMJ Open 8(4):e019231. https://doi.org/10.1136/bmjopen-2017-019231

Kullberg RFJ, Wikki I, Haak BW, Kauko A, Galenkamp H, Peters-Sengers H, Butler JM, Havulinna AS, Palmu J, McDonald D, Benchraka C, Abdel-Aziz MI, Prins M, Maitland van der Zee AH, van den Born BJ, Jousilahti P, de Vos WM, Salomaa V, Knight R et al (2024) Association between butyrate-producing gut bacteria and the risk of infectious disease hospitalisation: results from two observational, population-based microbiome studies. Lancet Microbe 5(9):100864. https://doi.org/10.1016/S2666-5247(24)00079-X

Lachmansingh DA, Lavelle A, Cryan JF, Clarke G (2023) Microbiota-Gut-Brain axis and antidepressant treatment. Curr Top Behav Neurosci. https://doi.org/10.1007/7854_2023_449

Lagomarsino VN, Kostic AD, Chiu IM (2021) Mechanisms of microbial-neuronal interactions in pain and nociception. Neurobiol Pain 9:100056. https://doi.org/10.1016/j.ynpai.2020.100056

Lai KP, Chung YT, Li R, Wan HT, Wong CK (2016) Bisphenol A alters gut microbiome: comparative metagenomics analysis. Environ Pollut 218:923–930. https://doi.org/10.1016/j.envpol.2016.08.039

Lane MM, Gamage E, Du S, Ashtree DN, McGuinness AJ, Gauci S, Baker P, Lawrence M, Rebholz CM, Srour B, Touvier M, Jacka FN, O'Neil A, Segasby T, Marx W (2024) Ultra-processed food exposure and adverse health outcomes: umbrella review of epidemiological meta-analyses. BMJ 384:e077310. https://doi.org/10.1136/bmj-2023-077310

Larraufie P, Martin-Gallausiaux C, Lapaque N, Dore J, Gribble FM, Reimann F, Blottiere HM (2018) SCFAs strongly stimulate PYY production in human enteroendocrine cells. Sci Rep 8(1):74. https://doi.org/10.1038/s41598-017-18259-0

Lathe R, Schultek NM, Balin BJ, Ehrlich GD, Auber LA, Perry G, Breitschwerdt EB, Corry DB, Doty RL, Rissman RA, Nara PL, Itzhaki R, Eimer WA, Tanzi RE, Intracell Research Group Consortium, C (2023) Establishment of a consensus protocol to explore the brain pathobiome in patients with mild cognitive impairment and Alzheimer's disease: research outline and call for collaboration. Alzheimers Dement 19(11):5209–5231. https://doi.org/10.1002/alz.13076

Lavelle A, Sokol H (2020) Gut microbiota-derived metabolites as key actors in inflammatory bowel disease. Nat Rev Gastroenterol Hepatol 17(4):223–237. https://doi.org/10.1038/s41575-019-0258-z

Lay C, Chu CW, Purbojati RW, Acerbi E, Drautz-Moses DI, de Sessions PF, Jie S, Ho E, Kok YJ, Bi X, Chen S, Mak SY, Chua MC, Goh AEN, Chiang WC, Rao R, Chaithongwongwatthana S, Khemapech N, Chongsrisawat V et al (2021) A synbiotic intervention modulates meta-omics signatures of gut redox potential and acidity in elective caesarean born infants. BMC Microbiol 21(1):191. https://doi.org/10.1186/s12866-021-02230-1

Leal Rodriguez C, Shah SA, Rasmussen MA, Thorsen J, Boulund U, Pedersen CT, Castro-Mejia JL, Poulsen CE, Poulsen CS, Deng L, Larsen FAN, Widdowson M, Zhang Y, Sorensen SJ, Moineau S, Petit MA, Chawes B, Bonnelykke K, Nielsen DS, Stokholm J (2024) The infant gut virome is associated with preschool asthma risk independently of bacteria. Nat Med 30(1):138–148. https://doi.org/10.1038/s41591-023-02685-x

LeDoux J (2003) The emotional brain, fear, and the amygdala [journal article]. Cell Mol Neurobiol 23(4–5):727–738. https://doi.org/10.1023/a:1025048802629

Lee JH, Wood TK, Lee J (2015) Roles of indole as an interspecies and interkingdom signaling molecule. Trends Microbiol 23(11):707–718. https://doi.org/10.1016/j.tim.2015.08.001

Lee JJ, Piras E, Tamburini S, Bu K, Wallach DS, Remsen B, Cantor A, Kong J, Goetz D, Hoffman KW, Bonner M, Joe P, Mueller BR, Robinson-Papp J, Lotan E, Gonen O, Malaspina D, Clemente JC (2023) Gut and oral microbiome modulate molecular and clinical markers of schizophrenia-related symptoms: a transdiagnostic, multilevel pilot study. Psychiatry Res 326:115279. https://doi.org/10.1016/j.psychres.2023.115279

Leigh SJ, Uhlig F, Wilmes L, Sanchez-Diaz P, Gheorghe CE, Goodson MS, Kelley-Loughnane N, Hyland NP, Cryan JF, Clarke G (2023) The impact of acute and chronic stress on gastrointestinal physiology and function: a microbiota-gut-brain axis perspective. J Physiol 601(20):4491–4538. https://doi.org/10.1113/JP281951

Lessard-Lord J, Roussel C, Lupien-Meilleur J, Genereux P, Richard V, Guay V, Roy D, Desjardins Y (2024) Short term supplementation with cranberry extract modulates gut microbiota in human and displays a bifidogenic effect. NPJ Biofilms Microbiomes 10(1):18. https://doi.org/10.1038/s41522-024-00493-w

Ley RE, Backhed F, Turnbaugh P, Lozupone CA, Knight RD, Gordon JI (2005) Obesity alters gut microbial ecology. Proc Natl Acad Sci USA 102(31):11070–11075. https://doi.org/10.1073/pnas.0504978102

Leyrolle Q, Decoeur F, Dejean C, Briere G, Leon S, Bakoyiannis I, Baroux E, Sterley TL, Bosch-Bouju C, Morel L, Amadieu C, Lecours C, St-Pierre MK, Bordeleau M, De Smedt-Peyrusse V, Sere A, Schwendimann L, Gregoire S, Bretillon L et al (2022) N-3 PUFA deficiency disrupts oligodendrocyte maturation and myelin integrity during brain development. Glia 70(1):50–70. https://doi.org/10.1002/glia.24088

Li D, Wu M (2021) Pattern recognition receptors in health and diseases. Signal Transduct Target Ther 6(1):291. https://doi.org/10.1038/s41392-021-00687-0

Li J, Jia H, Cai X, Zhong H, Feng Q, Sunagawa S, Arumugam M, Kultima JR, Prifti E, Nielsen T, Juncker AS, Manichanh C, Chen B, Zhang W, Levenez F, Wang J, Xu X, Xiao L, Liang S et al (2014) An integrated catalog of reference genes in the human gut microbiome. Nat Biotechnol 32(8):834–841. https://doi.org/10.1038/nbt.2942

Li H, Sun J, Du J, Wang F, Fang R, Yu C, Xiong J, Chen W, Lu Z, Liu J (2018) Clostridium butyricum exerts a neuroprotective effect in a mouse model of traumatic brain injury via the gut-brain axis. Neurogastroenterol Motil 30(5):e13260. https://doi.org/10.1111/nmo.13260

Lim MY, Nam YD (2023) Gut microbiome in healthy aging versus those associated with frailty. Gut Microbes 15(2):2278225. https://doi.org/10.1080/19490976.2023.2278225

Lin CH, Chen CC, Chiang HL, Liou JM, Chang CM, Lu TP, Chuang EY, Tai YC, Cheng C, Lin HY, Wu MS (2019) Altered gut microbiota and inflammatory cytokine responses in patients with Parkinson's disease. J Neuroinflammation 16(1):129. https://doi.org/10.1186/s12974-019-1528-y

Lin L, Zhang K, Xiong Q, Zhang J, Cai B, Huang Z, Yang B, Wei B, Chen J, Niu Q (2023) Gut microbiota in pre-clinical rheumatoid arthritis: from pathogenesis to preventing progression. J Autoimmun 141:103001. https://doi.org/10.1016/j.jaut.2023.103001

Link VM, Subramanian P, Cheung F, Han KL, Stacy A, Chi L, Sellers BA, Koroleva G, Courville AB, Mistry S, Burns A, Apps R, Hall KD, Belkaid Y (2024) Differential peripheral immune signatures elicited by vegan versus ketogenic diets in humans. Nat Med 30(2):560–572. https://doi.org/10.1038/s41591-023-02761-2

Liou AP, Paziuk M, Luevano JM Jr, Machineni S, Turnbaugh PJ, Kaplan LM (2013) Conserved shifts in the gut microbiota due to gastric bypass reduce host weight and adiposity. Sci Transl Med 5(178):178ra141. https://doi.org/10.1126/scitranslmed.3005687

Lippert RN, Bruning JC (2022) Maternal metabolic programming of the developing central nervous system: unified pathways to metabolic and psychiatric disorders. Biol Psychiatry 91(10):898–906. https://doi.org/10.1016/j.biopsych.2021.06.002

Liu B, Fang F, Pedersen NL, Tillander A, Ludvigsson JF, Ekbom A, Svenningsson P, Chen H, Wirdefeldt K (2017a) Vagotomy and Parkinson disease: a Swedish register-based matched-cohort study. Neurology 88(21):1996–2002. https://doi.org/10.1212/WNL.0000000000003961

Liu F, Li P, Chen M, Luo Y, Prabhakar M, Zheng H, He Y, Qi Q, Long H, Zhang Y, Sheng H, Zhou H (2017b) Fructooligosaccharide (FOS) and Galactooligosaccharide (GOS) increase bifidobacterium but reduce butyrate producing bacteria with adverse glycemic metabolism in healthy young population. Sci Rep 7(1):11789. https://doi.org/10.1038/s41598-017-10722-2

Liu H, Wang J, He T, Becker S, Zhang G, Li D, Ma X (2018a) Butyrate: a double-edged sword for health? Adv Nutr 9(1):21–29. https://doi.org/10.1093/advances/nmx009

Liu S, Marcelin G, Blouet C, Jeong JH, Jo YH, Schwartz GJ, Chua S Jr (2018b) A gut-brain axis regulating glucose metabolism mediated by bile acids and competitive fibroblast growth factor actions at the hypothalamus. Mol Metab 8:37–50. https://doi.org/10.1016/j.molmet.2017.12.003

Liu L, Ji G, Li G, Hu Y, Jin Q, Hu C, Zhao J, Meng Q, von Deneen KM, Chen A, Cui G, Wang H, Zhao Q, Wu K, Tian J, Shokri-Kojori E, Tomasi D, Volkow ND, Nie Y et al (2019) Structural changes in brain regions involved in executive-control and self-referential processing after sleeve gastrectomy in obese patients. Brain Imaging Behav 13(3):830–840. https://doi.org/10.1007/s11682-018-9904-2

Liu XF, Shao JH, Liao YT, Wang LN, Jia Y, Dong PJ, Liu ZZ, He DD, Li C, Zhang X (2023) Regulation of short-chain fatty acids in the immune system. Front Immunol 14:1186892. https://doi.org/10.3389/fimmu.2023.1186892

Llor C, Bjerrum L (2014) Antimicrobial resistance: risk associated with antibiotic overuse and initiatives to reduce the problem. Ther Adv Drug Saf 5(6):229–241. https://doi.org/10.1177/2042098614554919

Loh JS, Mak WQ, Tan LKS, Ng CX, Chan HH, Yeow SH, Foo JB, Ong YS, How CW, Khaw KY (2024) Microbiota-gut-brain axis and its therapeutic applications in neurodegenerative diseases. Signal Transduct

Target Ther 9(1):37. https://doi.org/10.1038/s41392-024-01743-1
Long SL, Gahan CGM, Joyce SA (2017) Interactions between gut bacteria and bile in health and disease. Mol Asp Med 56:54–65. https://doi.org/10.1016/j.mam.2017.06.002
Lonstein JS, Meinhardt TA, Pavlidi P, Kokras N, Dalla C, Charlier TD, Pawluski JL (2024) Maternal probiotic Lactocaseibacillus rhamnosus HN001 treatment alters postpartum anxiety, cortical monoamines, and the gut microbiome. Psychoneuroendocrinology 165:107033. https://doi.org/10.1016/j.psyneuen.2024.107033
Lopera-Maya EA, Kurilshikov A, van der Graaf A, Hu S, Andreu-Sanchez S, Chen L, Vila AV, Gacesa R, Sinha T, Collij V, Klaassen MAY, Bolte LA, Gois MFB, Neerincx PBT, Swertz MA, LifeLines Cohort S, Harmsen HJM, Wijmenga C, Fu J et al (2022) Effect of host genetics on the gut microbiome in 7,738 participants of the Dutch Microbiome Project. Nat Genet 54(2):143–151. https://doi.org/10.1038/s41588-021-00992-y
Lowe J, Briggs A, Whittle S, Hoon E, Stephenson M (2018) Effectiveness of probiotics in the management of inflammatory arthritis: a systematic review protocol. JBI Database System Rev Implement Rep 16(12):2295–2303. https://doi.org/10.11124/JBISRIR-2017-003692
Lu K, Abo RP, Schlieper KA, Graffam ME, Levine S, Wishnok JS, Swenberg JA, Tannenbaum SR, Fox JG (2014) Arsenic exposure perturbs the gut microbiome and its metabolic profile in mice: an integrated metagenomics and metabolomics analysis. Environ Health Perspect 122(3):284–291. https://doi.org/10.1289/ehp.1307429
Lynch CMK, Cowan CSM, Bastiaanssen TFS, Moloney GM, Theune N, van de Wouw M, Florensa Zanuy E, Ventura-Silva AP, Codagnone MG, Villalobos-Manriquez F, Segalla M, Koc F, Stanton C, Ross P, Dinan TG, Clarke G, Cryan JF (2023) Critical windows of early-life microbiota disruption on behaviour, neuroimmune function, and neurodevelopment. Brain Behav Immun 108:309–327. https://doi.org/10.1016/j.bbi.2022.12.008
Lynch CMK, Nagpal J, Luczynski P, McVey Neufeld K-A, Dinan TG, Clarke G, Cryan JF (2024) Germ-free animals. In: Hyland N, Stanton C (eds) The gut-brain axis. Academic Press, pp 401–454. https://doi.org/10.1016/b978-0-323-99971-7.00012-6
Lyte M, Daniels KM, Schmitz-Esser S (2019) Fluoxetine-induced alteration of murine gut microbial community structure: evidence for a microbial endocrinology-based mechanism of action responsible for fluoxetine-induced side effects. PeerJ 7:e6199. https://doi.org/10.7717/peerj.6199
Lyte JM, Gheorghe CE, Goodson MS, Kelley-Loughnane N, Dinan TG, Cryan JF, Clarke G (2020) Gut-brain axis serotonergic responses to acute stress exposure are microbiome-dependent. Neurogastroenterol Motil 32(11):e13881. https://doi.org/10.1111/nmo.13881
Ma J, Piao X, Mahfuz S, Long S, Wang J (2022) The interaction among gut microbes, the intestinal barrier and short chain fatty acids. Anim Nutr 9:159–174. https://doi.org/10.1016/j.aninu.2021.09.012
Macpherson AJ, Pachnis V, Prinz M (2023) Boundaries and integration between microbiota, the nervous system, and immunity. Immunity 56(8):1712–1726. https://doi.org/10.1016/j.immuni.2023.07.011
Madra M, Ringel R, Margolis KG (2020) Gastrointestinal issues and autism spectrum disorder. Child Adolesc Psychiatr Clin N Am 29(3):501–513. https://doi.org/10.1016/j.chc.2020.02.005
Mandelbaum N, Zhang L, Carasso S, Ziv T, Lifshiz-Simon S, Davidovich I, Luz I, Berinstein E, Gefen T, Cooks T, Talmon Y, Balskus EP, Geva-Zatorsky N (2023) Extracellular vesicles of the Gram-positive gut symbiont Bifidobacterium longum induce immune-modulatory, anti-inflammatory effects. npj Biofilms Microbiomes 9(1):30. https://doi.org/10.1038/s41522-023-00400-9
Manor O, Dai CL, Kornilov SA, Smith B, Price ND, Lovejoy JC, Gibbons SM, Magis AT (2020) Health and disease markers correlate with gut microbiome composition across thousands of people. Nat Commun 11(1):5206. https://doi.org/10.1038/s41467-020-18871-1
Manyi-Loh C, Mamphweli S, Meyer E, Okoh A (2018) Antibiotic use in agriculture and its consequential resistance in environmental sources: potential public health implications. Molecules 23(4). https://doi.org/10.3390/molecules23040795
Marco ML, Sanders ME, Ganzle M, Arrieta MC, Cotter PD, De Vuyst L, Hill C, Holzapfel W, Lebeer S, Merenstein D, Reid G, Wolfe BE, Hutkins R (2021) The International Scientific Association for Probiotics and Prebiotics (ISAPP) consensus statement on fermented foods. Nat Rev Gastroenterol Hepatol 18(3):196–208. https://doi.org/10.1038/s41575-020-00390-5
Martinez KB, Leone V, Chang EB (2017) Western diets, gut dysbiosis, and metabolic diseases: are they linked? Gut Microbes 8(2):130–142. https://doi.org/10.1080/19490976.2016.1270811
Marungruang N, Kovalenko T, Osadchenko I, Voss U, Huang F, Burleigh S, Ushakova G, Skibo G, Nyman M, Prykhodko O, Hallenius FF (2020) Lingonberries and their two separated fractions differently alter the gut microbiota, improve metabolic functions, reduce gut inflammatory properties, and improve brain function in ApoE−/− mice fed high-fat diet. Nutr Neurosci 23(8):600–612. https://doi.org/10.1080/1028415X.2018.1536423
Marx W, Lane MM, Hockey M, Aslam H, Walder K, Borsini A, Firth J, Pariante CM, Berding K, Cryan JF, Clarke G, Craig JM, Su KP, Mischoulon D, Gomez-Pinilla F, Foster JA, Cani PD, Thuret S, Staudacher HM et al (2022) Diet and depression: future needs to unlock the potential. Mol Psychiatry 27(2):778–780. https://doi.org/10.1038/s41380-021-01360-2

Marx W, Penninx B, Solmi M, Furukawa TA, Firth J, Carvalho AF, Berk M (2023) Major depressive disorder. Nat Rev Dis Primers 9(1):44. https://doi.org/10.1038/s41572-023-00454-1

Matsuzaki R, Gunnigle E, Geissen V, Clarke G, Nagpal J, Cryan JF (2023) Pesticide exposure and the microbiota-gut-brain axis. ISME J 17(8):1153–1166. https://doi.org/10.1038/s41396-023-01450-9

Mayer EA (2000) The neurobiology of stress and gastrointestinal disease. Gut 47(6):861–869. https://doi.org/10.1136/gut.47.6.861

Mayer EA, Ryu HJ, Bhatt RR (2023) The neurobiology of irritable bowel syndrome. Mol Psychiatry 28(4):1451–1465. https://doi.org/10.1038/s41380-023-01972-w

Mazzoli R, Pessione E (2016) The neuro-endocrinological role of microbial glutamate and GABA signaling. Front Microbiol 7:1934. https://doi.org/10.3389/fmicb.2016.01934

Mazzone L, Dooling SW, Volpe E, Uljarevic M, Waters JL, Sabatini A, Arturi L, Abate R, Riccioni A, Siracusano M, Pereira M, Engstrand L, Cristofori F, Adduce D, Francavilla R, Costa-Mattioli M, Hardan AY (2024) Precision microbial intervention improves social behavior but not autism severity: a pilot double-blind randomized placebo-controlled trial. Cell Host Microbe 32(1):106–116 e106. https://doi.org/10.1016/j.chom.2023.11.021

McColgan P, Tabrizi SJ (2018) Huntington's disease: a clinical review. Eur J Neurol 25(1):24–34. https://doi.org/10.1111/ene.13413

McDonald D, Hyde E, Debelius JW, Morton JT, Gonzalez A, Ackermann G, Aksenov AA, Behsaz B, Brennan C, Chen Y, DeRight Goldasich L, Dorrestein PC, Dunn RR, Fahimipour AK, Gaffney J, Gilbert JA, Gogul G, Green JL, Hugenholtz P et al (2018) American gut: an open platform for citizen science microbiome research. mSystems 3(3). https://doi.org/10.1128/mSystems.00031-18

McEvoy CT, Jennings A, Steves CJ, Macgregor A, Spector T, Cassidy A (2024) Diet patterns and cognitive performance in a UK Female Twin Registry (TwinsUK). Alzheimers Res Ther 16(1):17. https://doi.org/10.1186/s13195-024-01387-x

McGuinness AJ, Loughman A, Foster JA, Jacka F (2024) Mood disorders: the gut bacteriome and beyond. Biol Psychiatry 95(4):319–328. https://doi.org/10.1016/j.biopsych.2023.08.020

McKernan DP, Fitzgerald P, Dinan TG, Cryan JF (2010) The probiotic Bifidobacterium infantis 35624 displays visceral antinociceptive effects in the rat. Neurogastroenterol Motil 22(9):1029–1035, e1268. https://doi.org/10.1111/j.1365-2982.2010.01520.x

McNamara MP, Singleton JM, Cadney MD, Ruegger PM, Borneman J, Garland T (2021) Early-life effects of juvenile Western diet and exercise on adult gut microbiome composition in mice. J Exp Biol 224(Pt 4). https://doi.org/10.1242/jeb.239699

Medawar E, Beyer F, Thieleking R, Haange SB, Rolle-Kampczyk U, Reinicke M, Chakaroun R, von Bergen M, Stumvoll M, Villringer A, Witte AV (2024) Prebiotic diet changes neural correlates of food decision-making in overweight adults: a randomised controlled within-subject cross-over trial. Gut 73(2):298–310. https://doi.org/10.1136/gutjnl-2023-330365

Medzhitov R (2007) Recognition of microorganisms and activation of the immune response. Nature 449(7164):819–826. https://doi.org/10.1038/nature06246

Michaudel C, Danne C, Agus A, Magniez A, Aucouturier A, Spatz M, Lefevre A, Kirchgesner J, Rolhion N, Wang Y, Lavelle A, Galbert C, Da Costa G, Poirier M, Lapiere A, Planchais J, Nadvornik P, Illes P, Oeuvray C et al (2023) Rewiring the altered tryptophan metabolism as a novel therapeutic strategy in inflammatory bowel diseases. Gut 72(7):1296–1307. https://doi.org/10.1136/gutjnl-2022-327337

Millan MJ (2002) Descending control of pain. Prog Neurobiol 66(6):355–474. https://doi.org/10.1016/s0301-0082(02)00009-6

Miller JE, Goldacre R, Moore HC, Zeltzer J, Knight M, Morris C, Nowell S, Wood R, Carter KW, Fathima P, de Klerk N, Strunk T, Li J, Nassar N, Pedersen LH, Burgner DP (2020) Mode of birth and risk of infection-related hospitalisation in childhood: a population cohort study of 7.17 million births from 4 high-income countries. PLoS Med 17(11):e1003429. https://doi.org/10.1371/journal.pmed.1003429

Minnebo Y, De Paepe K, Raes J, Van de Wiele T (2023) Eating patterns contribute to shaping the gut microbiota in the mucosal simulator of the human intestinal microbial ecosystem. FEMS Microbiol Ecol 99(12). https://doi.org/10.1093/femsec/fiad149

Minter MR, Hinterleitner R, Meisel M, Zhang C, Leone V, Zhang X, Oyler-Castrillo P, Zhang X, Musch MW, Shen X, Jabri B, Chang EB, Tanzi RE, Sisodia SS (2017) Antibiotic-induced perturbations in microbial diversity during post-natal development alters amyloid pathology in an aged APP(SWE)/PS1(DeltaE9) murine model of Alzheimer's disease. Sci Rep 7(1):10411. https://doi.org/10.1038/s41598-017-11047-w

Mir H-D, Milman A, Monnoye M, Douard V, Philippe C, Aubert A, Castanon N, Vancassel S, Guérineau NC, Naudon L, Rabot S (2020) The gut microbiota metabolite indole increases emotional responses and adrenal medulla activity in chronically stressed male mice. Psychoneuroendocrinology 119:104750. https://doi.org/10.1016/j.psyneuen.2020.104750

Mir H-D, Yang Q, Maximin E, Montardy Q, Ji S, Cheng Q, Shan X, Wang L, Naudon L, Rabot S, Li L (2024) Indole induces anxiety-like behaviour in mice mediated by brainstem locus coeruleus activation. Neurobiol Dis 200:106606. https://doi.org/10.1016/j.nbd.2024.106606

Mischoulon D, Dunlop BW, Kinkead B, Schettler PJ, Lamon-Fava S, Rakofsky JJ, Nierenberg AA, Clain AJ, Mletzko Crowe T, Wong A, Felger JC, Sangermano L, Ziegler TR, Cusin C, Fisher LB, Fava M, Rapaport MH (2022) Omega-3 fatty acids for major depressive disorder with high inflammation: a randomized dose-finding clinical trial. J Clin Psychiatry 83(5). https://doi.org/10.4088/JCP.21m14074

Misiak B, Pawlak E, Rembacz K, Kotas M, Zebrowska-Rozanska P, Kujawa D, Laczmanski L, Piotrowski P, Bielawski T, Samochowiec J, Samochowiec A, Karpinski P (2024) Associations of gut microbiota alterations with clinical, metabolic, and immune-inflammatory characteristics of chronic schizophrenia. J Psychiatr Res 171:152–160. https://doi.org/10.1016/j.jpsychires.2024.01.036

Moeller AH, Foerster S, Wilson ML, Pusey AE, Hahn BH, Ochman H (2016) Social behavior shapes the chimpanzee pan-microbiome. Sci Adv 2(1):e1500997. https://doi.org/10.1126/sciadv.1500997

Mohle L, Mattei D, Heimesaat MM, Bereswill S, Fischer A, Alutis M, French T, Hambardzumyan D, Matzinger P, Dunay IR, Wolf SA (2016) Ly6C(hi) monocytes provide a link between antibiotic-induced changes in gut microbiota and adult hippocampal neurogenesis. Cell Rep 15(9):1945–1956. https://doi.org/10.1016/j.celrep.2016.04.074

Moloney GM, Viola MF, Hoban AE, Dinan TG, Cryan JF (2018) Faecal microRNAs: indicators of imbalance at the host-microbe interface? Benef Microbes 9(2):175–183. https://doi.org/10.3920/BM2017.0013

Moloney GM, Dinan TG, Clarke G, Cryan JF (2019) Microbial regulation of microRNA expression in the brain-gut axis. Curr Opin Pharmacol 48:120–126. https://doi.org/10.1016/j.coph.2019.08.005

Morais LH, Golubeva AV, Moloney GM, Moya-Perez A, Ventura-Silva AP, Arboleya S, Bastiaanssen TFS, O'Sullivan O, Rea K, Borre Y, Scott KA, Patterson E, Cherry P, Stilling R, Hoban AE, El Aidy S, Sequeira AM, Beers S, Moloney RD et al (2020) Enduring behavioral effects induced by birth by caesarean section in the mouse. Curr Biol 30(19):3761–3774 e3766. https://doi.org/10.1016/j.cub.2020.07.044

Morais S, Winkler S, Zorea A, Levin L, Nagies FSP, Kapust N, Lamed E, Artan-Furman A, Bolam DN, Yadav MP, Bayer EA, Martin WF, Mizrahi I (2024) Cryptic diversity of cellulose-degrading gut bacteria in industrialized humans. Science 383(6688):eadj9223. https://doi.org/10.1126/science.adj9223

Morales-Soto W, Gonzales J, Jackson WF, Gulbransen BD (2023) Enteric glia promote visceral hypersensitivity during inflammation through intercellular signaling with gut nociceptors. Sci Signal 16(812):eadg1668. https://doi.org/10.1126/scisignal.adg1668

Morin C, Faure F, Mollet J, Guenoun D, Heydari-Olya A, Sautet I, Diao S, Faivre V, Pansiot J, Tabet L, Hua J, Schwendimann L, Mokhtari A, Martin-Rosique R, Chadi S, Laforge M, Demene C, Delahaye-Duriez A, Diaz-Heijtz R et al (2025) C-section and systemic inflammation synergize to disrupt the neonatal gut microbiota and brain development in a model of prematurity. Brain Behav Immun 123:824–837. https://doi.org/10.1016/j.bbi.2024.10.023

Morton JT, Jin DM, Mills RH, Shao Y, Rahman G, McDonald D, Zhu Q, Balaban M, Jiang Y, Cantrell K, Gonzalez A, Carmel J, Frankiensztajn LM, Martin-Brevet S, Berding K, Needham BD, Zurita MF, David M, Averina OV et al (2023) Multi-level analysis of the gut-brain axis shows autism spectrum disorder-associated molecular and microbial profiles. Nat Neurosci 26(7):1208–1217. https://doi.org/10.1038/s41593-023-01361-0

Mueller NT, Differding MK, Ostbye T, Hoyo C, Benjamin-Neelon SE (2021a) Association of birth mode of delivery with infant faecal microbiota, potential pathobionts, and short chain fatty acids: a longitudinal study over the first year of life. BJOG 128(8):1293–1303. https://doi.org/10.1111/1471-0528.16633

Mueller NT, Differding MK, Zhang M, Maruthur NM, Juraschek SP, Miller ER 3rd, Appel LJ, Yeh HC (2021b) Metformin affects gut microbiome composition and function and circulating short-chain fatty acids: a randomized trial. Diabetes Care 44(7):1462–1471. https://doi.org/10.2337/dc20-2257

Mullish BH, Michael DR, Dabcheva M, Webberley TS, Coates N, John DA, Wang D, Luo Y, Plummer SF, Marchesi JR (2024) A double-blind, randomized, placebo-controlled study assessing the impact of probiotic supplementation on the symptoms of irritable bowel syndrome in females. Neurogastroenterol Motil 36(4):e14751. https://doi.org/10.1111/nmo.14751

Murray N, Al Khalaf S, Bastiaanssen TFS, Kaulmann D, Lonergan E, Cryan JF, Clarke G, Khashan AS, O'Connor K (2023) Compositional and functional alterations in intestinal microbiota in patients with psychosis or schizophrenia: a systematic review and meta-analysis. Schizophr Bull 49(5):1239–1255. https://doi.org/10.1093/schbul/sbad049

Muth AK, Park SQ (2021) The impact of dietary macronutrient intake on cognitive function and the brain. Clin Nutr 40(6):3999–4010. https://doi.org/10.1016/j.clnu.2021.04.043

Mysonhimer AR, Brown MD, Alvarado DA, Cornman E, Esmail M, Abdiel T, Gutierrez K, Vasquez J, Cannavale CN, Miller MJ, Khan NA, Holscher HD (2024) Honey added to yogurt with Bifidobacterium animalis subsp. lactis DN-173 010/CNCM I-2494 supports probiotic enrichment but does not reduce intestinal transit time in healthy adults: a randomized, controlled, crossover trial. J Nutr 154(8):2396–2410. https://doi.org/10.1016/j.tjnut.2024.05.028

Nagpal J, Cryan JF (2021a) Host genetics, the microbiome & behaviour-a 'Holobiont' perspective. Cell Res 31(8):832–833. https://doi.org/10.1038/s41422-021-00512-x

Nagpal J, Cryan JF (2021b) Microbiota-brain interactions: moving toward mechanisms in model organisms. Neuron 109(24):3930–3953. https://doi.org/10.1016/j.neuron.2021.09.036

Nguyen TT, Kosciolek T, Maldonado Y, Daly RE, Martin AS, McDonald D, Knight R, Jeste DV (2019) Differences in gut microbiome composition between persons with chronic schizophrenia and healthy comparison subjects. Schizophr Res 204:23–29. https://doi.org/10.1016/j.schres.2018.09.014

Nguyen TT, Kosciolek T, Daly RE, Vazquez-Baeza Y, Swafford A, Knight R, Jeste DV (2021) Gut microbiome in schizophrenia: altered functional pathways related to immune modulation and atherosclerotic risk. Brain Behav Immun 91:245–256. https://doi.org/10.1016/j.bbi.2020.10.003

Ni Lochlainn M, Bowyer RCE, Moll JM, Garcia MP, Wadge S, Baleanu AF, Nessa A, Sheedy A, Akdag G, Hart D, Raffaele G, Seed PT, Murphy C, Harridge SDR, Welch AA, Greig C, Whelan K, Steves CJ (2024) Effect of gut microbiome modulation on muscle function and cognition: the PROMOTe randomised controlled trial. Nat Commun 15(1):1859. https://doi.org/10.1038/s41467-024-46116-y

Nicolas S, Leime CSO, Hoban AE, Hueston CM, Cryan JF, Nolan YM (2022a) Enduring effects of an unhealthy diet during adolescence on systemic but not neurobehavioural measures in adult rats. Nutr Neurosci 25(4):657–669. https://doi.org/10.1080/1028415X.2020.1796041

Nicolas S, McGovern AJ, Hueston CM, O'Mahony SM, Cryan JF, O'Leary OF, Nolan YM (2022b) Prior maternal separation stress alters the dendritic complexity of new hippocampal neurons and neuroinflammation in response to an inflammatory stressor in juvenile female rats. Brain Behav Immun 99:327–338. https://doi.org/10.1016/j.bbi.2021.10.016

Nicolas S, Dohm-Hansen S, Lavelle A, Bastiaanssen TFS, English JA, Cryan JF, Nolan YM (2024) Exercise mitigates a gut microbiota-mediated reduction in adult hippocampal neurogenesis and associated behaviours in rats. Transl Psychiatry 14(1):195. https://doi.org/10.1038/s41398-024-02904-0

Nikolova VL, Smith MRB, Hall LJ, Cleare AJ, Stone JM, Young AH (2021) Perturbations in gut microbiota composition in psychiatric disorders: a review and meta-analysis. JAMA Psychiatry 78(12):1343–1354. https://doi.org/10.1001/jamapsychiatry.2021.2573

Noetel M, Sanders T, Gallardo-Gomez D, Taylor P, Del Pozo Cruz B, van den Hoek D, Smith JJ, Mahoney J, Spathis J, Moresi M, Pagano R, Pagano L, Vasconcellos R, Arnott H, Varley B, Parker P, Biddle S, Lonsdale C (2024) Effect of exercise for depression: systematic review and network meta-analysis of randomised controlled trials. BMJ 384:e075847. https://doi.org/10.1136/bmj-2023-075847

Nota MHC, Nicolas S, O'Leary OF, Nolan YM (2023) Outrunning a bad diet: interactions between exercise and a Western-style diet for adolescent mental health, metabolism and microbes. Neurosci Biobehav Rev 149:105147. https://doi.org/10.1016/j.neubiorev.2023.105147

O'Connor R, van De Wouw M, Moloney GM, Ventura-Silva AP, O'Riordan K, Golubeva AV, Dinan TG, Schellekens H, Cryan JF (2021) Strain differences in behaviour and immunity in aged mice: relevance to autism. Behav Brain Res 399:113020. https://doi.org/10.1016/j.bbr.2020.113020

Obata Y, Pachnis V (2016) The effect of microbiota and the immune system on the development and organization of the enteric nervous system. Gastroenterology 151(5):836–844. https://doi.org/10.1053/j.gastro.2016.07.044

O'Riordan KJ, Collins MK, Moloney GM, Knox EG, Aburto MR, Fulling C, Morley SJ, Clarke G, Schellekens H, Cryan JF (2022) Short chain fatty acids: microbial metabolites for gut-brain axis signalling. Mol Cell Endocrinol 546:111572. https://doi.org/10.1016/j.mce.2022.111572

Osman A, Mervosh NL, Strat AN, Euston TJ, Zipursky G, Pollak RM, Meckel KR, Tyler SR, Chan KL, Buxbaum Grice A, Drapeau E, Litichevskiy L, Gill J, Zeldin SM, Thaiss CA, Buxbaum JD, Breen MS, Kiraly DD (2023) Acetate supplementation rescues social deficits and alters transcriptional regulation in prefrontal cortex of Shank3 deficient mice. Brain Behav Immun 114:311–324. https://doi.org/10.1016/j.bbi.2023.08.020

Otaru N, Kourouma L, Pugin B, Constancias F, Braegger C, Mansuy IM, Lacroix C (2024) Transgenerational effects of early life stress on the fecal microbiota in mice. Commun Biol 7(1):670. https://doi.org/10.1038/s42003-024-06279-2

Palmer C, Bik EM, DiGiulio DB, Relman DA, Brown PO (2007) Development of the human infant intestinal microbiota. PLoS Biol 5(7):e177. https://doi.org/10.1371/journal.pbio.0050177

Palomar MM, Maldonado Galdeano C, Perdigon G (2014) Influence of a probiotic lactobacillus strain on the intestinal ecosystem in a stress model mouse. Brain Behav Immun 35:77–85. https://doi.org/10.1016/j.bbi.2013.08.015

Paoli A, Rubini A, Volek JS, Grimaldi KA (2013) Beyond weight loss: a review of the therapeutic uses of very-low-carbohydrate (ketogenic) diets. Eur J Clin Nutr 67(8):789–796. https://doi.org/10.1038/ejcn.2013.116

Park G, Kadyan S, Hochuli N, Pollak J, Wang B, Salazar G, Chakrabarty P, Efron P, Sheffler J, Nagpal R (2024) A modified Mediterranean-style diet enhances brain function via specific gut-microbiome-brain mechanisms. Gut Microbes 16(1):2323752. https://doi.org/10.1080/19490976.2024.2323752

Partty A, Kalliomaki M, Wacklin P, Salminen S, Isolauri E (2015) A possible link between early probiotic intervention and the risk of neuropsychiatric disorders later in childhood: a randomized trial [Article]. Pediatr Res 77(6):823–828. https://doi.org/10.1038/pr.2015.51

de Paula GC, Brunetta HS, Engel DF, Gaspar JM, Velloso LA, Engblom D, de Oliveira J, de Bem AF (2021) Hippocampal function is impaired by a short-term high-fat diet in mice: increased blood-brain barrier permeability and neuroinflammation as triggering events. Front Neurosci 15:734158. https://doi.org/10.3389/fnins.2021.734158

Pavlov I (1897) Lectures on activities of major gastrointestinal glands (Lekcji o rabotie glawnych piszcziewaritielnych zielez). St Petersburg

Pavlov VA, Tracey KJ (2005) The cholinergic anti-inflammatory pathway. Brain Behav Immun 19(6):493–499. https://doi.org/10.1016/j.bbi.2005.03.015

Perez-Burgos A, Wang L, McVey Neufeld KA, Mao YK, Ahmadzai M, Janssen LJ, Stanisz AM, Bienenstock J, Kunze WA (2015) The TRPV1 channel in rodents is a major target for antinociceptive effect of the probiotic Lactobacillus reuteri DSM 17938. J Physiol 593(17):3943–3957. https://doi.org/10.1113/JP270229

Petrican R, Fornito A, Boyland E (2024) Lifestyle factors counteract the neurodevelopmental impact of genetic risk for accelerated brain aging in adolescence. Biol Psychiatry 95(5):453–464. https://doi.org/10.1016/j.biopsych.2023.06.023

Pinto-Sanchez MI, Hall GB, Ghajar K, Nardelli A, Bolino C, Lau JT, Martin FP, Cominetti O, Welsh C, Rieder A, Traynor J, Gregory C, De Palma G, Pigrau M, Ford AC, Macri J, Berger B, Bergonzelli G, Surette MG et al (2017) Probiotic Bifidobacterium longum NCC3001 reduces depression scores and alters brain activity: a pilot study in patients with irritable bowel syndrome. Gastroenterology 153(2):448–459 e448. https://doi.org/10.1053/j.gastro.2017.05.003

Plovier H, Cani PD (2017) Enteroendocrine cells: metabolic relays between microbes and their host. In: Developmental biology of gastrointestinal hormones, vol 32. Karger Publishers, pp 139–164

Poewe W, Seppi K, Tanner CM, Halliday GM, Brundin P, Volkmann J, Schrag AE, Lang AE (2017) Parkinson disease. Nat Rev Dis Primers 3(1):17013. https://doi.org/10.1038/nrdp.2017.13

Potter HG, Kowash HM, Woods RM, Revill G, Grime A, Deeney B, Burgess MA, Aarons T, Glazier JD, Neill JC, Hager R (2023) Maternal behaviours and adult offspring behavioural deficits are predicted by maternal TNFalpha concentration in a rat model of neurodevelopmental disorders. Brain Behav Immun 108:162–175. https://doi.org/10.1016/j.bbi.2022.12.003

Prestinaci F, Pezzotti P, Pantosti A (2015) Antimicrobial resistance: a global multifaceted phenomenon. Pathog Glob Health 109(7):309–318. https://doi.org/10.1179/2047773215Y.0000000030

Prince N, Peralta Marzal LN, Markidi A, Ahmed S, Adolfs Y, Pasterkamp RJ, Kumar H, Roeselers G, Garssen J, Kraneveld AD, Perez-Pardo P (2024) Prebiotic diet normalizes aberrant immune and behavioral phenotypes in a mouse model of autism spectrum disorder. Acta Pharmacol Sin 45(8):1591–1603. https://doi.org/10.1038/s41401-024-01268-x

Pusceddu MM, Gareau MG (2018) Visceral pain: gut microbiota, a new hope? J Biomed Sci 25(1):73. https://doi.org/10.1186/s12929-018-0476-7

Quigley EMM (2018) The gut-brain axis and the microbiome: clues to pathophysiology and opportunities for novel management strategies in Irritable Bowel Syndrome (IBS). J Clin Med 7(1):6. https://doi.org/10.3390/jcm7010006

Radjabzadeh D, Bosch JA, Uitterlinden AG, Zwinderman AH, Ikram MA, van Meurs JBJ, Luik AI, Nieuwdorp M, Lok A, van Duijn CM, Kraaij R, Amin N (2022) Gut microbiome-wide association study of depressive symptoms. Nat Commun 13(1):7128. https://doi.org/10.1038/s41467-022-34502-3

Radulescu CI, Garcia-Miralles M, Sidik H, Bardile CF, Yusof N, Lee HU, Ho EXP, Chu CW, Layton E, Low D, De Sessions PF, Pettersson S, Ginhoux F, Pouladi MA (2019) Manipulation of microbiota reveals altered callosal myelination and white matter plasticity in a model of Huntington disease. Neurobiol Dis 127:65–75. https://doi.org/10.1016/j.nbd.2019.02.011

Ratsika A, Cruz Pereira JS, Lynch CMK, Clarke G, Cryan JF (2023) Microbiota-immune-brain interactions: a lifespan perspective. Curr Opin Neurobiol 78:102652. https://doi.org/10.1016/j.conb.2022.102652

Rea K, O'Mahony SM, Dinan TG, Cryan JF (2017) The role of the gastrointestinal microbiota in visceral pain. Handb Exp Pharmacol 239:269–287. https://doi.org/10.1007/164_2016_115

Rea K, SM OM, Cryan JF (2021) High and Mighty? Cannabinoids and the microbiome in pain. Neurobiol Pain 9:100061. https://doi.org/10.1016/j.ynpai.2021.100061

Rein B, Raymond K, Boustani C, Tuy S, Zhang J, St Laurent R, Pomrenze MB, Boroon P, Heifets B, Smith M, Malenka RC (2024) MDMA enhances empathy-like behaviors in mice via 5-HT release in the nucleus accumbens. Sci Adv 10(17):eadl6554. https://doi.org/10.1126/sciadv.adl6554

Ren L, Liang H, Zhu L, Yang X, Zhang H, Sun N, Huang D, Feng J, Wu Y, Xiong L, Ke X, Li M, Zhang A (2024) Dietary restriction improves perioperative neurocognitive disorders by inhibiting neuroinflammation and gut microbial dysbiosis. Neuroscience 540:48–67. https://doi.org/10.1016/j.neuroscience.2024.01.012

Ribeiro G, Camacho M, Fernandes AB, Cotovio G, Torres S, Oliveira-Maia AJ (2021) Reward-related gustatory and psychometric predictors of weight loss following bariatric surgery: a multicenter cohort study. Am J Clin Nutr 113(3):751–761. https://doi.org/10.1093/ajcn/nqaa349

Rinaman L (2010) Ascending projections from the caudal visceral nucleus of the solitary tract to brain regions involved in food intake and energy expenditure. Brain Res 1350:18–34. https://doi.org/10.1016/j.brainres.2010.03.059

Ritz NL, Bastiaanssen TFS, Cowan CSM, Smith L, Theune N, Brocka M, Myers EM, Moloney RD, Moloney GM, Shkoporov AN, Draper LA, Hill C, Dinan TG, Slattery DA, Cryan JF (2024a) Social fear extinction susceptibility is associated with Microbiota-Gut-Brain axis alterations. Brain Behav Immun 120:315–326. https://doi.org/10.1016/j.bbi.2024.06.009

Ritz NL, Draper LA, Bastiaanssen TFS, Turkington CJR, Peterson VL, van de Wouw M, Vlckova K, Fulling C, Guzzetta KE, Burokas A, Harris H, Dalmasso M, Crispie F, Cotter PD, Shkoporov AN, Moloney GM, Dinan TG, Hill C, Cryan JF (2024b) The gut virome is associated with stress-induced changes in behaviour and immune responses in mice. Nat Microbiol 9(2):359–376. https://doi.org/10.1038/s41564-023-01564-y

Roager HM, Licht TR (2018) Microbial tryptophan catabolites in health and disease. Nat Commun 9(1):3294. https://doi.org/10.1038/s41467-018-05470-4

Roager HM, Stanton C, Hall LJ (2023) Microbial metabolites as modulators of the infant gut microbiome and host-microbial interactions in early life. Gut Microbes 15(1):2192151. https://doi.org/10.1080/19490976.2023.2192151

Rodino-Janeiro BK, Vicario M, Alonso-Cotoner C, Pascua-Garcia R, Santos J (2018) A review of microbiota and irritable bowel syndrome: future in therapies. Adv Ther 35(3):289–310. https://doi.org/10.1007/s12325-018-0673-5

Rodrigues KB, Weng Z, Graham ZA, Lavin K, McAdam J, Tuggle SC, Peoples B, Seay R, Yang S, Bamman MM, Broderick TJ, Montgomery SB (2025) Exercise intensity and training alter the innate immune cell type and chromosomal origins of circulating cell-free DNA in humans. Proc Natl Acad Sci USA 122(3):e2406954122. https://doi.org/10.1073/pnas.2406954122

Rogers MAM, Aronoff DM (2016) The influence of non-steroidal anti-inflammatory drugs on the gut microbiome. Clin Microbiol Infect 22(2):178 e171–178 e179. https://doi.org/10.1016/j.cmi.2015.10.003

Romani-Perez M, Liebana-Garcia R, Flor-Duro A, Bonillo-Jimenez D, Bullich-Vilarrubias C, Olivares M, Sanz Y (2024a) Obesity and the gut microbiota: implications of neuroendocrine and immune signaling. FEBS J. https://doi.org/10.1111/febs.17249

Romani-Perez M, Lopez-Almela I, Bullich-Vilarrubias C, Evtoski Z, Benitez-Paez A, Sanz Y (2024b) Bacteroides uniformis CECT 7771 requires adaptive immunity to improve glucose tolerance but not to prevent body weight gain in diet-induced obese mice. Microbiome 12(1):103. https://doi.org/10.1186/s40168-024-01810-3

Romano S, Savva GM, Bedarf JR, Charles IG, Hildebrand F, Narbad A (2021) Meta-analysis of the Parkinson's disease gut microbiome suggests alterations linked to intestinal inflammation. NPJ Parkinsons Dis 7(1):27. https://doi.org/10.1038/s41531-021-00156-z

Rosin S, Xia K, Azcarate-Peril MA, Carlson AL, Propper CB, Thompson AL, Grewen K, Knickmeyer RC (2021) A preliminary study of gut microbiome variation and HPA axis reactivity in healthy infants. Psychoneuroendocrinology 124:105046. https://doi.org/10.1016/j.psyneuen.2020.105046

Ross FC, Mayer DE, Gupta A, Gill CIR, Del Rio D, Cryan JF, Lavelle A, Ross RP, Stanton C, Mayer EA (2024a) Existing and future strategies to manipulate the gut microbiota with diet as a potential adjuvant treatment for psychiatric disorders. Biol Psychiatry 95(4):348–360. https://doi.org/10.1016/j.biopsych.2023.10.018

Ross FC, Patangia D, Grimaud G, Lavelle A, Dempsey EM, Ross RP, Stanton C (2024b) The interplay between diet and the gut microbiome: implications for health and disease. Nat Rev Microbiol 22(11):671–686. https://doi.org/10.1038/s41579-024-01068-4

Rudzki L, Ostrowska L, Pawlak D, Malus A, Pawlak K, Waszkiewicz N, Szulc A (2019) Probiotic Lactobacillus Plantarum 299v decreases kynurenine concentration and improves cognitive functions in patients with major depression: a double-blind, randomized, placebo controlled study. Psychoneuroendocrinology 100:213–222. https://doi.org/10.1016/j.psyneuen.2018.10.010

Rueda GH, Causada-Calo N, Borojevic R, Nardelli A, Pinto-Sanchez MI, Constante M, Libertucci J, Mohan V, Langella P, Loonen LMP, Wells JM, Collins SM, Sokol H, Verdu EF, Bercik P (2024) Oral tryptophan activates duodenal aryl hydrocarbon receptor in healthy subjects: a crossover randomized controlled trial. Am J Physiol Gastrointest Liver Physiol 326(6):G687–G696. https://doi.org/10.1152/ajpgi.00306.2023

Rusch JA, Layden BT, Dugas LR (2023) Signalling cognition: the gut microbiota and hypothalamic-pituitary-adrenal axis. Front Endocrinol (Lausanne) 14:1130689. https://doi.org/10.3389/fendo.2023.1130689

Salehipour Z, Haghmorad D, Sankian M, Rastin M, Nosratabadi R, Soltan Dallal MM, Tabasi N, Khazaee M, Nasiraii LR, Mahmoudi M (2017) Bifidobacterium animalis in combination with human origin of Lactobacillus plantarum ameliorate neuroinflammation in experimental model of multiple sclerosis by altering CD4+ T cell subset balance. Biomed Pharmacother 95:1535–1548. https://doi.org/10.1016/j.biopha.2017.08.117

Salia S, Burke FF, Hinks ME, Randell AM, Matheson MA, Walling SG, Swift-Gallant A (2025) Gut microbiota transfer from the preclinical maternal immune activation model of autism is sufficient to induce sex-specific alterations in immune response and behavioural outcomes. Brain Behav Immun 123:813–823. https://doi.org/10.1016/j.bbi.2024.10.030

Salminen S, Collado MC, Endo A, Hill C, Lebeer S, Quigley EMM, Sanders ME, Shamir R, Swann JR, Szajewska H, Vinderola G (2021) The International Scientific Association of Probiotics and Prebiotics (ISAPP) consensus statement on the definition and scope of postbiotics. Nat Rev Gastroenterol Hepatol 18(9):649–667. https://doi.org/10.1038/s41575-021-00440-6

Samara J, Moossavi S, Alshaikh B, Ortega VA, Pettersen VK, Ferdous T, Hoops SL, Soraisham A, Vayalumkal J, Dersch-Mills D, Gerber JS, Mukhopadhyay S,

Puopolo K, Tompkins TA, Knights D, Walter J, Amin H, Arrieta MC (2022) Supplementation with a probiotic mixture accelerates gut microbiome maturation and reduces intestinal inflammation in extremely preterm infants. Cell Host Microbe 30(5):696–711 e695. https://doi.org/10.1016/j.chom.2022.04.005

Sampson TR, Debelius JW, Thron T, Janssen S, Shastri GG, Ilhan ZE, Challis C, Schretter CE, Rocha S, Gradinaru V, Chesselet MF, Keshavarzian A, Shannon KM, Krajmalnik-Brown R, Wittung-Stafshede P, Knight R, Mazmanian SK (2016) Gut microbiota regulate motor deficits and neuroinflammation in a model of Parkinson's disease. Cell 167(6):1469–1480 e1412. https://doi.org/10.1016/j.cell.2016.11.018

Sanchez M, Darimont C, Panahi S, Drapeau V, Marette A, Taylor VH, Dore J, Tremblay A (2017) Effects of a diet-based weight-reducing program with probiotic supplementation on satiety efficiency, eating behaviour traits, and psychosocial behaviours in obese individuals. Nutrients 9(3). https://doi.org/10.3390/nu9030284

Sanidad KZ, Rager SL, Carrow HC, Ananthanarayanan A, Callaghan R, Hart LR, Li T, Ravisankar P, Brown JA, Amir M, Jin JC, Savage AR, Luo R, Rowdo FM, Martin ML, Silver RB, Guo CJ, Krumsiek J, Inohara N, Zeng MY (2024) Gut bacteria-derived serotonin promotes immune tolerance in early life. Sci Immunol 9(93):eadj4775. https://doi.org/10.1126/sciimmunol.adj4775

Sanmarco LM, Chao CC, Wang YC, Kenison JE, Li Z, Rone JM, Rejano-Gordillo CM, Polonio CM, Gutierrez-Vazquez C, Piester G, Plasencia A, Li L, Giovannoni F, Lee HG, Faust Akl C, Wheeler MA, Mascanfroni I, Jaronen M, Alsuwailm M et al (2022) Identification of environmental factors that promote intestinal inflammation. Nature 611(7937):801–809. https://doi.org/10.1038/s41586-022-05308-6

Santangelo A, Corsello A, Spolidoro GCI, Trovato CM, Agostoni C, Orsini A, Milani GP, Peroni DG (2023) The influence of ketogenic diet on gut microbiota: potential benefits, risks and indications. Nutrients 15(17). https://doi.org/10.3390/nu15173680

Sarathy J, Detloff SJ, Ao M, Khan N, French S, Sirajuddin H, Nair T, Rao MC (2017) The Yin and Yang of bile acid action on tight junctions in a model colonic epithelium. Physiol Rep 5(10):e13294. https://doi.org/10.14814/phy2.13294

Sarnyai Z, Kraeuter AK, Palmer CM (2019) Ketogenic diet for schizophrenia: clinical implication. Curr Opin Psychiatry 32(5):394–401. https://doi.org/10.1097/YCO.0000000000000535

Savignac HM, Kiely B, Dinan TG, Cryan JF (2014) Bifidobacteria exert strain-specific effects on stress-related behavior and physiology in BALB/c mice. Neurogastroenterol Motil 26(11):1615–1627. https://doi.org/10.1111/nmo.12427

Schellekens H, Torres-Fuentes C, van de Wouw M, Long-Smith CM, Mitchell A, Strain C, Berding K, Bastiaanssen TFS, Rea K, Golubeva AV, Arboleya S, Verpaalen M, Pusceddu MM, Murphy A, Fouhy F, Murphy K, Ross P, Roy BL, Stanton C et al (2021) Bifidobacterium longum counters the effects of obesity: partial successful translation from rodent to human. EBioMedicine 63:103176. https://doi.org/10.1016/j.ebiom.2020.103176

Schellekens H, Ribeiro G, Cuesta-Marti C, Cryan JF (2023) The microbiome-gut-brain axis in nutritional neuroscience. Nutr Neurosci 26(11):1159–1171. https://doi.org/10.1080/1028415X.2022.2128007

Schneider KM, Thaiss CA (2022) When the brain feels the bugs. Immunity 55(6):976–978. https://doi.org/10.1016/j.immuni.2022.05.008

Schneider KM, Blank N, Alvarez Y, Thum K, Lundgren P, Litichevskiy L, Sleeman M, Bahnsen K, Kim J, Kardo S, Patel S, Dohnalova L, Uhr GT, Descamps HC, Kircher S, McSween AM, Ardabili AR, Nemec KM, Jimenez MT et al (2023) The enteric nervous system relays psychological stress to intestinal inflammation. Cell 186(13):2823–2838 e2820. https://doi.org/10.1016/j.cell.2023.05.001

Schneider E, Balasubramanian R, Ferri A, Cotter PD, Clarke G, Cryan JF (2024a) Conference on 'Diet and lifestyle strategies for prevention and management of multimorbidity' Symposium three: diet and the gut-brain-heart connection fibre & fermented foods: differential effects on the microbiota-gut-brain axis. Proc Nutr Soc 25:1–16. https://doi.org/10.1017/S0029665124004907

Schneider E, Balasubramanian R, Ferri A, Cotter PD, Clarke G, Cryan JF (2024b) Fibre & fermented foods: differential effects on the microbiota-gut-brain axis. Proc Nutr Soc:1–16. https://doi.org/10.1017/S0029665124004907

Schumacher SM, Doyle WJ, Hill K, Ochoa-Reparaz J (2024) Gut microbiota in multiple sclerosis and animal models. FEBS J. https://doi.org/10.1111/febs.17161

Selma-Royo M, Dubois L, Manara S, Armanini F, Cabrera-Rubio R, Valles-Colomer M, Gonzalez S, Parra-Llorca A, Escuriet R, Bode L, Martinez-Costa C, Segata N, Collado MC (2024) Birthmode and environment-dependent microbiota transmission dynamics are complemented by breastfeeding during the first year. Cell Host Microbe 32(6):996–1010 e1014. https://doi.org/10.1016/j.chom.2024.05.005

Sen P, Sherwin E, Sandhu K, Bastiaanssen TFS, Moloney GM, Golubeva A, Fitzgerald P, Paula Ventura Da Silva A, Chruscicka-Smaga B, Olavarria-Ramirez L, Druelle C, Campos D, Jayaprakash P, Rea K, Jeffery IB, Savignac H, Chetal S, Mulder I, Schellekens H et al (2022) The live biotherapeutic Blautia stercoris MRx0006 attenuates social deficits, repetitive behaviour, and anxiety-like behaviour in a mouse model relevant to autism. Brain Behav Immun 106:115–126. https://doi.org/10.1016/j.bbi.2022.08.007

Sens JP, Hofford RS, Kiraly DD (2023) Effect of germ-free status on transcriptional profiles in the nucleus accumbens and transcriptomic response to chronic morphine. Mol Cell Neurosci 126:103874. https://doi.org/10.1016/j.mcn.2023.103874

Serebrinsky-Duek K, Barra M, Danino T, Garrido D (2023) Engineered bacteria for short-chain-fatty-acid-repressed expression of biotherapeutic molecules. Microbiol Spectr 11(2):e0004923. https://doi.org/10.1128/spectrum.00049-23

Shao Y, Forster SC, Tsaliki E, Vervier K, Strang A, Simpson N, Kumar N, Stares MD, Rodger A, Brocklehurst P, Field N, Lawley TD (2019) Stunted microbiota and opportunistic pathogen colonization in caesarean-section birth. Nature 574(7776):117–121. https://doi.org/10.1038/s41586-019-1560-1

Sharkey KA, Mawe GM (2023) The enteric nervous system. Physiol Rev 103(2):1487–1564. https://doi.org/10.1152/physrev.00018.2022

Sheng JA, Bales NJ, Myers SA, Bautista AI, Roueinfar M, Hale TM, Handa RJ (2020) The hypothalamic-pituitary-adrenal axis: development, programming actions of hormones, and maternal-fetal interactions. Front Behav Neurosci 14:601939. https://doi.org/10.3389/fnbeh.2020.601939

Siciliano RA, Reale A, Mazzeo MF, Morandi S, Silvetti T, Brasca M (2021) Paraprobiotics: a new perspective for functional foods and nutraceuticals. Nutrients 13(4). https://doi.org/10.3390/nu13041225

Siegler Lathrop T, Perego S, Bastiaanssen TFS, van Hemert S, Chronakis IS, Diaz Heijtz R (2024) Multispecies probiotic intake during pregnancy modulates neurodevelopmental trajectories of offspring: aiming towards precision microbial intervention. Brain Behav Immun 122:547–554. https://doi.org/10.1016/j.bbi.2024.08.050

Silva YP, Bernardi A, Frozza RL (2020) The role of short-chain fatty acids from gut microbiota in gut-brain communication. Front Endocrinol (Lausanne) 11:25. https://doi.org/10.3389/fendo.2020.00025

Simren M, Tack J (2018) New treatments and therapeutic targets for IBS and other functional bowel disorders. Nat Rev Gastroenterol Hepatol 15(10):589–605. https://doi.org/10.1038/s41575-018-0034-5

Simren M, Ohman L, Olsson J, Svensson U, Ohlson K, Posserud I, Strid H (2010) Clinical trial: the effects of a fermented milk containing three probiotic bacteria in patients with irritable bowel syndrome - a randomized, double-blind, controlled study. Aliment Pharmacol Ther 31(2):218–227. https://doi.org/10.1111/j.1365-2036.2009.04183.x

Singh V, Sadler R, Heindl S, Llovera G, Roth S, Benakis C, Liesz A (2018) The gut microbiome primes a cerebroprotective immune response after stroke. J Cereb Blood Flow Metab 38(8):1293–1298. https://doi.org/10.1177/0271678X18780130

Slykerman RF, Davies N, Vlckova K, O'Riordan KJ, Bassett SA, Dekker J, Schellekens H, Hyland NP, Clarke G, Patterson E (2025) Precision psychobiotics for gut–brain axis health: advancing the discovery pipelines to deliver mechanistic pathways and proven health efficacy. Microb Biotechnol 18(1):e70079. https://doi.org/10.1111/1751-7915.70079

Smolensky I, Zajac-Bakri K, Odermatt TS, Bregere C, Cryan JF, Guzman R, Timper K, Inta D (2023) Sex-specific differences in metabolic hormone and adipose tissue dynamics induced by moderate low-carbohydrate and ketogenic diet. Sci Rep 13(1):16465. https://doi.org/10.1038/s41598-023-43587-9

Song X, Zhao Z, Zhao Y, Wang Z, Wang C, Yang G, Li S (2022) Lactobacillus plantarum DP189 prevents cognitive dysfunction in D-galactose/AlCl(3) induced mouse model of Alzheimer's disease via modulating gut microbiota and PI3K/Akt/GSK-3beta signaling pathway. Nutr Neurosci 25(12):2588–2600. https://doi.org/10.1080/1028415X.2021.1991556

Sonnenburg JL, Backhed F (2016) Diet-microbiota interactions as moderators of human metabolism. Nature 535(7610):56–64. https://doi.org/10.1038/nature18846

Sonnenburg JL, Sonnenburg ED (2019) Vulnerability of the industrialized microbiota. Science 366(6464). https://doi.org/10.1126/science.aaw9255

Stachulski AV, Knausenberger TB, Shah SN, Hoyles L, McArthur S (2023) A host-gut microbial amino acid co-metabolite, p-cresol glucuronide, promotes blood-brain barrier integrity in vivo. Tissue Barriers 11(1):2073175. https://doi.org/10.1080/21688370.2022.2073175

Stagaman K, Kmiecik MJ, Wetzel M, Aslibekyan S, Sonmez TF, Fontanillas P, Auton A, Babalola E, Bell RK, Bielenberg J, Bowes J, Bryc K, Chaudhary NS, Coker D, Das S, DelloRusso E, Elson SL, Eriksson N, Filshtein T et al (2024) Oral and gut microbiome profiles in people with early idiopathic Parkinson's disease. Commun Med 4(1):209. https://doi.org/10.1038/s43856-024-00630-8

Stilling RM, Bordenstein SR, Dinan TG, Cryan JF (2014) Friends with social benefits: host-microbe interactions as a driver of brain evolution and development? Front Cell Infect Microbiol 4:147. https://doi.org/10.3389/fcimb.2014.00147

Strandwitz P (2018) Neurotransmitter modulation by the gut microbiota. Brain Res 1693(Pt B):128–133. https://doi.org/10.1016/j.brainres.2018.03.015

Sudo N, Chida Y, Aiba Y, Sonoda J, Oyama N, Yu XN, Kubo C, Koga Y (2004) Postnatal microbial colonization programs the hypothalamic-pituitary-adrenal system for stress response in mice. J Physiol 558(Pt 1):263–275. https://doi.org/10.1113/jphysiol.2004.063388

Sun J, Wang F, Ling Z, Yu X, Chen W, Li H, Jin J, Pang M, Zhang H, Yu J, Liu J (2016) Clostridium butyricum attenuates cerebral ischemia/reperfusion injury in diabetic mice via modulation of gut microbiota. Brain Res 1642:180–188. https://doi.org/10.1016/j.brainres.2016.03.042

Sun J, Zhang Y, Kong Y, Ye T, Yu Q, Kumaran Satyanarayanan S, Su KP, Liu J (2022) Microbiota-derived metabolite indoles induced aryl hydrocarbon receptor activation and inhibited neuroinflammation in APP/PS1 mice. Brain Behav Immun 106:76–88. https://doi.org/10.1016/j.bbi.2022.08.003

Swanson KS, Gibson GR, Hutkins R, Reimer RA, Reid G, Verbeke K, Scott KP, Holscher HD, Azad MB, Delzenne NM, Sanders ME (2020) The International Scientific Association for Probiotics and Prebiotics (ISAPP) consensus statement on the definition and scope of synbiotics. Nat Rev Gastroenterol Hepatol 17(11):687–701. https://doi.org/10.1038/s41575-020-0344-2

Tabouy L, Getselter D, Ziv O, Karpuj M, Tabouy T, Lukic I, Maayouf R, Werbner N, Ben-Amram H, Nuriel-Ohayon M, Koren O, Elliott E (2018) Dysbiosis of microbiome and probiotic treatment in a genetic model of autism spectrum disorders. Brain Behav Immun 73:310–319. https://doi.org/10.1016/j.bbi.2018.05.015

Takeda K, Akira S (2004) Microbial recognition by Toll-like receptors. J Dermatol Sci 34(2):73–82. https://doi.org/10.1016/j.jdermsci.2003.10.002

Takeuchi T, Kubota T, Nakanishi Y, Tsugawa H, Suda W, Kwon AT, Yazaki J, Ikeda K, Nemoto S, Mochizuki Y, Kitami T, Yugi K, Mizuno Y, Yamamichi N, Yamazaki T, Takamoto I, Kubota N, Kadowaki T, Arner E et al (2023) Gut microbial carbohydrate metabolism contributes to insulin resistance. Nature 621(7978):389–395. https://doi.org/10.1038/s41586-023-06466-x

Tames H, Sabater C, Royo F, Margolles A, Falcon JM, Ruas-Madiedo P, Ruiz L (2024) Mouse intestinal microbiome modulation by oral administration of a GABA-producing Bifidobacterium adolescentis strain. Microbiol Spectr 12(1):e0258023. https://doi.org/10.1128/spectrum.02580-23

Tamtaji OR, Kouchaki E, Salami M, Aghadavod E, Akbari E, Tajabadi-Ebrahimi M, Asemi Z (2017) The effects of probiotic supplementation on gene expression related to inflammation, insulin, and lipids in patients with multiple sclerosis: a randomized, double-blind, placebo-controlled trial. J Am Coll Nutr 36(8):660–665. https://doi.org/10.1080/07315724.2017.1347074

Tanelian A, Nankova B, Miari M, Nahvi RJ, Sabban EL (2022) Resilience or susceptibility to traumatic stress: potential influence of the microbiome. Neurobiol Stress 19:100461. https://doi.org/10.1016/j.ynstr.2022.100461

Teichman EM, O'Riordan KJ, Gahan CGM, Dinan TG, Cryan JF (2020) When rhythms meet the blues: circadian interactions with the microbiota-gut-brain axis. Cell Metab 31(3):448–471. https://doi.org/10.1016/j.cmet.2020.02.008

Tette FM, Kwofie SK, Wilson MD (2022) Therapeutic anti-depressant potential of microbial GABA produced by Lactobacillus rhamnosus strains for GABAergic signaling restoration and inhibition of addiction-induced HPA Axis hyperactivity. Curr Issues Mol Biol 44(4):1434–1451. https://doi.org/10.3390/cimb44040096

Theis KR, Dheilly NM, Klassen JL, Brucker RM, Baines JF, Bosch TC, Cryan JF, Gilbert SF, Goodnight CJ, Lloyd EA, Sapp J, Vandenkoornhuyse P, Zilber-Rosenberg I, Rosenberg E, Bordenstein SR (2016) Getting the hologenome concept right: an eco-evolutionary framework for hosts and their microbiomes. mSystems 1(2). https://doi.org/10.1128/mSystems.00028-16

Thompson KN, Oulhote Y, Weihe P, Wilkinson JE, Ma S, Zhong H, Li J, Kristiansen K, Huttenhower C, Grandjean P (2022) Effects of lifetime exposures to environmental contaminants on the adult gut microbiome. Environ Sci Technol 56(23):16985–16995. https://doi.org/10.1021/acs.est.2c03185

Tian Z, Zhuang X, Luo M, Yin W, Xiong L (2020) The propionic acid and butyric acid in serum but not in feces are increased in patients with diarrhea-predominant irritable bowel syndrome. BMC Gastroenterol 20(1):73. https://doi.org/10.1186/s12876-020-01212-3

Tian YE, Di Biase MA, Mosley PE, Lupton MK, Xia Y, Fripp J, Breakspear M, Cropley V, Zalesky A (2023) Evaluation of brain-body health in individuals with common neuropsychiatric disorders. JAMA Psychiatry 80(6):567–576. https://doi.org/10.1001/jamapsychiatry.2023.0791

Ticinesi A, Milani C, Lauretani F, Nouvenne A, Mancabelli L, Lugli GA, Turroni F, Duranti S, Mangifesta M, Viappiani A, Ferrario C, Maggio M, Ventura M, Meschi T (2017) Gut microbiota composition is associated with polypharmacy in elderly hospitalized patients. Sci Rep 7(1):11102. https://doi.org/10.1038/s41598-017-10734-y

Tofani GSS, Leigh SJ, Gheorghe CE, Bastiaanssen TFS, Wilmes L, Sen P, Clarke G, Cryan JF (2025) Gut microbiota regulates stress responsivity via the circadian system. Cell Metab 37(1):138–153 e135. https://doi.org/10.1016/j.cmet.2024.10.003

Torres-Fuentes C, Schellekens H, Dinan TG, Cryan JF (2017) The microbiota-gut-brain axis in obesity. Lancet Gastroenterol Hepatol 2(10):747–756. https://doi.org/10.1016/S2468-1253(17)30147-4

Tremmel M, Gerdtham UG, Nilsson PM, Saha S (2017) Economic burden of obesity: a systematic literature review. Int J Environ Res Public Health 14(4). https://doi.org/10.3390/ijerph14040435

Tschop M, Weyer C, Tataranni PA, Devanarayan V, Ravussin E, Heiman ML (2001) Circulating ghrelin levels are decreased in human obesity. Diabetes 50(4):707–709. https://doi.org/10.2337/diabetes.50.4.707

Tuck CJ, Abu Omar A, De Palma G, Osman S, Jimenez-Vargas NN, Yu Y, Bennet SM, Lopez-Lopez C, Jaramillo-Polanco JO, Baker CC, Bennett AS, Guzman-Rodriguez M, Tsang Q, Alward T, Rolland S, Morissette C, Verdu EF, Bercik P, Vanner SJ et al (2022) Changes in signalling from faecal neuroactive metabolites following dietary modulation of IBS pain. Gut. https://doi.org/10.1136/gutjnl-2022-327260

Turna J, Grosman Kaplan K, Anglin R, Patterson B, Soreni N, Bercik P, Surette MG, Van Ameringen M (2020) The gut microbiome and inflammation in obsessive-compulsive disorder patients compared to age- and sex-matched controls: a pilot study. Acta Psychiatr Scand 142(4):337–347. https://doi.org/10.1111/acps.13175

Turnbaugh PJ, Backhed F, Fulton L, Gordon JI (2008) Diet-induced obesity is linked to marked but reversible alterations in the mouse distal gut microbiome. Cell Host Microbe 3(4):213–223. https://doi.org/10.1016/j.chom.2008.02.015

Turnbaugh PJ, Ridaura VK, Faith JJ, Rey FE, Knight R, Gordon JI (2009) The effect of diet on the human gut microbiome: a metagenomic analysis in humanized gnotobiotic mice. Sci Transl Med 1(6):6ra14. https://doi.org/10.1126/scitranslmed.3000322

Turner EH, Loftis JM, Blackwell AD (2006) Serotonin a la carte: supplementation with the serotonin precursor 5-hydroxytryptophan. Pharmacol Ther 109(3):325–338. https://doi.org/10.1016/j.pharmthera.2005.06.004

Ulker İ, Yildiran H (2019) The effects of bariatric surgery on gut microbiota in patients with obesity: a review of the literature. Biosci Microbiota Food Health 38(1):3–9. https://doi.org/10.12938/bmfh.18-018

Val-Laillet D, Guerin S, Coquery N, Nogret I, Formal M, Rome V, Le Normand L, Meurice P, Randuineau G, Guilloteau P, Malbert CH, Parnet P, Lalles JP, Segain JP (2018) Oral sodium butyrate impacts brain metabolism and hippocampal neurogenesis, with limited effects on gut anatomy and function in pigs. FASEB J 32(4):2160–2171. https://doi.org/10.1096/fj.201700547RR

Valles-Colomer M, Falony G, Darzi Y, Tigchelaar EF, Wang J, Tito RY, Schiweck C, Kurilshikov A, Joossens M, Wijmenga C, Claes S, Van Oudenhove L, Zhernakova A, Vieira-Silva S, Raes J (2019) The neuroactive potential of the human gut microbiota in quality of life and depression. Nat Microbiol 4(4):623–632. https://doi.org/10.1038/s41564-018-0337-x

Vasileva SS, Yang Y, Baker A, Siskind D, Gratten J, Eyles D (2024) Associations of the gut microbiome with treatment resistance in schizophrenia. JAMA Psychiatry 81(3):292–302. https://doi.org/10.1001/jamapsychiatry.2023.5371

Vermeulen W, De Man JG, Pelckmans PA, De Winter BY (2014) Neuroanatomy of lower gastrointestinal pain disorders. World J Gastroenterol 20(4):1005–1020. https://doi.org/10.3748/wjg.v20.i4.1005

Verstockt B, Ferrante M, Vermeire S, Van Assche G (2018) New treatment options for inflammatory bowel diseases. J Gastroenterol 53(5):585–590. https://doi.org/10.1007/s00535-018-1449-z

Viel JF, Rouget F, Warembourg C, Monfort C, Limon G, Cordier S, Chevrier C (2017) Behavioural disorders in 6-year-old children and pyrethroid insecticide exposure: the PELAGIE mother-child cohort. Occup Environ Med 74(4):275–281. https://doi.org/10.1136/oemed-2016-104035

Vogt NM, Kerby RL, Dill-McFarland KA, Harding SJ, Merluzzi AP, Johnson SC, Carlsson CM, Asthana S, Zetterberg H, Blennow K, Bendlin BB, Rey FE (2017) Gut microbiome alterations in Alzheimer's disease. Sci Rep 7(1):13537. https://doi.org/10.1038/s41598-017-13601-y

Voigt RM, Summa KC, Forsyth CB, Green SJ, Engen P, Naqib A, Vitaterna MH, Turek FW, Keshavarzian A (2016) The circadian clock mutation promotes intestinal dysbiosis. Alcohol Clin Exp Res 40(2):335–347. https://doi.org/10.1111/acer.12943

Wallace Fitzsimons SE, Chruscicka B, Druelle C, Stamou P, Nally K, Dinan TG, Cryan JF, Schellekens H (2019) A ghrelin receptor and oxytocin receptor heterocomplex impairs oxytocin mediated signalling. Neuropharmacology 152:90–101. https://doi.org/10.1016/j.neuropharm.2018.12.022

Walsh J, Olavarria-Ramirez L, Lach G, Boehme M, Dinan TG, Cryan JF, Griffin BT, Hyland NP, Clarke G (2020) Impact of host and environmental factors on beta-glucuronidase enzymatic activity: implications for gastrointestinal serotonin. Am J Physiol Gastrointest Liver Physiol 318(4):G816–G826. https://doi.org/10.1152/ajpgi.00026.2020

Walter J, Hornef MW (2021) A philosophical perspective on the prenatal in utero microbiome debate. Microbiome 9(1):5. https://doi.org/10.1186/s40168-020-00979-7

Wang LJ, Yang CY, Chou WJ, Lee MJ, Chou MC, Kuo HC, Yeh YM, Lee SY, Huang LH, Li SC (2020) Gut microbiota and dietary patterns in children with attention-deficit/hyperactivity disorder. Eur Child Adolesc Psychiatry 29(3):287–297. https://doi.org/10.1007/s00787-019-01352-2

Wang B, Zhou Y, Mao Y, Gong L, Li X, Xu S, Wang F, Guo Q, Zhang H, Li W (2021) Dietary supplementation with lactobacillus plantarum ameliorates compromise of growth performance by modulating short-chain fatty acids and intestinal dysbiosis in broilers under clostridium perfringens challenge. Front Nutr 8:706148. https://doi.org/10.3389/fnut.2021.706148

Wang M, Zhang H, Liang J, Huang J, Chen N (2023) Exercise suppresses neuroinflammation for alleviating Alzheimer's disease. J Neuroinflammation 20(1):76. https://doi.org/10.1186/s12974-023-02753-6

Wang G, Fan Y, Zhang G, Cai S, Ma Y, Yang L, Wang Y, Yu H, Qiao S, Zeng X (2024) Microbiota-derived indoles alleviate intestinal inflammation and modulate microbiome by microbial cross-feeding. Microbiome 12(1):59. https://doi.org/10.1186/s40168-024-01750-y

Warda AK, Clooney AG, Ryan F, de Almeida Bettio PH, Di Benedetto G, Ross RP, Hill C (2021) A postbiotic consisting of heat-treated lactobacilli has a bifidogenic effect in pure culture and in human fermented faecal communities. Appl Environ Microbiol 87(8). https://doi.org/10.1128/AEM.02459-20

Warren JL, Hunter GR, Gower BA, Bamman MM, Windham ST, Moellering DR, Fisher G (2020) Exercise effects on mitochondrial function and lipid metabolism during energy balance. Med Sci Sports Exerc 52(4):827–834. https://doi.org/10.1249/MSS.0000000000002190

Waters JL, Ley RE (2019) The human gut bacteria Christensenellaceae are widespread, heritable, and associated with health. BMC Biol 17(1):83. https://doi.org/10.1186/s12915-019-0699-4

Wehrwein EA, Orer HS, Barman SM (2016) Overview of the anatomy, physiology, and pharmacology of the autonomic nervous system. Compr Physiol 6(3):1239–1278. https://doi.org/10.1002/cphy.c150037

Wei S, Jespersen ML, Baunwall SMD, Myers PN, Smith EM, Dahlerup JF, Rasmussen S, Nielsen HB, Licht TR, Bahl MI, Hvas CL (2022) Cross-generational bacterial strain transfer to an infant after fecal microbiota transplantation to a pregnant patient: a case report. Microbiome 10(1):193. https://doi.org/10.1186/s40168-022-01394-w

Wei N, Ju M, Su X, Zhang Y, Huang Y, Rao X, Cui L, Lin Z, Dong Y (2024) Transplantation of gut microbiota derived from patients with schizophrenia induces schizophrenia-like behaviors and dysregulated brain transcript response in mice. Schizophrenia (Heidelb) 10(1):44. https://doi.org/10.1038/s41537-024-00460-6

Whelan K, Martin LD, Staudacher HM, Lomer MCE (2018) The low FODMAP diet in the management of irritable bowel syndrome: an evidence-based review of FODMAP restriction, reintroduction and personalisation in clinical practice. J Hum Nutr Diet 31(2):239–255. https://doi.org/10.1111/jhn.12530

Wiley NC, Cryan JF, Dinan TG, Ross RP, Stanton C (2021) Production of psychoactive metabolites by gut bacteria. Mod Trends Psychiatr 32:74–99. https://doi.org/10.1159/000510419

Wilmes L, Collins JM, O'Riordan KJ, O'Mahony SM, Cryan JF, Clarke G (2021) Of bowels, brain and behavior: a role for the gut microbiota in psychiatric comorbidities in irritable bowel syndrome. Neurogastroenterol Motil 33(3):e14095. https://doi.org/10.1111/nmo.14095

Woolf CJ, Bennett GJ, Doherty M, Dubner R, Kidd B, Koltzenburg M, Lipton R, Loeser JD, Payne R, Torebjork E (1998) Towards a mechanism-based classification of pain? Pain 77(3):227–229. https://doi.org/10.1016/S0304-3959(98)00099-2

van de Wouw M, Walsh CJ, Vigano GMD, Lyte JM, Boehme M, Gual-Grau A, Crispie F, Walsh AM, Clarke G, Dinan TG, Cotter PD, Cryan JF (2021) Kefir ameliorates specific microbiota-gut-brain axis impairments in a mouse model relevant to autism spectrum disorder. Brain Behav Immun 97:119–134. https://doi.org/10.1016/j.bbi.2021.07.004

Wu HJ, Wu E (2012) The role of gut microbiota in immune homeostasis and autoimmunity. Gut Microbes 3(1):4–14. https://doi.org/10.4161/gmic.19320

Wu WL, Adame MD, Liou CW, Barlow JT, Lai TT, Sharon G, Schretter CE, Needham BD, Wang MI, Tang W, Ousey J, Lin YY, Yao TH, Abdel-Haq R, Beadle K, Gradinaru V, Ismagilov RF, Mazmanian SK (2021) Microbiota regulate social behaviour via stress response neurons in the brain. Nature 595(7867):409–414. https://doi.org/10.1038/s41586-021-03669-y

Xu F, Xie Q, Kuang W, Dong Z (2023a) Interactions between antidepressants and intestinal microbiota. Neurotherapeutics 20(2):359–371. https://doi.org/10.1007/s13311-023-01362-8

Xu W, Rustenhoven J, Nelson CA, Dykstra T, Ferreiro A, Papadopoulos Z, Burnham CD, Dantas G, Fremont DH, Kipnis J (2023b) A novel immune modulator IM33 mediates a glia-gut-neuronal axis that controls lifespan. Neuron 111(20):3244–3254 e3248. https://doi.org/10.1016/j.neuron.2023.07.010

Yang D, Almanzar N, Chiu IM (2023) The role of cellular and molecular neuroimmune crosstalk in gut immunity. Cell Mol Immunol 20(11):1259–1269. https://doi.org/10.1038/s41423-023-01054-5

Yatsunenko T, Rey FE, Manary MJ, Trehan I, Dominguez-Bello MG, Contreras M, Magris M, Hidalgo G, Baldassano RN, Anokhin AP, Heath AC, Warner B, Reeder J, Kuczynski J, Caporaso JG, Lozupone CA, Lauber C, Clemente JC, Knights D et al (2012) Human gut microbiome viewed across age and geography. Nature 486(7402):222–227. https://doi.org/10.1038/nature11053

Yin J, Liao SX, He Y, Wang S, Xia GH, Liu FT, Zhu JJ, You C, Chen Q, Zhou L, Pan SY, Zhou HW (2015) Dysbiosis of gut microbiota with reduced Trimethylamine-N-Oxide level in patients with large-artery atherosclerotic stroke or transient ischemic attack. J Am Heart Assoc 4(11). https://doi.org/10.1161/JAHA.115.002699

Yolken R, Adamos M, Katsafanas E, Khushalani S, Origoni A, Savage C, Schweinfurth L, Stallings C, Sweeney K, Dickerson F (2016) Individuals hospitalized with acute mania have increased exposure to antimicrobial medications. Bipolar Disord 18(5):404–409. https://doi.org/10.1111/bdi.12416

Yoon HS, Cho CH, Yun MS, Jang SJ, You HJ, Kim JH, Han D, Cha KH, Moon SH, Lee K, Kim YJ, Lee SJ, Nam TW, Ko G (2021) Akkermansia muciniphila secretes a glucagon-like peptide-1-inducing protein that improves glucose homeostasis and ameliorates metabolic disease in mice. Nat Microbiol 6(5):563–573. https://doi.org/10.1038/s41564-021-00880-5

Yu Y, Daly DM, Adam IJ, Kitsanta P, Hill CJ, Wild J, Shorthouse A, Grundy D, Jiang W (2016) Interplay between mast cells, enterochromaffin cells, and sensory signaling in the aging human bowel. Neurogastroenterol Motil 28(10):1465–1479. https://doi.org/10.1111/nmo.12842

Yu L, Han X, Cen S, Duan H, Feng S, Xue Y, Tian F, Zhao J, Zhang H, Zhai Q, Chen W (2020) Beneficial effect of GABA-rich fermented milk on insomnia involving regulation of gut microbiota. Microbiol Res 233:126409. https://doi.org/10.1016/j.micres.2020.126409

Yue Y, Sun X, Tian S, Yan S, Sun W, Miao J, Huang S, Diao J, Zhou Z, Zhu W (2024) Multi-omics and gut microbiome: unveiling the pathogenic mechanisms of early-life pesticide exposure. Pestic Biochem Physiol 199:105770. https://doi.org/10.1016/j.pestbp.2024.105770

Zachariassen LF, Sorensen DB, Krych L, Hansen AK, Hansen CHF (2021) Effects of delivery mode on behavior in mouse offspring. Physiol Behav 230:113285. https://doi.org/10.1016/j.physbeh.2020.113285

Zelante T, Iannitti RG, Cunha C, De Luca A, Giovannini G, Pieraccini G, Zecchi R, D'Angelo C, Massi-Benedetti C, Fallarino F, Carvalho A, Puccetti P, Romani L (2013) Tryptophan catabolites from microbiota engage aryl hydrocarbon receptor and balance mucosal reactivity via interleukin-22. Immunity 39(2):372–385. https://doi.org/10.1016/j.immuni.2013.08.003

Zengeler KE, Shapiro DA, Bruch KR, Lammert CR, Ennerfelt H, Lukens JR (2023) SSRI treatment modifies the effects of maternal inflammation on in utero physiology and offspring neurobiology. Brain Behav Immun 108:80–97. https://doi.org/10.1016/j.bbi.2022.10.024

Zha X, Liu X, Wei M, Huang H, Cao J, Liu S, Bian X, Zhang Y, Xiao F, Xie Y, Wang W, Zhang C (2025) Microbiota-derived lysophosphatidylcholine alleviates Alzheimer's disease pathology via suppressing ferroptosis. Cell Metab 37(1):169–186 e169. https://doi.org/10.1016/j.cmet.2024.10.006

Zhai L, Huang C, Ning Z, Zhang Y, Zhuang M, Yang W, Wang X, Wang J, Zhang L, Xiao H, Zhao L, Asthana P, Lam YY, Chow CFW, Huang JD, Yuan S, Chan KM, Yuan CS, Lau JY et al (2023) Ruminococcus gnavus plays a pathogenic role in diarrhea-predominant irritable bowel syndrome by increasing serotonin biosynthesis. Cell Host Microbe 31(1):33–44 e35. https://doi.org/10.1016/j.chom.2022.11.006

Zhang X, Gerard P (2022) Diet-gut microbiota interactions on cardiovascular disease. Comput Struct Biotechnol J 20:1528–1540. https://doi.org/10.1016/j.csbj.2022.03.028

Zhang S, Jin Y, Zeng Z, Liu Z, Fu Z (2015) Subchronic exposure of mice to cadmium perturbs their hepatic energy metabolism and gut microbiome. Chem Res Toxicol 28(10):2000–2009. https://doi.org/10.1021/acs.chemrestox.5b00237

Zhang Z, Lin T, Meng Y, Hu M, Shu L, Jiang H, Gao R, Ma J, Wang C, Zhou X (2021b) FOS/GOS attenuates high-fat diet induced bone loss via reversing microbiota dysbiosis, high intestinal permeability and systemic inflammation in mice. Metabolism 119:154767. https://doi.org/10.1016/j.metabol.2021.154767

Zhang Y, Li Y, Barber AF, Noya SB, Williams JA, Li F, Daniel SG, Bittinger K, Fang J, Sehgal A (2023) The microbiome stabilizes circadian rhythms in the gut. Proc Natl Acad Sci USA 120(5):e2217532120. https://doi.org/10.1073/pnas.2217532120

Zhang S, Meng R, Jiang M, Qing H, Ni J (2024a) Emerging roles of microglia in blood-brain barrier integrity in aging and neurodegeneration. Curr Neuropharmacol 22(7):1189–1204. https://doi.org/10.2174/1570159X21666230203103910

Zhang YD, Shi DD, Liao BB, Li Y, Zhang S, Gao J, Lin LJ, Wang Z (2024b) Human microbiota from drug-naive patients with obsessive-compulsive disorder drives behavioral symptoms and neuroinflammation via succinic acid in mice. Mol Psychiatry 29(6):1782–1797. https://doi.org/10.1038/s41380-024-02424-9

Zheng Z, Zong Y, Ma Y, Tian Y, Pang Y, Zhang C, Gao J (2024) Glucagon-like peptide-1 receptor: mechanisms and advances in therapy. Signal Transduct Target Ther 9(1):234. https://doi.org/10.1038/s41392-024-01931-z

Zhernakova DV, Wang D, Liu L, Andreu-Sanchez S, Zhang Y, Ruiz-Moreno AJ, Peng H, Plomp N, Del Castillo-Izquierdo A, Gacesa R, Lopera-Maya EA, Temba GS, Kullaya VI, van Leeuwen SS, Lifelines Cohort S, Xavier RJ, de Mast Q, Joosten LAB, Riksen NP et al (2024) Host genetic regulation of human gut microbial structural variation. Nature 625(7996):813–821. https://doi.org/10.1038/s41586-023-06893-w

Zhou R, Qian S, Cho WCS, Zhou J, Jin C, Zhong Y, Wang J, Zhang X, Xu Z, Tian M, Chan LWC, Zhang H (2022) Microbiota-microglia connections in age-related cognition decline. Aging Cell 21(5):e13599. https://doi.org/10.1111/acel.13599

Zhou L, Qiu W, Wang J, Zhao A, Zhou C, Sun T, Xiong Z, Cao P, Shen W, Chen J, Lai X, Zhao LH, Wu Y, Li M, Qiu F, Yu Y, Xu ZZ, Zhou H, Jia W et al (2023) Effects of vaginal microbiota transfer on the neurodevelopment and microbiome of cesarean-born infants: a blinded randomized controlled trial. Cell Host Microbe 31(7):1232–1247 e1235. https://doi.org/10.1016/j.chom.2023.05.022

Zhu Y, Lin X, Zhao F, Shi X, Li H, Li Y, Zhu W, Xu X, Li C, Zhou G (2015) Meat, dairy and plant proteins alter bacterial composition of rat gut bacteria. Sci Rep 5:15220. https://doi.org/10.1038/srep15220

Zhu W, Romano KA, Li L, Buffa JA, Sangwan N, Prakash P, Tittle AN, Li XS, Fu X, Androjna C, DiDonato AJ, Brinson K, Trapp BD, Fischbach MA, Rey FE, Hajjar AM, DiDonato JA, Hazen SL (2021) Gut microbes impact stroke severity via the trimethylamine N-oxide pathway. Cell Host Microbe 29(7):1199–1208 e1195. https://doi.org/10.1016/j.chom.2021.05.002

Zhu L, Ming Y, Wu M, Zhang X, Wang X, Li H, Lin Z, Gao F, Zhu Y (2024a) Effect of fiber-rich diet and rope skipping on memory, executive function, and gut microbiota in young adults: a randomized controlled trial. Mol Nutr Food Res 68(3):e2300673. https://doi.org/10.1002/mnfr.202300673

Zhu Q, Chen B, Zhang F, Zhang B, Guo Y, Pang M, Huang L, Wang T (2024b) Toxic and essential metals: metabolic interactions with the gut microbiota and health implications. Front Nutr 11:1448388. https://doi.org/10.3389/fnut.2024.1448388

Zhuang ZQ, Shen LL, Li WW, Fu X, Zeng F, Gui L, Lu Y, Cai M, Zhu C, Tan YL, Zheng P, Li HY, Zhu J, Zhou HD, Bu XL, Wang YJ (2018) Gut microbiota is altered in patients with Alzheimer's disease. J Alzheimers Dis 63(4):1337–1346. https://doi.org/10.3233/JAD-180176

Zilber-Rosenberg I, Rosenberg E (2008) Role of microorganisms in the evolution of animals and plants: the hologenome theory of evolution. FEMS Microbiol Rev 32(5):723–735. https://doi.org/10.1111/j.1574-6976.2008.00123.x

Zuffa S, Schimmel P, Gonzalez-Santana A, Belzer C, Knol J, Bolte S, Falck-Ytter T, Forssberg H, Swann J, Diaz Heijtz R (2023) Early-life differences in the gut microbiota composition and functionality of infants at elevated likelihood of developing autism spectrum disorder. Transl Psychiatry 13(1):257. https://doi.org/10.1038/s41398-023-02556-6

An Emerging Role for Gut-Brain Signaling Involving Ghrelin in Chronic Stress

7

Alexis A. Salcido, Neftali F. Reyes, Andrea Y. Macias, Serina A. Batson, Dirk W. Beck, Alexander Friedman, and Ki A. Goosens

Abstract

Our internal and external environments are not stable; these ever-changing contexts produce stress on bodily systems. In response, the body recruits numerous peripheral hormones to bring those systems back within a desired homeostatic range. When our environments change in extreme ways and for prolonged periods of time, a different set of hormonal stress responses are recruited. These chronic stress responses produce adaptive changes but can also drive maladaptation. This chapter begins by reviewing the peripheral hormones that are recruited as part of the acute stress response and describing their adaptive impact on brain and peripheral function. We then examine new research describing the role of ghrelin, a hormone produced by the gut, in chronic stress. We review the role of ghrelin in hunger and consider how energy deficiency, a state shared by both hunger and stress, might explain why ghrelin is elevated by both. We consider how the unique recruitment of ghrelin during chronic stress mediates responses in the brain that can help an organism respond to future stressors, but also how chronic elevation of ghrelin can produce additional adaptations that contribute to stress-sensitive psychiatric disorders. Lastly, we identify important future areas for research on the biology of ghrelin.

A. A. Salcido · N. F. Reyes · A. Y. Macias · S. A. Batson
Department of Biological Sciences, University of Texas at El Paso, El Paso, TX, USA

D. W. Beck
Computational Science Program, University of Texas at El Paso, El Paso, TX, USA

A. Friedman
Department of Biological Sciences, University of Texas at El Paso, El Paso, TX, USA

Computational Science Program, University of Texas at El Paso, El Paso, TX, USA

K. A. Goosens (✉)
Departments of Psychiatry and Pharmacological Sciences, Center for Translational Medicine and Pharmacology, Friedman Brain Institute, Icahn School of Medicine at Mount Sinai, New York, NY, USA
e-mail: ki.goosens@mssm.edu

Keywords

Ghrelin · Stress · HPA axis · Sympathetic nervous system · Parasympathetic nervous system · Adaptation · Valenced behavior · Appetite · Metabolism · Brain-body

Abbreviation

ACTH	Adrenocorticotropic hormone
AUD	Alcohol use disorder
BChE	Butyrylcholinesterase

D. Tropea, E. Giacometti (eds.), *Brain-Body Connections*, Advances in Experimental Medicine and Biology 1477, https://doi.org/10.1007/978-3-031-89525-8_7

BLA	Basolateral amygdala
CRF	Corticotropin-releasing factor
DRD1	Dopamine receptor D1
DRD2	Dopamine receptor D2
DRD5	Dopamine receptor D5
GC	Glucocorticoid
GHSR	Growth-hormone secretagogue receptor
GOAT	Ghrelin *O*-acyltransferase
GR	Glucocorticoid receptor
GRE	Glucocorticoid response element
HPA	Hypothalamic-pituitary-adrenal
5HT2C	Serotonin 2c receptor
LEAP2	Liver-expressed antimicrobial peptide 2
MC3R	Melanocortin-3 receptor
MDD	Major depressive disorder
MR	Mineralocorticoid receptor
NAc	Nucleus accumbens
PFC	Prefrontal cortex
PTSD	Posttraumatic stress disorder
PVN	Paraventricular nucleus
SNP	Single nucleotide polymorphism
SST5	Somatostatin receptor-5
VTA	Ventral tegmental area

7.1 Defining Stress: Adaptation and Maladaptation

Stress is characterized by a set of bodily responses to a demand or challenge. In everyday language, "stress" typically has a negative connotation, referring to a state of worry or the feeling of being overloaded. Indeed, in medicine, there is clear evidence that prolonged stress exposure elevates the risk of a broad array of diseases (Hughes et al. 2017), including cardiovascular illnesses (Basu et al. 2017; Dong et al. 2004), cancer (Kelly-Irving et al. 2013; Bellis et al. 2015), and psychiatric illnesses (Daníelsdóttir et al. 2024), and conditions such as diarrhea (Zhang et al. 2025), sweating, sleep disturbance (Kalmbach et al. 2018) and increased body temperature (Oka 2015). Thus, some of the responses to stress must contribute to disease. Yet, in science, we have come to appreciate that not all bodily responses to a stressor are "bad" or drive disease risk. Some stress responses may compensate for others, preserving function in the face of biological perturbation (Nestler and Russo 2024). For example, studies of 'resilient' individuals, who are able to retain functionality of biological systems despite stress exposure, reveal that they do not return to a pre-stress state after stressor exposure; instead, they actively engage new mechanisms to preserve function. Such mechanisms include stress-associated enhancement of connectivity between the prefrontal cortex (PFC) and nucleus accumbens (NAc), which is linked to greater resilience to stressors (Francis et al. 2015; Bagot et al. 2015). Understanding whether a stress response is adaptive or maladaptive is especially important in clinical contexts, where we want to bolster resilience but blunt pathways that drive maladaptation.

7.2 A Canonical Stress Signaling Pathway in the Periphery: the HPA Axis

One of the best-studied systems for coordinating stress responses is the hypothalamic-pituitary-adrenal (HPA) axis (Fig. 7.1, left). When stressors occur, stress-related neural circuits of the paraventricular nucleus (PVN) of the hypothalamus are activated. Many types of signals can activate the PVN, including internal changes in temperature, blood glucose, or blood pressure and external stimuli such as loud noises or predator cues. A subset of parvocellular neurons in the PVN release corticotropin-releasing factor (CRF) locally within the PVN to induce central stress effects directly. Some CRF-positive PVN neurons send their axons into the median eminence and onward to the portal capillary system of the pituitary. The capillaries allow CRF to be transported to the anterior pituitary, where it binds to receptors on a subset of endocrine cells (corticotrophs) that then exocytose vesicles containing adrenocorticotropic hormone (ACTH) into the blood. ACTH acts at its receptor in the adrenal glands to drive the secretion of glucocorticoids (GCs; cortisol in humans, corticosterone in rodents) into the bloodstream. Glucocorticoid receptors are found throughout most tissues in

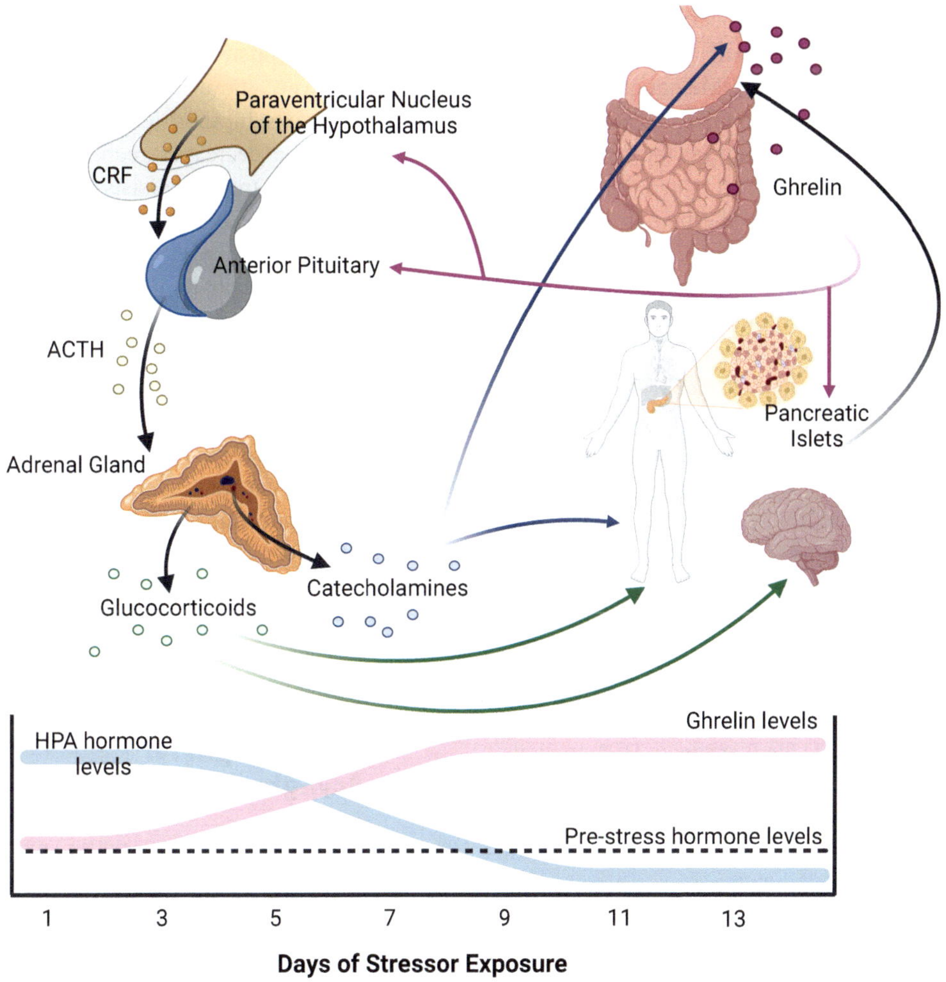

Fig. 7.1 The hormones of the HPA axis and ghrelin axis and their interactions are depicted. The green arrows indicate synergistic stimulatory actions between the two hormonal axes. The lower panel shows the differential recruitment of these axes over the course of a two-week stressor. While the HPA responses typically show habituation, eventually reaching slightly suppressed basal levels, ghrelin levels are not significantly enhanced early in stress exposure but reach roughly double the basal levels after approximately a week of stress exposure. (Created in BioRender. Goosens (2025) https://BioRender.com/w44p992)

the periphery and the brain, and thus, stress-induced elevation of GCs is poised to have broad effects.

There are two primary GC receptors: the mineralocorticoid receptor (MR) and the glucocorticoid receptor (GR). The two receptors can be co-expressed (Herman et al. 1989), but MR has a higher affinity for GCs than does GR (Joëls and de Kloet 2017). Thus, at lower levels of GCs, i.e., under non-stress conditions, MR signaling is expected to dominate. In contrast, when GCs are elevated, MR signaling reaches an asymptotic level and GR signaling pathways are increasingly recruited with increasing high levels of GCs (Reul et al. 1987).

Both MR and GR are ligand-activated transcription factors (Alvarez de la Rosa et al. 2024). Accordingly, there are many studies showing how activation of MR and GR by GCs causes the receptors to translocate to the nucleus and bind to glucocorticoid response elements (GREs) in open chromatin (John et al. 2011). MR is a transcriptional activator of many genes, including neuropeptides and growth factors (Pascual-Le Tallec and Lombès 2005; Meinel et al. 2014). In contrast, GR can bind to GREs to activate genes or to negative GREs to repress their transcription (Surjit et al. 2011); it is estimated that more than 1% of all transcripts in the human genome are regulated by GR (Wiley et al. 2016). GR can also bind to mRNA to induce degradation (Boo et al. 2024). However, membrane-bound MR can also mediate the rapid effects of GCs on neuronal excitability (Karst et al. 2010, 2005).

GRs also play an important role in terminating the acute HPA stress response, both in terms of ACTH secretion (Jacobson et al. 1988) as well as GC secretion (Herman et al. 2020). Within the PVN, GCs bind to GR to induce endocannabinoid synthesis in the CRF-containing neurons of the PVN, which leads to retrograde inhibition of the excitatory drive entering the PVN (Tasker and Herman 2011; Di et al. 2003). GC activation of the hippocampus (Herman et al. 2003) and PFC (Jones et al. 2011) also terminate HPA-mediated stress hormone responses (Radley and Sawchenko 2011). The multiple robust mechanisms by which GCs terminate their own secretion likely contribute to the observation that repeated stressor exposure produces fairly mild elevations of basal circulating GCs that persist for only a few weeks after stress terminates in rodents (Johnson et al. 2002; Sterlemann et al. 2008). In humans with posttraumatic stress disorder (PTSD), there are not clear links between altered GC levels and the disorder; a recent meta-analysis suggests that there is a trend towards observing mild hypocortisolism in PTSD (Sbisa et al. 2023). Interestingly, a recent meta-analysis also suggests that childhood stressors lead to blunted cortisol responses to stressors in adulthood (Brindle et al. 2022). Thus, the contributions of HPA activity to chronic stress responses are unclear.

As an acute stress response, HPA hormones drive multiple short-term adaptations that facilitate coping with stress. These include GC-dependent mobilization of energy stores (Swarbrick et al. 2021), increased cardiovascular tone (Yang and Zhang 2004), analgesia (Lewis et al. 1980), and inhibition of growth (Baxter 1978), immune function (Coutinho and Chapman 2011) and reproduction (Domes et al. 2024). Thus, collectively, the evidence supports a role for the HPA axis in short-term adaptations to acute stress. The decreased recruitment of the HPA axis when a stressor persists across days to weeks (Grissom and Bhatnagar 2009) (Fig. 7.1, lower panel), coupled with the observation that basal HPA activity is not dramatically altered following chronic stress, suggests that the HPA axis may not be the only system altered by chronic stress and that other mechanisms may contribute to changes following chronic stress. However, it is important to note that HPA activity can induce changes in behavior that emerge gradually, perhaps due to the genomic actions of GCs, and also sensitize the HPA axis itself so that a greater HPA response is mounted to subsequent novel stressors (Belda et al. 2015; Akana et al. 1992).

7.3 Additional Stress Signals in the Periphery: Acyl-ghrelin

The peptide hormone acyl-ghrelin, hereafter referred to as ghrelin, was discovered in the search for the ligand of the growth-hormone secretagogue receptor (GHSR) (Kojima et al. 1999). The posttranslational modification of ghrelin, an octanoylated serine, is necessary for it to bind and activate GHSR (Bednarek et al. 2000). GHSR was originally characterized as a receptor that regulated growth hormone secretion from the pituitary (Howard et al. 1996). It was surprising, then, that ghrelin was observed to be predominantly expressed in the stomach and has been subsequently confirmed to be almost exclusively in gastric endocrine cells (Kim et al. 2012; Date et al. 2000), with a smaller population of endocrine cells being in the small intestine (Wierup et al. 2007). A short time after its initial discovery, ghrelin was found to regulate energy balance: the administration of ghrelin promotes food consumption. Additionally, ghrelin is elevated by either short-term fasting (energy depletion) (Hollstein et al. 2022; Schéle et al. 2016) or chronic food restriction (Tezenas du Montcel et al. 2023; D'Cunha et al. 2020; Méquinion et al. 2013). Lastly, ghrelin increases sharply prior to expected meal times (Cummings et al. 2001) and decreases in a sated state (Tschöp et al. 2000).

Many additional studies confirmed and expanded the role of ghrelin in aspects of appetitive processing. Ghrelin influences the rewarding value of the food itself. In rodent studies using conditioned place preference, administration of ghrelin increased the amount of time rodents spent in the compartment previously paired with access to a high-fat diet (Perello et al. 2010). Consistent with these studies, ghrelin administration also enhances the activity of dopaminergic neurons in the ventral tegmental area (VTA) (Navarro et al. 2022; Cornejo et al. 2018) and enhances dopamine release into the downstream NAc (Jerlhag et al. 2007). Humans receiving intravenous ghrelin rate food cues as more pleasant (Han et al. 2018) and intravenous ghrelin also increases the neural response to food cues in multiple areas thought to encode the incentive value of food (Goldstone et al. 2014). Ghrelin also increases the preference for high-fat, sweet foods in both rodents (Perello et al. 2010; Disse et al. 2010; King et al. 2011; Chuang et al. 2011a; Shimbara et al. 2004) and humans (Zoon et al. 2018) and increases the intake of rewarding food (Egecioglu et al. 2010).

Elevated ghrelin also enhances the motivation to work for rewards. For example, mice who are bar-pressing for high-fat food pellets show a higher breakpoint in a progressive ratio schedule of reinforcement when given peripheral ghrelin injections compared to controls (Perello et al. 2010). Rats receiving peripheral ghrelin injections (Skibicka et al. 2012) or intra-VTA infusion of ghrelin (Skibicka et al. 2011) increase bar presses for sucrose pellets even when satiated. Intra-VTA ghrelin in rodents also increases cue-induced reinstatement of bar presses for high-fat food pellets (St-Onge et al. 2016).

Interestingly, the ability of ghrelin to mediate reward appears to extend well beyond food. Ghrelin receptor antagonism reduces drug-induced conditioned place preference for a number of drugs of abuse (Charalambous et al. 2021; Sustkova-Fiserova et al. 2020; Dunn et al. 2019; Jerlhag and Engel 2011). In male mice, impaired ghrelin signaling decreases interest in female mice in estrus, as well as sexual engagement with female mice (Egecioglu et al. 2016). Intravenous ghrelin enhances the craving for alcohol in alcohol-dependent humans (Leggio et al. 2014). Lastly, intravenous ghrelin administration decreases neural activity in response to anticipation of monetary losses in healthy human subjects (Pietrzak et al. 2023a). However, some types of reward do not seem to be related to ghrelin levels in humans, including social rewards (Sailer et al. 2023) and caressing touch (Pfabigan et al. 2024).

Collectively, the data suggest that elevated levels of ghrelin accompany hunger, a state of energy deficit, and that lower levels of ghrelin are observed with satiety, a state of energy excess (Mani et al. 2019). Interestingly, chronic stress represents another state of energy deficit. Chronic stress tends to produce elevated core body temperature (Marazziti et al. 1992; Oka 2018; Nakamura 2015) and also decreases body weight gain in rodents (Kuti et al. 2022; Shin et al. 2024). It should perhaps, then, not be surprising that ghrelin is elevated after chronic stress exposure in multiple species, including rodents (Lutter et al. 2008; Meyer et al. 2014; Harmatz et al. 2016), fish (Jönsson 2013), horses (Hemmann et al. 2012), and humans (Yousufzai et al. 2018; Jaremka et al. 2014; Malik et al. 2020; Wittekind et al. 2023) (Fig. 7.1), suggesting that elevated ghrelin may be a conserved response to chronic stress. Much like MR and GR, GHSR is found throughout the brain and body (Ueberberg et al. 2009; Mani et al. 2014; Zigman et al. 2006; Guan et al. 1997), which enables the stress-induced change in ghrelin levels to have a widespread impact. Consistent with the idea that stress-induced elevation of ghrelin may produce important behavioral adaptations to cope with the energy deficit produced during a chronic stressor, ghrelin receptor knockout mice do not have the same stress-induced changes in body weight and caloric intake that are observed in wild-type mice, and they also have different neurotransmitter alterations in the brain after stressor exposure (Patterson et al. 2010). Similarly, ghrelin receptor knockout mice do not exhibit the metabolic adaptions shown by wild-type mice during chronic stress, such as hyperleptinemia and hyperinsulinemia or changes in hypothalamic peptides associated with consummatory behaviors (Patterson et al. 2013).

In addition to the impact of ghrelin on reward processing, an increasingly large body of work supports the idea that ghrelin influences the processing of aversive stimuli (also called punishment or costs). The basolateral amygdala (BLA) is one brain region that is particularly tied to the processing of aversive memories (Perumal and Sah 2021), and it also happens to be a region where GHSR is highly expressed (Meyer et al. 2014; Alvarez-Crespo et al. 2012). Infusing ghrelin into the BLA inhibits the acquisition of conditioned taste aversion memories (Song et al. 2013). It also inhibits the formation of aversive Pavlovian fear conditioning memories (Harmatz et al. 2016). Consistent with these rodent studies, studies in healthy humans show that intravenous ghrelin decreased sensitivity to punishment (loss of monetary reward) in decision-making tasks (Pietrzak et al. 2024, 2023b) and also reduced neural activity during anticipation of monetary losses (Pietrzak et al. 2023b). This suggests that, in unstressed subjects, ghrelin inhibits aversive processing in multiple ways. However, rodents exposed to chronic stress display both higher levels of ghrelin and stronger fear memories than controls (Meyer et al. 2014; Harmatz et al. 2016). This suggests that ghrelin is no longer inhibiting aversive memories effectively after chronic stress exposure. In another aversive domain, chronically high ghrelin is positively associated with symptoms of physiological anxiety in otherwise healthy humans (Wittekind et al. 2022).

Further studies demonstrated that high levels of ghrelin were associated with a profound loss of GHSR binding sites in the BLA (Harmatz et al. 2016). This functional 'ghrelin resistance' is likely a compensatory mechanism by which receptors are downregulated in response to excessive ligand-dependent signaling. However, it should be noted that not all brain regions respond to elevated ghrelin by downregulating GHSR; in fact, in the VTA, GHSR levels are increased (Smith et al. 2024a). Thus, the elevated ghrelin observed after chronic stress reduces the ability of ghrelin to serve as an endogenous inhibitory signal for aversive processing.

7.4 The Adaptive Values of Persistently Elevated Ghrelin After Chronic Stress

Unlike the HPA axis, ghrelin remains approximately doubled long after stressors cease. Studies have shown elevated ghrelin in rodents for weeks (Lutter et al. 2008; Meyer et al. 2014; Harmatz et al. 2016; Smith et al. 2023; Kumar et al. 2013) to months (Yousufzai et al. 2018) after stressor cessation and there is increased brain penetrance of ghrelin after chronic stress (Smith et al. 2024b). In humans, elevated ghrelin has been demonstrated years after stressor exposure (Malik et al. 2020; Rossi et al. 2021). The increased ghrelin levels could be mediated by increased synthesis, increased release, increased posttranslational octanoylation, decreased breakdown, or a combination of these modulatory factors (see Fig. 7.2 for a summary of possible mechanisms). Excitation of the gastric afferent vagal nerve (Date et al. 2002) and repeated activation of β1-adrenergic receptors (Gupta et al. 2019), likely in the gastric ghrelin cells (Zhao et al. 2010a; Engelstoft et al. 2013), both increase circulating ghrelin, suggesting that increased vagal tone or enhanced catecholaminergic activity might be responsible for chronic stress-induced elevation of ghrelin. However, adrenalectomy, which eliminates the circulating catecholamines released by the adrenal glands, does not prevent stress-induced elevation of ghrelin (Meyer et al. 2014), suggesting that stress may elevate ghrelin via other pathways (for example, through sympathetic inputs to the gut). Ghrelin secretion is negatively regulated by increased blood glucose (Shiiya et al. 2002; Nakagawa et al. 2002). This raises the possibility that repeated hypoglycemia induced by chronic stress could also contribute to enhanced ghrelin secretion. One report demonstrated that gastric ghrelin cells express many types of G protein-

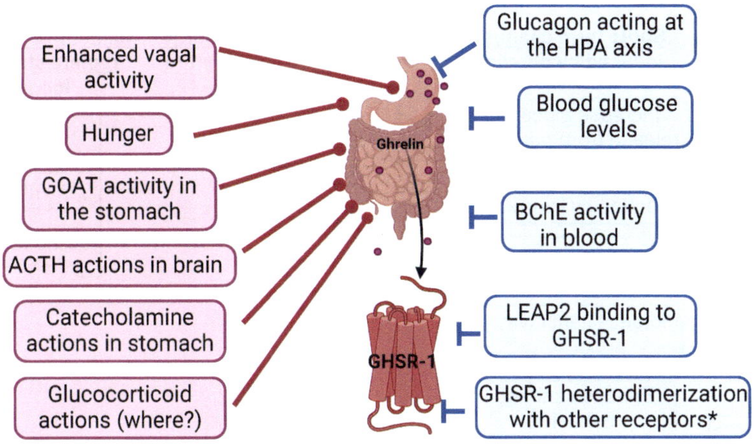

Fig. 7.2 Mechanisms by which stress could potentially elevate ghrelin or its signaling. Pink boxes indicate mechanisms that increase ghrelin. Blue boxes represent mechanisms that suppress ghrelin or its signaling. To enhance ghrelinergic activity, chronic stress would need to increase the mechanism(s) shown in pink or decrease the mechanism(s) shown in blue, or do both. *It is possible that heterodimerization with some receptors could elevate signaling through GHSR. (Created in BioRender. Goosens (2025) https://BioRender.com/w23c571)

coupled receptors (Engelstoft et al. 2013), suggesting that the control of ghrelin secretion is likely even more complex than described above. Regardless of the potential mechanisms, it is not understood how or why these mechanisms remain engaged after stress terminates (see Fig. 7.3 green boxes for a summary of adaptive changes). One potential adaptive value of the persistence of elevated ghrelin is that it renders the organism better prepared for the next encounter with stressors. This theory remains to be tested.

Another important role of elevated ghrelin is to maintain blood glucose levels during times of energy expenditure. Ghrelin may do this by promoting the release of glucagon in the short term (Chuang et al. 2011b), or enhancing growth hormone-mediated stimulation of glucose production in the liver and kidneys and driving insulin resistance (Kim and Park 2017; Zhao et al. 2010b; Zhang et al. 2015). In healthy subjects, glucagon exerts negative feedback to reduce ghrelin levels (Arafat et al. 2006). Administration of exogenous ghrelin is known to elevate blood glucose in both rodents (Chuang et al. 2011b) and humans (Broglio et al. 2001, 2004). It also promotes glucose intolerance (Tong et al. 2010; Page et al. 2018; Dezaki et al. 2004). In these studies, glucose intolerance was observed with acute elevation of ghrelin; it is tempting to speculate that chronically elevated ghrelin might produce even greater intolerance and contribute to elevated basal levels of blood glucose. Mice with either low blood levels of ghrelin or low levels of GHSR exhibit mild hypoglycemia after short-term caloric restriction (Longo et al. 2008; Sun et al. 2008) but display dangerously low blood glucose levels during prolonged caloric restriction (Zhao et al. 2010b; Li et al. 2012). It is possible that the energy deficit driven during chronic stress exposure enhances ghrelin levels to maintain blood glucose levels during stressor exposure, but it is not known why this mechanism would remain engaged after stress. It may be that these mechanisms evolved in response to environmental stressors, such as resource scarcity, that displayed less volatility than many contemporary stressors, and thus, having long-term elevations of ghrelin had more beneficial effects than detrimental effects when exposure to stressors was more consistent across time.

By promoting exploration or food-seeking behaviors, elevated ghrelin can also help indirectly to maintain blood glucose levels. Ghrelin signaling in the olfactory bulbs is important for promoting exploratory behavior, even in the absence of fasting, and also helps locate food (Stark et al. 2024). Ghrelin increases sensitivity to food odors in humans (Ginieis et al. 2022;

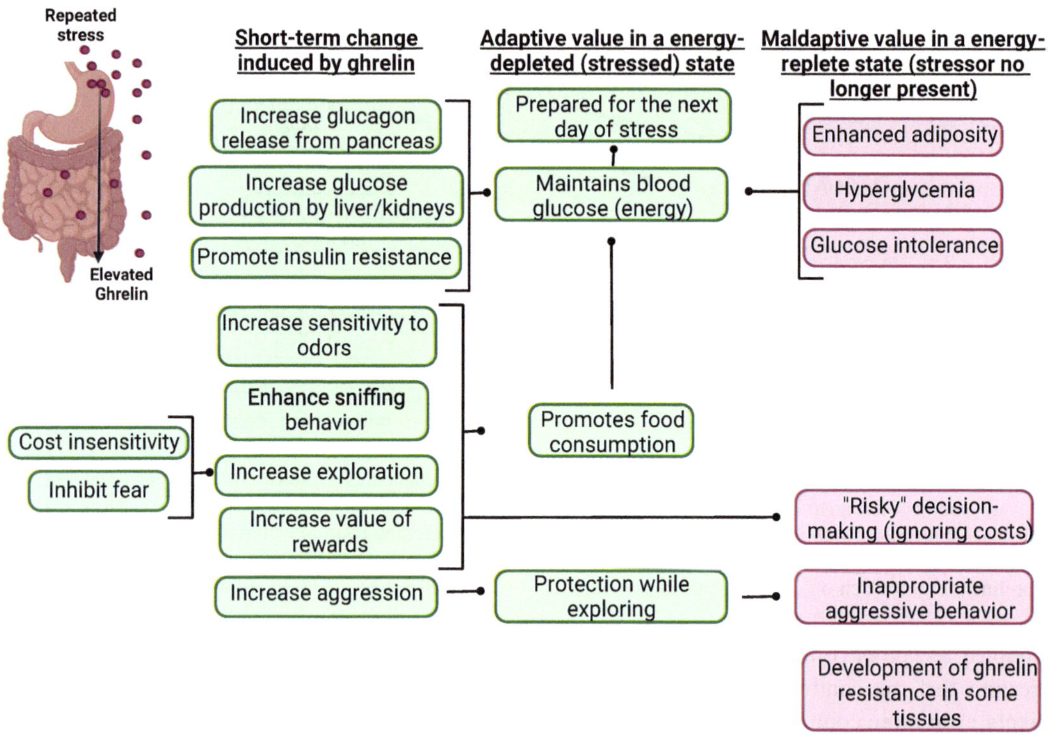

Fig. 7.3 Beneficial and maladaptive changes that can be mediated by increased ghrelin. The green boxes depict the changes that benefit an organism while remaining in a stressed state. The pink boxes depict the consequences of these changes when the stressor is no longer experienced. (Created in BioRender. Goosens (2025) https://BioRender.com/j66v932)

Tong et al. 2011) and rodents (Tong et al. 2011) and enhances the speed of responding to them (Han et al. 2018). These changes in behavior facilitate the ability to locate and identify foods, which indirectly facilitates the maintenance of blood glucose levels.

Lastly, for some types of stressors, like the presence of conspecifics encroaching on an organism's territory, elevated ghrelin may have an additional adaptive value of promoting aggressive behaviors. Male mice with persistently elevated ghrelin display enhanced aggression towards cagemates (Chen et al. 2015). Conversely, systemic administration of a GHSR antagonist reduced aggressive behaviors displayed by male mice towards conspecific intruder mice (Vestlund et al. 2019). People with antisocial personality disorder have elevated ghrelin (Tasci et al. 2022) and young men with polymorphisms in the ghrelin gene displayed significantly different scores on a questionnaire measuring overtly aggressive behaviors (Vestlund et al. 2019). To date, such studies have focused only on aggression in males; an important area of future research is to determine whether elevated ghrelin increases aggression in females. Regardless, a ghrelin-dependent increase in aggressive behaviors may help secure environmental resources when there is competition for such resources.

The ghrelin-dependent connection between stress and mechanisms controlling energy levels and growth positions ghrelin as a hub between metabolism and the stress-dependent exacerbation of neuropsychiatric conditions. The mechanisms connecting glucose metabolism and bioenergetics to psychiatric outcomes are *described in Chap. 3, Brain-body Communication in Glucose Metabolism*, and *Chap. 5, Neuronal Synaptic Communication and Mitochondrial Energetics in Human Health and Disease*, of this book.

7.5 Potential Maladaptive Tradeoffs of Persistently Elevated Ghrelin After Stress

As noted above, ghrelin can have many beneficial effects during stressor exposure. Thus, when a stressor is encountered repeatedly (for example, the organism lives in a food-scarce environment), then the behavioral and physiological changes produced by elevated ghrelin may be largely beneficial. However, when stressors are not reliably encountered, such as when an organism moves from a food-scarce (stressful) environment into a food-rich environment (no longer a source of stress), then the persistent physiological changes induced by high ghrelin may instead promote behaviors and physiological changes that are no longer needed (see Fig. 7.3, pink boxes for a summary of potential maladaptation). It is also possible that chronically elevated ghrelin may, itself, drive new adaptations to limit the impact of chronically elevated ghrelin.

There are several examples that one can consider where elevated ghrelin may no longer have beneficial effects in an energy-rich environment. First, ghrelin-induced elevation of blood glucose levels and insulin insensitivity are no longer adaptive when energetic demands return to normal levels following a stressor exposure. In this case, persistently heightened glucose levels and insulin insensitivity can contribute to adiposity and metabolic syndrome, two conditions that can ultimately contribute to diabetes. Indeed, there are strong links between stress exposure and the risk and severity of diabetes (Hackett and Steptoe 2017). Likewise, when ghrelin levels drive aggressive behaviors, it can help secure resources in a resource-scarce environment, but it also exposes the animal to greater potential for injury. In a resource-rich environment, aggression may expose an organism to injury without the positive benefit of increased resources. Also, for humans, who typically do not need to physically fight to gain resources, aggressive behaviors may only put one at risk for breaking the law. Stress exposure is known to increase the development of psychiatric disorders where aggression is a core feature (Veenema 2009). Lastly, while ghrelin can reduce sensitivity to costs, perhaps contributing to a willingness to explore and forage in new environments, ultimately leading to new food sources, a willingness to take on risk may not be beneficial when resources are not scarce. In humans, stress can increase risk-taking behavior (Reynolds et al. 2013; Pabst et al. 2013).

The adverse impact of ghrelin may be most apparent in enhanced risk for stress-sensitive human psychiatric conditions. In posttraumatic stress disorder (PTSD), which is the human disorder perhaps most closely linked to stress, prior lifetime stress exposure primes an individual for risk of PTSD following subsequent trauma (Catani et al. 2008; Gillespie et al. 2009). Consistent with the idea that elevated ghrelin drives an increased risk of PTSD, adolescents with PTSD have higher ghrelin levels than matched controls without PTSD (Yousufzai et al. 2018). One polymorphism in the ghrelin gene has been associated with PTSD symptom severity (Li et al. 2019).

Alcohol use disorder (AUD) is another stress-sensitive condition (Hughes et al. 2019) with compelling links to ghrelin. Ghrelin is generally positively correlated with alcohol craving in alcohol-dependent individuals (Koopmann et al. 2012, 2019), and intravenous delivery of ghrelin increased alcohol self-administration (Farokhnia et al. 2018) and craving (Leggio et al. 2014) in individuals with AUD. There are also multiple studies showing that polymorphisms in the ghrelin gene are associated with AUD (Landgren et al. 2008, 2010; Suchankova et al. 2016). Several studies suggest that antagonism of GHSR in mice would have beneficial effects on AUD. GHSR antagonism reduces binge-like alcohol drinking in mice (Richardson et al. 2024) and decreases alcohol intake in a two-bottle free choice test in dependent mice (Jerlhag et al. 2009; Kaur and Ryabinin 2010). GHSR antagonism also attenuates relapse consumption of alcohol after abstinence in rodents (Jerlhag et al. 2009; Suchankova et al. 2013), thought to reflect a reduction in alcohol craving. Disappointingly, the first study of a drug with GHSR antagonist activity (Kong et al. 2016) in humans with AUD observed no change in cue-elicited alcohol craving but did reduce the caloric content of food selected in a virtual food choice task (Faulkner et al. 2024). The role of ghrelin in *AUD is considered in greater depth in Chap. 8, Appetite-regulatory peptides ghrelin and GLP-1 in Alcohol Use Disorder*, of this book.

Major depressive disorder (MDD) is another psychiatric condition that is worsened by stressor exposure (Tafet and Nemeroff 2016), especially for childhood trauma (Heim and Nemeroff 2001), but the links to ghrelin are more equivocal for this disorder. Given that childhood trauma is shown to elevate ghrelin (Malik et al. 2020; Rossi et al. 2021), it is not surprising that some studies find elevated ghrelin in adults with MDD (Emül et al. 2007; Ozsoy et al. 2014; Kurt et al. 2007; Algul and Ozcelik 2018). On the other hand, some studies find no differences in ghrelin levels between people with MDD and healthy controls (Giménez-Palop et al. 2012; Matsuo et al. 2012; Schanze et al. 2008). One population that has yet to be examined is people with co-occurring MDD and PTSD. In rodents, the stressors used to induce depression-like behaviors elevate ghrelin reliably (Lutter et al. 2008; Kumar et al. 2013; Gupta et al. 2019), but it has been argued that elevated ghrelin actually buffers against depressive symptoms: stress-exposed GHSR knockout mice show greater stress-induced depressive behaviors than stress-exposed wild-type control mice (Chuang et al. 2011a; Lutter et al. 2008) and elevating ghrelin during stress reduces depression-like behaviors (Huang et al. 2017; Chang et al. 2024; Lu et al. 2019). Consistent with a positive role for ghrelin in depressive symptoms, short-term injections of ghrelin reduced depressive symptoms in men with MDD (Kluge et al. 2011). It is difficult to reconcile these disparate findings. One possibility is that GHSR knockout is functionally equivalent to 'ghrelin resistance' induced by high levels of ghrelin in some brain circuits and that loss of ghrelin signaling is what drives depressive behaviors. Another possibility is that the exploratory behaviors promoted by ghrelin may mask some of the depressive behaviors in mice. Yet another possibility is that GHSR knockout in mice is likely to impact signaling through other ligands in a manner that has nothing to do with ghrelin per se and is instead related to the complex modulatory effects of GHSR on other ligand systems (Shiimura et al. 2025; Ringuet et al. 2022). Further research will be needed to clarify whether elevated ghrelin is beneficial or detrimental to people with MDD and why.

One important reason for discrepancies between preclinical and clinical findings is that the drugs used to modify signaling through GHSR can have very diverse effects, depending on differences in how the compounds bind to GHSR (Shiimura et al. 2025). Both agonists and antagonists can bias G-protein coupling to GHSR (Shiimura et al. 2025). Some of the studies above showing that GHSR antagonism decreased alcohol consumption in preclinical rodent models (Jerlhag et al. 2009; Kaur and Ryabinin 2010) used pure GHSR antagonists, while the human study (Faulkner et al. 2024) used a drug with both inverse agonist (Bhattacharya et al. 2014) and competitive antagonist (Kong et al. 2016) properties. It is likely that the different compounds had distinct effects on GHSR signaling, which may have contributed to the different effects on the consumption of alcohol. Thus, a best practice is to compare the effects of the same drug in preclinical and clinical studies and not assume that drugs with similar activities (e.g., receptor antagonists) will exert their effects through the same mechanism.

7.6 Ghrelin: Interactions with the HPA Axis

Complicating the role of ghrelin in stress are multiple studies showing that ghrelin and the HPA axis can bidirectionally impact each other (Fig. 7.1, upper, green arrows). However, it should be noted that virtually all of these studies examine their interaction in healthy, unstressed subjects or subjects exposed to an acute stressor. It has been suggested that HPA-dependent elevation of ghrelin is GC-dependent: activation of the HPA axis with a single exogenous injection of ACTH elevates ghrelin, but this effect is blocked when metyrapone, a drug that blocks glucocorticoid synthesis, is administered (Azzam et al. 2017). On the other hand, adrenalectomy does not prevent chronic stress-induced elevation of ghrelin, suggesting that at least for chronic stress, glucocorticoids and circulating catecholamines released from the adrenal glands may not be the primary drivers of increased ghrelin.

In the other direction, ghrelin can modify the HPA axis. Infusion of ghrelin into the brain of chicks can increase ACTH and glucocorticoids (Gastón et al. 2017). Acute food deprivation, which elevates ghrelin, activates the HPA axis (Fernandez et al. 2022). Exogenous systemic ghrelin can activate CRF-positive neurons of the PVN and drive increases in plasma GCs (Fernandez et al. 2023). *Ghsr* mRNA is also observed in ACTH-expressing cells of the anterior pituitary (Reichenbach et al. 2012). In humans with AUD, intravenous administration of ghrelin elevated serum cortisol levels (Haass-Koffler et al. 2019). Thus, studies generally suggest that enhanced ghrelin activity drives increased HPA activity. In agreement with this, studies using methods to reduce ghrelin signaling generally find that a reduction in ghrelin signaling constrains HPA axis activation. For example, ghrelin receptor knockout mice show smaller increases of ACTH and GCs in response to acute stressor exposure compared to wild-type mice (Spencer et al. 2012).

Collectively, these studies suggest that ghrelin and the HPA axis both amplify each other. However, virtually nothing is known about these interactions in the context of chronic stress. In one relevant study, repeated injection of CRF into the PVN drove the elevation of ghrelin (Rayatpour et al. 2023). It is possible that repeated activation of CRF neurons by chronic stress may contribute to elevated ghrelin, especially considering that adrenalectomy enhances (rather than eliminates or reduces) activation of PVN neurons and the expression of CRF heteronuclear RNA in the PVN after acute stress exposure (Imaki et al. 1995; Pace et al. 2009). Yet, much remains to be understood about how the HPA and ghrelin systems interact in the face of chronic stress exposure, considering that the HPA stress response typically habituates for chronic stressors, particularly when the same stressor is experienced repeatedly (Grissom and Bhatnagar 2009; Belda et al. 2020). It is also important to note that the noradrenergic system is an important mediator of stress responses (*see Chap. 2, Noradrenaline Regulation of Brain-body Communication, for an overview of the role of this system in stress responses*), but even less is known about potential bidirectional interactions between this system and ghrelin; thus, we do not address this topic here.

7.7 New Frontiers for the Role of Ghrelin in Stress: Modulation of Ghrelin Signaling

One aspect of stress-associated ghrelin biology that is completely unknown is the mechanism by which ghrelin is persistently elevated. One enzyme that regulates the conversion of acyl-ghrelin to des-acyl-ghrelin is butyrylcholinesterase (BChE) (Chen et al. 2015; Schopfer et al. 2015). Single-nucleotide polymorphisms in the *BCHE* gene can considerably alter BChE enzymatic activity (La Du et al. 1990; Jensen et al. 1995; Dantas et al. 2011). While BChE is a ubiquitous enzyme whose levels are unlikely to be appreciably altered by chronic stress, it is possible that individuals with lower BChE activity may be more likely to have elevated ghrelin in response to chronic stress. Conversely, one might expect that individuals with higher BChE activity might be less likely to show elevated ghrelin after chronic stress exposure. These possibilities remain tantalizingly unexplored.

In the stomach, the cells that synthesize ghrelin also post-translationally activate it via the enzyme ghrelin *O*-acyltransferase (GOAT) in a process termed octanoylation (Yang et al. 2008; Gutierrez et al. 2008). In the gut, GOAT octanoylates proghrelin, the precursor molecule to ghrelin, so that proghrelin becomes acyl-ghrelin when cleaved (Yang et al. 2008; Gutierrez et al. 2008). In circulation, acyl-ghrelin is rapidly hydrolyzed into des-acyl-ghrelin (Schopfer et al. 2015), a form that does not act at GHSR (Kojima et al. 1999; Fernandez et al. 2016). In the brain, locally produced GOAT is capable of converting des-acyl-ghrelin back to acyl-ghrelin (Murtuza and Isokawa 2018). Consistent with the essential role of GOAT in producing acyl-ghrelin and the importance of the stomach in producing the acyl-ghrelin found in circulation, levels of acyl-ghrelin are correlated with GOAT expression in the stomach (Gahete et al. 2010). One possible explanation for the stress-associated increase in ghrelin is increased expression of GOAT in the gut, a possibility that also remains unexplored.

Liver-expressed antimicrobial peptide 2 (LEAP2) is a highly conserved peptide across mammals (Krause et al. 2003) and was originally named because of its structural similarity to other antimicrobial peptides in the liver, as well as its own antimicrobial properties (Henriques et al. 2010). However, it was suggested that LEAP2 likely had other functions (Henriques et al. 2010), and in 2018, it was discovered that LEAP2 is an endogenous inverse agonist and antagonist of GHSR (Ge et al. 2017; M'Kadmi et al. 2019), a role seemingly unrelated to its antimicrobial role. While most circulating LEAP2 comes from the liver, LEAP2 mRNA has been observed in other tissues including gut epithelium (Howard et al. 2010) and brain (Tufvesson-Alm et al. 2024; Islam et al. 2020). The expression of LEAP2 and ghrelin tend to inversely covary with each other (Mani et al. 2019; Islam et al. 2024), suggesting a shared, but opposing, mechanism of regulation. Consistent with inhibitory actions at GHSR, LEAP2 administration has been shown to attenuate many ghrelin-induced physiological changes, including food intake, blood glucose elevation, release of growth hormone, cFos expression in metabolic hypothalamus, and release of dopamine in the NAc (Tufvesson-Alm et al. 2024; Islam et al. 2020, 2024; Lugilde et al. 2022; Mustafá et al. 2021). Conversely, the reduction of endogenous LEAP2 actions enhances the impact of ghrelin and ghrelin-associated functions (Ge et al. 2017; Fei et al. 2024; Bhargava et al. 2023). Insofar as elevated ghrelin accompanies chronic stress, it seems likely that a decrease in LEAP2 accompanies chronic stress, but this has yet to be examined. Likewise, individual variability in LEAP2 expression may confer either vulnerability or resilience to the adaptations and maladaptation that accompany elevated ghrelin after chronic stress. Specifically, higher LEAP2 levels may protect against the stress-induced changes driven by elevated ghrelin, while lower LEAP2 levels may deepen the impact of elevated ghrelin post-stress.

Single nucleotide polymorphisms (SNPs) are another way that individual humans could have varied responses to chronic stress. SNPs in the genes encoding ghrelin and its receptor have been studied in the context of obesity (Gueorguiev et al. 2009; Mora et al. 2015; Vivenza et al. 2004), but not the context of stress responsivity; however, it is important to note that most studies do not show that ghrelin levels are impacted by studied variants in the ghrelin gene. Similarly, genetic variability in the LEAP2 gene (Andreoli et al. 2024a), some of which do predict LEAP2 levels (Andreoli et al. 2024b), would be interesting to study in the context of ghrelin-sensitive changes that accompany chronic stress.

Lastly, it would be remiss not to note that GHSR has a bidirectional modulatory effect on other types of signals. When GHSR forms heterodimers with other receptor types, it can impact the signaling of both the high level of constitutive activity observed in GHSR (Holst et al. 2003, 2004) as well as ligand-dependent signaling through GHSR. It has been suggested that the dimerization with other receptors may generally attenuate ghrelin-mediated signaling (Schellekens et al. 2013). For example, the serotonin 2c receptor (5HT2C) heterodimerizes with GHSR[217] and blockade of 5HT2C receptors potentiates the impact of ghrelin on food intake (Schellekens et al. 2015), suggesting that 5HT2C receptor activity limits signaling through GHSR. GHSR can heterodimerize with multiple types of prostanoid receptors, which reduces the constitutive activity of the GHSR (Chow et al. 2008). When GHSR forms heterodimers with the melanocortin-3 receptor (MC3R), it reduces both constitutive and ghrelin-induced signaling through GHSR (Rediger et al. 2011). In one case, instead of attenuating signaling through the GHSR, dimerization changes the type of signaling mediated by GHSR: dimerization with the somatostatin receptor-5 (SST5) shifts the ghrelin receptor-G-protein coupling from Gαq11 to Gαi/o in order to drive ghrelin-mediated inhibitory tone in pancreatic β-cells (Park et al. 2012).

GHSR heterodimers can also impact signaling through the receptors that are paired with GHSR, and these interactions are not always ghrelin-dependent. That is, it is clear that GHSR can act as a receptor-modifying protein for other receptors (Ringuet et al. 2022; Hedegaard and Holst 2020). For example, GHSR forms heterodimers with multiple dopamine receptors, including Dopamine D1, D2, and D5 (DRD1, DRD2 and DRD5) (Kern et al. 2015, 2012; Jiang et al. 2006). In hypothalamic neurons, GHSR-DRD2 heterodimerization switches dopamine signaling from inhibitory to excitatory (Kern et al. 2012). GHSR heterodimerizes with the melanocortin-3 receptor (MC3R), where it enhances melanocortin-induced intracellular cAMP accumulation compared to MC3R homodimers (Rediger et al. 2011). GHSR heterodimerization with the oxytocin receptor (OTR) attenuates OTR signaling (Wallace Fitzsimons et al. 2019). In this regard, stress-induced changes in ghrelin levels, which can modify GHSR expression (Harmatz et al. 2016; Smith et al. 2024a), can regulate the availability of GHSR for heterodimerization and thus indirectly impact other types of signaling. Differential heterodimer expression across brain regions may also explain why some brain regions show downregulation of GHSR after chronic stress (Harmatz et al. 2016), while other brain regions show upregulation of GHSR (Smith et al. 2024a). Future research will undoubtedly shed light on these possibilities.

7.8 Conclusions

Chronic stress produces long-lasting enhanced risk for multiple types of disease, suggesting that there are biological changes induced by stress that have maladaptive value. Here, we considered the role of ghrelin in such changes because large increases in ghrelin are observed in multiple species long after chronic stressors cease and ghrelin receptors are found throughout the periphery and brain. We considered the adaptive value of elevated ghrelin when organisms remain exposed to the stressor but also discussed how physiological changes driven by elevated ghrelin may contribute to disease when the stressor is no longer present. Because the vast majority of ghrelin that acts on the brain comes from the stomach, ghrelin represents a new frontier for connecting the brain and the body during times of energy deficit and energy excess, and the ghrelin system may represent a new frontier for tackling stress-sensitive disease.

Acknowledgments This project was supported by the following: NSF/CAREER (#2235858), NIH/NIDA (#R01DA058653), U-RISE T34GM145529 and G-RISE T32GM144919.

Bibliography

Akana SF, Dallman MF, Bradbury MJ, Scribner KA, Strack AM, Walker CD (1992) Feedback and facilitation in the adrenocortical system: unmasking facilitation by partial inhibition of the glucocorticoid response to prior stress. Endocrinology 131(1):57–68. PMID: 1319329

Algul S, Ozcelik O (2018) Evaluating the levels of Nesfatin-1 and ghrelin hormones in patients with moderate and severe major depressive disorders. Psychiatry Investig 15(2):214–218. PMCID: PMC5900400

Alvarez de la Rosa D, Ramos-Hernández Z, Weller-Pérez J, Johnson TA, Hager GL (2024) The impact of mineralocorticoid and glucocorticoid receptor interaction on corticosteroid transcriptional outcomes. Mol Cell Endocrinol 594:112389. PMID: 39423940

Alvarez-Crespo M, Skibicka KP, Farkas I, Molnár CS, Egecioglu E, Hrabovszky E, Liposits Z, Dickson SL (2012) The amygdala as a neurobiological target for ghrelin in rats: neuroanatomical, electrophysiological and behavioral evidence. PLoS One 7(10):e46321. PMCID: PMC3468604

Andreoli MF, Gentreau M, Rukh G, Perello M, Schiöth HB (2024a) Genetic variants of LEAP2 are associated with anthropometric traits and circulating insulin-like growth factor-1 concentration: a UK Biobank study. Diabetes Obes Metab 26(9):3565–3575. PMID: 38888057

Andreoli MF, Kruger AL, Sokolov AV, Rukh G, De Francesco PN, Perello M, Schiöth HB (2024b) LEAP2 is associated with impulsivity and reward sensitivity depending on the nutritional status and decreases with protein intake in humans. Diabetes Obes Metab 26(10):4734–4743. PMID: 39140219

Arafat AM, Perschel FH, Otto B, Weickert MO, Rochlitz H, Schöfl C, Spranger J, Möhlig M, Pfeiffer AFH (2006) Glucagon suppression of ghrelin secretion is exerted at hypothalamus-pituitary level. J Clin Endocrinol Metab 91(9):3528–3533. PMID: 16787987

Azzam I, Gilad S, Limor R, Stern N, Greenman Y (2017) Ghrelin stimulation by hypothalamic-pituitary-adrenal axis activation depends on increasing cortisol levels. Endocr Connect 6(8):847–855. PMCID: PMC5682420

Bagot RC, Parise EM, Peña CJ, Zhang HX, Maze I, Chaudhury D, Persaud B, Cachope R, Bolaños-Guzmán CA, Cheer JF, Deisseroth K, Han MH, Nestler EJ (2015) Ventral hippocampal afferents to the nucleus accumbens regulate susceptibility to depression. Nat Commun 6:7062. PMCID: PMC4430111

Basu A, McLaughlin KA, Misra S, Koenen KC (2017) Childhood maltreatment and health impact: the examples of cardiovascular disease and type 2 diabetes mellitus in adults. Clin Psychol 24(2):125–139. PMCID: PMC5578408

Baxter JD (1978) Mechanisms of glucocorticoid inhibition of growth. Kidney Int 14(4):330–333. PMID: 366226

Bednarek MA, Feighner SD, Pong SS, McKee KK, Hreniuk DL, Silva MV, Warren VA, Howard AD, Van Der Ploeg LH, Heck JV (2000) Structure-function studies on the new growth hormone-releasing peptide, ghrelin: minimal sequence of ghrelin necessary for activation of growth hormone secretagogue receptor 1a. J Med Chem 43(23):4370–4376. PMID: 11087562

Belda X, Fuentes S, Daviu N, Nadal R, Armario A (2015) Stress-induced sensitization: the hypothalamic-pituitary-adrenal axis and beyond. Stress Amst Neth 18(3):269–279. PMID: 26300109

Belda X, Fuentes S, Labad J, Nadal R, Armario A (2020) Acute exposure of rats to a severe stressor alters the circadian pattern of corticosterone and sensitizes to a novel stressor: relationship to pre-stress individual differences in resting corticosterone levels. Horm Behav 126:104865

Bellis MA, Hughes K, Leckenby N, Hardcastle KA, Perkins C, Lowey H (2015) Measuring mortality and the burden of adult disease associated with adverse childhood experiences in England: a national survey. J Public Health (Oxf) 37(3):445–454. PMCID: PMC4552010

Bhargava R, Luur S, Rodriguez Flores M, Emini M, Prechtl CG, Goldstone AP (2023) Postprandial increases in liver-gut hormone LEAP2 correlate with attenuated eating behavior in adults without obesity. J Endocr Soc 7(7):bvad061. PMCID: PMC10243873

Bhattacharya SK, Andrews K, Beveridge R, Cameron KO, Chen C, Dunn M, Fernando D, Gao H, Hepworth D, Jackson VM, Khot V, Kong J, Kosa RE, Lapham K, Loria PM, Londregan AT, McClure KF, Orr STM, Patel J, Rose C, Saenz J, Stock IA, Storer G, VanVolkenburg M, Vrieze D, Wang G, Xiao J, Zhang Y (2014) Discovery of PF-5190457, a potent, selective, and orally bioavailable ghrelin receptor inverse agonist clinical candidate. ACS Med Chem Lett 5(5):474–479. PMCID: PMC4027753

Boo SH, Shin MK, Ha H, Woo JS, Kim YK (2024) Transcriptome-wide analysis for glucocorticoid receptor-mediated mRNA decay reveals various classes of target transcripts. Mol Cells 47(11):100130

Brindle RC, Pearson A, Ginty AT (2022) Adverse childhood experiences (ACEs) relate to blunted cardiovascular and cortisol reactivity to acute laboratory stress: a systematic review and meta-analysis. Neurosci Biobehav Rev 134:104530. PMID: 35031343

Broglio F, Arvat E, Benso A, Gottero C, Muccioli G, Papotti M, van der Lely AJ, Deghenghi R, Ghigo E (2001) Ghrelin, a natural GH secretagogue produced by the stomach, induces hyperglycemia and reduces insulin secretion in humans. J Clin Endocrinol Metab 86(10):5083–5086. PMID: 11600590

Broglio F, Gottero C, Prodam F, Gauna C, Muccioli G, Papotti M, Abribat T, Van Der Lely AJ, Ghigo E (2004) Non-acylated ghrelin counteracts the metabolic but not the neuroendocrine response to acylated ghrelin in humans. J Clin Endocrinol Metab 89(6):3062–3065. PMID: 15181099

Catani C, Jacob N, Schauer E, Kohila M, Neuner F (2008) Family violence, war, and natural disasters: a study of the effect of extreme stress on children's mental health in Sri Lanka. BMC Psychiatry 8:33. PMCID: PMC2386780

Chang L, Niu F, Li B (2024) Ghrelin/GHSR signaling in the lateral septum ameliorates chronic stress-induced depressive-like behaviors. Prog Neuro-Psychopharmacol Biol Psychiatry 131:110953. PMID: 38278286

Charalambous C, Havlickova T, Lapka M, Puskina N, Šlamberová R, Kuchar M, Sustkova-Fiserova M (2021) Cannabinoid-induced conditioned place preference, intravenous self-administration, and behavioral stimulation influenced by ghrelin receptor antagonism in rats. Int J Mol Sci 22(5):2397. PMCID: PMC7957642

Chen VP, Gao Y, Geng L, Parks RJ, Pang YP, Brimijoin S (2015) Plasma butyrylcholinesterase regulates ghrelin to control aggression. Proc Natl Acad Sci U S A 112(7):2251–2256

Chow KBS, Leung PK, Cheng CHK, Cheung WT, Wise H (2008) The constitutive activity of ghrelin receptors is decreased by co-expression with vasoactive prostanoid receptors when over-expressed in human embryonic kidney 293 cells. Int J Biochem Cell Biol 40(11):2627–2637. PMID: 18573679

Chuang JC, Perello M, Sakata I, Osborne-Lawrence S, Savitt JM, Lutter M, Zigman JM (2011a) Ghrelin mediates stress-induced food-reward behavior in mice. J Clin Invest 121(7):2684–2692. PMCID: PMC3223843

Chuang JC, Sakata I, Kohno D, Perello M, Osborne-Lawrence S, Repa JJ, Zigman JM (2011b) Ghrelin directly stimulates glucagon secretion from pancreatic alpha-cells. Mol Endocrinol 25(9):1600–1611. PMCID: PMC3165914

Cornejo MP, Barrile F, De Francesco PN, Portiansky EL, Reynaldo M, Perello M (2018) Ghrelin recruits specific subsets of dopamine and GABA neurons of different ventral tegmental area sub-nuclei. Neuroscience 392:107–120. PMID: 30268780

Coutinho AE, Chapman KE (2011) The anti-inflammatory and immunosuppressive effects of glucocorticoids, recent developments and mechanistic insights. Mol Cell Endocrinol 335(1):2–13. PMCID: PMC3047790

Cummings DE, Purnell JQ, Frayo RS, Schmidova K, Wisse BE, Weigle DS (2001) A preprandial rise in plasma ghrelin levels suggests a role in meal initiation in humans. Diabetes 50(8):1714–1719

D'Cunha TM, Chisholm A, Hryhorczuk C, Fulton S, Shalev U (2020) A role for leptin and ghrelin in the augmentation of heroin seeking induced by chronic food restriction. Psychopharmacology 237(3):787–800. PMID: 31811350

Daníelsdóttir HB, Aspelund T, Shen Q, Halldorsdottir T, Jakobsdóttir J, Song H, Lu D, Kuja-Halkola R, Larsson H, Fall K, Magnusson PKE, Fang F, Bergstedt J, Valdimarsdóttir UA (2024) Adverse childhood experiences and adult mental health outcomes. JAMA Psychiatry 81(6):586–594

Dantas VGL, Furtado-Alle L, Souza RLR, Chautard-Freire-Maia EA (2011) Obesity and variants of the GHRL (ghrelin) and BCHE (butyrylcholinesterase) genes. Genet Mol Biol 34(2):205–207. PMCID: PMC3115310

Date Y, Kojima M, Hosoda H, Sawaguchi A, Mondal MS, Suganuma T, Matsukura S, Kangawa K, Nakazato M (2000) Ghrelin, a novel growth hormone-releasing acylated peptide, is synthesized in a distinct endocrine cell type in the gastrointestinal tracts of rats and humans. Endocrinology 141(11):4255–4261. PMID: 11089560

Date Y, Murakami N, Toshinai K, Matsukura S, Niijima A, Matsuo H, Kangawa K, Nakazato M (2002) The role of the gastric afferent vagal nerve in ghrelin-induced feeding and growth hormone secretion in rats. Gastroenterology 123(4):1120–1128. PMID: 12360474

Dezaki K, Hosoda H, Kakei M, Hashiguchi S, Watanabe M, Kangawa K, Yada T (2004) Endogenous ghrelin in pancreatic islets restricts insulin release by attenuating Ca2+ signaling in beta-cells: implication in the glycemic control in rodents. Diabetes 53(12):3142–3151. PMID: 15561944

Di S, Malcher-Lopes R, Halmos KC, Tasker JG (2003) Nongenomic glucocorticoid inhibition via endocannabinoid release in the hypothalamus: a fast feedback mechanism. J Neurosci 23(12):4850–4857. PMCID: PMC6741208

Disse E, Bussier AL, Veyrat-Durebex C, Deblon N, Pfluger PT, Tschöp MH, Laville M, Rohner-Jeanrenaud F (2010) Peripheral ghrelin enhances sweet taste food consumption and preference, regardless of its caloric content. Physiol Behav 101(2):277–281. PMID: 20515700

Domes G, Linnig K, von Dawans B (2024) Gonads under stress: a systematic review and meta-analysis on the effects of acute psychosocial stress on gonadal steroids secretion in humans. Psychoneuroendocrinology 164:107004. PMID: 38471257

Dong M, Giles WH, Felitti VJ, Dube SR, Williams JE, Chapman DP, Anda RF (2004) Insights into causal pathways for ischemic heart disease: adverse childhood experiences study. Circulation 110(13):1761–1766. PMID: 15381652

Dunn DP, Bastacky JMR, Gray CC, Abtahi S, Currie PJ (2019) Role of mesolimbic ghrelin in the acquisition of cocaine reward. Neurosci Lett 709:134367. PMID: 31278962

Egecioglu E, Jerlhag E, Salomé N, Skibicka KP, Haage D, Bohlooly-Y M, Andersson D, Bjursell M, Perrissoud D, Engel JA, Dickson SL (2010) Ghrelin increases intake of rewarding food in rodents. Addict Biol 15(3):304–311. PMCID: PMC2901520

Egecioglu E, Prieto-Garcia L, Studer E, Westberg L, Jerlhag E (2016) The role of ghrelin signalling for sexual behaviour in male mice. Addict Biol 21(2):348–359. PMID: 25475101

Emül HM, Serteser M, Kurt E, Ozbulut O, Guler O, Gecici O (2007) Ghrelin and leptin levels in patients with obsessive-compulsive disorder. Prog Neuro-Psychopharmacol Biol Psychiatry 31(6):1270–1274. PMID: 17597276

Engelstoft MS, Mee PW, Sakata I, Kristensen LV, Husted AS, Osborne-Lawrence S, Piper PK, Walker AK, Pedersen MH, Nøhr MK, Pan J, Sinz CJ, Carrington PE, Akiyama TE, Jones RM, Tang C, Ahmed K, Offermanns S, Egerod KL, Zigman JM, Schwartz TW (2013) Seven transmembrane G protein-coupled receptor repertoire of gastric ghrelin cells. Mol Metab 2(4):376–392

Farokhnia M, Grodin EN, Lee MR, Oot EN, Blackburn AN, Stangl BL, Schwandt ML, Farinelli LA, Momenan R, Ramchandani VA, Leggio L (2018) Exogenous ghrelin administration increases alcohol self-administration and modulates brain functional activity in heavy-drinking alcohol-dependent individuals. Mol Psychiatry 23(10):2029–2038. PMID: 29133954

Faulkner ML, Farokhnia M, Lee MR, Farinelli L, Browning BD, Abshire K, Daurio AM, Munjal V, Deschaine SL, Boukabara SR, Fortney C, Sherman G, Schwandt M, Akhlaghi F, Momenan R, Ross TJ, Persky S, Leggio L (2024) A randomized, double-blind, placebo-controlled study of a GHSR blocker in people with alcohol use disorder. JCI Insight 9(24):e182331. PMID: 39704175

Fei Y, Bao Z, Wang Q, Zhu Y, Lu J, Ouyang L, Hu Q, Zhou Y, Chen L (2024) CRISPR/Cas9-induced LEAP2 and GHSR1a knockout mutant zebrafish displayed abnormal growth and impaired lipid metabolism. Gen Comp Endocrinol 355:114563. PMID: 38830459

Fernandez G, Cabral A, Cornejo MP, De Francesco PN, Garcia-Romero G, Reynaldo M, Perello M (2016) Des-Acyl ghrelin directly targets the arcuate nucleus in a ghrelin-receptor independent manner and impairs the orexigenic effect of ghrelin. J Neuroendocrinol 28(2):12349. PMID: 26661382

Fernandez G, Cabral A, De Francesco PN, Uriarte M, Reynaldo M, Castrogiovanni D, Zubiría G, Giovambattista A, Cantel S, Denoyelle S, Fehrentz JA, Tolle V, Schiöth HB, Perello M (2022) GHSR controls food deprivation-induced activation of CRF neurons of the hypothalamic paraventricular nucleus in a LEAP2-dependent manner. Cell Mol Life Sci CMLS 79(5):277. PMCID: PMC11072678

Fernandez G, De Francesco PN, Cornejo MP, Cabral A, Aguggia JP, Duque VJ, Sayar N, Cantel S, Burgos JI, Fehrentz JA, Rorato R, Atasoy D, Mecawi AS, Perello M (2023) Ghrelin action in the PVH of male mice: accessibility, neuronal targets, and CRH neurons activation. Endocrinology 164(11):bqad154. PMCID: PMC11491828

Francis TC, Chandra R, Friend DM, Finkel E, Dayrit G, Miranda J, Brooks JM, Iñiguez SD, O'Donnell P, Kravitz A, Lobo MK (2015) Nucleus accumbens medium spiny neuron subtypes mediate depression-related outcomes to social defeat stress. Biol Psychiatry 77(3):212–222. PMCID: PMC5534173

Gahete MD, Córdoba-Chacón J, Salvatori R, Castaño JP, Kineman RD, Luque RM (2010) Metabolic regulation of ghrelin O-acyl transferase (GOAT) expression in the mouse hypothalamus, pituitary, and stomach. Mol Cell Endocrinol 317(1–2):154–160. PMCID: PMC2819060

Gastón MS, Cid MP, Salvatierra NA (2017) Bicuculline, a GABAA-receptor antagonist, blocked HPA axis activation induced by ghrelin under an acute stress. Behav Brain Res 320:464–472. PMID: 27780724

Ge X, Yang H, Bednarek MA, Galon-Tilleman H, Chen P, Chen M, Lichtman JS, Wang Y, Dalmas O, Yin Y, Tian H, Jermutus L, Grimsby J, Rondinone CM, Konkar A, Kaplan DD (2017) LEAP2 is an endogenous antagonist of the ghrelin receptor. Cell Metab 27(2):461–469.e6

Gillespie CF, Bradley B, Mercer K, Smith AK, Conneely K, Gapen M, Weiss T, Schwartz AC, Cubells JF, Ressler KJ (2009) Trauma exposure and stress-related disorders in inner city primary care patients. Gen Hosp Psychiatry 31(6):505–514. PMCID: PMC2785858

Giménez-Palop O, Coronas R, Cobo J, Gallart L, Barbero JD, Parra I, Fusté G, Vendrell J, Bueno M, González-Clemente JM, Caixàs A (2012) Fasting plasma peptide YY concentrations are increased in patients with major depression who associate weight loss. J Endocrinol Investig 35(7):645–648. PMID: 22183081

Ginieis R, Abeywickrema S, Oey I, Peng M (2022) Testing links of food-related olfactory perception to peripheral ghrelin and leptin concentrations. Front Nutr 9:888608. PMCID: PMC9130723

Goldstone AP, Prechtl CG, Scholtz S, Miras AD, Chhina N, Durighel G, Deliran SS, Beckmann C, Ghatei MA, Ashby DR, Waldman AD, Gaylinn BD, Thorner MO, Frost GS, Bloom SR, Bell JD (2014) Ghrelin mimics fasting to enhance human hedonic, orbitofrontal cortex, and hippocampal responses to food. Am J Clin Nutr 99(6):1319–1330. PMCID: PMC6410902

Grissom N, Bhatnagar S (2009) Habituation to repeated stress: get used to it. Neurobiol Learn Mem 92(2):215–224

Guan XM, Yu H, Palyha OC, McKee KK, Feighner SD, Sirinathsinghji DJ, Smith RG, Van der Ploeg LH, Howard AD (1997) Distribution of mRNA encoding the growth hormone secretagogue receptor in brain and peripheral tissues. Brain Res Mol Brain Res 48(1):23–29. PMID: 9379845

Gueorguiev M, Lecoeur C, Meyre D, Benzinou M, Mein CA, Hinney A, Vatin V, Weill J, Heude B, Hebebrand J, Grossman AB, Korbonits M, Froguel P (2009) Association studies on ghrelin and ghrelin receptor gene polymorphisms with obesity. Obesity (Silver Spring) 17(4):745–754. PMID: 19165163

Gupta D, Chuang JC, Mani BK, Shankar K, Rodriguez JA, Osborne-Lawrence S, Metzger NP, Zigman JM (2019) β1-adrenergic receptors mediate plasma acyl-ghrelin elevation and depressive-like behavior induced by chronic psychosocial stress. Neuropsychopharmacology 44(7):1319–1327. PMCID: PMC6785135

Gutierrez JA, Solenberg PJ, Perkins DR, Willency JA, Knierman MD, Jin Z, Witcher DR, Luo S, Onyia JE, Hale JE (2008) Ghrelin octanoylation mediated by an orphan lipid transferase. Proc Natl Acad Sci U A 105(17):6320–6325

Haass-Koffler CL, Long VM, Farokhnia M, Magill M, Kenna GA, Swift RM, Leggio L (2019) Intravenous administration of ghrelin increases serum cortisol and aldosterone concentrations in heavy-drinking alcohol-dependent individuals: results from a double-blind, placebo-controlled human laboratory study. Neuropharmacology 158:107711

Hackett RA, Steptoe A (2017) Type 2 diabetes mellitus and psychological stress - a modifiable risk factor. Nat Rev Endocrinol 13(9):547–560. PMID: 28664919

Han JE, Frasnelli J, Zeighami Y, Larcher K, Boyle J, McConnell T, Malik S, Jones-Gotman M, Dagher A (2018) Ghrelin enhances food odor conditioning in healthy humans: an fMRI study. Cell Rep 25(10):2643–2652.e4. PMID: 30517853

Harmatz ES, Stone L, Lim SH, Lee G, McGrath A, Gisabella B, Peng X, Kosoy E, Yao J, Liu E, Machado NJ, Weiner VS, Slocum W, Cunha RA, Goosens KA (2016) Central ghrelin resistance permits the overconsolidation of fear memory. Biol Psychiatry 81(12):1003–1013

Hedegaard MA, Holst B (2020) The complex signaling pathways of the ghrelin receptor. Endocrinology 161(4):bqaa020

Heim C, Nemeroff CB (2001) The role of childhood trauma in the neurobiology of mood and anxiety dis-

orders: preclinical and clinical studies. Biol Psychiatry 49(12):1023–1039. PMID: 11430844
Hemmann K, Raekallio M, Kanerva K, Hänninen L, Pastell M, Palviainen M, Vainio O (2012) Circadian variation in ghrelin and certain stress hormones in cribbiting horses. Vet J 193(1):97–102. PMID: 22040804
Henriques ST, Tan CC, Craik DJ, Clark RJ (2010) Structural and functional analysis of human liver-expressed antimicrobial peptide 2. Chembiochem Eur J Chem Biol 11(15):2148–2157. PMID: 20845358
Herman JP, Patel PD, Akil H, Watson SJ (1989) Localization and regulation of glucocorticoid and mineralocorticoid receptor messenger RNAs in the hippocampal formation of the rat. Mol Endocrinol 3(11):1886–1894. PMID: 2558306
Herman JP, Figueiredo H, Mueller NK, Ulrich-Lai Y, Ostrander MM, Choi DC, Cullinan WE (2003) Central mechanisms of stress integration: hierarchical circuitry controlling hypothalamo-pituitary-adrenocortical responsiveness. Front Neuroendocrinol 24(3):151–180. PMID: 14596810
Herman JP, Nawreen N, Smail MA, Cotella EM (2020) Brain mechanisms of HPA axis regulation: neurocircuitry and feedback in context Richard Kvetnansky lecture. Stress Amst Neth 23(6):617–632. PMCID: PMC8034599
Hollstein T, Basolo A, Unlu Y, Ando T, Walter M, Krakoff J, Piaggi P (2022) Effects of short-term fasting on Ghrelin/GH/IGF-1 axis in healthy humans: the role of ghrelin in the thrifty phenotype. J Clin Endocrinol Metab 107(9):e3769–e3780. PMCID: PMC9387714
Holst B, Cygankiewicz A, Jensen TH, Ankersen M, Schwartz TW (2003) High constitutive signaling of the ghrelin receptor--identification of a potent inverse agonist. Mol Endocrinol 17(11):2201–2210. PMID: 12907757
Holst B, Holliday ND, Bach A, Elling CE, Cox HM, Schwartz TW (2004) Common structural basis for constitutive activity of the ghrelin receptor family*. J Biol Chem 279(51):53806–53817
Howard AD, Feighner SD, Cully DF, Arena JP, Liberator PA, Rosenblum CI, Hamelin M, Hreniuk DL, Palyha OC, Anderson J, Paress PS, Diaz C, Chou M, Liu KK, McKee KK, Pong SS, Chaung LY, Elbrecht A, Dashkevicz M, Heavens R, Rigby M, Sirinathsinghji DJ, Dean DC, Melillo DG, Patchett AA, Nargund R, Griffin PR, DeMartino JA, Gupta SK, Schaeffer JM, Smith RG, Van der Ploeg LH (1996) A receptor in pituitary and hypothalamus that functions in growth hormone release. Science 273(5277):974–977
Howard A, Townes C, Milona P, Nile CJ, Michailidis G, Hall J (2010) Expression and functional analyses of liver expressed antimicrobial peptide-2 (LEAP-2) variant forms in human tissues. Cell Immunol 261(2):128–133. PMID: 20038463
Huang HJ, Zhu XC, Han QQ, Wang YL, Yue N, Wang J, Yu R, Li B, Wu GC, Liu Q, Yu J (2017) Ghrelin alleviates anxiety- and depression-like behaviors induced by chronic unpredictable mild stress in rodents. Behav Brain Res 326:33–43. PMID: 28245976
Hughes K, Bellis MA, Hardcastle KA, Sethi D, Butchart A, Mikton C, Jones L, Dunne MP (2017) The effect of multiple adverse childhood experiences on health: a systematic review and meta-analysis. Lancet Public Health 2(8):e356–e366. PMID: 29253477
Hughes K, Bellis MA, Sethi D, Andrew R, Yon Y, Wood S, Ford K, Baban A, Boderscova L, Kachaeva M, Makaruk K, Markovic M, Povilaitis R, Raleva M, Terzic N, Veleminsky M, Włodarczyk J, Zakhozha V (2019) Adverse childhood experiences, childhood relationships and associated substance use and mental health in young Europeans. Eur J Pub Health 29(4):741–747. PMCID: PMC6660110
Imaki T, Xiao-Quan W, Shibasaki T, Yamada K, Harada S, Chikada N, Naruse M, Demura H (1995) Stress-induced activation of neuronal activity and corticotropin-releasing factor gene expression in the paraventricular nucleus is modulated by glucocorticoids in rats. J Clin Invest 96(1):231–238. PMCID: PMC185193
Islam MN, Mita Y, Maruyama K, Tanida R, Zhang W, Sakoda H, Nakazato M (2020) Liver-expressed antimicrobial peptide 2 antagonizes the effect of ghrelin in rodents. J Endocrinol 244(1):13–23
Islam MN, Nabekura H, Ueno H, Nishida T, Nanashima A, Sakoda H, Zhang W, Nakazato M (2024) Liver-expressed antimicrobial peptide 2 is a hepatokine regulated by ghrelin, nutrients, and body weight. Sci Rep 14(1):24782. PMCID: PMC11494003
Jacobson L, Akana SF, Cascio CS, Shinsako J, Dallman MF (1988) Circadian variations in plasma corticosterone permit normal termination of adrenocorticotropin responses to stress. Endocrinology 122(4):1343–1348. PMID: 2831028
Jaremka LM, Belury MA, Andridge RR, Malarkey WB, Glaser R, Christian L, Emery CF, Kiecolt-Glaser JK (2014) Interpersonal stressors predict ghrelin and leptin levels in women. Psychoneuroendocrinology 48:178–188. PMCID: PMC4117712
Jensen FS, Skovgaard LT, Viby-Mogensen J (1995) Identification of human plasma cholinesterase variants in 6,688 individuals using biochemical analysis. Acta Anaesthesiol Scand 39(2):157–162. PMID: 7793180
Jerlhag E, Engel JA (2011) Ghrelin receptor antagonism attenuates nicotine-induced locomotor stimulation, accumbal dopamine release and conditioned place preference in mice. Drug Alcohol Depend 117(2–3):126–131. PMID: 21310553
Jerlhag E, Egecioglu E, Dickson SL, Douhan A, Svensson L, Engel JA (2007) Ghrelin administration into tegmental areas stimulates locomotor activity and increases extracellular concentration of dopamine in the nucleus accumbens. Addict Biol 12(1):6–16. PMID: 17407492
Jerlhag E, Egecioglu E, Landgren S, Salomé N, Heilig M, Moechars D, Datta R, Perrissoud D, Dickson SL, Engel JA (2009) Requirement of central ghrelin signaling for alcohol reward. Proc Natl Acad Sci USA 106(27):11318–11323. PMCID: PMC2703665

Jiang H, Betancourt L, Smith RG (2006) Ghrelin amplifies dopamine signaling by cross talk involving formation of growth hormone secretagogue receptor/dopamine receptor subtype 1 heterodimers. Mol Endocrinol Baltim Md 20(8):1772–1785. PMID: 16601073

Joëls M, de Kloet ER (2017) 30 years of the mineralocorticoid receptor: the brain mineralocorticoid receptor: a saga in three episodes. J Endocrinol 234(1):T49–T66. PMID: 28634266

John S, Sabo PJ, Thurman RE, Sung MH, Biddie SC, Johnson TA, Hager GL, Stamatoyannopoulos JA (2011) Chromatin accessibility pre-determines glucocorticoid receptor binding patterns. Nat Genet 43(3):264–268. PMCID: PMC6386452

Johnson JD, O'Connor KA, Deak T, Spencer RL, Watkins LR, Maier SF (2002) Prior stressor exposure primes the HPA axis. Psychoneuroendocrinology 27(3):353–365. PMID: 11818171

Jones KR, Myers B, Herman JP (2011) Stimulation of the prelimbic cortex differentially modulates neuroendocrine responses to psychogenic and systemic stressors. Physiol Behav 104(2):266–271. PMCID: PMC3640446

Jönsson E (2013) The role of ghrelin in energy balance regulation in fish. Gen Comp Endocrinol 187:79–85. PMID: 23557643

Kalmbach DA, Anderson JR, Drake CL (2018) The impact of stress on sleep: pathogenic sleep reactivity as a vulnerability to insomnia and circadian disorders. J Sleep Res 27(6):e12710. PMCID: PMC7045300

Karst H, Berger S, Turiault M, Tronche F, Schütz G, Joëls M (2005) Mineralocorticoid receptors are indispensable for nongenomic modulation of hippocampal glutamate transmission by corticosterone. Proc Natl Acad Sci USA 102(52):19204–19207. PMCID: PMC1323174

Karst H, Berger S, Erdmann G, Schütz G, Joëls M (2010) Metaplasticity of amygdalar responses to the stress hormone corticosterone. Proc Natl Acad Sci U S A 107(32):14449–14454. PMCID: PMC2922581

Kaur S, Ryabinin AE (2010) Ghrelin receptor antagonism decreases alcohol consumption and activation of perioculomotor urocortin-containing neurons. Alcohol Clin Exp Res 34(9):1525–1534. PMCID: PMC2929279

Kelly-Irving M, Lepage B, Dedieu D, Lacey R, Cable N, Bartley M, Blane D, Grosclaude P, Lang T, Delpierre C (2013) Childhood adversity as a risk for cancer: findings from the 1958 British birth cohort study. BMC Public Health 13:767. PMCID: PMC3765119

Kern A, Albarran-Zeckler R, Walsh HE, Smith RG (2012) Apo-ghrelin receptor forms heteromers with DRD2 in hypothalamic neurons and is essential for anorexigenic effects of DRD2 agonism. Neuron 73(2):317–332. PMCID: PMC3269786

Kern A, Mavrikaki M, Ullrich C, Albarran-Zeckler R, Brantley AF, Smith RG (2015) Hippocampal Dopamine/DRD1 signaling dependent on the ghrelin receptor. Cell 163(5):1176–1190. PMCID: PMC4937825

Kim SH, Park MJ (2017) Effects of growth hormone on glucose metabolism and insulin resistance in human. Ann Pediatr Endocrinol Metab 22(3):145–152. PMCID: PMC5642081

Kim HH, Jeon TY, Park DY, Kim YJ, Lee SY, Lee JY, Lee JG, Jeong DW, Yi YH, Cho YH, Im SJ, Bae MJ, Choi EJ (2012) Differential expression of ghrelin mRNA according to anatomical portions of human stomach. Hepato-Gastroenterology 59(119):2217–2221. PMID: 22366529

King SJ, Isaacs AM, O'Farrell E, Abizaid A (2011) Motivation to obtain preferred foods is enhanced by ghrelin in the ventral tegmental area. Horm Behav 60(5):572–580. PMID: 21872601

Kluge M, Schüssler P, Dresler M, Schmidt D, Yassouridis A, Uhr M, Steiger A (2011) Effects of ghrelin on psychopathology, sleep and secretion of cortisol and growth hormone in patients with major depression. J Psychiatr Res 45(3):421–426. PMID: 20888580

Kojima M, Hosoda H, Date Y, Nakazato M, Matsuo H, Kangawa K (1999) Ghrelin is a growth-hormone-releasing acylated peptide from stomach. Nature 402(6762):656–660

Kong J, Chuddy J, Stock IA, Loria PM, Straub SV, Vage C, Cameron KO, Bhattacharya SK, Lapham K, McClure KF, Zhang Y, Jackson VM (2016) Pharmacological characterization of the first in class clinical candidate PF-05190457: a selective ghrelin receptor competitive antagonist with inverse agonism that increases vagal afferent firing and glucose-dependent insulin secretion ex vivo. Br J Pharmacol 173(9):1452–1464. PMCID: PMC4831304

Koopmann A, von der Goltz C, Grosshans M, Dinter C, Vitale M, Wiedemann K, Kiefer F (2012) The association of the appetitive peptide acetylated ghrelin with alcohol craving in early abstinent alcohol dependent individuals. Psychoneuroendocrinology 37(7):980–986. PMID: 22172639

Koopmann A, Bach P, Schuster R, Bumb JM, Vollstädt-Klein S, Reinhard I, Rietschel M, Witt SH, Wiedemann K, Kiefer F (2019) Ghrelin modulates mesolimbic reactivity to alcohol cues in alcohol-addicted subjects: a functional imaging study. Addict Biol 24(5):1066–1076. PMID: 29984874

Krause A, Sillard R, Kleemeier B, Klüver E, Maronde E, Conejo-García JR, Forssmann WG, Schulz-Knappe P, Nehls MC, Wattler F, Wattler S, Adermann K (2003) Isolation and biochemical characterization of LEAP-2, a novel blood peptide expressed in the liver. Protein Sci Publ Protein Soc 12(1):143–152. PMCID: PMC2312392

Kumar J, Chuang JC, Na ES, Kuperman A, Gillman AG, Mukherjee S, Zigman JM, McClung CA, Lutter M (2013) Differential effects of chronic social stress and fluoxetine on meal patterns in mice. Appetite 64:81–88. PMCID: PMC3606634

Kurt E, Guler O, Serteser M, Cansel N, Ozbulut O, Altinbaş K, Alataş G, Savaş H, Gecici O (2007) The effects of electroconvulsive therapy on ghrelin, leptin and cholesterol levels in patients with mood disorders. Neurosci Lett 426(1):49–53. PMID: 17884293

Kuti D, Winkler Z, Horváth K, Juhász B, Szilvásy-Szabó A, Fekete C, Ferenczi S, Kovács KJ (2022) The metabolic stress response: adaptation to acute-, repeated- and chronic challenges in mice. iScience 25(8):104693. PMCID: PMC9307515

La Du BN, Bartels CF, Nogueira CP, Hajra A, Lightstone H, Van der Spek A, Lockridge O (1990) Phenotypic and molecular biological analysis of human butyrylcholinesterase variants. Clin Biochem 23(5):423–431. PMID: 2253336

Landgren S, Jerlhag E, Zetterberg H, Gonzalez-Quintela A, Campos J, Olofsson U, Nilsson S, Blennow K, Engel JA (2008) Association of pro-ghrelin and GHS-R1A gene polymorphisms and haplotypes with heavy alcohol use and body mass. Alcohol Clin Exp Res 32(12):2054–2061. PMID: 18828808

Landgren S, Jerlhag E, Hallman J, Oreland L, Lissner L, Strandhagen E, Thelle DS, Zetterberg H, Blennow K, Engel JA (2010) Genetic variation of the ghrelin signaling system in females with severe alcohol dependence. Alcohol Clin Exp Res 34(9):1519–1524. PMID: 20586762

Leggio L, Zywiak WH, Fricchione SR, Edwards SM, de la Monte SM, Swift RM, Kenna GA (2014) Intravenous ghrelin administration increases alcohol craving in alcohol-dependent heavy drinkers: a preliminary investigation. Biol Psychiatry 76(9):734–741. PMCID: PMC4176606

Lewis JW, Cannon JT, Liebeskind JC (1980) Opioid and nonopioid mechanisms of stress analgesia. Science 208(4444):623–625. PMID: 7367889

Li RL, Sherbet DP, Elsbernd BL, Goldstein JL, Brown MS, Zhao TJ (2012) Profound hypoglycemia in starved, ghrelin-deficient mice is caused by decreased gluconeogenesis and reversed by lactate or fatty acids. J Biol Chem 287(22):17942–17950. PMCID: PMC3365699

Li G, Zhang K, Wang L, Cao C, Fang R, Liu P, Luo S, Liberzon I (2019) The preliminary investigation of orexigenic hormone gene polymorphisms on posttraumatic stress disorder symptoms. Psychoneuroendocrinology 100:131–136. PMID: 30326460

Longo KA, Charoenthongtrakul S, Giuliana DJ, Govek EK, McDonagh T, Qi Y, DiStefano PS, Geddes BJ (2008) Improved insulin sensitivity and metabolic flexibility in ghrelin receptor knockout mice. Regul Pept 150(1–3):55–61. PMID: 18453014

Lu Y, Niu M, Qiu X, Cao H, Xing B, Sun Y, Zhou Z, Zhou Y (2019) Acute but not chronic calorie restriction defends against stress-related anxiety and despair in a GHS-R1a-Dependent manner. Neuroscience 412:94–104. PMID: 31185255

Lugilde J, Casado S, Beiroa D, Cuñarro J, Garcia-Lavandeira M, Álvarez CV, Nogueiras R, Diéguez C, Tovar S (2022) LEAP-2 counteracts ghrelin-induced food intake in a nutrient, growth hormone and age independent manner. Cells 11(3):324. PMCID: PMC8834077

Lutter M, Sakata I, Osborne-Lawrence S, Rovinsky SA, Anderson JG, Jung S, Birnbaum S, Yanagisawa M, Elmquist JK, Nestler EJ, Zigman JM (2008) The orexigenic hormone ghrelin defends against depressive symptoms of chronic stress. Nat Neurosci 11(7):752–753. PMCID: PMC2765052

M'Kadmi C, Cabral A, Barrile F, Giribaldi J, Cantel S, Damian M, Mary S, Denoyelle S, Dutertre S, Péraldi-Roux S, Neasta J, Oiry C, Banères JL, Marie J, Perello M, Fehrentz JA (2019) N-Terminal Liver-Expressed Antimicrobial Peptide 2 (LEAP2) region exhibits inverse agonist activity toward the ghrelin receptor. J Med Chem 62(2):965–973. PMID: 30543423

Malik MO, Shah M, Irfan Ul Akbar Yousufzai M, Ullah N, Burgess JA, Goosens KA (2020) Association of Acyl-Ghrelin with posttraumatic stress disorder in adolescents who experienced severe trauma. JAMA Netw Open 3(8):e2013946

Mani BK, Walker AK, Lopez Soto EJ, Raingo J, Lee CE, Perelló M, Andrews ZB, Zigman JM (2014) Neuroanatomical characterization of a growth hormone secretagogue receptor-green fluorescent protein reporter mouse. J Comp Neurol 522(16):3644–3666. PMCID: PMC4142102

Mani BK, Puzziferri N, He Z, Rodriguez JA, Osborne-Lawrence S, Metzger NP, Chhina N, Gaylinn B, Thorner MO, Thomas EL, Bell JD, Williams KW, Goldstone AP, Zigman JM (2019) LEAP2 changes with body mass and food intake in humans and mice. J Clin Invest 129(9):3909–3923. PMCID: PMC6715358

Marazziti D, Di Muro A, Castrogiovanni P (1992) Psychological stress and body temperature changes in humans. Physiol Behav 52(2):393–395. PMID: 1326118

Matsuo K, Nakano M, Nakashima M, Watanuki T, Egashira K, Matsubara T, Watanabe Y (2012) Neural correlates of plasma acylated ghrelin level in individuals with major depressive disorder. Brain Res 1473:185–192. PMID: 22819931

Meinel S, Gekle M, Grossmann C (2014) Mineralocorticoid receptor signaling: crosstalk with membrane receptors and other modulators. Steroids 91:3–10

Méquinion M, Langlet F, Zgheib S, Dickson S, Dehouck B, Chauveau C, Viltart O (2013) Ghrelin: central and peripheral implications in anorexia nervosa. Front Endocrinol 4:15. PMCID: PMC3581855

Meyer RM, Burgos-Robles A, Liu E, Correia SS, Goosens KA (2014) A ghrelin-growth hormone axis drives stress-induced vulnerability to enhanced fear. Mol Psychiatry 19(12):1284–1294

Mora M, Adam V, Palomera E, Blesa S, Díaz G, Buquet X, Serra-Prat M, Martín-Escudero JC, Palanca A, Chaves JF, Puig-Domingo M (2015) Ghrelin gene variants influence on metabolic syndrome components in aged spanish population. PLoS ONE 10(9):e0136931. PMCID: PMC4573319

Murtuza MI, Isokawa M (2018) Endogenous ghrelin-O-acyltransferase (GOAT) acylates local ghrelin in the hippocampus. J Neurochem 144(1):58–67. PMCID: PMC5832437

Mustafá ER, Cordisco González S, Damian M, Cantel S, Denoyelle S, Wagner R, Schiöth HB, Fehrentz JA, Banères JL, Perelló M, Raingo J (2021) LEAP2 impairs the capability of the growth hormone secretagogue receptor to regulate the Dopamine 2 receptor signaling. Front Pharmacol 12:712437. PMCID: PMC8383165

Nakagawa E, Nagaya N, Okumura H, Enomoto M, Oya H, Ono F, Hosoda H, Kojima M, Kangawa K (2002) Hyperglycaemia suppresses the secretion of ghrelin, a novel growth-hormone-releasing peptide: responses to the intravenous and oral administration of glucose. Clin Sci Lond Engl 103(3):325–328. PMID: 12193159

Nakamura K (2015) Neural circuit for psychological stress-induced hyperthermia. Temperature 2(3):352–361. PMCID: PMC4843917

Navarro G, Rea W, Quiroz C, Moreno E, Gomez D, Wenthur CJ, Casadó V, Leggio L, Hearing MC, Ferré S (2022) Complexes of Ghrelin GHS-R1a, GHS-R1b, and dopamine D1 receptors localized in the ventral tegmental area as main mediators of the dopaminergic effects of ghrelin. J Neurosci Off J Soc Neurosci 42(6):940–953. PMCID: PMC8824505

Nestler EJ, Russo SJ (2024) Neurobiological basis of stress resilience. Neuron 112(12):1911–1929. PMCID: PMC11189737

Oka T (2015) Psychogenic fever: how psychological stress affects body temperature in the clinical population. Temp Multidiscip Biomed J 2(3):368–378. PMCID: PMC4843908

Oka T (2018) Stress-induced hyperthermia and hypothermia. Handb Clin Neurol 157:599–621. PMID: 30459027

Ozsoy S, Besirli A, Abdulrezzak U, Basturk M (2014) Serum ghrelin and leptin levels in patients with depression and the effects of treatment. Psychiatry Investig 11(2):167–172. PMCID: PMC4023091

Pabst S, Schoofs D, Pawlikowski M, Brand M, Wolf OT (2013) Paradoxical effects of stress and an executive task on decisions under risk. Behav Neurosci 127(3):369–379. PMID: 23544598

Pace TWW, Gaylord RI, Jarvis E, Girotti M, Spencer RL (2009) Differential glucocorticoid effects on stress-induced gene expression in the paraventricular nucleus of the hypothalamus and ACTH secretion in the rat. Stress 12(5):400–411

Page LC, Gastaldelli A, Gray SM, D'Alessio DA, Tong J (2018) Interaction of GLP-1 and ghrelin on glucose tolerance in healthy humans. Diabetes 67(10):1976–1985. PMCID: PMC6152343

Park S, Jiang H, Zhang H, Smith RG (2012) Modification of ghrelin receptor signaling by somatostatin receptor-5 regulates insulin release. Proc Natl Acad Sci U S A 109(46):19003–19008. PMCID: PMC3503195

Pascual-Le Tallec L, Lombès M (2005) The mineralocorticoid receptor: a journey exploring its diversity and specificity of action. Mol Endocrinol 19(9):2211–2221

Patterson ZR, Ducharme R, Anisman H, Abizaid A (2010) Altered metabolic and neurochemical responses to chronic unpredictable stressors in ghrelin receptor-deficient mice. Eur J Neurosci 32(4):632–639. PMID: 20597975

Patterson ZR, Khazall R, Mackay H, Anisman H, Abizaid A (2013) Central ghrelin signaling mediates the metabolic response of C57BL/6 male mice to chronic social defeat stress. Endocrinology 154(3):1080–1091. PMID: 23341196

Perello M, Sakata I, Birnbaum S, Chuang JC, Osborne-Lawrence S, Rovinsky SA, Woloszyn J, Yanagisawa M, Lutter M, Zigman JM (2010) Ghrelin Increases the rewarding value of high-fat diet in an orexin-dependent manner. Biol Psychiatry 67(9):880–886

Perumal MB, Sah P (2021) Inhibitory circuits in the basolateral amygdala in aversive learning and memory. Front Neural Circuits 15:633235. PMCID: PMC8120102

Pfabigan DM, Frogner ER, Schéle E, Thorsby PM, Skålhegg BS, Dickson SL, Sailer U (2024) Ghrelin is related to lower brain reward activation during touch. Psychophysiology 61(2):e14443. PMID: 37737514

Pietrzak M, Yngve A, Hamilton JP, Kämpe R, Boehme R, Asratian A, Gauffin E, Löfberg A, Gustavson S, Persson E, Capusan AJ, Leggio L, Perini I, Tinghög G, Heilig M (2023a) A randomized controlled experimental medicine study of ghrelin in value-based decision making. J Clin Invest 133(12). Available from: https://www.jci.org/articles/view/168260. PMID: 0

Pietrzak M, Yngve A, Hamilton JP, Kämpe R, Boehme R, Asratian A, Gauffin E, Löfberg A, Gustavson S, Persson E, Capusan AJ, Leggio L, Perini I, Tinghög G, Heilig M (2023b) A randomized controlled experimental medicine study of ghrelin in value-based decision making. J Clin Invest 133(12):e168260. PMCID: PMC10266781

Pietrzak M, Yngve A, Hamilton JP, Asratian A, Gauffin E, Löfberg A, Gustavson S, Persson E, Capusan AJ, Leggio L, Perini I, Tinghög G, Heilig M, Boehme R (2024) Ghrelin decreases sensitivity to negative feedback and increases prediction-error related caudate activity in humans, a randomized controlled trial. Neuropsychopharmacol 49(6):1042–1049. PMCID: PMC11039644

Radley JJ, Sawchenko PE (2011) A common substrate for prefrontal and hippocampal inhibition of the neuroendocrine stress response. J Neurosci 31(26):9683–9695. PMCID: PMC3197245

Rayatpour A, Radahmadi M, Izadi MS, Ghasemi M (2023) Effects of sub-chronic CRH administration into the hypothalamic paraventricular and central amygdala nuclei in male rats with a focus on food intake biomarkers. An Acad Bras Cienc 95(4):e20200221. PMID: 38088701

Rediger A, Piechowski CL, Yi CX, Tarnow P, Strotmann R, Grüters A, Krude H, Schöneberg T, Tschöp MH, Kleinau G, Biebermann H (2011) Mutually opposite signal modulation by hypothalamic heterodimerization of ghrelin and melanocortin-3 receptors. J Biol Chem 286(45):39623–39631. PMCID: PMC3234785

Reichenbach A, Steyn FJ, Sleeman MW, Andrews ZB (2012) Ghrelin receptor expression and colocalization with anterior pituitary hormones using a GHSR-GFP mouse line. Endocrinology 153(11):5452–5466

Reul JM, van den Bosch FR, de Kloet ER (1987) Relative occupation of type-I and type-II corticosteroid receptors in rat brain following stress and dexamethasone treatment: functional implications. J Endocrinol 115(3):459–467. PMID: 3443807

Reynolds EK, Schreiber WM, Geisel K, MacPherson L, Ernst M, Lejuez CW (2013) Influence of social stress on risk-taking behavior in adolescents. J Anxiety Disord 27(3):272–277. PMCID: PMC3693744

Richardson RS, Kryszak LA, Vendruscolo JCM, Koob GF, Vendruscolo LF, Leggio L (2024) GHSR blockade, but not reduction of peripherally circulating ghrelin via β1-adrenergic receptor antagonism, decreases binge-like alcohol drinking in mice. Mol Psychiatry. PMID: 39232198

Ringuet MT, Furness JB, Furness SGB (2022) G protein-coupled receptor interactions and modification of signalling involving the ghrelin receptor, GHSR1a. J Neuroendocrinol 34(9):e13077

Rossi E, Cassioli E, Gironi V, Idrizaj E, Garella R, Squecco R, Baccari MC, Maggi M, Vignozzi L, Comeglio P, Ricca V, Castellini G (2021) Ghrelin as a possible biomarker and maintaining factor in patients with eating disorders reporting childhood traumatic experiences. Eur Eat Disord Rev 29(4):588–599. PMCID: PMC8251850

Sailer U, Riva F, Lieberz J, Campbell-Meiklejohn D, Scheele D, Pfabigan DM (2023) Hungry for compliments? Ghrelin is not associated with neural responses to social rewards or their pleasantness. Front Psychiatry 14:1104305. PMCID: PMC10106620

Sbisa AM, Madden K, Toben C, McFarlane AC, Dell L, Lawrence-Wood E (2023) Potential peripheral biomarkers associated with the emergence and presence of posttraumatic stress disorder symptomatology: a systematic review. Psychoneuroendocrinology 147:105954

Schanze A, Reulbach U, Scheuchenzuber M, Groschl M, Kornhuber J, Kraus T (2008) Ghrelin and eating disturbances in psychiatric disorders. Neuropsychobiology 57(3):126–130. PMID: 18552514

Schéle E, Bake T, Rabasa C, Dickson SL (2016) Centrally administered ghrelin acutely influences food choice in rodents. PLoS One 11(2):e0149456. PMCID: PMC4771210

Schellekens H, van Oeffelen WEPA, Dinan TG, Cryan JF (2013) Promiscuous dimerization of the growth hormone secretagogue receptor (GHS-R1a) attenuates ghrelin-mediated signaling. J Biol Chem 288(1):181–191. PMCID: PMC3537012

Schellekens H, De Francesco PN, Kandil D, Theeuwes WF, McCarthy T, van Oeffelen WEPA, Perelló M, Giblin L, Dinan TG, Cryan JF (2015) Ghrelin's orexigenic effect is modulated via a serotonin 2C receptor interaction. ACS Chem Neurosci 6(7):1186–1197. PMID: 25727097

Schopfer LM, Lockridge O, Brimijoin S (2015) Pure human butyrylcholinesterase hydrolyzes octanoyl ghrelin to desacyl ghrelin. Gen Comp Endocrinol 224:61–68

Shiimura Y, Im D, Tany R, Asada H, Kise R, Kurumiya E, Wakasugi-Masuho H, Yasuda S, Matsui K, Kishikawa JI, Kato T, Murata T, Kojima M, Iwata S, Masuho I (2025) The structure and function of the ghrelin receptor coding for drug actions. Nat Struct Mol Biol. PMID: 39833471

Shiiya T, Nakazato M, Mizuta M, Date Y, Mondal MS, Tanaka M, Nozoe SI, Hosoda H, Kangawa K, Matsukura S (2002) Plasma ghrelin levels in lean and obese humans and the effect of glucose on ghrelin secretion. J Clin Endocrinol Metab 87(1):240–244. PMID: 11788653

Shimbara T, Mondal MS, Kawagoe T, Toshinai K, Koda S, Yamaguchi H, Date Y, Nakazato M (2004) Central administration of ghrelin preferentially enhances fat ingestion. Neurosci Lett 369(1):75–79. PMID: 15380311

Shin HS, Lee SH, Moon HJ, So YH, Jang HJ, Lee KH, Ahn C, Jung EM (2024) Prolonged stress response induced by chronic stress and corticosterone exposure causes adult neurogenesis inhibition and astrocyte loss in mouse hippocampus. Brain Res Bull 208:110903

Skibicka KP, Hansson C, Alvarez-Crespo M, Friberg PA, Dickson SL (2011) Ghrelin directly targets the ventral tegmental area to increase food motivation. Neuroscience 180:129–137. PMID: 21335062

Skibicka KP, Hansson C, Egecioglu E, Dickson SL (2012) Role of ghrelin in food reward: impact of ghrelin on sucrose self-administration and mesolimbic dopamine and acetylcholine receptor gene expression. Addict Biol 17(1):95–107. PMCID: PMC3298643

Smith A, Hyland L, Al-Ansari H, Watts B, Silver Z, Wang L, Dahir M, Akgun A, Telfer A, Abizaid A (2023) Metabolic, neuroendocrine and behavioral effects of social defeat in male and female mice using the chronic non-discriminatory social defeat stress model. Horm Behav 155:105412. PMID: 37633226

Smith A, Rodrigues T, Wallace C, Mezher K, MacAulay B, Prowse R, Hyland L, Abizaid A (2024a) Growth Hormone Secretagogue Receptor (GHSR) Signaling in the Ventral Tegmental Area (VTA) mediates feeding produced by chronic social defeat stress in male mice. Neuroscience 547:17–27. PMID: 38583506

Smith A, MacAulay B, Scheufen J, Hudak A, Abizaid A (2024b) Chronic social defeat stress increases brain permeability to ghrelin in male mice. eNeuro 11(7):ENEURO.0093-24.2024. PMCID: PMC11253241

Song L, Zhu Q, Liu T, Yu M, Xiao K, Kong Q, Zhao R, Li GD, Zhou Y (2013) Ghrelin modulates lateral amygdala neuronal firing and blocks acquisition for conditioned taste aversion. PLoS One 8(6):e65422. PMCID: PMC3676403

Spencer SJ, Xu L, Clarke MA, Lemus M, Reichenbach A, Geenen B, Kozicz T, Andrews ZB (2012) Ghrelin regulates the hypothalamic-pituitary-adrenal axis and restricts anxiety after acute stress. Biol Psychiatry 72(6):457–465. PMID: 22521145

Stark R, Dempsey H, Kleeman E, Sassi M, Osborne-Lawrence S, Sheybani-Deloui S, Rushby HJ, Mirth CK, Austin-Muttitt K, Mullins J, Zigman JM, Davies JS, Andrews ZB (2024) Hunger signalling in the olfactory bulb primes exploration, food-seeking and peripheral metabolism. Mol Metab 89:102025. PMCID: PMC11471258

Sterlemann V, Ganea K, Liebl C, Harbich D, Alam S, Holsboer F, Müller MB, Schmidt MV (2008) Long-term behavioral and neuroendocrine alterations following chronic social stress in mice: implications for stress-related disorders. Horm Behav 53(2):386–394. PMID: 18096163

St-Onge V, Watts A, Abizaid A (2016) Ghrelin enhances cue-induced bar pressing for high fat food. Horm Behav 78:141–149. PMID: 26592452

Suchankova P, Steensland P, Fredriksson I, Engel JA, Jerlhag E (2013) Ghrelin receptor (GHS-R1A) antagonism suppresses both alcohol consumption and the alcohol deprivation effect in rats following long-term voluntary alcohol consumption. PLoS One 8(8):e71284s

Suchankova P, Nilsson S, von der Pahlen B, Santtila P, Sandnabba K, Johansson A, Jern P, Engel JA, Jerlhag E (2016) Genetic variation of the growth hormone secretagogue receptor gene is associated with alcohol use disorders identification test scores and smoking. Addict Biol 21(2):481–488. PMCID: PMC5033010

Sun Y, Butte NF, Garcia JM, Smith RG (2008) Characterization of adult ghrelin and ghrelin receptor knockout mice under positive and negative energy balance. Endocrinology 149(2):843–850. PMCID: PMC2219310

Surjit M, Ganti KP, Mukherji A, Ye T, Hua G, Metzger D, Li M, Chambon P (2011) Widespread negative response elements mediate direct repression by agonist-liganded glucocorticoid receptor. Cell 145(2):224–241. PMID: 21496643

Sustkova-Fiserova M, Puskina N, Havlickova T, Lapka M, Syslova K, Pohorala V, Charalambous C (2020) Ghrelin receptor antagonism of fentanyl-induced conditioned place preference, intravenous self-administration, and dopamine release in the nucleus accumbens in rats. Addict Biol 25(6):e12845. PMID: 31696597

Swarbrick M, Zhou H, Seibel M (2021) MECHANISMS IN ENDOCRINOLOGY: local and systemic effects of glucocorticoids on metabolism: new lessons from animal models. Eur J Endocrinol 185(5):R113–R129. PMID: 34478405

Tafet GE, Nemeroff CB (2016) The links between stress and depression: psychoneuroendocrinological, genetic, and environmental interactions. J Neuropsychiatry Clin Neurosci. American Psychiatric Publishing 28(2):77–88

Tasci G, Kaya S, Kalayci M, Atmaca M (2022) Increased ghrelin and decreased leptin levels in patients with antisocial personality disorder. J Affect Disord 317:22–28. PMID: 36028010

Tasker JG, Herman JP (2011) Mechanisms of rapid glucocorticoid feedback inhibition of the hypothalamic-pituitary-adrenal axis. Stress Amst Neth 14(4):398–406. PMCID: PMC4675656

Tezenas du Montcel C, Cao J, Mattioni J, Hamelin H, Lebrun N, Ramoz N, Gorwood P, Tolle V, Viltart O (2023) Chronic food restriction in mice and increased systemic ghrelin induce preference for running wheel activity. Psychoneuroendocrinology 155:106311. PMID: 37295225

Tong J, Prigeon RL, Davis HW, Bidlingmaier M, Kahn SE, Cummings DE, Tschöp MH, D'Alessio D (2010) Ghrelin suppresses glucose-stimulated insulin secretion and deteriorates glucose tolerance in healthy humans. Diabetes 59(9):2145–2151. PMCID: PMC2927935

Tong J, Mannea E, Aimé P, Pfluger PT, Yi CX, Castaneda TR, Davis HW, Ren X, Pixley S, Benoit S, Julliard K, Woods SC, Horvath TL, Sleeman MM, D'Alessio D, Obici S, Frank R, Tschöp MH (2011) Ghrelin enhances olfactory sensitivity and exploratory sniffing in rodents and humans. J Neurosci 31(15):5841–5846. PMCID: PMC3089941

Tschöp M, Smiley DL, Heiman ML (2000) Ghrelin induces adiposity in rodents. Nature. Nature Publishing Group 407(6806):908–913

Tufvesson-Alm M, Zhang Q, Aranäs C, Blid Sköldheden S, Edvardsson CE, Jerlhag E (2024) Decoding the influence of central LEAP2 on food intake and its effect on accumbal dopamine release. Prog Neurobiol 236:102615. PMID: 38641041

Ueberberg B, Unger N, Saeger W, Mann K, Petersenn S (2009) Expression of ghrelin and its receptor in human tissues. Horm Metab Res 41(11):814–821. PMID: 19670151

Veenema AH (2009) Early life stress, the development of aggression and neuroendocrine and neurobiological correlates: what can we learn from animal models? Front Neuroendocrinol 30(4):497–518

Vestlund J, Winsa-Jörnulf J, Hovey D, Lundström S, Lichtenstein P, Anckarsäter H, Studer E, Suchankova P, Westberg L, Jerlhag E (2019) Ghrelin and aggressive behaviours-evidence from preclinical and human genetic studies. Psychoneuroendocrinology 104:80–88. PMID: 30818255

Vivenza D, Rapa A, Castellino N, Bellone S, Petri A, Vacca G, Aimaretti G, Broglio F, Bona G (2004) Ghrelin gene polymorphisms and ghrelin, insulin, IGF-I, leptin and anthropometric data in children and adolescents. Eur J Endocrinol 151(1):127–133

Wallace Fitzsimons SE, Chruścicka B, Druelle C, Stamou P, Nally K, Dinan TG, Cryan JF, Schellekens H (2019) A ghrelin receptor and oxytocin receptor heterocomplex impairs oxytocin mediated signalling. Neuropharmacology 152:90–101. PMID: 30582955

Wierup N, Björkqvist M, Weström B, Pierzynowski S, Sundler F, Sjölund K (2007) Ghrelin and motilin are cosecreted from a prominent endocrine cell population in the small intestine. J Clin Endocrinol Metab 92(9):3573–3581. PMID: 17595255

Wiley JW, Higgins GA, Athey BD (2016) Stress and glucocorticoid receptor transcriptional programming in time and space: implications for the brain–gut axis. Neurogastroenterol Motil 28(1):12–25. PMCID: PMC4688904

Wittekind DA, Kratzsch J, Mergl R, Riedel-Heller S, Witte AV, Villringer A, Kluge M (2022) Serum ghrelin is positively associated with physiological anxiety but negatively associated with pathological anxiety in humans: data from a large community-based study. Psychoneuroendocrinology 140:105728. PMID: 35305404

Wittekind DA, Kratzsch J, Mergl R, Wirkner K, Baber R, Sander C, Witte AV, Villringer A, Kluge M (2023) Childhood sexual abuse is associated with higher total ghrelin serum levels in adulthood: results from a large, population-based study. Transl Psychiatry 13(1):219. PMCID: PMC10287627

Yang S, Zhang L (2004) Glucocorticoids and vascular reactivity. Curr Vasc Pharmacol 2(1):1–12. PMID: 15320828

Yang J, Brown MS, Liang G, Grishin NV, Goldstein JL (2008) Identification of the acyltransferase that octanoylates ghrelin, an appetite-stimulating peptide hormone. Cell 132(3):387–396. PMID: 18267071

Yousufzai M, Harmatz ES, Shah M, Malik MO, Goosens KA (2018) Ghrelin is a persistent biomarker for chronic stress exposure in adolescent rats and humans. Transl Psychiatry 8(1):74

Zhang Y, Fang F, Goldstein JL, Brown MS, Zhao TJ (2015) Reduced autophagy in livers of fasted, fat-depleted, ghrelin-deficient mice: reversal by growth hormone. Proc Natl Acad Sci USA 112(4):1226–1231. PMCID: PMC4313810

Zhang L, Wang HL, Zhang YF, Mao XT, Wu TT, Huang ZH, Jiang WJ, Fan KQ, Liu DD, Yang B, Zhuang MH, Huang GM, Liang Y, Zhu SJ, Zhong JY, Xu GY, Li XM, Cao Q, Li YY, Jin J (2025) Stress triggers irritable bowel syndrome with diarrhea through a spermidine-mediated decline in type I interferon. Cell Metab 37(1):87–103.e10. PMID: 39366386

Zhao TJ, Sakata I, Li RL, Liang G, Richardson JA, Brown MS, Goldstein JL, Zigman JM (2010a) Ghrelin secretion stimulated by {beta}1-adrenergic receptors in cultured ghrelinoma cells and in fasted mice. Proc Natl Acad Sci USA 107(36):15868–15873. PMCID: PMC2936616

Zhao TJ, Liang G, Li RL, Xie X, Sleeman MW, Murphy AJ, Valenzuela DM, Yancopoulos GD, Goldstein JL, Brown MS (2010b) Ghrelin O-acyltransferase (GOAT) is essential for growth hormone-mediated survival of calorie-restricted mice. Proc Natl Acad Sci USA 107(16):7467–7472. PMCID: PMC2867684

Zigman JM, Jones JE, Lee CE, Saper CB, Elmquist JK (2006) Expression of ghrelin receptor mRNA in the rat and the mouse brain. J Comp Neurol 494(3):528–548. PMCID: PMC4524499

Zoon HFA, de Bruijn SEM, Smeets PAM, de Graaf C, Janssen IMC, Schijns W, Aarts EO, Jager G, Boesveldt S (2018) Altered neural responsivity to food cues in relation to food preferences, but not appetite-related hormone concentrations after RYGB-surgery. Behav Brain Res 353:194–202. PMID: 30041007

The Connection Between the Appetite-Regulatory Peptides Ghrelin and GLP-1 and Alcohol Use Disorder

8

Elisabet Jerlhag

Abstract

The body-brain connection is well-established, further evidenced by studies on the orexigenic peptide ghrelin and the anorexigenic peptide glucagon-like peptide-1 (GLP-1). The ghrelin pathway consists of several substrates, which has been studied in relation to alcohol-related responses. Preclinical studies have found that central ghrelin infusions elevate alcohol intake, whereas suppression of the ghrelin receptors (GHSR) attenuates alcohol-related responses. On a similar note, the endogenous GHSR inverse agonist, LEAP2, and the precursor of ghrelin, DAG, block the stimulatory properties of alcohol and reduce alcohol intake. GLP-1 receptor agonists are currently approved for the treatment of type 2 diabetes and obesity. Recent advances in rodents have extended the pharmacological relevance of these GLP-1 receptor agonists as they mitigate alcohol-related responses. Specifically, in animal models reflecting alcohol use disorder (AUD) all tested GLP-1 receptor agonists reduce alcohol intake, suppress the motivation to consume alcohol, and prevent relapse drinking, effects most likely driven by an attenuation of alcohol-induced reward. Additional experiments were conducted in attempts to define areas and circuits responsible for ghrelin and GLP-1 to control alcohol-related responses. Specifically, areas associated with reward were defined as important modulatory areas. In summary, the ghrelin pathway and GLP-1 participate in the pathophysiology of AUD, thereby providing tentative treatment targets.

Keywords

Appetite-regulatory peptides · Gut-brain axis · Alcohol · Dopamine · Reward · Addiction · Dependence

Abbreviation

ADE	Alcohol deprivation effect
AUD	Alcohol use disorder
AUDIT	Alcohol Use Disorders Identification Test
CPP	Conditioned place preference
DAG	Des-acyl-ghrelin
DPP-IV	Dipeptidyl peptidase-IV
Ex4	Exendin-4

E. Jerlhag (✉)
Institute of Neuroscience and Physiology, Department of Pharmacology, The Sahlgrenska Academy at the University of Gothenburg, Gothenburg, Sweden
e-mail: Elisabet.Jerlhag@pharm.gu.se

D. Tropea, E. Giacometti (eds.), *Brain-Body Connections*, Advances in Experimental Medicine and Biology 1477, https://doi.org/10.1007/978-3-031-89525-8_8

GHSR	Growth hormone secretagogue receptor; Ghrelin receptor
GLP-1	Glucagon-like peptide-1
GLP-1R	Glucagon-like peptide-1 receptor
LEAP2	Liver-expressed antimicrobial peptide 2
LDTg	Laterodorsal tegmental area
NAc	Nucleus accumbens
NTS	Nucleus tractus solitarius
VTA	Ventral tegmental area

8.1 Ghrelin and Glucagon-Like Peptide-1

The connection between the brain and body is supported by studies on the gut-brain axis. Peptides associated with this axis are traditionally known for their ability to control feeding and appetite but have, over the last years, been attributed to an additional role in reward regulation. An extensive number of gut-brain peptides have been identified, where ghrelin and glucagon-like peptide-1 (GLP-1) appear to be central players.

Ghrelin is the first identified orexigenic peptide (Kojima et al. 1999), which is increased preprandially to signal appetite (Tschop et al. 2001). It is mainly produced by the stomach's endocrine cells (Kojima et al. 1999) and tentatively by neurons in the brain (Cowley et al. 2003; Lu et al. 2002; Mondal et al. 2005). Ghrelin exerts its effects via peripheral and centrally located ghrelin receptors (GHSR, Growth hormone secretagogue receptor) (for reviews, see, *e.g.*, Wolfe and Maisto (2000), Morganstern et al. (2011)). Besides being a traditional G-protein coupled receptor, GHSR has been identified to have intrinsic activity as well as a heterodimerization ability (for review see Cornejo et al. (2021)). Ghrelin was initially recognized for its orexigenic properties, as it elevates homeostatic feeding in both rodents and men (for review, see Cornejo et al. (2021)). Since these initial studies, it has been found that ghrelin also augments hedonic feeding via reward-related areas in the brain (for review, see Cornejo et al. (2021)). Although ghrelin regulates body weight and adiposity, no ghrelin receptor ligands exist clinically, and only experimental antagonists are used in research. Ghrelin has previously been referred to as the active form of ghrelin; however, the deacylated form of ghrelin, i.e., des-acyl-ghrelin (DAG), has, during later years, been attributed several physiological properties (De Vriese et al. 2004; Gauna et al. 2006; Hosoda et al. 2000; Delhanty et al. 2010; Inoue et al. 2013; Fernandez et al. 2016). An endogenous inverse agonist to the GHSR has been identified, namely the liver-expressed antimicrobial peptide 2 (LEAP2) (Ge et al. 2018; M'Kadmi et al. 2019). The production of LEAP2 is located in the liver and the small intestine (Krause et al. 2003), but given the central gene expression of LEAP2, central production may be possible (Islam et al. 2020; Tufvesson-Alm et al. 2024a). As ghrelin, LEAP2 controls the consumption of regular food (e.g., chow) (Ge et al. 2018) and palatable foods (e.g., high-fat diet intake) (Tufvesson-Alm et al. 2024a; Cornejo et al. 2019).

There are several anorexigenic peptides; however, glucagon-like peptide-1 (GLP-1) has gained extensive attention as agonists targeting its receptor (GLP-1R) are approved for type 2 diabetes and obesity. L-cells of the intestine account for the major GLP-1 production (Novak et al. 1987), whereas GLP-1 production has been found in neurons of the nucleus tractus solitarius (NTS) (Alvarez et al. 1996). After its production, GLP-1 binds to receptors expressed in the body as well as the brain (for reviews see, e.g., Wolfe and Maisto (2000), Morganstern et al. (2011)). The half-life of GLP-1 is approximately a few minutes as it is rapidly degraded by the dipeptidyl peptidase-IV (DPP-IV) enzyme, and following its inhibition, the GLP-1 levels in blood are increased. As for ghrelin, the physiological properties of GLP-1 are vast and include a reduction in hedonic and homeostatic food intake (for review see Srivastava and Apovian (2018)). Signals from the gut microbiota also control satiation, in fact, obesity has been linked to an altered microbiota composition, as discussed in the Chap. 6.

Additionally, GLP-1 modulates glucose-dependent insulin secretion, leading to the approval of GLP-1R agonists as type 2 diabetes treatment (Holst and Seino 2009). Besides treatment for diabetes type 2, GLP-1R agonists are used as anti-obesity medications as activation of GLP-1R reduces body weight (for review see Srivastava and Apovian (2018)). GLP-1R agonists exist as different trademarks with different durations, including short- (Exendin-4 (Ex4)/exenatide), moderate- (liraglutide), and long-acting (e.g., dulaglutide or semaglutide).

Although the ghrelin pathway and GLP-1 signaling system are well-known for controlling energy and glucose homeostasis, they have been attributed to an important regulatory role in reward and addiction, summarized below (Fig. 8.1).

8.1.1 Effects on Alcohol Intake: Preclinical Studies

In rodents, various drinking paradigms, such as intermittent access, continuous access, and drinking in the dark, are commonly used to investigate the impact of treatment on alcohol intake.

8.1.1.1 GHSR and LEAP2

The impact of GHSR antagonism on alcohol consumption has substantially been studied in rodent studies. This was initially demonstrated in 2009, as either genetic or pharmacological suppression of the GHSR lowers alcohol intake in male mice (Jerlhag et al. 2009). This reduction is further evident in male rats exposed to alcohol for long periods before treatment. As such, acute or repeated injections with GHSR antagonists decrease alcohol consumption in male high alcohol-preferring rats (both outbred and genetically modified AA) (Landgren et al. 2012; Suchankova et al. 2013). Clinical relevance may be provided, as GHSR antagonism's efficacy in male rats correlates with the duration and quantity of alcohol consumption, showing the most significant impact in cases of prolonged, heavy drinking (Suchankova et al. 2013). This study further revealed prolonged GHSR antagonist treatment does not cause tolerance development (Suchankova et al. 2013). A reduction in alcohol intake after GHSR antagonist treatment is further observed in several species (male mice/rats/prairie voles) with shorter alcohol access to alcohol before treatment (Bahi et al. 2013; Gomez et al. 2015; Kaur and Ryabinin 2010; Gomez and Ryabinin 2014; Stevenson et al. 2015). Notably, the lowered alcohol consumption is evident both after acute and repeated administrations of the GHSR antagonists (Bahi et al. 2013; Gomez et al. 2015; Kaur and Ryabinin 2010; Gomez and Ryabinin 2014; Stevenson et al. 2015). Male and female mice exposed to a binge drinking model exhibit reduced binge drinking behavior after GHSR antagonist treatment, demonstrating for the first time that GHSR antagonists effectively decrease alcohol consumption in females (Richardson et al. 2023). Further on, the reduced alcohol intake was similar between sexes, further demonstrating potential in a clinical setting (Richardson et al. 2023). While these initial studies used animals of regular weight, these findings extend to obese rats.

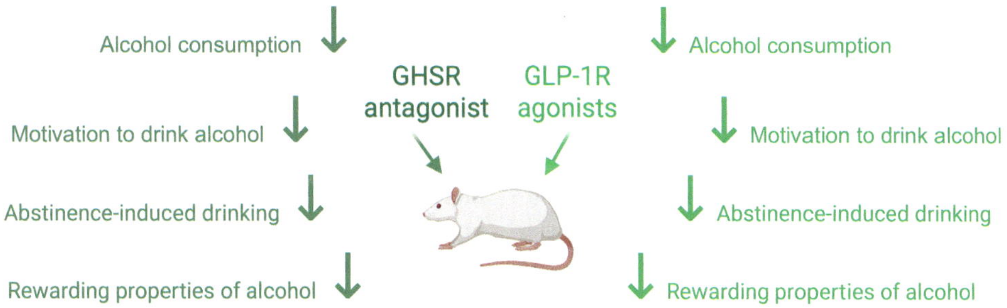

Fig. 8.1 Summary of preclinical studies. Ghrelin receptor (GHSR) antagonists suppress (↓) alcohol-related responses in rodents, effects also observed by GLP-1 receptor (GLP-1R) agonists. (The illustration was Created with BioRender.com)

Indeed, targeting GHSR pharmacologically led to a significant reduction in alcohol consumption among obese rats that underwent vertical sleeve gastrectomy or Roux-en-Y gastric bypass (Hajnal et al. 2012; Orellana et al. 2018). Moreover, inhibiting the GHSR may help prevent the development of AUD, as GHSR knockout rodents show lower alcohol consumption than their wild-type counterparts in a continuous access model (Jerlhag et al. 2009) and drinking in the dark model (Zallar et al. 2019). The initial study revealing that central GHSR appears to control alcohol intake (Jerlhag et al. 2009) was recently verified in a binge drinking model of AUD as central injection of GHSR antagonists reduced binge drinking in male and female mice (Richardson et al. 2023). Further support for central GHSR is provided by studies on LEAP2, the endogenous inverse GHSR agonist. Central, but not peripheral, LEAP2 infusion reduces alcohol intake in male and female rodents (Richardson et al. 2023; Tufvesson-Alm et al. 2024b). Central GHSR appears to modulate alcohol drinking, as repeated infusions of a GHSR antagonist reduce alcohol intake in male rats (Richardson et al. 2024a). Additional studies have found some of the circuits and brain regions central to alcohol consumption, where NAc appears to be one region of interest. Specifically, the expression of GHSR in NAc is significantly higher in rats that consumed high levels of alcohol compared to those with low alcohol consumption (Suchankova et al. 2013), and a lower alcohol intake is observed in female mice after local infusion of a GHSR antagonist into NAc (Abtahi et al. 2018). As the ventral tegmental area (VTA) projects dopaminergic neurons to NAc, as evidenced by dopamine release after activation, it may be of potential interest for the GHSR-alcohol connection. Indeed, GHSR on these dopaminergic-VTA neurons controls reward evoked by optogenetic stimulation (You et al. 2022), and rats that consumed high amounts of alcohol exhibit lower GHSR expression than those that consumed less (Suchankova et al. 2013). The amygdala is another region targeted by the dopaminergic projection of the VTA and GHSR protein levels, and immunoreactive cells in this region are elevated by high alcohol consumption (Yoshimoto et al. 2017). As evident from male mice with an intact vagus nerve GHSR antagonists cannot suppress alcohol intake, indicating that peripheral systems contribute to some extent (Godlewski et al. 2019).

Given that AUD encompasses a complex pathophysiology, several preclinical models that reflect other aspects of the AUD process can be used. One of these is the operant self-administration model that reflects the motivation or seeking to consume alcohol. Using this model in male rats, GHSR antagonists were found to reduce the motivation to consume alcohol (Landgren et al. 2012; Gomez et al. 2015). Supportively, the motivation to consume alcohol was lower in GHSR knockouts compared to wild types (Zallar et al. 2019). Another preclinical paradigm reflecting other aspects of AUD is the alcohol deprivation model, in which relapse drinking following alcohol abstinence is studied. In humans, this behavior has been suggested to reflect craving (Spanagel 2000), which cannot be measured in animals. Studies of male mice (Jerlhag et al. 2009) and male rats (Suchankova et al. 2013) reveal that pharmacological suppression of the GHSR prevents relapse-like drinking.

8.1.1.2 Ghrelin and Des-acyl-ghrelin

As ghrelin is the endogenous agonist of GHSR, ghrelin's impact on alcohol consumption warrants evaluation. Supportively, male mice with a complete ghrelin knockout exhibited lower alcohol intake than their wild-type littermates, further supporting the modulatory role of ghrelin (Bahi et al. 2013). Since ghrelin acts centrally and peripherally, the differentiating effects of ghrelin on alcohol intake after its central versus peripheral effects are important to decipher. In male mice with long-term alcohol exposure, a central infusion of ghrelin elevates alcohol intake after one acute injection (Jerlhag et al. 2009). This increase is further evident as ghrelin into the VTA or laterodorsal tegmental area (LDTg; an area projecting to the VTA) escalates alcohol drinking in male mice (Jerlhag et al. 2009). Supportively, an elevated alcohol intake is observed in male rats with short-time alcohol exposure after local infusion of ghrelin into the VTA (Colvin et al.

2022). Other areas where ghrelin appears to act to modulate alcohol consumption are the amygdala (Yoshimoto et al. 2017; Cruz et al. 2013), lateral hypothalamus (Yoshimoto et al. 2017), and Edinger-Westphal nucleus (Kaur and Ryabinin 2010). The effects of systemic ghrelin on alcohol consumption are inconclusive, as it elevates alcohol intake in male rats with short alcohol exposure (Cepko et al. 2014) as well as those that have undergone Roux-en-Y gastric bypass (Davis et al. 2012). Contrarily, the alcohol intake is unaffected by systemic ghrelin in another study using male mice (Lyons et al. 2008). Similarly, when neutralizing ghrelin, alcohol consumption is unaffected (Jerlhag et al. 2014), a finding replicated in a more recent study (Richardson et al. 2024b). This inconsistency may lay in different doses used and limited penetration of ghrelin from the blood to the brain (Pirnik et al. 2011; Furness et al. 2011; Grouselle et al. 2008; Sakata et al. 2009; Schaeffer et al. 2013). Specifically, the plasma levels of ghrelin are higher after systemic administration compared to attenuation of circulating levels, enabling more ghrelin to enter the brain in the former scenario. While ghrelin has been extensively studied in relation to alcohol intake, its counterpart, des-acyl-ghrelin (DAG)—the predominant form in plasma—has received significantly less attention. Recently, acute or repeated administration of DAG was demonstrated to reduce alcohol intake in both male and female rats, an effect tentatively mediated via attenuated dopaminergic signaling in the VTA (Witley et al. 2024).

8.1.1.3 GLP-1R Agonists

As mentioned, clinically approved GLP-1R agonists are approved for diabetes type 2 and obesity and therefore the active substance will be mentioned in relation to their impact on alcohol.

Ex4 (exenatide) was the first GLP-1R agonist investigated for its effects on alcohol intake. In male rats with long-term alcohol exposure before treatment, an acute systemic injection of Ex4 decreases the alcohol intake (Egecioglu et al. 2013), a decline foremost observed in those with high alcohol consumption (Davis et al. 2012; Shirazi et al. 2013). Supportively, not only does Ex4 lower the consumption of alcohol, but it also prevents ghrelin from escalating it (Brigande et al. 2023). While these studies show that Ex4 reduces alcohol intake in male rats, this treatment effect is evident across species. As such, the alcohol intake is lower after Ex4 treatment in group-housed male mice (Thomsen et al. 2017) and vervet monkeys (Thomsen et al. 2019). One factor contributing to the reduced alcohol consumption observed with Ex4 is its capacity to prolong the time until the first drink and decrease the frequency of drinking episodes (Thomsen et al. 2017). The findings that Ex4 lowers intravenous self-administration of alcohol in male mice exclude changed absorption or gastric emptying as confounding factors (Sorensen et al. 2016). Liraglutide was the second GLP-1R agonist identified to reduce alcohol consumption in several species. Indeed, a lower alcohol intake is observed in male rats (Vallof et al. 2015; Marty et al. 2020), male mice (Liu et al. 2024), and vervet monkeys (Thomsen et al. 2019) after both acute and repeated liraglutide injections. The disadvantage of both Ex4 and liraglutide is that they are short versus medium-acting, requiring more frequent injections. The long-acting GLP-1 receptor agonist dulaglutide, the third GLP-1R agonist investigated for its effects on alcohol consumption, offers the advantage of being administered once weekly. Treatment with dulaglutide for either 5 or 9 weeks significantly decreases alcohol consumption in male rats, and for the first time, this effect was also demonstrated in female rats (Vallof et al. 2020). While the reduction was somewhat more evident in males compared to females, this study provides clinically relevant data as there is no tolerance towards dulaglutide treatment and the treatment effect persists after treatment termination (Vallof et al. 2020). The fourth GLP-1R agonist with alcohol-reducing properties is semaglutide, another long-acting compound. Specifically, a lower alcohol intake was observed in male rats acutely treated with semaglutide (Marty et al. 2020). Furthermore, both acute and repeated injections of semaglutide dose-dependently reduce alcohol consumption in male and female rats that have been exposed to alcohol for extended periods (Aranas et al. 2023a, b). Although the reduction was observed in both sexes, the treatment effect was more pronounced in females than in males (Aranas et al. 2023a). On a similar note, repeated injections with semaglutide lower alcohol intake in alcohol-preferring male vervet monkeys (Fink-Jensen et al. 2024). However, the intake returned to baseline after treatment termination (Fink-Jensen et al. 2024). Besides reducing regular alcohol intake, semaglutide suppresses binge drinking in mice and rats (Chuong et al. 2023). As for dulaglutide, there does not appear to be any tolerance development towards semaglutide (Aranas et al. 2023a, b; Fink-Jensen et al. 2024; Chuong et al. 2023). While semaglutide significantly decreased alcohol consumption, its combination with anti-smoking cessation agents, such as bupropion or varenicline, did not enhance its effectiveness in reducing alcohol intake (Aranas et al. 2023b), raising the need for additional combination treatment studies. In retrospect, when visually evaluating the effectiveness of various GLP-1 receptor agonists, both dulaglutide and semaglutide demonstrate the most substantial reduction in alcohol consumption compared to Ex4 and liraglutide. GLP-1R agonists have been found to influence the motivation to consume alcohol, as well as relapse-like drinking. Ex4 dose-dependently (Bornebusch et al. 2019) diminishes the motivation to consume alcohol in male rats trained in the operant self-administration model (Egecioglu et al. 2013; Bornebusch et al. 2019). A similar treatment outcome was observed in male and female mice, whereas the effect was somewhat lower in females (Diaz-Megido and Thomsen 2023). Like Ex4, a lower motivation to consume alcohol is observed in alcohol-preferring rats (sP rats) after repeated liraglutide treatment (Vallof et al. 2015). When it comes to alcohol-seeking using the progressive ration test in the operant self-administration model, Ex4 prevents alcohol-seeking in male rats and mice (Egecioglu et al. 2013; Diaz-Megido and Thomsen 2023), but not in female mice (Diaz-Megido and Thomsen 2023). In the ADE model, acute systemic injection of Ex4 prevents relapse-like drinking in both individually-housed male rats (Vallof et al. 2015) and group-housed male mice (Thomsen et al. 2017). On a similar note, in male mice, both the GLP-1R agonist AC3174 (Suchankova et al. 2015) and liraglutide prevent relapse drinking (Liu et al. 2024). While studied to a lesser extent, abstinence systems during alcohol withdrawal are improved by liraglutide (Liu et al. 2024; Sharma et al. 2014a).

While GLP-1R is expressed throughout the body and brain, alcohol intake appears to involve central rather than systemic receptors (Sirohi et al. 2016; Vallof et al. 2019a). Supportively, both fluorescently labeled Ex4 and semaglutide are detected in NAc after their systemic administrations (Aranas et al. 2023a; Hernandez et al. 2019). On a similar note, a lower alcohol intake (Abtahi et al. 2018; Vallof et al. 2019b; Colvin et al. 2020), as well as alcohol self-administration (Allingbjerg et al. 2023), is observed after local infusion of Ex4 into NAc, and a positive correlation between alcohol intake and accumbal GLP-1R expression is observed (Vallof et al. 2019b). The ability of GLP-1R in the VTA to regulate alcohol intake is complex and may positively be linked to the heterogeneity of this area. As such, local infusion of Ex4 into the posterior part reduces alcohol consumption and alcohol self-administration in some (Shirazi et al. 2013; Colvin et al. 2020; Dixon et al. 2020), but not all studies (Vallof et al. 2019b). Neither is alcohol consumption affected after Ex4 into the anterior VTA (Vallof et al. 2019b). Although studied to a lesser extent, Ex4 into LDTg decreases alcohol intake in male rats (Vallof et al. 2019b). Together. These data indicate that similar brain regions contribute towards ghrelin's and GLP-1's ability to modulate alcohol intake. Furthermore, GLP-1R in areas like NTS (Vallof et al. 2019a), lateral hypothalamus (Colvin et al. 2020), lateral septum (Allingbjerg et al. 2023), or hippocampus (Liu et al. 2024; Allingbjerg et al. 2023) has been pinpointed to control alcohol intake. Some areas, such as the basolateral amygdala, arcuate nucleus, paraventricular nucleus (Colvin et al. 2020), and putamen (Allingbjerg et al. 2023), have been excluded as important for this interaction.

8.1.1.4 GLP-1

GLP-1 has been shown to induce a short-term reduction in alcohol intake, likely due to its short half-life (Shirazi et al. 2013). This effect may involve central GLP-1 mechanisms, as local infusion into the VTA has been found to decrease alcohol consumption (Shirazi et al. 2013; Marty et al. 2020). Additional studies further indicate that circulating GLP-1 plays a limited role in influencing alcohol consumption. Specifically, different pharmacological manipulations causing elevated endogenous GLP-1 levels did not result in any significant changes in alcohol intake (Marty et al. 2020), a finding replicated after both central and peripheral inhibition of GLP-1 production (Farokhina et al. 2025). While the role of GLP-1 in the motivation to consume alcohol and relapse-like drinking has not been studied, one study found that elevated GLP-1 levels by DPP-IV inhibition alleviates symptoms of abstinence during withdrawal (Sharma et al. 2014b).

8.1.2 Effects on Alcohol-Induced Reward: Preclinical Studies

In both humans and animals, alcohol triggers the release of dopamine in the NAc (for review see Jayaram-Lindström et al. (2016)), which in rodents can be measured by in vivo microdialysis. This increase can be further reflected in animal models, including the locomotor stimulation and conditioned place preference (CPP). This increase in dopamine levels is critically linked to the rewarding properties of alcohol and is associated with a heightened risk of developing AUD later in life (Boileau et al. 2003; King et al. 2021). Supportively, patinets with AUD report higher reward assoicated with alcohol compared to healthy controls (King et al. 2025).

8.1.2.1 GHSR and LEAP2

Both pharmacological and genetic suppression of the GHSR prevents the ability of alcohol to increase dopamine in NAc in male mice (Jerlhag et al. 2009). The link to suppressed alcohol-induced reward is further supported as GHSR antagonism attenuates the ability of alcohol to stimulate locomotion and cause a CPP in male mice (Jerlhag et al. 2009). An impact on alcohol's stimulatory properties is further supported as the alcohol-induced locomotor stimulation is attenuated by repeated GHSR antagonism treatment (Suchankova et al. 2016a). As for alcohol intake, central GHSR (into the third ventricle) appears important as a central infusion of a GHSR antagonist blocks the stimulatory and rewarding properties of alcohol as measured by dopamine release in NAc, locomotor stimulation and CPP (Jerlhag et al. 2009). A role for central GHSR in regulating alcohol-induced reward in male mice is provided by studies on LEAP2 as it prevents the alcohol-induced dopamine release in NAc, CPP and locomotor stimulation (Tufvesson-Alm et al. 2024b).

Interestingly, structures and mechanisms that control alcohol consumption also control glucose metabolism and obesity; hence, strategies that modulate such circuits have a possible role in regulating alcohol response. For a detailed description of glucose metabolism and its role in brain-body connections, *see the Chap. 3 in this book.*

8.1.2.2 Ghrelin and Des-acyl-ghrelin

The impact of ghrelin on alcohol's rewarding properties is provided in studies is provided by two studies in ghrelin knockout mice. These have found that the locomotor stimulatory and dopamine-releasing effects of alcohol are lower in ghrelin knockout mice compared to wild types (Jerlhag et al. 2011). On a similar note, ghrelin knockout mice exhibit reduced alcohol-induced CPP compared to their wild-type littermates (Bahi et al. 2013; Jerlhag et al. 2011). In addition to ghrelin, DAG-treated male mice display lower dopamine in NAc, hypermotion, and CPP after alcohol infusion than control-treated male mice (Witley et al. 2024).

8.1.2.3 GLP-1R Agonists

As for the ghrelin pathway, the different GLP-1R agonists appear to attenuate the rewarding properties of alcohol. Supportively, either Ex4, liraglutide, or semaglutide blocks the hypermotion and accumbal dopamine release observed after alcohol injections in male rodents (Egecioglu et al. 2013; Vallof et al. 2015; Aranas et al. 2023a). It should, however, be noted that the different GLP-1R agonists affect reward measured in the CPP paradigm differently, tentatively due to different brain penetrations (Gu et al. 2013; Salinas et al. 2018). Indeed, the expression of reward is suppressed by either Ex4 (Egecioglu et al. 2013; Shirazi et al. 2013) or liraglutide (Vallof et al. 2015), whereas only Ex4 attenuated the memory consolidation of alcohol reward (Egecioglu et al. 2013; Vallof et al. 2015). Furthermore, the ability of alcohol to impair memory function, as measured in the Morris Water Maze test, is blocked by liraglutide (Liu et al. 2024). As suggested for regulation of alcohol consumption, GLP-1R in NAc, LDTg, and NTS suppresses the ability of alcohol to cause a dopamine release in NAc, CPP, and locomotor stimulation (Vallof et al. 2019a, b). On the contrary, Ex4 into either the anterior or posterior part of the VTA altered these behaviors (Vallof et al. 2019b).

8.1.3 Effects on Alcohol Intake: Clinical Studies

8.1.3.1 GHSR and Ghrelin

As reviewed elsewhere, associations between ghrelin and alcohol-related events have been studied extensively in humans and rodents (for review see Kharbanda et al. (2022)). However, findings from human genetic studies reveal associations between polymorphisms on either the GHSR gene or pre-pro-ghrelin genes and alcohol-related events. Specifically, GHSR polymorphisms are associated with high alcohol intake, AUDIT scores, or AUD diagnosis and AUD-linked personality traits (Landgren et al. 2008, 2010, 2011; Suchankova et al. 2016b). Moreover, the diagnosis of AUD, average alcohol intake, subjective alcohol responses and personality traits are associated with pre-pro-ghrelin gene polymorphisms (Landgren et al. 2010, 2011; Suchankova, 2017). Furthermore in humans, the craving for alcohol and the number of alcohol bouts are enhanced by intravenous injection of ghrelin (Leggio et al. 2014;Farokhnia, 2017). Additionally, the anticipation of monetary loss was alleviated by ghrelin (Pietrzak et al. 2023). As expected, in contrast to ghrelin, a partial GHSR agonist decreases alcohol craving (Lee et al. 2020a) without affecting hangover symptoms (Farokhnia et al. 2018). While these studies do not provide a proof of concept that GHSR antagonism will reduce alcohol drinking in humans with AUD, they may offer some clinical relevance.

8.1.3.2 GLP-1R Agonists

As for the ghrelin pathway, initial reports providing clinical associations between GLP-1R and alcohol consumption were provided by human genetic studies. Supportively, the AUD diagnosis, intravenous alcohol self-administration (Suchankova et al. 2015) or AUDIT scores were associated with GLP-1R gene polymorphisms (Farokhnia et al. 2022). The first indication that GLP-1R agonists may reduce alcohol intake was provided by a subset of data from a randomized clinical trial, in which exenatide lowered alcohol consumption in overweight, but not normal-weight, AUD patients (Klausen et al. 2022). On a similar note, the alcohol intake was lowered by dulaglutide independent of smoking status (Probst et al. 2023) and semaglutide reduced alcohol intake in patients with AUD (Hendershot et al. 2025). Furthermore, semaglutide lowered self-reported intake (Quddos et al. 2023) and high AUDIT scores in overweight AUD patients (Quddos et al. 2023; Richards et al. 2023). On a similar note, the experience of stimulation/sedation after alcohol (Quddos et al. 2023), the alcohol-related posts (Quddos et al. 2023), and desire (Bremmer and Hendershot 2024) were reduced by semaglutide in overweight patients. The findings that GLP-1R agonists, but not DPP-IV inhibitors, lower the alcohol-related events (Wium-Andersen et al. 2022) or alcohol intake (Fakhorina et al. 2025) further support the preclinical studies indicating that the receptors rather than the endogenous ligand control alcohol intake. Together, these studies demonstrate the promising potential of GLP-1R agonists in reducing alcohol intake among individuals with both AUD and obesity, warranting further investigation through larger randomized clinical trials.

8.1.4 Future Directions and Conclusion

While these initial studies revealed that GHSR or GLP-1R expressed in reward-associated areas module alcohol-related responses, the mechanisms of actions and neurocircuits remain to be defined. The GABAergic system deserves some extra attention as these nerve terminals express GLP-1R presynaptically (Babateen et al. 2017; Korol et al. 2015a, b; Rebosio et al. 2018). Moreover, inflammation markers (Farokhnia et al. 2020), rather than other gut-brain peptides (Lee et al. 2020b), should be evaluated as downstream mechanisms. Although all the agonists of the GLP-1R lower alcohol intake, the efficacy varies. Moreover, they penetrate the brain differently, and their penetration may depend on previous alcohol exposure. Therefore, the mechanisms of action should be defined for each of the clinically GLP-1R agonists, and the impact on neurotransmission may depend on exposure to alcohol. Moreover, the sex-diverging differences following treatment deserve to be elucidated in future studies and may include sex hormones, different metabolism, distribution, brain penetration, or variations in underlying neurocircuits.

The possibility that treatments may have additional beneficial effects, including liver dysfunction, gut barrier function, or mucosal disruption, should be considered and studied further. Supportively, hepatic steatosis caused by long-term alcohol exposure is improved by Ex4 (Mahalingam et al. 2023).

Another area that deserves extra attention is the efficacy of agents targeting the GHSR or GLP-1R on multiple drug use in a choice situation. This is a health issue commonly observed in humans, and feeding experiments have revealed that GHSR substrates affect the most rewarding food choices (Tufvesson-Alm et al. 2024a; Egecioglu et al. 2010). For complex and heterogeneous disorders like AUD, combining multiple agents that each independently reduce alcohol consumption may lead to more effective treatment outcomes. Supportively, the beneficial effects of alcohol drinking have been found by the combination of bupropion and varenicline (Soderpalm et al. 2020), as well as naltrexone together with either varenicline or bupropion (Zhou et al. 2019; Grodin et al. 2021). However, this approach is unexplored when it comes to the impact of gut-brain peptide combinations on alcohol-related responses and upcoming studies should explore this potential interaction. On a similar note, the combination of the gut-brain peptides GLP-1 and amylin synergistically reduces feeding and body weight (Liberini et al. 2019), an approach that appears to be relevant for reducing alcohol intake (Aranas et al. 2025). Another way to boost the treatment outcome is to combine treatment with social intervention, exercise or social enrichment.

Either GHSR suppression or GLP-1R activation reduces alcohol consumption, tentatively by diminishing its rewarding effects, which are key drivers of consumption and relapse-like drinking. This suggests that targeting these pathways could be effective in treating AUD, a tentative subject of future randomized clinical trials.

Acknowledgments EH has no conflict of interest and has support from the Swedish Research Council (2023-2600) to write the article.

References

Abtahi S, Howell E, Currie PJ (2018) Accumbal ghrelin and glucagon-like peptide 1 signaling in alcohol reward in female rats. Neuroreport 29(12):1046–1053

Allingbjerg ML et al (2023) Glucagon-like peptide-1 receptors in nucleus accumbens, ventral hippocampus, and lateral septum reduce alcohol reinforcement in mice. Exp Clin Psychopharmacol 31(3):612–620

Alvarez E et al (1996) Expression of the glucagon-like peptide-1 receptor gene in rat brain. J Neurochem 66(3):920–927

Aranas C et al (2023a) Semaglutide reduces alcohol intake and relapse-like drinking in male and female rats. EBioMedicine 93:104642

Aranas C, Blid Skoldheden S, Jerlhag E (2023b) Antismoking agents do not contribute synergistically

to semaglutide's ability to reduce alcohol intake in rats. Front Pharmacol 14:1180512
Babateen O et al (2017) Liraglutide modulates GABAergic signaling in rat hippocampal CA3 pyramidal neurons predominantly by presynaptic mechanism. BMC Pharmacol Toxicol 18(1):83
Bahi A et al (2013) Ghrelin knockout mice show decreased voluntary alcohol consumption and reduced ethanol-induced conditioned place preference. Peptides 43:48–55
Boileau I et al (2003) Alcohol promotes dopamine release in the human nucleus accumbens. Synapse 49(4):226–231
Bornebusch AB et al (2019) Glucagon-Like peptide-1 receptor agonist treatment does not reduce abuse-related effects of opioid drugs. eNeuro 6(2)
Bremmer MP, Hendershot CS (2024) Social media as pharmacovigilance: the potential for patient reports to inform clinical research on Glucagon-Like peptide 1 (GLP-1) receptor agonists for substance use disorders. J Stud Alcohol Drugs 85(1):5–11
Brigande AM, Darwich JG, Currie PJ (2023) Mesolimbic exendin-4 attenuates reward salience evoked by neuropeptide Y and ghrelin. Behav Brain Res 440:114249
Cepko LC et al (2014) Ghrelin alters the stimulatory effect of cocaine on ethanol intake following mesolimbic or systemic administration. Neuropharmacology 85:224–231
Chuong V et al (2023) The glucagon-like peptide-1 (GLP-1) analogue semaglutide reduces alcohol drinking and modulates central GABA neurotransmission. JCI Insight 8(12)
Colvin KJ et al (2020) Brain site-specific inhibitory effects of the GLP-1 analogue Exendin-4 on alcohol intake and operant responding for palatable food. Int J Mol Sci 21(24):9710
Colvin KJ et al (2022) Differential effects of intra-ventral tegmental area ghrelin and glucagon-like peptide-1 on the stimulatory action of D-amphetamine and cocaine-induced ethanol intake in male Sprague Dawley rats. Behav Brain Res 421:113726
Cornejo MP et al (2019) Growth hormone secretagogue receptor signalling affects high-fat intake independently of plasma levels of ghrelin and LEAP2, in a 4-day binge eating model. J Neuroendocrinol 31(10):e12785
Cornejo MP et al (2021) The intriguing ligand-dependent and ligand-independent actions of the growth hormone secretagogue receptor on reward-related behaviors. Neurosci Biobehav Rev 120:401–416
Cowley MA et al (2003) The distribution and mechanism of action of ghrelin in the CNS demonstrates a novel hypothalamic circuit regulating energy homeostasis. Neuron 37(4):649–661
Cruz MT et al (2013) Ghrelin increases GABAergic transmission and interacts with ethanol actions in the rat central nucleus of the amygdala. Neuropsychopharmacology 38(2):364–375
Davis JF et al (2012) Gastric bypass surgery attenuates ethanol consumption in ethanol-preferring rats. Biol Psychiatry 72(5):354–360
De Vriese C et al (2004) Ghrelin degradation by serum and tissue homogenates: identification of the cleavage sites. Endocrinology 145(11):4997–5005
Delhanty PJ et al (2010) Unacylated ghrelin rapidly modulates lipogenic and insulin signaling pathway gene expression in metabolically active tissues of GHSR deleted mice. PLoS One 5(7):e11749
Diaz-Megido C, Thomsen M (2023) Sex-dependent divergence in the effects of GLP-1 agonist exendin-4 on alcohol reinforcement and reinstatement in C57BL/6J mice. Psychopharmacology 240(6):1287–1298
Dixon TN, McNally GP, Ong ZY (2020) Glucagon-Like Peptide-1 receptor signaling in the ventral tegmental area reduces alcohol self-administration in male rats. Alcohol Clin Exp Res 44(10):2118–2129
Egecioglu E et al (2010) Ghrelin increases intake of rewarding food in rodents. Addict Biol 15(3):304–311
Egecioglu E et al (2013) The glucagon-like peptide 1 analogue Exendin-4 attenuates alcohol mediated behaviors in rodents. Psychoneuroendocrinology 38(8):1259–1270
Farokhnia M et al (2018) Pharmacological manipulation of the ghrelin system and alcohol hangover symptoms in heavy drinking individuals: is there a link? Pharmacol Biochem Behav 172:39–49
Farokhnia M et al (2020) Effects of exogenous ghrelin administration and ghrelin receptor blockade, in combination with alcohol, on peripheral inflammatory markers in heavy-drinking individuals: results from two human laboratory studies. Brain Res 1740:146851
Farokhnia M et al (2022) Differential association between the GLP1R gene variants and brain functional connectivity according to the severity of alcohol use. Sci Rep 12(1):13027
Fernandez G et al (2016) Des-Acyl ghrelin directly targets the arcuate nucleus in a ghrelin-receptor independent manner and impairs the orexigenic effect of ghrelin. J Neuroendocrinol 28(2):12349
Fink-Jensen A et al (2024) Effect of the glucagon-like peptide-1 (GLP-1) receptor agonist semaglutide on alcohol consumption in alcohol-preferring male vervet monkeys. Psychopharmacology 242(1):63–70
Furness JB et al (2011) Investigation of the presence of ghrelin in the central nervous system of the rat and mouse. Neuroscience 193:1–9
Gauna C et al (2006) Unacylated ghrelin is active on the INS-1E rat insulinoma cell line independently of the growth hormone secretagogue receptor type. la and the corticotropin releasing factor 2 receptor. Mol Cell Endocrinol 251(1–2):103–111
Ge X et al (2018) LEAP2 is an endogenous antagonist of the ghrelin receptor. Cell Metab 27(2):461–469 e6
Godlewski G et al (2019) Targeting peripheral CB1 receptors reduces ethanol intake via a gut-brain axis. Cell Metab 29(6):1320–1333 e8
Gomez JL, Ryabinin AE (2014) The effects of ghrelin antagonists [D-Lys(3)]-GHRP-6 or JMV2959 on

ethanol, water, and food intake in C57BL/6J mice. Alcohol Clin Exp Res 38(9):2436–2444
Gomez JL et al (2015) Differential effects of ghrelin antagonists on alcohol drinking and reinforcement in mouse and rat models of alcohol dependence. Neuropharmacology 97:182–193
Grodin EN et al (2021) Combined varenicline and naltrexone attenuates alcohol cue-elicited activation in heavy drinking smokers. Drug Alcohol Depend 225:108825
Grouselle D et al (2008) Pulsatile cerebrospinal fluid and plasma ghrelin in relation to growth hormone secretion and food intake in the sheep. J Neuroendocrinol 20(10):1138–1146
Gu G et al (2013) Glucagon-like peptide-1 in the rat brain: distribution of expression and functional implication. J Comp Neurol 521(10):2235–2261
Hajnal A et al (2012) Alcohol reward is increased after Roux-en-Y gastric bypass in dietary obese rats with differential effects following ghrelin antagonism. PLoS One 7(11):e49121
Hernandez NS et al (2019) Activation of glucagon-like peptide-1 receptors in the nucleus accumbens attenuates cocaine seeking in rats. Addict Biol 24(2):170–181
Holst JJ, Seino Y (2009) GLP-1 receptor agonists: targeting both hyperglycaemia and disease processes in diabetes. Diabetes Res Clin Pract 85(1):1–3
Hosoda H et al (2000) Ghrelin and des-acyl ghrelin: two major forms of rat ghrelin peptide in gastrointestinal tissue. Biochem Biophys Res Commun 279(3):909–913
Inoue Y et al (2013) Central and peripheral des-acyl ghrelin regulates body temperature in rats. Biochem Biophys Res Commun 430(1):278–283
Islam MN et al (2020) Liver-expressed antimicrobial peptide 2 antagonizes the effect of ghrelin in rodents. J Endocrinol 244(1):13–23
Jayaram-Lindström N et al (2016) Dopamine and alcohol dependence: from bench to clinic. In: Meil W (ed) Recent advances in drug addiction research and clinical applications. IntechOpen, pp 81–114
Jerlhag E et al (2009) Requirement of central ghrelin signaling for alcohol reward. Proc Natl Acad Sci USA 106(27):11318–11323
Jerlhag E et al (2011) The alcohol-induced locomotor stimulation and accumbal dopamine release is suppressed in ghrelin knockout mice. Alcohol 45(4):341–347
Jerlhag E et al (2014) Peripherally circulating ghrelin does not mediate alcohol-induced reward and alcohol intake in rodents. Alcohol Clin Exp Res 38:959–968
Kaur S, Ryabinin AE (2010) Ghrelin receptor antagonism decreases alcohol consumption and activation of perioculomotor urocortin-containing neurons. Alcohol Clin Exp Res 34(9):1525–1534
Kharbanda KK et al (2022) Role of the ghrelin system in alcohol use disorder and alcohol-associated liver disease: a narrative review. Alcohol Clin Exp Res 46(12):2149–2159
King A et al (2021) Subjective responses to alcohol in the development and maintenance of alcohol use disorder. Am J Psychiatry 178(6):560–571
Klausen MK et al (2022) Exenatide once weekly for alcohol use disorder investigated in a randomized, placebo-controlled clinical trial. JCI Insight 7(19):e159863
Kojima M et al (1999) Ghrelin is a growth-hormone-releasing acylated peptide from stomach. Nature 402(6762):656–660
Korol SV et al (2015a) GLP-1 and exendin-4 transiently enhance GABAA receptor-mediated synaptic and tonic currents in rat hippocampal CA3 pyramidal neurons. Diabetes 64(1):79–89
Korol SV, Jin Z, Birnir B (2015b) The GLP-1 receptor agonist Exendin-4 and diazepam differentially regulate GABAA receptor-mediated tonic currents in rat hippocampal CA3 pyramidal neurons. PLoS One 10(4):e0124765
Krause A et al (2003) Isolation and biochemical characterization of LEAP-2, a novel blood peptide expressed in the liver. Protein Sci 12(1):143–152
Landgren S et al (2008) Association of pro-ghrelin and GHS-R1A gene polymorphisms and haplotypes with heavy alcohol use and body mass. Alcohol Clin Exp Res 32(12):2054–2061
Landgren S et al (2010) Genetic variation of the ghrelin signaling system in females with severe alcohol dependence. Alcohol Clin Exp Res 34(9):1519–1524
Landgren S et al (2011) Reward-related genes and personality traits in alcohol-dependent individuals: a pilot case control study. Neuropsychobiology 64(1):38–46
Landgren S et al (2012) Ghrelin receptor (GHS-R1A) antagonism suppresses both operant alcohol self-administration and high alcohol consumption in rats. Addict Biol 17(1):86–94
Lee MR et al (2020a) The novel ghrelin receptor inverse agonist PF-5190457 administered with alcohol: preclinical safety experiments and a phase 1b human laboratory study. Mol Psychiatry 25(2):461–475
Lee MR et al (2020b) Endocrine effects of the novel ghrelin receptor inverse agonist PF-5190457: results from a placebo-controlled human laboratory alcohol co-administration study in heavy drinkers. Neuropharmacology 170:107788
Leggio L et al (2014) Intravenous ghrelin administration increases alcohol craving in alcohol-dependent heavy drinkers: a preliminary investigation. Biol Psychiatry 76(9):734–741
Liberini CG et al (2019) Combined Amylin/GLP-1 pharmacotherapy to promote and sustain long-lasting weight loss. Sci Rep 9(1):8447
Liu W et al (2024) Liraglutide reduces alcohol consumption, anxiety, memory impairment, and synapse loss in alcohol dependent mice. Neurochem Res 49(4):1061–1075
Lu S et al (2002) Immunocytochemical observation of ghrelin-containing neurons in the rat arcuate nucleus. Neurosci Lett 321(3):157–160
Lyons AM et al (2008) Effects of food availability and administration of orexigenic and anorectic agents on elevated ethanol drinking associated with drinking in the dark procedures. Alcohol Clin Exp Res 32(11):1962–1968

M'Kadmi C et al (2019) N-Terminal Liver-Expressed Antimicrobial Peptide 2 (LEAP2) region exhibits inverse agonist activity toward the ghrelin receptor. J Med Chem 62(2):965–973
Mahalingam S et al (2023) Glucagon-like peptide 1 receptor agonist, exendin-4, reduces alcohol-associated fatty liver disease. Biochem Pharmacol 213:115613
Marty VN et al (2020) Long-acting glucagon-like peptide-1 receptor agonists suppress voluntary alcohol intake in male wistar rats. Front Neurosci 14:599646
Mondal MS et al (2005) Identification of ghrelin and its receptor in neurons of the rat arcuate nucleus. Regul Pept 126(1–2):55–59
Morganstern I, Barson JR, Leibowitz SF (2011) Regulation of drug and palatable food overconsumption by similar peptide systems. Curr Drug Abuse Rev 4(3):163–173
Novak U et al (1987) Identical mRNA for preproglucagon in pancreas and gut. Eur J Biochem 164(3): 553–558
Orellana ER et al (2018) Effect of vertical sleeve gastrectomy on alcohol consumption and preferences in dietary obese rats and mice: a plausible role for altered ghrelin signaling. Brain Res Bull 138:26–36
Pietrzak M et al (2023) A randomized controlled experimental medicine study of ghrelin in value-based decision making. J Clin Invest 133(12)
Pirnik Z et al (2011) Ghrelin agonists impact on Fos protein expression in brain areas related to food intake regulation in male C57BL/6 mice. Neurochem Int 59(6):889–895
Probst L et al (2023) Effects of dulaglutide on alcohol consumption during smoking cessation. JCI Insight 8(22)
Quddos F et al (2023) Semaglutide and Tirzepatide reduce alcohol consumption in individuals with obesity. Sci Rep 13(1):20998
Rebosio C et al (2018) Presynaptic GLP-1 receptors enhance the depolarization-evoked release of glutamate and GABA in the mouse cortex and hippocampus. Biofactors 44(2):148–157
Richards JR et al (2023) Significant decrease in alcohol use disorder symptoms secondary to semaglutide therapy for weight loss: a case series. J Clin Psychiatry 85(1)
Richardson RS et al (2023) Pharmacological GHSR (ghrelin receptor) blockade reduces alcohol binge-like drinking in male and female mice. Neuropharmacology 238:109643
Richardson RS et al (2024a) Centrally administered growth hormone secretagogue receptor antagonist DLys decreases alcohol intake and preference in male mice. Neuroreport 35(14):909–914
Richardson RS et al (2024b) GHSR blockade, but not reduction of peripherally circulating ghrelin via beta(1)-adrenergic receptor antagonism, decreases binge-like alcohol drinking in mice. Mol Psychiatry 30(3):1047–1056
Sakata I et al (2009) Characterization of a novel ghrelin cell reporter mouse. Regul Pept 155(1–3):91–98
Salinas CBG et al (2018) Integrated brain atlas for unbiased mapping of nervous system effects following liraglutide treatment. Sci Rep 8(1):10310
Schaeffer M et al (2013) Rapid sensing of circulating ghrelin by hypothalamic appetite-modifying neurons. Proc Natl Acad Sci USA 110(4):1512–1517
Sharma AN et al (2014a) Glucagon-like peptide-1 (GLP-1) receptor agonist prevents development of tolerance to anti-anxiety effect of ethanol and withdrawal-induced anxiety in rats. Metab Brain Dis 30(3):719–730
Sharma AN et al (2014b) Dipeptidyl-peptidase IV (DPP-IV) inhibitor delays tolerance to anxiolytic effect of ethanol and withdrawal-induced anxiety in rats. Metab Brain Dis 30(3):659–667
Shirazi RH, Dickson SL, Skibicka KP (2013) Gut peptide GLP-1 and its analogue, exendin-4, decrease alcohol intake and reward. PLoS One 8(4):e61965
Sirohi S et al (2016) Central & peripheral glucagon-like peptide-1 receptor signaling differentially regulate addictive behaviors. Physiol Behav 161: 140–144
Soderpalm B et al (2020) Combined administration of varenicline and bupropion produces additive effects on accumbal dopamine and abolishes the alcohol deprivation effect in rats. Addict Biol 25(5):e12807
Sorensen G, Caine SB, Thomsen M (2016) Effects of the GLP-1 agonist exendin-4 on intravenous ethanol self-administration in mice. Alcohol Clin Exp Res 40(10):2247–2252
Spanagel R (2000) Recent animal models of alcoholism. Alcohol Res Health 24(2):124–131
Srivastava G, Apovian C (2018) Future pharmacotherapy for obesity: new anti-obesity drugs on the horizon. Curr Obes Rep 7(2):147–161
Stevenson JR et al (2015) GHS-R1A antagonism reduces alcohol but not sucrose preference in prairie voles. Physiol Behav 147:23–29
Suchankova P et al (2013) Ghrelin receptor (GHS-R1A) antagonism suppresses both alcohol consumption and the alcohol deprivation effect in rats following long-term voluntary alcohol consumption. PLoS One 8(8):e71284
Suchankova P et al (2015) The glucagon-like peptide-1 receptor as a potential treatment target in alcohol use disorder: evidence from human genetic association studies and a mouse model of alcohol dependence. Transl Psychiatry 5:e583
Suchankova P, Engel JA, Jerlhag E (2016a) Sub-chronic ghrelin receptor blockade attenuates alcohol- and amphetamine-induced locomotor stimulation in mice. Alcohol Alcohol 51(2):121–127
Suchankova P et al (2016b) Genetic variation of the growth hormone secretagogue receptor gene is associated with alcohol use disorders identification test scores and smoking. Addict Biol 21(2):481–418
Thomsen M et al (2017) The glucagon-like peptide 1 receptor agonist Exendin-4 decreases relapse-like drinking in socially housed mice. Pharmacol Biochem Behav 160:14–20

Thomsen M et al (2019) Effects of glucagon-like peptide 1 analogs on alcohol intake in alcohol-preferring vervet monkeys. Psychopharmacology 236(2):603–611

Tschop M et al (2001) Post-prandial decrease of circulating human ghrelin levels. J Endocrinol Investig 24(6):RC19-21

Tufvesson-Alm M et al (2024a) Decoding the influence of central LEAP2 on food intake and its effect on accumbal dopamine release. Prog Neurobiol 236:102615

Tufvesson-Alm M et al (2024b) LEAP2, a ghrelin receptor inverse agonist, and its effect on alcohol-related responses in rodents. Transl Psychiatry 14(1):401

Vallof D et al (2015) The glucagon-like peptide 1 receptor agonist liraglutide attenuates the reinforcing properties of alcohol in rodents. Addict Biol 21(2):422–437

Vallof D, Vestlund J, Jerlhag E (2019a) Glucagon-like peptide-1 receptors within the nucleus of the solitary tract regulate alcohol-mediated behaviors in rodents. Neuropharmacology 149:124–132

Vallof D, Kalafateli AL, Jerlhag E (2019b) Brain region specific glucagon-like peptide-1 receptors regulate alcohol-induced behaviors in rodents. Psychoneuroendocrinology 103:284–295

Vallof D, Kalafateli AL, Jerlhag E (2020) Long-term treatment with a glucagon-like peptide-1 receptor agonist reduces ethanol intake in male and female rats. Transl Psychiatry 10(1):238

Witley S et al (2024) Des-acyl ghrelin reduces alcohol intake and alcohol-induced reward in rodents. Transl Psychiatry 14(1):277

Wium-Andersen IK et al (2022) Use of GLP-1 receptor agonists and subsequent risk of alcohol-related events. A nationwide register-based cohort and self-controlled case series study. Basic Clin Pharmacol Toxicol 131(5):372–379

Wolfe WL, Maisto SA (2000) The relationship between eating disorders and substance use: moving beyond co-prevalence research. Clin Psychol Rev 20(5):617–631

Yoshimoto K et al (2017) Enhanced alcohol-drinking behavior associated with active ghrelinergic and serotoninergic neurons in the lateral hypothalamus and amygdala. Pharmacol Biochem Behav 153:1–11

You ZB et al (2022) Involvement of the ghrelin system in the maintenance of oxycodone self-administration: converging evidence from endocrine, pharmacologic and transgenic approaches. Mol Psychiatry 27(4):2171–2181

Zallar LJ et al (2019) Ghrelin receptor deletion reduces binge-like alcohol drinking in rats. J Neuroendocrinol 31(7):e12663

Zhou Y et al (2019) Sex differences in the effect of bupropion and naltrexone combination on alcohol drinking in mice. Pharmacol Biochem Behav 181:28–36

Rett Syndrome: Thinking Beyond Brain Borders

9

Yi Liu, Ruisi Guo, and Rudolf Jaenisch

Abstract

Rett Syndrome (RTT) is a severe neurological disorder caused by mutations in the MECP2 gene. This disorder primarily affects brain development, leading to profound cognitive and motor impairments. Conventionally, research on RTT has concentrated on the neurological impact of MECP2 mutations. However, other evidence suggests that the influence of RTT extends beyond neural boundaries, impacting a broad spectrum of other tissues and organ systems. This chapter provides an overview of identified abnormalities in respiratory, skeletal, cardiac, liver, endocrine, urinary, immune, and gastrointestinal systems in patients with RTT or related disease models. We further explore the potential pathways involved in the brain-body circuit and their impact on the development of RTT. These insights not only enhance our understanding of the multi-system involvement in RTT, but also reveal the necessity for a more comprehensive approach to investigating the systemic courses of neurological disorders.

Keywords

Rett syndrome · Autonomic nervous system · Endocrine system · Immune system · Metabolism · MECP2 · Transcriptional regulation

Abbreviations

RTT	Rett Syndrome
ANS	Autonomic nervous system
MECP2	Methyl-CpG-binding protein 2
MDS	MECP2 duplication syndrome
AD	Alzheimer's disease
PD	Parkinson's disease
ASD	Autism spectrum disorder
BDNF	Brain-derived neurotrophic factor
GABA	γ-aminobutyric acid
nTS	Nucleus tractus solitarius
LQT	Long QT syndrome
iPSCs	Induced pluripotent stem cells
GI	Gastrointestinal
ENS	Enteric nervous system
HSP	Homeostatic synaptic plasticity

Yi Liu and Ruisi Guo contributed equally with all other contributors.

Y. Liu · R. Guo
Whitehead Institute for Biomedical Research, Cambridge, MA, USA

R. Jaenisch (✉)
Whitehead Institute for Biomedical Research, Cambridge, MA, USA

Department of Biology, Massachusetts Institute of Technology, Cambridge, MA, USA
e-mail: jaenisch@wi.mit.edu

D. Tropea, E. Giacometti (eds.), *Brain-Body Connections*, Advances in Experimental Medicine and Biology 1477, https://doi.org/10.1007/978-3-031-89525-8_9

CNS	Central nervous system
nNOS	Nitric oxide synthase
BAT	Brown adipose tissue
GSH	Glutathione
ROS	Reactive oxygen species
CRH	Corticotropin-releasing hormone
HPA	Hypothalamic-pituitary-adrenal
BMD	Bone mineral density
BM-MSCs	Bone marrow-derived mesenchymal stem cells
GH/IGF	Growth hormone/insulin-like growth factor
TNF	Tumor necrosis factor
Tregs	Regulatory T cells
ER alpha	Estrogen receptor alpha
AE2	Alveolar epithelial 2
scRNA-seq	Single-cell RNA sequencing
5-HT	Serotonin

9.1 Introduction

Rett syndrome (RTT) is a severe neurological disorder predominantly observed in females, with an estimated incidence of 1 in 10,000. Approximately 95% of cases are attributed to mutations in the X-linked gene methyl-CpG-binding protein 2 (*MECP2*) (Amir et al. 1999), which encodes a critical epigenetic factor. This protein is highly expressed in neurons at levels comparable to histone octamers (Skene et al. 2010) and loss of function in MECP2 is mainly associated with a progressive regression in early neurodevelopment (Hagberg et al. 1983). Conversely, overexpression of MECP2 can be equally detrimental, as an extra copy of *MECP2* leads to MECP2 duplication syndrome (MDS), characterized by intellectual impairment and significant motor impairments (Peters et al. 2013).

Initially, MECP2 was identified as a transcriptional repressor, binding to methylated DNA to inhibit gene expression by its association with the NCoR/SMRT co-repressor complex (Nan et al. 1993, 1997; Lyst et al. 2013; Gabel et al. 2015; Boxer et al. 2020). Later studies have expanded our understanding of MECP2, revealing its activator functions (Chahrour et al. 2008; Li et al. 2013) by interacting with RNA polymerase II, CREB1, TCF20 as well as the super elongation complex to facilitate their recruitment to target genes (Chahrour et al. 2008; Zhou et al. 2022; Liu et al. 2024; Sonn et al. 2024).

The primary clinical phenotypes of RTT are associated with severe intellectual disability, seizures, autistic features, loss of hand skills and speech, and microcephaly (Chahrour and Zoghbi 2007). Most research on MECP2 has been focused on brain tissues or cultured neurons. However, there is growing evidence of systemic impairment in non-neuronal tissues in RTT patients, including respiratory, skeletal, cardiac, liver, endocrine, urinary, immune, and gastrointestinal systems (Fig. 9.1). In particular, symptoms affecting peripheral systems can impact patients' quality of life and increase the incidence of sudden death (Singh et al. 2020; Borloz et al. 2021). Further clarification of the origins of these symptoms, whether in the brain or peripheral systems, is still required to advance our understanding of RTT.

Researchers have explored peripheral mechanisms and brain-body connections in various neurological disorders. One notable area of interest is the gut-brain axis, well-documented for its implications in conditions such as Alzheimer's disease (AD), Parkinson's disease (PD), autism spectrum disorder (ASD), and amyotrophic lateral sclerosis, although direct evidence linking the gut-brain axis to RTT remains elusive (Chen et al. 2024; Kowalski and Mulak 2019; Tan et al. 2022; Taniya et al. 2022). This axis is linked to interactions among the enteric microbiota, the enteric nervous system, and the central nervous system (Fried et al. 2021). For example, gut microbiota may cause the accumulation of amyloid proteins, lipopolysaccharides, and biotoxins that disrupt physiological barriers and provoke systemic inflammation, potentially triggering the onset of AD (Kowalski and Mulak 2019). Another example is the autonomic dysfunction, which is observed in cognitive disorders including AD and PD, and can lead to symptoms including anxiety, urinary retention, constipation, and dysphagia (Beishon et al. 2022; Allan 2019). Peripheral symptoms of nervous system disorders also frequently correlate with metabolic disorders, endo-

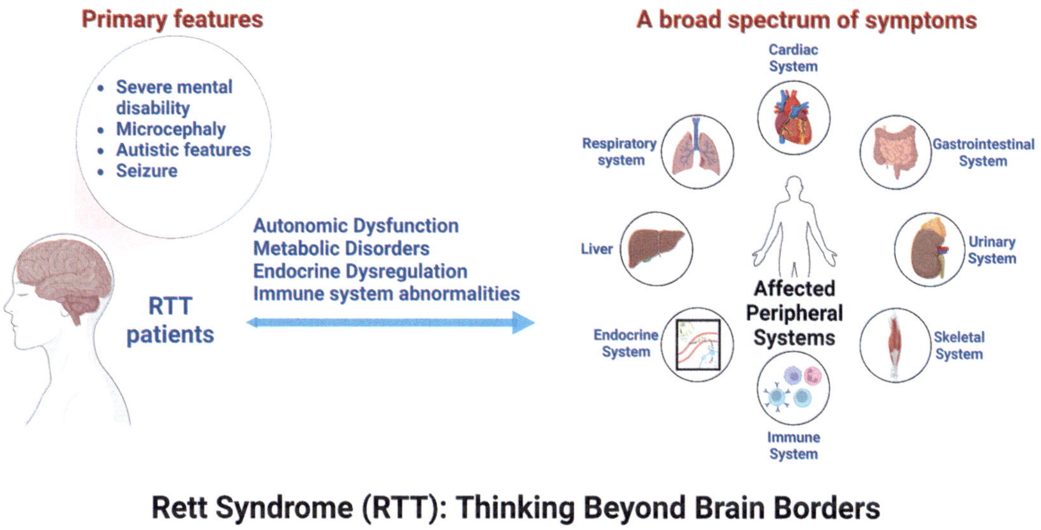

Fig. 9.1 Rett Syndrome (RTT): Thinking Beyond Brain Borders. Although most symptoms of RTT occur in the brain, there is a growing awareness of the importance of peripheral systems (respiratory, skeletal, cardiac, liver, endocrine, urinary, immune, and gastrointestinal systems) in this disease. This review features the brain-body connection in RTT from three different perspectives: autonomic dysfunction, autonomic-independent dysfunction, and understudied dysfunction. Autonomic dysfunction affects the respiratory system, the cardiac system, and the gastrointestinal system. Autonomic-independent dysfunction includes metabolic disorders affecting the liver, lungs, gastrointestinal system, adipose tissue, and skeletal muscles, as well as endocrine dysregulation impacting the endocrine, skeletal, and immune systems. Additionally, defects in the urinary system in RTT are notably understudied

crine dysregulation, and immune system abnormalities. For instance, multiple hormone dysfunctions within the endocrine system are commonly seen in ASD (Tareen and Kamboj 2012). Depression is often associated with hyperactivity of the hypothalamic-pituitary-adrenal (HPA) axis (Tichomirowa et al. 2005) and systemic immune activation, with a notable correlation between depression severity and increased pro-inflammatory cytokines (Howren et al. 2009; Beurel et al. 2020). Notably, in RTT, while most autonomic dysfunction is thought to be secondary to neuronal deficits, the expression of MECP2 in all human tissues suggests potential regulatory mechanisms independent of the nervous system. Mouse models with conditional *Mecp2* knockout (KO) in specific tissues have been used to study the origins of the corresponding symptoms.

Overall, it is of great interest to explore the molecular basis of the peripheral symptoms associated with RTT and their connections to the central nervous system. Understanding these nuances could provide deeper insights into the disease mechanisms and therapeutic development in a context-dependent manner. In this chapter, we will review the abnormalities in RTT across key systems outside the brain, including autonomic nervous system (ANS)-dependent regulation in respiratory, cardiac, and gastrointestinal systems, ANS-independent regulation in metabolism, the endocrine system, and the immune system, as well as other understudied areas, as illustrated in Fig. 9.1, to explore how MECP2's broad regulatory functions impact various non-neuronal systems, highlighting its systemic effects beyond the brain.

9.2 ANS-Dependent Peripheral Dysfunction

ANS dysfunctions in RTT are mainly caused by neuronal dysfunction within the brain. These dysfunctions manifest as sympathetic overactivity, parasympathetic underactivity, and sympathovagal imbalance, primarily caused by immaturity of the brainstem (Julu et al. 2008; Kumar et al. 2017) (Fig. 9.2). Dysfunction of the autonomic nervous system can lead to abnormalities in multiple systems, including respiratory, cardiac, and gastrointestinal (Fig. 9.2), further complicating the clinical presentation and management of RTT. Therefore, understanding the extent and mechanisms of autonomic dysfunction is crucial for developing effective therapeutic strategies to improve patient outcomes.

9.2.1 Respiratory System

Breathing abnormalities are observed in nearly all RTT patients, with hyperventilation and breath-holding being the most prevalent dysfunctions (Fig. 9.2). These symptoms typically manifest early in childhood and persist throughout the patient's life (Tarquinio et al. 2018). Additionally, shallow breathing tends to increase with age and is more prevalent among teenagers and adults (Julu et al. 2001; Pini et al. 2016). Difficulties with breathing and swallowing can increase the risk of aspiration pneumonia (Ramirez et al. 2020) (Fig. 9.2).

RTT mouse models, including the male *Mecp2*$^{-/y}$ and female *Mecp2*$^{+/-}$ mice, recapitulate similar phenotypes observed in humans. These models exhibit hyperventilation, breath-holding, and breathing arrhythmias, mirroring the respiratory dysfunctions seen in RTT patients (Viemari et al. 2005; Johnson et al. 2015). *Mecp2*$^{-/y}$ mice also exhibit abnormal pulmonary structure, particularly in the right anterior lobe. The elevated number of infiltrating cells in the alveolar epithelial layer and alveolar spaces indicates severe lung inflammation. Additionally, the presence of polarizable foreign material in the injured lungs of *Mecp2*-null mice strongly suggests the occurrence of aspiration pneumonia (Kida et al. 2017).

Researchers subsequently developed conditional KO mouse models to identify which regions of the brain stem are responsible for respiratory system abnormalities in RTT. Since multiple anatomical regions and neuron categories contribute to the respiratory network, these studies first focused on the mouse hindbrain, which contains several respiratory and autonomic centers. Conditional *Mecp2* KO in this area was sufficient to induce abnormal breathing patterns, indicating the critical role of the hindbrain in respiratory regulation in RTT (Ward et al. 2011). Additionally, conditional *Mecp2* KO in excitatory, inhibitory, and modulatory neuronal populations can contribute to a decline in hypoxic ventilatory response (Ward et al. 2020). Consistently, rescue models where MECP2 function is restored to these neuronal groups demonstrate significant improvements in these phenotypes (Ward et al. 2020). These findings reinforce the connection between autonomic nervous system dysfunctions and respiratory abnormalities in RTT pathology.

Neurochemical alterations and misregulation of brain-derived neurotrophic factor (BDNF) are key molecular mechanisms contributing to those respiratory network dysfunctions. MECP2 deficiency in the brainstem disrupts the normal production of critical neurotransmitters, including substance P, γ-aminobutyric acid (GABA), glutamate, monoamines, norepinephrine, serotonin (5-HT), leading to respiratory irregularities (Viemari et al. 2005; Kida et al. 2017) (Fig. 9.3). Substance P is involved in cardiorespiratory regulation, gastrointestinal regulation, and the cough reflex, indicative of the significant impact of autonomic dysregulation on the fatal aspiration pneumonia observed in *Mecp2*$^{-/y}$ mice (Kida et al. 2017). In the ventrolateral medulla of *Mecp2*$^{-/y}$ mice, there is diminished GABA synaptic inhibition of expiratory neurons, which is thought to be a fundamental cause of the apneas observed in RTT (Medrihan et al. 2008). Increasing $GABA_A$ inhibition substantially reduces the occurrence of apnea and periodic breathing in *Mecp2*$^{+/-}$ mice (Abdala et al. 2010). Altered levels of norepinephrine and serotonin (5-HT) in the medulla also contribute to the irregular breathing rhythms (Viemari et al. 2005). In *Mecp2*$^{+/-}$ mice, BDNF expres-

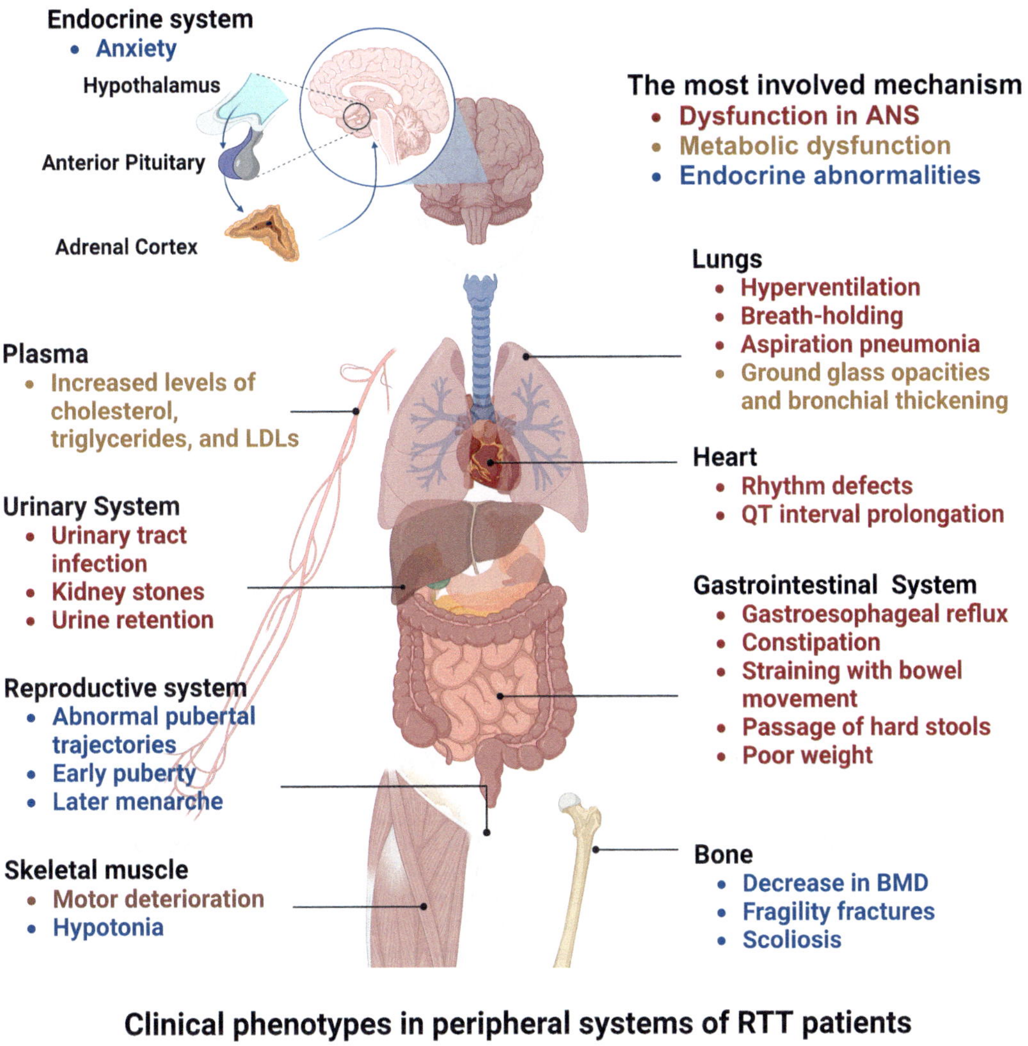

Fig. 9.2 Clinical phenotypes in RTT patients. The potential contributing mechanism is indicated by the text color: ANS-dependent (red), ANS-independent (yellow), and understudied (blue)

sion is reduced within the (nucleus tractus solitarius) nTS (Kline et al. 2010), a crucial area where signals from the cardiorespiratory and gastrointestinal systems are received and integrated (Travagli 2007). Specifically, downregulation of BDNF leads to an excitatory/inhibitory synaptic imbalance in *Mecp2*$^{+/-}$ nTS (Fig. 9.3), while exogenous BDNF expression can rescue this phenotype (Kline et al. 2010). These findings highlight the critical role of BDNF in maintaining synaptic balance and proper autonomic function in RTT.

9.2.2 Heart

Autonomic dysregulation in RTT also significantly impacts the cardiovascular system, causing rhythm defects that can lead to sudden death in patients (Guideri and Acampa 2005), and approximately 20% of RTT patients exhibit a severe cardiac conduction disorder known as long QT syndrome (LQT) (McCauley et al. 2011) (Fig. 9.2).

Both male and female *Mecp2*-null mice have reproduced the LQT phenotype observed in RTT (McCauley et al. 2011). Conditional *Mecp2* KO

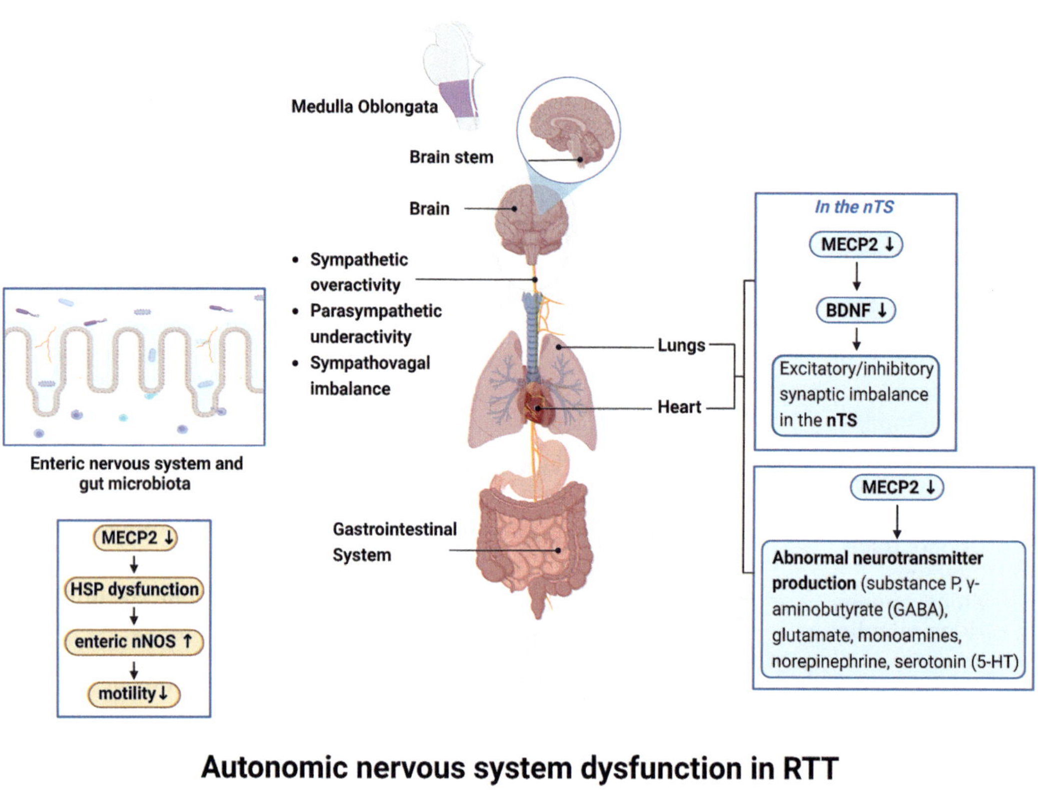

Fig. 9.3 Autonomic nervous system dysfunction in RTT. RTT patients exhibit phenotypes of sympathetic overactivity, parasympathetic underactivity, and sympathovagal imbalance (Julu et al. 2008; Kumar et al. 2017), primarily affecting the lungs, heart, and gastrointestinal systems. These disorders originate from the medulla oblongata, where MECP2 deficiency leads to abnormal production of key neurotransmitters involved in cardiorespiratory regulation (Viemari et al. 2005; Kida et al. 2017). MECP2 deficiency also causes decreased expression of BDNF in the nTS, leading to an excitatory/inhibitory synaptic imbalance (Kline et al. 2010). In the enteric nervous system, MECP2 deficiency affects homeostatic synaptic plasticity, leading to elevated nNOS levels and reduced motility (Wahba et al. 2016)

in the nervous system has been shown to induce both LQT and pacing-induced arrhythmia (McCauley et al. 2011). Furthermore, the conditional removal of *Mecp2* from cholinergic neurons, which play a crucial role in autonomic functions, can replicate the cardiac rhythm defects observed in *Mecp2* null mice (Herrera et al. 2016). These findings confirm that the abnormalities in the autonomic nervous system can affect the cardiovascular system. Misregulation of BDNF in the nTS may also lead to cardiovascular dysfunction, similar to its effects on the respiratory system (Kline et al. 2010) (Fig. 9.3). Additionally, alterations in neurotransmitters such as substance P may further exacerbate these cardiac issues (Singh et al. 2020) (Fig. 9.3). Collectively, the cardiac defects observed in RTT patients are likely caused by the misregulation of BDNF or other neurotransmitters in the autonomic nervous system.

In contrast, MECP2 deficiency may affect cardiac development and function by directly regulating the expression of cardiac-specific genes. For example, ChIP analysis has shown that MECP2 binds to the CpG islands of *Tbx5*, and the expression of *Tbx5* is upregulated in *Mecp2*-null heart, leading to cardiac dysfunction, including arrhythmias (Hara et al. 2015). Consistently, overexpression of *Mecp2* in the mouse heart results in the downregulation of *Tbx5* (Alvarez-

Saavedra et al. 2010). Additionally, in cardiomyocytes derived from induced pluripotent stem cells (iPSCs) of a 3-year-old female patient with the *MECP2*^{1155del32}, there is an upregulation of multiple sodium, potassium, and calcium ion channel genes as well as several WNT family genes, which disrupt normal electrophysiological properties (Ng et al. 2022). However, due to the lack of in vivo evidence, we do not fully understand how these transcriptional dysregulations are related to RTT symptoms. Additionally, it remains elusive which transcriptional dysregulations and structural changes originate from the ANS and which are directly caused by the absence of MECP2 in the heart, which requires further investigation using conditional knockout mouse models.

9.2.3 Gastrointestinal (GI) System and Enteric Nervous System (ENS)

Gastrointestinal and nutritional issues are frequently reported in RTT patients. In a study of 983 females with RTT, 92% had gastrointestinal issues, including gastroesophageal reflux, constipation, straining with bowel movements, and hard stools, while 47% experienced nutritional problems, primarily characterized by poor weight (Motil et al. 2012) (Fig. 9.2).

Mecp2$^{-/y}$ mice also exhibit a marked increase in transit time and reduced body weight (Wahba et al. 2016; Guy et al. 2001), with the latter occurs in the C57BL/6 genetic background (Guy et al. 2001). These mice also display shorter colons with severe epithelial changes similar to colitis, along with abnormal localization of key membrane proteins (Millar-Buchner et al. 2016). The gastrointestinal phenotype observed in *Mecp2*-null models is not replicated by conditional *Mecp2* KO in intestinal tissue alone. This indicates that the GI abnormalities seen in RTT may stem from defects in the nervous system that innervate or regulate the intestinal tissue rather than originate within the gut itself.

MECP2 deficiency in mice disrupts homeostatic synaptic plasticity (HSP) in the ENS, mirroring effects observed in the central nervous system (CNS) (Wahba et al. 2016). This dysfunction results in elevated neuronal nitric oxide synthase (nNOS) levels, causing defective nitric oxide production and impaired peristalsis in the GI system (Wahba et al. 2016) (Fig. 9.3). Interestingly, the expression levels of substance P, tyrosine hydroxylase, and vasoactive intestinal peptide remain normal in the GI tracts of RTT patients (Deguchi et al. 2000). These findings contrast with those in the brain stem, suggesting that the autonomic nervous system, alongside the ENS, plays a role in RTT-related GI abnormalities (Deguchi et al. 2000). Additionally, dysfunctional MECP2 may promote a dysbiotic intestinal microbial community, exacerbating RTT gastrointestinal physiopathology (Strati et al. 2016) (Fig. 9.3). This dysbiosis can affect the production of short-chain fatty acids, leading to persistent constipation, inflammation, and cytokine dysregulation (Strati et al. 2016). Moreover, dysbiotic gut microbiota might affect nervous system function through various mechanisms, including cytokine release from mucosal immune cells, hormone secretion from enteroendocrine cells, and signals via afferent neural pathways such as the vagus nerve (Collins et al. 2012). In summary, abnormalities in the ENS, the ANS, and gut microbiota collectively impair GI motility and contribute to other GI symptoms in RTT patients, while the gut microbiota may, in turn, influence CNS function.

9.3 ANS-Independent Peripheral Dysfunction

9.3.1 Metabolism Disorders

Peripheral metabolic dysfunction is a notable aspect of RTT, affecting a range of tissues such as the liver, lungs, adipose tissues, and skeletal muscles (Fig. 9.4). Studying these metabolic parameters provides insights into the underlying pathological mechanisms and paves the way for identifying biomarkers indicative of disease severity.

9.3.1.1 Lipid Metabolism

9.3.1.1.1 Liver

The liver, a critical metabolic organ, is affected in RTT. In female RTT patients, alterations in key proteins involved in cholesterol regulation lead to abnormal cholesterol metabolism (Segatto et al. 2014). Additionally, these patients typically exhibit an abnormal plasma lipid profile characterized by elevated levels of both HDL and LDL (Sticozzi et al. 2013; Justice et al. 2013) (Fig. 9.2).

Mecp2$^{-/y}$ null mice also exhibit symptoms of dyslipidemia, fatty liver disease, and insulin resistance (Kyle et al. 2016; Buchovecky et al. 2013). Furthermore, conditional *Mecp2* KO in the liver is sufficient to cause dyslipidemia and fatty liver disease in these mouse models, highlighting the liver-specific effects of MECP2 deficiency (Kyle et al. 2016). Mechanistically, MECP2 regulates the expression of liver lipogenic enzymes such as *Sqle* through its interaction with the NCoR1/HDAC3 corepressor complex (Kyle et al. 2016) (Fig. 9.4). Loss of

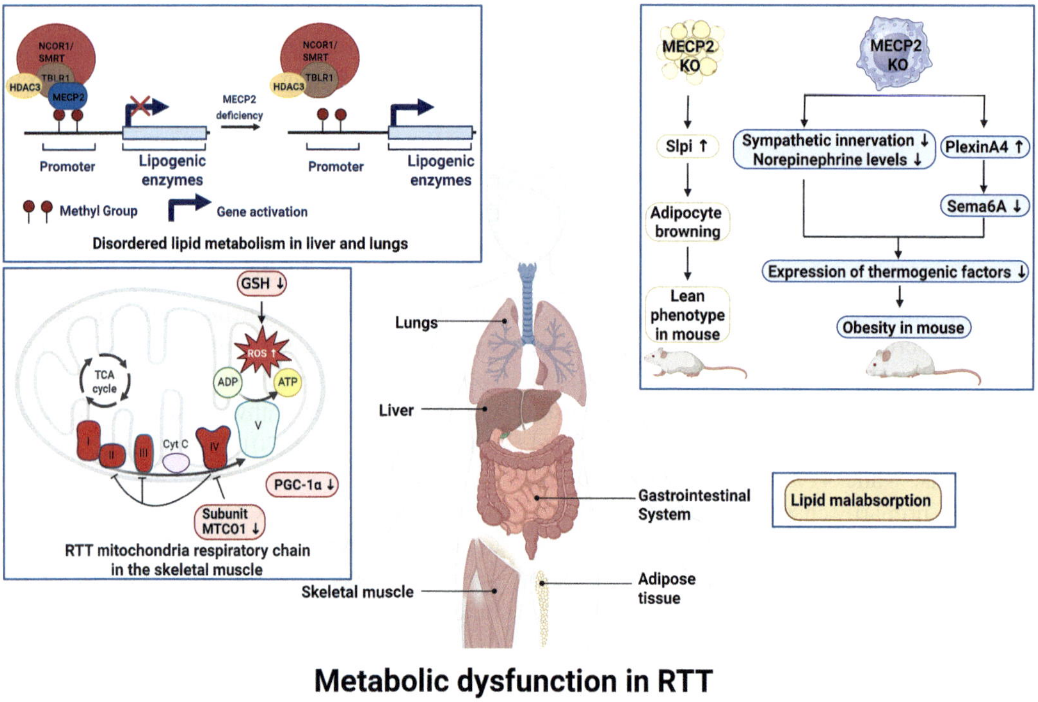

Fig. 9.4 Metabolic dysfunction in RTT. MECP2 deficiency leads to metabolic dysfunction that specifically affects the liver, lungs, gastrointestinal system, skeletal muscle, and adipose tissue. Disordered lipid metabolism is observed in the liver and lungs. MECP2 binds to promoters of lipogenic enzymes, recruiting the NCOR1/HDAC3 co-repressor complex to suppress their expression (Vashi et al. 2021; Kyle et al. 2016). Consequently, the loss of MECP2 results in the overexpression of these enzymes and disrupts lipid metabolism (Vashi et al. 2021; Kyle et al. 2016). Furthermore, MECP2 deficiency impairs the mitochondrial respiratory chain, particularly in skeletal muscle. This impairment is characterized by reduced expression of critical Complex IV (COIV) subunit MTCO1, while decreased COIV activity further affects COII and COIII activity (Hargreaves et al. 2007; Gold et al. 2014). A reduction in PGC-1α, a master regulator of mitochondrial biogenesis (Li et al. 2017; Finck and Kelly 2006), has been observed (Gold et al. 2014). Additionally, a deficiency in the non-enzymatic antioxidant GSH leads to elevated ROS production and disrupted oxidative metabolism (Gold et al. 2014). MECP2 KO in adipocytes results in increased Slpi expression, promoting adipocyte browning and a lean phenotype (Liu et al. 2020). Conversely, MECP2 knockout in macrophages decreases sympathetic innervation and norepinephrine levels, reduces the expression of thermogenic factors, and alters signaling pathways involving PlexinA4 and Sema6A, ultimately contributing to obesity in mice (Wolf et al. 2017)

MECP2 interferes with the binding of this corepressor complex to promoter regions, leading to aberrant transcription of lipogenic enzymes and increased production of hepatic triglycerides and cholesterol (Kyle et al. 2016). Overall, the disruption of MECP2 plays a critical role in regulating liver lipogenic enzyme activity, contributing to increased production of triglycerides and cholesterol.

9.3.1.1.2 Lungs

The lungs, which contribute to lipid metabolism, are also implicated in RTT. Computed tomography scans of RTT patients' lungs reveal ground-glass opacities and bronchial thickening (Fig. 9.2), indicative of surfactant dysfunction (De Felice et al. 2010). Furthermore, studies on *Mecp2*$^{-/y}$ mice have noted abnormal lipid production and emphysema-like changes in the lungs (Kida et al. 2017; Vashi et al. 2021). Conditional *Mecp2* KO from lung alveolar cells is sufficient to alter the expression of lipogenic genes, indicating that the lungs exhibit autonomous phenotypes (Vashi et al. 2021).

Previous studies using single-cell RNA sequencing (scRNA-seq) have identified the misregulation of many lipid metabolism genes in *Mecp2*$^{-/y}$ alveolar epithelial 2 (AE2) cells (Vashi et al. 2021). ChIP assays further confirm that the lipogenic genes *Acot1* and *Hmgcs2* are direct targets of MECP2 and the NCoR1/2 co-repressor complex (Vashi et al. 2021) (Fig. 9.4). In addition, the number of ABCA3-positive AE2 cells is significantly reduced in *Mecp2*-deficient lungs (Kida et al. 2017). ABCA3, a crucial lipid transport protein, is essential for the normal synthesis and storage of pulmonary surfactants in AE2 cells (Whitsett et al. 2010). The reduction in ABCA3-positive AE2 cells can disrupt normal alveolar development and injury repair, potentially leading to emphysema-like changes (Kaminski et al. 2006). In summary, MECP2 deficiency significantly affects lung health in RTT patients by altering the expression of key lipogenic genes.

9.3.1.1.3 GI System

Alterations in the microbiome and metabolome are evident in RTT patients and animal models. Notably, the phenotype in female *Mecp2*$^{+/-}$ mice is more pronounced than in male *Mecp2*$^{-/y}$ mice, implying a potential sex-specific mechanism influencing disease progression in females (Strati et al. 2016; Neier et al. 2021). In female *Mecp2*$^{+/-}$ mice, elevated fecal lipid levels are observed at nine weeks of age, just before the onset of neurological and motor symptoms at 10–11 weeks. This increase in fecal lipids, indicative of lipid malabsorption in the GI system (Fig. 9.4), is often associated with lipidome alterations in brain (Neier et al. 2021). These findings suggest that metabolic abnormalities may serve as a new molecular pathway to precede disease onset in females, indicating their critical roles in the pathophysiology of RTT.

9.3.1.2 Energy Metabolism and Adipose Tissue

Abnormalities in energy metabolism have been observed in RTT mouse models. Specifically, *Mecp2*$^{+/-}$ mice exhibit an increased fat mass compared to that of WT mice, despite similar body weight, which may be due to a reduced capacity for fat oxidation (Park et al. 2014). In adipocyte-specific *Mecp2* KO mice, a lean phenotype is observed, indicating a significant role of MECP2 in adipose tissue regulation (Liu et al. 2020). Mechanistically, the deletion of MECP2 in adipocytes leads to the upregulation of *Slpi*, a gene known to enhance browning of white adipose tissue and reduce obesity (Liu et al. 2020) (Fig. 9.4). This suggests that MECP2's influence on adipocytes may directly affect energy storage and expenditure processes, promoting a leaner body composition through enhanced metabolic activity. In contrast, mice with macrophage-specific MECP2 deficiency exhibit obesity and impaired brown adipose tissue (BAT) function (Wolf et al. 2017). The dysfunction in BAT observed in these pre-obese mice is associated with decreased sympathetic innervation and lower local norepinephrine levels, which are crucial for the expression of thermogenic factors by adipocytes (Wolf et al. 2017). Moreover, these mutant macrophages exhibit increased expression of *PlexinA4*, a signaling receptor that interacts with the transmembrane semaphorin Sema6A to repel sympathetic axons (Wolf et al. 2017). This reduction in sympathetic input to BAT diminishes its thermogenic capacity, thereby altering body composition and contributing to an obese phenotype (Fig. 9.4). These findings highlight a complex interplay where MECP2 regulates energy metabolism and body composition through its distinct effects on different cell types within the adipose tissue.

Interestingly, elevated levels of leptin, a hormone primarily secreted by white adipocytes, have been noted in both RTT patients and animal models (Park et al. 2014; Acampa et al. 2008). Its increase in RTT is associated with disruptions in the ANS, particularly affecting the respiratory network (Acampa et al. 2008; Belaidouni et al. 2023). Leptin receptors are located in respiratory control centers, including the NTS and the rostro ventrolateral medulla (Caron et al. 2010). Furthermore, antagonizing leptin has been shown to restore the synaptic excitation/inhibition imbalance and plasticity impairment in hippocampal neurons in *Mecp2*$^{-/y}$ mice (Belaidouni et al. 2023). These observations not only illustrate the profound connection between peripheral disturbances and central nervous system dysfunctions in RTT, but also suggest a potential new therapeutic approach by targeting leptin.

9.3.1.3 Skeletal Muscles

Movement disorders are a critical clinical feature of RTT, potentially linked to mitochondrial abnormalities in the nervous system and skeletal muscles (Li et al. 2013; Kriaucionis et al. 2006; Gibson et al. 2010; Forlani et al. 2010; Singh et al. 2021; Tomasello et al. 2024). Skeletal muscles rely on the substantial energy produced by their abundant mitochondria and their ability to maintain oxidative metabolism homeostasis (Dong and Tsai 2023). Electron microscopy analyses of skin and muscle biopsies from RTT patients have revealed abnormal mitochondrial characteristics, including swelling, dumbbell shapes, and vacuolization (Wakai et al. 1990; Eeg-Olofsson et al. 1988). Additionally, elongated mitochondria with vacuolization and concentric membranous profiles have been noted (Ruch et al. 1989). Similar mitochondrial abnormalities have been observed in *Mecp2*$^{-/y}$ mice, characterized by misshapen, enlarged mitochondria with electron-lucent central matrices and disorganized non-parallel cristae (Park et al. 2014).

Gold and colleagues reported reduced levels of proteins critical for mitochondrial function in the skeletal muscles of *Mecp2*$^{-/y}$ mice, leading to compromised mitochondrial respiratory chain activity and increased oxidative stress (Gold et al. 2014). Specifically, the absence of MECP2 results in decreased expression of MTCO1, a subunit of complex IV, which diminishes complex IV activity (Gold et al. 2014) (Fig. 9.4). This reduction triggers a secondary decrease in the activity of complexes II and III (Hargreaves et al. 2007; Gold et al. 2014) (Fig. 9.4). Additionally, there is a decrease in the expression of PGC-1α, potentially disrupting mitochondrial biogenesis and the cellular oxidant-antioxidant balance (Gold et al. 2014; Kang and Li Ji 2012) (Fig. 9.4). The levels of glutathione (GSH), a crucial non-enzymatic antioxidant, are also reduced in mitochondrial lysates from *Mecp2*$^{-/y}$ mouse muscle samples, leading to excess production of reactive oxygen species (ROS) (Gold et al. 2014) (Fig. 9.4). Overall, mitochondrial dysfunction caused by the reduced expression of these key genes results in ROS accumulation and subsequent motor deterioration, significantly contributing to the progression of RTT.

9.3.2 Endocrine Dysregulation

9.3.2.1 Endocrine System

RTT patients exhibit abnormal pubertal trajectories, characterized by early onset of puberty with delayed menarche (Killian et al. 2014) (Fig. 9.2). Additionally, these individuals experience heightened anxiety with frequent spells of panic and episodes of body rigidity (Fig. 9.2), more frequently than those with severe to profound mental retardation (Mount et al. 2002). Similarly, *Mecp2*$^{308/Y}$ mice, which express a truncated form of MECP2 (amino acids 1-308), demonstrate increased anxiety-like behavior in behavioral assays, such as the open field test and the elevated plus maze, reflecting the heightened anxiety observed in RTT patients (McGill et al. 2006). While this behavior likely reflects nervous system alterations, it may also be associated with dysfunctions in the endocrine system, indicating a complex interplay between these systems in RTT patients.

Endocrine system dysfunction has been implicated in the anxiety-related behaviors and abnormal pubertal trajectories observed in patients with RTT. Preliminary evidence in *Mecp2*$^{308/Y}$ mice suggests a molecular link where MECP2 may bind to the promoter of corticotropin-releasing hormone (CRH) to repress its gene expression (McGill et al. 2006) (Fig. 9.5). CRH is critical for regulating behavioral and physiological stress responses (Bale and Vale 2004), as it initiates HPA (hypothalamic-pituitary-adrenal) axis activation and glucocorticoid release from the adrenal cortex (Bruhn et al. 1984; Herman et al. 2016) (Fig. 9.5). Mutations in MECP2 that lead to CRH overexpression could result in anxiety and HPA axis hyperactivity (McGill et al. 2006) (Fig. 9.5). In contrast, studies in *Mecp2*$^{-/y}$ mice demonstrate decreased CRH levels and reduced anxiety-like behaviors, which may be related to enhanced whisker sensitivity (Abellan-Alvaro et al. 2021; Flores Gutierrez et al. 2020). Other findings highlight that MECP2 suppresses estrogen receptor alpha (ER alpha) mRNA expression in the cortex during postnatal development through methylation of its promoters (Westberry et al. 2010) (Fig. 9.5). MECP2 deficiency also correlates with the overexpression of FXYD1 in the frontal cortex, implicated in abnormal female sexual development (Deng et al. 2007; Garcia-Rudaz et al. 2009) (Fig. 9.5). In summary, these studies suggest that MECP2 mutations may induce HPA axis hyperactivity, contributing to anxiety-like behaviors, and affect pubertal trajectories by altering proteins essential for female sexual development.

9.3.2.2 Musculoskeletal System

Females with RTT often suffer from reduced bone mineral density (BMD), which predisposes them to fragility fractures (Roende et al. 2011) (Fig. 9.2). Markers indicative of bone metabolism, such as osteocalcin and bone-specific alkaline phosphatase, are consistently reported to be lower in RTT patients (Killian et al. 2017). Additionally, the prevalence of scoliosis is significantly high (Fig. 9.2), affecting approximately 64% of patients (Killian et al. 2017), with over 80% experiencing orthopedic comorbidities (Neul et al. 2010).

Previous studies using *Mecp2*$^{-/Y}$ models have revealed that low bone mass is a defining characteristic of RTT, as evidenced by reduced cortical thickness, diminished spinal bone density, and decreased mineralization within the medullary cavities of long bones (Shapiro et al. 2010). Vitamin D deficiency, characterized by low serum 25-hydroxyvitamin D (25(OH)D) levels, is observed in RTT patients (Motil et al. 2011; Sarajlija et al. 2013). This deficiency leads to decreased calcium absorption and elevated bone turnover, increasing the risk of fractures (Lips 2001) (Fig. 9.5). In the liver of *Mecp2*$^{+/-}$ mice, decreased levels of CYP27A1, an enzyme critical for synthesizing 25(OH)D and maintaining vitamin D homeostasis, have been noted, suggesting a potential mechanism underlying the vitamin D deficiency associated with RTT (Ribeiro and MacDonald 2022). Human RTT bone marrow-derived mesenchymal stem cells (BM-MSCs) exhibit early senescence and reduced apoptosis, also contributing to the abnormal development and regeneration of bone (Squillaro et al. 2008). In summary, disruptions in vitamin D homeostasis and MSCs may play crucial roles in the development of bone abnormalities in RTT.

Pronounced hypotonia, commonly observed in individuals with RTT, further complicates their physical health challenges within the skeletal system (Neul et al. 2010) (Fig. 9.5). Hypotonia is consistently observed in *Mecp2*$^{-/y}$ mice, characterized by disorganization and atrophy of skeletal muscle (Conti et al. 2015). Interestingly, experiments involving mice with conditional *Mecp2* KO in myoblasts and muscle fibers failed to rep-

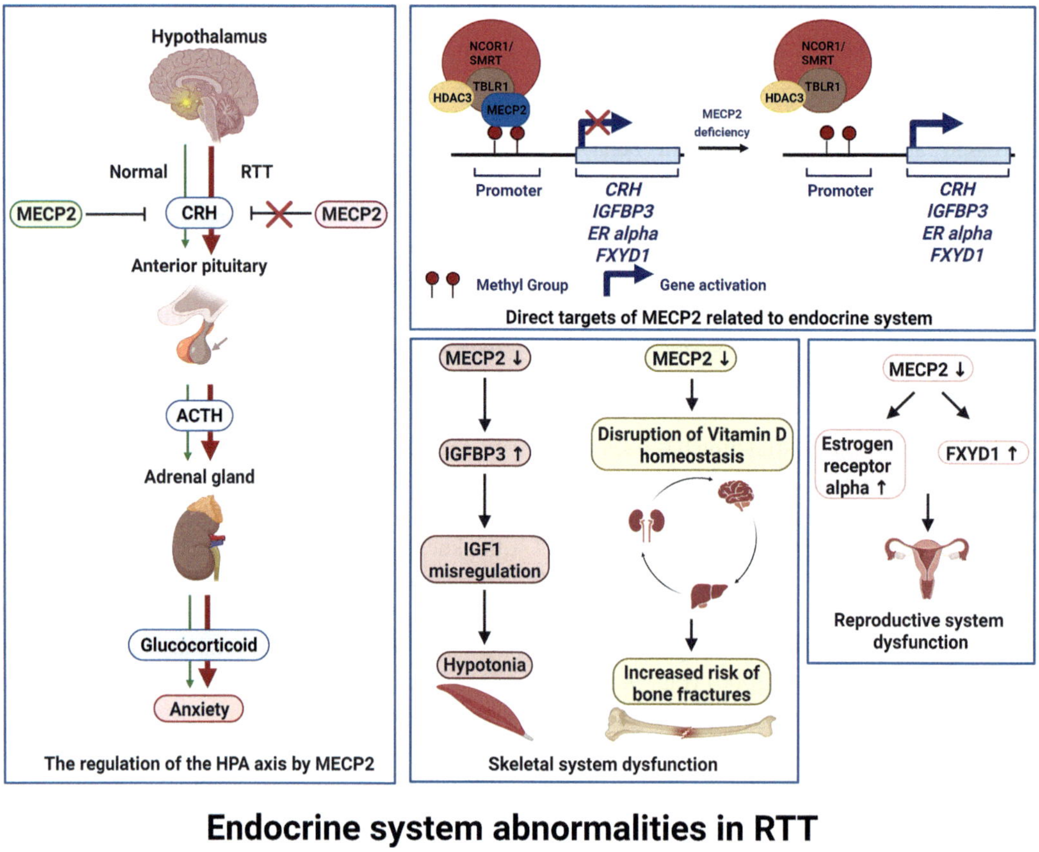

Fig. 9.5 Endocrine system abnormalities in RTT. Endocrine system abnormalities in RTT distinctly impact the HPA axis, skeletal system, and reproductive system. MECP2 can bind to the promoters of key genes such as CRH, IGFBP3, ER alpha, and FXYD1, and facilitate the recruitment of the NCOR1/SMRT/TBLR1 co-repressor complex to suppress their expression (Bale and Vale 2004; Westberry et al. 2010; Deng et al. 2007; Modric et al. 2001). MECP2 loss is associated with the overexpression of these genes, contributing to endocrine system dysfunction. Specifically, MECP2 deficiency disrupts CRH expression in the hypothalamus, leading to abnormal HPA axis activation, increased glucocorticoid production, and heightened anxiety (McGill et al. 2006). Furthermore, this deficiency contributes to skeletal system issues by altering IGFBP3 expression and vitamin D homeostasis; increased IGFBP3 levels disrupt IGF1 regulation, contributing to hypotonia and impaired skeletal system (Modric et al. 2001). Alteration in IGF1 concentration lead to other disfunctions *as discussed in Chap. 4* "IGF1 as brain body connector." Impaired vitamin D metabolism increases the risk of bone fractures (Ribeiro and MacDonald 2022; Motil et al. 2011; Sarajlija et al. 2013). In the reproductive system, elevated levels of estrogen receptor alpha and FXYD1 due to MECP2 deficiency are linked to reproductive dysfunctions (Westberry et al. 2010; Deng et al. 2007; Garcia-Rudaz et al. 2009)

licate the hypotonia phenotype (Conti et al. 2015), suggesting that non-cell-autonomous effects might underpin this symptom in RTT (Conti et al. 2015). Notably, abnormalities in paracrine/endocrine signaling pathways, especially within the growth hormone/insulin-like growth factor (GH/IGF) axis, are considered crucial for maintaining the structural integrity and functional capacity of the skeletal system (Conti et al. 2015; Tropea et al. 2009). The mRNA and protein levels of IGF-1 are decreased in skeletal muscles of the *Mecp2*$^{-/y}$ mice, but not in mice with conditional *Mecp2* KO in the muscle, indicating a consistent non-cell-autonomous role in this regulation (Conti et al. 2015). These initial observations highlight the necessity to further investigate their underlying mechanism. Some evidence suggests that MECP2 can directly target IGFBP3, and its overexpression in *Mecp2* null mice and RTT patients may inhibit IGF-1 signaling (Modric et al. 2001; Chang et al. 2004; Itoh et al. 2007) (Fig. 9.4), potentially impairing skeletal muscle growth and development. Additionally, BDNF, another direct target of MECP2 (Martinowich et al. 2003), plays a crucial role in paracrine/endocrine signaling and is involved in muscle regeneration (Iizuka et al. 2014; Clow and Jasmin 2010); its downregulation in *Mecp2*$^{-/y}$ mice may adversely impact skeletal health (Conti et al. 2015). Overall, the paracrine/endocrine signaling pathways involving BDNF, IGF-1, and IGFBP3 are likely to be the potential mechanisms contributing to the hypotonia observed in RTT; however, the role of neuronal functions in this context remains unclear.

9.3.3 Immune System Abnormalities

Immune response dysfunctions are common in patients with RTT, often indicated by an elevated erythrocyte sedimentation rate that typically suggests a subclinical chronic inflammatory condition (Cortelazzo et al. 2014; Brigden 1999). However, it's important to recognize that this measurement can also be influenced by other medical conditions such as anemia, macrocytosis, and infection. Additionally, there is significant upregulation of cytokines that correlates with both the clinical severity and inflammatory status of these patients (Leoncini et al. 2015). In mice, MECP2 is widely expressed in macrophages, and its deficiency results in transcriptional impairments and/or the loss of macrophages across multiple tissues, potentially contributing to the systemic pathologies observed in RTT (Cronk et al. 2015) (Fig. 9.6).

MECP2 has been identified as a critical regulator of glucocorticoids, hypoxia, and tumor necrosis factor (TNF) responses in microglia and peritoneal macrophages in *Mecp2*-null mouse (Cronk et al. 2015) (Fig. 9.6). Increased expression of glucocorticoid-induced transcriptional signature genes is observed in these cells (Cronk et al. 2015), suggesting that MECP2 functions as a repressor of this pathway. Specifically, the MECP2 recruits NCoR2 and HDAC3 to suppress the expression of *Fkbp5*, a key target gene in the glucocorticoid signaling pathway, which plays a role in chronic stress and metabolic regulation (Cronk et al. 2015; Balsevich et al. 2014, 2017). Furthermore, elevated expression of canonical hypoxia-responsive genes—Hif3a, Ddit4, and Cyr61, has been observed in microglia under hypoxic conditions in *Mecp2*-null mice and confirmed in cultured *Mecp2*-null macrophages in vitro (Cronk et al. 2015). Additionally, an increase in TNF-induced transcriptional signature genes was noted in *Mecp2*-null microglia, which may amplify inflammatory responses and contribute to neurotoxicity (Cronk et al. 2015). Overall, these findings indicate that MECP2 deficiency leads to alternations in inflammatory responses in microglia and macrophages (Cronk et al. 2015), which may affect broader systemic functions in RTT (Fig. 9.6).

Another type of immune cell affected in RTT includes T cells, specifically CD8+ T cells and regulatory T cells (Tregs) (Fig. 9.6). $CD8^+$ T cells in *Mecp2*-null mice exhibit increased expression of *Ifng*, which encodes IFN-γ, a critical regulator in innate and adaptive immunity (Li et al. 2022) (Fig. 9.6). Additionally, these cells demonstrate enhanced NF-κB activity and elevated levels of pro-inflammatory cytokines such as TNFα, IL-6, and IL-3, signaling an inflammatory response in Mecp2-deficient immune cells (O'Driscoll et al. 2013, 2015) (Fig. 9.6). MECP2 also plays a crucial role in regulating the immunosuppressive activity of Treg by enhancing the expression of the transcription factor *Foxp3* (Li et al. 2014), which is essential for the development of Treg within the thymus and its lineage identity maintenance in the peripheral systems (Williams and Rudensky 2007) (Fig. 9.6). In summary, these findings highlight that MECP2 deficiency contributes to immune system dysregulation and increased inflammation in RTT patients.

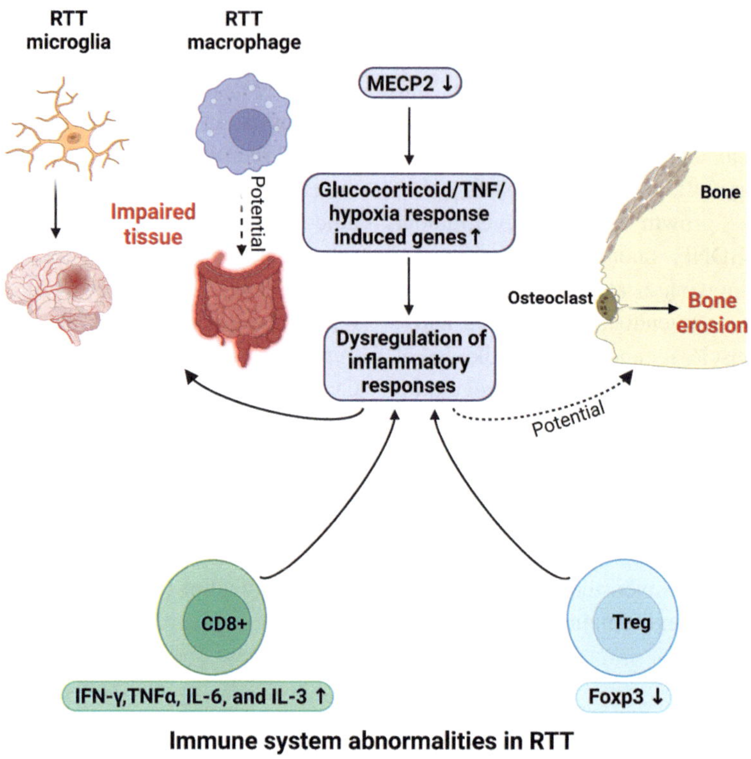

Fig. 9.6 Immune system abnormalities. Immune system abnormalities resulting from MECP2 deficiency specifically affect microglia, macrophages, CD8+ T cells, and Tregs. MECP2 deficiency leads to the upregulation of glucocorticoid/TNF/hypoxia response-induced genes in microglia and macrophages, which may lead to inflammatory responses in various tissues (Cronk et al. 2015). Defects of microglia could impair brain function (Paolicelli and Ferretti 2017). In addition, this dysregulation may also affect the intestine, since it contains a large number of macrophages (Zigmond and Jung 2013), although further research is needed. Another potential mechanism is the increased activity of osteoclasts, contributing to bone erosion (Davies et al. 2013). Furthermore, MECP2 deficiency in CD8+ T cells results in the overproduction of pro-inflammatory cytokines such as IFN-γ, TNFα, IL-6, and IL-3, contributing to immune dysregulation (Li et al. 2022; O'Driscoll et al. 2013, 2015). On the other hand, the reduction of Foxp3 expression in Tregs due to MECP2 deficiency impairs their regulatory function, further contributing to the overall immune system abnormalities observed in RTT (Li et al. 2014)

9.4 Understudied Dysfunction

9.4.1 Urinary System

Urological dysfunction is not a common phenotype in RTT patients, with only 8% experiencing symptoms such as urinary tract infections, kidney stones, or urine retention (Ward et al. 2016) (Fig. 9.2). However, this dysfunction, observed in both *Mecp2*$^{-/y}$ and *Mecp2*$^{+/-}$ mouse models, can lead to abnormalities in micturition, causing urine reflux from the bladder back into the kidneys and potentially resulting in kidney damage (Ward et al. 2016).

The mechanism behind RTT urinary system abnormalities has not been well-studied, although the ANS is likely involved in regulating urinary bladder function (Ochodnicky et al. 2013). Also, MECP2 is known to influence water homeostasis by repressing vasopressin expression through binding to its enhancer region, while miR-132 antagonizes this regulation (Bijkerk et al. 2018). These insights suggest pathways by which MECP2 mutations could impact the urinary system in RTT patients.

9.5 Discussion

In this review, we use RTT as a case study to highlight the significance of the brain-body connection in neurological disorders. Previous studies suggest that MECP2 may drive tissue-specific regulation, with certain peripheral symptoms stemming from localized systems, while others are closely linked to the nervous system. Specifically, dysfunction in the ANS and the ENS can lead to impairments in the cardiopulmonary and gastrointestinal systems (Singh et al. 2020; Viemari et al. 2005; Kida et al. 2017; Kline et al. 2010; Wahba et al. 2016). In contrast, the loss of MECP2 in the liver or lungs can directly lead to alterations in lipid metabolism, resulting in disrupted cholesterol levels and impaired surfactant function (Vashi et al. 2021; Kyle et al. 2016). MECP2 deficiency also influences the energy balance in adipocytes and the mitochondrial oxidative stress of skeletal muscle, affecting body weight and motor abilities (Gold et al. 2014; Liu et al. 2020; Wolf et al. 2017). Other MECP2-induced tissue-specific regulations, including effects on endocrine hormones and immune cells, may contribute to a spectrum of RTT symptoms such as hypotonia, fragile bones, anxiety-related behaviors, reproductive system irregularities, and abnormal inflammatory responses (Westberry et al. 2010; Deng et al. 2007; McGill et al. 2006; Motil et al. 2011; Sarajlija et al. 2013; Conti et al. 2015; Cronk et al. 2015; Li et al. 2022; O'Driscoll et al. 2013, 2015; Li et al. 2014). Conversely, symptoms in the peripheral systems of RTT may also influence normal brain functions. For example, dysregulated inflammatory factors and hormones in the gut might impact the central nervous system, impaired lipid absorption might result in brain lipid deficiencies, and elevated plasma leptin levels could disrupt the respiratory network controlled by the ANS (Collins et al. 2012; Neier et al. 2021; Acampa et al. 2008; Belaidouni et al. 2023). Additionally, MECP2 deficiency has been associated with defective cilia formation in RTT patients' fibroblasts, cultured neurons, and mouse brains (Liu et al. 2024; Frasca et al. 2020; Lopes et al. 2024; D'Mello 3rd. 2021; Gupta et al. 2015), implying that ciliopathy, which affects a variety of human tissues, may potentially contribute to the broad spectrum of RTT symptoms, although RTT is not classified as a ciliopathy. To better understand the origins of symptoms associated with MECP2 dysfunction, future studies might utilize mouse models with conditional MECP2 knockouts alongside induced pluripotent stem cell (iPSC)-derived human cells or organoids in specific systems. Such models will allow for detailed studies of MECP2's roles in different tissues, offering insights into its diverse effects on physiological and pathological processes.

These findings also highlight the potential of targeting the brain-body connection as a therapeutic approach for RTT patients. Sarizotan, a full 5-HT agonist, has demonstrated a reduction in breathing apneas by 15–30% in *Mecp2*$^{-/Y}$ mouse models (Abdala et al. 2014). Additionally, the sodium channel blocker phenytoin has been effective in partially alleviating symptoms of long QT syndrome in these models (McCauley et al. 2011). Regarding metabolism, statin drugs, which lower lipid levels, have been shown to improve motor symptoms and extend lifespan in RTT mice (Buchovecky et al. 2013). BDNF and IGF-1 are critical targets for RTT therapies (Vashi and Justice 2019). For example, fingolimod and an ampakine compound CX456 has been shown to enhance motor activity and improve respiratory function in *Mecp2*$^{-/Y}$ mice by restoring BDNF expression, respectively (Deogracias et al. 2012; Ogier et al. 2007).The small molecule LM22A-4, which is a mimetic of the BDNF loop domain, can partially activate BDNF receptor TrkB and reverse the respiratory abnormalities of *Mecp2*$^{+/-}$[140,141]. In addition, direct supplementation of IGF-1, either its tri-peptide form or human recombinant IGF-1, has been shown to improve the respiratory pattern and motor abilities in RTT mouse models (Tropea et al. 2009; Castro et al. 2014). Meanwhile, Trofinetide, a synthetic analog of IGF-1, has received FDA approval as a therapeutic option for treating RTT, offering new hope for improving overall health and motor function in patients (Glaze et al. 2019; Neul et al. 2023).

The pathogenic mechanisms of MECP2 in regulating peripheral systems require further investigation, particularly to determine if MECP2 directly influences transcriptional and epigenetic regulation in peripheral tissues. Currently, the field lacks a comprehensive genetic and epigenetic profiling of MECP2 in non-neuronal human tissues. As evidenced by recent studies, advanced techniques like CUT&Tag or CUT&RUN, coupled with RNA-Seq, are essential for distinguishing MECP2's functional versus non-functional DNA binding, providing superior mapping resolution compared to traditional ChIP-Seq methods (Liu et al. 2024; Kaya-Okur et al. 2019). Moreover, single-cell sequencing or MERFISH spatial transcriptomics will be instrumental in revealing the cell-type-specific epigenome and transcriptional signature of MECP2 across diverse peripheral systems (Liu et al. 2023; Vandereyken et al. 2023). Additionally, considering the physiological differences between humans and mice, alternative models such as those derived from iPSCs or primate models like monkeys should be considered (Ng et al. 2022; Chen et al. 2017). For example, when studying Long QT syndrome in RTT, mice might not fully recapitulate the condition due to their substantial electrophysiological differences in cardiac ion channel profiles (Tanner and Beeton 2018), indicating the requirement for more representative biological models in RTT research to address human-specific pathologies.

Overall, our review has systematically outlined the origins of peripheral system symptoms in RTT, elucidating that while many symptoms stem from the nervous system, significant contributions also arise from abnormalities in metabolism, the endocrine system, and the immune system across different tissues. Comprehensive research is required to further delineate the molecular mechanisms driven by MECP2 within these peripheral systems. Such studies are crucial not only for developing targeted interventions that address both central and peripheral dysfunctions in RTT, but also for enhancing our understanding of the brain-body interplay, which may have implications for other neurological disorders.

Acknowledgments The authors thank Daniela Tropea and Emanuela Giacometti for helpful discussions and resources. Figures 9.1, 9.2, 9.3, 9.4, 9.5 and 9.6 were created using BioRender.com. This work was supported by NIH grant 5R01MH104610 (R.J.) and International Rett Syndrome Foundation (IRSF) Research Independence Award (Y.L.). This research was supported by a generous gift from The Owens Family Foundation (Y.L., R.J.).

Author Contributions Y.L., R.G., and R.J. contributed to all aspects of the manuscript, including drafting figures and writing the text.

Declaration of Interests R.J. is an advisor and co-founder of Fate Therapeutics and Fulcrum Therapeutics. All other authors declare no competing interests.

References

Abdala AP, Dutschmann M, Bissonnette JM, Paton JF (2010) Correction of respiratory disorders in a mouse model of Rett syndrome. Proc Natl Acad Sci USA 107:18208–18213. https://doi.org/10.1073/pnas.1012104107

Abdala AP et al (2014) Effect of Sarizotan, a 5-HT1a and D2-like receptor agonist, on respiration in three mouse models of Rett syndrome. Am J Respir Cell Mol Biol 50:1031–1039. https://doi.org/10.1165/rcmb.2013-0372OC

Abellan-Alvaro M, Stork O, Agustin-Pavon C, Santos M (2021) MeCP2 haplodeficiency and early-life stress interaction on anxiety-like behavior in adolescent female mice. J Neurodev Disord 13:59. https://doi.org/10.1186/s11689-021-09409-7

Acampa M et al (2008) Sympathetic overactivity and plasma leptin levels in Rett syndrome. Neurosci Lett 432:69–72. https://doi.org/10.1016/j.neulet.2007.12.030

Allan LM (2019) Diagnosis and management of autonomic dysfunction in dementia syndromes. Curr Treat Options Neurol 21:38. https://doi.org/10.1007/s11940-019-0581-2

Alvarez-Saavedra M et al (2010) Elevated expression of MeCP2 in cardiac and skeletal tissues is detrimental for normal development. Hum Mol Genet 19:2177–2190. https://doi.org/10.1093/hmg/ddq096

Amir RE et al (1999) Rett syndrome is caused by mutations in X-linked MECP2, encoding methyl-CpG-binding protein 2. Nat Genet 23:185–188. https://doi.org/10.1038/13810

Bale TL, Vale WW (2004) CRF and CRF receptors: role in stress responsivity and other behaviors. Annu Rev Pharmacol Toxicol 44:525–557. https://doi.org/10.1146/annurev.pharmtox.44.101802.121410

Balsevich G et al (2014) Interplay between diet-induced obesity and chronic stress in mice: potential role of FKBP51. J Endocrinol 222:15–26. https://doi.org/10.1530/JOE-14-0129

Balsevich G et al (2017) Stress-responsive FKBP51 regulates AKT2-AS160 signaling and metabolic function. Nat Commun 8:1725. https://doi.org/10.1038/s41467-017-01783-y

Beishon LC et al (2022) The role of the autonomic nervous system in cerebral blood flow regulation in dementia: A review. Auton Neurosci 240:102985. https://doi.org/10.1016/j.autneu.2022.102985

Belaidouni Y et al (2023) Leptin antagonism improves Rett syndrome phenotype in symptomatic male Mecp2-null mice. bioRxiv. https://doi.org/10.1101/2023.02.03.526251

Beurel E, Toups M, Nemeroff CB (2020) The bidirectional relationship of depression and inflammation: double trouble. Neuron 107:234–256. https://doi.org/10.1016/j.neuron.2020.06.002

Bijkerk R et al (2018) MicroRNA-132 controls water homeostasis through regulating MECP2-mediated vasopressin synthesis. Am J Physiol Renal Physiol 315:F1129–F1138. https://doi.org/10.1152/ajprenal.00087.2018

Borloz E, Villard L, Roux JC (2021) Rett syndrome: think outside the (skull) box. Fac Rev 10:59. https://doi.org/10.12703/r/10-59

Boxer LD et al (2020) MeCP2 represses the rate of transcriptional initiation of highly methylated long genes. Mol Cell 77:294–309 e299. https://doi.org/10.1016/j.molcel.2019.10.032

Brigden ML (1999) Clinical utility of the erythrocyte sedimentation rate. Am Fam Physician 60:1443–1450

Bruhn TO, Plotsky PM, Vale WW (1984) Effect of paraventricular lesions on corticotropin-releasing factor (CRF)-like immunoreactivity in the stalk-median eminence: studies on the adrenocorticotropin response to ether stress and exogenous CRF. Endocrinology 114:57–62. https://doi.org/10.1210/endo-114-1-57

Buchovecky CM et al (2013) A suppressor screen in Mecp2 mutant mice implicates cholesterol metabolism in Rett syndrome. Nat Genet 45:1013–1020. https://doi.org/10.1038/ng.2714

Caron E, Sachot C, Prevot V, Bouret SG (2010) Distribution of leptin-sensitive cells in the postnatal and adult mouse brain. J Comp Neurol 518:459–476. https://doi.org/10.1002/cne.22219

Castro J et al (2014) Functional recovery with recombinant human IGF1 treatment in a mouse model of Rett syndrome. Proc Natl Acad Sci USA 111:9941–9946. https://doi.org/10.1073/pnas.1311685111

Chahrour M, Zoghbi HY (2007) The story of Rett syndrome: from clinic to neurobiology. Neuron 56:422–437. https://doi.org/10.1016/j.neuron.2007.10.001

Chahrour M et al (2008) MeCP2, a key contributor to neurological disease, activates and represses transcription. Science 320:1224–1229. https://doi.org/10.1126/science.1153252

Chang YS et al (2004) Mechanisms underlying lack of insulin-like growth factor-binding protein-3 expression in non-small-cell lung cancer. Oncogene 23:6569–6580. https://doi.org/10.1038/sj.onc.1207882

Chen Y et al (2017) Modeling Rett syndrome using TALEN-edited MECP2 mutant cynomolgus monkeys. Cell 169:945–955 e910. https://doi.org/10.1016/j.cell.2017.04.035

Chen S et al (2024) Brain-gut-microbiota Axis in amyotrophic lateral sclerosis: A historical overview and future directions. Aging Dis 15:74–95. https://doi.org/10.14336/AD.2023.0524

Clow C, Jasmin BJ (2010) Brain-derived neurotrophic factor regulates satellite cell differentiation and skeltal muscle regeneration. Mol Biol Cell 21:2182–2190. https://doi.org/10.1091/mbc.e10-02-0154

Collins SM, Surette M, Bercik P (2012) The interplay between the intestinal microbiota and the brain. Nat Rev Microbiol 10:735–742. https://doi.org/10.1038/nrmicro2876

Conti V et al (2015) MeCP2 affects skeletal muscle growth and morphology through non cell-autonomous mechanisms. PLoS One 10:e0130183. https://doi.org/10.1371/journal.pone.0130183

Cortelazzo A et al (2014) Subclinical inflammatory status in Rett syndrome. Mediat Inflamm 2014:480980. https://doi.org/10.1155/2014/480980

Cronk JC et al (2015) Methyl-CpG binding protein 2 regulates microglia and macrophage gene expression in response to inflammatory stimuli. Immunity 42:679–691. https://doi.org/10.1016/j.immuni.2015.03.013

D'Mello SR 3rd. (2021) MECP2 and the biology of MECP2 duplication syndrome. J Neurochem 159:29–60. https://doi.org/10.1111/jnc.15331

Davies LC, Jenkins SJ, Allen JE, Taylor PR (2013) Tissue-resident macrophages. Nat Immunol 14:986–995. https://doi.org/10.1038/ni.2705

De Felice C et al (2010) Unrecognized lung disease in classic Rett syndrome. Chest 138:386–392. https://doi.org/10.1378/chest.09-3021

Deguchi K et al (2000) Substance P immunoreactivity in Rett syndrome. Pediatr Neurol 22:259–266. https://doi.org/10.1016/s0887-8994(00)00120-x

Deng V et al (2007) FXYD1 is an MeCP2 target gene overexpressed in the brains of Rett syndrome patients and Mecp2-null mice. Hum Mol Genet 16:640–650. https://doi.org/10.1093/hmg/ddm007

Deogracias R et al (2012) Fingolimod, a sphingosine-1 phosphate receptor modulator, increases BDNF levels and improves symptoms of a mouse model of Rett syndrome. Proc Natl Acad Sci USA 109:14230–14235. https://doi.org/10.1073/pnas.1206093109

Dong H, Tsai SY (2023) Mitochondrial properties in skeletal muscle fiber. Cells 12. https://doi.org/10.3390/cells12172183

Eeg-Olofsson O, al-Zuhair AG, Teebi AS, al-Essa MM (1988) Abnormal mitochondria in the Rett syndrome.

Brain Dev 10:260–262. https://doi.org/10.1016/s0387-7604(88)80010-x

Finck BN, Kelly DP (2006) PGC-1 coactivators: inducible regulators of energy metabolism in health and disease. J Clin Invest 116:615–622. https://doi.org/10.1172/JCI27794

Flores Gutierrez J et al (2020) Protective role of mirtazapine in adult female Mecp2(+/−) mice and patients with Rett syndrome. J Neurodev Disord 12:26. https://doi.org/10.1186/s11689-020-09328-z

Forlani G et al (2010) The MeCP2/YY1 interaction regulates ANT1 expression at 4q35: novel hints for Rett syndrome pathogenesis. Hum Mol Genet 19:3114–3123. https://doi.org/10.1093/hmg/ddq214

Frasca A et al (2020) MECP2 mutations affect ciliogenesis: a novel perspective for Rett syndrome and related disorders. EMBO Mol Med 12:e10270. https://doi.org/10.15252/emmm.201910270

Fried S, Wemelle E, Cani PD, Knauf C (2021) Interactions between the microbiota and enteric nervous system during gut-brain disorders. Neuropharmacology 197:108721. https://doi.org/10.1016/j.neuropharm.2021.108721

Gabel HW et al (2015) Disruption of DNA-methylation-dependent long gene repression in Rett syndrome. Nature 522:89–93. https://doi.org/10.1038/nature14319

Garcia-Rudaz C et al (2009) FXYD1, a modulator of Na,K-ATPase activity, facilitates female sexual development by maintaining gonadotrophin-releasing hormone neuronal excitability. J Neuroendocrinol 21:108–122. https://doi.org/10.1111/j.1365-2826.2008.01812.x

Gibson JH et al (2010) Downstream targets of methyl CpG binding protein 2 and their abnormal expression in the frontal cortex of the human Rett syndrome brain. BMC Neurosci 11:53. https://doi.org/10.1186/1471-2202-11-53

Glaze DG et al (2019) Double-blind, randomized, placebo-controlled study of trofinetide in pediatric Rett syndrome. Neurology 92:e1912–e1925. https://doi.org/10.1212/WNL.0000000000007316

Gold WA et al (2014) Mitochondrial dysfunction in the skeletal muscle of a mouse model of Rett syndrome (RTT): implications for the disease phenotype. Mitochondrion 15:10–17. https://doi.org/10.1016/j.mito.2014.02.012

Guideri F, Acampa M (2005) Sudden death and cardiac arrhythmias in Rett syndrome. Pediatr Cardiol 26:111. https://doi.org/10.1007/s00246-004-0701-x

Gupta GD et al (2015) A dynamic protein interaction landscape of the human centrosome-cilium Interface. Cell 163:1484–1499. https://doi.org/10.1016/j.cell.2015.10.065

Guy J, Hendrich B, Holmes M, Martin JE, Bird A (2001) A mouse Mecp2-null mutation causes neurological symptoms that mimic Rett syndrome. Nat Genet 27:322–326. https://doi.org/10.1038/85899

Hagberg B, Aicardi J, Dias K, Ramos O (1983) A progressive syndrome of autism, dementia, ataxia, and loss of purposeful hand use in girls: Rett's syndrome: report of 35 cases. Ann Neurol 14:471–479. https://doi.org/10.1002/ana.410140412

Hara M et al (2015) Disturbance of cardiac gene expression and cardiomyocyte structure predisposes Mecp2-null mice to arrhythmias. Sci Rep 5:11204. https://doi.org/10.1038/srep11204

Hargreaves IP et al (2007) Inhibition of mitochondrial complex IV leads to secondary loss complex II-III activity: implications for the pathogenesis and treatment of mitochondrial encephalomyopathies. Mitochondrion 7:284–287. https://doi.org/10.1016/j.mito.2007.02.001

Herman JP et al (2016) Regulation of the hypothalamic-pituitary-adrenocortical stress response. Compr Physiol 6:603–621. https://doi.org/10.1002/cphy.c150015

Herrera JA, Ward CS, Wehrens XH, Neul JL (2016) Methyl-CpG binding-protein 2 function in cholinergic neurons mediates cardiac arrhythmogenesis. Hum Mol Genet 25:4983–4995. https://doi.org/10.1093/hmg/ddw326

Howren MB, Lamkin DM, Suls J (2009) Associations of depression with C-reactive protein, IL-1, and IL-6: a meta-analysis. Psychosom Med 71:171–186. https://doi.org/10.1097/PSY.0b013e3181907c1b

Iizuka K, Machida T, Hirafuji M (2014) Skeletal muscle is an endocrine organ. J Pharmacol Sci 125:125–131. https://doi.org/10.1254/jphs.14r02cp

Itoh M et al (2007) Methyl CpG-binding protein 2 (a mutation of which causes Rett syndrome) directly regulates insulin-like growth factor binding protein 3 in mouse and human brains. J Neuropathol Exp Neurol 66:117–123. https://doi.org/10.1097/nen.0b013e3180302078

Johnson CM, Cui N, Zhong W, Oginsky MF, Jiang C (2015) Breathing abnormalities in a female mouse model of Rett syndrome. J Physiol Sci 65:451–459. https://doi.org/10.1007/s12576-015-0384-5

Julu PO et al (2001) Characterisation of breathing and associated central autonomic dysfunction in the Rett disorder. Arch Dis Child 85:29–37. https://doi.org/10.1136/adc.85.1.29

Julu PO et al (2008) Cardiorespiratory challenges in Rett's syndrome. Lancet 371:1981–1983. https://doi.org/10.1016/S0140-6736(08)60849-1

Justice MJ, Buchovecky CM, Kyle SM, Djukic A (2013) A role for metabolism in Rett syndrome pathogenesis: new clinical findings and potential treatment targets. Rare Dis 1:e27265. https://doi.org/10.4161/rdis.27265

Kaminski WE, Piehler A, Wenzel JJ (2006) ABC A-subfamily transporters: structure, function and disease. Biochim Biophys Acta 1762:510–524. https://doi.org/10.1016/j.bbadis.2006.01.011

Kang C, Li Ji L (2012) Role of PGC-1alpha signaling in skeletal muscle health and disease. Ann N Y Acad Sci 1271:110–117. https://doi.org/10.1111/j.1749-6632.2012.06738.x

Kaya-Okur HS et al (2019) CUT&tag for efficient epigenomic profiling of small samples and single cells.

Nat Commun 10:1930. https://doi.org/10.1038/s41467-019-09982-5
Kida H et al (2017) Pathogenesis of lethal aspiration pneumonia in Mecp2-null mouse model for Rett syndrome. Sci Rep 7:12032. https://doi.org/10.1038/s41598-017-12293-8
Killian JT et al (2014) Pubertal development in Rett syndrome deviates from typical females. Pediatr Neurol 51:769–775. https://doi.org/10.1016/j.pediatrneurol.2014.08.013
Killian JT et al (2017) Scoliosis in Rett syndrome: progression, comorbidities, and predictors. Pediatr Neurol 70:20–25. https://doi.org/10.1016/j.pediatrneurol.2017.01.032
Kline DD, Ogier M, Kunze DL, Katz DM (2010) Exogenous brain-derived neurotrophic factor rescues synaptic dysfunction in Mecp2-null mice. J Neurosci 30:5303–5310. https://doi.org/10.1523/JNEUROSCI.5503-09.2010
Kowalski K, Mulak A (2019) Brain-gut-microbiota Axis in Alzheimer's disease. J Neurogastroenterol Motil 25:48–60. https://doi.org/10.5056/jnm18087
Kriaucionis S et al (2006) Gene expression analysis exposes mitochondrial abnormalities in a mouse model of Rett syndrome. Mol Cell Biol 26:5033–5042. https://doi.org/10.1128/MCB.01665-05
Kron M et al (2014) A BDNF loop-domain mimetic acutely reverses spontaneous apneas and respiratory abnormalities during behavioral arousal in a mouse model of Rett syndrome. Dis Model Mech 7:1047–1055. https://doi.org/10.1242/dmm.016030
Kumar A et al (2017) Cardiovascular autonomic dysfunction in children and adolescents with Rett syndrome. Pediatr Neurol 70:61–66. https://doi.org/10.1016/j.pediatrneurol.2017.01.010
Kyle SM, Saha PK, Brown HM, Chan LC, Justice MJ (2016) MeCP2 co-ordinates liver lipid metabolism with the NCoR1/HDAC3 corepressor complex. Hum Mol Genet 25:3029–3041. https://doi.org/10.1093/hmg/ddw156
Leoncini S et al (2015) Cytokine dysregulation in MECP2- and CDKL5-related Rett syndrome: relationships with aberrant redox homeostasis, inflammation, and omega-3 PUFAs. Oxidative Med Cell Longev 2015:421624. https://doi.org/10.1155/2015/421624
Li Y et al (2013) Global transcriptional and translational repression in human-embryonic-stem-cell-derived Rett syndrome neurons. Cell Stem Cell 13:446–458. https://doi.org/10.1016/j.stem.2013.09.001
Li C et al (2014) MeCP2 enforces Foxp3 expression to promote regulatory T cells' resilience to inflammation. Proc Natl Acad Sci USA 111:E2807–E2816. https://doi.org/10.1073/pnas.1401505111
Li PA, Hou X, Hao S (2017) Mitochondrial biogenesis in neurodegeneration. J Neurosci Res 95:2025–2029. https://doi.org/10.1002/jnr.24042
Li H et al (2022) A single-cell atlas reveals the heterogeneity of meningeal immunity in a mouse model of methyl CpG binding protein 2 deficiency. Front Immunol 13:1056447. https://doi.org/10.3389/fimmu.2022.1056447
Lips P (2001) Vitamin D deficiency and secondary hyperparathyroidism in the elderly: consequences for bone loss and fractures and therapeutic implications. Endocr Rev 22:477–501. https://doi.org/10.1210/edrv.22.4.0437
Liu C et al (2020) Fat-specific knockout of Mecp2 upregulates Slpi to reduce obesity by enhancing browning. Diabetes 69:35–47. https://doi.org/10.2337/db19-0502
Liu S et al (2023) Cell-type-specific 3D-genome organization and transcription regulation in the brain. bioRxiv. https://doi.org/10.1101/2023.12.04.570024
Liu Y et al (2024) MECP2 directly interacts with RNA polymerase II to modulate transcription in human neurons. Neuron 112:1943. https://doi.org/10.1016/j.neuron.2024.04.007
Lopes AG, Loganathan SK, Caliaperumal J (2024) Rett syndrome and the role of MECP2: signaling to clinical trials. Brain Sci 14. https://doi.org/10.3390/brainsci14020120
Lyst MJ et al (2013) Rett syndrome mutations abolish the interaction of MeCP2 with the NCoR/SMRT co-repressor. Nat Neurosci 16:898–902. https://doi.org/10.1038/nn.3434
Martinowich K et al (2003) DNA methylation-related chromatin remodeling in activity-dependent BDNF gene regulation. Science 302:890–893. https://doi.org/10.1126/science.1090842
McCauley MD et al (2011) Pathogenesis of lethal cardiac arrhythmias in Mecp2 mutant mice: implication for therapy in Rett syndrome. Sci Transl Med 3:113ra125. https://doi.org/10.1126/scitranslmed.3002982
McGill BE et al (2006) Enhanced anxiety and stress-induced corticosterone release are associated with increased Crh expression in a mouse model of Rett syndrome. Proc Natl Acad Sci USA 103:18267–18272. https://doi.org/10.1073/pnas.0608702103
Medrihan L et al (2008) Early defects of GABAergic synapses in the brain stem of a MeCP2 mouse model of Rett syndrome. J Neurophysiol 99:112–121. https://doi.org/10.1152/jn.00826.2007
Millar-Buchner P et al (2016) Severe changes in colon epithelium in the Mecp2-null mouse model of Rett syndrome. Mol Cell Pediatr 3:37. https://doi.org/10.1186/s40348-016-0065-3
Modric T et al (2001) Phenotypic manifestations of insulin-like growth factor-binding protein-3 overexpression in transgenic mice. Endocrinology 142:1958–1967. https://doi.org/10.1210/endo.142.5.8165
Motil KJ et al (2011) Vitamin D deficiency is prevalent in girls and women with Rett syndrome. J Pediatr Gastroenterol Nutr 53:569–574. https://doi.org/10.1097/MPG.0b013e3182267a66
Motil KJ et al (2012) Gastrointestinal and nutritional problems occur frequently throughout life in girls and women with Rett syndrome. J Pediatr Gastroenterol Nutr 55:292–298. https://doi.org/10.1097/MPG.0b013e31824b6159

Mount RH, Charman T, Hastings RP, Reilly S, Cass H (2002) The Rett syndrome behaviour questionnaire (RSBQ): refining the behavioural phenotype of Rett syndrome. J Child Psychol Psychiatry 43:1099–1110. https://doi.org/10.1111/1469-7610.00236

Nan X, Meehan RR, Bird A (1993) Dissection of the methyl-CpG binding domain from the chromosomal protein MeCP2. Nucleic Acids Res 21:4886–4892. https://doi.org/10.1093/nar/21.21.4886

Nan X, Campoy FJ, Bird A (1997) MeCP2 is a transcriptional repressor with abundant binding sites in genomic chromatin. Cell 88:471–481. https://doi.org/10.1016/s0092-8674(00)81887-5

Neier K et al (2021) Sex disparate gut microbiome and metabolome perturbations precede disease progression in a mouse model of Rett syndrome. Commun Biol 4:1408. https://doi.org/10.1038/s42003-021-02915-3

Neul JL et al (2010) Rett syndrome: revised diagnostic criteria and nomenclature. Ann Neurol 68:944–950. https://doi.org/10.1002/ana.22124

Neul JL et al (2023) Trofinetide for the treatment of Rett syndrome: a randomized phase 3 study. Nat Med 29:1468–1475. https://doi.org/10.1038/s41591-023-02398-1

Ng KM et al (2022) Isogenic human-induced pluripotent stem-cell-derived cardiomyocytes reveal activation of Wnt signaling pathways underlying intrinsic cardiac abnormalities in Rett syndrome. Int J Mol Sci 23. https://doi.org/10.3390/ijms232415609

Ochodnicky P, Uvelius B, Andersson KE, Michel MC (2013) Autonomic nervous control of the urinary bladder. Acta Physiol (Oxf) 207:16–33. https://doi.org/10.1111/apha.12010

O'Driscoll CM, Kaufmann WE, Bressler JP (2013) MeCP2 deficiency enhances glutamate release through NF-kappaB signaling in myeloid derived cells. J Neuroimmunol 265:61–67. https://doi.org/10.1016/j.jneuroim.2013.09.002

O'Driscoll CM, Lima MP, Kaufmann WE, Bressler JP (2015) Methyl CpG binding protein 2 deficiency enhances expression of inflammatory cytokines by sustaining NF-kappaB signaling in myeloid derived cells. J Neuroimmunol 283:23–29. https://doi.org/10.1016/j.jneuroim.2015.04.005

Ogier M et al (2007) Brain-derived neurotrophic factor expression and respiratory function improve after ampakine treatment in a mouse model of Rett syndrome. J Neurosci 27:10912–10917. https://doi.org/10.1523/JNEUROSCI.1869-07.2007

Paolicelli RC, Ferretti MT (2017) Function and dysfunction of microglia during brain development: consequences for synapses and neural circuits. Front Synaptic Neurosci 9:9. https://doi.org/10.3389/fnsyn.2017.00009

Park MJ et al (2014) Anaplerotic triheptanoin diet enhances mitochondrial substrate use to remodel the metabolome and improve lifespan, motor function, and sociability in MeCP2-null mice. PLoS One 9:e109527. https://doi.org/10.1371/journal.pone.0109527

Peters SU et al (2013) The behavioral phenotype in MECP2 duplication syndrome: a comparison with idiopathic autism. Autism Res 6:42–50. https://doi.org/10.1002/aur.1262

Pini G et al (2016) Rett syndrome: a wide clinical and autonomic picture. Orphanet J Rare Dis 11:132. https://doi.org/10.1186/s13023-016-0499-7

Ramirez JM et al (2020) The pathophysiology of Rett syndrome with a focus on breathing dysfunctions. Physiology (Bethesda) 35:375–390. https://doi.org/10.1152/physiol.00008.2020

Ribeiro MC, MacDonald JL (2022) Vitamin D modulates cortical transcriptome and behavioral phenotypes in an Mecp2 heterozygous Rett syndrome mouse model. Neurobiol Dis 165:105636. https://doi.org/10.1016/j.nbd.2022.105636

Roende G et al (2011) Patients with Rett syndrome sustain low-energy fractures. Pediatr Res 69:359–364. https://doi.org/10.1203/PDR.0b013e31820bc6d3

Ruch A, Kurczynski TW, Velasco ME (1989) Mitochondrial alterations in Rett syndrome. Pediatr Neurol 5:320–323. https://doi.org/10.1016/0887-8994(89)90027-1

Sarajlija A, Djuric M, Tepavcevic DK, Grkovic S, Djordjevic M (2013) Vitamin D deficiency in Serbian patients with Rett syndrome. J Clin Endocrinol Metab 98:E1972–E1978. https://doi.org/10.1210/jc.2013-2627

Schmid DA et al (2012) A TrkB small molecule partial agonist rescues TrkB phosphorylation deficits and improves respiratory function in a mouse model of Rett syndrome. J Neurosci 32:1803–1810. https://doi.org/10.1523/JNEUROSCI.0865-11.2012

Segatto M et al (2014) Cholesterol metabolism is altered in Rett syndrome: a study on plasma and primary cultured fibroblasts derived from patients. PLoS One 9:e104834. https://doi.org/10.1371/journal.pone.0104834

Shapiro JR et al (2010) Bone mass in Rett syndrome: association with clinical parameters and MECP2 mutations. Pediatr Res 68:446–451. https://doi.org/10.1203/PDR.0b013e3181f2edd2

Singh J, Lanzarini E, Santosh P (2020) Autonomic dysfunction and sudden death in patients with Rett syndrome: a systematic review. J Psychiatry Neurosci 45:150–181. https://doi.org/10.1503/jpn.190033

Singh J, Lanzarini E, Nardocci N, Santosh P (2021) Movement disorders in patients with Rett syndrome: A systematic review of evidence and associated clinical considerations. Psychiatry Clin Neurosci 75:369–393. https://doi.org/10.1111/pcn.13299

Skene PJ et al (2010) Neuronal MeCP2 is expressed at near histone-octamer levels and globally alters the chromatin state. Mol Cell 37:457–468. https://doi.org/10.1016/j.molcel.2010.01.030

Sonn JY et al (2024) MeCP2 interacts with the super elongation complex to regulate transcription. bioRxiv. https://doi.org/10.1101/2024.06.30.601446

Squillaro T et al (2008) A case report: bone marrow mesenchymal stem cells from a Rett syndrome patient are prone to senescence and show a lower degree of apoptosis. J Cell Biochem 103:1877–1885. https://doi.org/10.1002/jcb.21582

Sticozzi C et al (2013) Scavenger receptor B1 post-translational modifications in Rett syndrome. FEBS Lett 587:2199–2204. https://doi.org/10.1016/j.febslet.2013.05.042

Strati F et al (2016) Altered gut microbiota in Rett syndrome. Microbiome 4:41. https://doi.org/10.1186/s40168-016-0185-y

Tan AH, Lim SY, Lang AE (2022) The microbiome-gut-brain axis in Parkinson disease—from basic research to the clinic. Nat Rev Neurol 18:476–495. https://doi.org/10.1038/s41582-022-00681-2

Taniya MA et al (2022) Role of gut microbiome in autism spectrum disorder and its therapeutic regulation. Front Cell Infect Microbiol 12:915701. https://doi.org/10.3389/fcimb.2022.915701

Tanner MR, Beeton C (2018) Differences in ion channel phenotype and function between humans and animal models. Front Biosci (Landmark Ed) 23:43–64. https://doi.org/10.2741/4581

Tareen RS, Kamboj MK (2012) Role of endocrine factors in autistic spectrum disorders. Pediatr Clin N Am 59:75–88. https://doi.org/10.1016/j.pcl.2011.10.013

Tarquinio DC et al (2018) The course of awake breathing disturbances across the lifespan in Rett syndrome. Brain Dev 40:515–529. https://doi.org/10.1016/j.braindev.2018.03.010

Tichomirowa MA et al (2005) Endocrine disturbances in depression. J Endocrinol Investig 28:89–99. https://doi.org/10.1007/BF03345535

Tomasello DL et al (2024) Mitochondrial dysfunction and increased reactive oxygen species production in MECP2 mutant astrocytes and their impact on neurons. Sci Rep 14:20565. https://doi.org/10.1038/s41598-024-71040-y

Travagli RA (2007) The nucleus tractus solitarius: an integrative Centre with 'task-matching' capabilities. J Physiol 582:471. https://doi.org/10.1113/jphysiol.2007.137091

Tropea D et al (2009) Partial reversal of Rett syndrome-like symptoms in MeCP2 mutant mice. Proc Natl Acad Sci USA 106:2029–2034. https://doi.org/10.1073/pnas.0812394106

Vandereyken K, Sifrim A, Thienpont B, Voet T (2023) Methods and applications for single-cell and spatial multi-omics. Nat Rev Genet 24:494–515. https://doi.org/10.1038/s41576-023-00580-2

Vashi N, Justice MJ (2019) Treating Rett syndrome: from mouse models to human therapies. Mamm Genome 30:90–110. https://doi.org/10.1007/s00335-019-09793-5

Vashi N, Ackerley C, Post M, Justice MJ (2021) Aberrant lung lipids cause respiratory impairment in a Mecp2-deficient mouse model of Rett syndrome. Hum Mol Genet 30:2161–2176. https://doi.org/10.1093/hmg/ddab182

Viemari JC et al (2005) Mecp2 deficiency disrupts norepinephrine and respiratory systems in mice. J Neurosci 25:11521–11530. https://doi.org/10.1523/JNEUROSCI.4373-05.2005

Wahba G et al (2016) Activity and MeCP2-dependent regulation of nNOS levels in enteric neurons. Neurogastroenterol Motil 28:1723–1730. https://doi.org/10.1111/nmo.12873

Wakai S et al (1990) Rett syndrome: findings suggesting axonopathy and mitochondrial abnormalities. Pediatr Neurol 6:339–343. https://doi.org/10.1016/0887-8994(90)90028-y

Ward CS et al (2011) MeCP2 is critical within HoxB1-derived tissues of mice for normal lifespan. J Neurosci 31:10359–10370. https://doi.org/10.1523/JNEUROSCI.0057-11.2011

Ward CS et al (2016) Loss of MeCP2 causes urological dysfunction and contributes to death by kidney failure in mouse models of Rett syndrome. PLoS One 11:e0165550. https://doi.org/10.1371/journal.pone.0165550

Ward CS et al (2020) Loss of MeCP2 function across several neuronal populations impairs breathing response to acute hypoxia. Front Neurol 11:593554. https://doi.org/10.3389/fneur.2020.593554

Westberry JM, Trout AL, Wilson ME (2010) Epigenetic regulation of estrogen receptor alpha gene expression in the mouse cortex during early postnatal development. Endocrinology 151:731–740. https://doi.org/10.1210/en.2009-0955

Whitsett JA, Wert SE, Weaver TE (2010) Alveolar surfactant homeostasis and the pathogenesis of pulmonary disease. Annu Rev Med 61:105–119. https://doi.org/10.1146/annurev.med.60.041807.123500

Williams LM, Rudensky AY (2007) Maintenance of the Foxp3-dependent developmental program in mature regulatory T cells requires continued expression of Foxp3. Nat Immunol 8:277–284. https://doi.org/10.1038/ni1437

Wolf Y et al (2017) Brown-adipose-tissue macrophages control tissue innervation and homeostatic energy expenditure. Nat Immunol 18:665–674. https://doi.org/10.1038/ni.3746

Zhou J et al (2022) Disruption of MeCP2-TCF20 complex underlies distinct neurodevelopmental disorders. Proc Natl Acad Sci USA 119. https://doi.org/10.1073/pnas.2119078119

Zigmond E, Jung S (2013) Intestinal macrophages: well educated exceptions from the rule. Trends Immunol 34:162–168. https://doi.org/10.1016/j.it.2013.02.001

Brain-Resident Immune Cells in Neurodevelopmental Disorders

10

Luca Pangrazzi, Birgit Weinberger, and Yuri Bozzi

Abstract

Brain-resident immune cells are involved in the protection against microbes invading the brain. In addition, their essential role in regulating physiological processes, including neurogenesis and neuronal migration, has been documented. In this chapter, we will provide a general introduction to immune cell types that reside and infiltrate the brain. Furthermore, the impact of brain-resident immune cells on brain physiology will be discussed. The phenotype of immune cells is known to be impaired in neurodevelopmental disorders. Thus, we will describe how the accumulation of dysfunctional immune cell types contributes to the pathogenesis of this group of diseases, with a particular emphasis on the connection between immune dysfunction, inflammation, and oxidative stress. Finally, possible interventions to counteract neurodevelopmental disorders by targeting oxidative stress and inflammation will be summarized.

Keywords

Autism spectrum disorder · Microglia · Pharmacological treatment · Mouse model

Abbreviations

ASD	autism spectrum disorder
BBB	blood-brain barrier
CAT	catalase
CD	cluster of differentiation
CNS	central nervous system
CNTNAP2	contactin associated protein like 2
DCs	dendritic cells
GABA	γ-aminobutyric acid
GPX	glutathione peroxidase
GSH	glutathione
GWAS	genome-wide association studies
H_2O	water
IFN	interferon
IL	interleukin
IL-R	interleukin receptor
MIA	maternal immune activation
mTOR	mammalian target of rapamycin
NAC	N-acetylcysteine

L. Pangrazzi
Institute for Biomedical Aging Research, University of Innsbruck, Innsbruck, Austria

CIMeC – Center for Mind/Brain Sciences, University of Trento, Rovereto, Italy

B. Weinberger
Institute for Biomedical Aging Research, University of Innsbruck, Innsbruck, Austria

Y. Bozzi (✉)
CIMeC – Center for Mind/Brain Sciences, University of Trento, Rovereto, Italy

CNR Institute of Neuroscience, Pisa, Italy
e-mail: yuri.bozzi@unitn.it

D. Tropea, E. Giacometti (eds.), *Brain-Body Connections*, Advances in Experimental Medicine and Biology 1477, https://doi.org/10.1007/978-3-031-89525-8_10

NADPH	nicotinamide adenine dinucleotide phosphate
NOX	NADPH oxidase
NK	natural killer cells
$O_2^{\cdot-}$	superoxide anion
H_2O_2	hydrogen peroxide
O_2	oxygen
OXPHOS	oxidative phosphorylation
PD-L1	programmed death-ligand 1
RORγ	receptor-related orphan receptor γ
ROS	reactive oxygen species
S1	primary somatosensory cortex
SCID	severe combined immunodeficiency
SHANK3	SH3 and multiple ankyrin repeat domains 3
SOD	superoxide dismutase
TNF	tumor necrosis factor

10.1 Brain-Resident Immune Cells in Brain Physiology

The immune system is essential for maintaining body homeostasis and overall health. In particular, it detects invading pathogens, malignancies, and damaged tissues and coordinates wound healing and tissue remodelling. The immune system is strategically present throughout the body to fulfil its functions. Historically, research has primarily studied immune cells located in accessible immune hubs such as the blood, spleen and lymph nodes, and reduced cell numbers were identified in other anatomical sites. More recently, it become more evident that immune cells exhibit distinct phenotypes, differentiation states and functions based on their anatomical compartment of residence (Gray and Farber 2022; Barros et al. 2022). Among the various tissues, immune resident brain cells have long remained underexplored. Indeed, the prevailing concept was that the brain might be immune privileged, protected by the blood-brain barrier (BBB) from the infiltration of circulating immune cells, with microglia being the only recognized immune cell type present in this organ (Ransohoff and Engelhardt 2012; Louveau et al. 2015a, b). However, in the last two decades, several studies have described the critical role of neuroimmune communication in sustaining brain homeostasis and function. Under homeostatic conditions, immune cells are mainly located within the choroid plexus and meninges, with rare populations resident in the brain parenchyma (Rustenhoven and Kipnis 2019). After peripheral priming, non-resident immune cells can migrate into and survive within the brain parenchyma (Mrdjen et al. 2018). The importance of these immune cell subsets was documented, as they were described to substantially impact brain functions, including learning, memory and social behavior (Schepanski et al. 2022; Pasciuto et al. 2020).

The immune system has classically been divided into innate and adaptive immunity. The innate immune system represents the first line of defence against invading pathogens. As it is based on recognizing cell structures conserved among different species, it can be activated very quickly, and pathogens may be destroyed within hours. It comprises physical and chemical barriers, a humoral response consisting of proteins interacting with invaders, and cellular defences, which neutralize the pathogens and activate the adaptive immune system. Adaptive immunity requires more time to get activated, as its activation may take days or even weeks to develop after the first encounter. Unlike innate immunity, it is specific to antigens and can retain long-term memory to enable faster responses if pathogens are encountered again.

It is now well-recognized that cells of both the innate and adaptive immune system are present in the central nervous system (CNS) under homeostatic conditions. In 2015, it was described that the brain possesses functional lymphatic vessels in the meninges, which can carry immune cells from the cerebrospinal fluid (Louveau et al. 2015a, b). Microglia are the most frequent immune cell type in the brain, as they represent approximately 10–15% of total brain cells (Dos Santos et al. 2020). Microglia cells are known to constantly surveil the brain parenchyma. In addition, neutrophils, monocytes, macrophages, and adaptive immune T and B cells can enter the meninges and the brain, affecting neural development and learning (Engelhardt et al. 2017; Filiano et al. 2016; Herz et al. 2021; Pasciuto et al. 2020; Radjavi et al. 2014; Ribeiro et al. 2019; Chen and Holtzman 2022). Moreover, both innate and adaptive immune cells inhabit the dura through skull channels that link the skull bone marrow to the dura. (Brioschi et al. 2021; Cugurra et al. 2021). Under homeostatic conditions, immune cells mainly access the brain's border regions, establishing a potential link with other cell populations within the parenchyma. Consequently, since most adaptive immune cells in the CNS are located at its borders, they can both respond to disturbances and support neuronal function through immunological protection (Rustenhoven and Kipnis 2019; Chen and Holtzman 2022).

10.1.1 Innate Immune Cells in the Brain: Microglia and Macrophages

Microglia cells derive from yolk sac macrophages and arise during the first wave of primitive hematopoiesis. After establishing embryonic circulation, they populate the developing CNS via the bloodstream (Ginhoux et al. 2010; Gomez Perdiguero et al. 2015). After the formation of the blood-brain barrier, the influx of yolk sac macrophages into the CNS stops (McGrath et al. 2003), and the microglia population self-maintains throughout life by continuous turnover (Réu et al. 2017). All tissues contain parenchymal macrophages specialized to support the homeostasis of their home organ. Each tissue-resident population shows distinct gene expression profiles reflecting their function within a particular niche (Gautier et al. 2012; Gosselin et al. 2014). Similarly, microglia express distinct signature genes that are absent or expressed at lower levels by other macrophage populations (Butovsky et al. 2014; Gautier et al. 2012).

In the brain, microglia serve as a first line of defence against pathogens and are crucial for maintaining homeostasis. Microglia cells constantly survey their environment and play an essential role in regulating neuronal activity and modulating synaptic connections (Badimon et al. 2020; Weinhard et al. 2018). Furthermore, they engage in universal macrophage functions, including removing apoptotic cells and cell debris, which prevent inflammation and autoimmunity (Arandjelovic and Ravichandran 2015). In addition, microglia clear apoptotic cells in developing white matter tracts, where many differentiating oligodendrocyte progenitor cells undergo apoptosis, and in the cerebellum (Ayata et al. 2018). This cell type is also required for synapse pruning, an essential process in which excess synapses are eliminated (Bohlen et al. 2019). In addition to microglia meningeal macrophages, involved in learning and memory (Derecki et al. 2010), as well as perivascular macrophages, regulating BBB function and neutrophil recruitment (Willis et al. 2007; Polfliet et al. 2001), can be found in the brain.

10.1.2 Other Innate Immune Cells in the Brain

In addition to microglia, other innate immune cells such as dendritic cells, natural killer (NK) cells, and neutrophils are present in the brain (D'Agostino et al. 2012; Ning et al. 2023; Salken et al. 2023). These cells can migrate from peripheral tissues and contribute to immune surveillance and response. Together with meningeal macrophages, dendritic cells are involved in antigen presentation and can influence T cell activation and differentiation. While the numbers of DCs within parenchyma are low under steady-state conditions, functional dendritic cells can be identified in the presence of severe injury or pro-inflammatory conditions (Pachter et al. 2003). In mice, NK cells were found in the healthy brain and made up a small fraction ($1.1 \pm 0.14\%$) of the total immune cells in CNS compartments (Korin et al. 2017; Mrdjen et al. 2018). NK cells reside primarily in the meninges and the choroid plexus but are absent in the parenchyma. A higher degree of NK cell migration to the CNS occurs in case of BBB breakdown (Lepennetier et al. 2019). In the CNS, NK cells might function as gatekeepers, allowing immune cell infiltration and initiating inflammation by their interferon (IFN)-γ production (Kunis et al. 2013). The understanding of the role of brain neutrophils in healthy conditions remains limited. Studies have often described how these cells infiltrate and damage CNS functions in many pathologies (Salken et al. 2023).

10.1.3 Adaptive Immune Cells (T and B Cells) in the Brain

In addition to innate immune cells, cell populations belonging to the adaptive immune system were identified in the brain in both physiological and pathological conditions.

T cells, although traditionally associated with peripheral immune responses, have been found in the brain in physiological conditions, particularly in the meninges and choroid plexus (Abbaoui et al. 2023; Baruch and Schwartz 2013; Strominger et al. 2018). In particular, this cell type plays a crucial role in CNS homeostasis. In addition, T cells can enter the brain parenchyma under pathological conditions. $CD4^+$ T helper cells and $CD8^+$ cytotoxic T cells were the main types observed. These cells can exert effector functions, such as producing cytokines that modulate microglial activity and influence neuroinflammation.

Kipnis and colleagues described that deficiency of mature T cells impairs the performance of mice in cognitive tasks (Kipnis et al. 2004, 2012). Severe combined immune deficiency (SCID) and nude mice (lacking mature T cells) showed restored cognitive functions and increased hippocampal neurogenesis after T cell repopulation (Kipnis et al. 2004; Brynskikh et al. 2008; Wolf et al. 2009). However, B cell-deficient mice do not show any changes in cognitive tasks compared to their wild-type littermates (Brynskikh et al. 2008). Furthermore, the adoptive transfer of $CD4^+$ T cells, but not $CD8^+$ T cells, improved spatial learning and memory performances and reverted anxiety behaviour (Rattazzi et al. 2013; Radjavi et al. 2014). IFN-γ-expressing T cells are known to regulate neuronal connectivity and social behaviors (Filiano et al. 2015). Cognitive tasks increase the number of T cells in the meningeal space. In particular, interleukin (IL)-4-producing T cells regulate learning and memory (Derecki et al. 2010). Furthermore, SCID mice showed neuronal hyperconnectivity between the prefrontal cortex and insular regions, and T cell repopulation during the juvenile period partially rescued hyperconnectivity (Filiano et al. 2015). Thus, these studies show that T cells mediate neuronal plasticity during brain development. Within the brain, $CD8^+$ and $CD4^+$ T cells were described to show the typical phenotype of tissue-resident memory cells, expressing the canonical marker CD69. These populations are indispensable for local protection against pathogens invading the brain, as they trigger neuroinflammatory responses after CNS viral infections (Ayasoufi et al. 2023).

B cells are rarely present in the healthy CNS parenchyma. Small numbers can be found in cerebrospinal fluid and the choroid plexus, and they are constitutively present in the subdural meninges (Korin et al. 2017; Schafflick et al. 2021). A large fraction of meningeal B cells includes immature B cells generated in the skull's bone marrow (Hwang et al. 2023). The capacity of peripheral B cells to enter the brain in healthy conditions is limited in young mice but increases as mice age and under pathological conditions (Jain and Yong 2022).

Also microbial metabolites such as SCFAs can modulate immune cell function *as mentioned in the Chap. 6*, such can modulate immune cell function and promote anti-inflammatory responses maintaining gut barrier integrity. For example, microbial metabolites can stimulate the production of regulatory T cells, which help suppress excessive immune responses and prevent inflammation. This modulation of the immune system by the gut microbiota is essential for preventing chronic inflammation and maintaining overall health.

10.2 The Impact of Brain-Resident Immune Cells on Neurodevelopmental Disorders

10.2.1 Immune Dysfunction in Neurodevelopmental Disorders

Recent studies on ASD patients reported strong pro-inflammatory states in the brain and periphery (Brigida et al. 2017; Siniscalco et al. 2018; Croonenberghs et al. 2002). Indeed, increased levels of pro-inflammatory cytokines IL-1, IL-6, IL-8, IL-12p40, IL-17, and tumor necrosis factor (TNF) were found in the plasma of 97 medication-free ASD children. Some of these molecules were additionally overexpressed in post-mortem samples from ASD individuals. (Li et al. 2009; Morgan et al. 2010; Vargas et al. 2004). Note that there might be a protective effect of IL-6 in the brain *as reported in Chap. 11*.

In addition, the over-activation of immune cells, imbalanced production of cytokines and chemokines, and increased permeability of the blood-brain barrier were documented (Gottfried et al. 2015; Masi et al. 2017; Pangrazzi et al. 2020a). Thus, evidence exists that immune dysregulation may contribute to ASD-related behaviors.

Dysfunctions in the interplay between different cell types during development result in the onset of neurodevelopmental and psychological disorders. These disorders were promoted by maternal immune activation (MIA) during pregnancy. This condition mimics autism spectrum disorder (ASD), is induced by infections during pregnancy, and therefore is accompanied by pro-inflammatory conditions (Careaga et al. 2017; Scola and Duong 2017). Furthermore, genome-wide association studies (GWAS) have shown that molecules related to the immune system are implicated in autism and schizophrenia. These observations suggest that the immune system plays a determinant role in the development of the CNS and the pathogenesis of neurodevelopmental disorders (Schizophrenia Working Group 2014; Lintas et al. 2012). Under pathological conditions, the balance between innate and adaptive immune responses shifts, leading to the onset of neuroinflammatory processes. While short-term inflammation is beneficial, as it facilitates repair and recovery, chronic inflammation is associated with the onset of brain diseases.

10.2.2 Microglia

The importance of microglia in brain physiology suggests that this cell type may play a key role in neurodevelopmental disorders. In the developing brain, the generation of neurons and glial cells is controlled by the immune system. Therefore, immunological impairments strongly contribute to the onset of this group of disorders, including ASD. This aspect is further supported by the fact that microglia are the most numerous immune cell type in the brain. In addition, pro-inflammatory cytokines, highly produced by microglia, can impact neurodevelopment and neural function. Indeed, evidence exists that a direct link may be present between changes in microglia and ASD pathophysiology. Increased microglia activation was described in brain tissues of ASD patients compared to healthy individuals (Vargas et al. 2004; Suzuki et al. 2013; Morgan et al. 2010). Furthermore, the density of resting (ramified) microglia was reduced in the white and grey matter of ASD patients compared to neurotypical controls (Lee et al. 2017). Furthermore, higher numbers of reactive rod microglia were observed in patients aged 2–3 years, and increased primed microglia were described at later ages. Furthermore, microglia clusters were more frequent in the patients than in healthy controls. Impairments in microglia from patients have additionally been detected on a transcriptional level. In a single-nucleus RNA sequencing experiment, microglia from ASD patients showed increased activity of genes associated with microglial activation and transcriptional factors controlling developmental processes (Velmeshev et al. 2019).

10.2.3 T and B Cells

GWAS analysis has linked B-cell-related genes CD19 and CD20 to the onset of schizophrenia (Schizophrenia Working Group 2014). Elevated levels of CD5+ B cells have been detected in the blood of schizophrenia patients (Printz et al. 1999; Seiner et al. 2010), suggesting that dysregulated B-cell immunity may contribute to the pathogenesis of the disorder. B cells play a role in oligodendrogenesis and myelination during brain development, and alterations in the expression of myelin and oligodendrocyte-related genes have been observed in schizophrenia patients (Katsel et al. 2005). Therefore, evidence suggests that disruptions in B-cell function during brain development may impair myelin formation, increasing susceptibility to schizophrenia. In the mouse brain, CD4+ T cells are present as early as embryonic day 16 (E16), with their proportion remaining stable throughout development (Tanabe and Yamashita 2018a). IFN-γ secreted by T cells promotes inhibitory activity in GABAergic neurons within the prefrontal cortex (Filiano et al. 2016). Additionally, IFN-γ influences dendrite morphology and suppresses the formation of excitatory synapses (Fujita et al. 2017). Conversely, during synaptogenesis, IFN-γ enhances inhibition in hippocampal neurons (Brask et al. 2004). Since T-cell dysfunction leads to altered IFN-γ production, abnormalities in this immune cell subset may significantly impact brain physiology. The involvement of T cells in neurodevelopmental disorders has been investigated using a MIA model, in which the Toll-like receptor 3 agonist polyinosinic:polycytidylic acid [poly (I:C)] is administered to a pregnant mouse at E12.5 (Smith et al. 2007).

After this treatment, pups exhibit abnormal behaviors, including hyperactivity, anxiety, and social impairments. MIA has been shown to promote the development of Th17 cells (Choi et al. 2016). Specifically, this condition leads to increased IL-17A secretion into the serum and upregulated IL-17Ra expression in cortical neurons. Blocking IL-17A with antibodies or knocking out retinoic acid receptor-related orphan receptor γt (RORγt), a transcription factor essential for Th17 cell development, prevents the manifestation of MIA (Choi et al. 2016). Following poly (I:C) injection, bacterial-specific T cells in the maternal gut secrete IL-17A, resulting in the loss of parvalbumin+ cortical interneurons. This disruption contributes to neuronal hyperactivation and MIA-induced behavioral abnormalities (Shin Yim et al. 2017; Tanabe and Yamashita 2018b). Thus, a connection is present between pro-inflammatory impairments in maternal T cells and the onset of ASD (Fig. 10.1).

10.3 The Interplay Between Inflammation and Oxidative Stress in Neurodevelopmental Disorders

Alongside pro-inflammatory impairments, the accumulation of reactive oxygen species (ROS) in the brain is associated with pathological conditions, including neurodevelopmental disorders (Pangrazzi et al. 2020a). The imbalance between the production of ROS and the organism's ability to block their harmful effects through the antioxidative systems supports the onset of oxidative stress. This condition is often accompanied by mitochondrial abnormalities and is present under pathological conditions. Indeed, mitochondrial dysfunction leads to increased ROS levels in the cells (Swerdlow 2011; Reddy and Reddy 2011). Mitochondrial dysfunction has been associated with the accumulation of ROS, which, when not adequately scavenged, may lead to oxidative stress (Georgieva et al. 2017).

Superoxide anion ($O_2{\cdot}-$) is a major cellular ROS. It is generated as a byproduct of oxidative phosphorylation (OXPHOS) in mitochondria, as well as through the activity of nicotinamide adenine dinucleotide phosphate (NADPH) oxidase (NOX) located at the plasma membrane (Jiang et al. 2011; Terzi and Suter 2020). $O_{2{\cdot}-}$ is converted to hydrogen peroxide (H_2O_2), oxygen (O_2) and water (H_2O) by enzymatic antioxidants, including superoxide dismutase (SOD), catalase (CAT), and glutathione peroxidase (GPX), and by other nonenzymatic antioxidants such as glu-

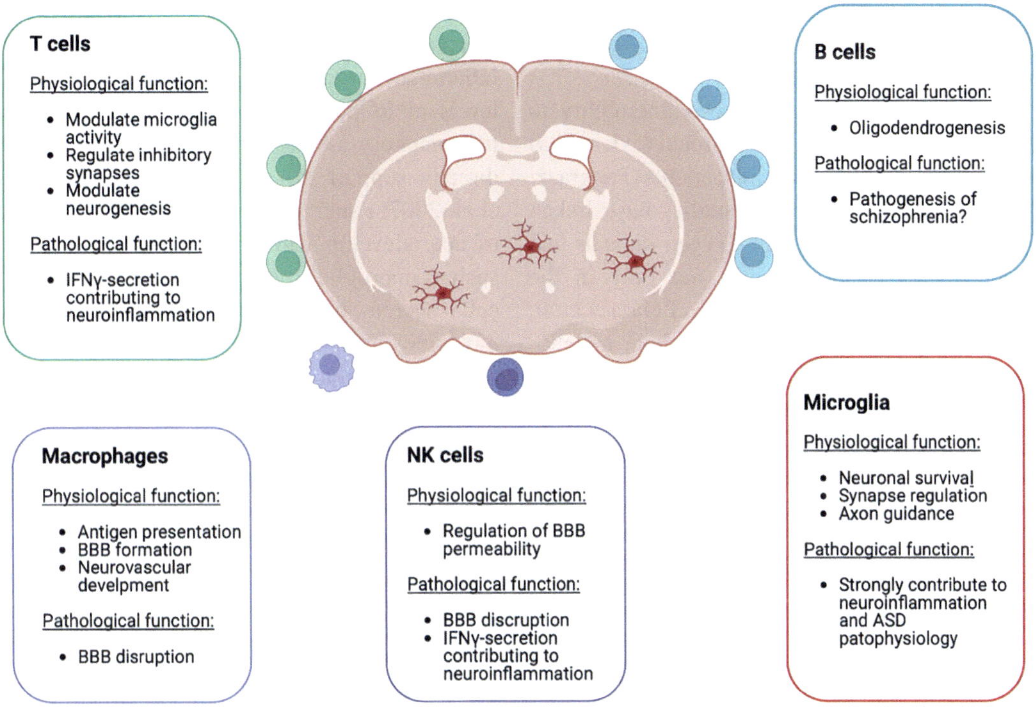

Fig. 10.1 Brain-resident immune cells and their impact on neurodevelopmental disorders. The role of most characterized brain-resident immune cell types in physiological conditions, as well as in neurodevelopmental disorders, is reported. (Figure created with Biorender)

tathione (GSH). The correct control of ROS production and activity is fundamental for physiological functions, including neuronal development (Wilson et al. 2018). Indeed, excessive production of ROS leads to damage of DNA, RNA, proteins, and lipids (Schieber and Chandel 2014). When this occurs in brain cells, the damage can alter neuronal functions, impairing cognitive processes such as memory, learning, and cognition (Spinu et al. 2019).

A large body of evidence has shown that ROS may play a role in schizophrenia. Indeed, oxidative stress may represent an important mechanism contributing to the pathogenesis of this disease (Armstrong and Stratton 2016). Altered mechanisms in schizophrenia may be caused by oxidative damage of proteins, lipids, and nucleic acids, which leads to impaired neuronal functions. Deficits in synaptic circuitry may represent the primary deficit in schizophrenia, and these mechanisms are observed after the first onset of psychosis when antioxidant protection against oxidative stress is low (Armstrong and Stratton 2016). Oxidative stress was additionally observed at a biochemical level, particularly in the dopaminergic system, due to the excessive production of free radicals leading to oxidative damage in brain structures. It has been proposed that schizophrenia may show a progressive route, therefore supporting the neurodegenerative origin of this disorder. Alongside oxidative stress, increased production of pro-inflammatory molecule IL-6 and elevated cyclooxygenase-2 have been observed (Armstrong and Stratton 2016).

A link between oxidative stress, inflammation, and brain disorders has also been *discussed in the Chap. 5* where it is noted that schizophrenia disease effects on mitochondrial OXPHOS gene expression in cortical gray matter. It is further proposed that, unlike schizophrenia, the multifunctional aspects of mitochondria may be more engaged with the bipolar disease process.

Consistent with the above, alteration in mitochondrial-related processes is correlated to neuropsychiatric conditions.

Oxidative stress, as well as vulnerability to ROS-mediated damage and neuronal toxicity, has been widely described in ASD patients (Pangrazzi et al. 2020a, 2020b). Some studies have linked ASD to increased levels of oxygen radicals and reduced antioxidant capacity, not only in the brain but also at the systemic level (James et al. 2006). Impairments in the glutathione metabolism (including increased levels of oxidized glutathione, low reduced glutathione and diminished glutathione redox ratios) were described in the temporal cortices and cerebella of autistic patients (Rose et al. 2012). In parallel, the plasma concentration of reduced GSH was reduced, and the ratio of oxidized GSH/reduced GSH was increased in autistic patients in comparison to healthy controls, suggesting that systemic oxidative stress may be present with ASD (Swerdlow 2011). The transcriptional pattern of 84 genes related to oxidative stress was measured in the peripheral blood of ASD patients and healthy controls. Notably, eight genes coding for key proteins involved in ROS metabolism (GCLM, SOD2, NCF2, PRNP, PTGS2, TXN, and FTH1) were downregulated in autistic persons. These molecules may be useful as biomarkers of autism, therefore being relevant for early diagnostic and therapeutic purposes. Furthermore, red blood cells were found to be damaged by ROS in ASD individuals (Bolotta et al. 2018).

Accumulation of ROS is known to be associated with pro-inflammatory conditions. Whether high levels of oxygen radicals may induce inflammation or whether pro-inflammatory molecules may support the onset of oxidative stress is unclear. Despite this, a relationship between oxidative stress and inflammation has been documented, as evidence indicates that oxidative stress supports chronic inflammatory diseases. Indeed, in many cell types present in the brain or the periphery, high concentrations of ROS can activate signalling pathways, including NF-κB, a downstream mediator of TNF signalling. This leads to the production of pro-inflammatory molecules and creates vicious cycles, which maintain a high secretion of pro-inflammatory cytokines and chemokines (Yu et al. 2020). In parallel, pro-inflammatory cytokines directly lead to increased levels of ROS, and it was shown that the pro-inflammatory cytokines TNF, IFN-γ, and IL-1β directly support the production of ROS (Yang et al. 2007; Pangrazzi et al. 2017). In the context of neurodevelopmental disorders, immune cells such as microglia or other brain-resident immune cells secrete high levels of pro-inflammatory molecules, which may support the vicious cycle of inflammation-oxidative stress-inflammation (Popa-Wagner et al. 2013). This situation supports cell death, therefore leading to neuron degeneration. In addition, other studies have shown that microglia and astrocytes secrete pro-inflammatory cytokines in response to oxidative stress (Fischer et al. 2012; Voet et al. 2019). Taken together, the interplay between oxidative stress and inflammation may contribute to the severity, as well as to the pathogenesis of neurodevelopmental disorders.

10.4 Targeting Oxidative Stress and Immune Dysfunction to Counteract Neurodevelopmental Disorders

Several studies, summarized in the previous paragraphs, clearly indicated that oxidative stress and immune dysfunction markedly contribute to the pathogenesis of neurodevelopmental disorders, including ASD. Thus, research in recent years has investigated the feasibility of targeting oxidative stress and immune dysfunction as potential therapeutic approaches in ASD.

10.4.1 Targeting Oxidative Stress

As mentioned above, several studies reported the presence of oxidative stress markers in individuals with ASD, including increased levels of lipid peroxidation and reduced activity of antioxidant enzymes (for additional references, the reader is referred to reviews by Varun and Singh 2019, and

Pangrazzi et al. 2020a). In addition, polymorphisms in genes related to glutathione metabolism have been linked to increased oxidative stress levels in ASD (Ghanizadeh and Moghimi 2013), suggesting a genetic predisposition to oxidative stress, at least in some individuals with ASD. This evidence led to testing the hypothesis that dietary interventions, including supplementation with antioxidants and diets rich in fruits and vegetables, might contribute to ameliorating AS symptoms. A couple of clinical trials explored the efficacy of antioxidant treatments, such as N-acetylcysteine (NAC), in improving behavioral and cognitive symptoms in children with ASD. One study indicated that this approach could help reduce oxidative stress markers and improve behavioral outcomes (Adams et al. 2011), and the second one reported reduced irritability and improved social skills (Hardan et al. 2012) in ASD subjects. However, other trials showed no beneficial effects (Wink et al. 2016; Dean et al. 2017), indicating that more rigorous trials are needed (see also references in Pangrazzi et al. 2020b, and Curpan et al. 2021). A possible explanation of these contradictive results may rely on the high heterogeneity of the ASD cohorts investigated in the different studies, which might have included ASD subjects with very different levels of oxidative stress.

Notwithstanding the high variability of human studies reported above, our recent work on mouse models of ASD contributed to shed light on the possible mechanisms through which targeting oxidative stress by NAC might exert its beneficial effects. Specifically, we showed that both *Cntnap2* and *Shank3b* knockout mice display high levels of oxidative stress and inflammation in the cerebellum but not in other brain areas (Pangrazzi et al. 2025, 2024). The cerebellum has long been recognized as a critical region for ASD pathogenesis, as shown by the extensive series of neuropathological alterations (including neuroinflammation) and circuit dysfunctions that have been detected in ASD subjects (Tsai 2016; Kelly et al. 2021). Treatment of adult *Cntnap2* and *Shank3b* knockout mice with NAC was able to rescue cerebellar neuroinflammation, altered microglia branching, and ASD-like behavioral manifestations in mutant mice (Pangrazzi et al. 2025, 2024). Our results suggest that, at least in some forms of genetic syndromes, microglia dysfunction might contribute to ASD pathogenesis and that targeting microglia might represent a valid therapeutic strategy.

10.4.2 Targeting Microglia

Among the different brain-resident immune cell types, microglia dysfunction recently attracted attention in the context of the pathogenesis of neurodevelopmental disorders, including ASD (Fleiss and Gressens 2024; Mastenbroek et al. 2024; Meng et al. 2024).

An interesting hypothesis on the role of microglia in the genesis of these disorders implicates epigenetic reprogramming of microglia function during early phases of CNS development in response to immune activation. Microglia start populating the embryonic CNS very early, modulating immune responses and controlling neurodevelopmental processes. Immune activation during early CNS development may disrupt normal microglial activity, potentially resulting in lasting effects on brain function in postnatal life (Mastenbroek et al. 2024).

Recent studies performed on mice lacking ASD-associated genes confirmed the role of microglia as mediators of key neurodevelopmental processes, whose derangement may lead to pathological outcomes. For example, haploinsufficiency of Cyfip1 leads to increased adult hippocampal neurogenesis as a consequence of the failure of microglia to induce neuronal apoptosis through the secretion of specific factors (Haan et al. 2021). The absence of microglia proliferation following brain trauma was detected in *Shank3*-deficient mice, suggesting that promoting microglia proliferation might counteract neuropathology in neurodevelopmental disorders (Urrutia-Ruiz et al. 2022). Moreover, microglia showed an increased activated morphology and phagocytosis of synaptic structures in male but not female *Cntnap2* knockout mice, suggesting that ASD neuropathology may result from sexually dimorphic microglial activity (Dawson et al.

Table 10.1 Major outcomes of recent studies targeting oxidative stress or microglia activity in ASD human cohorts and mouse models

Treatment	Human cohort or animal model	Outcome	Reference
NAC	Human ASD	Reduced oxidative stress, improved behavior	Adams et al. (2011)
NAC	Human ASD	Reduced irritability and improved social skills	Hardan et al. (2012)
NAC	Human ASD	No beneficial effects	Wink et al. (2016) and Dean et al. (2017)
NAC	*Cntnap2* Knockout mice	Rescued cerebellar inflammation/ microglia branching and ASD-like behavior	Pangrazzi et al. (2025)
NAC	*Shank3b* Knockout mice	Rescues cerebellar inflammation/ microglia gene expression and ASD-like behavior	Pangrazzi et al. (2024)
Histamine H3 receptor antagonist ciproxifan (CPX)	Prenatal exposure to valproic acid in mice	Rescue of social and repetitive behaviors	Baronio et al. (2015)
Histamine H3 receptor antagonists (CPX, pitolisant)	BTBR mice	Rescue of motor but not repetitive behaviors	Molenhuis et al. (2022)
Pharmacological ablation of microglia by PLX3397 administration	*Scn2a* Knockout mice	Partially restored synaptic transmission and spine density	Wu et al. (2024)
PD-L1Fc administration (intraperitoneal in MIA pregnant females; intracerebral in MIA offsprings)	Poly(I:C)-induced MIA in pregnant mice	Suppressed neuroinflammation and partially rescued ASD-like behaviors	Zeng et al. (2024)
Rapamycin (mTOR inhibitor)	*IL-1R* Knockout mice	Rescued behavioral and microglia defects	Borreca et al. (2024)

Abbreviations are described in the text

2023). As mentioned above, we also detected neuroinflammation and altered microglia morphology/gene expression in the cerebellum of *Cntnap2* and *Shank3b* knockout mice, features that were rescued by NAC treatment in adult animals (Pangrazzi et al. 2025, 2024). In line with these findings, several recent studies indicate that targeting microglia dysfunction and modulating microglia activity might represent a promising strategy against ASD.

Compounds that modulate histamine microglial receptors may help restore normal microglia function and reduce inflammation. Studies performed on preclinical ASD mouse models showed that treatment with histamine H3 receptor antagonists (which are known to enhance microglial activity and reduce neuroinflammation) could result in a partial rescue of ASD-like behaviours (Baronio et al. 2015; Molenhuis et al. 2022; see also Thomas et al. 2024 for an extensive review).

Proper microglial function is critical for synaptic pruning during development, and both increased and decreased pruning (resulting from hyperactive and hypoactive microglia, respectively) are thought to contribute to ASD. Mice lacking the ASD-associated gene *Scn2a* display activated microglia and excessive synaptic pruning via phagocytosis. In these mice, microglia ablation by PLX3397 partially restores synaptic transmission and spine density, indicating that directly targeting the amount of microglia in the brain can effectively restore function in genetic forms of ASD (Wu et al. 2024).

Finally, targeting immune pathways has revealed an exciting potential for treatment in ASD. A recent study investigated the effects of the administration of PD-L1Fc, a fusion protein combining the extracellular domain of Programmed Death-Ligand 1 (PD-L1) with the IgG Fc region, in the poly(I:C) induced maternal immune activation (MIA) model of ASD (Zeng et al. 2024). MIA resulted in a marked increase of pro-inflammatory proteins (including PD-L1) in the brain of offspring mice, which exhibited typical ASD-like behaviors accompanied by activation of microglial cells. Intracerebral injections of PD-L1Fc to MIA offspring mice and intraperitoneal injections to MIA pregnant mice were able to suppress neuroinflammation in offspring mice, partially rescuing ASD-like behaviours (Zeng et al. 2024). Another recent study targeted microglia IL-1 signalling to counteract ASD pathology. IL-1 receptor (IL-1R) knockout mice display autistic-like defects associated with an excessive number of synapses (Borreca et al. 2024). The authors of the study showed that microglia lacking IL-1R could not properly eliminate synapses due to excessive activation of the mammalian target of rapamycin (mTOR) signalling. Notably, rapamycin treatment in IL-1R knockout mice rescued microglia function and behavioral defects (Borreca et al. 2024).

Table 10.1 summarizes strategies recently developed to target oxidative stress or microglia activity in ASD human cohorts and mouse models. Together, these studies indicate that targeting specific pro-inflammatory pathways in microglia has considerable potential for treatment in ASD. Further research into microglia biology, using both animal models and human-derived cells and organoids (Mesci et al. 2024), will be essential for understanding the critical role of microglia in neurodevelopment, oxidative stress, and neuroinflammation as well as for testing new therapies.

Acknowledgements The authors thank the technical and administrative staff of CIMeC (University of Trento) and Institute for Biomedical Aging Research (University of Innsbruck) for excellent support.

This work was supported by the Autism Research Institute (ARI) 2023 Research Award to YB.

References

Abbaoui A, Fatoba O, Yamashita T (2023) Meningeal T cells function in the central nervous system homeostasis and neurodegenerative diseases. Front Cell Neurosci 17:1181071

Adams JB, Audhya T, McDonough-Means S, Rubin RA, Quig D, Geis E, Gehn E, Loresto M, Mitchell J, Atwood S, Barnhouse S, Lee W (2011) Effect of a vitamin/mineral supplement on children and adults with autism. BMC Pediatr 11:111

Arandjelovic S, Ravichandran KS (2015) Phagocytosis of apoptotic cells in homeostasis. Nat Immunol 16(9):907–917

Armstrong D, Stratton RD (2016) Oxidative stress and antioxidant protection the science of free radical biology and disease. Wiley

Ayasoufi K, Wolf DM, Namen SL, Jin F, Tritz ZP, Pfaller CK, Zheng J, Goddery EN, Fain CE, Gulbicki LR, Borchers AL, Reesman RA, Yokanovich LT, Maynes MA, Bamkole MA, Khadka RH, Hansen MJ, Wu LJ, Johnson AJ (2023) Brain resident memory T cells rapidly expand and initiate neuroinflammatory responses following CNS viral infection. Brain Behav Immun 112:51–76

Ayata P, Badimon A, Strasburger HJ, Duff MK, Montgomery SE et al (2018) Epigenetic regulation of brain region-specific microglia clearance activity. Nat Neurosci 21(8):1049–1060

Badimon A, Strasburger HJ, Ayata P, Chen X, Nair A, Ikegami A, Hwang P, Chan AT, Graves SM, Uweru JO, Ledderose C, Kutlu MG, Wheeler MA, Kahan A, Ishikawa M, Wang YC, Loh YE, Jiang JX, Surmeier DJ, Robson SC, Junger WG, Sebra R, Calipari ES, Kenny PJ, Eyo UB, Colonna M, Quintana FJ, Wake H, Gradinaru V, Schaefer A (2020) Negative feedback control of neuronal activity by microglia. Nature 586(7829):417–423

Baronio D, Castro K, Gonchoroski T, deMelo GM, Nunes GDF, Bambini-Junior V, Gottfried C, Riesgo R (2015) Effects of an H3R antagonist on the animal model of autism induced by prenatal exposure to valproic acid. PLoS One 10:e0116363

Barros L, Ferreira C, Veldhoen M (2022) The fellowship of regulatory and tissue-resident memory cells. Mucosal Immunol 15(1):64–73

Baruch K, Schwartz M (2013) CNS-specific T cells shape brain function via the choroid plexus. Brain Behav Immun 34:11–16. https://doi.org/10.1016/j.bbi.2013.04.002

Bohlen CJ, Friedman BA, Dejanovic B, Sheng M (2019) Microglia in brain development, homeostasis, and neurodegeneration. Annu Rev Genet 53:263–288. https://doi.org/10.1146/annurev-genet-112618-043515

Bolotta A, Battistelli M, Falcieri E, Ghezzo A, Manara MC, Manfredini S, Marini M, Posar A, Visconti P, Abruzzo PM (2018) Oxidative stress in autistic children alters erythrocyte shape in the absence of quantitative protein alterations and of loss of membrane

phospholipid asymmetry. Oxidative Med Cell Longev 2018:6430601
Borreca A, Mantovani C, Desiato G, Corradini I, Filipello F, Elia CA, D'Autilia F, Santamaria G, Garlanda C, Morini R, Pozzi D, Matteoli M (2024) Loss of interleukin 1 signaling causes impairment of microglia- mediated synapse elimination and autistic-like behaviour in mice. Brain Behav Immun 117:493–509
Brask J, Kristensson K, Hill RH (2004) Exposure to interferon-gamma during synaptogenesis increases inhibitory activity after a latent period in cultured rat hippocampal neurons. Eur J Neurosci 19(12):3193–3201
Brigida AL, Schultz S, Cascone M, Antonucci N, Siniscalco D (2017) Endocannabinod signal dysregulation in autism spectrum disorders: a correlation link between inflammatory state and neuro-immune alterations. Int J Mol Sci 18(7):1425
Brioschi S, Wang WL, Peng V, Wang M, Shchukina I et al (2021) Heterogeneity of meningeal B cells reveals a lymphopoietic niche at the CNS borders. Science 373(6553):eabf9277
Brynskikh A, Warren T, Zhu J, Kipnis J (2008) Adaptive immunity affects learning behavior in mice. Brain Behav Immun 22(6):861–869
Butovsky O, Jedrychowski MP, Moore CS, Cialic R, Lanser AJ, Gabriely G, Koeglsperger T, Dake B, Wu PM, Doykan CE, Fanek Z, Liu L, Chen Z, Rothstein JD, Ransohoff RM, Gygi SP, Antel JP, Weiner HL (2014) Identification of a unique TGF-β-dependent molecular and functional signature in microglia. Nat Neurosci 17(1):131–143
Careaga M, Murai T, Bauman MD (2017) Maternal immune activation and autism spectrum disorder: from rodents to nonhuman and human primates. Biol Psychiatry 81(5):391–401
Chen X, Holtzman DM (2022) Emerging roles of innate and adaptive immunity in Alzheimer's disease. Immunity 55(12):2236–2254
Choi GB, Yim YS, Wong H, Kim S, Kim H, Kim SV, Hoeffer CA, Littman DR, Huh JR (2016) The maternal interleukin-17a pathway in mice promotes autism-like phenotypes in offspring. Science 351(6276):933–939
Croonenberghs J, Bosmans E, Deboutte D, Kenis G, Maes M (2002) Activation of the inflammatory response system in autism. Neuropsychobiology 45(1):1–6
Cugurra A, Mamuladze T, Rustenhoven J, Dykstra T, Beroshvili G, Greenberg ZJ, Baker W, Papadopoulos Z, Drieu A, Blackburn S, Kanamori M, Brioschi S, Herz J, Schuettpelz LG, Colonna M, Smirnov I, Kipnis J (2021) Skull and vertebral bone marrow are myeloid cell reservoirs for the meninges and CNS parenchyma. Science 373(6553):eabf7844
Curpan AS, Luca AC, Ciobica A (2021) Potential novel therapies for neurodevelopmental diseases targeting oxidative stress. Oxidative Med Cell Longev 2021:6640206
D'Agostino PM, Gottfried-Blackmore A, Anandasabapathy N, Bulloch K (2012) Brain dendritic cells: biology and pathology. Acta Neuropathol 124(5):599–614
Dawson MS, Gordon-Fleet K, Yan L, Tardos V, He H, Mui K, Nawani S, Asgarian Z, Catani M, Fernandes C, Drescher U (2023) Sexual dimorphism in the social behaviour of Cntnap2-null mice correlates with disrupted synaptic connectivity and increased microglial activity in the anterior cingulate cortex. Commun Biol 6(1):846
Dean OM, Gray KM, Villagonzalo KA, Dodd S, Mohebbi M, Vick T, Tonge BJ, Berk M (2017) A randomised, double blind, placebo-controlled trial of a fixed dose of N-acetyl cysteine in children with autistic disorder. Aust N Z J Psychiatry 51(3):241–249
Derecki NC, Cardani AN, Yang CH, Quinnies KM, Crihfield A, Lynch KR et al (2010) Regulation of learning and memory by meningeal immunity: a key role for IL-4. J Exp Med 207:1067–1080
Dos Santos SE, Medeiros M, Porfirio J, Tavares W, Pessôa L, Grinberg L, Leite REP, Ferretti-Rebustini REL, Suemoto CK, Filho WJ, Noctor SC, Sherwood CC, Kaas JH, Manger PR, Herculano-Houzel S (2020) Similar microglial cell densities across brain structures and mammalian species: implications for brain tissue function. J Neurosci 40(24):4622–4643
Engelhardt B, Vajkoczy P, Weller RO (2017) The movers and shapers in immune privilege of the CNS. Nat Immunol 18(2):123–131
Filiano AJ, Gadani SP, Kipnis J (2015) Interactions of innate and adaptive immunity in brain development and function. Brain Res 1617:18–27
Filiano AJ, Xu Y, Tustison NJ, Marsh RL, Baker W, Smirnov I, Overall CC, Gadani SP, Turner SD, Weng Z, Peerzade SN, Chen H, Lee KS, Scott MM, Beenhakker MP, Litvak V, Kipnis J (2016) Unexpected role of interferon-γ in regulating neuronal connectivity and social behaviour. Nature 535(7612):425–429
Fischer MT, Sharma R, Lim JL, Haider L, Frischer JM, Drexhage J, Mahad D, Bradl M, van Horssen J, Lassmann H (2012) NADPH oxidase expression in active multiple sclerosis lesions in relation to oxidative tissue damage and mitochondrial injury. Brain 135(Pt 3):886–899
Fleiss B, Gressens P (2024) Role of microglial modulation in therapies for perinatal brain injuries leading to neurodevelopmental disorders. Adv Neurobiol 37:591–606
Fujita Y, Masuda K, Bando M, Nakato R, Katou Y et al (2017) Decreased cohesin in the brain leads to defective synapse development and anxiety-related behavior. J Exp Med 214(5):1431–1452
Gautier EL, Shay T, Miller J, Greter M, Jakubzick C, Ivanov S, Helft J, Chow A, Elpek KG, Gordonov S, Mazloom AR, Ma'ayan A, Chua WJ, Hansen TH, Turley SJ, Merad M, Randolph GJ, Immunological Genome Consortium (2012) Gene-expression profiles and transcriptional regulatory pathways that underlie the identity and diversity of mouse tissue macrophages. Nat Immunol 13(11):1118–1128

Georgieva E, Ivanova D, Zhelev Z, Bakalova R, Gulubova M, Aoki I (2017) Mitochondrial dysfunction and redox imbalance as a diagnostic marker of "free radical diseases". Anticancer Res 37(10):5373–5381

Ghanizadeh A, Moghimi M (2013) The relationship between genetic polymorphisms of antioxidant enzymes and autism spectrum disorders. J Neural Transm 120(11):1545–1551

Ginhoux F, Greter M, Leboeuf M, Nandi S, See P, Gokhan S et al (2010) Fate mapping analysis reveals that adult microglia derive from primitive macrophages. Science 330:841–845

Gomez Perdiguero E, Klapproth K, Schulz C, Busch K, Azzoni E, Crozet L et al (2015) Tissue-resident macrophages originate from yolk-sac-derived erythromyeloid progenitors. Nature 518:547–551

Gosselin D, Link VM, Romanoski CE, Fonseca GJ, Eichenfield DZ et al (2014) Environment drives selection and function of enhancers controlling tissue-specific macrophage identities. Cell 159:1327–1340

Gottfried C, Bambini-Junior V, Francis F, Riesgo R, Savino W (2015) The impact of neuroimmune alterations in autism Spectrum disorder. Front Psych 6:121

Gray JI, Farber DL (2022) Tissue-resident immune cells in humans. Annu Rev Immunol 40:195–220

Haan N, Westacott LJ, Carter J, Owen MJ, Gray WP, Hall J, Wilkinson LS (2021) Haploinsufficiency of the schizophrenia and autism risk gene Cyfip1 causes abnormal postnatal hippocampal neurogenesis through microglial and Arp2/3 mediated actin dependent mechanisms. Transl Psychiatry 11(1):313

Hardan AY, Fung LK, Libove RA, Obukhanych TV, Nair S, Herzenberg LA, Frazier TW, Tirouvanziam R (2012) A randomized controlled pilot trial of oral N-acetylcysteine in children with autism. Biol Psychiatry 71(11):956–961

Herz J, Fu Z, Kim K, Dykstra T, Wall M, Li H, Salvador AF, Zou B, Yan N, Blackburn SM, Andrews PH, Goldman DH, Papadopoulos Z, Smirnov I, Xie XS, Kipnis J (2021) GABAergic neuronal IL-4R mediates T cell effect on memory. Neuron 109(22):3609–3618.e9

Hwang S, Jang J, Park K, Yim YS (2023) Unveiling the enigma of Brain-resident immune cells. bioRxiv [Preprint]. 2023 Sep 28:2023.09.26.559602. https://doi.org/10.1101/2023.09.26.559602

Jain RW, Yong VW (2022) B cells in central nervous system disease: diversity, locations and pathophysiology. Nat Rev Immunol 22(8):513–524. https://doi.org/10.1038/s41577-021-00652-6

James SJ, Melnyk S, Jernigan S, Cleves MA, Halsted CH, Wong DH, Cutler P, Bock K, Boris M, Bradstreet JJ, Baker SM, Gaylor DW (2006) Metabolic endophenotype and related genotypes are associated with oxidative stress in children with autism. Am J Med Genet B Neuropsychiatr Genet 141B(8):947–956

Jiang F, Zhang Y, Dusting GJ (2011) NADPH oxidase-mediated redox signaling: roles in cellular stress response, stress tolerance, and tissue repair. Pharmacol Rev 63(1):218–242

Katsel P, Davis KL, Haroutunian V (2005) Variations in myelin and oligodendrocyte-related gene expression across multiple brain regions in schizophrenia: a gene ontology study. Schizophr Res 79(2–3):157–173

Kelly E, Escamilla CO, Tsai PT (2021) Cerebellar dysfunction in autism Spectrum disorders: deriving mechanistic insights from an internal model framework. Neuroscience 462:274–287

Kipnis J, Cohen H, Cardon M, Ziv Y, Schwartz M (2004) T cell deficiency leads to cognitive dysfunction: implications for therapeutic vaccination for schizophrenia and other psychiatric conditions. Proc Natl Acad Sci USA 101(21):8180–8185

Kipnis J et al (2012) T cell deficiency and brain function. Nat Rev Immunol 12(3):176–187

Korin B, Ben-Shaanan TL, Schiller M, Dubovik T, Azulay-Debby H et al (2017) High-dimensional, single-cell characterization of the brain's immune compartment. Nat Neurosci 20:1300–1309

Kunis G, Baruch K, Rosenzweig N, Kertser A, Miller O et al (2013) IFN-γ-dependent activation of the brain's choroid plexus for CNS immune surveillance and repair. Brain 136:3427–3440

Lee A, Azmitia E, Whitaker-Azmitia P (2017) Developmental microglial priming in postmortem autism spectrum disorder temporal cortex. Brain Behav Immun 62:193–202

Lepennetier G, Hracsko Z, Unger M, Van Griensven M, Grummel V et al (2019) Cytokine and immune cell profiling in the cerebrospinal fluid of patients with neuro-inflammatory diseases. J Neuroinflammation 16:219

Li X, Chauhan A, Sheikh AM, Patil S, Chauhan V, Li XM, Ji L, Brown T, Malik M (2009) Elevated immune response in the brain of autistic patients. J Neuroimmunol 207(1–2):111–116

Lintas C, Sacco R, Persico AM (2012) Genome-wide expression studies in autism spectrum disorder, Rett syndrome, and down syndrome. Neurobiol Dis 45(1):57–68

Louveau A, Harris TH, Kipnis J (2015a) Revisiting the mechanisms of CNS immune privilege. Trends Immunol 36(10):569–577

Louveau A, Smirnov I, Keyes TJ, Eccles JD, Rouhani SJ, Peske JD, Derecki NC, Castle D, Mandell JW, Lee KS, Harris TH, Kipnis J (2015b) Structural and functional features of central nervous system lymphatic vessels. Nature 523(7560):337–341

Masi A, Glozier N, Dale R, Guastella AJ (2017) The immune system, cytokines, and biomarkers in autism spectrum disorder. Neurosci Bull 33(2):194–204

Mastenbroek LJM, Kooistra SM, Eggen BJL, Prins JR (2024) The role of microglia in early neurodevelopment and the effects of maternal immune activation. Semin Immunopathol 46(1–2):1

McGrath KE, Koniski AD, Malik J, Palis J (2003) Circulation is established in a stepwise pattern in the mammalian embryo. Blood 101:1669–1676
Meng J, Zhang L, Zhang YW (2024) Microglial dysfunction in autism spectrum disorder. Neuroscientist:10738584241252576
Mesci P, LaRock CN, Jeziorski JJ, Nakashima H, Chermont N, Ferrasa A, Herai RH, Ozaki T, Saleh A, Snethlage CE, Sanchez S, Goldberg G, Trujillo CA, Nakashima K, Nizet V, Muotri AR (2024) Human microglial cells as a therapeutic target in a neurodevelopmental disease model. Stem Cell Rep 19(8):1074–1091
Molenhuis RT, Hutten L, MJH K (2022) Histamine H3 receptor antagonism modulates autism-like hyperactivity but not repetitive behaviors in BTBR T+Itpr3tf/J inbred mice. Pharmacol Biochem Behav 212:173304
Morgan J, Chana G, Pardo C, Achim C, Semendeferi K, Buckwalter J, Courchesne E, Everall IP (2010) Microglial activation and increased microglial density observed in the dorsolateral prefrontal cortex in autism. Biol Psych 68(4):368–376
Mrdjen D, Pavlovic A, Hartmann FJ, Schreiner B, Utz SG et al (2018) High-dimensional single-cell mapping of central nervous system immune cells reveals distinct myeloid subsets in health, aging, and disease. Immunity 48:380–395. e386
Ning Z, Liu Y, Guo D, Lin WJ, Tang Y (2023) Natural killer cells in the central nervous system. Cell Commun Signal 21(1):341
Pachter JS, de Vries HE, Fabry Z (2003) The blood–brain barrier and its role in immune privilege in the central nervous system. J Neuropathol Exp Neurol 62(6):593–604
Pangrazzi L, Meryk A, Naismith E, Koziel R, Lair J, Krismer M, Trieb K, Grubeck-Loebenstein B (2017) "Inflamm-aging" influences immune cell survival factors in human bone marrow. Eur J Immunol 47(3):481–492. https://doi.org/10.1002/eji.201646570
Pangrazzi L, Balasco L, Bozzi Y (2020a) Oxidative stress and immune system dysfunction in autism spectrum disorders. Int J Mol Sci 21:3288
Pangrazzi L, Balasco L, Bozzi Y (2020b) Natural antioxidants: a novel therapeutic approach to autism spectrum disorders? Antioxidants (Basel) 9(12):1186
Pangrazzi L, Cerilli E, Balasco L, Khurshid C, Tobia C, Dall'O' GM, Chelini G, Perini S, Filosi M, Barbieri A, Ravizza T, Vezzani A, Provenzano G, Pastore A, Weinberger B, Rubert J, Domenici E, Bozzi Y (2025) The interplay between oxidative stress and inflammation supports autistic-related behaviors in Cntnap2 knockout mice. Brain Behav Immun 127:57–71. https://doi.org/10.1016/j.bbi.2025.02.030
Pangrazzi P, Cerilli E, Balasco L, Dall'O' GM, Chelini G, Pastore A, Weinberger B, Bozzi Y (2024) N-acetylcysteine counteracts immune dysfunction and autistic-related behaviors in the Shank3b mouse model of autism Spectrum disorders. bioRxiv 2024.03.13.584809; https://doi.org/10.1101/2024.03.13.584809
Pasciuto E, Burton OT, Roca CP, Lagou V, Rajan WD, Theys T et al (2020) Microglia require CD4 T cells to complete the fetal-to-adult transition. Cell 182(3):625–640.e24
Polfliet MM, Zwijnenburg PJ, van Furth AM, van der Poll T, Döpp EA, Renardel de Lavalette C et al (2001) Meningeal and perivascular macrophages of the central nervous system play a protective role during bacterial meningitis. J Immunol 167:4644–4650
Popa-Wagner A, Mitran S, Sivanesan S, Chang E, Buga AM (2013) ROS and brain diseases: the good, the bad, and the ugly. Oxidative Med Cell Longev 2013:963520
Printz DJ, Strauss DH, Goetz R, Sadiq S, Malaspina D, Krolewski J, Gorman JM (1999) Elevation of CD5+ B lymphocytes in schizophrenia. Biol Psychiatry 46(1):110–118
Radjavi A, Smirnov I, Kipnis J (2014) Brain antigen-reactive CD4+ T cells are sufficient to support learning behavior in mice with limited T cell repertoire. Brain Behav Immun 35:58–63
Ransohoff RM, Engelhardt B (2012) The anatomical and cellular basis of immune surveillance in the central nervous system. Nat Rev Immunol 12(9):623–635
Rattazzi L, Piras G, Ono M, Deacon R, Pariante CM, D'Acquisto F (2013) $CD4^+$ but not $CD8^+$ T cells revert the impaired emotional behavior of immunocompromised RAG-1-deficient mice. Transl Psychiatry 3(7):e280
Reddy PH, Reddy TP (2011) Mitochondria as a therapeutic target for aging and neurodegenerative diseases. Curr Alzheimer Res 8(4):393–409
Réu P, Khosravi A, Bernard S, Mold JE, Salehpour M, Alkass K et al (2017) The lifespan and turnover of microglia in the human brain. Cell Rep 20:779–784
Ribeiro M, Brigas HC, Temido-Ferreira M, Pousinha PA, Regen T, Santa C et al (2019) Meningeal gammadelta T cell-derived IL-17 controls synaptic plasticity and short-term memory. Sci Immunol 4. https://doi.org/10.1126/sciimmunol.aay5199
Rose S, Melnyk S, Pavliv O, Bai S, Nick TG, Frye RE, James SJ (2012) Evidence of oxidative damage and inflammation associated with low glutathione redox status in the autism brain. Transl Psychiatry 2(7):e134
Rustenhoven J, Kipnis J (2019) Bypassing the blood-brain barrier. Science 366(6472):1448–1449
Salken I, Provencio JJ, Coulibaly AP (2023) A potential therapeutic target: the role of neutrophils in the central nervous system. Brain Behav Immun Health 33:100688
Schafflick D, Wolbert J, Heming M, Thomas C, Hartlehnert M, Börsch AL et al (2021) Single-cell profiling of CNS border compartment leukocytes reveals that B cells and their progenitors reside in non-diseased meninges. Nat Neurosci 24(9):1225–1234

Schepanski S, Chini M, Sternemann V, Urbschat C, Thiele K, Sun T et al (2022) Pregnancy-induced maternal microchimerism shapes neurodevelopment and behavior in mice. Nat Commun 13(1):4571

Schieber M, Chandel NS (2014) ROS function in redox signaling and oxidative stress. Curr Biol 24(10):R453–R462

Schizophrenia Working Group of the Psychiatric Genomics Consortium (2014) Biological insights from 108 schizophrenia-associated genetic loci. Nature 511:421

Scola G, Duong A (2017) Prenatal maternal immune activation and brain development with relevance to psychiatric disorders. Neuroscience 346:403–408

Seiner J, Jacobs R, Panteli B, Brauner M, Schiltz K, Bahn S, Herberth M, Westphal S, Gos T, Walter M, Bernstein HG, Myint AM, Bogerts B (2010) Acute schizophrenia is accompanied by reduced T cell and increased B cell immunity. Eur Arch Psychiatry Clin Neurosci 260(7):509–518

Shin Yim Y, Park A, Berrios J, Lafourcade M, Pascual LM, Soares N et al (2017) Reversing behavioural abnormalities in mice exposed to maternal inflammation. Nature 549(7673):482–487

Siniscalco D, Schultz S, Brigida AL, Antonucci N (2018) Inflammation and neuro-immune dysregulations in autism Spectrum disorders. Pharmaceuticals (Basel) 11(2):56

Smith SE, Li J, Garbett K, Mirnics K, Patterson PH (2007) Maternal immune activation alters fetal brain development through interleukin-6. J Neurosci 27(40):10695–10702

Spinu N, Bal-Price A, Cronin MTD, Enoch SJ, Madden JC, Worth AP (2019) Development and analysis of an adverse outcome pathway network for human neurotoxicity. Arch Toxicol 93(10):2759–2772

Strominger I, Elyahu Y, Berner O, Reckhow J, Mittal K, Nemirovsky A, Monsonego A (2018) The choroid plexus functions as a niche for T-cell stimulation within the central nervous system. Front Immunol 9:1066. https://doi.org/10.3389/fimmu.2018.01066

Suzuki K, Sugihara G, Ouchi Y, Nakamura K, Futatsubashi M, Takebayashi K, Yoshihara Y, Omata K, Matsumoto K, Tsuchiya KJ, Iwata Y, Tsujii M, Sugiyama T, Mori N (2013) Microglial activation in young adults with autism Spectrum disorder. JAMA Psych 70(1):49–58

Swerdlow RH (2011) Brain aging, Alzheimer's disease, and mitochondria. Biochim Biophys Acta 1812:1630–1639

Tanabe S, Yamashita T (2018a) B-1a lymphocytes promote oligodendrogenesis during brain development. Nat Neurosci 21(4):506–516

Tanabe S, Yamashita T (2018b) The role of immune cells in brain development and neurodevelopmental diseases. Int Immunol 30(10):437–444

Terzi A, Suter DM (2020) The role of NADPH oxidases in neuronal development. Free Radic Biol Med 154:33–47

Thomas SD, Abdalla S, Eissa N, Akour A, Jha NK, Ojha S, Sadek B (2024) Targeting microglia in neuroinflammation: H3 receptor antagonists as a novel therapeutic approach for Alzheimer's disease, Parkinson's disease, and autism spectrum disorder. Pharmaceuticals (Basel) 17(7):831

Tsai PT (2016) Autism and cerebellar dysfunction: evidence from animal models. Semin Fetal Neonatal Med 21(5):349–355

Urrutia-Ruiz C, Rombach D, Cursano S, Gerlach-Arbeiter S, Schoen M, Bockmann J, Demestre M, Boeckers TM (2022) Deletion of the autism-associated protein SHANK3 abolishes structural synaptic plasticity after brain trauma. Int J Mol Sci 23(11):6081

Vargas D, Nascimbene C, Krishnan C, Zimmerman A, Pardo C (2004) Neuroglial activation and neuroinflammation in the brain of patients with autism. Ann Neurol 57(1):67–81

Varun S, Singh A (2019) Role of oxidative stress in the pathophysiology of autism spectrum disorder. J Neuroinflammation 16(1):1–12

Velmeshev D, Schirmer L, Jung D, Haeussler M, Perez Y, Mayer S, Bhaduri A, Goyal N, Rowitch D, Kriegstein A (2019) Single-cell genomics identifies cell type–specific molecular changes in autism. Science 364(6441):685–689

Voet S, Prinz M, van Loo G (2019) Microglia in central nervous system inflammation and multiple sclerosis pathology. Trends Mol Med 25(2):112–123

Weinhard L, di Bartolomei G, Bolasco G, Machado P, Schieber NL, Neniskyte U, Exiga M, Vadisiute A, Raggioli A, Schertel A, Schwab Y, Gross CT (2018) Microglia remodel synapses by presynaptic trogocytosis and spine head filopodia induction. Nat Commun 9(1):1228

Willis CL, Garwood CJ, Ray DE (2007) A size selective vascular barrier in the rat area postrema formed by perivascular macrophages and the extracellular matrix. Neuroscience 150:498–409

Wilson C, Muñoz-Palma E, González-Billault C (2018) From birth to death: a role for reactive oxygen species in neuronal development. Semin Cell Dev Biol 80:43–49

Wink LK, Adams R, Wang Z, Klaunig JE, Plawecki MH, Posey DJ, McDougle CJ, Erickson CA (2016) A randomized placebo-controlled pilot study of N-acetylcysteine in youth with autism spectrum disorder. Mol Autism 7:26

Wolf SA, Steiner B, Akpinarli A, Kammertoens T, Nassenstein C, Braun A, Blankenstein T, Kempermann G (2009) CD4-positive T lymphocytes provide a neuroimmunological link in the control of adult hippocampal neurogenesis. J Immunol 182(7):3979–3984

Wu J, Zhang J, Chen X, Wettschurack K, Que Z, Deming BA, Olivero-Acosta MI, Cui N, Eaton M, Zhao Y, Li SM, Suzuki M, Chen I, Xiao T, Halurkar MS, Mandal P, Yuan C, Xu R, Koss WA, Du D, Chen F, Wu LJ, Yang Y (2024) Microglial over-pruning of synapses during development in autism-associated SCN2A-deficient mice and human cerebral organoids. Mol Psychiatry 29(8):2424–2437

Yang D, Elner SG, Bian ZM, Till GO, Petty HR, Elner VM (2007) Pro-inflammatory cytokines increase reactive oxygen species through mitochondria and NADPH oxidase in cultured RPE cells. Exp Eye Res 85(4):462–472

Yu Y, Liu S, Yang L, Song P, Liu Z, Liu X, Yan X, Dong Q (2024) Roles of reactive oxygen species in inflammation and cancer. MedComm (2020) 5(4):e519

Zeng X, Fan L, Qin Q, Zheng D, Wang H, Li M, Jiang Y, Wang H, Liu H, Liang S, Wu L, Liang S (2024) Exogenous PD-L1 binds to PD-1 to alleviate and prevent autism-like behaviors in maternal immune activation-induced male offspring mice. Brain Behav Immun 122:527–546

11 Beyond Infections: Exploring Immune-Mediated Pathways Linking Cannabis and Emerging Environmental Contaminants to Neurodevelopmental Outcomes

Taylor Doolittle and Natalia Duque-Wilckens

Abstract

The development of the brain is a highly coordinated process that begins early in gestation and relies on intricate interactions between maternal and fetal immune systems. Disruptions to this delicate prenatal immune environment can significantly impact fetal brain development, increasing the risk of a spectrum of neurological and behavioral disorders, including autism spectrum disorder (ASD), schizophrenia, depression, and anxiety. While maternal exposure to viral and bacterial infections has been extensively studied as a driver of these disruptions, emerging research highlights the potential role of non-infectious exposures—such as drugs of abuse and environmental contaminants—in shaping neurodevelopmental outcomes. This issue has gained urgency with the rising prevalence of neurodevelopmental disorders; in the United States, for instance, ASD diagnoses have surged from 1 in 150 children in the early 2000s to 1 in 36 by 2020. Within this context, the increasing detection of cannabis, micro- and nanoplastics (MPs/NPs), and flame retardants in maternal and fetal tissues warrants close scrutiny. These substances, which are growing in prevalence due to changing societal norms, widespread environmental pollution, and industrial practices, may disrupt neurodevelopment through immune-mediated mechanisms. Cannabis use during pregnancy, for example, has increased significantly with legalization, while MPs/NPs and flame retardants are now frequently detected in maternal blood, placenta, and breast milk, raising concerns about their impacts on fetal health. In this chapter, we summarize human and preclinical evidence to explore how perinatal exposure to these substances may alter neurodevelopment by disrupting maternal and/or fetal immunity. We begin with an overview of central nervous system development and the critical role of immune interactions in ensuring a healthy pregnancy. We then review evidence linking perinatal exposure to cannabis, MPs/NPs, and flame retardants to neurodevelopmental outcomes, emphasizing immune-mediated pathways such as alterations in cytokine production, microglial activation, and adaptive immune cell function. Finally, we identify key gaps in the literature and

T. Doolittle
Department of Biological Sciences, College of Sciences, North Carolina State University, Raleigh, NC, USA

N. Duque-Wilckens (✉)
Department of Biological Sciences, College of Sciences, North Carolina State University, Raleigh, NC, USA

Center for Human Health and the Environment, North Carolina State University, Raleigh, NC, USA
e-mail: nduquew@ncsu.edu

D. Tropea, E. Giacometti (eds.), *Brain-Body Connections*, Advances in Experimental Medicine and Biology 1477, https://doi.org/10.1007/978-3-031-89525-8_11

propose future research directions to better understand the complex interplay between environmental exposures, immune dysregulation, and neurodevelopmental outcomes.

Keywords

Maternal-fetal immunity · Neurodevelopment · Cannabis exposure pregnancy · Microplastics and immunity · Flame retardants brain development · Neuroimmune interactions · Non-infectious exposures

11.1 Introduction

The prenatal environment plays a pivotal role in shaping fetal brain development, with maternal and fetal immune responses serving as critical regulators during this period (Bilbo and Schwarz 2012). Perinatal immune dysregulation—arising from maternal exposure to infections or stress—has been consistently linked to an increased risk of neurodevelopmental and psychiatric disorders later in life, including autism spectrum disorder (ASD), schizophrenia, and depression (Jiang et al. 2016; Brown et al. 2004, 2005; Murphy et al. 2017; Quagliato et al. 2021). This association likely stems from immune-mediated disruptions in brain formation and synaptic connectivity, which can result in lasting effects on brain function and behavior (Bilbo and Schwarz 2012; Han et al. 2021; Mastenbroek et al. 2024).

Amid these established risks, increasing attention is being drawn to the growing prevalence of substances in maternal tissues—such as cannabis, micro- and nanoplastics (MPs/NPs), and flame retardants, among others—that are known to affect neurodevelopment and may exert their effects, in part, by altering maternal and/or fetal immunity. Cannabis use during pregnancy, driven by shifting perceptions and increasing accessibility following legalization, has risen significantly (Marchetti et al. 2017; Chang et al. 2019). Cannabinoids readily cross the placenta and are secreted in breast milk, potentially reaching offspring tissues where they may interfere with immune-regulated processes essential for neurodevelopment (Marchetti et al. 2017). Meanwhile, micro- and nanoplastics (MPs/NPs), generated from the degradation of plastic products, are now widely detected across environments and in human tissues, including the placenta and amniotic fluid (Prata et al. 2020; Ragusa et al. 2021, 2022; Sangkham et al. 2022; Xue et al. 2024). Similarly, flame retardants, once considered confined to consumer products, are being detected with alarming frequency in maternal blood, breast milk, and children's urine (Huang et al. 2022; Chen et al. 2023; Zheng et al. 2021), reflecting their widespread and growing environmental presence.

This rise in exposures also coincides with a marked increase in the prevalence of neurodevelopmental disorders. In the United States, ASD diagnoses rose from approximately 1 in 150 children in the early 2000s to about 1 in 36 by 2020 (CDC 2024). Similarly, global schizophrenia-related health impacts have escalated, with incident cases reaching 1.13 million and disability-adjusted life years (DALYs) increasing by over 60% between 1990 and 2017 (He et al. 2020). Although multiple factors likely contribute to these trends, the immune-disrupting properties of these contaminants could be a significant, yet underexplored, driver.

To address this gap, this chapter explores the growing body of human and preclinical evidence on how immune disruptions may mediate the neurodevelopmental impacts of emerging non-infectious exposures. We begin with an overview of central nervous system development, emphasizing the critical role of the immune system in processes such as synaptic pruning, neuronal migration, and axonogenesis. Following this, we review the neurodevelopmental effects of cannabis, MPs/NPs, and flame retardants, focusing on their potential to disrupt maternal and/or fetal immunity. While these three substances are the focus of this chapter, it is important to acknowledge that they represent only a subset of the myriad environmental exposures potentially impacting neurodevelopment. Finally, we identify key gaps in the literature and propose future research directions to deepen our understanding and inform strategies to mitigate these risks.

11.2 Brief Overview of Central Nervous System Development and the Role of the Immune System

The complexity of the nervous system arises from the precise coordination of numerous cellular and molecular processes, which begin early in embryonic development and follow conserved spatiotemporal patterns across mammals (Fig. 11.1). These processes ensure that the brain forms highly interconnected networks capable of integrating information from the body and environment to generate appropriate responses.

A pivotal event in this development is the formation of the neural tube, an early embryonic structure that will give rise to the brain, brainstem, and spinal cord (He et al. 2020; Stiles and Jernigan 2010). In the innermost layer of the neural tube, known as the ventricular zone, progenitor cells undergo symmetric division to self-renew and expand the progenitor pool. As development progresses, progenitor cells shift to asymmetric division to generate neurons. These newly generated neurons then begin migrating in successive waves to specific regions of the brain, where they further differentiate into specialized neuronal subtypes, extend axons and dendrites, and ultimately form synaptic connections in a process called synaptogenesis (Stiles and Jernigan 2010; Tau and Peterson 2010). Although synaptogenesis continues throughout life (Zito and Svoboda 2002), it peaks during early brain development, establishing the foundational neural circuits necessary for sensory processing, motor control, and cognitive function.

After most neurogenesis has occurred, progenitor cells in the ventricular zone begin differentiating into macroglia, including astrocytes and

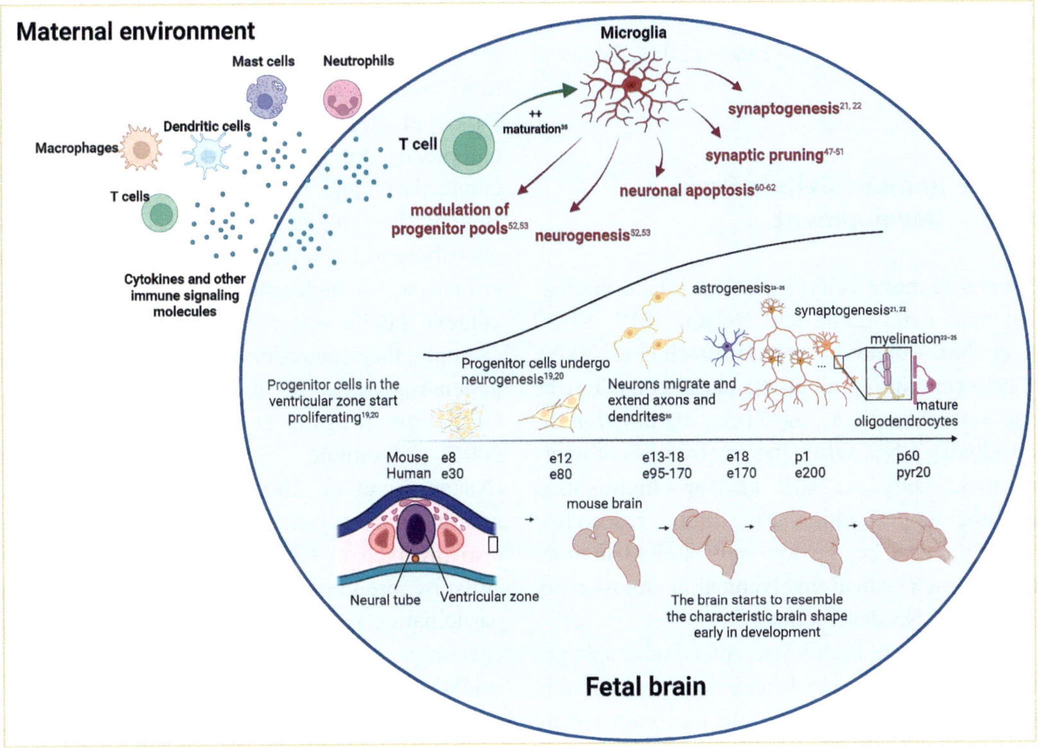

Fig. 11.1 Development of fetal brain and role of the immune system

cells of the oligodendroglial lineage, a process that continues into the postnatal period (Akdemir et al. 2020; Vivi and Di Benedetto 2024; van Tilborg et al. 2018). While research into the diverse roles and subtypes of macroglia is ongoing (Werkman et al. 2020), several key functions of astrocytes and oligodendrocytes in CNS physiology are well-established. Astrocytes, for instance, provide crucial modulation of synaptic transmission, maintain neuronal homeostasis, and contribute to the formation of the blood-brain barrier (Gradisnik and Velnar 2023; Valles et al. 2023), while oligodendrocytes are essential for axon myelination and metabolic support to neurons (Philips and Rothstein 2017).

Crucially, throughout these intricate stages of brain development—from neuronal differentiation to synaptogenesis and macroglial maturation—the immune system plays an indispensable role. In this section, we will provide a brief overview of the immune cells and molecules that contribute to brain development; however, for a more comprehensive exploration of this topic, we recommend consulting the available excellent reviews (Mastenbroek et al. 2024; Dziabis and Bilbo 2022; Zengeler and Lukens 2021; Lenz and Nelson 2018).

11.2.1 Immune Cells in Brain Development

Several immune cells, including innate microglia, mast cells (Lenz and Nelson 2018; Khalil et al. 2007), dendritic cells (Bulloch et al. 2008), T cells (Pasciuto et al. 2020), and B cells (Tanabe and Yamashita 2018), have been identified in the developing CNS. While the precise roles of many of these cells are still unclear—highlighting exciting opportunities for future research—microglia have been extensively studied, demonstrating their critical involvement in almost every aspect of CNS development.

Microglia are highly specialized macrophages that originate from erythromyeloid precursors in the yolk sac and migrate into the brain during early embryonic development—well before the blood-brain barrier forms and neurogenesis begins (Alliot et al. 1999; Kierdorf et al. 2013). Once established in the brain, microglia proliferate locally, undergo specialized differentiation and maturation (Bennett et al. 2016; Nikodemova et al. 2015), and continue to self-renew throughout life (Tay et al. 2017; Ajami et al. 2007). While they are best known for their role in neuroinflammatory processes, microglia represent a highly heterogeneous population (Bennett and Bennett 2020) that plays critical roles in brain development and homeostasis through complex interactions with both neuronal and non-neuronal cells (Mehl et al. 2022).

Among their many roles in brain development, one of the primary ways microglia contribute is by regulating cell populations through the selective pruning of neurons and synapses. During early embryonic stages, microglia migrate to proliferative zones and actively phagocytose neural (Cunningham et al. 2013) and oligodendrocyte (Nemes-Baran et al. 2020) progenitor cells, helping to control the size of the progenitor pool. Similarly, microglia play a key role in eliminating weak or underutilized synaptic connections (Schafer et al. 2012; Filipello et al. 2018; Park et al. 2021; Scott-Hewitt et al. 2020), a crucial process for the normal development of brain connectivity and behavior (Zhan et al. 2014). Beyond their role in synaptic pruning, microglia contribute to brain development in several other critical ways, highlighting the complexity and context-specific nature of their influence. For example, they can promote neurogenesis and oligodendrogenesis in the subventricular zone (Shigemoto-Mogami et al. 2014; Walton et al. 2006), facilitate astrocyte differentiation (Nakanishi et al. 2007), and enhance cortical neuron survival (Ueno et al. 2013). Microglia are also involved in forming functional neural circuits by promoting synaptic remodeling, thus participating in brain wiring. For instance, they encourage dendritic spine formation in the somatosensory cortex (Miyamoto et al. 2016) and modulate axonal outgrowth of dopaminergic neurons and neocortical interneuron positioning in the forebrain (Squarzoni et al. 2014), among many other functions.

The mechanisms by which microglia contribute to these diverse functions are varied and depend heavily on brain region, developmental timing, and specific signaling pathways. For instance, microglial synaptic pruning in the developing mouse cortex and the retinogeniculate system are mediated by the complement system (Deivasigamani et al. 2023; Stevens et al. 2007), a set of innate immune proteins that help tag cells for removal. In addition, the modulation of dopaminergic axon outgrowth and neocortical interneuron positioning involves CX3CR1 (Squarzoni et al. 2014)—the receptor for neuron-secreted fractalkine, a chemokine that guides microglia to specific synaptic sites for pruning and maintenance. Microglia can also induce neuronal death by secreting noxious factors such as reactive oxygen species and nerve growth factor (Marín-Teva et al. 2004; Wakselman et al. 2008; Frade and Barde 1998).

Although the role of other immune cells in prenatal brain development has been less studied compared to that of microglia, emerging evidence suggests they also play important roles. For instance, a study found that B-1a cells—a subtype of B lymphocytes involved in innate immunity through the production of antibodies that recognize antigens with low specificity—contribute to oligodendrogenesis. Specifically, by producing IgM antibody, B-1a cells promote the development of oligodendrocyte precursor cells during brain development, supporting the formation of myelinating cells essential for efficient neural signaling (Tanabe and Yamashita 2018). Similarly, recent research has shown that CD4 T cells—a subtype of T lymphocytes integral to adaptive immunity, known for recognizing antigens presented by MHC class II molecules and releasing cytokines that regulate other immune cells—play a crucial role in promoting microglial maturation and enabling microglia to acquire synaptic pruning functions (Pasciuto et al. 2020).

11.2.2 Immune Molecules in Brain Development

A main way in which immune and non-immune CNS-resident cells interact to influence brain development is through the action of immune-related molecules. One of the most studied groups of these molecules are cytokines, a major class of signaling proteins. In peripheral immunity, cytokines are primarily produced by immune cells to orchestrate immune responses, guide immune cell differentiation, modulate inflammation, and promote tissue repair (Turner et al. 2014; Fang et al. 2018; Moon et al. 2023). Within the CNS, cytokines and their receptors are not only expressed in microglia but also in neurons, astrocytes, and oligodendrocytes, where, among other functions, they play critical roles in brain development (Borsini et al. 2015; Deverman and Patterson 2009).

The specific functions of cytokines will largely vary depending on the developmental stage and the surrounding microenvironment. For instance, IL-1β and IL-1α, both members of the IL-1 family of cytokines—traditionally considered proinflammatory mediators—have been shown to either promote (Park et al. 2018; de la Mano et al. 2007) or inhibit (Wang et al. 2007; Zhang et al. 2013) neuronal differentiation, depending on the developmental stage and/or brain area. They also play roles in guiding neuronal migration (Ma et al. 2014) and inhibiting axonal growth (Nolan et al. 2011). Similarly, IL-33, which is typically released during tissue damage in peripheral immunity, is expressed postnatally by synapse-associated astrocytes, where it signals to microglia to promote synaptic engulfment (Vainchtein et al. 2018).

Another important cytokine is IL-6, a multifunctional cytokine involved in both immune and neural processes. In peripheral immunity, IL-6 is produced by various immune cells to regulate inflammation, acute-phase responses, and immune cell differentiation. Within the CNS, although the physiological roles of IL-6 in brain development have not been extensively studied, research in adult brains and progenitor cell cultures suggests that IL-6 contributes to neuronal progenitor expansion, differentiation, and survival (Kummer et al. 2021). Through these actions, IL-6 may support the growth and maturation of neural circuits during early brain development.

11.2.3 Immune Function Dysregulation and Brain Development

Given the vast and heterogeneous roles of the immune system during brain development, it is not surprising that impairments in immune function during this critical stage are linked to neurological dysfunctions that may only manifest later in life. For example, human epidemiological studies support an association between maternal inflammation and an increased risk of offspring neurodevelopmental disorders, including autism spectrum disorder (Jiang et al. 2016), attention-deficit/hyperactivity disorder (Mann and McDermott 2011; Pineda et al. 2007), and schizophrenia spectrum disorder (Brown et al. 2004, 2005). These findings increasingly support the hypothesis that environmental factors interact with genetic influences to shape both the likelihood and severity of these conditions (Han et al. 2021).

Preclinical studies using maternal immune activation (MIA) models provide substantial evidence supporting the role of impaired maternal immunity in driving neurodevelopmental abnormalities in offspring. In these models, the maternal immune system is artificially stimulated during pregnancy, often through exposure to immune activators such as poly(I:C), a viral mimic, or bacterial components like lipopolysaccharide (LPS). These agents activate pattern recognition receptors on both peripheral immune cells and central nervous system (CNS) cells (Akira and Takeda 2004), initiating a maternal inflammatory response that disrupts the balance of pro-inflammatory and anti-inflammatory cytokines reaching the fetal brain. This immune imbalance can interfere with critical neurodevelopmental processes, leading to lasting neurobehavioral effects. Cytokines such as IL-6, IL-17, tumor necrosis factor (TNF), and IL-1β are key mediators in transmitting MIA effects to the fetus, resulting in altered immune signaling within the fetal brain, disrupted microglial functions, changes in neuronal signaling, and abnormal behaviors (for comprehensive reviews, see (Dziabis and Bilbo 2022; Zengeler and Lukens 2021)).

While preclinical models using infectious stimuli have been instrumental in establishing the link between immune system disruption and prenatal brain development, other environmental factors have also been shown to increase susceptibility to neurodevelopmental abnormalities. These include diet (Edlow 2016), socioeconomic status (Hackman et al. 2010), use of drugs, and exposure to pollutants (Rock and Patisaul 2018; Tartaglione et al. 2024). Among these, we will focus on some which use and/or prevalence in the environment has particularly risen in the past decade and are predicted to continue to do so: cannabis, micro and nanoplastics, and flame retardant contaminants (Fuller et al. 2022; Afifa et al. 2024; Thompson et al. 2024).

11.3 Impact of Emerging Contaminants on Neurodevelopment: Exploring the Role of the Immune System

11.3.1 Cannabis

Cannabis refers to the dried leaves, flowers, stems, and seeds of the *Cannabis sativa* plant. It contains delta-9-tetrahydrocannabinol (Δ9THC), the primary psychoactive compound, along with other related compounds collectively known as phytocannabinoids. These compounds exert many of their effects by interacting with the endogenous cannabinoid system, a complex signaling network that includes the endogenous ligands N-arachidonoylethanolamine (anandamide, AEA) and 2-arachidonoylglycerol

(2-AG), the proteins responsible for their transport, synthesis, and degradation, as well as the G protein-coupled cannabinoid receptors CB1 and CB2 (Lu and Mackie 2016). However, emerging evidence reveals that phytocannabinoids can also influence a diverse range of other receptors, transporters, enzymes, cellular structures, membranes, and ion channels, significantly broadening their impact across multiple physiological systems (Watkins 2019; Soderstrom et al. 2017; Ghovanloo et al. 2018; Franzen et al. 2023).

For decades, cannabis has been the most commonly reported illicit drug used by pregnant women (Martin et al. 2015; Beatty et al. 2012; Moore et al. 2009; Passey et al. 2014). However, its use has surged dramatically in recent years, likely due to increasing legalization for medicinal and recreational purposes, shifting public perceptions that downplay its risks, and its appeal as a remedy for common pregnancy symptoms like nausea (Mark et al. 2017; Chang et al. 2019; Latuskie et al. 2019; Levy et al. 2021). A recent study using data from the National Survey on Drug Use and Health (Volkow et al. 2019) reported that the prevalence of past-month cannabis use among pregnant women increased from 3.4% to 7.0%, with the most significant rise occurring during the first trimester, where usage jumped from 5.7% to 12.1%. Similarly, a cross-sectional time-series study conducted in Northern California assessed prenatal substance use during the first trimester through self-administered questionnaires and/or positive urine toxicology tests (Young-Wolff et al. 2022). This study found that the use of only cannabis during pregnancy increased substantially from 2.39% in 2009 to 6.30% in 2018, whereas the use of cannabis with one other substance rose more modestly, from 1.19% to 1.92%, suggesting that the sharp increase in cannabis use during pregnancy is primarily driven by its growing perception as a safe option.

This trend is concerning, as cannabinoids readily cross the placenta (Marchetti et al. 2017; Concheiro et al. 2021) and are secreted in breast milk (Castro-Navarro et al. 2024), where they can potentially impact fetal development and neonatal health. Indeed, ample evidence has linked prenatal cannabis exposure to a range of adverse outcomes, with some, such as low birth weight and premature delivery (Corsi et al. 2019; Gunn et al. 2016; Gabrhelík et al. 2021), evident at birth, while others, like neurobehavioral abnormalities (see below) (Bolhuis et al. 2018; Avalos et al. 2024), emerging later in life.

The specific mechanisms underlying the effects of prenatal cannabis exposure on offspring health, however, remain poorly understood (Mulligan and Hamre 2023). This is partly due to the fact that these effects may result from interactions with the endocannabinoid system—critical for reproductive health, placental function, and fetal development (Walker et al. 2019; Correa et al. 2016; Bukiya 2019; Maccarrone et al. 2014; Bara et al. 2021)—as well as its broader actions on other signaling pathways, adding complexity to their study. Interpreting data from human studies is further complicated by confounding factors. For example, the detrimental effects of prenatal cannabis exposure on children's cognition, such as decreased attention span, have been primarily reported in low socioeconomic groups (Braillon and Bewley 2017), suggesting that factors like access to prenatal care, maternal nutrition, or exposure to additional environmental stressors may contribute to these outcomes. Additionally, women who use cannabis during pregnancy are less likely to take folic acid supplements and more likely to experience domestic abuse (Braillon and Bewley 2017)—both of which can independently affect fetal health (Irvine et al. 2022).

This complexity underscores the importance of animal studies, which allow researchers to isolate and directly examine the effects of prenatal cannabis exposure on maternal and fetal systems. In this section, we will first review findings from human studies on perinatal cannabis exposure, with a focus on its impact on brain development and behavior. We will then explore insights gained from animal research and conclude by examining the potential role of the immune system in mediating these effects.

11.3.1.1 Perinatal Cannabis Exposure on Brain Development and Behavior

Long-term effects of prenatal cannabis exposure in humans have been investigated through several population-based longitudinal cohort studies. These studies typically recruit healthy mothers during pregnancy and track both maternal and offspring outcomes at regular intervals from childhood through early adolescence. Maternal cannabis use is primarily determined through self-reported data, though some studies have incorporated objective measures, such as urinary or blood cannabinoid levels, to enhance accuracy. While findings across studies are sometimes conflicting (Mulligan and Hamre 2023), most have identified associations between prenatal cannabis exposure and a variety of neurodevelopmental outcomes, including cognitive and attention deficits (Day et al. 1994; Fried et al. 1992, 1998; Fried and Watkinson 2000, 2001; Goldschmidt et al. 2008; Leech et al. 1999; Richardson et al. 2002), increased impulsivity (Leech et al. 1999; Richardson et al. 2002; Smith et al. 2004), psychotic-like experiences (Bolhuis et al. 2018), and depressive symptoms (Gray et al. 2005) at various stages of postnatal life.

Emerging evidence has begun to reveal specific brain alterations that may underlie these neurobehavioral outcomes, including disruptions in connectivity and potential changes in specific regions and systems. A recent study using data from the Adolescent Brain Cognitive Development (ABCD) Study® examined brain structural and functional connectivity through diffusion and resting-state functional magnetic resonance imaging (MRI) and correlated it with parent-reported child behavioral checklist scores. While the study did not confirm direct anatomical or functional differences between unexposed children and children prenatal cannabis exposure, it identified significant interactions between prenatal cannabis exposure, brain network organization, and behavioral outcomes (Vishnubhotla et al. 2024). For example, in the right lateral orbitofrontal cortex (OFC), graph network measures such as local efficiency and clustering coefficient were linked to externalizing behavior problems, but the relationship differed between exposed and unexposed children. Specifically, prenatal cannabis exposure appeared to reverse the typical association, suggesting that altered network properties in the OFC—a region critical for decision-making and emotional regulation—may underlie the behavioral differences observed in exposed children.

Importantly, the timing of prenatal cannabis exposure may significantly influence its effects on brain connectivity. For instance, another study using the ABCD database (Faraj et al. 2022) found that exposure during early gestation (i.e., before pregnancy awareness) was associated with altered connectivity in the salience and sensorimotor networks, which are critical for processing external stimuli and coordinating movement. In contrast, exposures occurring later in pregnancy did not show similar effects. These findings may be linked to the specific roles of the endogenous cannabinoid system in early pregnancy processes, such as embryo transport, implantation, and placentation (Kozakiewicz et al. 2021) —where even mild disruptions could create suboptimal conditions for fetal development and potentially affect long-term brain connectivity—as well as its critical role in early neurodevelopmental events, including neurogenesis and neuronal migration (Díaz-Alonso et al. 2012). Alternatively, the observed differences could reflect variations in cannabis doses or methods of consumption, as well as biases in self-reporting, with mothers potentially less likely to disclose cannabis use after becoming aware of their pregnancy.

Prenatal cannabis exposure has further been associated with structural and functional changes within specific brain regions and/or neurotransmitter systems. Among these, the prefrontal cortex (PFC), essential for executive functions such as decision-making, impulse control, and emotional regulation (Pizzagalli and Roberts 2022; Friedman and Robbins 2022), seems particularly vulnerable. For instance, prenatal cannabis exposure combined with tobacco use, as opposed to tobacco use alone, has been associated with increased PFC thickness in young children (El Marroun et al. 2016). Similarly, prenatal cannabis exposure has been linked to heightened neural activity in the adolescent PFC during both response inhibition tasks—which assess the ability to suppress inappropriate or automatic responses—and visuospatial working memory tasks—which involve maintaining and manipulating visual and spatial information (Smith et al. 2004; Smith et al. 2006). These results could potentially explain the deleterious effects of prenatal cannabis exposure on cognitive processing.

In addition to the PFC, disruptions in dopaminergic signaling may further contribute to the effects of prenatal cannabis exposure. The dopaminergic system—a neurotransmitter network heavily implicated in various psychiatric and substance use disorders—plays a critical role in cognitive and emotional processes. Dopamine neurons primarily originate in the ventral tegmental area (VTA) and substantia nigra, projecting to brain regions such as the PFC, amygdala, and dorsal striatum, where they modulate functions including reward learning, motivation, motor control, and cognitive functioning (Money and Stanwood 2013; Schultz 2007; Spencer et al. 1998; Liu et al. 2021). Dopamine exerts its effects by binding to two primary types of receptors: the excitatory D1-like receptors and the inhibitory D2-like receptors (DR2). Studies in midgestation human fetal specimens have shown that cannabis use during pregnancy reduces the expression of mRNA for DR2 in both the striatum (DiNieri et al. 2011) and the amygdala (Wang et al. 2004). In the ventral striatum, DR2 play a critical role in modulating motivated behaviors (Gallo et al. 2018), while in the amygdala, they exert anxiolytic effects (Hernández-Mondragón et al. 2023) and optimize appetitive and defensive behaviors strategies (Valjent et al. 2023), among others. Therefore, these disruptions could help explain the effects of prenatal cannabis exposure on emotional regulation and stress-related behaviors in exposed offspring.

Animal studies further support these findings and highlight significant sex differences, emphasizing the critical need to incorporate analyses of sex and gender influences when investigating cannabis-induced effects. For example, a study in rats exposed pregnant dams to 5 mg/kg THC from gestation day 5 to gestation day 20—equivalent to the first trimester and early second trimester in humans—found that cannabinoid exposure heightened excitability of PFC pyramidal neurons in males but not females (Bara et al. 2018). Likewise, another study where dams were randomly assigned to either vehicle or 3 mg/kg THC daily from gestation day 7–22 observed hyperactive PFC pyramidal neuron activity in both male and female offspring. However, this study also reported distinct effects in the ventral hippocampus, with males exhibiting hypoactivity and females showing hyperactivity (Sarikahya et al. 2023). In addition to these findings, studies in rodents have also reported effects on dopaminergic neurotransmission. For instance, two separate studies in which rats were exposed to 2 mg/kg THC from gestation day 5–20 found that male, but not female offspring exhibited dysregulated dopaminergic neuron excitability (Frau et al. 2019; Sagheddu et al. 2021). Similarly, supporting their findings in human tissues, the study by Di Nieri et al. (2011) demonstrated that THC exposure (0.15 mg/kg) from gestation day 5 to postnatal day 2 reduced DR2 mRNA expression in offspring, although sex differences were not reported. An important caveat of these studies, however, is their reliance on injection-based THC delivery, which does not fully replicate human cannabis use methods like smoking, vaping, or edibles (Young-Wolff et al. 2022; Baggio et al. 2014). Differences in THC metabolism and exposure levels between these methods could impact outcomes, highlighting the need for future studies to use exposure models that better reflect human use patterns for improved translational relevance.

To summarize, prenatal cannabis exposure has been associated with a range of neurodevelopmental and behavioral outcomes in both human and animal studies. While many of these effects on neuronal function are likely due to prenatal THC acting directly on cannabinoid receptors in neurons (Maccarrone et al. 2014; de Salas-Quiroga et al. 2015; Hillard 2014; Moreira et al. 2008) and/or macroglia (Navarrete and Araque 2010; Mato et al. 2010), THC-induced changes in maternal and fetal immunity may also play a role, as cannabinoids play an integral role in modulating innate and adaptive immune functions (Cabral et al. 2015; Almogi-Hazan and Or 2020).

11.3.1.2 Perinatal Cannabinoids as Immunomodulators

Immune cells from both the adaptive and innate immune systems not only express classical cannabinoid receptors CB1 and CB2, but also possess the necessary machinery for endocannabinoid biosynthesis, uptake, and degradation. Beyond these classical receptors, "non-classical" cannabinoid targets, such as transient receptor potential (TRP) channels and peroxisome proliferator-activated receptors (PPARs), have emerged as important regulators of immune function, highlighting the central role of the endocannabinoid system in immune modulation (for reviews, see (Almogi-Hazan and Or 2020; Oláh et al. 2017)). Reflecting this broad involvement, exogenous cannabinoids can have profound effects on immune activity. Among the most well-documented effects of cannabis are its robust anti-inflammatory and immunosuppressive activities. For example, multiple studies have shown that cannabis inhibit pro-inflammatory cytokines, including TNF-α, IL-1β, and IL-6, an effect largely driven by cannabinoids such as cannabidiol (CBD) and cannabigerol (CBG) (Henshaw et al. 2021). Additionally, cannabinoids have been shown to prevent proliferation and induce apoptosis of effector T cells (McKallip et al. 2005) and B cells (Lampron et al. 2023), an effect that can be beneficial in the context of inflammatory diseases associated with dysregulated T-cell (Hegde et al. 2008) or B-cell (Melén et al. 2019) activity. However, this immunosuppressive activity may also be deleterious in certain contexts, such as certain cancers, where it can suppress antitumor immune responses (McKallip et al. 2005), or infections, where it can reduce resistance to bacterial and viral pathogens (Hernández-Cervantes et al. 2017).

In the context of cannabis use during pregnancy, human studies have shown that concurrent use of cannabis and tobacco during the first trimester, as opposed to tobacco alone, significantly reduces maternal circulating levels of TNF-α (Ashford et al. 2019). Furthermore, a recent study analyzing placental biopsies collected at birth and processed for RNA sequencing found that maternal cannabis use was associated with reduced expression of immune-related genes, including proinflammatory cytokines and immune cell–type markers. Importantly, using gene coexpression analyses, the authors further demonstrated that immune-related gene networks in the placenta significantly correlated with elevated early childhood anxiety problems and hyperactivity (Rompala et al. 2021).

While these studies suggest that cannabinoids exert largely comparable effects—primarily anti-inflammatory activity—in both non-pregnant and pregnant individuals, caution is warranted when extrapolating findings from non-pregnant individuals. Animal studies indicate that pregnancy itself can modify the effects of cannabinoids on immune responses. For example, one study found that pregnant mice exhibited increased CB1 and decreased CB2 expression in splenic B cells compared to non-pregnant mice. Furthermore, the downstream effects were distinct: in B cells isolated from non-pregnant mice, CB1 agonist treatment reduced IL-10 production following lipopolysaccharide stimulation, whereas in B cells from pregnant mice, the same treatment significantly increased IL-10 production (Wolfson et al. 2016).

Beyond their effects on maternal and placental tissues, animal studies suggest that prenatal cannabis exposure can also impact offspring immunity. For example, just one intraperitoneal injection of 20 or 50 mg/kg THC exposure on gestational day 16 in mice led to significant thymic atrophy and altered T cell subpopulations in fetal thymuses, with effects persisting in pups through postnatal day 11 (Lombard et al. 2011). Intriguingly, these immune effects may be long-lasting; in a study where pregnant rats were exposed to the synthetic cannabinoid HU-210, adult male offspring exhibited a reduction in T-helper cells in the spleen and a dose-dependent decrease in the T-helper/cytotoxic T cell ratio in peripheral blood (del Arco et al. 2000). Given the essential roles of cytokines and T cells in early brain development and function, these findings suggest that at least some long-term effects of prenatal cannabis exposure on brain function may be mediated by immune system alterations. For instance, reduced expression of IL-1β caused by prenatal cannabis exposure could disrupt neuronal differentiation, migration, and axonal growth, affecting connectivity (Accogli et al. 2020). Similarly, the inhibition of effector T cell functions might impair microglial maturation and their critical role in synaptic pruning, potentially contributing to the enlarged PFC observed in offspring (Kolk and Rakic 2022). However, these mechanisms remain speculative and require further study.

The effects of prenatal cannabis could also be mediated by its direct actions on microglia. While the specific effects of cannabis exposure on microglial function during prenatal brain development have, to our knowledge, not yet been directly studied, evidence from in vitro studies and animal models in adults suggests that cannabinoids, through activation of CB1 and/or CB2 receptors, generally suppress microglial activation and promote an anti-inflammatory (M2-like) phenotype (Araujo et al. 2019; Duffy et al. 2021; Ferrisi et al. 2023). This raises important questions about how prenatal cannabis exposure might influence microglia during early brain development—a critical period when microglia, often through pro-inflammatory actions, play essential roles in processes such as synaptic pruning, neuronal migration, and axonal growth (Thion et al. 2018). However, caution is warranted when extrapolating findings from adult models, as the effects of cannabinoids on microglia appear to vary depending on the developmental stage. For example, rather than promoting an anti-inflammatory phenotype, a recent study found that adolescent THC exposure induces microglial apoptosis in the medial PFC (Hasegawa et al. 2023). This, in turn, triggered *reduced* excitability of pyramidal-tract neurons in the PFC—a particularly intriguing observation given that prenatal cannabis exposure has been associated with *increased* neuronal activity in the PFC (Smith et al. 2004, 2006), suggesting that the timing of exposure may critically influence the direction and nature of these outcomes.

Collectively, these findings highlight the complex and multifaceted effects of prenatal cannabis exposure on maternal and offspring health, particularly through its interactions with the immune system. However, it is important to note that the effects of cannabis on immune responses appear highly dependent on the method of administration. For instance, a recent study (Black et al. 2023) comparing cytokine profiles in the placenta and fetal brains following maternal cannabis exposure via injection versus inhalation revealed significant differences. Pro-inflammatory cytokines such as TNF-α and IL-1β were elevated in fetal brains after maternal THC injection but decreased following high-THC smoke exposure. These findings underscore the importance of addressing variations in administration methods to better understand how cannabis impacts immune and developmental outcomes—a critical step toward improving the translational relevance of animal research.

Lastly, we want to highlight that, to the best of our knowledge, no studies have yet explored the potential role of pathogen exposure in mediating the long-term effects of cannabis use during pregnancy. Given the immunosuppressive properties of cannabinoids (Hernández-Cervantes et al. 2017), it is plausible that cannabis use could increase maternal susceptibility to infections. Such heightened vulnerability might influence prenatal development by altering maternal immune responses and increasing the exposure of maternal and fetal tissues to pathogens, which themselves can disrupt developmental processes. Addressing this gap in knowledge is essential to fully understand the interplay between cannabis use, immune modulation, and pathogen exposure, and how these factors collectively shape maternal and offspring health outcomes.

In sum, the evidence underscores the urgent need for a deeper understanding of how cannabis affects immune-mediated processes critical for neurodevelopment. Cannabinoids exert diverse effects on immune cells, including anti-inflammatory and immunosuppressive activities, but these effects can vary significantly depending on physiological states such as pregnancy, as well as the method of administration. Future research must systematically investigate these variables to elucidate the mechanisms by which prenatal cannabis exposure shapes maternal and offspring health.

11.3.2 Micro- and Nanoplastics

Plastic products are used worldwide due to their durability, light weight, resistance to degradation, strength, affordability, and insulating properties against heat and electricity. Today, around 45 types of plastics, including polypropylene, polyethylene, polyethylene terephthalate, polystyrene, polyurethane, polyvinyl chloride, and polycarbonate, are commercially available (Sangkham et al. 2022; Thompson et al. 2024). The extensive use of plastics has introduced serious environmental challenges, as plastic waste accumulates in ecosystems with limited recycling success. In 2019, only 9% of the 460 million tons produced globally were recycled, and annual production is projected to keep growing and reach 1.2 billion tons by 2060 (OECD 2022).

A growing concern related to plastic pollution is the formation of microplastics (MPs; 1 μm–5 mm) and nanoplastics (NPs; 1 nm–1 μm), which primarily emerge as larger plastic items break down over time (Thompson et al. 2004). However, other sources, including textile fibers (Acharya et al. 2021), cosmetic products (Kukkola et al. 2024; Zhou et al. 2023), paints (Gaylarde et al. 2021), infill from sports fields (Armada et al. 2022), mechanical recycling processes (Suzuki et al. 2024), plastic-coated fertilizers and agricultural mulch films (Hofmann et al. 2023), also contribute to microplastic contamination. As a result, MP/NP pollution is now widespread across environments around the globe (Thompson et al. 2024)—even those that are generally considered more remote and pristine (Bergmann et al. 2022; Allen et al. 2019; Kukkola et al. 2022). Because of their size, these particles can easily be ingested, inhaled, and even be absorbed via the skin (Prata et al. 2020), leading to MP/NP accumulating across human and animal tissues (Thompson et al. 2024). MP/NP have now been detected in over 1300 species, spanning aquatic and terrestrial environments, including fish, mammals, birds, and insects (Santos et al. 2021; Kühn and van Franeker 2020; Edo et al. 2021). In recent years, MP/NP have also been reported across human tissues, including blood (Leslie et al. 2022), liver (Horvatits et al. 2022), kidneys (Porta et al. 2023), and brain (Amato-Lourenço et al. 2024; Campen et al. 2024), indicating their ability to cross biological barriers and distribute across human bodily systems.

Importantly, recent findings have shown that MP/NP accumulate in the human amniotic fluid and the placenta (Ragusa et al. 2021, 2022; Halfar et al. 2023), raising critical questions about the potential impacts on maternal and fetal health. As the developing brain is particularly vulnerable to environmental exposures during pregnancy, concerns about the effects of MP/NP on neurodevelopment have intensified (Sharma et al. 2024). Although evidence remains limited, the following section reviews current findings, primarily from animal and in vitro studies, to shed light on how plastic-related pollutants may influence neurodevelopment and neurobehavioral outcomes.

11.3.2.1 Perinatal MP/NP Exposure on Brain Development and Behavior

Studies examining the effects of adult exposure to MP and NP span a range of species, including invertebrates (Lei et al. 2018; Chen et al. 2020; Gambardella et al. 2017), fish (Kashiwada 2006; Mattsson et al. 2017; Mak et al. 2019), and rodents (Liang et al. 2022; Lee et al. 2022; Shan et al. 2022). This body of research has highlighted their neurotoxic potential, revealing their ability to reach brain tissues and disrupt normal brain function. Notably, under physiological conditions, NP seem to penetrate the brain more effectively than MP, likely due to their smaller size and ability to cross biological barriers more readily. Reported neurotoxic effects include oxidative stress, disruptions in neurotransmitter levels and energy metabolism, neuroinflammation, neuronal apoptosis, excitotoxicity, and neurodegeneration. Additionally, exposure to MPs and NPs in adult animals has been associated with alterations in social, motor, and anxiety-like behaviors, as well as impairments in cognitive function (for review, see (Prüst et al. 2020)).

In contrast, evidence on the effects of perinatal MP/NP exposure remains limited, though existing studies consistently indicate significant impacts. Most research has focused on polystyrene NP (PSNPs) administered throughout pregnancy and lactation. Using high-resolution 3D ex vivo magnetic resonance imaging, Harvey et al. (2023) demonstrated that maternal PSNP exposure via drinking water during gestation and lactation in mice led to sex-specific structural alterations across multiple brain regions in mice by postnatal day 23. Affected areas included the olfactory bulbs, motor cortex, hippocampus, hypothalamus, and medulla. These structural changes may relate to oxidative damage-induced neurotoxicity: Chen et al. (2024) reported that oral PSNP administration from gestation to PN21 in rats caused an increase in reactive oxygen species (ROS) within the hippocampus. This ROS elevation activated the P53 pathway, which subsequently triggered NCOA4-mediated ferritinophagy, ultimately leading to neuronal death. Notably, these effects resulted in long-term cognitive impairments.

Interestingly, while MPs and NPs have been detected in placental tissues, a study by Jeong et al. (2022) suggests that PSNPs administered to mothers may not reach the offspring's brain until after birth. By using PSNPs conjugated with yellow-green fluorescence, they tracked maternal transfer, finding that PSNPs were passed to offspring through breast milk rather than directly during gestation. Once in the postnatal brain, PSNP exposure significantly reduced cell proliferation and decreased neuronal layer thickness in hippocampal subregions in one-week-old pups. RNA sequencing analysis linked these effects to the downregulation of genes involved in cell division and proliferation. Notably, only female offspring exhibited persistent cognitive deficits, with increased GABA levels and reduced estrogen receptor α expression in the hippocampus, suggesting a potential sex-specific impact on brain maturation.

However, the effects of prenatal MP and NP exposure on brain development and behavior may occur even without direct brain infiltration. For example, Mercer et al (2023) exposed mice to PSNPs via drinking water throughout gestation and analyzed fetal brains as early as embryonic day 17.5—prior to the PSNPs reaching the brain, as suggested by Jeong et al. (2022) While PSNPs were not directly measured in brain tissues, the study employed magic angle spinning nuclear magnetic resonance spectroscopy and found significant reductions in gamma-aminobutyric acid (GABA), creatine, and glucose levels—decreased by 40%, 21%, and 30%, respectively. Supporting a role for MPs and NPs in affecting brain function without necessarily reaching brain tissue, Zaheer et al (2022) reported that prenatal exposure to PSMPs in mice led to social deficits and repetitive-like behaviors at 5–6 weeks of age. However, while this study assessed brain infiltration in animals exposed during adulthood, it did not evaluate brain infiltration following prenatal exposure. Similarly, a study by So et al (2023) investigating PSMPs found that chronic PSMPs exposure via drinking water from the embryonic stage to adulthood in mice led to impaired social behaviors, particularly in females. Interestingly, the study reported that the PSMPs did not reach the brain. Instead, they primarily accumulated in the digestive and excretory organs, suggesting peripherally induced changes in brain physiology. Nonetheless, it's worth noting that this study assessed PSMPs in the brains of adult animals only, leaving open the question of whether PSMPs might reach the brain transiently during earlier developmental stages. Other studies have shown that when NP are administered at PN1, but not at PN28, they can be detected in the brain by day 28, but are cleared by day 60, suggesting a possible mechanism of early susceptibility to brain penetration followed by subsequent elimination (Zou et al. 2024).

While more studies are needed to determine whether similar effects occur in humans, these studies in rodents collectively highlight the possibility that MPs and NPs may influence brain development and behavior, and that this may occur through indirect mechanisms. Below we will explore the possibility that the immune system plays a mediating role.

Before delving into the impacts specifically of MPs and NPs on the immune system, it is important to note that evidence suggests that additives commonly found in MPs and NPs may independently contribute to neurodevelopmental conditions in humans. In fact, some studies argue that health concerns may stem more from these additives than from the plastics themselves (Han et al. 2020). One such additive, bisphenol A (BPA), has been linked to neurodevelopmental conditions such as autism. Increased maternal BPA levels have been associated with higher ASD symptom severity (Hansen et al. 2021; Lim et al. 2017), and children with autism have been found to have 2–3 times higher urinary BPA levels compared to neurotypical children (Stein et al. 2015). Similarly, phthalates—another common additive widely present in the environment (Henkel et al. 2022)—have been associated with increased maternal anxiety and depression, as well as social abnormalities in children (Cohen-Eliraz et al. 2023).

11.3.2.2 Perinatal MP/NP as Immunomodulators

Although still in its early stages, research on the immune effects of MPs and NPs suggests a consistent trend of immune system disruption across species and model systems (for review, see (Yang et al. 2022)). In adult animals, the primary immune effects of MPs and NPs have been observed through interactions with innate immune cells, such as macrophages and neutrophils, leading to increased production of inflammatory cytokines—including IL-1α, IL-1β, TNF-α, and IL-6—along with heightened oxidative stress and immunotoxicity (Jin et al. 2018; Xuan et al. 2024; Li et al. 2020a; 2024; Park et al. 2024; Florance et al. 2021). Although less studied, accumulating evidence indicates that the adaptive immune system is also affected, with the main findings being reductions in CD8+ and CD4+ T cell populations and altered B lymphocyte function (Li et al. 2020a, 2022, 2024).

However, the precise mechanisms by which MPs and NPs affect immune cells remain incompletely understood. This complexity arises, in part, because factors such as particle size, surface charge, and plastic type significantly influence how MPs and NPs interact with the immune system (Yang et al. 2022). Additionally, MPs and NPs can bind to various biomolecules due to their high surface-area-to-mass ratio and small size. In particular, interactions with proteins can form polystyrene-protein complexes, resulting in a protein coat, commonly referred to as a "corona" (Lundqvist et al. 2008). This biomolecular corona provides MPs and NPs with a new biological identity, modifying how they are recognized by the immune system—sometimes masking them from immune cells or, conversely, making them more immunogenic (Gopinath et al. 2019). Moreover, in addition to plastic additives, that can themselves affect immune function (Kharrazian 2014; Xu et al. 2016; Zhang et al. 2022a), MPs and NPs often act as carriers for environmental toxins, heavy metals, and other pollutants (Mozafarjalali et al. 2023; Rafa et al. 2024), which could further modify their immune impact.

Within the context of prenatal exposures, emerging evidence indicates that MP/NP can affect both maternal and fetal immune health. For example, Park et al. (2020) found that, in mice, daily oral administration of 2 mg of polyethylene MPs (PE-MPs) by gavage throughout pregnancy and lactation led to observable immune changes in both dams and their offspring. Compared to the vehicle group, PE-MP-exposed dams exhibited increased neutrophil but reduced lymphocyte counts in the blood, along with heightened degranulation of stomach mast cells, indicating an overall increase in mucosal and systemic inflammatory responses. Additionally, although the proportions of T cells, B cells, and NK cells in the dams' spleens were not significantly different between groups, the ratio of helper T cells to cytotoxic T cells (CD4+/CD8+) increased with dose, suggesting a shift in T cell balance. In the offspring, this study found sex-specific immune effects at PN21. The total number T cells within the spleens of female pups significantly decreased compared to controls, while the proportion of helper T cells increased in both male and female pups. Furthermore, the maturation of dendritic cells (CD11b−/CD11c+) was notably inhibited in male pups but enhanced in female pups, indicating differential immune impacts based on sex.

Intriguingly, recent studies suggest that microglia may be particularly vulnerable to perinatal NP/MP exposure, with effects on microglial function depending on the timing of exposure. For example, Tian et al. (2024) found that in rats, compared to vehicle controls, oral PSNP exposure during pregnancy led to reductions in dopamine and its primary metabolite, dihydroxyphenylacetic acid (DOPAC), alongside increases in norepinephrine metabolites in the cortex of offspring. These neurochemical changes—absent in mice exposed exclusively postnatally—were accompanied by cortical thinning, elevated cell apoptosis, and an increase in microglia numbers. These findings suggest that microglia are particularly sensitive to activation resulting from prenatal PS-NP exposure, which may drive excessive microglia-induced cortical cell death (Lull and Block 2010; Madrigal et al. 2005).

In contrast, postnatal exposures appear to exert distinct effects on microglia, which may be associated with the findings by Jeong et al. (2022) that PSNPs administered to mothers may not reach the offspring's brain until after birth. For example, Zou et al. (2024) demonstrated that PSMP and PSNP exposure at postnatal days 1 or 14 impaired microglia energy metabolism. This impairment was attributed to the active phagocytosis of plastic particles by microglia, as these were detected within microglia but not in other brain cell types. These effects were further accompanied by increased synaptic density in the cortex, hippocampus, and striatum, suggesting that altered microglia metabolism induced by postnatal PSMP and PSNP exposure impairs critical microglia synaptic pruning functions (Dziabis and Bilbo 2022).

Collectively, these findings underscore the complex and potentially profound effects of MPs and NPs on both immune and neural systems, particularly during critical developmental periods. The evidence suggests that MPs and NPs may influence neurodevelopment not only through direct interactions with brain tissue but also by modulating immune responses in both maternal and fetal contexts. However, the variability in outcomes based on exposure timing, particle size, and type highlights the need for further research to clarify the underlying mechanisms and determine how these factors influence susceptibility.

A significant challenge in advancing this research lies in controlling plastic exposure in laboratory settings, which can confound experimental findings. Materials such as cages, water bottles, and other equipment may unintentionally introduce additional plastic exposure, complicating interpretations of experimental results. Unfortunately, many studies fail to report these details, hindering the standardization of methodologies and limiting the reliability of conclusions.

Given the increasing prevalence of plastic pollution and the widespread detection of MPs and NPs in human and environmental samples, it is crucial to extend this work to human contexts. Elucidating these pathways is essential for assessing potential risks to human health and shaping effective public health policies that address the growing concerns associated with plastic exposure. By prioritizing transparent reporting in experimental studies and bridging the gap between animal models and real-world human scenarios, future research can provide actionable insights into mitigating the impact of plastic pollution on neurodevelopment and overall health.

11.3.3 Flame Retardants

Flame retardants are chemicals extensively used in consumer products—including textiles, plastics, furnishings, electronics, building materials, and transportation products—to inhibit flame spread and prevent combustion. Their use surged in the twentieth century, driven by increasingly stringent fire safety regulations (Flame Retardants n.d.; Flammable Fabrics Act n.d.). Initially, the most prevalent flame retardants were polybrominated biphenyls (PBBs) and polychlorinated biphenyls (PCBs). However, these chemicals were banned in many industrialized nations due to strong evidence of their toxicity, persistence in both the environment and human tissues, and incidents of accidental poisoning (Birnbaum and Staskal 2004; Carpenter 2006; Fonnum et al. 2006; Silberhorn et al. 1990). In response, industries shifted to other brominated compounds, notably polybrominated diphenyl ethers (PBDEs), hexabromocyclododecane (HBCD), and tetrabromobisphenol A (TBBPA). Since the 1970s, PBDEs became the most widely used flame retardants, but they were phased out in the U.S. market starting in 2004 due to their bioaccumulation and confirmed neurotoxic, thyrotoxic, estrogenic, and carcinogenic effects (Siddiqi et al. 2003; Dodson et al. 2012; Costa and Giordano 2007). Despite the phase-out, PBDEs continue to persist in the environment, with ongoing detections in human tissues that raise both health and environmental concerns (Altarawneh et al. 2020; Akintunde et al. 2023; Marchitti et al. 2017).

With the decline in PBDE use, industries have increasingly adopted organophosphate flame retardants (OPFRs)—including compounds such as tris(2-chloroethyl) phosphate (TCEP), triphenyl phosphate (TPP), and tris(1,3-dichloro-2-propyl) phosphate (TDCPP)—and newer brominated flame retardants (BFRs), particularly decabromodiphenyl ethane (DBDPE).

OPFRs are used widely across various industries, including textiles, chemicals, glues, and printed circuit boards, with global production volumes estimated at approximately 598,422 metric tons annually (Huang et al. 2022). OPFRs can easily enter the environment through volatilization, dissolution, and wear (Marklund et al. 2005; Lai et al. 2015), leading to widespread across environments (Hou et al. 2021; Zhang et al. 2022b; Pantelaki and Voutsa 2020; Xie et al. 2022; Kung et al. 2022) and species (Huang et al. 2019; Bekele et al. 2019; Sala et al. 2019), including human milk, placenta, and blood (Zheng et al. 2021; Ma et al. 2019; Ding et al. 2016; Hou et al. 2020), which is raising further health risk concerns. Accumulating research increasingly reports adverse effects, including neurotoxicity (Sun et al. 2016; Wang and Song 2024), carcinogenesis (Saquib et al. 2022), and endocrine disruption (Meeker and Stapleton 2010; Zhang et al. 2020).

Among the newer BFRs, DBDPE has emerged as one of the most widely used, valued for its stable flame-retardant properties, UV resistance, and cost-effectiveness. It is extensively added to plastics, electronics, textiles, and building materials. Correspondingly, DBDPE production has risen sharply and is projected to continue increasing (Chen et al. 2023; Shen et al. 2019). DBDPE can volatilize from these products, leading to its release into the environment; consequently, DBDPE has been detected globally in air (Yadav et al. 2017; Esplugas et al. 2022), house dust (Stapleton et al. 2008; Wang et al. 2018), water (Wang et al. 2023), and food products (Shi et al. 2016; Zheng et al. 2012). It has also been found in animal and human tissues, including breast milk (Wemken et al. 2020; Wang et al. 2020) and blood (Chen et al. 2019). Emerging studies are raising concerns about DBDPE's potential impact on health, with documented links to hepatotoxicity (Sun et al. 2020), reproductive toxicity (Xue et al. 2022), cardiovascular toxicity (Jing et al. 2019), thyroid disruption (Chen et al. 2019), and developmental impairments (Sun et al. 2016).

Due to the extensive and growing use of OPFRs and newer BFRs, this chapter will focus specifically on these high-production substitutes for PBDEs. We will explore the current evidence regarding their neurodevelopmental effects, including how immune system modulation may play a critical role in mediating these impacts.

11.3.3.1 Perinatal Exposure to OPFRs and Newer BFRs on Brain Development and Behavior

Research on the neurobehavioral impacts of perinatal exposure to contemporary flame retardants is still in its early stages, yet emerging evidence indicates that these chemicals may carry long-term risks for brain development and behavior (see (Doherty et al. 2019a; Vuong et al. 2020) for review). Current human studies have primarily focused on OPFRs, revealing concerning associations with child development outcomes. These studies typically assess maternal exposure by measuring OPFR metabolites in urine during pregnancy and then tracking neurobehavioral outcomes in children at various developmental stages.

For instance, Castorina et al. (2017) found that higher maternal urinary levels of the OPFR metabolite diphenyl phosphate were inversely associated with children's Full-Scale IQ and Working Memory scores at age seven. Similarly, studies by Doherty et al. (2019b, c) showed that elevated maternal urinary concentrations of isopropyl-phenyl phenyl phosphate during the third trimester correlated with lower fine motor skills and early language abilities in children at ages two and three. Additionally, increased levels of the metabolite bis(1,3-dichloro-2-propyl) phosphate were linked to behavioral issues, including heightened Withdrawal, Attention Problems, Depression, Hyperactivity, and Aggression at 36 months. Building on these findings, a recent study by Cheng et al. (2024) revealed that maternal urine levels of bis(1-chloro-2-propyl) phosphate and di-n-butyl phosphate were associated with poorer gross motor, adaptive, and social skills in one-year-old infants.

Animal studies have begun to elucidate possible mechanisms by which these flame retardants impact neurodevelopment, often affecting multiple brain regions. For example, maternal exposure to OPFRs has been shown to affect the development of monoaminergic systems. In one study, exposure to FM550—a commercial flame-retardant mixture containing both brominated and organophosphate compounds—resulted in lengthened serotonergic axons in the forebrain of rats at embryonic day 14, with effects more pronounced in males than females (Rock et al. 2020). Newell et al. (2022) further demonstrated that maternal exposure to OPFRs disrupted the formation of dopaminergic cell clusters, resulting in increased height but reduced width and length at the same embryonic stage in both male and female rats.

Other findings indicate that these neurodevelopmental disruptions may be driven by changes in neurogenesis, axonogenesis, and sex-specific differences. For example, Newell and Patisaul (2023) reported that OPFR exposure reduced the volume of the hippocampal dentate gyrus and decreased the proliferation of neural progenitor cells in male rats. Witchy et al. (2022) found that FM550 exposure in rats altered the expression of genes related to mitochondrial function, synaptic transmission, and axonogenesis in the cortex, with effects more prominent in males.

The mechanisms underlying the increased vulnerability of male offspring to neurodevelopmental disruptions from OPFR and FM550 exposure are not yet fully understood. However, one possibility is that these effects may be linked to the endocrine-disruptive properties of flame retardants. OPFRs have been shown to disrupt hormonal signaling by interfering with androgen synthesis and receptor pathways (Zhang et al. 2020; Chen et al. 2015). Male neurodevelopment is particularly influenced by the organizational effects of androgens during critical prenatal windows, such as the establishment of sexually dimorphic neural circuits and the shaping of synaptic connectivity (McCarthy and Arnold 2011). Thus, disruption of androgen signaling during these periods could leave male brains more vulnerable to exposure to endocrine disrupting chemicals.

Regarding DBDPE, studies remain limited; however, emerging evidence in zebrafish models suggests possible neurotoxic effects. For instance, a study by Yang et al. (2023) in which zebrafish embryos were exposed to DBDPE until 120 h post-fertilization (hpf), resulted in increased swimming speed, altered neurotransmitter levels, and disrupted transcription of neurodevelopment-related genes in larvae, indicating neurodevelopmental toxicity. Intriguingly, the neurotoxic effects of DBDPE in zebrafish may extend across generations. Sun et al. (2023) revealed that when only the parental (F0) zebrafish were exposed to DBDPE, subsequent generations, which were never directly exposed, displayed significant alterations. Specifically, the F2 generation exhibited changes in methylation patterns and transcription of genes involved in calcium signaling, neurotransmission, and glutathione metabolism.

Studies in mammals remain limited, but preliminary findings suggest potential neurodevelopmental consequences. One study in mice found that maternal DBDPE exposure did not affect the growth and development of offspring but did result in behavioral alterations (Shi et al. 2022). Female offspring from DBDPE-exposed mothers exhibited increased motor activity and social behavior deficits, while both male and female offspring demonstrated cognitive memory impairments. These findings highlight the potential for DBDPE to exert lasting neurodevelopmental effects across species, warranting further investigation.

Together, these studies underscore the neurodevelopmental risks posed by perinatal exposure to contemporary flame retardants, including OPFRs and newer BFRs. Despite growing evidence of their detrimental effects on brain structure and function, significant knowledge gaps remain, particularly regarding the specific mechanisms driving these effects. While endocrine disruption is a suspected mediator (Belcher et al. 2014; Liu et al. 2013; Sun et al. 2024), immune system modulation may also play a role, which we explore further in the following section.

11.3.3.2 Perinatal OPFRs and Newer BFRs as Immunomodulators

Emerging research suggests that organophosphate flame retardants (OPFRs) and newer brominated flame retardants (BFRs), like their predecessors, may disrupt immune function. Reported effects include altered cytokine production and signaling across tissues (Almughamsi and Whalen 2016; Dunnick et al. 2017; Meng et al. 2022; Win-Shwe et al. 2022), as well as changes in innate immune cell (Li et al. 2020b; Giles et al. 2024; Canbaz et al. 2017). However, studies specifically investigating their developmental immune impacts remain scarce, with current data focusing primarily on OPFRs and their effects on microglia.

Marinello et al. (2022) demonstrated that exposure to FM550 via subcutaneous injections to dams during gestation, combined with direct postnatal exposure of pups from birth until weaning, resulted in sex-dependent effects on microglial density. Male prairie voles exposed to FM550 exhibited reduced microglial density in the hippocampus, amygdala, and cerebellum, while microglial density in the prefrontal cortex was reduced in both sexes compared to controls. Furthermore, FM550-exposed males, but not females, displayed decreased microglial ramification and fewer processes, suggesting a more reactive or potentially dystrophic state. Although more studies are needed to isolate the effects of prenatal exposure, evidence suggests that postnatal exposures may play a dominant role in microglial alterations. Zhou et al. (Zhong et al. 2020) reported that daily exposure to the OPFR TDCPP via oral gavage from postnatal day 10–38 increased microglial activation in the hippocampus. This activation, marked by an amoeboid-like phenotype, was accompanied by elevated levels of inflammatory markers, including IL-1β and TNFα, as well as hippocampal apoptosis. These findings indicate that neonatal exposure to TDCPP may induce neuronal damage through microglia-mediated inflammation.The observed alterations in microglial activity could contribute to disruptions in OPFR-induced dopaminergic and serotonergic signaling (Thion et al. 2018; She et al. 2024) and impairments in hippocampal neurogenesis (Ueno and Yamashita 2014).

Although no studies have specifically examined DBDPE's effects on the immune system, its structural similarity to other brominated flame retardants, such as PBDEs—known to impact maternal and fetal immune systems (Liu et al. 2012)—suggests that DBDPE might elicit similar immunotoxic effects. Further research is critical to elucidate the immune-mediated pathways through which perinatal flame-retardant exposure influences neurodevelopment.

11.4 Conclusions

In this chapter, we examined the growing body of evidence on the neurodevelopmental impacts of substances such as cannabis, micro- and nanoplastics, and flame retardants, increasingly detected in maternal and fetal tissues—particularly through their potential to disrupt immune processes. Cannabis use during pregnancy, increasingly common due to shifting social and legal norms, has been linked to altered brain connectivity and behavioral impairments in offspring. Micro- and nanoplastics, now widely detected in the environment and even in human tissues, present additional risks by potentially interfering with cellular function and promoting neuroinflammation. Similarly, newer flame retardants have been associated with cognitive and behavioral deficits in children exposed in utero. Collectively, these findings underscore the need for further research to clarify the mechanisms by which these substances impact neurodevelopment and to assess their full implications for human health.

Cannabis, MP/NP, and flame retardants such as OPFRs and newer BFRs represent distinct classes of non-infectious environmental exposures with unique chemical properties, yet emerging evidence suggests areas of intersection in their mechanisms of action on the immune system and brain development. All three have been shown to disrupt immune function, which may act as a critical intermediary in their neurodevelopmental impacts. Cannabinoids modulate immune responses through cannabinoid receptors (CB1 and CB2), largely leading to immunosuppressive effects, such as reduced pro-inflammatory cytokines (e.g., IL-1β, TNF-α) and altered T-cell and B-cell activity. In contrast, MPs/NPs and flame retardants are associated with mostly pro-inflammatory responses and changes in cytokine production but can also impair adaptive immune functions, such as altering T-cell populations or inducing microglial activation. This immune dysregulation could, in turn, influence neurodevelopmental processes such as synaptic pruning, neurogenesis, and neuronal migration.

Particularly, cannabis, plastics, and flame retardants appear to affect microglia—albeit through varying pathways. Cannabis generally promotes an anti-inflammatory (M2-like) microglial phenotype, potentially dampening microglial activation. Conversely, MPs/NPs and flame retardants have been shown to activate microglia, leading to neuroinflammation. Despite these differences, the net result of microglial dysregulation from these contaminants can manifest as impaired synaptic pruning and abnormal neurodevelopmental outcomes, including alterations in connectivity, neurotransmitter systems (e.g., dopaminergic and serotonergic pathways), and behavioral deficits.

An interesting observation is the persistent sex-specific effects across substance exposures. For example, cannabis has shown sex-dependent impacts on PFC activity in rodents, with males exhibiting heightened excitability of PFC pyramidal neurons, while females are relatively unaffected or show opposing trends. Flame retardants like FM550 also exhibit sex-specific effects, with male prairie voles showing reduced microglial density in multiple brain regions, including the hippocampus, amygdala, and cerebellum, as well as decreased microglial ramification and fewer processes, indicative of a more reactive or dystrophic state. In contrast, female offspring exhibit fewer or less pronounced alterations. MPs/NPs further display sex-specific impacts, as demonstrated in studies reporting immune disruptions, such as differential T-cell maturation and dendritic cell activity, as well as persistent cognitive and behavioral deficits primarily in females. These results highlight the pivotal need to always include both sexes in every study...

Additionally, the potential for synergistic effects among contaminants raises important questions. In real-world settings, exposures rarely occur in isolation—combined exposures to various contaminants may amplify their developmental impacts. For example, simultaneous exposure to flame retardants and cannabis could potentially create additive or synergistic effects on immune dysregulation, compounding the risks to the developing brain. Moreover, early-life exposure to these substances may prime or sensitize the brain to other stressors later in life, increasing susceptibility to neurodevelopmental and psychiatric conditions. Future studies employing approaches such as the "two-hit" model—in which initial exposure to environmental contaminants is followed by exposures to additional stressors later in life—could offer valuable insights into these complex, long-term effects.

In conclusion, while we have made strides in identifying the hazards associated with flame retardants, cannabis, and micro/nanoplastics, significant knowledge gaps remain. Addressing these gaps is essential for advancing public health protections, particularly for populations most vulnerable to toxic exposures, such as pregnant women and developing fetuses. Effective regulation and public health interventions will require a robust evidence base that reflects the multifactorial, dynamic nature of environmental exposures and their impacts on neurodevelopment. By expanding research efforts in these directions, we can move toward a future where policy and practice are more attuned to safeguarding the developmental health of future generations in an increasingly polluted world.

References

Accogli A, Addour-Boudrahem N, Srour M (2020) Chapter 4: Neurogenesis, neuronal migration, and axon guidance. In: Gallagher A, Bulteau C, Cohen D, Michaud JL (eds) Handbook of clinical neurology, vol 173. Elsevier, pp 25–42

Acharya S, Rumi SS, Hu Y, Abidi N (2021) Microfibers from synthetic textiles as a major source of microplastics in the environment: a review. Text Res J 91:2136–2156

Afifa, Arshad K, Hussain N, Ashraf MH, Saleem MZ (2024) Air pollution and climate change as grand challenges to sustainability. Sci Total Environ 928:172370

Ajami B, Bennett JL, Krieger C, Tetzlaff W, Rossi FMV (2007) Local self-renewal can sustain CNS microglia maintenance and function throughout adult life. Nat Neurosci 10:1538–1543

Akdemir ES, Huang AY-S, Deneen B (2020) Astrocytogenesis: where, when, and how. F1000Res 9:F1000 Faculty Rev-233

Akintunde ME et al (2023) Ex vivo exposure to polybrominated diphenyl ether (PBDE) selectively affects the immune response in autistic children. Brain Behav Immun Health 34:100697

Akira S, Takeda K (2004) Toll-like receptor signalling. Nat Rev Immunol 4:499–511

Allen S et al (2019) Atmospheric transport and deposition of microplastics in a remote mountain catchment. Nat Geosci 12:339–344

Alliot F, Godin I, Pessac B (1999) Microglia derive from progenitors, originating from the yolk sac, and which proliferate in the brain. Dev Brain Res 117:145–152

Almogi-Hazan O, Or R (2020) Cannabis, the endocannabinoid system and immunity—the journey from the bedside to the bench and Back. Int J Mol Sci 21:4448

Almughamsi H, Whalen MM (2016) Hexabromocyclododecane and tetrabromobisphenol A alter secretion of interferon gamma (IFN-γ) from human immune cells. Arch Toxicol 90:1695–1707

Altarawneh M, Saeed A, Siddique K, Jansson S, Dlugogorski BZ (2020) Formation of polybrominated dibenzofurans (PBDFs) and polybrominated diphenyl ethers (PBDEs) from oxidation of brominated flame retardants (BFRs). J Hazard Mater 400:123166

Amato-Lourenço LF et al (2024) Microplastics in the olfactory bulb of the human brain. JAMA Netw Open 7:e2440018

Araujo DJ, Tjoa K, Saijo K (2019) The endocannabinoid system as a window into microglial biology and its relationship to autism. Front Cell Neurosci 13:424

Armada D et al (2022) Global evaluation of the chemical hazard of recycled tire crumb rubber employed on worldwide synthetic turf football pitches. Sci Total Environ 812:152542

Ashford K et al (2019) Associations of first trimester co-use of tobacco and Cannabis with prenatal immune response and psychosocial well-being. Neurotoxicol Teratol 73:42–48

Avalos LA et al (2024) Early maternal prenatal cannabis use and child developmental delays. JAMA Netw Open 7:e2440295

Baggio S et al (2014) Routes of administration of cannabis used for nonmedical purposes and associations with patterns of drug use. J Adolesc Health 54:235–240

Bara A et al (2018) Sex-dependent effects of in utero cannabinoid exposure on cortical function. elife 7:e36234

Bara A, Ferland J-MN, Rompala G, Szutorisz H, Hurd YL (2021) Cannabis and synaptic reprogramming of the developing brain. Nat Rev Neurosci 22:423–438

Beatty JR, Svikis DS, Ondersma SJ (2012) Prevalence and perceived financial costs of marijuana versus tobacco use among urban low-income pregnant women. J Addict Res Ther 3:1000135

Bekele TG, Zhao H, Wang Q, Chen J (2019) Bioaccumulation and trophic transfer of emerging organophosphate flame retardants in the marine food webs of Laizhou Bay, North China. Environ Sci Technol 53:13417–13426

Belcher S, Cookman CJ, Patisaul H, Stapleton H (2014) In vitro assessment of human nuclear hormone receptor activity and cytotoxicity of the flame retardant mixture FM 550 and its triarylphosphate and brominated components. Toxicol Lett 228(2):93–102

Bennett ML, Bennett FC (2020) The influence of environment and origin on brain resident macrophages and implications for therapy. Nat Neurosci 23:157–166

Bennett ML et al (2016) New tools for studying microglia in the mouse and human CNS. Proc Natl Acad Sci USA 113:E1738–E1746

Bergmann M et al (2022) Plastic pollution in the Arctic. Nat Rev Earth Environ 3:323–337

Bilbo SD, Schwarz JM (2012) The immune system and developmental programming of brain and behavior. Front Neuroendocrinol 33:267–286

Birnbaum LS, Staskal DF (2004) Brominated flame retardants: cause for concern? Environ Health Perspect 112:9

Black T et al (2023) Characterization of cannabinoid plasma concentration, maternal health, and cytokine levels in a rat model of prenatal cannabis smoke exposure. Sci Rep 13:21070

Bolhuis K et al (2018) Maternal and paternal cannabis use during pregnancy and the risk of psychotic-like experiences in the offspring. Schizophr Res 202:322–327

Borsini A, Zunszain PA, Thuret S, Pariante CM (2015) The role of inflammatory cytokines as key modulators of neurogenesis. Trends Neurosci 38:145–157

Braillon A, Bewley S (2017) Committee Opinion No. 722: marijuana use during pregnancy and lactation. Obstet Gynecol 130:e205

Brown AS et al (2004) Serologic evidence of prenatal influenza in the etiology of schizophrenia. Arch Gen Psychiatry 61:774–780

Brown AS et al (2005) Maternal exposure to toxoplasmosis and risk of schizophrenia in adult offspring. Am J Psychiatry 162:767–773

Bukiya AN (2019) Physiology of the endocannabinoid system during development. In: Bukiya AN (ed) Recent advances in cannabinoid physiology and pathology. Springer International Publishing, Cham, pp 13–37. https://doi.org/10.1007/978-3-030-21737-2_2

Bulloch K et al (2008) CD11c/EYFP transgene illuminates a discrete network of dendritic cells within the embryonic, neonatal, adult, and injured mouse brain. J Comp Neurol 508:687–710

Cabral GA, Rogers TJ, Lichtman AH (2015) Turning over a new leaf: cannabinoid and endocannabinoid modulation of immune function. J Neuroimmune Pharmacol 10:193

Campen M et al (2024) Bioaccumulation of microplastics in decedent human brains assessed by pyrolysis gas chromatography-mass spectrometry. Res Sq. https://doi.org/10.21203/rs.3.rs-4345687/v1

Canbaz D, Logiantara A, van Ree R, van Rijt LS (2017) Immunotoxicity of organophosphate flame retardants TPHP and TDCIPP on murine dendritic cells in vitro. Chemosphere 177:56–64

Carpenter DO (2006) Polychlorinated biphenyls (PCBs): routes of exposure and effects on human health. Rev Environ Health 21:1–23

Castorina R et al (2017) Current-use flame retardants: maternal exposure and neurodevelopment in children of the CHAMACOS cohort. Chemosphere 189:574–580

Castro-Navarro I et al (2024) Maternal cannabis use during lactation and potential effects on human Milk composition and production: a narrative review. Adv Nutr 15:100196

CDC (2024) Data and statistics on autism spectrum disorder. Autism Spectrum Disorder (ASD) https://www.cdc.gov/autism/data-research/index.html

Chang JC et al (2019) Beliefs and attitudes regarding prenatal marijuana use: perspectives of pregnant women who report use. Drug Alcohol Depend 196:14–20

Chen G, Jin Y, Wu Y, Liu L, Fu Z (2015) Exposure of male mice to two kinds of organophosphate flame retardants (OPFRs) induced oxidative stress and endocrine disruption. Environ Toxicol Pharmacol 40:310–318

Chen T et al (2019) Thyroid function and decabromodiphenyl ethane (DBDPE) exposure in Chinese adults from a DBDPE manufacturing area. Environ Int 133:105179

Chen Y, Liu X, Leng Y, Wang J (2020) Defense responses in earthworms (Eisenia fetida) exposed to low-density polyethylene microplastics in soils. Ecotoxicol Environ Saf 187:109788

Chen A et al (2023) Emission and environmental distribution of decabromodiphenyl ethane (DBDPE) in China from 2006 to 2026: retrospection, forecasting, and implications for assessment and management. Environ Pollut 327:121536

Chen J et al (2024) Maternal exposure to nanopolystyrene induces neurotoxicity in offspring through P53-mediated ferritinophagy and ferroptosis in the rat hippocampus. J Nanobiotechnol 22:651

Cheng X et al (2024) Prenatal exposure to a mixture of organophosphate flame retardants and infant neurodevelopment: a prospective cohort study in Shandong, China. Int J Hyg Environ Health 258:114336

Cohen-Eliraz L et al (2023) Prenatal exposure to phthalates and emotional/behavioral development in young children. Neurotoxicology 98:39–47

Concheiro M et al (2021) Assessment of biological matrices for the detection of in utero cannabis exposure. Drug Test Anal 13:1371–1382

Correa F, Wolfson ML, Valchi P, Aisemberg J, Franchi AM (2016) Endocannabinoid system and pregnancy. Reproduction 152:R191–R200

Corsi DJ et al (2019) Association between self-reported prenatal cannabis use and maternal, perinatal, and neonatal outcomes. JAMA 322:145–152

Costa LG, Giordano G (2007) Developmental neurotoxicity of Polybrominated Diphenyl Ether (PBDE) flame retardants. Neurotoxicology 28:1047–1067

Cunningham CL, Martínez-Cerdeño V, Noctor SC (2013) Microglia regulate the number of neural precursor cells in the developing cerebral cortex. J Neurosci 33:4216–4233

Day NL et al (1994) Effect of prenatal marijuana exposure on the cognitive development of offspring at age three. Neurotoxicol Teratol 16:169–175

de la Mano A et al (2007) Role of interleukin-1beta in the control of neuroepithelial proliferation and differentiation of the spinal cord during development. Cytokine 37:128–137

de Salas-Quiroga A et al (2015) Prenatal exposure to cannabinoids evokes long-lasting functional alterations by targeting CB1 receptors on developing cortical neurons. Proc Natl Acad Sci USA 112:13693–13698

Deivasigamani S et al (2023) Microglia complement signaling promotes neuronal elimination and normal

brain functional connectivity. Cereb Cortex (New York, NY) 33:10750
del Arco I et al (2000) Maternal exposure to the synthetic cannabinoid HU-210: effects on the endocrine and immune systems of the adult male offspring. Neuroimmunomodulation 7:16–26
Deverman BE, Patterson PH (2009) Cytokines and CNS development. Neuron 64:61–78
Díaz-Alonso J, Guzmán M, Galve-Roperh I (2012) Endocannabinoids via CB1 receptors act as neurogenic niche cues during cortical development. Philos Trans R Soc Lond Ser B Biol Sci 367:3229–3241
Ding J, Xu Z, Huang W, Feng L, Yang F (2016) Organophosphate ester flame retardants and plasticizers in human placenta in Eastern China. Sci Total Environ 554–555:211–217
DiNieri JA et al (2011) Maternal cannabis use alters ventral striatal dopamine D2 gene regulation in the offspring. Biol Psychiatry 70:763–769
Dodson RE et al (2012) After the PBDE phase-out: a broad suite of flame retardants in repeat house dust samples from California. Environ Sci Technol 46:13056–13066
Doherty B, Hammel S, Daniels J, Stapleton H, Hoffman K (2019a) Organophosphate esters: are these flame retardants and plasticizers affecting children's health? Curr Environ Health Rep 6:201–213
Doherty B et al (2019b) Prenatal exposure to organophosphate esters and cognitive development in young children in the Pregnancy, Infection, and Nutrition Study. Environ Res 169:33–40
Doherty B et al (2019c) Prenatal exposure to organophosphate esters and behavioral development in young children in the Pregnancy, Infection, and Nutrition Study. Neurotoxicology 73:150–160
Duffy SS, Hayes JP, Fiore NT, Moalem-Taylor G (2021) The cannabinoid system and microglia in health and disease. Neuropharmacology 190:108555
Dunnick JK et al (2017) Tetrabromobisphenol A activates the hepatic interferon pathway in rats. Toxicol Lett 266:32–41
Dziabis JE, Bilbo SD (2022) Microglia and sensitive periods in brain development. Curr Top Behav Neurosci 53:55–78
Edlow AG (2016) Maternal obesity and neurodevelopmental and psychiatric disorders in offspring. Prenat Diagn 37:95
Edo C et al (2021) Honeybees as active samplers for microplastics. Sci Total Environ 767:144481
El Marroun H et al (2016) Prenatal cannabis and tobacco exposure in relation to brain morphology: a prospective neuroimaging study in young children. Biol Psychiatry 79:971–979
Esplugas R et al (2022) Emerging and legacy flame retardants in indoor air and dust samples of Tarragona Province (Catalonia, Spain). Sci Total Environ 806:150494
Fang P et al (2018) Immune cell subset differentiation and tissue inflammation. J Hematol Oncol 11:97
Faraj MM et al (2022) Impact of prenatal cannabis exposure on functional connectivity of the salience network in children. J Neurosci Res 101:162
Ferrisi R et al (2023) The interplay between cannabinoid receptors and microglia in the pathophysiology of Alzheimer's disease. J Clin Med 12:7201
Filipello F et al (2018) The microglial innate immune receptor TREM2 is required for synapse elimination and normal brain connectivity. Immunity 48:979–991.e8
Flame Retardants (n.d.) National Institute of Environmental Health Sciences. https://www.niehs.nih.gov/health/topics/agents/flame_retardants
Flammable Fabrics Act (n.d.) U.S. Consumer Product Safety Commission. https://www.cpsc.gov/Regulations-Laws%2D%2DStandards/Statutes/Flammable-Fabrics-Act
Florance I, Ramasubbu S, Mukherjee A, Chandrasekaran N (2021) Polystyrene nanoplastics dysregulate lipid metabolism in murine macrophages in vitro. Toxicology 458:152850
Fonnum F, Mariussen E, Reistad T (2006) Molecular mechanisms involved in the toxic effects of polychlorinated biphenyls (PCBs) and brominated flame retardants (BFRs). J Toxicol Environ Health A 69:21–35
Frade JM, Barde YA (1998) Microglia-derived nerve growth factor causes cell death in the developing retina. Neuron 20:35–41
Franzen JM et al (2023) Cannabidiol attenuates fear memory expression in female rats via hippocampal 5-HT1A but not CB1 or CB2 receptors. Neuropharmacology 223:109316
Frau R et al (2019) Prenatal THC exposure produces a hyperdopaminergic phenotype rescued by pregnenolone. Nat Neurosci 22:1975
Fried PA, Watkinson B (2000) Visuoperceptual functioning differs in 9- to 12-year olds prenatally exposed to cigarettes and marihuana. Neurotoxicol Teratol 22:11–20
Fried PA, Watkinson B (2001) Differential effects on facets of attention in adolescents prenatally exposed to cigarettes and marihuana. Neurotoxicol Teratol 23:421–430
Fried PA, Watkinson B, Gray R (1992) A follow-up study of attentional behavior in 6-year-old children exposed prenatally to marihuana, cigarettes, and alcohol. Neurotoxicol Teratol 14:299–311
Fried PA, Watkinson B, Gray R (1998) Differential effects on cognitive functioning in 9- to 12-year olds prenatally exposed to cigarettes and marihuana. Neurotoxicol Teratol 20:293–306
Friedman NP, Robbins TW (2022) The role of prefrontal cortex in cognitive control and executive function. Neuropsychopharmacology 47:72–89
Fuller R et al (2022) Pollution and health: a progress update. Lancet Planet Health 6:e535–e547
Gabrhelík R et al (2021) Cannabis use during pregnancy and risk of adverse birth outcomes: a longitudinal cohort study. Eur Addict Res 27:131–141

Gallo EF et al (2018) Accumbens dopamine D2 receptors increase motivation by decreasing inhibitory transmission to the ventral pallidum. Nat Commun 9:1086
Gambardella C et al (2017) Effects of polystyrene microbeads in marine planktonic crustaceans. Ecotoxicol Environ Saf 145:250–257
Gaylarde CC, Neto JAB, da Fonseca EM (2021) Paint fragments as polluting microplastics: a brief review. Mar Pollut Bull 162:111847
Ghovanloo M-R et al (2018) Inhibitory effects of cannabidiol on voltage-dependent sodium currents. J Biol Chem 293:16546–16558
Giles BH, Kukolj N, Mann KK, Robaire B (2024) Phenotypic and functional outcomes in macrophages exposed to an environmentally relevant mixture of organophosphate esters in vitro. Environ Health Perspect 132:87002
Goldschmidt L, Richardson GA, Willford J, Day NL (2008) Prenatal marijuana exposure and intelligence test performance at age 6. J Am Acad Child Adolesc Psychiatry 47:254–263
Gopinath PM et al (2019) Assessment on interactive prospectives of nanoplastics with plasma proteins and the toxicological impacts of virgin, coronated and environmentally released-nanoplastics. Sci Rep 9:8860
Gradisnik L, Velnar T (2023) Astrocytes in the central nervous system and their functions in health and disease: a review. World J Clin Cases 11:3385–3394
Gray KA, Day NL, Leech S, Richardson GA (2005) Prenatal marijuana exposure: effect on child depressive symptoms at ten years of age. Neurotoxicol Teratol 27:439–448
Gunn JKL et al (2016) Prenatal exposure to cannabis and maternal and child health outcomes: a systematic review and meta-analysis. BMJ Open 6:e009986
Hackman DA, Farah MJ, Meaney MJ (2010) Socioeconomic status and the brain: mechanistic insights from human and animal research. Nat Rev Neurosci 11:651
Halfar J et al (2023) Microplastics and additives in patients with preterm birth: the first evidence of their presence in both human amniotic fluid and placenta. Chemosphere 343:140301
Han S et al (2020) Surface pattern analysis of microplastics and their impact on human-derived cells. ACS Appl Polym Mater 2:4541–4550
Han VX, Patel S, Jones HF, Dale RC (2021) Maternal immune activation and neuroinflammation in human neurodevelopmental disorders. Nat Rev Neurol 17:564–579
Hansen JB et al (2021) Prenatal exposure to bisphenol A and autistic- and ADHD-related symptoms in children aged 2 and5 years from the Odense Child Cohort. Environ Health 20:24
Harvey NE et al (2023) Maternal exposure to polystyrene nanoplastics impacts developmental milestones and brain structure in mouse offspring. Environ Sci Adv 2:622–628
Hasegawa Y et al (2023) Microglial cannabinoid receptor type 1 mediates social memory deficits in mice produced by adolescent THC exposure and 16p11.2 duplication. Nat Commun 14:6559
He H et al (2020) Trends in the incidence and DALYs of schizophrenia at the global, regional and national levels: results from the Global Burden of Disease Study 2017. Epidemiol Psychiatr Sci 29:e91
Hegde VL et al (2008) Attenuation of experimental autoimmune hepatitis by exogenous and endogenous cannabinoids: involvement of regulatory T cells. Mol Pharmacol 74:20
Henkel C, Hüffer T, Hofmann T (2022) Polyvinyl chloride microplastics leach phthalates into the aquatic environment over decades. Environ Sci Technol 56:14507–14516
Henshaw FR, Dewsbury LS, Lim CK, Steiner GZ (2021) The effects of cannabinoids on pro- and anti-inflammatory cytokines: a systematic review of in vivo studies. Cannabis Cannabinoid Res 6:177
Hernández-Cervantes R, Méndez-Díaz M, Prospéro-García Ó, Morales-Montor J (2017) Immunoregulatory role of cannabinoids during infectious disease. Neuroimmunomodulation 24:183–199
Hernández-Mondragón JC et al (2023) Evidence for the existence of facilitatory interactions between the dopamine D2 receptor and the oxytocin receptor in the amygdala of the rat. Relevance for anxiolytic actions. Front Pharmacol 14:1251922
Hillard CJ (2014) Stress regulates endocannabinoid-CB1 receptor signaling. Semin Immunol 26:380–388
Hofmann T et al (2023) Plastics can be used more sustainably in agriculture. Commun Earth Environ 4:1–11
Horvatits T et al (2022) Microplastics detected in cirrhotic liver tissue. EBioMedicine 82:104147
Hou M, Shi Y, Jin Q, Cai Y (2020) Organophosphate esters and their metabolites in paired human whole blood, serum, and urine as biomarkers of exposure. Environ Int 139:105698
Hou M, Shi Y, Na G, Cai Y (2021) A review of organophosphate esters in indoor dust, air, hand wipes and silicone wristbands: implications for human exposure. Environ Int 146:106261
Huang Q et al (2019) Organophosphate flame retardants in heron eggs from upper Yangtze River basin, Southwest China. Chemosphere 236:124327
Huang J, Ye L, Fang M, Su G (2022) Industrial production of organophosphate flame retardants (OPFRs): big knowledge gaps need to be filled? Bull Environ Contam Toxicol 108:809–818
Irvine N, England-Mason G, Field CJ, Dewey D, Aghajafari F (2022) Prenatal folate and choline levels and brain and cognitive development in children: a critical narrative review. Nutrients 14:364
Jeong B et al (2022) Maternal exposure to polystyrene nanoplastics causes brain abnormalities in progeny. J Hazard Mater 426:127815
Jiang H-Y et al (2016) Maternal infection during pregnancy and risk of autism spectrum disorders: a systematic review and meta-analysis. Brain Behav Immun 58:165–172

Jin Y et al (2018) Polystyrene microplastics induce microbiota dysbiosis and inflammation in the gut of adult zebrafish. Environ Pollut 235:322–329

Jing L et al (2019) Cardiovascular toxicity of decabrominated diphenyl ethers (BDE-209) and decabromodiphenyl ethane (DBDPE) in rats. Chemosphere 223:675–685

Kashiwada S (2006) Distribution of nanoparticles in the see-through Medaka (Oryzias latipes). Environ Health Perspect 114:1697–1702

Khalil M et al (2007) Brain mast cell relationship to neurovasculature during development. Brain Res 1171:18–29

Kharrazian D (2014) The potential roles of Bisphenol A (BPA) pathogenesis in autoimmunity. Autoimmune Dis 2014:743616

Kierdorf K et al (2013) Microglia emerge from erythromyeloid precursors via Pu.1- and Irf8-dependent pathways. Nat Neurosci 16:273

Kolk SM, Rakic P (2022) Development of prefrontal cortex. Neuropsychopharmacology 47:41–57

Kozakiewicz ML, Grotegut CA, Howlett AC (2021) Endocannabinoid system in pregnancy maintenance and labor: a mini-review. Front Endocrinol (Lausanne) 12:699951

Kühn S, van Franeker JA (2020) Quantitative overview of marine debris ingested by marine megafauna. Mar Pollut Bull 151:110858

Kukkola AT et al (2022) A large-scale study of microplastic abundance in sediment cores from the UK continental shelf and slope. Mar Pollut Bull 178:113554

Kukkola A, Chetwynd AJ, Krause S, Lynch I (2024) Beyond microbeads: examining the role of cosmetics in microplastic pollution and spotlighting unanswered questions. J Hazard Mater 476:135053

Kummer KK, Zeidler M, Kalpachidou T, Kress M (2021) Role of IL-6 in the regulation of neuronal development, survival and function. Cytokine 144:155582

Kung H-C, Hsieh Y-K, Huang B-W, Cheruiyot NK, Chang-Chien G-P (2022) An overview: organophosphate flame retardants in the atmosphere. Aerosol Air Qual Res 22:220148

Lai S et al (2015) Occurrence and dry deposition of organophosphate esters in atmospheric particles over the northern South China Sea. Chemosphere 127:195–200

Lampron M-C, Paré I, Al-Zharani M, Semlali A, Loubaki L (2023) Cannabinoid mixture affects the fate and functions of B cells through the modulation of the caspase and MAP kinase pathways. Cells 12:588

Latuskie KA et al (2019) Reasons for substance use continuation and discontinuation during pregnancy: a qualitative study. Women Birth 32:e57–e64

Lee C-W et al (2022) Exposure to polystyrene microplastics impairs hippocampus-dependent learning and memory in mice. J Hazard Mater 430:128431

Leech SL, Richardson GA, Goldschmidt L, Day NL (1999) Prenatal substance exposure: effects on attention and impulsivity of 6-year-olds. Neurotoxicol Teratol 21:109–118

Lei L et al (2018) Polystyrene (nano)microplastics cause size-dependent neurotoxicity, oxidative damage and other adverse effects in Caenorhabditis elegans. Environ Sci Nano 5:2009–2020

Lenz KM, Nelson LH (2018) Microglia and beyond: innate immune cells as regulators of brain development and behavioral function. Front Immunol 9:698

Leslie HA et al (2022) Discovery and quantification of plastic particle pollution in human blood. Environ Int 163:107199

Levy NS, Mauro PM, Mauro CM, Segura LE, Martins SS (2021) Joint perceptions of the risk and availability of cannabis in the United States, 2002–2018. Drug Alcohol Depend 226:108873

Li B et al (2020a) Polyethylene microplastics affect the distribution of gut microbiota and inflammation development in mice. Chemosphere 244:125492

Li X, Li N, Rao K, Huang Q, Ma M (2020b) In vitro immunotoxicity of organophosphate flame retardants in human THP-1-derived macrophages. Environ Sci Technol 54:8900–8908

Li Y et al (2022) In vitro study on the toxicity of nanoplastics with different charges to murine splenic lymphocytes. J Hazard Mater 424:127508

Li L et al (2024) Chronic exposure to polystyrene nanoplastics induces intestinal mechanical and immune barrier dysfunction in mice. Ecotoxicol Environ Saf 269:115749

Liang B et al (2022) Brain single-nucleus transcriptomics highlights that polystyrene nanoplastics potentially induce Parkinson's disease-like neurodegeneration by causing energy metabolism disorders in mice. J Hazard Mater 430:128459

Lim Y-H et al (2017) Prenatal and postnatal bisphenol A exposure and social impairment in 4-year-old children. Environ Health 16:79

Liu X, Zhan H, Zeng X, Zhang C, Chen D (2012) The PBDE-209 exposure during pregnancy and lactation impairs immune function in rats. Mediat Inflamm 2012:692467

Liu X, Ji K, Jo A, Moon H-B, Choi K (2013) Effects of TDCPP or TPP on gene transcriptions and hormones of HPG axis, and their consequences on reproduction in adult zebrafish (Danio rerio). Aquat Toxicol 134–135:104–111

Liu C, Goel P, Kaeser PS (2021) Spatial and temporal scales of dopamine transmission. Nat Rev Neurosci 22:345–358

Lombard C, Hegde VL, Nagarkatti M, Nagarkatti PS (2011) Perinatal exposure to Δ9-tetrahydrocannabinol triggers profound defects in T cell differentiation and function in fetal and postnatal stages of life, including decreased responsiveness to HIV antigens. J Pharmacol Exp Ther 339:607–617

Lu H-C, Mackie K (2016) An introduction to the endogenous cannabinoid system. Biol Psychiatry 79:516–525

Lull ME, Block ML (2010) Microglial activation and chronic neurodegeneration. Neurotherapeutics 7:354–365

Lundqvist M et al (2008) Nanoparticle size and surface properties determine the protein corona with possible implications for biological impacts. Proc Natl Acad Sci 105:14265–14270
Ma L et al (2014) Interleukin-1 beta guides the migration of cortical neurons. J Neuroinflammation 11:114
Ma J, Zhu H, Kannan K (2019) Organophosphorus flame retardants and plasticizers in breast milk from the United States. Environ Sci Technol Lett 6:525
Maccarrone M, Guzman M, Mackie K, Doherty P, Harkany T (2014) Programming and reprogramming neural cells by (endo-) cannabinoids: from physiological rules to emerging therapies. Nat Rev Neurosci 15:786
Madrigal JLM, Feinstein DL, Russo CD (2005) Norepinephrine protects cortical neurons against microglial-induced cell death. J Neurosci Res 81:390–396
Mak CW, Ching-Fong Yeung K, Chan KM (2019) Acute toxic effects of polyethylene microplastic on adult zebrafish. Ecotoxicol Environ Saf 182:109442
Mann JR, McDermott S (2011) Are maternal genitourinary infection and pre-eclampsia associated with ADHD in school-aged children? J Atten Disord 15:667–673
Marchetti D, Di Masi G, Cittadini F, La Monaca G, De Giovanni N (2017) Placenta as alternative specimen to detect in utero cannabis exposure: a systematic review of the literature. Reprod Toxicol 73:250–258
Marchitti SA, Fenton SE, Mendola P, Kenneke JF, Hines EP (2017) Polybrominated diphenyl ethers in human milk and serum from the U.S. EPA MAMA Study: modeled predictions of infant exposure and considerations for risk assessment. Environ Health Perspect 125:706–713
Marinello WP et al (2022) Effects of developmental exposure to FireMaster® 550 (FM 550) on microglia density, reactivity and morphology in a prosocial animal model. Neurotoxicology 91:140–154
Marín-Teva JL et al (2004) Microglia promote the death of developing Purkinje cells. Neuron 41:535–547
Mark K, Gryczynski J, Axenfeld E, Schwartz RP, Terplan M (2017) Pregnant Women's current and intended cannabis use in relation to their views toward legalization and knowledge of potential harm. J Addict Med 11:211
Marklund A, Andersson B, Haglund P (2005) Traffic as a source of organophosphorus flame Retardants and plasticizers in snow. Environ Sci Technol 39:3555–3562
Martin CE, Longinaker N, Mark K, Chisolm MS, Terplan M (2015) Recent trends in treatment admissions for marijuana use during pregnancy. J Addict Med 9:99–104
Mastenbroek LJM, Kooistra SM, Eggen BJL, Prins JR (2024) The role of microglia in early neurodevelopment and the effects of maternal immune activation. Semin Immunopathol 46:1
Mato S, Victoria Sánchez-Gómez M, Matute C (2010) Cannabidiol induces intracellular calcium elevation and cytotoxicity in oligodendrocytes. Glia 58:1739–1747
Mattsson K et al (2017) Brain damage and behavioural disorders in fish induced by plastic nanoparticles delivered through the food chain. Sci Rep 7:11452
McCarthy MM, Arnold AP (2011) Reframing sexual differentiation of the brain. Nat Neurosci 14:677–683
McKallip RJ, Nagarkatti M, Nagarkatti PS (2005) Delta-9-tetrahydrocannabinol enhances breast cancer growth and metastasis by suppression of the antitumor immune response. J Immunol 174:3281–3289
Meeker JD, Stapleton HM (2010) House dust concentrations of organophosphate flame retardants in relation to hormone levels and semen quality parameters. Environ Health Perspect 118:318–323
Mehl LC, Manjally AV, Bouadi O, Gibson EM, Leng Tay T (2022) Microglia in brain development and regeneration. Development 149:dev200425
Melén CM et al (2019) A clinical trial of cannabis as targeted therapy for indolent leukemic lymphoma. Blood 134:5487
Meng Y et al (2022) Alkyl organophosphate flame retardants (OPFRs) induce lung inflammation and aggravate OVA-simulated asthmatic response via the NF-κB signaling pathway. Environ Int 163:107209
Mercer GV et al (2023) Maternal exposure to polystyrene nanoplastics alters fetal brain metabolism in mice. Metabolomics 19:96
Miyamoto A et al (2016) Microglia contact induces synapse formation in developing somatosensory cortex. Nat Commun 7:12540
Money KM, Stanwood GD (2013) Developmental origins of brain disorders: roles for dopamine. Front Cell Neurosci 7:260
Moon S, Hong J, Go S, Kim B-S (2023) Immunomodulation for tissue repair and regeneration. Tissue Eng Regen Med 20:389–409
Moore DG et al (2009) During pregnancy, recreational drug-using women stop taking ecstasy (3,4-methyl enedioxy-N-methylamphetamine) and reduce alcohol consumption, but continue to smoke tobacco and cannabis: initial findings from the Development and Infancy Study. J Psychopharmacol (Oxford, England) 24:1403
Moreira FA et al (2008) Antiaversive effects of cannabinoids: is the periaqueductal Gray involved? Neural Plast 2009:625469
Mozafarjalali M, Hamidian AH, Sayadi MH (2023) Microplastics as carriers of iron and copper nanoparticles in aqueous solution. Chemosphere 324:138332
Mulligan MK, Hamre KM (2023) Influence of prenatal cannabinoid exposure on early development and beyond. Adv Drug Alcohol Res 3:10981
Murphy SK et al (2017) Maternal infection and stress during pregnancy and depressive symptoms in adolescent offspring. Psychiatry Res 257:102–110
Nakanishi M et al (2007) Microglia-derived interleukin-6 and leukaemia inhibitory factor promote astrocytic differentiation of neural stem/progenitor cells. Eur J Neurosci 25:649–658

Navarrete M, Araque A (2010) Endocannabinoids potentiate synaptic transmission through stimulation of astrocytes. Neuron 68:113–126

Nemes-Baran AD, White DR, DeSilva TM (2020) Fractalkine-dependent microglial pruning of viable oligodendrocyte progenitor cells regulates myelination. Cell Rep 32:108047

Newell AJ, Patisaul HB (2023) Developmental organophosphate flame retardant exposure disrupts adult hippocampal neurogenesis in Wistar rats. Neurotoxicology 99:104–114

Newell AJ et al (2022) Maternal organophosphate flame retardant exposure alters the developing mesencephalic dopamine system in fetal rat. Toxicol Sci 191:357

Nikodemova M et al (2015) Microglial numbers attain adult levels after undergoing a rapid decrease in cell number in the third postnatal week. J Neuroimmunol 278:280–288

Nolan AM, Nolan YM, O'Keeffe GW (2011) IL-1β inhibits axonal growth of developing sympathetic neurons. Mol Cell Neurosci 48:142–150

OECD (2022) Global plastics outlook: policy scenarios to 2060. Organisation for Economic Co-operation and Development, Paris

Oláh A, Szekanecz Z, Bíró T (2017) Targeting cannabinoid signaling in the immune system: "high"-ly exciting questions, possibilities, and challenges. Front Immunol 8:1487

Pantelaki I, Voutsa D (2020) Occurrence, analysis and risk assessment of organophosphate esters (OPEs) in biota: a review. Mar Pollut Bull 160:111547

Park S-Y, Kang M-J, Han J-S (2018) Interleukin-1 beta promotes neuronal differentiation through the Wnt5a/RhoA/JNK pathway in cortical neural precursor cells. Mol Brain 11:39

Park E-J et al (2020) Repeated-oral dose toxicity of polyethylene microplastics and the possible implications on reproduction and development of the next generation. Toxicol Lett 324:75–85

Park J et al (2021) Microglial MERTK eliminates phosphatidylserine-displaying inhibitory post-synapses. EMBO J 40:e107121

Park K-M, Kim B, Woo W, Kim LK, Hyun Y-M (2024) Polystyrene microplastics induce activation and cell death of neutrophils through strong adherence and engulfment. J Hazard Mater 480:136100

Pasciuto E et al (2020) Microglia require CD4 T cells to complete the fetal-to-adult transition. Cell 182:625–640.e24

Passey ME, Sanson-Fisher RW, D'Este CA, Stirling JM (2014) Tobacco, alcohol and cannabis use during pregnancy: clustering of risks. Drug Alcohol Depend 134:44–50

Philips T, Rothstein JD (2017) Oligodendroglia: metabolic supporters of neurons. J Clin Invest 127:3271

Pineda DA et al (2007) Environmental influences that affect attention deficit/hyperactivity disorder: study of a genetic isolate. Eur Child Adolesc Psychiatry 16:337–346

Pizzagalli DA, Roberts AC (2022) Prefrontal cortex and depression. Neuropsychopharmacology 47:225–246

Porta EL et al (2023) Microplastics and kidneys: an update on the evidence for deposition of plastic microparticles in human organs, tissues and fluids and renal toxicity concern. Int J Mol Sci 24:14391

Prata JC, da Costa JP, Lopes I, Duarte AC, Rocha-Santos T (2020) Environmental exposure to microplastics: an overview on possible human health effects. Sci Total Environ 702:134455

Prüst M, Meijer J, Westerink RHS (2020) The plastic brain: neurotoxicity of micro- and nanoplastics. Part Fibre Toxicol 17:24

Quagliato LA, de Matos U, Nardi AE (2021) Maternal immune activation generates anxiety in offspring: a translational meta-analysis. Transl Psychiatry 11:1–6

Rafa N et al (2024) Microplastics as carriers of toxic pollutants: source, transport, and toxicological effects. Environ Pollut 343:123190

Ragusa A et al (2021) Plasticenta: first evidence of microplastics in human placenta. Environ Int 146:106274

Ragusa A et al (2022) Deeply in plasticenta: presence of microplastics in the intracellular compartment of human placentas. Int J Environ Res Public Health 19:11593

Richardson GA, Ryan C, Willford J, Day NL, Goldschmidt L (2002) Prenatal alcohol and marijuana exposure: effects on neuropsychological outcomes at 10 years. Neurotoxicol Teratol 24:309–320

Rock KD, Patisaul HB (2018) Environmental mechanisms of neurodevelopmental toxicity. Curr Environ Health Rep 5:145

Rock KD et al (2020) Effects of prenatal exposure to a mixture of organophosphate flame retardants on placental gene expression and serotonergic innervation in the fetal rat brain. Toxicol Sci 176:203–223

Rompala G, Nomura Y, Hurd YL (2021) Maternal cannabis use is associated with suppression of immune gene networks in placenta and increased anxiety phenotypes in offspring. Proc Natl Acad Sci 118:e2106115118

Sagheddu C et al (2021) Mesolimbic dopamine dysregulation as a signature of information processing deficits imposed by prenatal THC exposure. Prog Neuro-Psychopharmacol Biol Psychiatry 105:110128

Sala B, Giménez J, de Stephanis R, Barceló D, Eljarrat E (2019) First determination of high levels of organophosphorus flame retardants and plasticizers in dolphins from Southern European waters. Environ Res 172:289–295

Sangkham S et al (2022) A review on microplastics and nanoplastics in the environment: their occurrence, exposure routes, toxic studies, and potential effects on human health. Mar Pollut Bull 181:113832

Santos RG, Machovsky-Capuska GE, Andrades R (2021) Plastic ingestion as an evolutionary trap: toward a holistic understanding. Science 373:56–60

Saquib Q et al (2022) Organophosphorus flame retardant TDCPP displays genotoxic and carcinogenic risks in human liver cells. Cells 11:195

Sarikahya MH et al (2023) Prenatal THC exposure induces long-term, sex-dependent cognitive dysfunction associated with lipidomic and neuronal pathology in the prefrontal cortex-hippocampal network. Mol Psychiatry 28:4234–4250

Schafer DP et al (2012) Microglia sculpt postnatal neural circuits in an activity and complement-dependent manner. Neuron 74:691–705

Schultz W (2007) Multiple dopamine functions at different time courses. Annu Rev Neurosci 30:259–288

Scott-Hewitt N et al (2020) Local externalization of phosphatidylserine mediates developmental synaptic pruning by microglia. EMBO J 39:e105380

Shan S, Zhang Y, Zhao H, Zeng T, Zhao X (2022) Polystyrene nanoplastics penetrate across the blood-brain barrier and induce activation of microglia in the brain of mice. Chemosphere 298:134261

Sharma RK, Kumari U, Kumar S (2024) Impact of microplastics on pregnancy and fetal development: a systematic review. Cureus 16:e60712

She K et al (2024) Emerging role of microglia in the developing dopaminergic system: perturbation by early life stress. Neural Regen Res. https://doi.org/10.4103/NRR.NRR-D-24-00742

Shen K, Li L, Liu J, Chen C, Liu J (2019) Stocks, flows and emissions of DBDPE in China and its international distribution through products and waste. Environ Pollut 250:79–86

Shi Z et al (2016) Novel brominated flame retardants in food composites and human milk from the Chinese Total Diet Study in 2011: concentrations and a dietary exposure assessment. Environ Int 96:82–90

Shi F et al (2022) Decabromodiphenyl ethane affects embryonic development by interfering with nuclear F-actin in zygotes and leads to cognitive and social disorders in offspring mice. FASEB J 36:e22445

Shigemoto-Mogami Y, Hoshikawa K, Goldman JE, Sekino Y, Sato K (2014) Microglia enhance neurogenesis and oligodendrogenesis in the early postnatal subventricular zone. J Neurosci 34:2231–2243

Siddiqi MA, Laessig RH, Reed KD (2003) Polybrominated diphenyl ethers (PBDEs): new pollutants-old diseases. Clin Med Res 1:281

Silberhorn EM, Glauert HP, Robertson LW (1990) Carcinogenicity of polyhalogenated biphenyls: PCBs and PBBs. Crit Rev Toxicol 20:440–496

Smith AM, Fried PA, Hogan MJ, Cameron I (2004) Effects of prenatal marijuana on response inhibition: an fMRI study of young adults. Neurotoxicol Teratol 26:533–542

Smith AM, Fried PA, Hogan MJ, Cameron I (2006) Effects of prenatal marijuana on visuospatial working memory: an fMRI study in young adults. Neurotoxicol Teratol 28:286–295

So YH et al (2023) Maternal exposure to polystyrene microplastics impairs social behavior in mouse offspring with a potential neurotoxicity. Neurotoxicology 99:206–216

Soderstrom K, Soliman E, Van Dross R (2017) Cannabinoids modulate neuronal activity and cancer by CB1 and CB2 receptor-independent mechanisms. Front Pharmacol 8:720

Spencer GE, Klumperman J, Syed NI (1998) Neurotransmitters and neurodevelopment. Role of dopamine in neurite outgrowth, target selection and specific synapse formation. Perspect Dev Neurobiol 5:451–467

Squarzoni P et al (2014) Microglia modulate wiring of the embryonic forebrain. Cell Rep 8:1271–1279

Stapleton HM et al (2008) Alternate and new brominated flame Retardants detected in U.S. House Dust. Environ Sci Technol 42:6910–6916

Stein TP, Schluter MD, Steer RA, Guo L, Ming X (2015) Bisphenol A exposure in children with autism Spectrum disorders. Autism Res 8:272–283

Stevens B et al (2007) The classical complement Cascade mediates CNS synapse elimination. Cell 131:1164–1178

Stiles J, Jernigan TL (2010) The basics of brain development. Neuropsychol Rev 20:327–348

Sun L et al (2016) Developmental exposure of zebrafish larvae to organophosphate flame retardants causes neurotoxicity. Neurotoxicol Teratol 55:16–22

Sun Y et al (2020) Hepatotoxicity of decabromodiphenyl ethane (DBDPE) and decabromodiphenyl ether (BDE-209) in 28-day exposed Sprague-Dawley rats. Sci Total Environ 705:135783

Sun Y et al (2023) Multi- and transgenerational developmental impairments are induced by Decabromodiphenyl ethane (DBDPE) in zebrafish larvae. Environ Sci Technol 57:2887–2897

Sun Y et al (2024) Life-time exposure to decabromodiphenyl ethane (DBDPE) caused transgenerational epigenetic alterations of thyroid endocrine system in zebrafish. Sci Total Environ 950:175337

Suzuki G et al (2024) Global discharge of microplastics from mechanical recycling of plastic waste. Environ Pollut 348:123855

Tanabe S, Yamashita T (2018) B-1a lymphocytes promote oligodendrogenesis during brain development. Nat Neurosci 21:506–516

Tartaglione AM, Camoni L, Calamandrei G, Chiarotti F, Venerosi A (2024) The contribution of environmental pollutants to the risk of autism and other neurodevelopmental disorders: a systematic review of case-control studies. Neurosci Biobehav Rev 164:105815

Tau GZ, Peterson BS (2010) Normal development of brain circuits. Neuropsychopharmacology 35:147–168

Tay TL et al (2017) A new fate mapping system reveals context-dependent random or clonal expansion of microglia. Nat Neurosci 20:793–803

Thion MS, Ginhoux F, Garel S (2018) Microglia and early brain development: an intimate journey. Science 362:185–189

Thompson RC et al (2004) Lost at sea: where is all the plastic? Science 304:838

Thompson RC et al (2024) Twenty years of microplastic pollution research—what have we learned? Science 386:eadl2746

Tian L et al (2024) Effects of nanoplastic exposure during pregnancy and lactation on neurodevelopment of rat offspring. J Hazard Mater 474:134800
Turner MD, Nedjai B, Hurst T, Pennington DJ (2014) Cytokines and chemokines: at the crossroads of cell signalling and inflammatory disease. Biochim Biophys Acta (BBA) – Mol Cell Res 1843:2563–2582
Ueno M, Yamashita T (2014) Bidirectional tuning of microglia in the developing brain: from neurogenesis to neural circuit formation. Curr Opin Neurobiol 27:8–15
Ueno M et al (2013) Layer V cortical neurons require microglial support for survival during postnatal development. Nat Neurosci 16:543–551
Vainchtein ID et al (2018) Astrocyte-derived Interleukin-33 promotes microglial synapse engulfment and neural circuit development. Science 359:1269–1273
Valjent E et al (2023) Dopamine D2 receptors in the extended amygdala scale the optimization of seeking and avoidance behaviors. https://doi.org/10.21203/rs.3.rs-2595678/v1
Valles SL et al (2023) Functions of astrocytes under Normal conditions and after a brain disease. Int J Mol Sci 24:8434
van Tilborg E et al (2018) Origin and dynamics of oligodendrocytes in the developing brain: implications for perinatal white matter injury. Glia 66:221–238
Vishnubhotla RV, Ahmad ST, Zhao Y, Radhakrishnan R (2024) Impact of prenatal marijuana exposure on adolescent brain structural and functional connectivity and behavioural outcomes. Brain Commun 6:fcae001
Vivi E, Di Benedetto B (2024) Brain stars take the lead during critical periods of early postnatal brain development: relevance of astrocytes in health and mental disorders. Mol Psychiatry 29:2821–2833
Volkow ND, Han B, Compton WM, McCance-Katz EF (2019) Self-reported medical and nonmedical cannabis use among pregnant women in the United States. JAMA 322:167–169
Vuong AM et al (2020) Flame retardants and neurodevelopment: an updated review of epidemiological literature. Curr Epidemiol Rep 7:220–236
Wakselman S et al (2008) Developmental neuronal death in hippocampus requires the microglial CD11b integrin and DAP12 Immunoreceptor. J Neurosci 28:8138–8143
Walker OS, Holloway AC, Raha S (2019) The role of the endocannabinoid system in female reproductive tissues. J Ovarian Res 12:3
Walton NM et al (2006) Microglia instruct subventricular zone neurogenesis. Glia 54:815–825
Wang X, Song F (2024) The neurotoxicity of organophosphorus flame retardant tris (1,3-dichloro-2-propyl) phosphate (TDCPP): Main effects and its underlying mechanisms. Environ Pollut 346:123569
Wang X, Dow-Edwards D, Anderson V, Minkoff H, Hurd YL (2004) In utero marijuana exposure associated with abnormal amygdala dopamine D2 gene expression in the human fetus. Biol Psychiatry 56:909–915
Wang X et al (2007) Interleukin-1beta mediates proliferation and differentiation of multipotent neural precursor cells through the activation of SAPK/JNK pathway. Mol Cell Neurosci 36:343–354
Wang J, Wang Y, Shi Z, Zhou X, Sun Z (2018) Legacy and novel brominated flame retardants in indoor dust from Beijing, China: occurrence, human exposure assessment and evidence for PBDEs replacement. Sci Total Environ 618:48–59
Wang J, Li J, Shi Z (2020) Dietary exposure assessment of a nursing mother-infant cohort to legacy and novel brominated flame retardants: results of a 3-day duplicate diet study in Beijing, China. Chemosphere 254:126843
Wang R, Cheng H, Gong Y, Huang T (2023) New brominated flame retardant decabromodiphenyl ethane (DBDPE) in water sediments: a review of contamination characteristics, exposure pathways, ecotoxicological effects and health risks. Environ Pollut 334:122121
Watkins AR (2019) Cannabinoid interactions with ion channels and receptors. Channels 13:162
Wemken N et al (2020) Emerging and legacy brominated flame retardants in the breast milk of first time Irish mothers suggest positive response to restrictions on use of HBCDD and Penta- and Octa-BDE formulations. Environ Res 180:108805
Werkman IL, Lentferink DH, Baron W (2020) Macroglial diversity: white and grey areas and relevance to remyelination. Cell Mol Life Sci 78:143–171
Win-Shwe T-T et al (2022) Dietary exposure to flame retardant Tris (2-Butoxyethyl) phosphate altered Neurobehavior and Neuroinflammatory responses in a mouse model of allergic asthma. Int J Mol Sci 23:655
Witchey SK et al (2022) Impacts of gestational FireMaster 550 exposure on the neonatal cortex are sex specific and largely attributable to the organophosphate esters. Neuroendocrinology 113:1262–1282
Wolfson ML et al (2016) Expression analysis of cannabinoid receptors 1 and 2 in B cells during pregnancy and their role on cytokine production. J Reprod Immunol 116:23–27
Xie Z et al (2022) Organophosphate ester pollution in the oceans. Nat Rev Earth Environ 3:309–322
Xu J, Huang G, Guo TL (2016) Developmental Bisphenol A exposure modulates immune-related diseases. Toxics 4:23
Xuan L et al (2024) Exposure to polystyrene nanoplastics induces abnormal activation of innate immunity via the cGAS-STING pathway. Ecotoxicol Environ Saf 275:116255
Xue J et al (2022) Decabromodiphenyl ethane induces male reproductive toxicity by glycolipid metabolism imbalance and meiotic failure. Ecotoxicol Environ Saf 246:114165
Xue J et al (2024) Microplastics in maternal amniotic fluid and their associations with gestational age. Sci Total Environ 920:171044

Yadav IC, Devi NL, Li J, Zhang G (2017) Occurrence and source apportionment of halogenated flame retardants in the indoor air of Nepalese cities: implication on human health. Atmos Environ 161:122–131
Yang W et al (2022) Impacts of microplastics on immunity. Front Toxicol 4:956885
Yang L et al (2023) Mitochondrial dysfunction was involved in Decabromodiphenyl ethane-induced Glucolipid metabolism disorders and neurotoxicity in zebrafish larvae. Environ Sci Technol 57:11043–11055
Young-Wolff KC et al (2022) Trends in cannabis polysubstance use during early pregnancy among patients in a large health care system in Northern California. JAMA Netw Open 5:e2215418
Zaheer J et al (2022) Pre/post-natal exposure to microplastic as a potential risk factor for autism spectrum disorder. Environ Int 161:107121
Zengeler KE, Lukens JR (2021) Innate immunity at the crossroads of healthy brain maturation and neurodevelopmental disorders. Nat Rev Immunol 21:454–468
Zhan Y et al (2014) Deficient neuron-microglia signaling results in impaired functional brain connectivity and social behavior. Nat Neurosci 17:400–406
Zhang K, Xu H, Cao L, Li K, Huang Q (2013) Interleukin-1β inhibits the differentiation of hippocampal neural precursor cells into serotonergic neurons. Brain Res 1490:193–201
Zhang Q, Yu C, Fu L, Gu S, Wang C (2020) New insights in the endocrine disrupting effects of three primary metabolites of organophosphate flame retardants. Environ Sci Technol 54:4465–4474
Zhang Y, Lyu L, Tao Y, Ju H, Chen J (2022a) Health risks of phthalates: a review of immunotoxicity. Environ Pollut 313:120173
Zhang Q et al (2022b) A review of organophosphate esters in soil: implications for the potential source, transfer, and transformation mechanism. Environ Res 204:112122
Zheng X-B et al (2012) Halogenated flame retardants in home-produced eggs from an electronic waste recycling region in South China: levels, composition profiles, and human dietary exposure assessment. Environ Int 45:122–128
Zheng G et al (2021) Organophosphate esters and their metabolites in breast Milk from the United States: breastfeeding is an important exposure pathway for infants. Environ Sci Technol Lett 8:224–230
Zhong X et al (2020) Neonatal exposure to organophosphorus flame retardant TDCPP elicits neurotoxicity in mouse hippocampus via microglia-mediated inflammation in vivo and in vitro. Arch Toxicol 94:541–552
Zhou Y et al (2023) Current research trends on cosmetic microplastic pollution and its impacts on the ecosystem: a review. Environ Pollut 320:121106
Zito K, Svoboda K (2002) Activity-dependent synaptogenesis in the adult mammalian cortex. Neuron 35:1015–1017
Zou L et al (2024) Neonatal exposure to polystyrene Nanoplastics impairs microglia-mediated synaptic pruning and causes social behavioral defects in adulthood. Environ Sci Technol 58:11945–11957

MicroRNAs Fine-Tune Brain and Body Communication in Health and Disease

12

MicroRNAs in Brain-Body Communication

David C. Henshall

Abstract

MicroRNAs are short noncoding RNAs that post-transcriptionally regulate gene expression. Their function is to fine-tune protein levels, thereby reducing gene expression noise. MicroRNAs serve critical functions during brain development, sculpting the transcript landscapes that direct cell proliferation, migration, and differentiation. In the mature brain, microRNAs impose stability on gene expression patterns to protect the constancy of network behavior while also facilitating the mechanics of synaptic plasticity. Brain diseases feature alterations in levels of specific microRNAs which influence injury and repair processes including changes to neuronal and glia morphology, immune and inflammatory responses, and cellular metabolism. MicroRNAs are also found within the extracellular space, including circulating in the cerebrospinal fluid and plasma. Characterization of biofluid microRNAs indicates some have CNS origins leading to the idea that microRNAs function in local and long-distance signalling. This chapter reviews the microRNA pathway and functions in the brain, the physical forms of extracellular microRNAs including encapsulation within extracellular vesicles, the changes to circulating microRNAs in brain diseases and evidence that circulating microRNAs come from the brain, and the effects, if any, of extracellular microRNA on recipient cells. The chapter also reviews the clinical applications of these findings, including diagnostic point-of-care testing and as therapeutic targets.

Keywords

Biomarker · Epigenetic · Epileptogenesis · Exosome · Medicine · Hippocampus

Abbreviations

ALS	Amyotrophic lateral sclerosis
AAV	Adeno-associated virus
Ago	Argonaute
BBB	Blood-brain barrier
CSF	Cerebrospinal fluid
CNS	Central nervous system
EEG	Electroencephalogram
EVs	Extracellular vesicles

D. C. Henshall (✉)
Department of Physiology & Medical Physics, RCSI University of Medicine & Health Sciences, Dublin, Ireland

FutureNeuro Research Ireland Centre for Translational Brain Science, RCSI University of Medicine & Health Sciences, Dublin, Ireland
e-mail: dhenshall@rcsi.ie

D. Tropea, E. Giacometti (eds.), *Brain-Body Connections*, Advances in Experimental Medicine and Biology 1477, https://doi.org/10.1007/978-3-031-89525-8_12

ESCRT	Endosomal sorting complexes required for transport
FLAG	Epitope for tagging a protein (AspTyrLysAspAspAspAspLys)
MRI	Magnetic resonance imaging
mRNA	Messenger RNA
miRNA	MicroRNAs
Pol II	RNA polymerase II
Pri-miRNA	Primary miRNA transcript
Pre-miRNA	Precursor miRNA (processed by Drosha)
RNA	Ribonucleic acid
RNases	Ribonucleases
RISC	RNA-induced silencing complex
RBPs	RNA binding proteins
TBI	Traumatic brain injury
UTR	Untranslated region
YFP	Yellow fluorescent protein

12.1 Introduction

Understanding, diagnosing, and treating brain diseases require analysis of cell function and communication. Increasingly, this is being achieved by the sampling, integration and experimental interrogation of multimodal data spanning from single molecule through to single cell, network, tissue and high-order organization (Buccitelli and Selbach 2020). Ultimately, cell function and dysfunction are a result of the activity of genes and the programmes and pathways they cause and serve. Thus, organism physiology depends on correctly executed programs of gene expression. Gene expression is, however, a stochastic process, noise being generated due to intrinsic and extrinsic variation in the components that control each step from transcription through to translation and protein turnover (Raj and Van Oudenaarden 2008). Negative feedback is a principle mechanism to impose stability and limit noisy gene expression. One of the key mediators of the negative control of gene expression are microRNA (miRNA) (Ebert and Sharp 2012; Henshall 2024).

miRNA are short noncoding RNAs that negatively regulate protein levels mainly via posttranscriptional base-pairing to complementary sites in the 3′ untranslated region (UTR) of protein-coding mRNAs (Shang et al. 2023). Since their discovery, just over 30 years ago, miRNA have emerged as a major class of non-protein coding gene with critical roles in multicellular life. Indeed, miRNAs are found in all branches of eukaryotic life, and the number of miRNA genes tracks closely with organism complexity, evidence that the system supports or contributed to the evolution of cellular diversity in higher organisms including humans (Hornstein and Shomron 2006; Berezikov 2011). A number of miRNAs have been found that are unique to primates, unique to the great apes, and unique to *homo sapiens*. Most of the unique-to-human miRNA are expressed in the brain (Qureshi and Mehler 2012; Kosik and Nowakowski 2018).

The brain is the most miRNA-rich organ in the body and miRNA regulate every step during brain development (Kosik 2006; Kosik and Nowakowski 2018). After maturation is complete, miRNA continue to enforce the gene expression programs that ensure cells such as neurons retain the appropriate gene cohorts to support their physiology, while also conferring the ability of neurons to remain plastic and reactive to incoming information and to strengthen or weaken synaptic connections. Expression of miRNAs is disturbed in brain disease, leading to under- and over-suppression of gene pathways and even the re-directing of miRNAs to new targets, for example through RNA editing (Nishikura 2016; Li et al. 2018). Differences in levels of circulating miRNAs have been found in the body fluids of people with brain disease and this has been recapitulated in animal models. Such signals may serve as a basis for the diagnosis of disease but their value may be far greater than simply a window into a brain condition. The distribution of miRNA from their sites of origin to other cells within their locality and beyond the brain may have a role in long-distance intercellular communication (Budnik et al. 2016; Thompson et al. 2016). Evidence has emerged that miRNAs enclosed in vesicles can influence the physiological function of cells to which they make contact. This idea has revolutionized our thinking about brain-body communication and represents a new approach to the treatment as well as diagnosis and monitoring of brain health. This chapter reviews the role of miRNA as master controllers of gene expression within cells, and the evidence that extracellular miRNAs function as agents in communication between the brain and body in health and disease.

12.2 miRNAs: Post-Transcriptional Control of Gene Expression in the Brain

12.2.1 Discovery, Biogenesis, and Mechanism of Action of miRNA

The discovery of miRNA was made during studies of the development of *C. elegans*, by the labs of Ruvkun and Ambros. The importance of this discovery was recognized with the award of the 2024 Nobel prize in physiology or medicine. Both teams were studying genes in the worm called *lin-4* and *lin-14* and were trying to uncover how *lin-4* was blocking the translation of the LIN-14 protein during progression from late larval to adult forms (Lee et al. 1993; Wightman et al. 1993). Through elegant experiments, they demonstrated that the *lin-4* gene product was not a protein-coding transcript but in fact a unique 22 nucleotide short RNA that could bind via Watson-Crick base-pairing to a region of the 3′UTR of the *lin-14* transcript, acting within the cytoplasm to limit translation of LIN-14. A few years later, Ruvkun's team discovered another miRNA in *C. elegans*, *let-7*, but this time found other animals, including humans made *let-7* and the mature miRNA sequence was identical to that in *C. elegans* (Pasquinelli et al. 2000). Within a few short years, hundreds of miRNA genes had been identified in humans and the main events in their biogenesis and mechanism of action had been resolved (Ambros 2004). Researchers have continued to work toward agreement on the exact number of miRNAs in the human genome, with databases such as miRBase listing about 2700 mature miRNAs (Kozomara et al. 2019), but a consensus has not yet been reached, and depending on the stringency of criteria for what is a bona fide miRNA, the range may be as few as 550 (Alles et al. 2019; Fromm et al. 2022) to upper estimates of several thousand (Londin et al. 2015). Regardless of the exact number, miRNA are understood to regulate the levels of more than half of all protein-coding genes, influencing almost every aspect of cellular physiology (Bartel 2018).

The transcription of a miRNA begins in the nucleus via the actions of RNA polymerase II (Pol II) (Ha and Kim 2014) (Fig. 12.1). This generates a long (~1 kB or more) transcript called the pri-miRNA that bears a distinctive hairpin loop generated by an RNA fold-back where a sequence of nucleotides (e.g., AUGC) pair to their complementary opposites (i.e., UACG) via Watson-Crick base-pairing. Close to the loop lies what will eventually become the mature miRNA. The pri-miRNA undergoes processing within the nucleus via the microprocessor complex. This comprises the RNase III enzyme Drosha and a co-factor, DiGeorge Critical Region 8 (DGCR8) (Ruiz-Arroyo and Nam 2022). The product of this cleavage event is a pre-miRNA which still contains the loop but is now ~70 nucleotides in length. Specific features in the pre-miRNA are recognized by Exportin 5 (XPO5) which takes the RNA from the nucleus out to the cytoplasm. There, the pre-miRNA is further cleaved by the actions of Dicer, another RNase III enzyme. This produces a two-stranded RNA duplex from which one strand (either the 3′ or 5′ arm) is selected for uptake to form the RNA-induced silencing complex (RISC) (Gebert and Macrae 2019). The core protein within the RISC and into which the miRNA strand is lodged, is Argonaute (Ago). The duplex arm selected for loading is determined by the chemistry of nucleotides as well as the binding pocket of the Ago protein. Specifically, the thermodynamic base-pairing stability at the putative 5′ end of each arm is the main influence, with a U or A having weaker bond pairing and therefore favoring incorporation over a 5′ arm with a G or C terminal nucleotide (Schwarz et al. 2003; Gebert and Macrae 2019). Often, the 5′ end of mature miRNAs begins with a U (Shang et al. 2023). Once a miRNA is within the binding

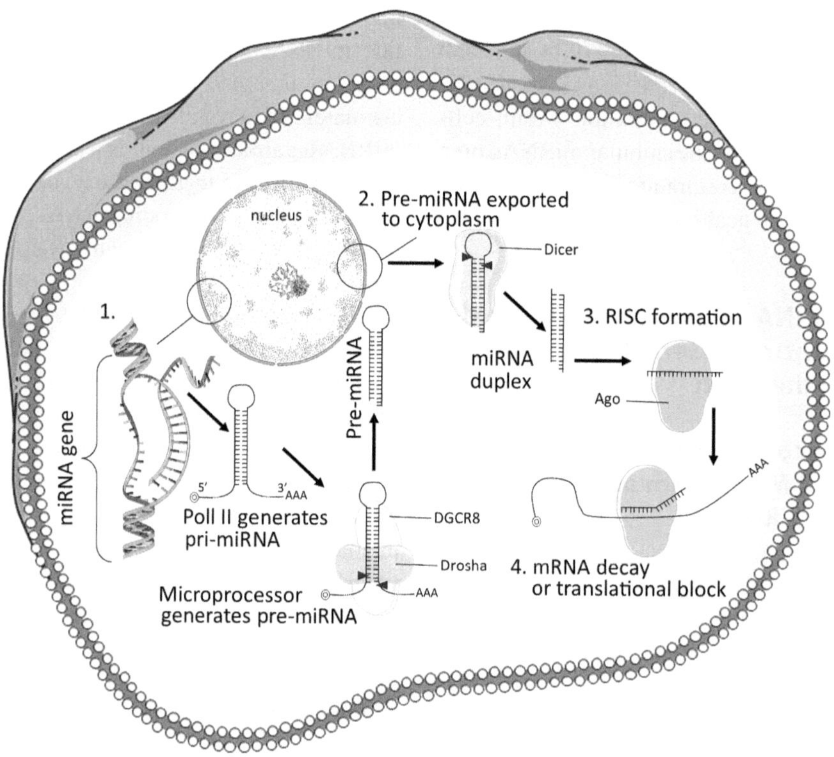

Fig. 12.1 The miRNA pathway. Schematic showing a simple overview of the biogenesis of miRNA. (1) miRNA genes are transcribed generating a pri-miRNA. This is then processed by Drosha-DGCR8 to generate the pre-miRNA. (2) The pre-miRNA is exported from the nucleus via Exportin 5 (not shown) and cleaved a second time by Dicer. One strand of the duplex miRNA is taken up by an Ago protein to form the RISC. The RISC traffics along mRNAs searching for potential sites of complementarity within the 3′UTR. Once a sequence of 7–8 nucleotides at the 5′ end of the miRNA is found to match the mRNA, the RISC locks in place. (4) The final events are either mRNA decay or translational block. Grey triangles indicate cut points by the two RNase III enzymes. AAA, polyA tail. Round shape on the 5′ end of the pri-miRNA is the cap. Figure created using images from Servier Medical Art. Servier Medical Art by Servier is licensed under a Creative Commons Attribution 4.0 International License (CC BY 4.0) (https://creativecommons.org/licenses/by/4.0/)

pocket of an Ago it begins to "search" for potential targets, typically along the 3′UTR of the mRNA (Chandradoss et al. 2015). Once a region of 6–8 nucleotides counting from the 5′ end of the miRNA is found to match, the RISC locks down. This results in either translational inhibition or mRNA decay (Gebert and Macrae 2019). Both processes cause a reduction in protein levels of the target but involve recruitment of different components to the RISC. The strength of the suppression depends on several factors and stronger complementarity and mRNA degradation can result in stronger target suppression (Baek et al. 2008).

Since miRNA:target matching requires only a short seed sequence, individual miRNAs can have dozens and often hundreds of potential targets. The implication is that miRNA are multi-targeting regulators of gene expression (Mcgeary et al. 2019). Single miRNAs can target several mRNAs within the same or a different pathway. Likewise, single mRNAs may be targeted by a few or even many miRNAs (Gebert and Macrae 2019). Generally, transcripts with longer 3′UTRs are more susceptible to miRNA regulation. Notably, the brain expresses the highest number of transcripts with long 3′UTRs and therefore miRNA control is particular exquisite for brain-expressed transcripts (Schieweck et al. 2021). The collective actions of miRNAs for organisms include (Henshall 2024):

1. Buffering against variation in gene expression, to reduce fluctuations in target protein number.
2. Opposing transient fluctuations in gene expression (which should be counteracted) and permanent changes (which should be maintained).
3. Suppressing residual transcripts during stages of development and directing and maintaining cell phenotypes.
4. Creating "tunability" of cell functions by the mixing of combinations of miRNA sites with cell type-specific transcript milieus.
5. Providing a means to regulate translation in local compartments (e.g., synapses).

12.2.2 miRNA and the Brain

Early studies in mice indicated a wide variety of miRNA genes switch on during embryonic and later postnatal brain development (Krichevsky et al. 2003; Sempere et al. 2004). Later, researchers confirmed this occurs on a larger and more complex scale in the developing human brain (Ziats and Rennert 2014; Nowakowski et al. 2018). Target predictions and functional studies indicate these miRNAs adjust the expression of genes during key events and transition phases including neural precursor division, differentiation, migration, and maturation (Kosik 2006; Soutschek and Schratt 2023). This also includes the growth and contact of axons and dendrites and the subsequent adjustment of such contacts (Soutschek and Schratt 2023). The brain expresses a rich diversity of unique miRNAs such as miR-9, miR-128, and miR-134 which each steer and shape the gene expression programs required for proper brain formation. Interruption of the biogenesis of miRNAs or even the loss of certain individual miRNAs can have devastating effects on brain development, restricting growth and leading to neurodegeneration (Bartel 2018). For example, deletion of various components of the miRNA biogenesis pathway in adult animals results in neurodegeneration and often seizures and death (Schaefer et al. 2007; Tao et al. 2011). Selective knockout of specific miRNAs is also sufficient to trigger brain disease. This was elegantly demonstrated for miR-128 which, upon knockout, resulted in lethal seizures in mice (Tan et al. 2013) and for miR-218, the loss of which produces a mouse with phenotypic similarities to amyotrophic lateral sclerosis (ALS) (Amin et al. 2015).

In the mature brain, various miRNAs continue to be expressed at high levels, indicating an ongoing need for miRNA-based tuning of gene expression. In this setting, many miRNAs serve to stabilize and maintain normal patterns of gene expression that define the features of the cell, including neurons. This includes miR-124, the expression of which suppresses non-neuronal genes in neurons (Lim et al. 2005). Other miRNAs are important for synaptic plasticity. This includes sets of miRNAs which regulate the translation of proteins that underpin dendritic spine growth and shrinkage in response to incoming synaptic signals (Soutschek and Schratt 2023). Notably, many miRNAs are shuttled beyond the soma to be co-located at sites where there is local translation of their target mRNAs (Leung 2015). Sequences within precursors of miRNA attract RNA binding proteins (RBPs) which then shuttle the pre-miRNA to dendritic spines to await local processing (Bicker et al. 2013). Indeed, elegant studies have shown that pre-miRNAs sit close to dendritic spines and are locally matured by Dicer upon calcium entry downstream of glutamatergic input (Sambandan et al. 2017). Here, they act on co-located mRNAs such as *CaMKIIα* (calcium-calmodulin kinase IIα) (Sambandan et al. 2017) and transcripts that encode proteins that adjust the physical size of dendrites such as *Limk1* (LIM kinase 1) and *Apt1* (acyl protein thioesterase 1) (Schratt et al. 2006; Siegel et al. 2009). Thus, miRNA function is carefully controlled in the brain and tuned to react to and further adjust synaptic communication. The functions of glia are also regulated by miRNA, including pathways that control their metabolic support services and responses to injury by shaping the neuroinflammatory response (Mor et al. 2011; Ponomarev et al. 2013; Wang et al. 2014). Additional methods of miRNA trafficking within cells have recently been reported, including via endosomes to traffic miRNAs to axonal growth cones (Corradi et al. 2020). Other noncoding RNA species also support the localization of miRNAs to synapses. For example, *circDlc1(2)* is required for the correct postsynaptic localization of miR-130b at glutamatergic synapses (Silenzi et al. 2024).

12.2.3 miRNA and Brain Diseases

The identification of various brain-enriched miRNAs in the early 2000s was shortly followed by the discovery of altered miRNA levels in various brain diseases. Using microarray- and RNA sequencing-based technologies, researchers revealed hundreds of miRNAs were dysregulated in brain diseases, including chronic neurologic disorders such as epilepsy (Kan et al. 2012; Mckiernan et al. 2012; Mills et al. 2020) and multiple sclerosis (Junker et al. 2009; Keller et al. 2014), neuropsychiatric diseases such as schizophrenia (Perkins et al. 2007; Miller et al. 2012; Van Erp et al. 2014), and neurodegenerative diseases including Alzheimer's (Cogswell et al. 2008; Hebert et al. 2008), ALS (Amin et al. 2015), Huntington's (Packer et al. 2008; Lee et al. 2011), and Parkinson's (Kim et al. 2007; Junn et al. 2009). Researchers reported changes to miRNAs enriched in neurons such as miR-9, miR-124, miR-134 as well as miRNAs that are expressed by glia and other cell types such as miR-146a and miR-155. The findings prompted studies exploring whether miRNAs could be targeted to adjust the gene landscape in brain diseases. Since most brain diseases feature either complex causes or a rich diversity of perturbed pathways, the multi-targeting actions of miRNAs offered a new therapeutic strategy—a novel way to collectively adjust multiple proteins to treat complex brain disease. Two broad approaches have been taken in the field of miRNA-based therapies; either over-expressing a miRNA, for example using a viral vector such as an adeno-associated virus (AAV) or depleting a miRNA, for example using a designed antisense oligonucleotide (termed antimir or antagomir) (Van Rooij and Kauppinen 2014; Morris et al. 2021). The outcomes of deploying miRNA-based approaches for brain diseases, at least in preclinical models, include neuroprotective and disease-modifying effects in models of Alzheimer's (Zovoilis et al. 2011; Koval et al. 2013), ALS (Amin et al. 2021), Huntington's (Valles et al. 2021), multiple sclerosis (Du et al. 2009; Guerau-De-Arellano et al. 2011), Parkinson's (Zhang et al. 2021; Stein et al. 2022), psychosis (Kocerha et al. 2009; Siegert et al. 2015), (Amin et al. 2021; Valles et al. 2021), and epilepsy (Jimenez-Mateos et al. 2012; Tan et al. 2013; Bekenstein et al. 2017). In the latter case, more than 20 different miRNAs have been identified that, when targeted, produce effects on evoked or spontaneous seizures in animal models (Brennan and Henshall 2020). This has led, in some cases, to clinical trials using miRNA-based approaches for brain diseases (Mueller et al. 2020).

12.3 Brain Injury Results in Changes to the miRNA Composition of Biofluids

12.3.1 Brain miRNAs Found in Biofluids

The brain is not the only place where changes in miRNA levels have been found in experimental and human brain disease. Levels of miRNAs also change within various biofluids in patients with brain diseases and experimental models of these conditions (Enright et al. 2018; Van Den Berg et al. 2020). This includes in cerebrospinal fluid (CSF) and blood (plasma and serum) and has been documented for all major brain diseases, including both acquired and genetic conditions. This includes Alzheimer's (reviewed in (Yoon et al. 2022)), ALS (reviewed in (Casado Gama et al. 2024)), Huntington's (reviewed in (Zhang et al. 2023)), multiple sclerosis (reviewed in (Pietrasik et al. 2021)), Parkinson's (reviewed in

(Roser et al. 2018)), psychosis (reviewed in (Liu et al. 2024b)), and epilepsy (reviewed in (Enright et al. 2018)). Taking epilepsy as an example, blood-profiling studies in patients with epilepsy and animal models of epilepsy including both acquired and genetic forms have identified changes to dozens of miRNAs including several enriched in the brain and relevant to mechanisms of neuronal excitability such as miR-9 (brain development), miR-134 (neuronal microstructure), miR-106b (potassium control), miR-335 (sodium channels), and miR-146a (neuroinflammation) (Gorter et al. 2014; Roncon et al. 2015; Wang et al. 2015a, b; An et al. 2016; Surges et al. 2016; Trelinska et al. 2016; Wang et al. 2016; Avansini et al. 2017; Martins-Ferreira et al. 2020; Heiland et al. 2023). In several studies, the same miRNA was demonstrated to be dysregulated in the same direction in patient biofluids and the same biofluid obtained from animals with epilepsy (Raoof et al. 2018; Brennan et al. 2020). Some clinical studies have also found blood levels of certain miRNA correlate with seizure frequency (An et al. 2016; Che et al. 2017). Notably, deployment of treatments that attenuate seizures in rodents or humans has been shown to correct circulating miRNA levels, providing evidence for an association between the varying amounts of biofluid miRNAs and active pathophysiology (Trelinska et al. 2016; Raoof et al. 2018; Brennan et al. 2020).

Circulating miRNAs have been identified that distinguish Alzheimer's disease patients from controls (Leidinger et al. 2013). This included higher levels of miR-112, miR-161, let-7d, miR-5010, miR-26a, miR-1285, and miR-151a and lower levels of miR-103a, miR-107, miR-532, miR-26b, and let-7 f. The panel did not, however, distinguish samples from patients with mild versus moderate symptoms although did have some accuracy for distinguishing patients with mild cognitive impairment from AD. Running the 12-panel miRNA with samples from other brain diseases found the panel also distinguished patients with psychosis and depression from healthy controls (Leidinger et al. 2013). The accumulation of individual studies makes it possible to perform meta-analyses where the most consistent circulating miRNA profiles can be identified. This was recently performed for psychosis, using data from over 30 circulating miRNA studies, and converged on a set of ten miRNAs which included miR-210, miR-146a, miR-150, miR-155, miR-222, miR-16, and miR-20a (Han et al. 2024). Notably, these miRNAs were linked to processes implicated in schizophrenia such as prenatal infection and epigenetic functions but changes to blood levels of several (e.g., miR-146a) are common to many brain diseases. These findings indicate body-wide disruption to the normal pattern of miRNA expression is a common feature of brain diseases. Notably, the nature and extent of the miRNA changes varies according to the disease. While some miRNA changes in biofluids appear in many or most brain diseases, others appear to be more selective. Since each brain disease is characterized by specific contributions of genetics, neuronal versus glial involvement, inflammation, and degrees of immune response along with many other variables, these biofluid-based changes in miRNA could represent a molecular signature of each brain disease. This has clear diagnostic implications and various studies have explored the value of measuring biofluid-based miRNA profiles to distinguish between patients and controls or longitudinally, to track disease progression or remission (discussed later in this chapter).

A variety of biofluids appear to hold miRNA signatures of brain disease. Most focus has been on commonly obtained fluids such as plasma and serum. However, the dilution effect of any brain-released miRNA in the blood has led other groups to explore alternate sources of circulating miRNAs, including CSF. Although levels of RNA are extremely low in CSF, miRNA can be detected and levels have been reported to change in various brain diseases (Gallego et al. 2012). Obtaining CSF requires, however, withdrawal of spinal fluid, for example by lumbar puncture. This has motivated researchers to explore other biofluids which has led to the discovery that less-invasive sources of brain disease-relevant miRNA exist, including tears (Kenny et al. 2019; Wijesinghe et al. 2023).

While the diagnostic implications of circulating miRNAs in biofluids became an active area of study, researchers began to ask new questions. Could the circulating miRNAs be more than just a trace of a brain disease and perhaps a deliberate and controlled process? That is, biofluid-based miRNAs might also be a signal and communication system, a way in which events within the brain can exert effects locally and at long distance. Once outside of its host brain cell, what functions might extracellular miRNA serve?

12.3.2 The Origins of Extracellular RNAs as Signals in Cell-to-Cell Communication

Extracellular RNA, that is RNA generated from within a cell but then released into the outside environment, was first proposed as a form of signalling mechanism in the 1980s (Gruner and Mcmanus 2021). The idea developed further with studies in the mid-2000s on cancer and stem cells when RNA was found in the media surrounding cells. After that, interest in this topic quickly increased, culminating in the emergence of a new field of study with implications for organism physiology, mechanisms of disease, biomarkers, and therapeutics (Kalluri and Lebleu 2020). Extracellular RNA has been proposed as an important class of cell-to-cell signalling process, a mechanism by which distant communication occurs, with recipient cells engulfing the material followed by biological effects. In the case of the extracellular RNA being mRNA, evidence emerged this could be translated in the recipient cell. If the received RNA was miRNA, the effect in the receiver cell was the silencing of target transcripts. These biological effects remain, however, controversial (Witwer and Halushka 2016). Indeed, studies suggest the majority of RNA material engulfed by recipient cells ends up in the lysosome (Gruner and Mcmanus 2021). What is not contested is that there is extracellular RNA present in various forms—protein-bound, encapsulated in vesicles—in all body fluids.

How do RNAs become extracellular? Naked extracellular RNA does not survive long, being rapidly degraded by RNases. For extracellular RNA to persist requires a cloaking mechanism. One of the pathways by which RNAs become protected is via secretion within extracellular vesicles (EVs), for example generated through the endosomal sorting complexes required for transport (ESCRT) pathway (Thompson et al. 2016; Kalluri and Lebleu 2020). The terminology for EVs has recently been reviewed but they are defined as lipid bilayer-enclosed, cell-derived particles (Kalluri and Lebleu 2020; Buzas 2023). There are two main classes of EVs. Exosomes originate from the endosomal pathway and average ~ 100 nm in diameter (Fig. 12.2). The second category, ectosomes, is formed by the outward budding of the plasma membrane, thereby generating vesicles of sizes ranging from as small as 50 nm to as large as 1 μm (Kalluri and Lebleu 2020). The pathways by which EVs are generated have been extensively reviewed and will not be covered in this chapter (Budnik et al. 2016; Kalluri and Lebleu 2020). The biochemical and biophysical properties of EVs and their relative abundance vary according to organism, tissue, and context. In 2007, a landmark study by Lotvall's group at Göteburg University reported that exosomes represented a new mechanism of cell-cell communication (Valadi et al. 2007). They demonstrated that many different cell types generate exosomes, and these contain nucleic acids including various miRNAs. Some of the enclosed miRNAs were found at concentrations that implied active loading and the specific miRNAs detected were derived from a restricted set of genes; the presence of miRNAs in exosomes was coordinated and non-random. The team hypothesized this might represent an information-shuttling mechanism between cells.

Extracellular RNA survives via encapsulation, membrane-enclosed, protein-bound or via lipid carriers. In the case of extracellular miRNA, there is evidence that both EVs and protein-bound forms are present in most body fluids, although the relative amounts and their specific

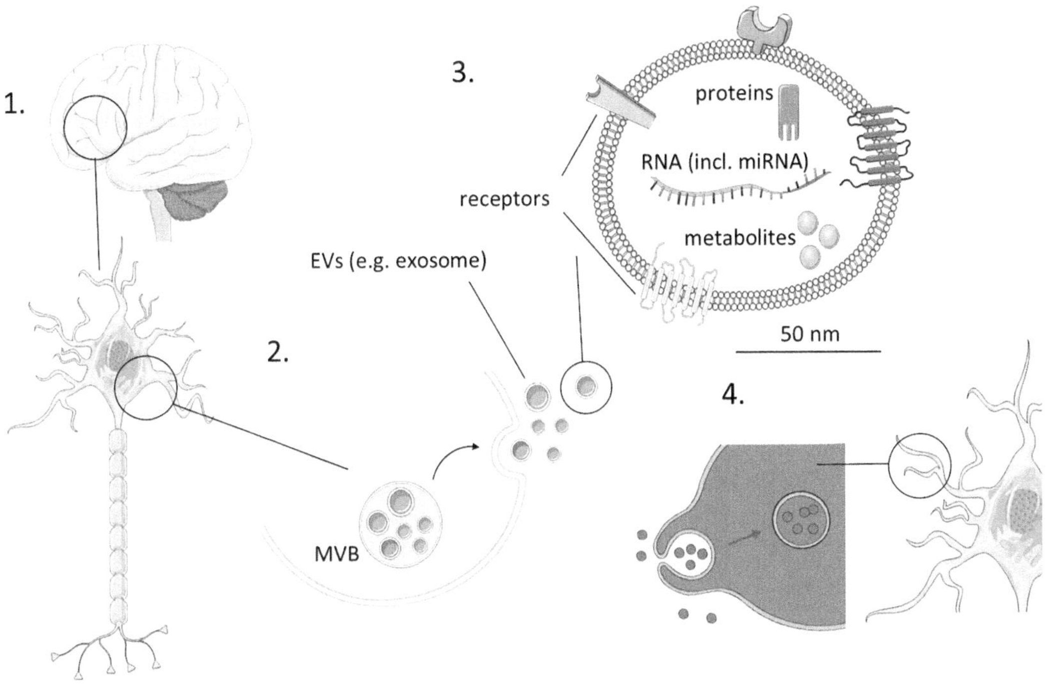

Fig. 12.2 Production, characteristics, and engulfment of exosomes. Schematic showing a simple overview of the production of EVs. (1) Neurons and other neural cell types generate extracellular vesicles such as exosomes via the (2) endosomal pathway, which involves formation of multivesicular bodies (MVB). The MVBs then fuse with the plasma membrane and release EVs. (3) EVs such as exosomes comprise a double-membrane containing a variety of molecules including miRNA. Released EVs can be (4) engulfed by recipient cells, depending on surface chemistry and, once inside, the released contents may bring about changes in gene expression. Figure created using images from Servier Medical Art. Servier Medical Art by Servier is licensed under a Creative Commons Attribution 4.0 International License(CC BY 4.0) (https://creativecommons.org/licenses/by/4.0/)

contents vary (O'Brien et al. 2020). An extensive range of studies have made claims about effects of extracellular RNAs, including miRNAs at sites and tissues distant from origin. Elegant studies include showing that a phenotype in mice bearing a deletion of a noncoding RNA gene could be rescued by blood transferred from a donor wild-type animal, a process termed parabiosis (Gruner and Mcmanus 2021). Studies on the immune system have reported that miRNAs can be transferred via EVs between T and B cells (Mittelbrunn et al. 2011) and miRNA transfer via EVs has been reported between macrophages and endothelial cells (Zhang et al. 2010). This and other work seem to support biological functions of this material, but contaminants are difficult to fully exclude. Indeed, the surface chemistry of EVs is a known contributor to the effect they have on recipient cells (Buzas 2023). Research continues to be needed to prove the biological functions of extracellular RNA, including miRNA (O'Brien et al. 2020).

12.3.3 Signals that Direct miRNA Loading for Export

If cells use EVs to perform intercellular communication, including via their contained miRNAs, and these are distributed between brain cells and throughout the body, then mechanisms likely exist which allow the EV production to change, contents to be controlled and to vary this according to context (e.g., inflammatory or metabolic status). Some of the mechanisms have been uncovered, although further studies are needed to properly understand the selection of contents, the trafficking and release and uptake into recipient cells (O'Brien et al. 2020). Among the known factors which influence packaging are sequence motifs within RNAs, association with RNA binding proteins, and posttranscriptional and posttranslational modifications thereof (O'Brien et al. 2020). One of the early breakthroughs in understanding the mechanisms which influence the selection of miRNAs for incorporation into EVs was reported by Koppers-Lalic and colleagues (Koppers-Lalic et al. 2014). The team analyzed exosomes produced by B cells, finding that overall the miRNA contents of EVs reflected the more general abundance of miRNAs in the host cell. There were, however, a subset of miRNAs that were enriched in EVs, which included miR-451 and miR-127 (Koppers-Lalic et al. 2014). Other researchers have found selective loading of donor cell miRNAs into the EVs released by activated microglia as well as evidence the miRNA contents varies according to the type of EV (e.g., exosomes versus ectosomes) (Prada et al. 2018). When the Koppers-Lalic team looked for biochemical features among the miRNAs enriched in Evs, they identified a posttranscriptional process, namely 3′ uridylation (the addition of U nucleotides to the 3′ end of the miRNA). They also found that nontemplated 3′ adenylated modifications were enriched among miRNAs that were retained within the cell and not EV-loaded. Their findings demonstrated that the chemistry of the added nucleotide at the 3′ end of a miRNA influences the loading of miRNAs into EVs. The observations may reflect broader principles of miRNA sorting into EVs because the same nontemplated 3′ uridylated changes were detected among miRNAs in EVs in urine samples (Koppers-Lalic et al. 2014). Whether the same or different features influence miRNA loading into brain-generated EVs remains unclear. Thus, miRNAs are not simply subsumed into EVs from the surrounding milieu but instead reflect planned or directed release by cells via specific posttranscriptional pathways.

12.3.4 miRNA-Containing EVs Are Generated in the Brain

There is accumulating evidence that EVs are produced within the brain and serve as a mechanism of communication and information transfer under healthy and pathological conditions. Studies show miRNAs, contained in EVs, are released from neurons to produce effects on local as well as distant, non-CNS cell types (Simeoli et al. 2017; Xu et al. 2017; Men et al. 2019; Hermann et al. 2024). The production of EVs in the brain and their entry to biofluids occurs under physiological conditions because both neuron- and glia-specific proteins have been detected in EVs extracted from healthy human CSF (Chiasserini et al. 2014). Research suggests EVs and the constituents thereof as contributing to a wide range of functions including trophic support of neurons, such as the maintenance of myelination around axons, synaptic plasticity, as well as being a route to extrude unwanted material (Thompson et al. 2016).

Studies in cell culture demonstrate that neurons release EVs, including with the biophysical features of exosomes, when depolarized and upon other forms of stimulation (Budnik et al. 2016). Using centrifugation and other techniques, EVs that display the surface epitopes and physical features of exosomes have also been isolated from the mouse and human brain. Studies in non-neuronal cell types suggest substantial numbers of exosomes are generated by cells each hour (Chiu et al. 2016). This rate of production leads, over time, to significant accumulation of EVs. Specific amounts vary, but studies have reported conditioned media from neurons to contain from ~2×10^8 particles/mL (Antoniou et al. 2023), to as much as ~2×10^{12} particles/mL (Men et al. 2019). The rate of exosome production from activated rat microglia was estimated at ~1×10^8 per hour (Prada et al. 2018). Approximately half of the EVs generated by neurons, at least in culture, appear to be via the ceramide-dependent biogenesis pathway (Antoniou et al. 2023). Recent studies using genetically encoded labelling techniques to mark EVs revealed they are generated within neurons, astrocytes, oligodendrocytes, and microglia and their presence is widespread in the mouse brain, including in the hippocampus (Men et al. 2019). In neurons, EVs can be observed within the soma, consistent with generation via the endosomal pathway, but are also found in dendrites of neurons. EVs are found at much lower levels in axons, indicating the shuttling of EVs around neurons is selective and directed. At least 10% of EVs are found extracellularly and the uptake of neuronally generated EVs and uptake by astrocytes has been demonstrated, confirming intercellular EV-based signalling occurs in the adult brain in vivo (Men et al. 2019). Transfer of glial-generated miRNAs to neurons via EVs has also been reported (Prada et al. 2018).

Estimates of the relative amount of miRNA in EVs compared to other RNAs vary widely, from 15% to as much as 80% of the RNA content (O'Brien et al. 2020). Research has shown that the miRNA contents of brain exosomes differ from the ambient milieu in which the EVs are generated. That is, the miRNAs found in EVs have been selectively accumulated. Examples of miRNAs found enriched in EVs generated within neurons include the brain-abundant miR-124 but also less-highly expressed miRNAs such as miR-669, miR-297a, miR-3082, and miR-7004 (Men et al. 2019). EVs generated by activated glial cells contain a distinct miRNA pool that includes miR-146a (Prada et al. 2018).

12.3.5 Signals That Adjust miRNA Composition of EVs

Some of the upstream mechanisms that link changes in cell state to changes to the miRNA contents of EVs have also been uncovered. For example, inflammatory stimuli that converge to activate the inflammasome trigger caspase-1 activation which then cleaves Rab-interacting lysosomal protein (RILP) (Wozniak et al. 2020). This promotes an increase in EV production via the ESCRT pathway which include a set of miRNAs carrying a specific motif, chaperoned along via the fragile X messenger ribonucleoprotein 1 (FMRP). The motif, AAUGC, was present in a number of miRNAs known to be elevated by inflammatory signalling such as miR-155 (Wozniak et al. 2020). Notably, the same sequence is found in miR-124, which is highly enriched in the brain. Thus, inflammatory stimuli directly influence the amount and contents of miRNAs loaded into and released within EVs (Buzas 2023).

Studies have also explored the mechanisms by which neuronal signalling affects the miRNA composition of EVs. There is evidence that signals that shape the strength of synapses such as brain derived neurotrophic factor (BDNF) also influence the miRNA content of neuronal EVs (Antoniou et al. 2023). For example, neuronal EVs from cells treated with BDNF were shown to stimulate the outgrowth of dendrites in cultured neurons, promoting network connectivity, and this effect was mediated via a change in the miRNA composition of the EVs (Antoniou et al. 2023).

12.3.6 Uptake of EVs by Recipient Cells

The mechanisms by which EVs that contain miRNAs are taken up by recipient cells remain incompletely understood (O'Brien et al. 2020). Both selective and nonselective mechanisms have been reported and the surface chemistry of the EV and physiologic state and nature of the recipient cell appear to influence whether and how uptake occurs. Proteoglycans and extracellular matrix components, phosphatidylserine, and integrins have all been linked to the process of recognition-internalization (Prada et al. 2018; O'Brien et al. 2020; Hermann et al. 2024). Vesicle uptake appears to occur following contact with the recipient cell by means of clathrin-dependent endocytosis and clathrin-independent mechanisms including some exotic processes such as tunnelling nanotubes (Rustom et al. 2004;

O'Brien et al. 2020). Recent in vitro studies using imaging-based markers on EVs indicate uptake by neurons is mediated via dynamin-dependent endocytosis (Antoniou et al. 2023). In order to exert effects in the recipient cell, the contents of EVs must be released into the cytosol which requires endosomal escape (Hermann et al. 2024). The efficiency of that release, including for miRNA contents, is uncertain. The final destination of internalized contents may include the nucleus of recipient cells (Hermann et al. 2024), and it is interesting that the sites of EV uptake are not necessarily their final destination in neurons. Recent studies, for example, found internalized EVs on neuronal extensions such as dendrites that may be later trafficked to the soma (Antoniou et al. 2023).

12.4 How EV-Loaded miRNAs Communicate Between Brain and Body

12.4.1 Biological Effects of Extracellular miRNA and the Effects of Disease on EV miRNAs

There is evidence that miRNAs within EVs released by neurons have biological effects in recipient cells and some of the mechanisms by which EV-conveyed miRNAs exert functional effects in recipient cells are understood. This includes effects of neuronal EVs containing miRNAs on other neurons (Antoniou et al. 2023) but neuronal miRNAs such as miR-124 have been demonstrated to traffic to astrocytes under certain conditions (Men et al. 2019). Experiments have demonstrated that EV-delivered miRNA cargoes can interact with complementary sites within 3′UTRs in the recipient cell (Antoniou et al. 2023). EV release from glia including oligodendroglia has also been demonstrated and found to occur in an activity-dependent manner and these EVs are taken up by neighboring neurons and glia (Fruhbeis et al. 2013). The effect of received EVs ranges but can include changes to metabolic function and differentiation status suggestive of roles in feedback control of network excitability and maturation. For example, the glutamate transporter GLT1 (encoded by the *Slc1a2* transcript) is upregulated in astrocytes upon neuronal stimulation. This effect appears to be mediated by miR-124 conveyed in EVs which, upon entry to astrocytes, downregulates the transcription of *Slc1a2*-targeting miRNAs in astrocytes. This suggests that miRNA conveyed in EVs serve physiological functions, fine-tuning neuron-glia communication and neurotransmitter recycling according to prevailing activity levels (Men et al. 2019). More recently, a set of EV-enclosed miRNAs that included miR-132 were found to produce effects in recipient neurons that included changes to transcripts that encode proteins involved in excitatory synapse development and maturation (Antoniou et al. 2023). The transfer of miR-146a within EVs from activated glia to neurons has also been reported to target transcripts encoding synaptic proteins, resulting in reductions in dendritic spine density and the strength of excitatory synapses (Prada et al. 2018). Together, the findings reveal EV-delivered miRNAs as a distinct communication mechanism that can adjust neuron-neuron and neuron-glial function in the brain.

While there remains conflicting evidence on whether neuronal activity causes an increase in EV production (Lachenal et al. 2011; Antoniou et al. 2023), both the quantity and biochemical composition of brain EVs have been reported to change in brain diseases (Budnik et al. 2016; Kalluri and Lebleu 2020). This includes Alzheimer disease (Liu et al. 2014; Peng et al. 2019), ALS (Ferrara et al. 2018), and Parkinson's disease (Cerri et al. 2018). Several studies have also linked EVs to the mechanisms of disease, in particular propagation of pathologic protein aggregates. Currently, EVs have been proposed to serve dual and opposing functions, as an agent of disease propagation but potentially helping clear neurotoxic waste. Indeed, inhibiting EV production has been reported to exacerbate pathogenicity in some neurodegenerative conditions (Iguchi et al. 2016).

In the context of epilepsy, there is emerging evidence that the production and miRNA contents of EVs is altered. This includes changes to the abundance of EV-associated proteins and EV-enclosed miRNAs in brain tissue from experimental or human epilepsy (Batool et al. 2019; Gitai et al. 2020; Cukovic et al. 2021). Research has also reported functional effects of miRNAs conveyed in EVs in models of epilepsy. Notably, among reported miRNAs are miR-142 and miR-21 (Cukovic et al. 2021), which are miRNAs found to be upregulated across multiple models of epilepsy and their inhibition produces anti-seizure effects in mice (Veno et al. 2020). EV-conveyed miRNAs have been suggested to reflect and influence mainly inflammatory processes. Indeed, a preparation of EVs made from surgically obtained human epileptogenic brain tissue was reported to alter inflammatory signalling when exposed to human SHSy5y cells (Cukovic et al. 2021). Thus, EV production, distribution, and effects on local cells within the brain appear to occur under physiological and pathological conditions and may perform functions that modulate and coordinate intercellular signalling.

12.4.2 Mechanisms of Transfer of Extracellular miRNA Beyond the Brain

The release of miRNAs from the brain into the circulation could occur via a number of different mechanisms (Fig. 12.3). One route is via simple passive transfer from the releasing cell (e.g., a neuron) into a neighboring capillary or venule or passing into the CSF via the interstitial fluid (Thompson et al. 2016). The glymphatic system may provide another route for EVs and other material, perhaps including protein-bound miRNAs, to transition into the lymph (Bohr et al. 2022). Alternatively, miRNA contents may be disbursed via controlled release of EVs. If brain-generated EVs contain miRNAs and if the contents are present in amounts that can produce biological effects (i.e., engage mRNA targets), then this has the basis for serving as a route between brain and body, but for miRNA to be a source of brain-body information exchange, there must be mechanisms for the transfer of EVs. A key impediment is the blood-brain barrier (BBB). The BBB is a formidable barrier to the passage of entities as large as EVs and studies have challenged whether EVs can pass through an intact BBB. However, this view is changing, with an increasing awareness that material of this size can exchange across the BBB under physiological as well as pathologic conditions (Budnik et al. 2016). Under pathological conditions, a major mechanism for brain miRNA distribution to the rest of the body is probably lytic release, whereby damaged cells passively release material which then moves into the circulation, for example because of breaches in the integrity of the BBB or via the glymphatic drainage system. The integrity of the BBB is compromised in many and perhaps most brain diseases, which would provide a route by which brain-generated EVs could pass into the circulation or immune and periphery-generated EVs could pass into the brain (Obermeier et al. 2013) (*see Chap. 1 by Garcia-Gallardo and Campbell "Understanding the Blood-Brain Barrier: From Physiology to Pathology"* for a comprehensive overview of the BBB in health and disease). Nevertheless, most research proposing EVs as biological signal carriers within the nervous system and outside has been based on cell culture studies rather than in vivo work and the challenges of standardization of purification, physiologically relevant quantities, and the criteria for assessing composition, continue to create interpretative challenges. Regardless and as discussed elsewhere, EVs are present within the CSF and EVs with chemical signatures of the brain are present within biofluids such as blood (Budnik et al. 2016). Finally, the microbiota produces EVs. Bacterial membrane vesicles (currently regarded as microbiota-derived EVs) may cross the BBB and contribute to neuroinflammation and cognitive impairment linked with neurodegeneration disease such as Alzheimer's disease, Parkinson's disease, and dementia (Zhao et al. 2021). *Chapter 6*

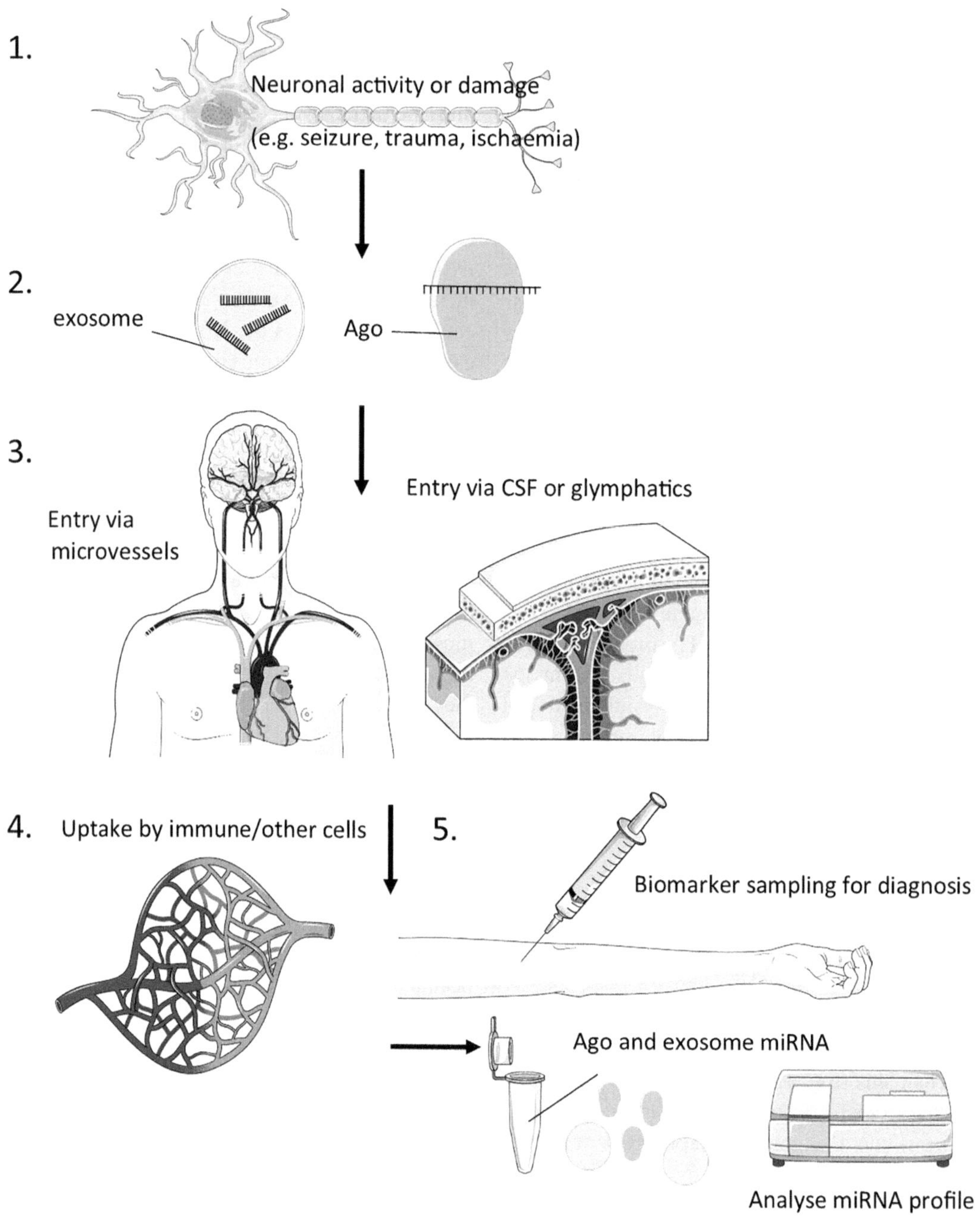

Fig. 12.3 Dispersal and destination of brain-derived extracellular miRNA. Schematic showing a scenario by which brain-generated extracellular miRNA may escape the CNS and enter the circulation. (1) miRNA are released under physiological or pathological conditions in forms such as (2) enclosed in EVs or bound to Ago. (3) Extracellular miRNA exits the CNS via vascular supply or other routes (e.g. glymphatics). (4) Extracellular miRNA encounter target cells (e.g., macrophages). (5) Circulating extracellular miRNA may be sampled and analyzed to support the diagnosis of a brain condition. Figure created using images from Servier Medical Art. Servier Medical Art by Servier is licensed under a Creative Commons Attribution 4.0 International License (CC BY 4.0) (https://creativecommons.org/licenses/by/4.0/)

"Microbiome—A Key Regulator of Body-Brain Interactions" covers in details the gut microbiota-brain connections.

12.4.3 The Biology of Circulating miRNAs as Paracrine Signals

What do miRNA released from the brain do in the body? There is evidence that circulating miRNAs may inform other systems within the body about injury and act as a form of molecular primer for inflammatory and immune responses. There is certainly evidence that circulating EVs are important to immune system function (Buzas 2023). For example, the materials in EVs have been shown to possess chemotactic properties, alter metabolic states, and exert both pro- and anti-inflammatory effects. That is, when encountering recipient cells or tissues, the circulating miRNA trigger responses that serve a purpose. Circulating miRNAs are, therefore, a potential warning system and coordinator of repair and restitution, but they may also convey pathology. There is some evidence to support EVs as having a role in intercellular propagation of disease, at least in the area of neurodegenerations. This includes Alzheimer, ALS, and Parkinson diseases, where the pathogenic material in question is aggregated proteins such as β-amyloid (Gabrielli et al. 2022) and α-synuclein (Guo et al. 2020). Whether miRNA within EVs propagate disease or convey pathogenicity from the CNS to elsewhere in the body is unclear. Within the immune system, EVs have been reported to have effects on recipient cells via transfer of contents including bioactive lipids, danger signals and cytokines. There is evidence that EVs can affect T lymphocyte development, contributing to their specificity for antigens, as well as B cell development and signalling including antigen presentation (Buzas 2023). EVs and their miRNA contents are exchanged at the so-called immune synapse (from T cell to antigen presenting cell) (Mittelbrunn et al. 2011) and miRNA contained within EVs are exchanged between T cells and B cells. Stem cell-derived EVs also contain miRNA and these EVs have been reported to affect a wide range of immune cell types. There is accumulating evidence that the packaged miRNA in immune cell EVs produce gene-silencing effects in the recipient cell and this may influence adaptive immunity and inflammatory signalling in the body. Examples include EV-delivered miR-146a, miR-155, and miR-182 suppressing components within the Fas, toll-like signalling and PI3kinase-Akt pathways (Buzas 2023). Both cardiovascular and cancer-derived EVs have also been found to contain miRNAs that produce effects on target cells and tissues. For example, cardiomyocyte-derived EVs convey miR-320 that affects fibrosis in target tissues and glioblastoma-derived EVs containing brain-enriched miR-9 have been reported to have effects on microglia function (O'Brien et al. 2020). Notably, several of the identified EV-based miRNAs regulating immune function and inflammation are among those repeatedly altered in experimental and human neurological disorders, including epilepsy. While further research is required, brain EVs and their miRNA contents may serve as sentinels of disease and their presence in the circulation may alert the immune system and trigger engagement.

12.4.4 Extracellular RNA Signals Are bi-Directional

While most of the interest on extracellular RNA and the brain relates to the movement of RNA material out and into the circulation, the direction of this signalling can also be from body to brain. That is, RNAs and their carriage material also travel the other way, from the circulation into the CNS. There is extensive evidence that the immune system communicates with the brain in health and disease, via molecules such as cytokines as well as direct neural mechanisms (Kipnis 2016; Rustenhoven and Kipnis 2022). Some of this communication is achieved via EVs. An elegant study by Momma's team used a transgenic mouse model that uses the Cre-LoxP system to trace the movement of RNA within extracellular cargo into the brain (Kur et al. 2020). The mouse expresses Cre mRNA restricted to hematopoetic cells, a portion of which gets loaded into EVs. In recipient cells the Cre mRNA is released and a fraction gets translated to protein which then triggers expression of yellow fluorescent protein (YFP). The presence of YFP-positive cells is evidence that RNA was transferred from one cell type to another via EVs. The team found virtually no YFP-positive brain cells in the brain of healthy mice. However, a large number of neurons as well as some microglia within the hippocampus and other brain areas became YFP positive following exposure of mice to the seizure-triggering excitotoxin kainic acid. The findings indicate that increased or pathologic brain activity triggers the transfer of the RNA content from immune cell EVs to the brain. Notably, glial cells rarely or never became YFP positive suggesting the EV uptake is specific to neurons. Further experiments demonstrated this was largely due to the cell that originated the EV rather than any inability of glia to absorb peripherally circulating blood cell-derived EVs (Kur et al. 2020). Finally, the team demonstrated that exposure of mice to a novel environment was also sufficient to trigger YFP in hippocampal neurons, indicating that the transfer of EVs from immune cells can even occur under non-disease states. Although the study did not investigate miRNA, the ubiquity of miRNA within immune cell-derived EVs makes it likely that miRNA was also transferred in the process. These findings are important for at least two reasons. First, they reveal an expanding view of the interaction between the brain and the rest of the body, with the door open to information being submitted to the brain by cells of the immune system or elsewhere that may influence local or larger-scale network actions and, potentially, the behavior of the organism or its response to an injury. A second implication is that anytime EV contents are detected within the brain, a possibility exist that they did not originate there but in fact arrived from an external, non-brain source such as the immune system. Last, we are yet to know the biological value of the molecular information being transferred from the immune system to the brain. However, we may speculate the effect is to prime, enhance, or attenuate neuroinflammatory responds which may be relevant to most brain diseases. Finally, the immune system is unlikely to be the only external source of EVs containing miRNA that could affect brain function. Notably, recent studies have found skeletal muscle cells release miRNA-containing EVs to adjust the neuronal inputs they receive (Huang et al. 2024).

12.4.5 Other Trafficking Mechanisms Challenge the Hypothesis of EVs as Information-Containing Signals

While there is much enthusiasm for the notion of EV-conveyed miRNAs as vehicles for cell-to-cell communication over distance, there remain three significant challenges. First, with rare exceptions, Ago is not found within EVs (O'Brien et al. 2020). Ago is an essential component for miRNAs to mediate effects on targets so this raises the question of whether and how any miRNA delivered from an EV might become functional. Perhaps, once inside a recipient cell, released miRNA can become Ago-bound and then begin to search for targets. The second challenge is around the amount of miRNA within EVs within biofluids. Studies suggest that the majority of circulating miRNA is not within EVs but in fact bound to Ago proteins (Arroyo et al. 2011; Turchinovich et al. 2011). The third, related challenge is on the basis of the amount of miRNA contained within individual EVs. Research has suggested this is extremely low. Specifically, studies by teams at the Fred Hutchinson Cancer Research Center in Seattle performed extensive analysis of the miRNA composition of EVs. Their findings indicated that, on average, each EV (exosome) contained far less than one copy per miRNA (Chevillet et al. 2014). This was the case across various sources of exosomes from different biofluids. Since miRNA binding sites in targets are usually not fully saturated, this would challenge the ability of such a small change in the overall miRNA contingent to produce a biological effect in a recipient cell. This represents a significant challenge to the plausibility of EV-delivered miRNAs as a long-distance communication system. Where claims have been made around effects of EV-delivered miRNAs in recipient cells, this may be accounted for by supra-physiological amounts of EVs or their content.

12.5 Diagnostic and Therapeutic Applications of Circulating miRNAs

12.5.1 Measuring Extracellular miRNA to Support a Diagnosis

There is an urgent and unmet need for biomarkers of brain diseases. More specifically, a simple, cheap, and easy-to-measure fluid biomarker would revolutionize diagnosis, clinical trials, and treatment. This is particularly true for the field of epilepsy where there is often a diagnostic "odyssey" that begins when someone first presents with an unprovoked seizure (Simonato et al. 2021). While there are diagnostic tools available to clinicians such as MRI and EEG, the results are often inconclusive. That is, many people who develop epilepsy have normal MRI scans at the time of first seizure and routine EEG recordings lack any tell-tale spikes and sharp waves. The presence of a molecule in a convenient biofluid such as blood that is sensitive and specific for epilepsy (or a recent seizure) would provide a critical support for clinical decision-making. A biomarker of seizures would also have applications in tracking the effectiveness of new drugs or changes to existing regimens and could provide more objective outcome measures than seizure diaries in clinical trials (Simonato et al. 2021). Likewise, the potential use of circulating miRNA signatures for diagnosis, prognosis, and monitoring outcomes in clinical trials is being explored for other brain diseases (Roser et al. 2018; Pietrasik et al. 2021; Yoon et al. 2022; Zhang et al. 2023; Casado Gama et al. 2024; Liu et al. 2024b).

As reviewed above, extracellular miRNAs possess various qualities that support their use as circulating biomarkers for the diagnosis of brain diseases. Similar to EVs, which bear a biochemical signature of the contents of the cells that produce them, many miRNAs such as miR-124, miR-128, and miR-181a are uniquely expressed within the brain. There are also glia-specific miRNAs (Jovicic et al. 2013). The presence of any of these within a biofluid such as blood would indicate some cerebral event had occurred. We also know that specific cell types express specific miRNAs. For example, there are differences in miRNA expression between excitatory and inhibitory neurons (Nowakowski et al. 2018). The presence of a brain cell type-specific miRNA in the blood or another biofluid would provide more specific information, perhaps in terms of the balance of cell types involved in the event or condition (Fig. 12.3). Importantly, miRNA levels change in the brain in disease and changes in miRNAs can cause brain disease. As a result, the physiology and pathophysiologic processes occurring within the brain generate a miRNA signature, some of which may be conveyed—via EVs or otherwise—beyond the location of their source to other parts of the body. The detection and decoding of that signal has diagnostic value. The link between miRNAs and pathophysiology also makes them potential therapeutic targets (Brennan and Henshall 2020; Morris et al. 2021). The changes in levels of miRNAs in brain diseases are driven by a combination of transcriptional changes in specific cells (up/down-regulation) as well as degeneration and cell death processes triggering the release of material. Production of EVs or ejection of protein-bound miRNAs would therefore carry a molecular signature of tissue pathology. If this were preserved upon entering the circulation, this would mean that specific brain diseases have specific circulating miRNA profiles. There is evidence to support this idea. An analysis of CSF has shown that the amount of miR-19 within EVs versus Ago-bound is different between samples from patients with Alzheimer disease, multiple sclerosis and epilepsy (Raoof et al. 2017). Ago-bound miRNAs in plasma have also been shown to display more of a change in level after a seizure than EV-measured miRNAs (Raoof et al. 2018). Recent work on other brain diseases has emphasized how measuring only the EV-carried miRNAs in the circulation enhances specificity. An eight-miRNA panel comprising miRs -10b, −4454, −199a, −151a (both 3p and 5p arms), −199a, 146a, and 29b showed high specificity for ALS when compared to profiles in other control groups and Parkinson's patients (Banack et al. 2024). The biological functions of the identified miRNAs suggested potential mechanistic links, including roles in oxidative stress, neurodegeneration, synaptic transmission, and inflammation.

There are plenty of mRNAs that are highly specific to the brain and to specific brain cell types, so why not use those as circulating biomarkers for diagnosis? Here, miRNAs appear once again to be more favourable. Regardless of whether they are carried within EVs or protein-bound, miRNAs circulate in highly stable forms. There are one or two further arguments in favor of miRNAs as disease biomarkers. Levels of miRNA can easily be assayed using standard PCR-type assays (Dave et al. 2019). This makes it possible to measure them within any decent laboratory or clinical setting. There are also a number of technological developments that suggest these measurements can be performed within a short period of time (e.g., 1 hour) and with minimal pre-processing of the sample (Mcardle et al. 2017). Taken together, circulating miRNAs offer a strong candidate for biomarkers. Large numbers of studies in different brain diseases appear to back up the potential of measuring circulating miRNAs to diagnose disease and meta-analyses are beginning to refine the selection of diagnostic "'panels"' of miRNAs. Whether measured individually or as a panel, the sensitivity and specificity of circulating miRNA for brain diseases has the potential to reach a clinically practical standard. Nevertheless, while miRNA-based blood tests have entered the market for cancers (e.g., GASTROclear from Mirxes), we do not yet have a clinical miRNA-based test for a brain disease. Clinical trials are ongoing, however, including to validate in larger and diverse patient cohorts, refine and real-world test circulating miRNAs as diagnostic biomarkers of brain disease.

12.5.2 Limitations in the Value of Circulating miRNA Profiles for Predicting Outcomes

Despite the enthusiasm and promise of circulating miRNAs as biomarkers of brain disease, recent studies have identified limitations. While circulating miRNAs appear suitable biomarkers for brain injury, they may not help predict some of the functional outcomes of that injury. Studies led by Pitkanen and colleagues found a set of miRNAs were consistently altered in blood samples after traumatic brain injury (TBI) in rats and this included known brain-enriched miRNAs such as miR-9, miR-124, and miR-136 (Heiskanen et al. 2023). Measuring plasma levels of these, alone or in combination, could easily distinguish animals with TBI compared to controls. They also found that levels of these and other miRNAs correlated with the degree of cortical damage assessed using brain imaging (MRI). Notably, some of the best correlations were for the miRNAs with highest abundance in the brain (e.g., miR-9, miR-124). Thus, brain injury causes the release of miRNAs into the circulation and the composition and quantity of that released miRNA reflects injury severity. However, no circulating miRNA was found to be at significantly higher (or lower) levels in the TBI animals that went on to develop epilepsy compared to those animals that did not develop epilepsy (Heiskanen et al. 2023). This means that plasma miRNAs assayed shortly after TBI could not predict the risk of epilepsy. There were also no plasma miRNAs that had predictive value for high versus low rates of spontaneous seizures in those animals that did develop epilepsy. Thus, circulating miRNAs can confirm a TBI has taken place and even help infer how severe an injury it was, but cannot predict the risk of epilepsy, at least in this model and at the timepoints chosen. The findings are disappointing because there is evidence that the severity of TBI is a risk factor for epilepsy development (Christensen et al. 2009). There is also evidence from models of status epilepticus that while plasma miRNA changes can be detected after an epilepsy-triggering event and before spontaneous seizure occur, the same miRNAs often remain dysregulated into the period of chronic epilepsy (Brennan et al. 2020). This would limit any clinical applications of these miRNAs as biomarkers of epileptogenesis which remains a critical unmet need (Simonato et al. 2021). Further studies are needed to determine the opportunities for using circulating miRNAs as biomarkers of brain injury and epilepsy risk. Efforts are also ongoing to use circulating miRNA panels to predict conversion from mild cognitive impairment to AD and diagnose relapsing-remitting multiple sclerosis, among other applications.

12.5.3 Improving Mechanistic Links Between Circulating miRNA and Brain Pathology

How sure are we that any of the circulating miRNA biomarkers of brain diseases *really* come from the brain? Evidence supports the presence of CNS-enriched miRNAs in the circulation as having biological significance in terms of reflecting tissue damage and acting as a signalling system, and diagnostic applications, in terms of being useful to confirm the presence of a particular pathologic state or brain damage, but does the circulating miRNA actually come from the brain? This is important to establish because if circulating miRNA changes do not reflect material released from the brain, then this may compromise the validity of these as signalling mechanisms and undermine confidence in circulating miRNAs as mechanistic biomarkers of brain disease states. There is strong circumstantial evidence that a portion of the circulating pool has come from the brain, established by studies which found known brain-enriched miRNAs in biofluids such as blood and CSF after injuries to the CNS. For example, circulating levels of miR-9, normally largely undetectable in plasma, are increased in the hours and days after status epilepticus (Roncon et al. 2015) and TBI (Heiskanen et al. 2023). Brindley and colleagues also reported changes in the miRNA composition of plasma was greatest in the first hours following status epilepticus in mice, and then declined over time (Brindley et al. 2023). Several of the miRNAs with high levels in plasma collected within a few hours of the insult were brain-enriched, including miR-181a (Brindley et al. 2023). Chronic brain diseases also show evidence for elevated levels of known CNS-enriched miRNAs. For example, higher plasma levels of miR-134 are found in patients with drug-resistant epilepsy (Spain et al. 2015; Leontariti et al. 2020; Yakovleva et al. 2022), but many of the miRNAs found to be altered in the circulation after brain injuries are those associated with inflammation. This includes miR-146a, miR-155, and miR-132 which have been reported to be upregulated in blood samples from people with epilepsy (Yakovleva et al. 2022). It is unclear whether the source of these miRNAs is the brain or circulating immune and/or vascular components. Indeed, many of the miRNAs which display changes in the plasma following brain injuries are more enriched in other organs. This may reflect that while an insult such as status epilepticus primarily damages the brain, the consequences of the event are experienced in other body systems. Prolonged seizures, for example, produce a range of systemic effects including raised blood pressure, arrhythmia, acidosis, and altered body temperature (Hocker 2015). Thus, changes to the circulating pool of miRNAs after a CNS injury reflect brain injury but also pathophysiologic effects on non-brain systems that vary according to the insult.

The presence of brain-enriched miRNAs in biofluids is inadequate on its own to prove the source was the brain. To confirm the transit of miRNAs from the brain to the circulation, Brindley and colleagues recently reported a tracing method to detect brain-generated RISC in the circulation. The team expressed a FLAG-tagged Ago2 in mice under a neuron (Thy1) or microglial (Cx3cr1) promoter (Brindley et al. 2023). Characterization of samples from the transgenic mice confirmed the presence of FLAG-Ago2 within the brain and no other organs. The cell type restriction was also demonstrated with immunohistochemistry showing the FLAG tag in scattered neuronal nuclear protein (NeuN)-positive hippocampal neurons from the Thy1 line and in ionized calcium binding adaptor molecule 1 (IBA1)-positive microglia in the hippocampus of the Cx3cr1 line. Status epilepticus was then evoked in the animals and then blood samples taken two weeks later, a time when chronic epilepsy has usually developed. The researchers were able to elute FLAG from the plasma for both mouse lines and detect bound miRNAs (Brindley et al. 2023). This provides strong evidence that some of the circulating pool of miRNAs has come from the brain and indicates that both neurons and glia are capable of contributing to this signal.

12.5.4 Therapeutic Applications of EV-Based miRNAs

Many of the brain diseases which have been found to display changes in circulating miRNAs might be treatable using miRNA-based therapies. Indeed, in parallel with the growing evidence that miRNAs are released from various brain cells to communicate over varying distances and exert effects on recipient cells, therapeutic applications of these findings have emerged (Hermann et al. 2024). The approach in most of these studies has been, however, to centrally deliver the miRNA therapy, but what if there were other ways to deliver miRNAs to the brain? EVs have emerged as promising cargo agents in this regard (O'Brien et al. 2020). The surface composition and contents of EVs can be manipulated to adjust bioavailability. While challenges still remain, proteins can be engineered to be expressed on the surface that may improve longevity (reduce clearance) and guide selective uptake by recipient cells and the EVs can be pre-loaded with one or more therapeutic miRNA (or indeed other therapeutic modalities) (Kalluri and Lebleu 2020; O'Brien et al. 2020). This versatility could enable a variety of minimally invasive delivery routes to be used for EV-based therapeutics, such as via intranasal application (Hermann et al. 2024). Direct brain entry of EVs may not be required for all applications. For example, central effects of EVs on brain injury or repair might be achieved by EV transfer and modulation of the function of endothelial cells of the BBB or non-resident immune cells (Hermann et al. 2024). Remarkably, the delivery of RNA via EVs has entered clinical testing. One of the sources of EVs are mesenchymal stem cells, and there are clinical trials that are testing cargoes including miR-124 to treat ischemic brain damage (O'Brien et al. 2020). There are some promising preclinical studies to suggest the approach of EVs to deliver therapeutic miRNA may have applications in epilepsy (Liu et al. 2024a; Yang et al. 2024). Since reduced levels of prominent miRNAs such as miR-128 have been associated with epilepsy, this represents a promising strategy that might avoid the need for a viral gene therapy-based approach.

12.6 Conclusion

This chapter has summarized the evidence that miRNAs generated within the brain serve functions in local and longer distance intercellular communication. Released miRNAs protected and chaperoned by proteins or enclosed in EVs also represent sentinels of brain disease. Examples are accumulating that neurons and other brain cell types generate EVs that contain specific combinations of miRNAs. Upon receipt by a target cell, such material may be engulfed, whereupon the released miRNA can adjust the protein landscape within the recipient cell. This may result in changes in cell physiology and behavior that include the remodelling of synapses and the broader neuronal network as well as adjust the responses of cells that coordinate neuroinflammation. Challenges to this hypothesis persist, however. There remains quite limited in vivo evidence that miRNAs within EVs generated by brain cells serve physiologic and pathologic communication functions, particularly outside the CNS. Resolving these questions requires further research and improved ways to isolate and characterize EVs, trace the cellular origins of EVs, track their movement in vivo, uptake by recipient cells, and demonstrate effects upon their targets. As technologies and experimental methodologies are brought to bear on these challenges, we can expect to see the promise of this line of study for applications in the fields of diagnostic medicine and therapy and for a general improvement in our understanding of brain-body connections.

Acknowledgements The author would like to thank colleagues and the operations team at FutureNeuro and acknowledges funding for miRNA and biomarker research from Science Foundation Ireland and industry partners under grants 16/RC/3948 and 21/RC/10294_P2 (FutureNeuro) as well as 13/IA/1891 and 14/ADV/RC2721, the Wellcome Trust (222648/Z/21/Z), the European Union's "Seventh Framework" Programme (FP7) under Grant Agreement 602130 (EpimiRNA), the EU's Horizon 2020 program PRIME project (No. 964712) and the Higher Education Authority (HEA) Ireland (SeeDeepER). The author reports patent US9803200 B2 Inhibition of microRNA-134 for the treatment of seizure-related disorders and neurologic injuries.

References

Alles J, Fehlmann T, Fischer U, Backes C, Galata V, Minet M, Hart M, Abu-Halima M, Grasser FA, Lenhof HP, Keller A, Meese E (2019) An estimate of the total number of true human miRNAs. Nucleic Acids Res 47:3353–3364

Ambros V (2004) The functions of animal microRNAs. Nature 431:350–355

Amin ND, Bai G, Klug JR, Bonanomi D, Pankratz MT, Gifford WD, Hinckley CA, Sternfeld MJ, Driscoll SP, Dominguez B, Lee KF, Jin X, Pfaff SL (2015) Loss of motoneuron-specific microRNA-218 causes systemic neuromuscular failure. Science 350:1525–1529

Amin ND, Senturk G, Costaguta G, Driscoll S, O'leary B, Bonanomi D, Pfaff SL (2021) A hidden threshold in motor neuron gene networks revealed by modulation of miR-218 dose. Neuron 109(3252–3267):e3256

An N, Zhao W, Liu Y, Yang X, Chen P (2016) Elevated serum miR-106b and miR-146a in patients with focal and generalized epilepsy. Epilepsy Res 127:311–316

Antoniou A, Auderset L, Kaurani L, Sebastian E, Zeng Y, Allahham M, Cases-Cunillera S, Schoch S, Grundemann J, Fischer A, Schneider A (2023) Neuronal extracellular vesicles and associated microRNAs induce circuit connectivity downstream BDNF. Cell Rep 42:112063

Arroyo JD, Chevillet JR, Kroh EM, Ruf IK, Pritchard CC, Gibson DF, Mitchell PS, Bennett CF, Pogosova-Agadjanyan EL, Stirewalt DL, Tait JF, Tewari M (2011) Argonaute2 complexes carry a population of circulating microRNAs independent of vesicles in human plasma. Proc Natl Acad Sci USA 108:5003–5008

Avansini SH, De Sousa Lima BP, Secolin R, Santos ML, Coan AC, Vieira AS, Torres FR, Carvalho BS, Alvim MK, Morita ME, Yasuda CL, Pimentel-Silva LR, Dogini DB, Rogerio F, Cendes F, Lopes-Cendes I (2017) MicroRNA hsa-miR-134 is a circulating biomarker for mesial temporal lobe epilepsy. PLoS One 12:e0173060

Baek D, Villen J, Shin C, Camargo FD, Gygi SP, Bartel DP (2008) The impact of microRNAs on protein output. Nature 455:64–71

Banack SA, Dunlop RA, Mehta P, Mitsumoto H, Wood SP, Han M, Cox PA (2024) A microRNA diagnostic biomarker for amyotrophic lateral sclerosis. Brain Commun 6:fcae268

Bartel DP (2018) Metazoan MicroRNAs. Cell 173:20–51

Batool A, Hill TDM, Nguyen NT, Langa E, Diviney M, Mooney C, Brennan GP, Connolly NMC, Sanz-Rodriguez A, Cavanagh BL, Henshall DC (2019) Altered biogenesis and microRNA content of hippocampal exosomes following experimental status epilepticus. Front Neurosci 13:1404

Bekenstein U, Mishra N, Milikovsky DZ, Hanin G, Zelig D, Sheintuch L, Berson A, Greenberg DS, Friedman A, Soreq H (2017) Dynamic changes in murine forebrain miR-211 expression associate with cholinergic imbalances and epileptiform activity. Proc Natl Acad Sci USA 114:E4996–E5005

Berezikov E (2011) Evolution of microRNA diversity and regulation in animals. Nat Rev Genet 12:846–860

Bicker S, Khudayberdiev S, Weiss K, Zocher K, Baumeister S, Schratt G (2013) The DEAH-box helicase DHX36 mediates dendritic localization of the neuronal precursor-microRNA-134. Genes Dev 27:991–996

Bohr T, Hjorth PG, Holst SC, Hrabetova S, Kiviniemi V, Lilius T, Lundgaard I, Mardal KA, Martens EA, Mori Y, Nagerl UV, Nicholson C, Tannenbaum A, Thomas JH, Tithof J, Benveniste H, Iliff JJ, Kelley DH, and Nedergaard M (2022). The glymphatic system: current understanding and modeling. iScience 25:104987

Brennan GP, Henshall DC (2020) MicroRNAs as regulators of brain function and targets for treatment of epilepsy. Nat Rev Neurol 16:506–519

Brennan GP, Bauer S, Engel T, Jimenez-Mateos EM, Del Gallo F, Hill TDM, Connolly NMC, Costard LS, Neubert V, Salvetti B, Sanz-Rodriguez A, Heiland M, Mamad O, Brindley E, Norwood B, Batool A, Raoof R, El-Naggar H, Reschke CR, Delanty N, Prehn JHM, Fabene P, Mooney C, Rosenow F, Henshall DC (2020) Genome-wide microRNA profiling of plasma from three different animal models identifies biomarkers of temporal lobe epilepsy. Neurobiol Dis 144:105048

Brindley E, Heiland M, Mooney C, Diviney M, Mamad O, Hill TDM, Yan Y, Venø MT, Reschke CR, Batool A, Langa E, Sanz-Rodriguez A, Heller JP, Morris G, Conboy K, Kjems J, Brennan GP, Henshall DC (2023)

Brain cell-specific origin of circulating microRNA biomarkers in experimental temporal lobe epilepsy. Front Mol Neurosci 16
Buccitelli C, Selbach M (2020) mRNAs, proteins and the emerging principles of gene expression control. Nat Rev Genet 21:630–644
Budnik V, Ruiz-Canada C, Wendler F (2016) Extracellular vesicles round off communication in the nervous system. Nat Rev Neurosci 17:160–172
Buzas EI (2023) The roles of extracellular vesicles in the immune system. Nat Rev Immunol 23:236–250
Casado Gama H, Amoros MA, Andrade De Araujo M, Sha CM, Vieira MPS, Torres RGD, Souza GF, Junkes JA, Dokholyan NV, Leite Goes Gitai D, Duzzioni M (2024) Systematic review and meta-analysis of dysregulated microRNAs derived from liquid biopsies as biomarkers for amyotrophic lateral sclerosis. Noncoding RNA Res 9:523–535
Cerri S, Ghezzi C, Sampieri M, Siani F, Avenali M, Dornini G, Zangaglia R, Minafra B, Blandini F (2018) The exosomal/total alpha-synuclein ratio in plasma is associated with glucocerebrosidase activity and correlates with measures of disease severity in PD patients. Front Cell Neurosci 12:125
Chandradoss SD, Schirle NT, Szczepaniak M, Macrae IJ, Joo C (2015) A dynamic search process underlies microRNA targeting. Cell 162:96–107
Che N, Zu G, Zhou T, Wang X, Sun Y, Tan Z, Liu Y, Wang D, Luo X, Zhao Z, Zhang Y, Wei M, Yin J (2017) Aberrant expression of miR-323a-5p in patients with refractory epilepsy caused by focal cortical dysplasia. Genet Test Mol Biomarkers 21:3–9
Chevillet JR, Kang Q, Ruf IK, Briggs HA, Vojtech LN, Hughes SM, Cheng HH, Arroyo JD, Meredith EK, Gallichotte EN, Pogosova-Agadjanyan EL, Morrissey C, Stirewalt DL, Hladik F, Yu EY, Higano CS, Tewari M (2014) Quantitative and stoichiometric analysis of the microRNA content of exosomes. Proc Natl Acad Sci USA 111:14888–14893
Chiasserini D, Van Weering JR, Piersma SR, Pham TV, Malekzadeh A, Teunissen CE, De Wit H, Jimenez CR (2014) Proteomic analysis of cerebrospinal fluid extracellular vesicles: a comprehensive dataset. J Proteome 106:191–204
Chiu YJ, Cai W, Shih YR, Lian I, Lo YH (2016) A single-cell assay for time lapse studies of exosome secretion and cell behaviors. Small 12:3658–3666
Christensen J, Pedersen MG, Pedersen CB, Sidenius P, Olsen J, Vestergaard M (2009) Long-term risk of epilepsy after traumatic brain injury in children and young adults: a population-based cohort study. Lancet 373:1105–1110
Cogswell JP, Ward J, Taylor IA, Waters M, Shi Y, Cannon B, Kelnar K, Kemppainen J, Brown D, Chen C, Prinjha RK, Richardson JC, Saunders AM, Roses AD, Richards CA (2008) Identification of miRNA changes in Alzheimer's disease brain and CSF yields putative biomarkers and insights into disease pathways. J Alzheimers Dis 14:27–41
Corradi E, Dalla Costa I, Gavoci A, Iyer A, Roccuzzo M, Otto TA, Oliani E, Bridi S, Strohbuecker S, Santos-Rodriguez G, Valdembri D, Serini G, Abreu-Goodger C, Baudet ML (2020) Axonal precursor miRNAs hitchhike on endosomes and locally regulate the development of neural circuits. EMBO J 39:e102513
Cukovic D, Bagla S, Ukasik D, Stemmer PM, Jena BP, Naik AR, Sood S, Asano E, Luat A, Chugani DC, Dombkowski AA (2021) Exosomes in epilepsy of tuberous sclerosis complex: carriers of pro-inflammatory microRNAs. Noncoding. RNA 7
Dave VP, Ngo TA, Pernestig AK, Tilevik D, Kant K, Nguyen T, Wolff A, Bang DD (2019) MicroRNA amplification and detection technologies: opportunities and challenges for point of care diagnostics. Lab Investig 99:452–469
Du C, Liu C, Kang J, Zhao G, Ye Z, Huang S, Li Z, Wu Z, Pei G (2009) MicroRNA miR-326 regulates TH-17 differentiation and is associated with the pathogenesis of multiple sclerosis. Nat Immunol 10:1252–1259
Ebert MS, Sharp PA (2012) Roles for microRNAs in conferring robustness to biological processes. Cell 149:515–524
Enright N, Simonato M, Henshall DC (2018) Discovery and validation of blood microRNAs as molecular biomarkers of epilepsy—ways to close current knowledge gaps, vol 3. Epilepsia Open, p 427
Ferrara D, Pasetto L, Bonetto V, Basso M (2018) Role of extracellular vesicles in amyotrophic lateral sclerosis. Front Neurosci 12:574
Fromm B, Zhong X, Tarbier M, Friedlander MR, Hackenberg M (2022) The limits of human microRNA annotation have been met. RNA 28:781–785
Fruhbeis C, Frohlich D, Kuo WP, Amphornrat J, Thilemann S, Saab AS, Kirchhoff F, Mobius W, Goebbels S, Nave KA, Schneider A, Simons M, Klugmann M, Trotter J, Kramer-Albers EM (2013) Neurotransmitter-triggered transfer of exosomes mediates oligodendrocyte-neuron communication. PLoS Biol 11:e1001604
Gabrielli M, Prada I, Joshi P, Falcicchia C, D'arrigo G, Rutigliano G, Battocchio E, Zenatelli R, Tozzi F, Radeghieri A, Arancio O, Origlia N, Verderio C (2022) Microglial large extracellular vesicles propagate early synaptic dysfunction in Alzheimer's disease. Brain 145:2849–2868
Gallego JA, Gordon ML, Claycomb K, Bhatt M, Lencz T, Malhotra AK (2012) In vivo microRNA detection and quantitation in cerebrospinal fluid. J Mol Neurosci 47:243–248
Gebert LFR, Macrae IJ (2019) Regulation of microRNA function in animals. Nat Rev Mol Cell Biol 20:21–37
Gitai DLG, Dos Santos YDR, Upadhya R, Kodali M, Madhu LN, Shetty AK (2020) Extracellular vesicles in the forebrain display reduced miR-346 and miR-331-3p in a rat model of chronic temporal lobe epilepsy. Mol Neurobiol 57:1674–1687
Gorter JA, Iyer A, White I, Colzi A, Van Vliet EA, Sisodiya S, Aronica E (2014) Hippocampal subregion-specific microRNA expression during epileptogenesis

in experimental temporal lobe epilepsy. Neurobiol Dis 62:508–520

Gruner HN, Mcmanus MT (2021) Examining the evidence for extracellular RNA function in mammals. Nat Rev Genet 22:448–458

Guerau-De-Arellano M, Smith KM, Godlewski J, Liu Y, Winger R, Lawler SE, Whitacre CC, Racke MK, Lovett-Racke AE (2011) Micro-RNA dysregulation in multiple sclerosis favours pro-inflammatory T-cell-mediated autoimmunity. Brain 134:3578–3589

Guo M, Wang J, Zhao Y, Feng Y, Han S, Dong Q, Cui M, Tieu K (2020) Microglial exosomes facilitate alpha-synuclein transmission in Parkinson's disease. Brain 143:1476–1497

Ha M, Kim VN (2014) Regulation of microRNA biogenesis. Nat Rev Mol Cell Biol 15:509–524

Han S, Li Y, Gao J (2024) Peripheral blood microRNAs as biomarkers of schizophrenia: expectations from a meta-analysis that combines deep learning methods. World J Biol Psychiatry 25:65–81

Hebert SS, Horre K, Nicolai L, Papadopoulou AS, Mandemakers W, Silahtaroglu AN, Kauppinen S, Delacourte A, De Strooper B (2008) Loss of microRNA cluster miR-29a/b-1 in sporadic Alzheimer's disease correlates with increased BACE1/beta-secretase expression. Proc Natl Acad Sci USA 105:6415–6420

Heiland M, Connolly NMC, Mamad O, Nguyen NT, Kesavan JC, Langa E, Fanning K, Sanfeliu A, Yan Y, Su J, Venø MT, Costard LS, Neubert V, Engel T, Hill TD, Freiman T, Mahesh A, Tiwari VK, Rosenow F, Bauer S, Kjems J, Morris G, Henshall DC (2023) MicroRNA-335-5p suppresses voltage-gated sodium channel expression and may be a target for seizure control. Proc Natl Acad Sci USA 120:e2216658120

Heiskanen M, Das Gupta S, Mills JD, Van Vliet EA, Manninen E, Ciszek R, Andrade P, Puhakka N, Aronica E, Pitkanen A (2023) Discovery and validation of circulating microRNAs as biomarkers for epileptogenesis after experimental traumatic brain injury-the EPITARGET cohort. Int J Mol Sci 24

Henshall DC (2024) Fine-tuning life: a guide to microRNAs, your genome's master regulators. Cambridge University Press

Hermann DM, Peruzzotti-Jametti L, Giebel B, Pluchino S (2024) Extracellular vesicles set the stage for brain plasticity and recovery by multimodal signalling. Brain 147:372–389

Hocker S (2015) Systemic complications of status epilepticus--An update. Epilepsy Behav 49:83–87

Hornstein E, Shomron N (2006) Canalization of development by microRNAs. Nat Genet 38(Suppl):S20–S24

Huang KY, Upadhyay G, Ahn Y, Sakakura M, Pagan-Diaz GJ, Cho Y, Weiss AC, Huang C, Mitchell JW, Li J, Tan Y, Deng YH, Ellis-Mohr A, Dou Z, Zhang X, Kang S, Chen Q, Sweedler JV, Im SG, Bashir R, Chung HJ, Popescu G, Gillette MU, Gazzola M, Kong H (2024) Neuronal innervation regulates the secretion of neurotrophic myokines and exosomes from skeletal muscle. Proc Natl Acad Sci USA 121:e2313590121

Iguchi Y, Eid L, Parent M, Soucy G, Bareil C, Riku Y, Kawai K, Takagi S, Yoshida M, Katsuno M, Sobue G, Julien JP (2016) Exosome secretion is a key pathway for clearance of pathological TDP-43. Brain 139:3187–3201

Jimenez-Mateos EM, Engel T, Merino-Serrais P, Mckiernan RC, Tanaka K, Mouri G, Sano T, O'tuathaigh C, Waddington JL, Prenter S, Delanty N, Farrell MA, O'brien DF, Conroy RM, Stallings RL, Defelipe J, Henshall DC (2012) Silencing microRNA-134 produces neuroprotective and prolonged seizure-suppressive effects. Nat Med 18:1087–1094

Jovicic A, Roshan R, Moisoi N, Pradervand S, Moser R, Pillai B, Luthi-Carter R (2013) Comprehensive expression analyses of neural cell-type-specific miRNAs identify new determinants of the specification and maintenance of neuronal phenotypes. J Neurosci 33:5127–5137

Junker A, Krumbholz M, Eisele S, Mohan H, Augstein F, Bittner R, Lassmann H, Wekerle H, Hohlfeld R, Meinl E (2009) MicroRNA profiling of multiple sclerosis lesions identifies modulators of the regulatory protein CD47. Brain 132:3342–3352

Junn E, Lee KW, Jeong BS, Chan TW, Im JY, Mouradian MM (2009) Repression of alpha-synuclein expression and toxicity by microRNA-7. Proc Natl Acad Sci USA 106:13052–13057

Kalluri R, Lebleu VS (2020) The biology, function, and biomedical applications of exosomes. Science 367

Kan AA, Van Erp S, Derijck AA, De Wit M, Hessel EV, O'duibhir E, De Jager W, Van Rijen PC, Gosselaar PH, De Graan PN, Pasterkamp RJ (2012) Genome-wide microRNA profiling of human temporal lobe epilepsy identifies modulators of the immune response. Cell Mol Life Sci 69:3127–3145

Keller A, Leidinger P, Steinmeyer F, Stahler C, Franke A, Hemmrich-Stanisak G, Kappel A, Wright I, Dorr J, Paul F, Diem R, Tocariu-Krick B, Meder B, Backes C, Meese E, Ruprecht K (2014) Comprehensive analysis of microRNA profiles in multiple sclerosis including next-generation sequencing. Mult Scler 20:295–303

Kenny A, Jimenez-Mateos EM, Zea-Sevilla MA, Rabano A, Gili-Manzanaro P, Prehn JHM, Henshall DC, Avila J, Engel T, Hernandez F (2019) Proteins and microRNAs are differentially expressed in tear fluid from patients with Alzheimer's disease. Sci Rep 9:15437

Kim J, Inoue K, Ishii J, Vanti WB, Voronov SV, Murchison E, Hannon G, Abeliovich A (2007) A MicroRNA feedback circuit in midbrain dopamine neurons. Science 317:1220–1224

Kipnis J (2016) Multifaceted interactions between adaptive immunity and the central nervous system. Science 353:766–771

Kocerha J, Faghihi MA, Lopez-Toledano MA, Huang J, Ramsey AJ, Caron MG, Sales N, Willoughby D, Elmen J, Hansen HF, Orum H, Kauppinen S, Kenny

PJ, Wahlestedt C (2009) MicroRNA-219 modulates NMDA receptor-mediated neurobehavioral dysfunction. Proc Natl Acad Sci USA 106:3507–3512
Koppers-Lalic D, Hackenberg M, Bijnsdorp IV, Van Eijndhoven MA, Sadek P, Sie D, Zini N, Middeldorp JM, Ylstra B, De Menezes RX, Wurdinger T, Meijer GA, Pegtel DM (2014) Nontemplated nucleotide additions distinguish the small RNA composition in cells from exosomes. Cell Rep 8:1649–1658
Kosik KS (2006) The neuronal microRNA system. Nat Rev Neurosci 7:911–920
Kosik KS, Nowakowski T (2018) Evolution of new miRNAs and cerebro-cortical development. Annu Rev Neurosci 41:119–137
Koval ED, Shaner C, Zhang P, Du Maine X, Fischer K, Tay J, Chau BN, Wu GF, Miller TM (2013) Method for widespread microRNA-155 inhibition prolongs survival in ALS-model mice. Hum Mol Genet 22:4127–4135
Kozomara A, Birgaoanu M, Griffiths-Jones S (2019) miRBase: from microRNA sequences to function. Nucleic Acids Res 47:D155–D162
Krichevsky AM, King KS, Donahue CP, Khrapko K, Kosik KS (2003) A microRNA array reveals extensive regulation of microRNAs during brain development. RNA 9:1274–1281
Kur IM, Prouvot PH, Fu T, Fan W, Muller-Braun F, Das A, Das S, Deller T, Roeper J, Stroh A, Momma S (2020) Neuronal activity triggers uptake of hematopoietic extracellular vesicles in vivo. PLoS Biol 18:e3000643
Lachenal G, Pernet-Gallay K, Chivet M, Hemming FJ, Belly A, Bodon G, Blot B, Haase G, Goldberg Y, Sadoul R (2011) Release of exosomes from differentiated neurons and its regulation by synaptic glutamatergic activity. Mol Cell Neurosci 46:409–418
Lee RC, Feinbaum RL, Ambros V (1993) The C. elegans heterochronic gene lin-4 encodes small RNAs with antisense complementarity to lin-14. Cell 75:843–854
Lee ST, Chu K, Im WS, Yoon HJ, Im JY, Park JE, Park KH, Jung KH, Lee SK, Kim M, Roh JK (2011) Altered microRNA regulation in Huntington's disease models. Exp Neurol 227:172–179
Leidinger P, Backes C, Deutscher S, Schmitt K, Mueller SC, Frese K, Haas J, Ruprecht K, Paul F, Stahler C, Lang CJ, Meder B, Bartfai T, Meese E, Keller A (2013) A blood based 12-miRNA signature of Alzheimer disease patients. Genome Biol 14:R78
Leontariti M, Avgeris M, Katsarou MS, Drakoulis N, Siatouni A, Verentzioti A, Alexoudi A, Fytraki A, Patrikelis P, Vassilacopoulou D, Gatzonis S, Sideris DC (2020) Circulating miR-146a and miR-134 in predicting drug-resistant epilepsy in patients with focal impaired awareness seizures. Epilepsia 61:959–970
Leung AKL (2015) The whereabouts of microRNA actions: cytoplasm and beyond. Trends Cell Biol 25:601–610
Li L, Song Y, Shi X, Liu J, Xiong S, Chen W, Fu Q, Huang Z, Gu N, Zhang R (2018) The landscape of miRNA editing in animals and its impact on miRNA biogenesis and targeting. Genome Res 28:132–143
Lim LP, Lau NC, Garrett-Engele P, Grimson A, Schelter JM, Castle J, Bartel DP, Linsley PS, Johnson JM (2005) Microarray analysis shows that some microRNAs downregulate large numbers of target mRNAs. Nature 433:769–773
Liu CG, Song J, Zhang YQ, Wang PC (2014) MicroRNA-193b is a regulator of amyloid precursor protein in the blood and cerebrospinal fluid derived exosomal microRNA-193b is a biomarker of Alzheimer's disease. Mol Med Rep 10:2395–2400
Liu T, Liu H, Xue S, Xiao L, Xu J, Tong S, Wei X (2024a) MiR129-5p-loaded exosomes suppress seizure-associated neurodegeneration in status epilepticus model mice by inhibiting HMGB1/TLR4-mediated neuroinflammation. Mol Biol Rep 51:292
Liu X, Dong L, Jiang Z, Song M, Yan P (2024b) Identifying the differentially expressed peripheral blood microRNAs in psychiatric disorders: a systematic review and meta-analysis. Front Psych 15:1390366
Londin E, Loher P, Telonis AG, Quann K, Clark P, Jing Y, Hatzimichael E, Kirino Y, Honda S, Lally M, Ramratnam B, Comstock CE, Knudsen KE, Gomella L, Spaeth GL, Hark L, Katz LJ, Witkiewicz A, Rostami A, Jimenez SA, Hollingsworth MA, Yeh JJ, Shaw CA, Mckenzie SE, Bray P, Nelson PT, Zupo S, Van Roosbroeck K, Keating MJ, Calin GA, Yeo C, Jimbo M, Cozzitorto J, Brody JR, Delgrosso K, Mattick JS, Fortina P, Rigoutsos I (2015) Analysis of 13 cell types reveals evidence for the expression of numerous novel primate- and tissue-specific microRNAs. Proc Natl Acad Sci USA 112:E1106–E1115
Martins-Ferreira R, Chaves J, Carvalho C, Bettencourt A, Chorao R, Freitas J, Samoes R, Boleixa D, Lopes J, Ramalheira J, Da Silva BM, Martins Da Silva A, Costa PP, Leal B (2020) Circulating microRNAs as potential biomarkers for genetic generalized epilepsies: a three microRNA panel. Eur J Neurol 27:660–666
Mcardle H, Jimenez-Mateos EM, Raoof R, Carthy E, Boyle D, Elnaggar H, Delanty N, Hamer H, Dogan M, Huchtemann T, Krtvelyessy P, Rosenow F, Forster RJ, Henshall DC, Spain E (2017) "TORNADO"—Theranostic one-step RNA detector; microfluidic disc for the direct detection of microRNA-134 in plasma and cerebrospinal fluid. Sci Rep 7:1750
Mcgeary SE, Lin KS, Shi CY, Pham TM, Bisaria N, Kelley GM, Bartel DP (2019) The biochemical basis of microRNA targeting efficacy. Science 366
Mckiernan RC, Jimenez-Mateos EM, Bray I, Engel T, Brennan GP, Sano T, Michalak Z, Moran C, Delanty N, Farrell M, O'brien D, Meller R, Simon RP, Stallings RL, Henshall DC (2012) Reduced mature microRNA levels in association with Dicer loss in human temporal lobe epilepsy with hippocampal sclerosis. PLoS One 7:e35921
Men Y, Yelick J, Jin S, Tian Y, Chiang MSR, Higashimori H, Brown E, Jarvis R, Yang Y (2019) Exosome reporter mice reveal the involvement of exosomes in

mediating neuron to astroglia communication in the CNS. Nat Commun 10:4136

Miller BH, Zeier Z, Xi L, Lanz TA, Deng S, Strathmann J, Willoughby D, Kenny PJ, Elsworth JD, Lawrence MS, Roth RH, Edbauer D, Kleiman RJ, Wahlestedt C (2012) MicroRNA-132 dysregulation in schizophrenia has implications for both neurodevelopment and adult brain function. Proc Natl Acad Sci USA 109:3125–3130

Mills JD, Van Vliet EA, Chen BJ, Janitz M, Anink JJ, Baayen JC, Idema S, Devore S, Friedman D, Diehl B, Thom M, Scott C, Thijs R, Aronica E, Devinsky O (2020) Coding and non-coding transcriptome of mesial temporal lobe epilepsy: critical role of small non-coding RNAs. Neurobiol Dis 134:104612

Mittelbrunn M, Gutierrez-Vazquez C, Villarroya-Beltri C, Gonzalez S, Sanchez-Cabo F, Gonzalez MA, Bernad A, Sanchez-Madrid F (2011) Unidirectional transfer of microRNA-loaded exosomes from T cells to antigen-presenting cells. Nat Commun 2:282

Mor E, Cabilly Y, Goldshmit Y, Zalts H, Modai S, Edry L, Elroy-Stein O, Shomron N (2011) Species-specific microRNA roles elucidated following astrocyte activation. Nucleic Acids Res 39:3710–3723

Morris G, O'brien D, Henshall DC (2021) Opportunities and challenges for microRNA-targeting therapeutics for epilepsy. Trends Pharmacol Sci 42:605–616

Mueller C, Berry JD, Mckenna-Yasek DM, Gernoux G, Owegi MA, Pothier LM, Douthwright CL, Gelevski D, Luppino SD, Blackwood M, Wightman NS, Oakley DH, Frosch MP, Flotte TR, Cudkowicz ME, Brown RH Jr (2020) SOD1 suppression with adeno-associated virus and microRNA in familial ALS. N Engl J Med 383:151–158

Nishikura K (2016) A-to-I editing of coding and non-coding RNAs by ADARs. Nat Rev Mol Cell Biol 17:83–96

Nowakowski TJ, Rani N, Golkaram M, Zhou HR, Alvarado B, Huch K, West JA, Leyrat A, Pollen AA, Kriegstein AR, Petzold LR, Kosik KS (2018) Regulation of cell-type-specific transcriptomes by microRNA networks during human brain development. Nat Neurosci 21:1784–1792

O'brien K, Breyne K, Ughetto S, Laurent LC, Breakefield XO (2020) RNA delivery by extracellular vesicles in mammalian cells and its applications. Nat Rev Mol Cell Biol 21:585–606

Obermeier B, Daneman R, Ransohoff RM (2013) Development, maintenance and disruption of the blood-brain barrier. Nat Med 19:1584–1596

Packer AN, Xing Y, Harper SQ, Jones L, Davidson BL (2008) The bifunctional microRNA miR-9/miR-9* regulates REST and CoREST and is downregulated in Huntington's disease. J Neurosci 28:14341–14346

Pasquinelli AE, Reinhart BJ, Slack F, Martindale MQ, Kuroda MI, Maller B, Hayward DC, Ball EE, Degnan B, Muller P, Spring J, Srinivasan A, Fishman M, Finnerty J, Corbo J, Levine M, Leahy P, Davidson E, Ruvkun G (2000) Conservation of the sequence and temporal expression of let-7 heterochronic regulatory RNA. Nature 408:86–89

Peng KY, Perez-Gonzalez R, Alldred MJ, Goulbourne CN, Morales-Corraliza J, Saito M, Saito M, Ginsberg SD, Mathews PM, Levy E (2019) Apolipoprotein E4 genotype compromises brain exosome production. Brain 142:163–175

Perkins DO, Jeffries CD, Jarskog LF, Thomson JM, Woods K, Newman MA, Parker JS, Jin J, Hammond SM (2007) microRNA expression in the prefrontal cortex of individuals with schizophrenia and schizoaffective disorder. Genome Biol 8:R27

Pietrasik S, Dziedzic A, Miller E, Starosta M, Saluk-Bijak J (2021) Circulating miRNAs as potential biomarkers distinguishing relapsing-remitting from secondary progressive multiple sclerosis. A review. Int J Mol Sci 22

Ponomarev ED, Veremeyko T, Weiner HL (2013) MicroRNAs are universal regulators of differentiation, activation, and polarization of microglia and macrophages in normal and diseased CNS. Glia 61:91–103

Prada I, Gabrielli M, Turola E, Iorio A, D'arrigo G, Parolisi R, De Luca M, Pacifici M, Bastoni M, Lombardi M, Legname G, Cojoc D, Buffo A, Furlan R, Peruzzi F, Verderio C (2018) Glia-to-neuron transfer of miRNAs via extracellular vesicles: a new mechanism underlying inflammation-induced synaptic alterations. Acta Neuropathol 135:529–550

Qureshi IA, Mehler MF (2012) Emerging roles of non-coding RNAs in brain evolution, development, plasticity and disease. Nat Rev Neurosci 13:528–541

Raj A, Van Oudenaarden A (2008) Nature, nurture, or chance: stochastic gene expression and its consequences. Cell 135:216–226

Raoof R, Jimenez-Mateos EM, Bauer S, Tackenberg B, Rosenow F, Lang J, Onugoren MD, Hamer H, Huchtemann T, Kortvelyessy P, Connolly NMC, Pfeiffer S, Prehn JHM, Farrell MA, O'brien DF, Henshall DC, Mooney C (2017) Cerebrospinal fluid microRNAs are potential biomarkers of temporal lobe epilepsy and status epilepticus. Sci Rep 7:3328

Raoof R, Bauer S, El Naggar H, Connolly NMC, Brennan GP, Brindley E, Hill T, Mcardle H, Spain E, Forster RJ, Prehn JHM, Hamer H, Delanty N, Rosenow F, Mooney C, Henshall DC (2018) Dual-center, dual-platform microRNA profiling identifies potential plasma biomarkers of adult temporal lobe epilepsy. EBioMedicine 38:127–141

Roncon P, Soukupova M, Binaschi A, Falcicchia C, Zucchini S, Ferracin M, Langley SR, Petretto E, Johnson MR, Marucci G, Michelucci R, Rubboli G, Simonato M (2015) MicroRNA profiles in hippocampal granule cells and plasma of rats with pilocarpine-induced epilepsy—comparison with human epileptic samples. Sci Rep 5:14143

Roser AE, Caldi Gomes L, Schunemann J, Maass F, Lingor P (2018) Circulating miRNAs as diagnostic biomarkers for Parkinson's disease. Front Neurosci 12:625

Ruiz-Arroyo VM, Nam Y (2022) Dynamic protein-RNA recognition in primary microRNA processing. Curr Opin Struct Biol 76:102442
Rustenhoven J, Kipnis J (2022) Brain borders at the central stage of neuroimmunology. Nature 612:417–429
Rustom A, Saffrich R, Markovic I, Walther P, Gerdes HH (2004) Nanotubular highways for intercellular organelle transport. Science 303:1007–1010
Sambandan S, Akbalik G, Kochen L, Rinne J, Kahlstatt J, Glock C, Tushev G, Alvarez-Castelao B, Heckel A, Schuman EM (2017) Activity-dependent spatially localized miRNA maturation in neuronal dendrites. Science 355:634–637
Schaefer A, O'carroll D, Tan CL, Hillman D, Sugimori M, Llinas R, Greengard P (2007) Cerebellar neurodegeneration in the absence of microRNAs. J Exp Med 204:1553–1558
Schieweck R, Ninkovic J, Kiebler MA (2021) RNA-binding proteins balance brain function in health and disease. Physiol Rev 101:1309–1370
Schratt GM, Tuebing F, Nigh EA, Kane CG, Sabatini ME, Kiebler M, Greenberg ME (2006) A brain-specific microRNA regulates dendritic spine development. Nature 439:283–289
Schwarz DS, Hutvagner G, Du T, Xu Z, Aronin N, Zamore PD (2003) Asymmetry in the assembly of the RNAi enzyme complex. Cell 115:199–208
Sempere LF, Freemantle S, Pitha-Rowe I, Moss E, Dmitrovsky E, Ambros V (2004) Expression profiling of mammalian microRNAs uncovers a subset of brain-expressed microRNAs with possible roles in murine and human neuronal differentiation. Genome Biol 5:R13
Shang R, Lee S, Senavirathne G, Lai EC (2023) microRNAs in action: biogenesis, function and regulation. Nat Rev Genet 24:816
Siegel G, Obernosterer G, Fiore R, Oehmen M, Bicker S, Christensen M, Khudayberdiev S, Leuschner PF, Busch CJ, Kane C, Hubel K, Dekker F, Hedberg C, Rengarajan B, Drepper C, Waldmann H, Kauppinen S, Greenberg ME, Draguhn A, Rehmsmeier M, Martinez J, Schratt GM (2009) A functional screen implicates microRNA-138-dependent regulation of the depalmitoylation enzyme APT1 in dendritic spine morphogenesis. Nat Cell Biol 11:705–716
Siegert S, Seo J, Kwon EJ, Rudenko A, Cho S, Wang W, Flood Z, Martorell AJ, Ericsson M, Mungenast AE, Tsai LH (2015) The schizophrenia risk gene product miR-137 alters presynaptic plasticity. Nat Neurosci 18:1008–1016
Silenzi V, D'ambra E, Santini T, D'uva S, Setti A, Salvi N, Nicoletti C, Scarfo R, Cordella F, Mongiardi B, Cavezza D, Liessi N, Ferrucci L, Ragozzino D, Armirotti A, Di Angelantonio S, De Leonibus E, Bozzoni I, Morlando M (2024) A tripartite circRNA/mRNA/miRNA interaction regulates glutamatergic signaling in the mouse brain. Cell Rep 43:114766
Simeoli R, Montague K, Jones HR, Castaldi L, Chambers D, Kelleher JH, Vacca V, Pitcher T, Grist J, Al-Ahdal H, Wong LF, Perretti M, Lai J, Mouritzen P, Heppenstall P, Malcangio M (2017) Exosomal cargo including microRNA regulates sensory neuron to macrophage communication after nerve trauma. Nat Commun 8:1778
Simonato M, Agoston DV, Brooks-Kayal A, Dulla C, Fureman B, Henshall DC, Pitkanen A, Theodore WH, Twyman RE, Kobeissy FH, Wang KK, Whittemore V, Wilcox KS (2021) Identification of clinically relevant biomarkers of epileptogenesis—a strategic roadmap. Nat Rev Neurol 17:231–242
Soutschek M, Schratt G (2023) Non-coding RNA in the wiring and remodeling of neural circuits. Neuron 111:2140
Spain E, Jimenez-Mateos EM, Raoof R, El Naggar H, Delanty N, Forster RJ, Henshall DC (2015) Direct, non-amplified detection of microRNA-134 in plasma from epilepsy patients. RSC Adv 5:90071–90078
Stein CS, Mclendon JM, Witmer NH, Boudreau RL (2022) Modulation of miR-181 influences dopaminergic neuronal degeneration in a mouse model of Parkinson's disease. Mol Ther Nucleic Acids 28:1–15
Surges R, Kretschmann A, Abnaof K, Van Rikxoort M, Ridder K, Frohlich H, Danis B, Kaminski RM, Foerch P, Elger CE, Weinsberg F, Pfeifer A (2016) Changes in serum miRNAs following generalized convulsive seizures in human mesial temporal lobe epilepsy. Biochem Biophys Res Commun 481:13–18
Tan CL, Plotkin JL, Veno MT, Von Schimmelmann M, Feinberg P, Mann S, Handler A, Kjems J, Surmeier DJ, O'carroll D, Greengard P, Schaefer A (2013) MicroRNA-128 governs neuronal excitability and motor behavior in mice. Science 342:1254–1258
Tao J, Wu H, Lin Q, Wei W, Lu XH, Cantle JP, Ao Y, Olsen RW, Yang XW, Mody I, Sofroniew MV, Sun YE (2011) Deletion of astroglial dicer causes non-cell-autonomous neuronal dysfunction and degeneration. J Neurosci 31:8306–8319
Thompson AG, Gray E, Heman-Ackah SM, Mager I, Talbot K, Andaloussi SE, Wood MJ, Turner MR (2016) Extracellular vesicles in neurodegenerative disease—pathogenesis to biomarkers. Nat Rev Neurol 12:346–357
Trelinska J, Fendler W, Dachowska I, Kotulska K, Jozwiak S, Antosik K, Gnys P, Borowiec M, Mlynarski W (2016) Abnormal serum microRNA profiles in tuberous sclerosis are normalized during treatment with everolimus: possible clinical implications. Orphanet J Rare Dis 11:129
Turchinovich A, Weiz L, Langheinz A, Burwinkel B (2011) Characterization of extracellular circulating microRNA. Nucleic Acids Res 39:7223–7233
Valadi H, Ekstrom K, Bossios A, Sjostrand M, Lee JJ, Lotvall JO (2007) Exosome-mediated transfer of mRNAs and microRNAs is a novel mechanism of genetic exchange between cells. Nat Cell Biol 9:654–659
Valles A, Evers MM, Stam A, Sogorb-Gonzalez M, Brouwers C, Vendrell-Tornero C, Acar-Broekmans S, Paerels L, Klima J, Bohuslavova B, Pintauro R,

Fodale V, Bresciani A, Liscak R, Urgosik D, Starek Z, Crha M, Blits B, Petry H, Ellederova Z, Motlik J, Van Deventer S, Konstantinova P (2021) Widespread and sustained target engagement in Huntington's disease minipigs upon intrastriatal microRNA-based gene therapy. Sci Transl Med 13:13

Van Den Berg MMJ, Krauskopf J, Ramaekers JG, Kleinjans JCS, Prickaerts J, Briede JJ (2020) Circulating microRNAs as potential biomarkers for psychiatric and neurodegenerative disorders. Prog Neurobiol 185:101732

Van Erp TG, Guella I, Vawter MP, Turner J, Brown GG, Mccarthy G, Greve DN, Glover GH, Calhoun VD, Lim KO, Bustillo JR, Belger A, Ford JM, Mathalon DH, Diaz M, Preda A, Nguyen D, Macciardi F, Potkin SG (2014) Schizophrenia miR-137 locus risk genotype is associated with dorsolateral prefrontal cortex hyperactivation. Biol Psychiatry 75:398–405

Van Rooij E, Kauppinen S (2014) Development of microRNA therapeutics is coming of age. EMBO Mol Med 6:851–864

Veno MT, Reschke CR, Morris G, Connolly NMC, Su J, Yan Y, Engel T, Jimenez-Mateos EM, Harder LM, Pultz D, Haunsberger SJ, Pal A, Heller JP, Campbell A, Langa E, Brennan GP, Conboy K, Richardson A, Norwood BA, Costard LS, Neubert V, Del Gallo F, Salvetti B, Vangoor VR, Sanz-Rodriguez A, Muilu J, Fabene PF, Pasterkamp RJ, Prehn JHM, Schorge S, Andersen JS, Rosenow F, Bauer S, Kjems J, Henshall DC (2020) A systems approach delivers a functional microRNA catalog and expanded targets for seizure suppression in temporal lobe epilepsy. Proc Natl Acad Sci USA 117:15977–15988

Wang CY, Yang SH, Tzeng SF (2014) MicroRNA-145 as one negative regulator of astrogliosis. Glia 63:194

Wang J, Tan L, Tan L, Tian Y, Ma J, Tan CC, Wang HF, Liu Y, Tan MS, Jiang T, Yu JT (2015a) Circulating microRNAs are promising novel biomarkers for drug-resistant epilepsy. Sci Rep 5:10201

Wang J, Yu JT, Tan L, Tian Y, Ma J, Tan CC, Wang HF, Liu Y, Tan MS, Jiang T, Tan L (2015b) Genome-wide circulating microRNA expression profiling indicates biomarkers for epilepsy. Sci Rep 5:9522

Wang X, Sun Y, Tan Z, Che N, Ji A, Luo X, Sun X, Li X, Yang K, Wang G, Luan L, Liu Y, Wei M, Yin J (2016) Serum microRNA-4521 is a potential biomarker for focal cortical dysplasia with refractory epilepsy. Neurochem Res 41:905–912

Wightman B, Ha I, Ruvkun G (1993) Posttranscriptional regulation of the heterochronic gene lin-14 by lin-4 mediates temporal pattern formation in C. elegans. Cell 75:855–862

Wijesinghe P, Xi J, Cui J, Campbell M, Pham W, Matsubara JA (2023) MicroRNAs in tear fluids predict underlying molecular changes associated with Alzheimer's disease. Life Sci Alliance 6

Witwer KW, Halushka MK (2016) Toward the promise of microRNAs—enhancing reproducibility and rigor in microRNA research. RNA Biol 13:1103–1116

Wozniak AL, Adams A, King KE, Dunn W, Christenson LK, Hung WT, Weinman SA (2020) The RNA binding protein FMR1 controls selective exosomal miRNA cargo loading during inflammation. J Cell Biol 219

Xu B, Zhang Y, Du XF, Li J, Zi HX, Bu JW, Yan Y, Han H, Du JL (2017) Neurons secrete miR-132-containing exosomes to regulate brain vascular integrity. Cell Res 27:882–897

Yakovleva KD, Dmitrenko DV, Panina IS, Usoltseva AA, Gazenkampf KA, Konovalenko OV, Kantimirova EA, Novitsky MA, Nasyrova RF, Shnayder NA (2022) Expression profile of miRs in mesial temporal lobe epilepsy: systematic review. Int J Mol Sci 23

Yang X, Yang X, Sun A, Chen S, Wang X, Zhao X (2024) The miR-23b-3p from adipose-derived stem cell exosomes alleviate inflammation in mice experiencing kainic acid-induced epileptic seizures. Neuroreport 35:612–620

Yoon S, Kim SE, Ko Y, Jeong GH, Lee KH, Lee J, Solmi M, Jacob L, Smith L, Stickley A, Carvalho AF, Dragioti E, Kronbichler A, Koyanagi A, Hong SH, Thompson T, Oh H, Salazar De Pablo G, Radua J, Shin JI, Fusar-Poli P (2022) Differential expression of MicroRNAs in Alzheimer's disease: a systematic review and meta-analysis. Mol Psychiatry 27:2405–2413

Zhang Y, Liu D, Chen X, Li J, Li L, Bian Z, Sun F, Lu J, Yin Y, Cai X, Sun Q, Wang K, Ba Y, Wang Q, Wang D, Yang J, Liu P, Xu T, Yan Q, Zhang J, Zen K, Zhang CY (2010) Secreted monocytic miR-150 enhances targeted endothelial cell migration. Mol Cell 39:133–144

Zhang J, Zhao M, Yan R, Liu J, Maddila S, Junn E, Mouradian MM (2021) MicroRNA-7 protects against neurodegeneration induced by alpha-synuclein preformed fibrils in the mouse brain. Neurotherapeutics 18:2529–2540

Zhang S, Cheng Y, Shang H (2023) The updated development of blood-based biomarkers for Huntington's disease. J Neurol 270:2483–2503

Zhao L, Ye Y, Gu L, Jian Z, Stary CM, Xiong X (2021) Extracellular vesicle-derived miRNA as a novel regulatory system for bi-directional communication in gut-brain-microbiota axis. J Transl Med 19:202

Ziats MN, Rennert OM (2014) Identification of differentially expressed microRNAs across the developing human brain. Mol Psychiatry 19:848–852

Zovoilis A, Agbemenyah HY, Agis-Balboa RC, Stilling RM, Edbauer D, Rao P, Farinelli L, Delalle I, Schmitt A, Falkai P, Bahari-Javan S, Burkhardt S, Sananbenesi F, Fischer A (2011) microRNA-34c is a novel target to treat dementias. EMBO J 30:4299–4308

Conclusions

Daniela Tropea, Giulia Guasoni,
and Emanuela Giacometti

The study of the connections between brain and body systems is an emerging topic in neuroscience and it promises to grow in the coming years. The connections presented in this book are only a part of the interactions existing between body and brain, and they show that many of these mechanisms overlap also across different body systems (see Fig. 1). These heterogeneous mechanisms—including hormones, metabolic systems, microRNA, and others—force us to reconsider the hierarchical communications and decision check points in the organism, and open for a new interpretation of systems in health and disease.

In fact, basic mechanisms linked to vital functions can ultimately control cognitive systems and have an impact on disorders of the brain. Hence, the identification of peripheral systems impacting on brain function may have applications in diagnosis, prognosis, and therapeutic interventions.

In addition, understanding how the brain responds to the signals coming from the periphery will clarify the mechanisms involved in brain modulation and feedback and will provide further insights on how the brain works. It is also worth remembering that the organization of brain connectivity changes over time in response to an intrinsic plan and external stimuli. It is reasonable to speculate that also the interactions between the brain and other body systems may change and present some degree of plasticity depending on the specific interaction.

All considered, this short paragraph, instead to represent the conclusion of this volume, can be looked as the introduction to a new branch in neuroscience: "The Science of the Changing Brain Body Connections."

D. Tropea, E. Giacometti (eds.), *Brain-Body Connections*, Advances in Experimental Medicine and Biology 1477, https://doi.org/10.1007/978-3-031-89525-8

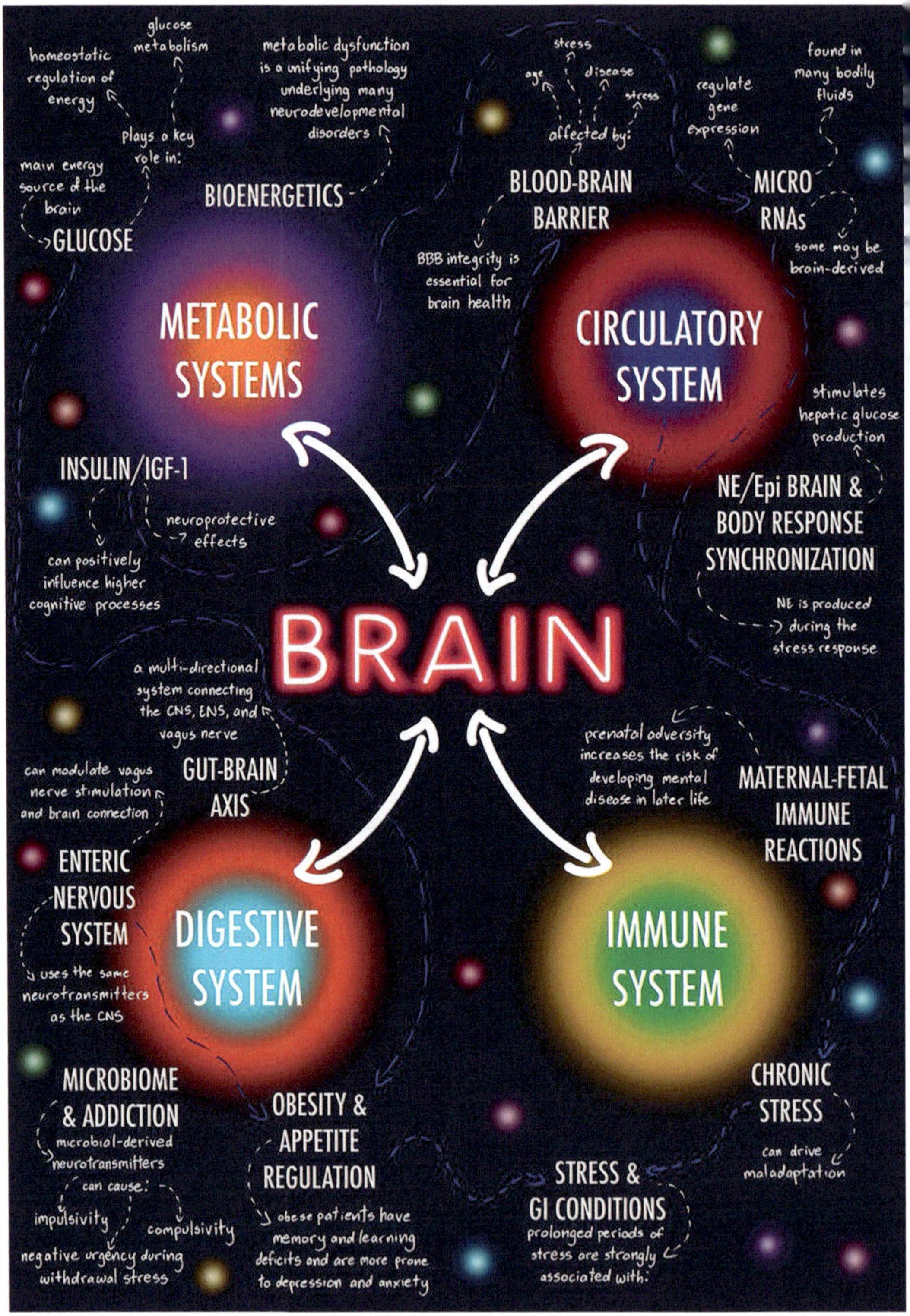

Fig. 1 A graphical representation of the connections between the brain and major systems of the body (arranged in clockwise sequence from the top-right): circulatory system, immune system, digestive system, and metabolic systems. Core concepts from each chapter are shown and interconnections between chapters are highlighted

GPSR Compliance

The European Union's (EU) General Product Safety Regulation (GPSR) is a set of rules that requires consumer products to be safe and our obligations to ensure this.

If you have any concerns about our products, you can contact us on ProductSafety@springernature.com

In case Publisher is established outside the EU, the EU authorized representative is:

Springer Nature Customer Service Center GmbH
Europaplatz 3
69115 Heidelberg, Germany

Batch number: 10731395

Printed by Printforce, the Netherlands